Polymers in Energy Conversion and Storage

Polymers in Energy Conversion and Storage

Edited by
Inamuddin
Mohd Imran Ahamed
Rajender Boddula
Tariq Altalhi

CRC Press
Taylor & Francis Group
Boca Raton London New York

CRC Press is an imprint of the
Taylor & Francis Group, an **informa** business

First edition published [2022]
by CRC Press
6000 Broken Sound Parkway NW, Suite 300, Boca Raton, FL 33487-2742

and by CRC Press
2 Park Square, Milton Park, Abingdon, Oxon, OX14 4RN

CRC Press is an imprint of Taylor & Francis Group, LLC

Library of Congress Cataloging-in-Publication Data
Names: Inamuddin, 1980- editor. | Ahamed, Mohd Imran, editor. | Boddula, Rajender, editor. | Altalhi, Tariq, editor.
Title: Polymers in energy conversion and storage / edited by Inamuddin, Mohd Imran Ahamed, Rajender Boddula, Tariq A Altalhi.
Description: First edition. | Boca Raton : Taylor and Francis, 2022. | Includes bibliographical references and index.
Identifiers: LCCN 2021055547 | ISBN 9780367770815 (hardback) | ISBN 9780367770891 (paperback) | ISBN 9781003169727 (ebook)
Subjects: LCSH: Electric batteries--Materials. | Energy storage--Materials. | Capacitors--Materials. | Energy conversion. | Polymers--Electric properties. | Polyelectrolytes. | Conducting polymers.
Classification: LCC TK2945.P65 P66 2022 | DDC 621.31/242--dc23/eng/20220105
LC record available at https://lccn.loc.gov/2021055547

ISBN: 978-0-367-77081-5 (hbk)
ISBN: 978-0-367-77089-1 (pbk)
ISBN: 978-1-003-16972-7 (ebk)

DOI: 10.1201/9781003169727

Typeset in Times
by SPi Technologies India Pvt Ltd (Straive)

Contents

Preface

Research and development activities in energy conversion and storage play a significant role in our daily lives owing to the rising interest in clean energy technologies to alleviate the fossil-fuel crisis. A variety of materials, including polymers, inorganic materials, and carbons, have been studied in energy conversion and storage applications. Among them, polymers have been extensively utilized due to their low cost, softness, ductility, and flexibility compared to carbon and inorganic materials. The polymers can also be conveniently used as binders, separators, electrolytes, and so on for additional applications in energy fields compared to inorganic materials and carbons. This book explores the wide range of applicability of polymers in energy conversion and storage. The potential applications include hydrogen production, solar cells, photovoltaics, water splitting, fuel cells, supercapacitors, and batteries. This book provides in-depth literature on the applicability of polymers in energy conversion and storage, its history and progress, fabrication techniques, and potential applications. It brings together panels of highly accomplished experts working in the field of polymeric materials for energy conversion and storage and will be an invaluable guide to students, professors, scientists, and R&D industrial experts working in the field of energy conversion and storage.

Chapter 1 covers the history and progress of polymer materials used in several energy applications divided into energy storage and conversion. Polymers are used in supercapacitors, batteries, fuel cells, and solar cells. Conducting polymers exhibit semiconductor-like properties, and they have emerged as fascinating materials for the fabrication of electronic devices.

Chapter 2 highlights the recent advancements of various polymer-based electrolytes employed in supercapacitor applications, namely: solid polymer electrolytes; gel polymer electrolytes including hydrogels and organogels; ionogel-based polymer electrolytes, proton-conducting polymer electrolytes, and polyelectrolytes. In addition,, the observations have been analyzed to interpret their practicability in designing smart, flexible, and multifunctional supercapacitors.

Chapter 3 discusses recently developed polyaniline-based electrode materials used for supercapacitor applications. To overcome, the short cycle life of polyaniline and the low specific capacity of carbon materials, metal compounds have been included as the third component to obtain ternary nanocomposites for interesting applications in the development of new supercapacitors.

Chapter 4 discusses different covalent and non-covalent approaches utilized for synthesizing self-healing gel electrolytes. The autonomous self-restoration abilities, ionic conductivities, and elongation efficiencies of various gel electrolytes, and their application in integrated flexible supercapacitors, are summarized. Also, the abilities of redox mediator doped gel electrolytes for overall augmentation in capacitance are highlighted.

Chapter 5 provides a detailed discussion of the operating mechanism of different types of polymer-based nanogenerators. The focus of the study has been on energy-harvesting device designs with polymer materials and combinations of polymer nanogenerators employed in a wide range of applications. The advantages and importance of polymer nanogenerators are highlighted with their performance.

Chapter 6 presents the concepts of piezo/pyroelectric effects, their figures of merit (FOM), measurement tools, and the materials that exhibit these properties. The various types of piezo/pyroelectric polymers, co-polymers, fillers, and composites are discussed along with an emphasis on their applications in energy harvesting, its principles, methods, and FOM.

Chapter 7 reports on the recent advances in polymers and their composites for solar cells. Polymers are used to fabricate flexible substrates, or pore-forming for photoanodes and cathodes, as well as solidify the electrolytes. Polymers showed significantly enhanced long-term stability of solar cell devices which opens the way for their practical applications.

Chapter 8 overviews various aspects of the polymer-based systems used both as electrode materials as well as ion exchange membranes in different fuel cells. It gives a detailed account of polymers and their composites utilized to accomplish a high power and energy density while also overcoming the shortcomings of currently used materials.

Chapter 9 describes the various characteristics of a solid polymer electrolyte including ionic conductivity, glass transition temperature, degree of crystallinity, and crystal growth. Further, various recent works implementing solid polymer electrolytes in rechargeable batteries are highlighted by discussing the existing issues and their possible solutions.

Chapter 10 details polymer electrolytes based on ionic conductivity. Also, the role of plasticizers, that is, organic solvents and ionic liquids in polymer electrolytes, as well as types of batteries and their parameters, are discussed in detail. Finally, the major focus is on the electrochemical measurements of polymer electrolytes for efficient polymer batteries.

Chapter 11 describes the synthesis and properties (electronic, optical, mechanical, and physical) of polymer semiconductors. It also collects the main techniques employed to study these materials (physicochemical, electro, and optical characterization). The significant applications and their highlighted features are discussed in detail.

Chapter 12 discusses the fabrication of organic polymer-based photovoltaic devices for the aim of converting sunlight into electricity. The working principle of the device,

J-V characteristics, device architecture, the effect of various factors in the operating process, future possibilities in the enhancement of power conversion efficiency, and the scope for commercialization are discussed in detail.

Chapter 13 discusses the new field of polymer application in wearable devices. Their most important properties such as stretchability, electrical conductivity, and thermoelectric properties are discussed. Conductive polymers, carbon materials, piezoelectric elastomer polymers, chitin and chitosan, rare earth nanoparticles, and composites of these materials are focused on.

Chapter 14 details various polymers and their conjugate classes that allow the manufacturing of cost-effective, lightweight, and flexible devices with remarkable applications in organic photovoltaics, organic field-effect transistors, organic light-emitting diodes, and so on. It discusses the integration of various conjugated polymers in organic electronics.

Chapter 15 summarizes recent progress in polymer thermoelectrics. It highlights the strategies for improving the thermoelectric properties of polymers by engineering the Seebeck coefficient, the electrical conductivity, and the thermal conductivity of polymers and polymer-based composites.

Chapter 16 discusses various types of polymer-based hydrogen storage systems with a focus on their mechanism of action. The gas sorption measurement techniques are also briefly covered. The primary goal is to outline the synthesis methods, benefits, and drawbacks of several polymeric materials utilized in hydrogen storage.

Chapter 17 reports on how to use water splitting applications to generate hydrogen and oxygen gases. Since materials used in water-splitting applications are not very effective, this chapter focuses on recent advances and increasing the efficiency of water-splitting processes with polymers and their composites.

Editors

Inamuddin is Assistant Professor at the Department of Applied Chemistry, Aligarh Muslim University, Aligarh, India. He obtained his Master of Science degree in Organic Chemistry from Chaudhary Charan Singh University, Meerut, India, in 2002. He received his Master of Philosophy and Doctor of Philosophy degrees in Applied Chemistry from Aligarh Muslim University, India, in 2004 and 2007, respectively. He has extensive research experience in the multidisciplinary fields of Analytical Chemistry, Materials Chemistry, and Electrochemistry and, more specifically, Renewable Energy and the Environment. He has worked on different research projects as a project fellow and senior research fellow funded by the University Grants Commission, Government of India, and the Council of Scientific and Industrial Research, Government of India. He received the Fast Track Young Scientist Award from the Department of Science and Technology, India, to work in the area of bending actuators and artificial muscles. He completed four major research projects sanctioned by the University Grant Commission, the Department of Science and Technology, the Council of Scientific and Industrial Research, and the Council of Science and Technology, India. He has published 196 research articles in international journals of repute and 19 book chapters published by renowned international publishers. He has published 150 edited books. He also serves as: Associate Editor for the journals *Environmental Chemistry Letters*, *Applied Water Science*, and the *Euro-Mediterranean Journal for Environmental Integration*; Frontiers Section Editor for *Current Analytical Chemistry*; Editorial Board Member for *Scientific Reports—Nature*; Editor for the *Eurasian Journal of Analytical Chemistry*; and Review Editor for *Frontiers in Chemistry*. He has worked as a postdoctoral fellow, leading a research team at the Creative Research Initiative Center for Bio-Artificial Muscle, Hanyang University, South Korea, in the field of renewable energy, especially biofuel cells. He has also worked as a postdoctoral fellow at the Center of Research Excellence in Renewable Energy, King Fahd University of Petroleum and Minerals, Saudi Arabia, in the field of computational fluid dynamics of polymer electrolyte membrane fuel cells. He is a life member of the *Journal of the Indian Chemical Society*. His research interests include ion exchange materials, sensors for heavy metal ions, biofuel cells, supercapacitors, and bending actuators.

Mohd Imran Ahamed received his Ph.D. degree on the topic "Synthesis and Characterization of Inorganic-organic Composite Heavy Metal Selective Cation-exchangers and Their Analytical Applications" from Aligarh Muslim University, Aligarh, India in 2019. He has published several research and review articles in journals of international recognition. He completed his B.Sc. (Hons) in Chemistry from Aligarh Muslim University, Aligarh, India, and his M.Sc. (Organic Chemistry) from Dr. Bhimrao Ambedkar University, Agra, India. He has co-edited more than 20 books. His research work includes ion-exchange chromatography, wastewater treatment and analysis, bending actuators, and electrospinning.

Rajender Boddula is currently working with the Chinese Academy of Sciences—President's International Fellowship Initiative (CAS-PIFI) at the National Center for Nanoscience and Technology (NCNST) Beijing. He obtained his Master of Science in Organic Chemistry from Kakatiya University, Warangal, India, in 2008. He received his Doctor of Philosophy in Chemistry with the highest honors in 2014 for the work entitled "Synthesis and Characterization of Polyanilines for Supercapacitor and Catalytic Applications" at the CSIR-Indian Institute of Chemical Technology and Kakatiya University (India). Before joining NCNST as a CAS-PIFI research fellow, he worked as a senior research associate and postdoctoral fellow at the National Tsing-Hua University (Taiwan) in the fields of bio-fuel and CO_2 reduction applications, respectively. His academic honors include a University Grants Commission National Fellowship and many merit scholarships, as well as study-abroad fellowships including an Australian Endeavour Research Fellowship and a CAS-PIFI Fellowship. He has published many scientific articles in international peer-reviewed journals and has authored around 20 book chapters; he also serves as an editorial board member and a referee for international journals. He has also published edited books. His specialized areas of research are energy conversion and storage, including sustainable nanomaterials, graphene, polymer composites, heterogeneous catalysis for organic transformation, environmental remediation technologies, photoelectrochemical water-splitting devices, biofuel cells, batteries, and supercapacitors.

Tariq Altalhi, PhD, joined the Department of Chemistry at Taif University, Saudi Arabia as Assistant Professor in 2014. He received his doctorate degree from the University of Adelaide, Australia in the year 2014 with the Dean's Commendation for Doctoral Thesis Excellence. He was promoted to the position of Head of the Chemistry Department at Taif University in 2017 and as Vice Dean of the Science College in 2019. In 2015, one of his works was nominated for the Green Tech awards in Germany, Europe's largest environmental and business prize, and was amongst the top 10 entries. He has co-edited various scientific books. His group is involved in fundamental multidisciplinary research in nanomaterial synthesis and engineering, characterization, and its application in molecular separation, desalination, membrane systems, drug delivery, and biosensing. In addition, he has established key contacts with major industries in the Kingdom of Saudi Arabia.

Contributors

Sofía Abad-Sojos
Institute of Biology
ELTE Eötvös Loránd University
Budapest, Hungary

Adnan
Department of Chemical Engineering, Faculty of Engineering and Technology,
Aligarh Muslim University
Aligarh, India

Abdelaal S. A. Ahmed
Chemistry Department
Al-Azhar University
Assiut, Egypt

Mohammad Faraz Ahmer
Department of Electrical and Electronic Engineering
Mewat Engineering College
Nuh, Haryana, India

Gomaa A. M. Ali
Chemistry Department
Al–Azhar University
Assiut, Egypt

Sandeep Arya
Department of Physics
University of Jammu
Jammu and Kashmir, India

Swapan Kumar Bhattacharya
Department of Chemistry
Physical Chemistry Division
Jadavpur University
Kolkata, India

Rutuja S. Bhoje
Department of Chemical Engineering
Institute of Chemical Technology
Mumbai, India

Emilio Bucio
Institute of Nuclear Sciences
National Autonomous University of Mexico
Mexico City, Mexico

Moises Bustamante-Torres
Biomedical Engineering Department
School of Biological and Engineering
Yachay Tech University
Urcuqui City, Ecuador
and
Department of Radiation Chemistry and Radiochemistry
Institute of Nuclear Sciences
National Autonomous University of Mexico
Mexico City, Mexico

Bryan Chiguano-Tapia
School of Chemical and Engineering
Yachay Tech University
Urcuqui City, Ecuador

Kwok Feng Chong
Faculty of Industrial Sciences & Technology
Universiti Malaysia Pahang
Gambang, Malaysia

Mir Saeed Seyed Dorraji
Applied Chemistry Research Laboratory
Department of Chemistry
University of Zanjan
Zanjan, Iran

Dimple Dutta
Chemistry Division
Bhabha Atomic Research Centre
Mumbai, India

Jocelyne Estrella-Nuñez
School of Chemical and Engineering
Yachay Tech University
Urcuqui City, Ecuador

Yu Guo
Mechanical & Industrial Engineering Department
University of Massachusetts Amherst
Massachusetts, USA

Zahra Heidari
Shahrood University of Technology
Shahrood, Semnan
Iran

Seyyedeh Fatemeh Hosseini
Applied Chemistry Research Laboratory
Department of Chemistry
University of Zanjan
Zanjan, Iran

Sahidul Islam
Department of Chemistry
Trivenidevi Bhalotia College
Paschim Burdwan, West Bengal, India
and
Department of Chemistry
The University of Burdwan
Burdwan, West Bengal, India

Abdullah Jan
Laboratory of New Materials and Conversion Devices
Wuhan University of Technology
Wuhan, P. R. China

Mathiyarasu Jayaraman
Electrodics and Electrocatalysis Division
CSIR-Central Electrochemical Research Institute
Karaikudi, Tamil Nadu, India

Dragana Jovanović
Vinča Institute of Nuclear Sciences, University of Belgrade,
National Institute of the Republic of Serbia
Belgrade, Serbia

Svetlana Jovanović
Vinča Institute of Nuclear Sciences, University of Belgrade,
National Institute of the Republic of Serbia
Belgrade, Serbia

Pragati Kumar
Department of Nanosciences and Materials
Central University of Jammu
Jammu, J&K, India

Dipanwita Majumdar
Department of Chemistry
Chandernagore College
West Bengal, India

Mandira Majumder
Indian Institute of Technology (Indian School of Mines)
Dhanbad, India

Shiva Mohajer
Applied Chemistry Research Laboratory
Department of Chemistry
University of Zanjan
Zanjan, Iran

A. M. Vinu Mohan
Electrodics and Electrocatalysis Division
CSIR-Central Electrochemical Research Institute
Karaikudi, Tamil Nadu, India

Ujjwal Mandal
Department of Chemistry
The University of Burdwan
Burdwan, West Bengal, India

S. K. Nataraj
Centre for Nano and Material Sciences
Jain University
Jain Global Campus
Bangalore, India

Parag R. Nemade
Department of Chemical Engineering
Institute of Chemical Technology
Mumbai, India
and
Department of Oils, Oleochemicals and Surfactants Technology
Institute of Chemical Technology
Mumbai, India

Bhavya Padha
Department of Physics
University of Jammu
Jammu and Kashmir, India

Archana Patole
Khalifa University of Science and Technology
Abu Dhabi, United Arab Emirates

Shashikant P. Patole
Applied Quantum Materials Laboratory
Department of Physics
Khalifa University of Science and Technology
Abu Dhabi, United Arab Emirates

Zahra Pezeshki
Faculty of Electrical and Robotic Engineering
Shahrood University of Technology
Shahrood, Semnan, Iran

Prerna
Department of Physics
University of Jammu
Jammu and Kashmir, India

Vinoth Rajendran
Electrodics and Electrocatalysis Division
CSIR-Central Electrochemical Research Institute
Karaikudi, Tamil Nadu, India

Mashallah Rezakazemi
Faculty of Chemical and Materials Engineering
Shahrood University of Technology
Shahrood, Semnan, Iran

Anita Samage
Centre for Nano and Material Sciences
Jain University
Jain Global Campus
Bangalore, India

Rajendra Kumar Singh
Ionic Liquid and Solid State Ionics Laboratory
Department of Physics
Institute of Science
Banaras Hindu University
Varanasi, India

Shishir Kumar Singh
Ionic Liquid and Solid State Ionics Laboratory
Department of Physics, Institute of Science
Banaras Hindu University
Varanasi, India

Anukul K. Thakur
Department of Advanced Components and Materials Engineering
Sunchon National University
Chonnam, Republic of Korea

Mohammad R. Thalji
Independent researcher
Amman, Jordan

Odalys Torres
Institute of Biology
ELTE Eötvös Loránd University
Budapest, Hungary

Lakshmi Unnikrishnan
School for Advanced Research in Petrochemicals (SARP)—APDDRL
Central Institute of Petrochemicals Engineering & Technology
Bengaluru, Karnataka, India

Sonali Verma
Department of Physics
University of Jammu
Jammu and Kashmir, India

Yanfei Xu
Mechanical & Industrial Engineering Department
University of Massachusetts Amherst
Massachusetts, USA
and
Chemical Engineering Department
University of Massachusetts Amherst
Massachusetts, USA

Mohammad Younas
Department of Chemical Engineering
University of Engineering and Technology
Peshawar, Pakistan

1 History and Progress of Polymers for Energy Applications

Mohammad R. Thalji
Independent Researcher, Amman, Jordan

Gomaa A. M. Ali
Al-Azhar University, Assiut, Egypt

Kwok Feng Chong
Universiti Malaysia Pahang, Gambang, Malaysia

CONTENTS

1.1 INTRODUCTION: HISTORICAL PERSPECTIVE OF POLYMERS IN THE ENERGY FIELD

Polymer materials have been considered the golden gateway to the future, which deals with developing many novel materials to suit our daily lives. Polymer materials are essential in energy concepts with a wide range of applications, including semiconductors [1, 2], light-emitting diodes (LEDs) [3], flexible supercapacitors [4], flexible solar cells [5], and fuel cells [6]. Berzelius (in 1930) coined the term “polymer” to define molecules that contained the repeatedly arranged atomic groups [7]. Over time, the term came to refer to larger molecules when it was applied to long macromolecules composed of various entities (monomers) [8]. The period 1920–1940 was regarded as a golden age in the development of synthetic polymers, during which new monomers were synthesized from abundant raw materials [9]. Simultaneously, the polymerization and polycondensation processes were refined to improve their efficiency. New synthetic methods have been developed to enhance the characterization of polymer macromolecules. This enabled the creation of polymers with distinct physico-chemical characteristics by modifying the arrangement of their chains.

Polymer material has long been considered to be an insulating material. Typically, it was employed in insulating cables in electronic systems, and it was uncommon to use it as an electrode material for conducting [10]. As a result, since the discovery of polyacetylene in 1977, the growth of conducting polymers has quickly piqued the interest of both academia and industry [11]. Conducting polymers are promising energy materials due to their electrical conductivity and reversible

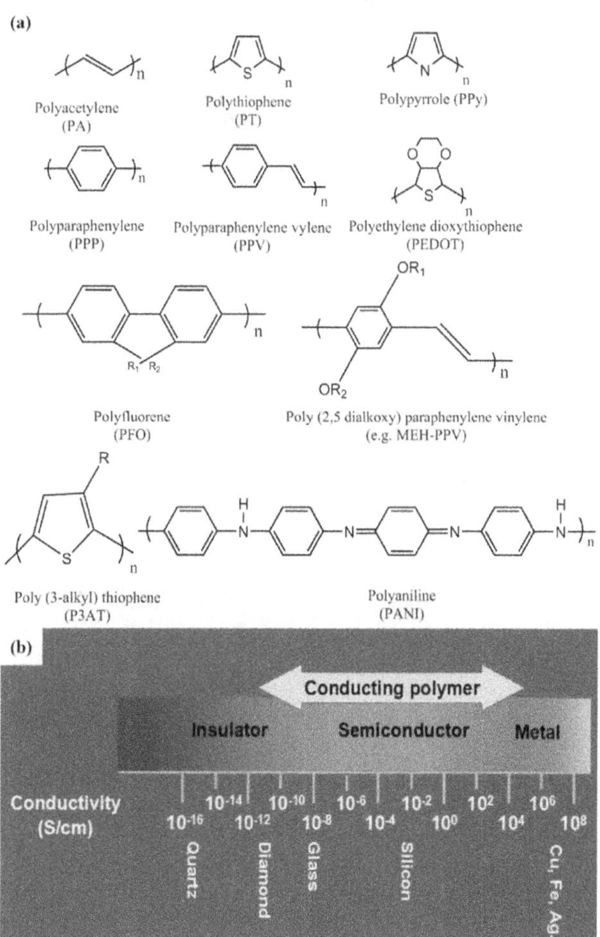

FIGURE 1.1 (a) Chemical formulas of some conducting polymers. (b) Electrical conductivity range of conducting polymers [12].

DOI: 10.1201/9781003169727-1

electrochemical performance [12]. Polyaniline (PANI), polythiophene (PT), polypyrrole (PPY), poly(ethylene dioxythiophene) (PEDOT), and poly(3-hexylthiophene) (P3HT) are some examples of the conjugated double bonds conducting polymers (Figure 1.1a). They have sp^2 in their chemical structure, facilitating charge transport and improving their electrochemical and conductivity properties (Figure 1.1b).

These polymers exhibit high potentials as electrodes for various energy devices. Although polymer materials have broad applications, this chapter will focus on their energy (conversion and storage).

1.2 POLYMER MATERIALS FOR ENERGY STORAGE APPLICATIONS

The development of a novel electrode material with a large electrochemically active surface area [13], excellent porosity [14], high conductivity [15], and pseudocapacitive properties [16] is the primary key to improving the electrochemical performance of energy storage systems. With the development of supercapacitors, some new materials have appeared gradually. Research interest has expanded beyond numerous conducting polymer materials in recent years, focusing on developing distinct electrode materials for electrochemical capacitors and batteries.

Although batteries are the most common energy source in each field, they have some limitations in the life cycle and power performance [17, 18]. Supercapacitors (SCs), known as electrochemical capacitors and ultracapacitors, are classified as energy storage devices with a high power density and long charge–discharge cycles [15, 19–21]. SCs, as shown in Figure 1.2, have a distinct Ragone plot position that bridges the gap between batteries and capacitors [22]. Also, when compared to a conventional capacitor, SCs have a higher specific energy density. Furthermore, SCs have a higher specific power density than batteries because of their unique charge storage mechanism.

Electrical double-layer capacitors (EDLCs) and faradaic pseudocapacitors (PCs) are the two main types of SCs [23]. EDLCs store electrical charges at the electrode–electrolyte interface using electrostatic force rather than electrochemical reactions on the electrode surface (Figure 1.3a). Carbon-based materials like carbon nanotubes (CNTs), carbon fibers, carbide-derived carbons, activated carbon, and graphene are commonly used as electrode materials for EDLCs [13–15, 24, 25]. Despite having a high power density and fantastic charge–discharge cycling stability, carbon-based materials have a low energy density [26]. Metal oxides, metal sulfides, and conducting polymers have been investigated as the electrode for pseudocapacitors to improve the specific capacitance and energy density of PCs [4, 20, 21, 27–32]. In PCs, energy storage is derived from reversible redox reactions at the electrolyte and electroactive interface. Because metal oxides have multiple oxidation states for redox charge transfer reactions, PCs can typically yield higher energy density than EDLCs.

Conducting polymers belong to pseudocapacitive electrode material that has been widely investigated for various energy storage applications like supercapacitors and batteries. This is associated with high energy density, redox-storage capability, relatively high conductivity, and large voltage windows [29, 33, 34]. PANI is an intrinsic polymer commonly used as electrode material in electrochemical energy storage applications. It has many distinguishing characteristics, including high conductivity and excellent electrochemical performance. Despite these benefits, it is prone to rapid performance degradation due to repeated charge–discharge cycles. To address this issue, carbon-based materials were

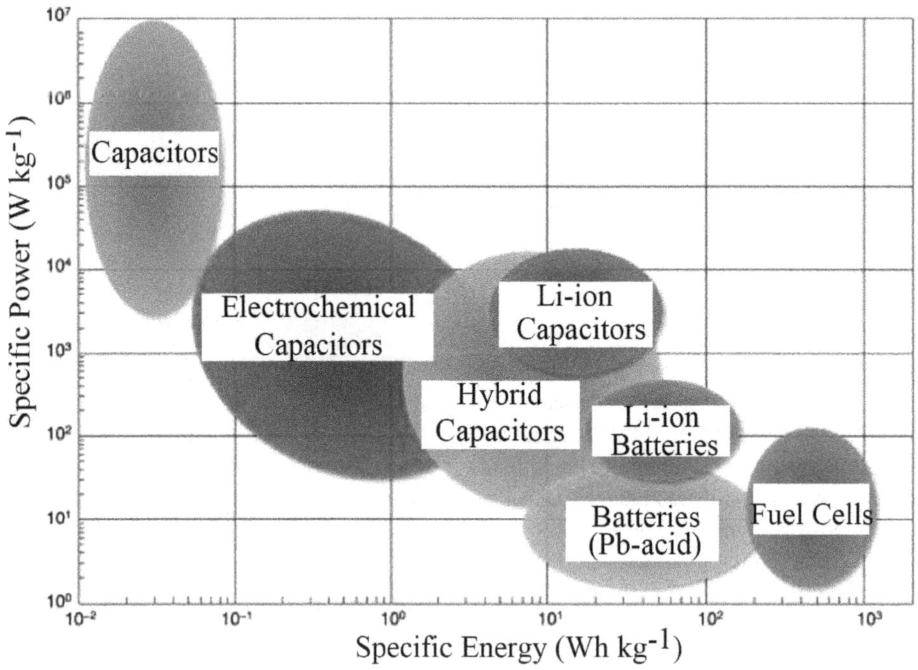

FIGURE 1.2 Ragone plot for energy storage systems [22].

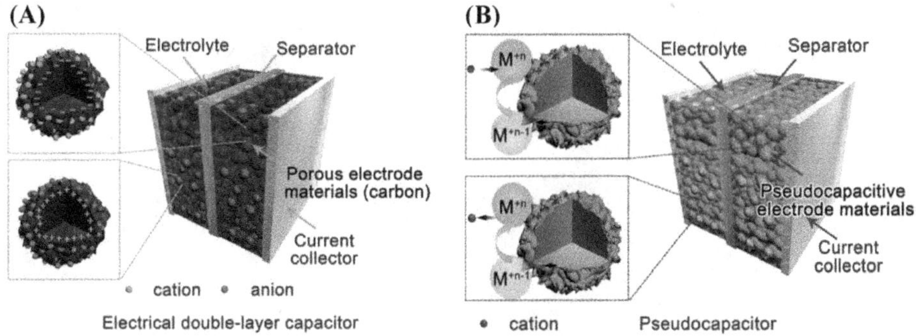

FIGURE 1.3 Representative diagrams for (a) EDLCs and (b) PCs [23].

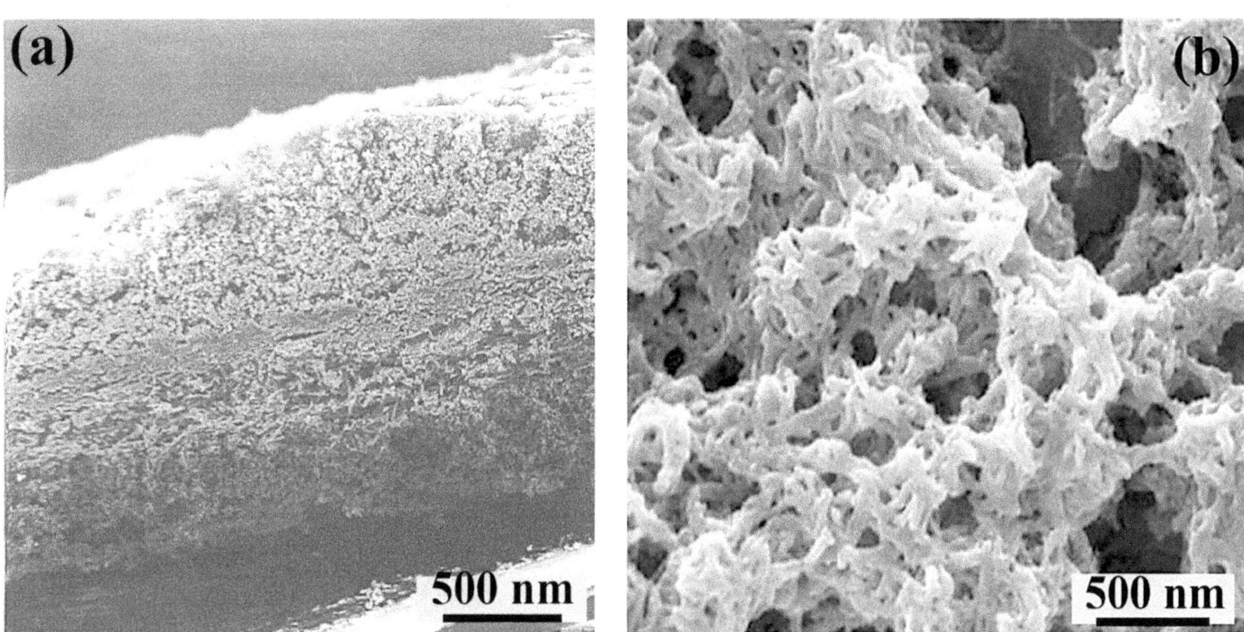

FIGURE 1.4 Scanning electron microscopy image of (a) a PANI-coated carbon paper electrode in cross-section, and (b) a PANI film fabricated by the pulse galvanostatic method (PGM) in 0.5 M H_2SO_4 and 0.2 M aniline [35].

combined with PANI material to form a novel composite that contributed to the reinforced PANI stability and enhanced the specific capacitance. For instance, Fusalba et al. synthesized a highly porous PANI film 250 mm thick (Figure 1.4a), and used it as the active material in a supercapacitor device with a nonaqueous electrolyte (1 M Et_4NBF_4/ACN) [35]. This increased the potential window in an aqueous solution from 0.75 to 1.0 V.

A low-frequency capacitance of 3 F cm^{-2} (150 F g^{-1}) was achieved in the as-prepared film. Furthermore, the PANI–PANI capacitor's cycling stability was low for the first 60 cycles (the loss of charge accounts for approximately 25% of the initial charge). In another study, Zhou et al. obtained a capacitance value of 609 F g^{-1} and an energy density of 26.8 W h kg^{-1} for a nanofibrous PANI capacitor at 1.5 mA cm^{-2} (Figure 1.4b) [36]. According to Zhou et al., the outstanding capacitance is due to a highly porous nanofibrous architecture that provides a high surface area and a great charge-transfer rate, allowing it to be a promising electrode material supercapacitor. Liu and his co-workers [37] used *in situ* aqueous polymerization to create porous PANI, and they compared its performance to that of as-prepared nonporous PANI. As shown in Figure 1.5a and b, the porous PANI had a more random pore arrangement than the nonporous PANI.

Furthermore, under 10 mA g^{-1}, the porous PANI had a specific capacitance of 837 F g^{-1}, much higher than the nonporous ones (519 F g^{-1}) (Figure 1.5c). The porous electrode exhibits high long-cycle stability and high-rate capability (Figure 1.5d). The porous PANI's notable electrochemical performance is assigned to its meso/macropores with a remarkable surface area (211 m^2 g^{-1}), compared to nonporous ones (6.0 m^2 g^{-1}), which facilitates ion transport to entire surfaces.

A flexible supercapacitor was made from polypyrrole as electrode material by creating hollow polypyrrole/cellulose hybrid hydrogels using a simple electrochemical deposition strategy on an Ni current collector (Figure 1.6a) [38]. The symmetric supercapacitor device assembly (Figure 1.6b) has a high capacitance, energy density, and capacitance retention of 255 F g^{-1}, 20.4 W h kg^{-1}, and 80%, respectively.

FIGURE 1.5 Scanning electron microscopy images of the (a) nonporous PANI and (b) porous PANI. (c) Specific capacitances and (d) cycle stabilities of the porous and the nonporous PANI (Liu et al. [37] with permission).

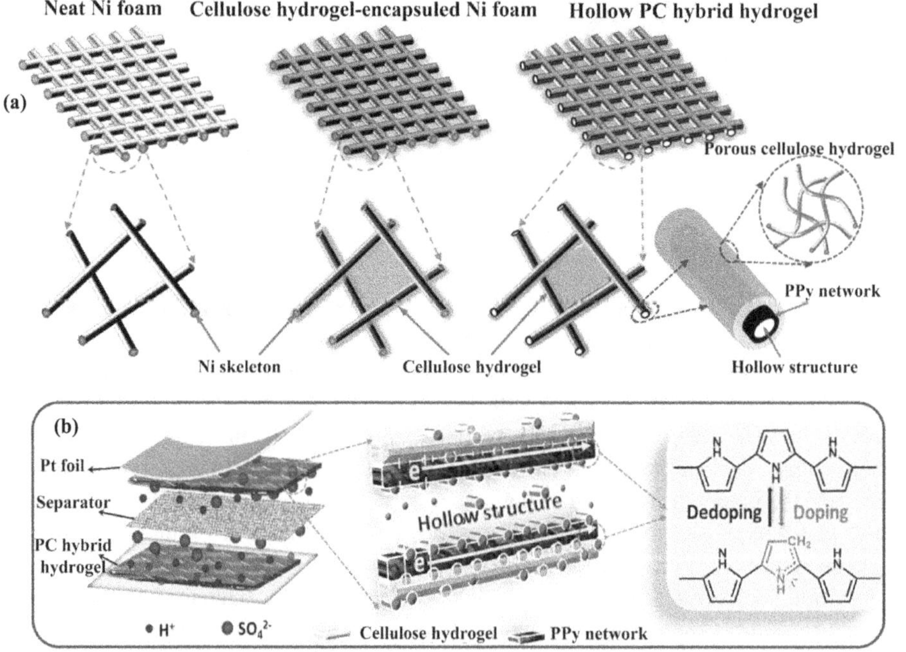

FIGURE 1.6 (a) The synthetic path and (b) the proposed electrochemical mechanism for the hollow polypyrrole/cellulose hybrid hydrogel (Zhang et al. [38] with permission).

This improvement can be attributed to cellulose hydrogel, which prevents PPy volume shrinkage and expansion during the electrochemical processes. Furthermore, its biphase porous and hollow structure can improve electrolyte ion accessibility and efficient interfacial electrochemical reaction kinetics, resulting in enhanced charge storage. As shown in Figure 1.6b, the inner phase of the polypyrrole network provides a fast electron transfer throughout the hybrid hydrogel, while the outer phase of the cellulose hydrogel is accountable for efficient ion transport and superior mechanical performance.

The insertion of carbon-based materials within conducting polymers plays a critical part in improving overall electrochemical performance. Special electrode capacity and cycling stability can be enhanced synergistically by integrating all materials for effective energy storage applications. Feng et al., for instance, reported that graphite oxide had been modified to produce composites of polypyrrole/graphite oxide using *in situ* polymerization [30]. The specific capacitance of the as-fabricated electrode was 202 F g^{-1} (a three-electrode system) and 87 F g^{-1} (a two-electrode system) at 1 A g^{-1}. The retention rate of capacity was 83.8% after 1000 cycles.

Polyindole/carbon black (PIn/CB) composite has been used as a supercapacitor electrode. Incorporating CB into the PIn improves its capacitive performance while increasing energy and power densities. Electrodes with PIn, CB, and polyvinylidene difluoride ratios of 80 wt.%, 10 wt.%, and 10 wt.%, respectively were prepared [39]. The PIn has a continuously interlinked globular morphology, as seen in the field emission scanning electron microscopy (FESEM) image (Figure 1.7a). Furthermore, as shown in Figure 1.7b, the PIn/CB composite has similar globular structures with tiny particles that are identified as CB particles. At 1.0 A g^{-1}, the authors reported a Coulombic efficiency of 89.1 and 91.1% of polyindole and polyindole/carbon black, respectively. Compared with pure polyindole (112 F g^{-1}), the composite showed an improved specific capacitance of 193 F g^{-1} at 1 A g^{-1}. Carbon's amorphous and conducting nature facilitates charge transport within the electrode material. The designable functional groups able to interact chemically with the polyindole chains facilitate the synergy of the polyindole/carbon black composite.

In another work, graphene oxide (GO) was premixed with Poly(3,4-ethylenedioxythiophene):poly(styrenesulfonate) (PEDOT:PSS) before being bar-coated on a flexible polyvinylidene fluoride current collector to create a highly flexible rGO-PEDOT/PSS electrode (Figure 1.8a) that achieved an areal capacitance of 448 mF cm^{-2} at 10 mV s^{-1} [41]. However, the GO material was agglomerated by this strategy, which was detrimental to the composite's performance (Figure 1.8b). To address this issue, a PPy/graphene composite was developed to prevent graphene sheet agglomeration (Figures 1.8c and d), in which the PPy works as gaps between graphene sheets and improves the electrolyte/electrode interface for ion transport and charge storage performance [40]. Due to the two components' synergistic effect, the resulting supercapacitor exhibited a high mechanical flexibility film (Figure 1.8e).

The carbon-based supercapacitors and conducting polymers have shown improvements in electrochemical performance, as well as flexibility. However, their performance for practical applications remains far from what is expected. Transitional metal oxides were recently admitted as an effective strategy based on their high capacities, and their electrical conductivity is relatively high. For instance, a remarkable pseudocapacitive performance of 2055 F g^{-1} at 1 A g^{-1} with outstanding cycling stability (90% retention after 5000 long-term cycles) was obtained by a supercapacitor of the $NiCo_2O_4$/PPy nanostructure that was grown on a hemp-derived carbon fiber (HDC) [42]. Polypyrrole was employed to improve the conductivity by hastening the electron transfer of $NiCo_2O_4$. As shown in Figure 1.9c, the symmetric (wire-type) supercapacitor was built utilizing NaOH/PVA as a solid electrolyte and HDC@$NiCo_2O_4$@PPy fiber composites as the current electrode. The symmetric supercapacitor had a power density of 0.5 kW kg^{-1} and a considerable energy density of 17.5 W h kg^{-1}.

Polyaniline was also employed in the presence of $CuCo_2S_4$ nanosheets on carbon cloth, as shown in Figure 1.10a, to improve the electrical performance of $CuCo_2S_4$ nanosheets via chemical polymerization of aniline [43]. At 1 A g^{-1}, the pure $CuCo_2S_4$ electrode achieved a capacitance value of 780 F g^{-1}. The electrode significantly improved after compositing ($CuCo_2S_4$/polyaniline) to the capacitance value of 920 F g^{-1}. Furthermore, the capacitance retention

FIGURE 1.7 FESEM images of (a) polyindole and (b) polyindole/carbon black composite (Majumder et al. [39] with permission).

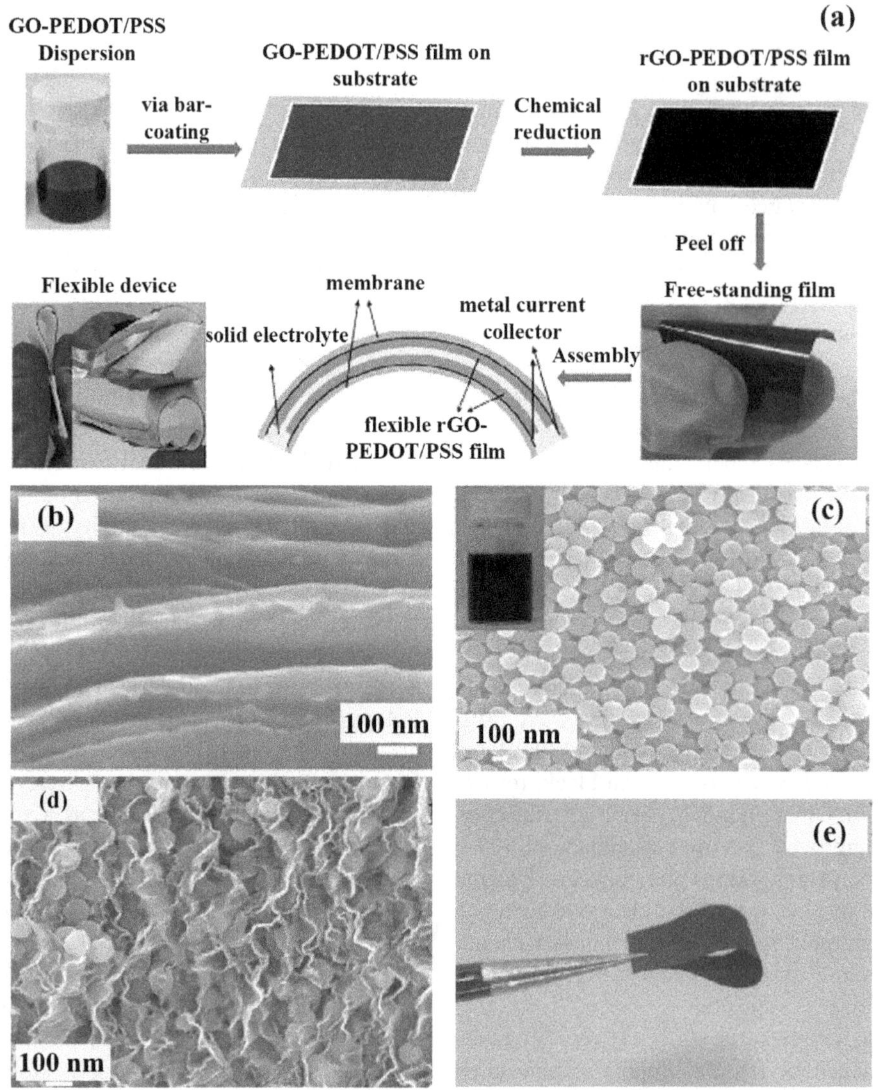

FIGURE 1.8 (a) A graphical illustration of rGO-PEDOT/PSS film preparation and assembled supercapacitor device structure. (b) rGO-PEDOT/PSS film scanning electron microscopy (SEM) cross-section image. SEM images for (c) PPy NPs (the inset is a PPy NPs aqueous dispersion image at 0.5 mg mL^{-1}), and (d) PPy/graphene film. (e) Image of a flexible PPy/graphene film (Ge et al. [40] with permission.

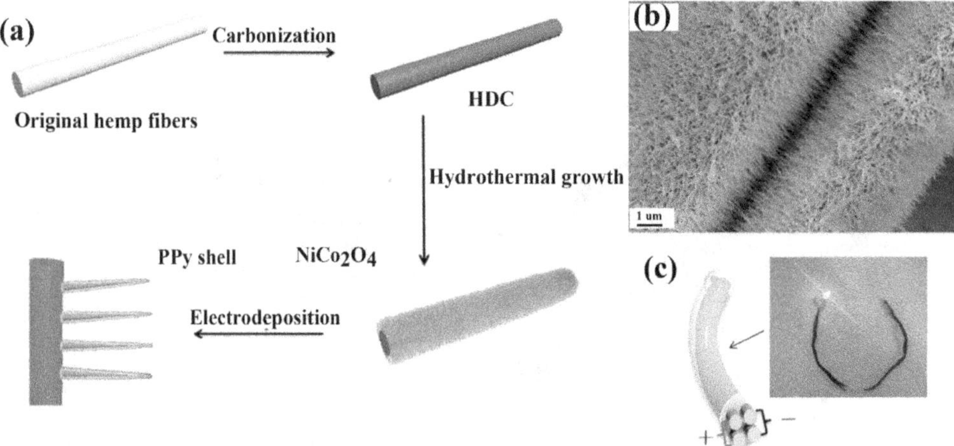

FIGURE 1.9 (a) Schematic representation of the preparation process of HDC@$NiCo_2O_4$@polypyrrole. (b) SEM image of the HDC@$NiCo_2O_4$ composite. (c) Schematic design of symmetric supercapacitor (the inset is a red LED powered by two supercapacitors connected in series).

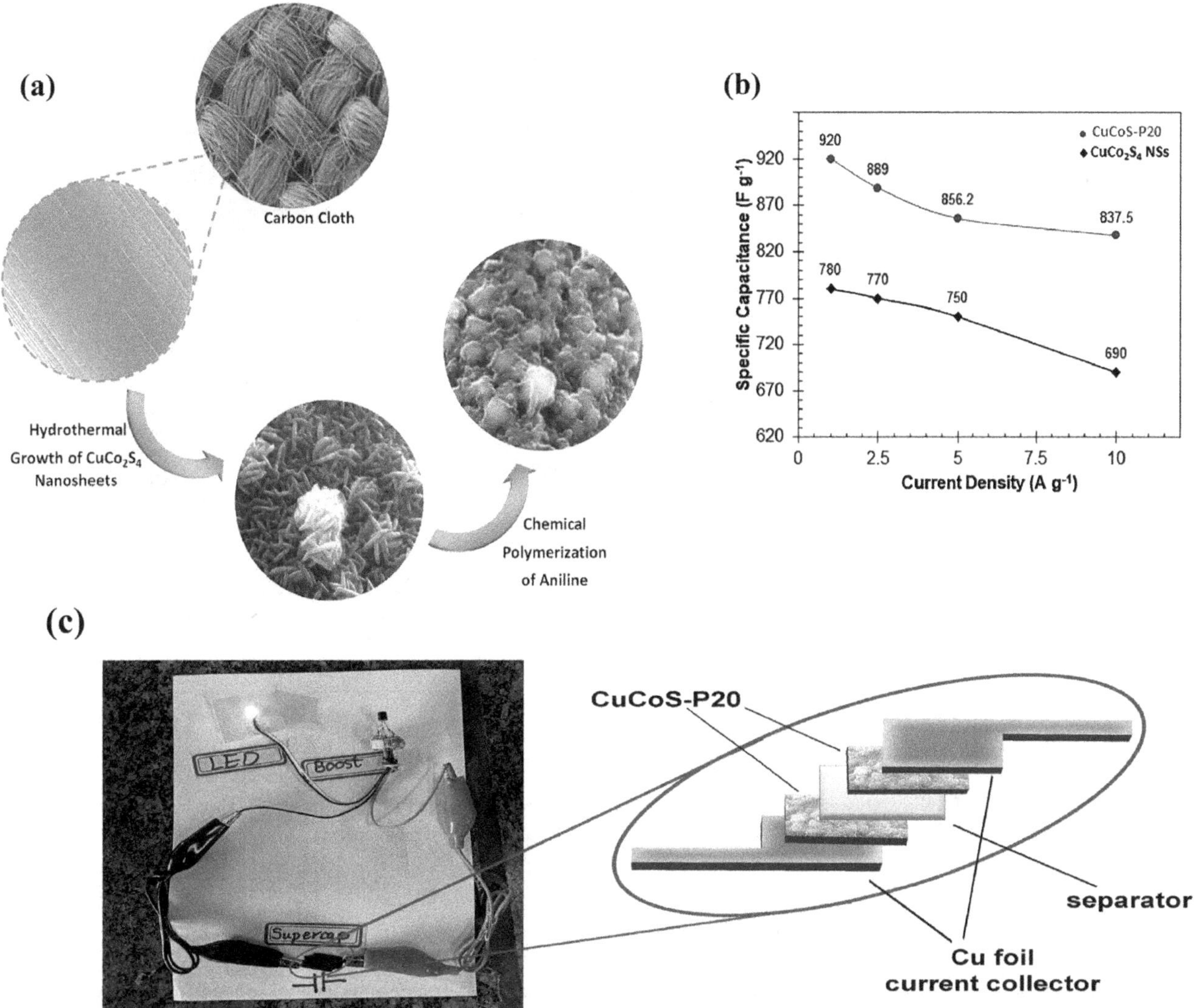

FIGURE 1.10 (a) Synthesis scheme of $CuCo_2S_4$ on carbon cloth and subsequent aniline polymerization. (b) Specific capacitances for pure and composite electrodes. (c) A fabricated device and schematic illustration of a symmetric supercapacitor assembled with $CuCo_2S_4$/polyaniline. Abuali et al. [43] with permission.

of the pure $CuCo_2S_4$ electrode and the $CuCo_2S_4$/polyaniline electrode was 88.46 and 91.03%, respectively, from 1 to 10 A g^{-1} (Figure 1.10b). This indicates that after composing with polyaniline, the rate capability of the $CuCo_2S_4$ electrode was improved, and the prepared nanocomposite performed better during rapid charge transfer. Such notable electrochemical performance was attributed to the enhancement of electrical conductivity rendered by the excellent coupling between $CuCo_2S_4$ and polyaniline.

In a 1 M Na_2SO_4 electrolyte, the $CuCo_2S_4$/polyaniline composite was used to assemble a symmetric supercapacitor. Figure 1.10c shows that the completely charged (1.8 V) device has been connected with the DC converter and can light up a 3.2 V white LED for a period of 12 min.

1.3 POLYMER MATERIALS FOR ENERGY CONVERSION APPLICATIONS

High efficient energy conversion systems such as fuel cells and solar cells can convert chemical to electrical energy. During the last decade, they have been actively developed. Considerable research has been undertaken on fuel cells such as the direct fuel cell of methanol, the polymeric electrolyte membrane fuel cell, the proton membrane exchange cell, and the alkaline cell. Due to the excellent conductivity and high flexibility of conducting polymers, which typically display high catalytic activity and good stability, they promise use as electrocatalysts of fuel cells. Several studies have shown that PANI-based electrocatalysts provide excellent catalytic activity in hydrogen evolution (HER), hydrogen oxidation (HOR), oxygen reduction (ORR), and methanol oxidation reaction (MOR). This section summarizes the recent research on metal electrocatalysts supported by PANI and metal-free electrocatalysts derived by PANI.

In 2011, Yuan et al. created an effective ORR electrocatalyst from a PANI/carbon black composite assisted iron phthalocyanine (PANI/FePc/C composite) for ORR in an air-cathode microbial fuel cell [44]. The resulting catalyst outperformed on bare Pt, PANI/C, and FePc/C in terms of catalytic activity, indicating that the activity of C in ORR is

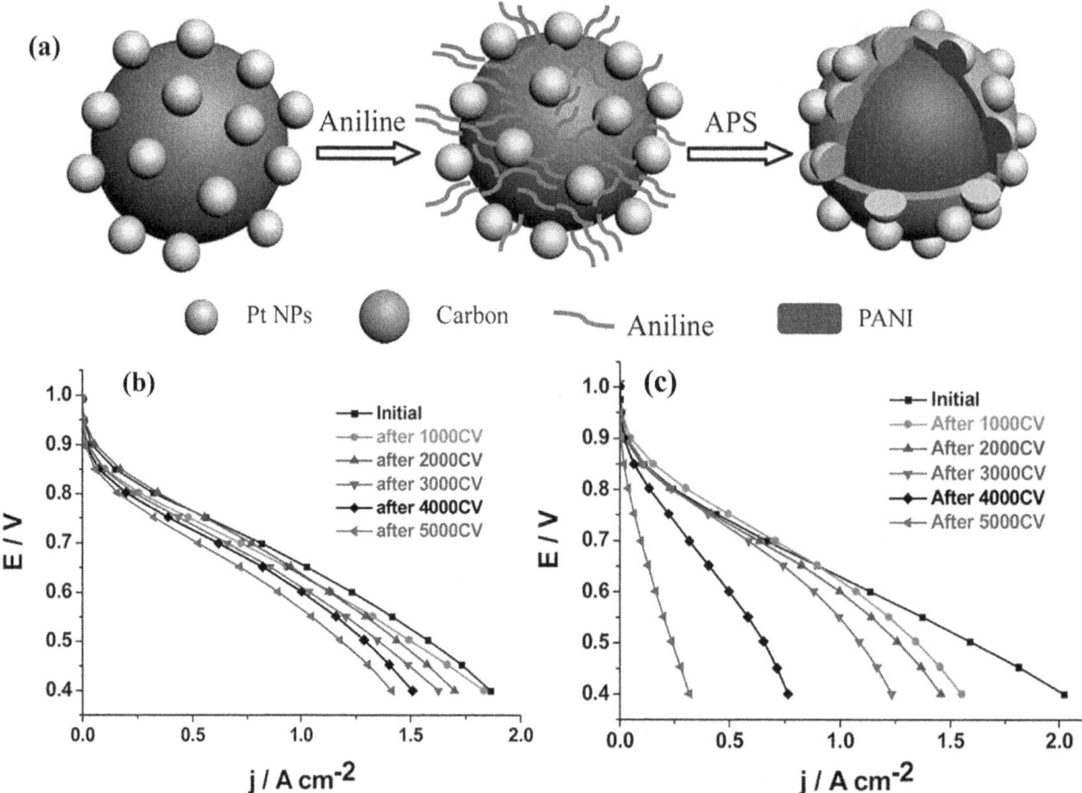

FIGURE 1.11 (a) Pt/C@PANI catalyst preparation process. Polarizing curves of single PEM cell fuel cells with (b) PANI@Pt/C (30%) and (c) Pt/C catalysts, following the indicated CV cycle numbers (Chen et al. [45] with permission).

improved by the PANI addition. Furthermore, a PANI/C/FePc catalyst had a power per cost that was much higher than the bare Pt catalyst. Chen et al. developed a PANI-decorated Pt/C@PANI core–shell catalyst in 2012, as shown in Figure 1.11a, with a 5 nm thickness PANI layer covering the Pt/C catalyst surface [45]. This core–shell structure caused electron delocalization between Pt and PANI-conjugated ligand d-orbitals, facilitating the electron transfer to PANI. As a result, it outperformed nondecorated Pt/C catalysts in terms of catalytic activity and durability (Figure 1.11b and c).

Zhang et al. prepared an MoS_2/PANI 3D catalyst, which achieved catalytic functionality with HER [46]. The flexible PANI inhibits MoS_2 aggregation, confirming that PANI branches with MoS_2 nanosheets are uniformly and vertically dispersed. In addition to other HER electrocatalysts based on polymer-MoS_2, the as-prepared MoS_2/PANI has achieved excellent stability.

Osmieri and co-workers used pyrrole to synthesize N-doped carbon-based material for ORR using a pyrolysis method with iron and cobalt salts [47]. In alkaline conditions, an excellent ORR activity, excellent surface stability, and well-developed current-restraint diffusion was achieved. The use of a 3D macroporous graphene/PANI anode in microbial fuel cells has been investigated [48]. SEM was used to verify the active deposition of PANI on the graphene surface by *in situ* polymerization (Figure 1.12a). The output of the as-prepared anode was compared to that of commercially available carbon cloth.

Bacteria attached to the graphene surface went deep within the 3D graphene/PANI (Figure 1.12b). The overall power density achieved from this microbial fuel cell (3D graphene/PANI) was 768 mW m^{-2} (four times higher than obtained from the carbon cloth microbial fuel cell (158 mW m^{-2})) [18]. Therefore, the 3D graphene/PANI electrode is a good candidate for high-performance and efficient microbial fuel cells. Yousfi et al. suggested a physical model for the analysis in a GO/poly (3-hexylthiophene) (P3HT) nanocomposite transistor thin film of portable concentration and mobility parameters as shown in Figure 1.13 [49]. This model has produced excellent results and has explained that the P3HT increase in the nanocomposite changed the interspace of the GO by the oxygen groups, which altered the energy of its bandgap.

By combining CdS/CdSe quantum dots with a P3HT/PCBM bulk heterojunction, Grissom et al. created flexible 3D CNT-based dye-sensitized solar cells [50]. The cells had good morphological properties and a high photon conversion rate of 7.6%. The cells are well-suited for extended multi-cell systems as they are performed in series and parallel configurations. As shown by the study findings, these novel three-dimensional optoelectronic-electrodes, which are low cost and innovative, have stable photovoltaic (PV) performance when enhanced by specific hybridization with

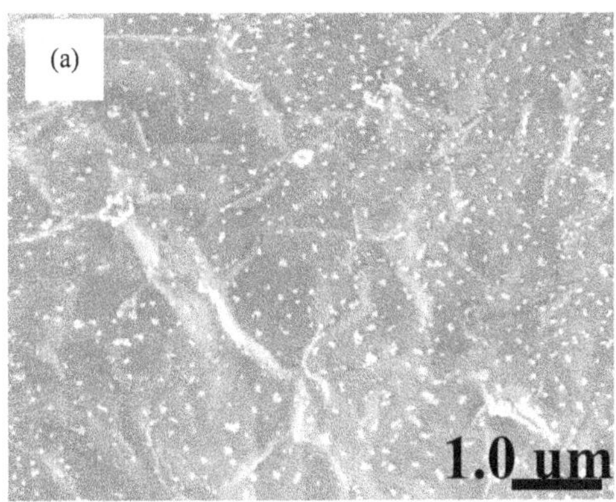

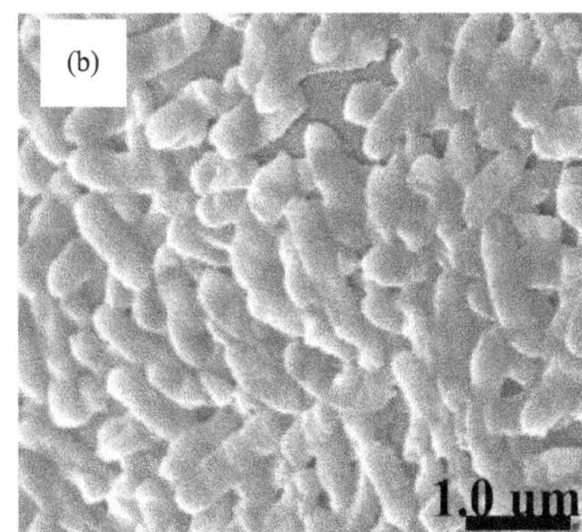

FIGURE 1.12 SEM images of (a) PANI/graphene composite and (b) bacteria cells on graphene/PANI surface. Yong et al. [48] with permission.

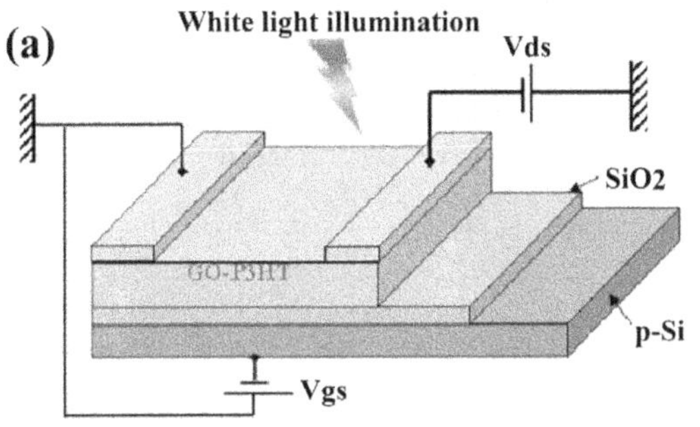

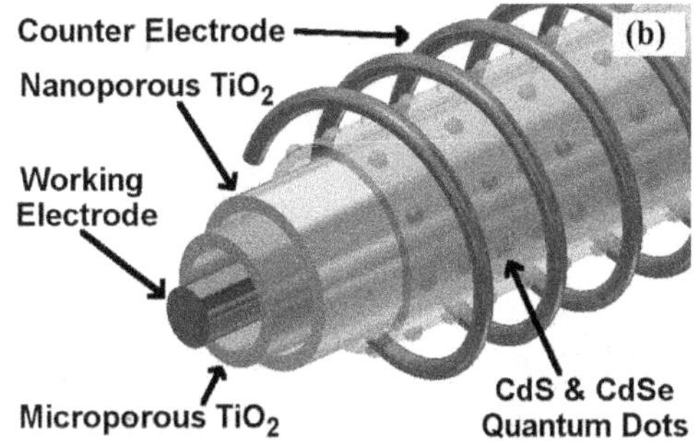

FIGURE 1.13 (a) GO/P3HT nanocomposite transistor scheme. (b) The structure of the photovoltaic cell. Yousfi et al. [49] with permission.

P3HT-PCBM. The hybrid interface formed by the inorganic CdS–CdSe QD layer and the organic P3HT-PCBM layer prevents the interfacial recombination of the photoinjected electrons from TiO_2 into the electrolyte from occurring.

1.4 CONCLUSION

Polymer materials have been used in various energy applications, including supercapacitors, batteries, solar cells, fuel cells, and electrochromic devices. Conducting polymers play an important role in electronic device fabrication. Carbon-based materials inserted into conducting polymers play a critical role in improving overall electrochemical efficiency. The electrodes' basic capacitance and cycling stability are enhanced synergistically for efficient energy storage applications by integrating both materials.

REFERENCES

1. Shi, L., Guo, Y., Hu, W., and Liu, Y. 2017. Design and effective synthesis methods for high-performance polymer semiconductors in organic field-effect transistors. *Materials Chemistry Frontiers* 1(12): 2423–2456.
2. Meng, Q., Cai, K., Chen, Y., and Chen, L. 2017. Research progress on conducting polymer based supercapacitor electrode materials. *Nano Energy* 36: 268–285.
3. Low, J.Y., Aljunid Merican, Z.M., and Hamza, M.F. 2019. Polymer light emitting diodes (PLEDs): An update review on current innovation and performance of material properties. *Materials Today: Proceedings* 16: 1909–1918.
4. Han, Y., and Dai, L. 2019. Conducting polymers for flexible supercapacitors. *Macromolecular Chemistry and Physics* 220(3): 1–14.
5. Hou, W., Xiao, Y., Han, G., and Lin, J.Y. 2019. The applications of polymers in solar cells: A review. *Polymers* 11(1): 1–46.
6. Chandran, P., Ghosh, A., and Ramaprabhu, S. 2018. High-performance platinum-free oxygen reduction reaction and hydrogen oxidation reaction catalyst in polymer electrolyte membrane fuel cell. *Scientific Reports* 8(1): 1–11.
7. Wisniak, J. 2000. Jöns Jacob Berzelius a guide to the perplexed chemist. *The Chemical Educator* 5(6): 343–350.
8. Feldman, D. 2008. Polymer history. *Designed Monomers and Polymers* 11(1): 1–15.
9. Frey, H., and Johann, T. 2019. Celebrating 100 years of "polymer science": Hermann Staudinger's 1920 manifesto. *Polymer Chemistry* 11(1): 8–14.
10. Ismail, N.H., and Mustapha, M. 2018. A review of thermoplastic elastomeric nanocomposites for high voltage insulation applications. *Polymer Engineering and Science* 58: E36–E63.
11. Shirakawa, H., Louis, E.J., MacDiarmid, A.G., Chiang, C.K., and Heeger, A.J. 1977. Synthesis of electrically conducting organic polymers: Halogen derivatives of polyacetylene, (CH)x. *Journal of the Chemical Society, Chemical Communications* (16): 578–580.
12. Tajik, S., Beitollahi, H., Nejad, F.G., Shoaie, I.S., Khalilzadeh, M.A., Asl, M.S., Van Le, Q., Zhang, K., Jang, H.W., and Shokouhimehr, M. 2020. Recent developments in conducting polymers: Applications for electrochemistry. *RSC Advances* 10(62): 37834–37856

13. Ali, G.A.M., Megiel, E., Cieciórski, P., Thalji, M.R., Romański, J., Algarni, H., and Chong, K.F. 2020. Ferrocene functionalized multi-walled carbon nanotubes as supercapacitor electrodes. *Journal of Molecular Liquids* 318: 114064.
14. Ali, G.A.M., Habeeb, O.A., Algarni, H., and Chong, K.F. 2019. CaO impregnated highly porous honeycomb activated carbon from agriculture waste: symmetrical supercapacitor study. *Journal of Materials Science* 54(1): 683–692.
15. Ali, G.A.M., Yusoff, M.M., Algarni, H., and Chong, K.F. 2018. One-step electrosynthesis of MnO_2/rGO nanocomposite and its enhanced electrochemical performance. *Ceramics International* 44(7): 7799–7807.
16. Thalji, M.R., Ali, G.A.M., Algarni, H., and Chong, K.F. 2019. Al^{3+} ion intercalation pseudocapacitance study of $W_{18}O_{49}$ nanostructure. *Journal of Power Sources* 438: 227028.
17. Han, X., Lu, L., Zheng, Y., Feng, X., Li, Z., Li, J., and Ouyang, M. 2019. A review on the key issues of the lithium ion battery degradation among the whole life cycle. *eTransportation* 1: 100005.
18. Ali, G.A.M. 2020. Recycled MnO_2 nanoflowers and graphene nanosheets for low-cost and high performance asymmetric supercapacitor. *Journal of Electronic Materials* 49(9): 5411–5421.
19. Thalji, M.R., Ali, G.A.M., Poh, S., and Chong, K.F. 2019. Solvothermal synthesis of reduced graphene oxide as electrode material for supercapacitor application. *Chemistry of Advanced Materials* 4(3): 17–26.
20. Ali, G.A.M., Fouad, O.A., Makhlouf, S.A., Yusoff, M.M., and Chong, K.F. 2014. Co_3O_4/SiO_2 nanocomposites for supercapacitor application. *Journal of Solid State Electrochemistry* 18(9): 2505–2512.
21. Ali, G.A.M., Yusoff, M.M., Shaaban, E.R., and Chong, K.F. 2017. High performance MnO_2 nanoflower supercapacitor electrode by electrochemical recycling of spent batteries. *Ceramics International* 43(11): 8440–8448.
22. Collins, J., Gourdin, G., Foster, M., and Qu, D. 2015. Carbon surface functionalities and SEI formation during Li intercalation. *Carbon* 92: 193–244.
23. Abbas, Q., Raza, R., Shabbir, I., and Olabi, A.G. 2019. Heteroatom doped high porosity carbon nanomaterials as electrodes for energy storage in electrochemical capacitors: A review. *Journal of Science: Advanced Materials and Devices* 4(3): 341–352
24. Vijayakumar, M., Santhosh, R., Adduru, J., Rao, T.N., and Karthik, M. 2018. Activated carbon fibres as high performance supercapacitor electrodes with commercial level mass loading. *Carbon* 140: 465–476.
25. Thalji, M.R., Ali, G.A.M., Liu, P., Zhong, Y.L., and Chong, K.F. 2021. $W_{18}O_{49}$ nanowires-graphene nanocomposite for asymmetric supercapacitors employing $AlCl_3$ aqueous electrolyte. *Chemical Engineering Journal* 409: 128216.
26. He, S., Wang, S., Chen, H., Hou, X., and Shao, Z. 2020. A new dual-ion hybrid energy storage system with energy density comparable to that of ternary lithium ion batteries. *Journal of Materials Chemistry A* 8(5): 2571–2580.
27. Ali, G.A.M., Thalji, M.R., Soh, W.C., Algarni, H., and Chong, K.F. 2020. One-step electrochemical synthesis of MoS_2/graphene composite for supercapacitor application. *Journal of Solid State Electrochemistry* 24(1): 25–34.
28. Aboelazm, E.A.A., Ali, G.A.M., Algarni, H., and Chong, K.F. 2018. Flakes size-dependent optical and electrochemical properties of MoS_2. *Current Nanoscience* 14(2): 1–5.
29. Da-Wei, W., Feng, L., Jinping, Z., Wencai, R., Zhi-Gang, C., Jun, T., Zhong-Shuai, W., Ian, G., Gao Qing, L., and Hui-Ming, C. 2009. Fabrication of graphene/polyaniline composite paper via in situ anodic electropolymerization for high-performance flexible electrode. *ACS Nano* 3(7): 1745–1752.
30. Feng, H., Wang, B., Tan, L., Chen, N., Wang, N., and Chen, B. 2014. Polypyrrole/hexadecylpyridinium chloride-modified graphite oxide composites: Fabrication, characterization, and application in supercapacitors. *Journal of Power Sources* 246: 621–628.
31. Ali, G.A.M., Yusoff, M.M., Ng, Y.H., Lim, N.H., and Chong, K.F. 2015. Potentiostatic and galvanostatic electrodeposition of MnO_2 for supercapacitors application: A comparison study. *Current Applied Physics* 15(10): 1143–1147.
32. Ali, G.A.M., Wahba, O.A.G., Hassan, A.M., Fouad, O.A., and Chong, K.F. 2015. Calcium-based nanosized mixed metal oxides for supercapacitor application. *Ceramics International* 41(6): 8230–8234.
33. Kim, J., Kim, J.H., and Ariga, K. 2017. Redox-active polymers for energy storage nanoarchitectonics. *Joule* 1(4): 739–768.
34. Shehata, N., Sayed, E.T., Abdelkareem, M.A., Ali, G.A.M., and Olabi, A.G. 2021. Smart Electronic Materials, in *Reference Module in Materials Science and Materials Engineering*, Elsevier.
35. Fusalba, F., Gouérec, P., Villers, D., and Bélanger, D. 2001. Electrochemical characterization of polyaniline in nonaqueous electrolyte and its evaluation as electrode material for electrochemical supercapacitors. *Journal of The Electrochemical Society* 148(1): A1.
36. Zhou, H., Chen, H., Luo, S., Lu, G., Wei, W., and Kuang, Y. 2005. The effect of the polyaniline morphology on the performance of polyaniline supercapacitors. *Journal of Solid State Electrochemistry* 9(8): 574–580.
37. Liu, J., Zhou, M., Fan, L.Z., Li, P., and Qu, X. 2010. Porous polyaniline exhibits highly enhanced electrochemical capacitance performance. *Electrochimica Acta* 55(20): 5819–5822.
38. Zhang, X., Zhao, J., Xia, T., Li, Q., Ao, C., Wang, Q., Zhang, W., Lu, C., and Deng, Y. 2020. Hollow polypyrrole/cellulose hydrogels for high-performance flexible supercapacitors. *Energy Storage Materials* 31: 135–145.
39. Majumder, M., Bilash, R., Koiry, S.P., and Thakur, A.K. 2017. Gravimetric and volumetric capacitive performance of polyindole/carbon black/MoS_2 hybrid electrode material for supercapacitor applications. *Electrochimica Acta* 248: 98–111.
40. Ge, Y., Wang, C., Shu, K., Zhao, C., Jia, X., Gambhir, S., and Wallace, G.G. 2015. A facile approach for fabrication of mechanically strong graphene/polypyrrole films with large areal capacitance for supercapacitor applications. *RSC Advances* 5(124): 102643–102651.
41. Liu, Y., Weng, B., Razal, J.M., Xu, Q., Zhao, C., Hou, Y., Seyedin, S., Jalili, R., Wallace, G.G., and Chen, J. 2015. High-performance flexible all-solid-state supercapacitor from large free-standing graphene-PEDOT/PSS films. *Scientific Reports* 5: 1–11.
42. Xiong, W., Hu, X., Wu, X., Zeng, Y., Wang, B., He, G., and Zhu, Z. 2015. A flexible fiber-shaped supercapacitor utilizing hierarchical $NiCo_2O_4$@polypyrrole core-shell nanowires on hemp-derived carbon. *Journal of Materials Chemistry A* 3(33): 17209–17216
43. Abuali, M., Arsalani, N., and Ahadzadeh, I. 2020. Investigation of electrochemical performance of a new nanocomposite: $CuCo_2S_4$/polyaniline on carbon cloth. *Journal of Energy Storage* 32: 101694–101694.

44. Yuan, Y., Ahmed, J., and Kim, S. 2011. Polyaniline/carbon black composite-supported iron phthalocyanine as an oxygen reduction catalyst for microbial fuel cells. *Journal of Power Sources* 196(3): 1103–1106.
45. Chen, Siguo, Wei, Zidong, Qi, XueQiang, Dong, Lichun, Guo, Yu-Guo, Wan, Lijun, Shao, Zhigang, and Li, L. 2012. Nanostructured polyaniline-decorated Pt/C@PANI core–shell catalyst with enhanced durability and activity. *Journal of the American Chemical Society* 134: 13252–13255.
46. Zhang, N., Ma, W., Wu, T., Wang, H., Han, D., and Niu, L. 2015. Electrochimica acta edge-rich MoS_2 naonosheets rooting into polyaniline nano fi bers as effective catalyst for electrochemical hydrogen evolution. *Electrochimica Acta* 180: 155–163.
47. Osmieri, A.L., Zafferoni, C., Wang, L., Videla, H.A.M., and Lavacchi, A. 2018. Polypyrrole-derived Fe-Co-N-C catalyst for oxygen reduction reaction: Performance in alkaline hydrogen and ethanol fuel cells. *ChemElectroChem* 5(14): 1954–1965.
48. Yong, Y.C., Dong, X.C., Chan-Park, M.B., Song, H., and Chen, P. 2012. Macroporous and monolithic anode based on polyaniline hybridized three-dimensional Graphene for high-performance microbial fuel cells. *ACS Nano* 6(3): 2394–2400.
49. Yousfi, Y., Jouili, A., Mansouri, S., El Mir, L., Al-Ghamdi, A., and Yakuphanoglu, F. 2020. Studies of the dirac point in a GO/P3HT nanocomposite thin-film phototransistor. *Journal of Electronic Materials* 49(10): 5808–5815.
50. Grissom, G., Jaksik, J., McEntee, M., Durke, E.M., Aishee, S.T.J., Cua, M., Okoli, O., Touhami, A., Moore, H.J., and Uddin, M.J. 2018. Three-dimensional carbon nanotube yarn based solid state solar cells with multiple sensitizers exhibit high energy conversion efficiency. *Solar Energy* 171: 16–22.

2 Polymer Electrolytes for Supercapacitor Applications

Dipanwita Majumdar
Chandernagore College, West Bengal, India

Swapan Kumar Bhattacharya
Jadavpur University, Kolkata, India

CONTENTS

2.1 INTRODUCTION

Critical issues like sky-high global energy challenges, environmental pollution, and fast-depleting fossil fuel resources have driven worldwide motivation for active renewable and sustainable energy stratagems (Dincer 2000; Baños et al. 2011; Ellabban et al. 2014; Kim, Oh, and Kim 2015; Sorrell 2015). However, such renewable energy technologies require dynamic energy storage and rapid power output on instant demand (Winter and Brodd 2004; Kim, Oh, and Kim 2015; Raza et al. 2018). Batteries and electrochemical capacitors or supercapacitors have been amongst the popular electrochemical energy storage (EES) systems that show an inspiring electrochemical to electrical energy conversion proficiency (Salanne et al. 2016; Wang, Song, and Xia 2016; Gür 2018; Poonam et al. 2019; Staffell et al. 2019). The device performances of these EES are principally assessed by their energy density (energy accumulated per unit volume) or specific energy (energy stored per unit mass) and specific power (energy delivered per unit time per unit mass) or power density (energy delivered per unit time per unit volume) values (Gidwani et al. 2014; Chee et al. 2016; Dong et al. 2016). In general, batteries record higher energy storage characteristics but face practical limitations in respect to their poor portability features, high installation expenditures,

DOI: 10.1201/9781003169727-2

short life period, and above all substandard power delivering efficiencies. Supercapacitors, on the other hand, exhibit high power applications in addition to possessing attractive features of easy transportability, durability, and least maintenance although they largely suffer from marketable energy storage deficiencies (Khomenko et al. 2006; Majumdar 2016; Choudhary et al. 2017; Shao et al. 2018; Majumdar, Maiyalagan, and Jiang 2019) Thus, the latter currently play the role of supporting accessories to the batteries for providing uninterrupted power to different technological sectors (Kim, Sy, et al. 2015). Nevertheless, their potency for adequately addressing the existing difficulties and providing continuous energy back-ups to therapeutic machinery, wireless devices, and portable electronics employed in various sectors of daily life including everyday electronic appliances, traffic management and transportation, industry, and biomedical, astronomical, and defense areas, and so on, have motivated scientists to explore the innovative designing of smart and flexible electrochemical supercapacitors to substitute the existing bulky, low-lasting batteries (Conway 1991; Kondrat et al. 2012; Chen 2017; Yassine and Fabri 2017; Zhao et al. 2017; Zuo et al. 2017).

2.1.1 Effect of the Electrolyte on Supercapacitor Performance

The basic operational set-up of supercapacitors includes two opposite polarity electrodes composed of highly porous electroactive materials ionically connected via electrolytes. Supercapacitors generally store energy via charge accumulation arising either from rapid ion adsorption/desorption processes at the electrolyte/electrode interfaces during charging/discharging procedures, typically observed for electrical double layer capacitances (EDLCs), or via a faradaic, capacitive charge transfer process driven by the diffusion-controlled movement of electrolyte ions, characteristic of pseudocapacitors, respectively, as illustrated in Figure 2.1(a) (Wang et al. 2017; Majumdar, Mandal, and Bhattacharya 2019). Irrespective of the operating mechanisms, charge storage occurs at the electrode/electrolyte interface and thus the interfacial chemistry needs a better understanding for upgrading the energy performances of these gadgets (Majumdar and Bhattacharya 2017; Majumdar et al. 2017). Several research reviews have been penned that have excellently described in detail the different kinds of charge storage mechanisms prevailing in different electrode materials employed for supercapacitor applications (Wang, Zhang, and Zhang 2012; Wang, Xiao, et al. 2013; Augustyn et al. 2014; Wang, Guo, et al. 2015). Systematic research investigation has well established the fact that the appropriate choice of electrode material and its dimensions, surface nature, and crystal forms often promote modifications in capacitive charge storage processes, although the vital role of electrolytes was initially overlooked (Lin et al. 2009; Wang and Pan 2017; Majumdar 2021a; Majumdar 2021b). Consequently, detailed investigation of charge storage kinetics, the thorough interpretation of charging/discharging dynamics, and the analysis of capacitive responses prompted an equivalent importance to the electrolytes employed in device arrangements (Akinwolemiwa and Chen 2018; Mondal et al. 2019; Majumdar et al. 2020; Majumdar 2021a; Majumdar and Ghosh 2021). This deep realization, as demonstrated by the schematic diagram in Figure 2.1(b), shows the gradual development of advanced generations of supercapacitor technology that have promoted the judicious designing of both electrode materials as well as electrolytes so as to ensure not only faster electron/ion transportation to personalize exceptional energy storage/delivery efficiencies but also to promote multifunctional integrated electronics that can fruitfully develop portable self-powering systems as well as tackle the intense and comprehensive energy crisis (Zhong, Deng, et al. 2015; Luo, Ye, and Wang 2017; Majumdar 2021b). Accordingly, sincere efforts are being channeled in selecting electrolyte materials that can satisfactorily complement the charge storing capacity of the targeted electrodes. Thus, it is very crucial to fundamentally comprehend the influence of electrolytes on the electrochemical performance of these energy storage devices (Khomenko et al. 2006; Akinwolemiwa et al. 2015; Yi et al. 2018; Pal et al. 2019).

The fundamental criteria for accessing the electrochemical performances of supercapacitors are primarily based on two key parameters, namely specific energy (***E***) and specific power (***P***), as shown in Equations (2.1) and (2.2) (Zhao et al. 2017):

$$E = \frac{1}{2} C V^2 \tag{2.1}$$

$$P = \frac{V^2}{2R} \tag{2.2}$$

where C is the cell's specific capacitance, V is the operating voltage range, and R stands for the internal resistance.

The principal challenge in the development of diligent supercapacitors involves elevating the specific energy value, which can be accomplished by increasing the capacitance as well as expanding the operating voltage range of the device. Thus, for enhancing **E** and **P**, both **C** and **V** need to be improved, while R has to be minimized as much as possible. Since the voltage is related in the form of a square term, it will obviously have a higher impact for amplifying the energy characteristics of supercapacitors (Zhao et al. 2017; Majumdar et al. 2020). The functional limits of the voltage window are strongly controlled by the electrochemical characteristics of the electrolyte employed for the device fabrication in contrast to the deployed electrode materials and therein lies the importance of electrolytes in determining the charge performance of supercapacitors (Ma et al. 2018). Nonetheless, the electrolytes employed significantly influence all the parameters related to **E** and **P** and thus the proper selection of the same is extremely crucial to accomplish the desired results.

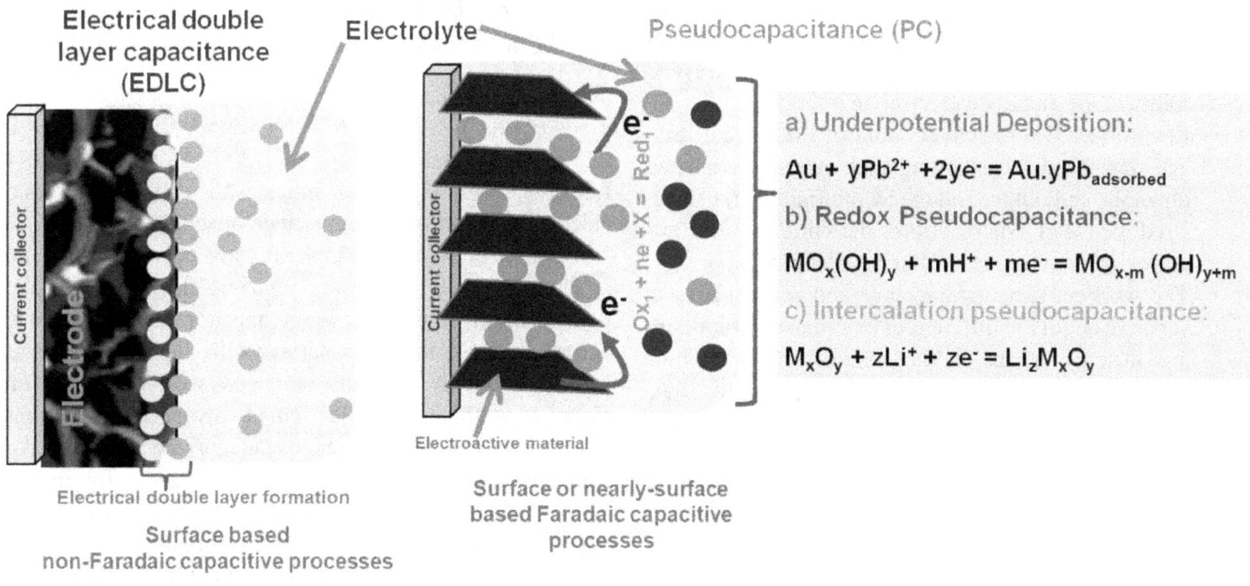

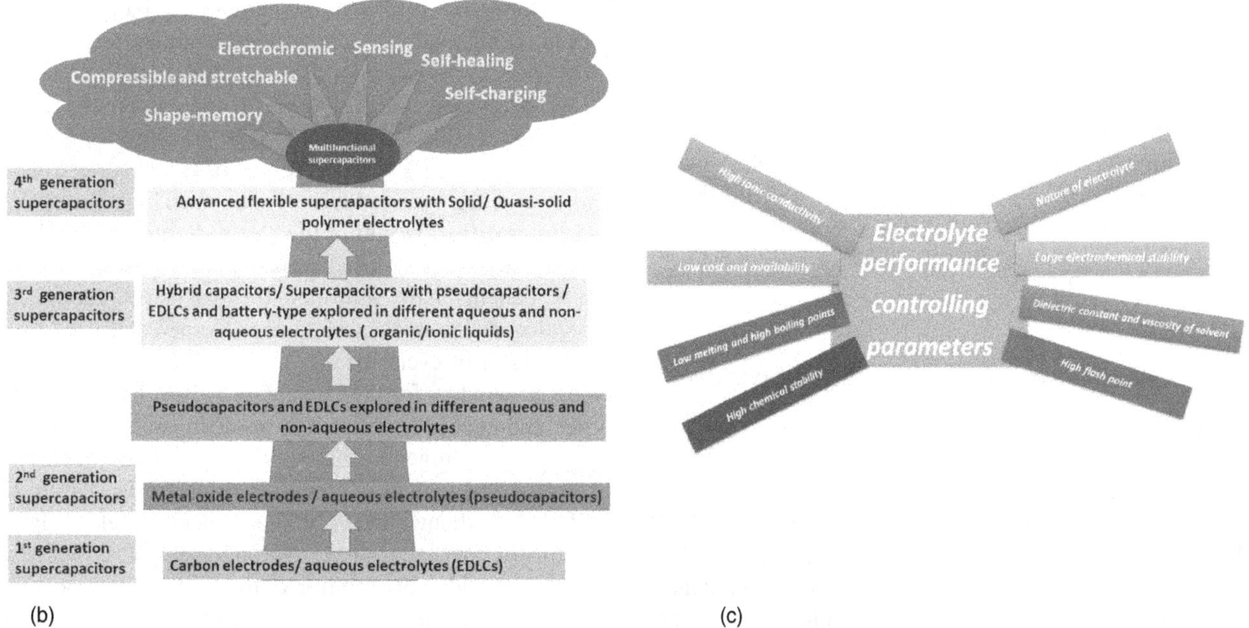

FIGURE 2.1 (a) Schematic representation of a charge storage mechanism in electrical double layer capacitors and pseudocapacitors. (b) Schematic tree diagram showing gradual development of different generations of supercapacitor technology. (c) Schematic representation of the determining parameters controlling an electrolyte towards achieving high energy performance.

2.1.2 Essential Electrochemical Performance Parameters Controlled by the Electrolytes

A number of electrolytic parameters play a decisive role in guiding the energy efficiency of the supercapacitors (Zhong, Deng, et al. 2015; Yi et al. 2018):

- A low mass and a reduced volume of the electrolyte in the device are highly desirable to make the system portable without compromising cell performance.
- Maximum utilization of the electrode materials through better wettability and interfacial interactions increase electrode capacitance.
- Lowering the resistances, i.e., the equivalent series resistance (ESR), of the bulk electrolyte solution and the electrolyte/electrode interface as well as within the electrode's pores dictates the device's power density.
- Ionic conductivity of the electrolytes varies with ionic size and significantly affects the ESR.

- The appropriate choice of electrolyte for the working electrodes can considerably improve cyclic stability by minimizing the probable electrochemical irreversibility at the electrode/electrolyte interfaces.
- Electrolytes can effectively control the current leakage due to self-discharging that otherwise narrows down the cell output voltage by minimizing the undesired localized side reactions, the current collector's corrosion course, and similar parallel processes.
- The working temperature range and performance of supercapacitors in different environmental conditions are strongly reliant on the electrolyte: the thermal properties of the solute, and explicit properties such as the freezing point, boiling point, flash point, and viscosity of the solvent, enable the resisting of fire/explosions often caused by uncontrolled heat release during charging–discharging sequences.

Much research with sophisticated spectroscopic techniques and computational methodologies have accomplished both tuning of the electrode materials as well as the electrolytes' characteristics which can lead to high conductivity and a smoother and better interface chemistry for faster ion/charge transport for outstanding capacitance values (Khomenko et al. 2006; Akinwolemiwa et al. 2015; Pal et al. 2019). Hence, the judicious choice of electrolytes along with that of the electrodes determines the overall EES device performance. The determining parameters that guide an electrolyte towards achieving superior energy performances are shown in Figure 2.1c.

2.1.3 Characteristics of an Ideal Electrolyte

Systematic and rigorous research efforts have resulted in the continuous revision of strategies to improve electrolyte functioning for improved energy efficiency (Brandon et al. 2007; Kim, Oh, and Kim 2015; Hui et al. 2019; Panda et al. 2020). An ideal electrolyte for a supercapacitor has been proposed to satisfy the following fundamental criteria. It should have (i) a wide operating voltage window; (ii) high ionic conductivity; (iii) electrochemical workability over a wide range of operating temperatures; (iv) high biocompatibility; (v) low viscosity; (vi) low pricing and availability; (vii) low inflammability; and (viii) a low volatile property and chemical stability towards other cell components apart from electrode materials. An ideal electrolyte has not yet been achieved as it is difficult to satisfy all the desirable qualities and thus the choice is restricted and subjected to a compromise between performance and other essentialities based on device application requirements. Different types of electrolytes have been under trial and each electrolyte class has its own merits and limitations, and thus nonstop efforts are involved in the research for the development of a unique electrolyte satisfying all the above criteria (Hong et al. 2002; Béguin et al. 2014).

2.2 DIFFERENT CLASSES OF ELECTROLYTES FOR SUPERCAPACITORS

To meet the desired electrochemical performances, various kinds of electrolytes have been developed and reported in the scientific database (Hong et al. 2002; Béguin et al. 2014; Lee et al. 2019). Figure 2.2 indicates the different classes of electrolytes used in devising supercapacitors till date.

Liquid electrolytes, in general, have been classified into aqueous electrolytes and non-aqueous electrolytes respectively. Initial studies of supercapacitor technology with carbon-based electrodes started with liquid electrolytes involving widely available aqueous solutions that were modified with the introduction of advanced generations. Aqueous electrolytes are again categorized as acid, alkaline, and neutral salt solutions. They generally involve dilute solutions of H_2SO_4, HCl, KOH, NaOH, LiOH, Na_2SO_4, K_2SO_4, Li_2SO_4, KCl, and so on, depending on the nature of the electrode systems designed for the fabrication of the supercapacitor (Akinwolemiwa et al. 2015; Yi et al. 2018; Pal et al. 2019). Water is a popular universal solvent that solubilizes a substantial number of inorganic and organic compounds in appreciable quantities to purposefully serve as an essential "active component" in various EES systems (Pal et al. 2019). Further, it exists in the liquid phase over a wide temperature range, having insignificant vapor pressure and reasonable viscosity. The electrolyte ions even in the hydrated form render much higher mobility and faster diffusion within the electrode pore channels to significantly cause raised capacitances (Ohtaki 2002). These benefits have informed the extensive practice of designing aqueous electrolyte-based supercapacitors over the years. However, the thermodynamically recognized potential window for liquid water is 1.23 V, beyond which water undergoes electrolysis to result in oxygen evolution (OER) and hydrogen evolution reactions (HER), due to its oxidation and reduction processes respectively (Pal et al. 2019). Even though the neutral salt aqueous electrolytes contain low concentrations of hydronium as well as hydroxyl ions that considerably minimizes the HER/OER probabilities, the former have low ionic conductivities due to larger hydrated ionic radii than the hydronium/hydroxyl ions which in turn increases the cell ESR as well as significantly reduces the observed capacitances (Ohtaki 2002; Fic et al. 2012; He et al. 2016). Thus, the constricted working potential range, and possible leakages of corrosive liquids, severely restrict the extensive usage of aqueous electrolytes in commercial supercapacitors (Gao, Virya, et al. 2014; Merrill et al. 2014). Nevertheless, in the recent past, the formal potential of a water splitting course has been judiciously pushed above 2.5 V by employing super-concentrations of salts in a liquid water solvent to form the so-called "water-in-salt" electrolytes (Chen et al. 2020). Accordingly, Sundaram and Appadoo (2020) carried out comparative work between conventional water-in-salt electrolytes and salt-in-water electrolytes in the presence of binary metal oxide symmetric supercapacitors that

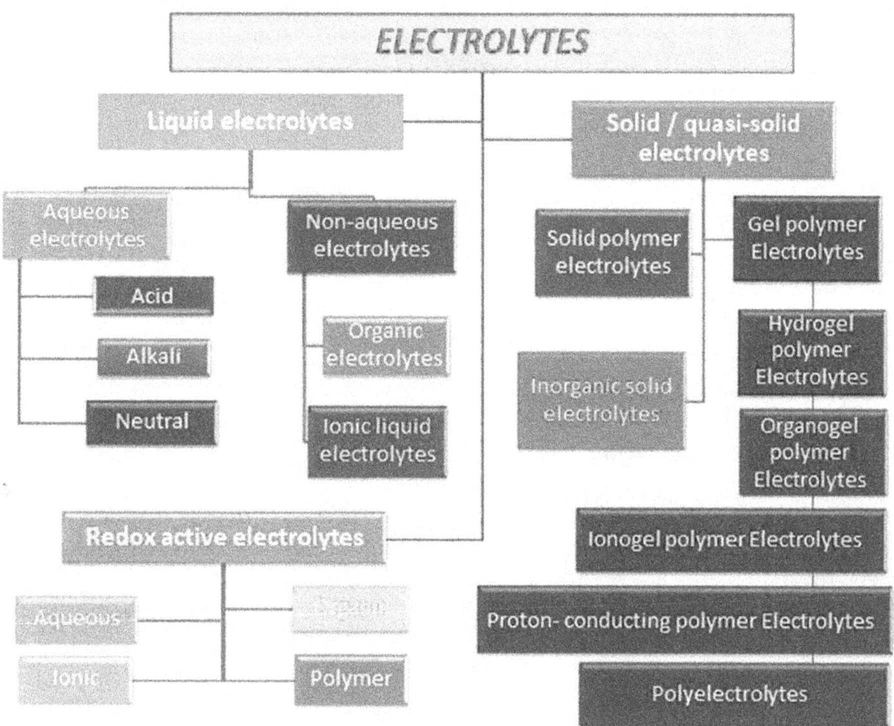

FIGURE 2.2 Classification of electrolytes employed in various supercapacitors.

indicated the superior electrochemical stability with enhanced storage capacitance for water-in-salt electrolyte-based devices owing to substantial modifications in the ion-solvation structure and solubility that suppressed the water decomposition reactions remarkably and correspondingly extended the operating voltage window. However, to date the recorded voltage window does not satisfy the practicability of aqueous supercapacitors and hence many scientists have suggested various non-aqueous electrolytes as serving the purpose successfully (Xia et al. 2012).

Non-aqueous electrolytes essentially involve salt solutions in organic solvents or the usage of fused ions that exhibit extended electrochemical stability over a wide voltage range without allowing any decomposition/degradation of the involved electroactive components (Fic et al. 2012). Non-aqueous electrolytes are basically classified as: organic electrolytes, ionic salt electrolytes, and their mixed electrolytes (Pal et al. 2019). Organic electrolytes comprising of acetonitrile-based systems are less viscous and work better at low temperatures, illustrating improved ionic conductivity and thus recording a higher power density compared to the other popularly recommended electrolytes such as propylene carbonate (Ishimoto et al. 2009; Lewandowski et al. 2010; Orita, Kamijima, and Yoshida 2010). Nonetheless, devices assembled using propylene carbonate-based electrolytes record wider operative voltages, as high as 4 V, besides being more eco-friendly and having a higher flash point (less inflammable) compared to acetonitrile (Ruch et al. 2007). Notably, a number of important constrictions severely oppose their practicability. EDLCs or pseudocapacitors employing organic electrolytes record inferior values of specific capacitance, mainly due to the larger radii of solvated ions that results in restricted accessibility to interfaces and then only to the very few available larger diameter pore channels of the electrodes (Ghosh and Lee 2012; Frackowiak et al. 2013). Further, low ionic conductivity (~20 mS cm^{-1}) in organic electrolytes as compared to aqueous electrolytes (1 S cm^{-1}) results in a higher ESR and, hence, a power fall in the devices (Pal et al. 2019). Again, most of the organic solvents are extremely volatile and have high fire risks; they also undergo decomposition on repeated charging/discharging cycles to liberate toxic products and thus additional safety procedures are required for their proper handling. Moreover, traces of water in these organic solvents can spoil the overall electrochemical performances of the devices to a massive extent. Even the ionic mobilities are rather low in the organic medium which leads to low capacitances and further a high cost of the organic solvents, reduced availability, and rigorous purification procedures, which altogether severely restrict their employment in practical gadgets (Ding et al. 2014). Consequently, to cope with the worrying issues of the decomposition of organic solvents, high flammability, and the narrow potential window of aqueous electrolytes, another class of non-aqueous electrolytes in the form of ionic liquids have been targeted (Zhu et al. 2007). Ionic liquid electrolytes are usually composed of room temperature melting salts having an asymmetric carbon skeleton-based structural cation coordinated to an inorganic/carbon-based anion and bestowed with substantial thermal, chemical, electrochemical, and non-volatile properties (Orita, Kamijima, and Yoshida 2010; Chen et al. 2011; Huang, Luo, et al. 2015). The hydrocarbon functional moieties assist them to become readily soluble in organic solvents and also to facilitate their usage in a neat form as electrolytes so as to effectively evade the ion-solvation consequences (Wang, Lu, et al. 2012; Huang, Luo, et al. 2015). Often combinations of ionic liquid mixtures have resulted in increased ionic conductivity as well as a widening of the voltage window in order to enhance the energy efficiency of ionic liquid-based supercapacitors (Orita, Kamijima, et al. 2010). However, the proper composition of mixtures is to be optimized for superior electrochemical performances (Vraneš et al. 2014). The acute need for devising flexible electronics has been the driving motivation for engineering the cell components, namely the electrodes and the electrolytes, with specially designed electroactive materials having exquisite mechanical properties, so that they can restore their utmost electrochemical performances even under structurally distorted conditions and thus can be readily tailored to fit any desired shape (Huang, Zhang, et al. 2016; Na et al. 2016; Cao et al. 2017).

Polymers have been renowned for a long time for their high mechanical flexibility as well as tunable electronic and charge transport properties (Majumdar and Saha 2010; Majumdar and Saha 2015; Roy et al. 2019). Conducting polymers and various types of polymer-based composites have been widely employed as electrode materials with high mechanical flexibility as well as charge storage efficiencies (Majumdar 2016; Han and Dai 2019; Majumdar 2019a; Majumdar 2019b). This idea led to the advent of polymer electrolytes in supercapacitor technology to cope with the unwanted spillage and safe handling complications related to liquid electrolytes that otherwise severely restrict their applications in miniature, portable, and wearable electronic gadgets. These issues have motivated extensive scientific investigation into exploring polymer based solid/quasi-solid types of electrolytes for constructing advanced supercapacitors that can well meet the top-priority needs of the highly flexible, miniature, as well as cost-effective electronics of the near future (Meng et al. 2010; Yadav et al. 2017). A comparative study of the different features of various types of electrolytes are illustrated in Table 2.1 highlighting their advantageous features as well as limitations associated (Lu et al. 2014; Aziz et al. 2019, 2021; Deng et al. 2019; Pal et al. 2019).

2.3 DIFFERENT SOLID AND QUASI-SOLID TYPES OF ELECTROLYTES USED IN SUPERCAPACITOR TECHNOLOGY

Solid or quasi-solid electrolytes are increasing in importance owing to better life-spans, minimum charging rates, low chances of hazardous contamination, safe handling, and being tailorable towards designing reliable and robust, miniature, portable, and wearable electronic devices. Usage of these electrolytes is further advantageous in respect to their ease of production, minimal internal corrosion problems, simpler and mechanically conducive modes of device construction, and packaging flexibilities that have seen highly stimulated worldwide exploration in recent years (Na et al. 2016; Cao et al. 2017). Solid or quasi-solid electrolytes have been popularly classified as solid polymer electrolytes, gel polymer electrolytes, and inorganic solid electrolytes, depending on the nature of the solvents employed and the function of the polymer used. The following section also discusses the advantages and limitations faced in the usage of solid polymer electrolytes and various types of gel polymer electrolytes in designing flexible multifunctional supercapacitors. The present chapter restricts discussion to polymer-based electrolytes, hence inorganic solid electrolytes are not elaborated here.

2.3.1 Solid Polymer Electrolytes

Solid polymer electrolytes are extremely encouraging energy storage devices that support thinner, lighter, portable, and wearable electronics, highly compact micro-electronic devices, and printable and flexible technologies (Kaempgen et al. 2009; Huang, Zhang, et al. 2016). Undesired leakages, ease of design, and packaging advantages of solid electrolyte-based supercapacitors over liquid-based supercapacitors have found noticeably widespread applications in recent years (Lu et al. 2014). Popular solid polymer electrolytes bestowed with the admirable qualities of mechanical flexibility, biocompatibility, thermal stability, and chemical and electrochemical stability are usually employed in flexible, multifaceted supercapacitors. Such solid polymers include examples like chitosan, poly(aryl ether ketone), and polyethylene glycol (Meng et al. 2010; Yadav et al. 2017; Aziz et al. 2019). Chitosan is a renowned biopolymer comprising of glucosamine/N-acetyl glusamine units containing glucoside bonds that have the several advantages of safe-handling, hydrophilicity, biodegradability, as well as exceptional mechanical flexibility and film-forming ability and which can be specifically employed in designing resilient film-supercapacitors (Sudhakar and Selvakumar 2012; Shukur et al. 2014). However, chitosan displays an exceptionally low ionic conductivity ($\sigma \sim 10^{-5}$ to 10^{-6} S cm^{-1} under standard ambient conditions) that severely reduces its practical applicability (Hamsan et al. 2020). Poly(aryl ether ketone) is another ecofriendly semi-crystalline macromolecule that possesses splendid chemical, electrochemical, thermal, and mechanical properties (Dominici et al. 2019). Similarly, polyethylene glycol (PEG) is also a cost effective, eco-compatible, and highly processable electrolyte reported to have inspiring electrochemical efficiency records (Karaman et al. 2019). However, the major drawback lies in low ionic conductivity at ambient temperatures owing to the poor-conductive behavior of its crystalline phase (Golodnitsky et al. 2002). Thus, to eradicate the problem, combinations of polyethylene glycol–borate ester plasticizer with poly(ethylene glycol) methacrylate and LiTFSI as a solid polymer electrolyte have displayed exceptional ionic conductivity and thermal stability in addition to a wide electrochemical window of 4.5 V (Chakrabarti et al.

TABLE 2.1
Comparative Study of Features of Different Electrolytes Used in Supercapacitors

Electrolytes	Electrode Materials	Cell Voltage (V)	Temperature (°C)	Specific Capacitance (F g^{-1})	Energy Density	Power Density	Manufacturing Cost	Safety Features	Advantages	Limitations
Aqueous	Carbon-based materials and pseudocapacitive materials	~2.5	5 to 80	100–1800	Low	High	Cost depends on salt	Non-flammable	High ionic conductivity	Narrow voltage output, leakage problems
Organic	Carbon-based and pseudocapacitive materials	2.5–2.9	−40 to 60	50–600	Medium	Low	Cost depends on salt and solvent	Less corrosive	Wider operating voltage, cyclic stability	High flammability, leakage and cost
Ionic liquids	Mostly carbon-based materials	~2.5–4.0	−100 to 25	<100	High	Medium	High cost	Safe to handle	Wide functioning temperature range, non-corrosive, non-flammable, cyclic stability	Low conductivity, cost, and power density
Polymers	Carbon-based and pseudocapacitive materials	~2.0–4.0	25 to 90	~100	High	Medium	Cost depends on constituents	Non-flammability and safe to handle	Miniaturized portable and wearable electronics, high cyclic stability, no leakage problems, multifaceted properties like self-charging, self-healing, shape memory, electrochromism	Low ionic conductivity and fabrication complexities

2008). Likewise, a novel solid polymer electrolyte designed by mixing poly(methyl methacrylate) with suitable proportions of KOH have enhanced the much-needed polymer amorphicity that highly promotes rapid ion transportation and conductivity enhancement for upgrading the overall electrochemical competency of the designed energy storage system (Ibrahim et al. 2020).

Nonetheless, up to today, the fabrication of solid-state polymers experiences several technical problems. One of the prime issues involves electrode/electrolyte layer-delaminating due to poor cohesion that often shortens the lifespan of the devices. Moreover, optimum ionic conductivity can only be accomplished if the electrode material forms an interpenetrating network-like structure with the solid polymer electrolyte matrix to facilitate easy ion transportation, which is an imperfect prospect with dissimilar polymers. Thus, these issues need further systematic exploration to adequately address the emerging challenges and seek further industrial upgrades as far as their practicability is concerned (Lee, Lee, et al. 2008).

2.3.2 Gel Polymer Electrolytes

A gel polymer electrolyte (GPE), also recognized as a plasticized polymer electrolyte (PPE), was first presented by Feuillade and Perche in 1975 (Kotobuki 2020). A characteristic gel electrolyte comprises a polymer matrix embedded in a liquid electrolyte explicitly designed to hold its flexible features and tailorable morphology (to be employed in bendable, wearable, and portable devices) (Duay et al. 2012). The liquid electrolyte used can be aqueous, organic, an ionic liquid, or a redox active solvent mixture. GPE can be prepared easily by mixing the appropriate polymer matrix, an alkali metal salt, and a solvent, and then subjecting it to controlled heating. The resultant viscous liquid is cast under hot conditions and then chilled under the pressure of the electrodes to produce a thin film. Different strategies have been adopted for the physical and chemical cross-linking of polymer chains to form the resultant gels, as illustrated in Figure 2.3 (Yang, Zhang, et al. 2013; Ngai et al. 2016).

Gel polymer electrolytes are also popularly classified as quasi-solid electrolytes owing to the existence of adequate liquid trapped in the structure (Yu et al. 2019). Unlike common solid polymer electrolytes (SPEs) that mainly consist of the host polymer and inorganic salt with zero liquid content, GPE steals large quantities of liquid electrolytes entrapped within host polymers that impart better ionic conductivity than those of SPEs and accordingly dominate the solid electrolyte-based supercapacitor devices of recent days (Wang and Zhong 2015). Such entrapment of liquid also enhances the amorphous characteristics of the host polymer and facilitates considerable ionic mobility within the matrix. Several varieties of polymers like poly(vinyl alcohol) (PVA), poly(methylmethacrylate) (PMMA), polyacrylamide (PAM), sodium/potassium salts of polyacrylic acid (sodium polyacrylate/potassium polyacrylate) (SPA/PAAK), poly(ether-ether-ketone) (PEEK), and poly(vinylidene fluoride-*co*-hexafluoropropylene) (PVDF-HFP), as shown in Figure 2.4, have been investigated for their possible application as a suitable host polymer medium to form the required GPE compositions owing to their respectable mechanical flexibility, thermal stability, interfacial contact, and ionic conductivity (Choudhury et al. 2009; Chen, Li, et al. 2014; Wang and Zhong 2015; Ngai et al. 2016; Peng et al. 2019; Han et al. 2020; Mondal et al. 2020).

2.3.2.1 Hydrogel Polymer Electrolytes

Water is one of the most commonly used solvents for the preparation of gel polymer electrolytes. Besides water,

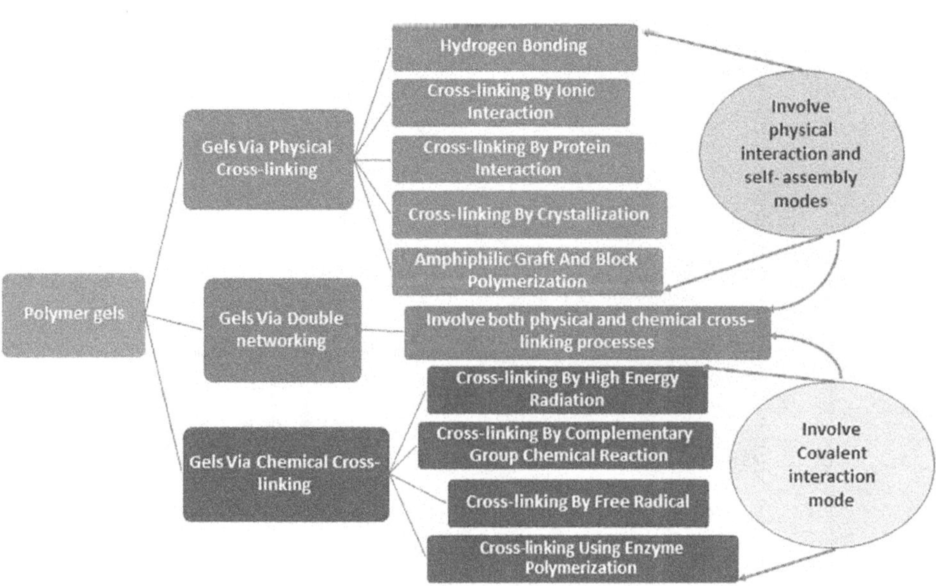

FIGURE 2.3 Representation of different cross-linking modes for the preparation of polymer gels.

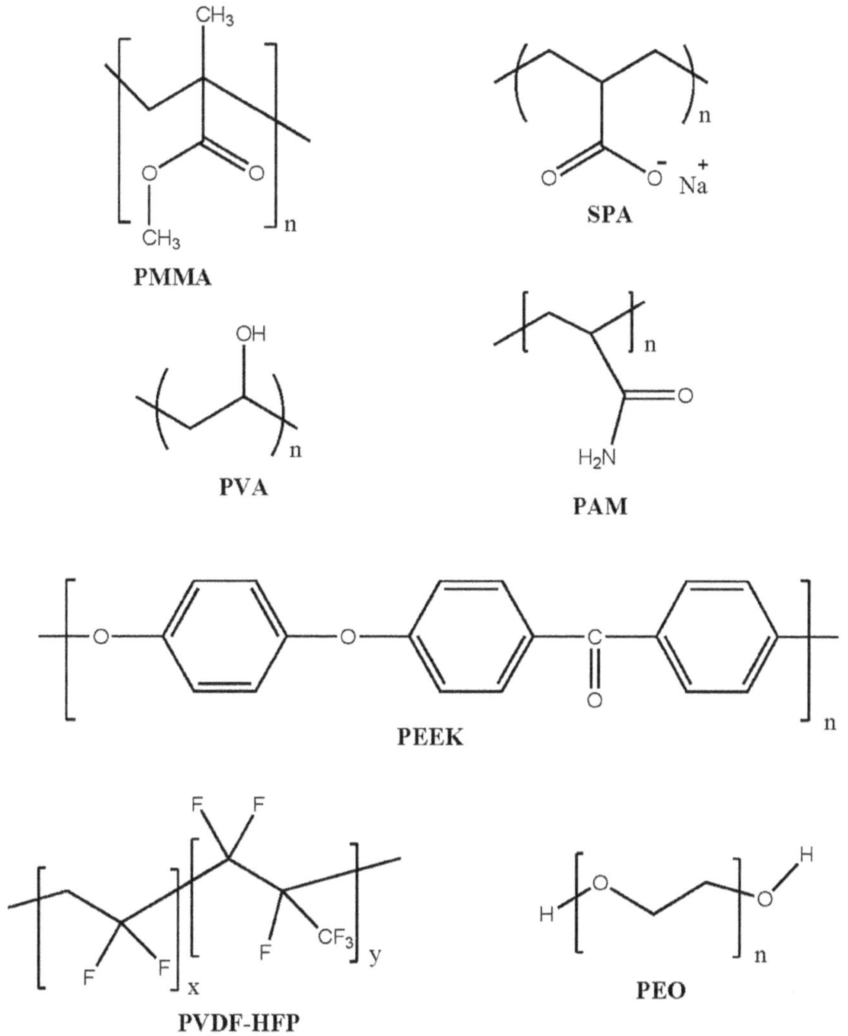

FIGURE 2.4 Chemical structures of some polymer hosts commonly used as electrolyte components in supercapacitors.

other common organic solvents like propylene carbonate (PC), ethylene carbonate (EC), and dimethylformamide (DMF) or their mixtures are often employed as the plasticizers in GPEs (Mondal et al. 2020). The extent of plasticization in a characteristic GPE is governed by the relative proportions of the host polymer and the liquid electrolyte present in order to tune the following features for superior supercapacitor devices: (i) high ionic conductivity; (ii) greater chemical, thermal, and electrochemical stability; and (iii) appropriate mechanical flexibility and dimensional constancy (Choudhury et al. 2009). The polymer hydrogel electrolytes can be further sub-categorized as natural and artificial or synthesized, purely on the basis of their existence and availability, as illustrated in the subsequent sections.

2.3.2.1.1 Synthesized Polymer Hydrogel Electrolytes

In synthesized polymer hydrogel electrolytes, water is the plasticizer which readily remains trapped within the complex polymer matrices because of surface tension, leading to the formation of fascinating 3D-network assemblies with high mechanical flexibility. In general, depending on the nature of the polymer chain interactions that lead to the interconnected network structure, reserve tunable quantities of water can get trapped and reserved, which can also dissolve ions to enhance ionic conductivity, thus rendering hydrogels sufficiently ionic conducting (Choudhury et al. 2009; Mondal et al. 2020). Moreover, grafting the polymer chains with polar functionalities would reinforce improved ionic conductivity and water retentivity. One of the premier approaches to strengthen the structural features of hydrogels has been carried out by introducing cross-linking via different non-covalent interactions including hydrogen-bonding and dipolar or electrostatic forces, so as to restore the electrode/electrolyte interactions particularly during extreme mechanical deformations (Huang et al. 2012; Ramasamy, del Vel, and Anderson 2014; Na et al. 2018). Furthermore, strong interfacial bonding via introduction of suitable surface functional moities are necessary to enhance the pastiness of the hydrogel electrolytes for better adherence to the electrodes' surfaces and thus to avoid the unwanted collapsing of the device on subjecting it to structural deformations

TABLE 2.2
Polymer Hydrogel Electrolytes for Supercapacitors

Electrolyte Materials	Electrode Materials	Gravimetric/Volumetric Capacitance @ Current Density/Potential Sweep Rate	Cyclic Stability	Reference
PVA/H_3PO_4	SWCNT film	110 F g^{-1} @ current density of 40 μA cm^{-2}	—	Kaempgen et al. (2009)
PVA/H_3PO_4	Graphene	2.6–3.7 mF cm^{-2} @ potential sweep rate of 100 mV s^{-1}	~90% capacitance retention after 5000 cycles	Chen, Li, et al. (2014)
Cross-linked porous PVA/ H_2SO_4	Graphene/CB	144.5 F g^{-1} @ current density of 0.5 A g^{-1}	78% capacitance retention after 1000 cycles	Fei et al. (2014)
Fe^{3+} cross-linked polyacrylic acid	Graphene foam/PPy	87.4 F g^{-1} @ current density of 0.5 A g^{-1}	89% capacitance retention after 5000 cycles	Guo et al. (2016a)
Potassium polyacrylate/KCl	3D-MnO_2/graphene and CNT/graphene electrodes	69.4 F g^{-1} @ current density of 0.5 A g^{-1} & 38.2 F g^{-1} @ current density of 10 A g^{-1}	84.4% capacitance retention after 10,000 cycles	Lee and Wu (2008)
Cross-linked potassium polyacrylate/KOH	$Ni(OH)_2$//AC	0.97 F g^{-1} @ potential sweep rate of 10 mV s^{-1}	90% capacitance retention after 20,000 cycles	Nohara et al. (2006)
Poly(propylsulfonate dimethylammonium propylmethacrylamide)/LiCl	Graphene	300.8 F cm^{-3} @ current density of 0.8 A cm^{-3}	103% capacitance retention after 10,000 cycles	Peng et al. (2016)
Agarose/NaCl	MnO_2	286 F g^{-1} @ current density of 0.5 A g^{-1}	80% capacitance retention after 1200 cycles	Moon et al. (2015)
Guar gum/$LiClO_4$/glycerol (plasticizer)	Activated carbon	164 mF cm^{-2} @ current density of 2 mA cm^{-2}	96 % capacitance retention after 2000 cycles	Sudhakar et al. (2014)

and/or a large number of repetitive charging/discharging tests. The electrochemical performances of some interesting polymer hydrogel electrolyte-based supercapacitors are summarized in Table 2.2.

PVA based hydrogel electrolytes, owing to their good hydrophilic property, simplistic synthesis methodology, low toxicity, and cost-effectiveness, have been by far the commonest electrolytes used for flexible, stretchable, and micro-supercapacitor gadgets so far reported (Fei et al. 2014). They are mostly based on H_2SO_4, H_3PO_4 (acidic), KOH, NaOH (alkaline), and NaCl, LiCl (neutral) as dopants (Li, Zang, et al. 2014; Ma et al. 2014; Wu et al. 2013; Subramani et al. 2017). A large number of various carbon-based electrodes have shown improved electrochemical response with these gel electrolytes, as illustrated in Table 2.2. For instance, Wang, Lu, et al. (2012) fabricated graphene-based in-plane micro-supercapacitors that withstood high charging/discharging characteristics even at a voltage sweep rate of 1000 mV s^{-1} using an H_2SO_4/PVA electrolyte. Further, most of the pseudocapacitive electrode materials (e.g., VO_x and VN), which undergo rapid dissolution in pure aqueous media, have been found to experience significantly improved electrochemical stability in hydrogels, as observed in Figure 2.5(a) (Wang, Lu, et al. 2012). Further explorations were carried out on the capacitive behavior of the highly porous activated carbon-based symmetric supercapacitors using PVA and KOH hydrogel electrolytes mixed with various additives like carbon black (CB) and a conducting polymer as well as PANI; the observations were compared. Figure 2.5(b–d) shows cyclic voltammograms, specific capacitance variations with current density, and the Ragone plots (energy density versus power density plots) for the as-prepared cells in the presence of these different electrolytes (Barzegar et al. 2015). The observations clearly point out that, although the specific capacitance is lower in polymer electrolytes, higher operating voltage ranges in them offer a larger weightage to the ultimate energy and power densities as per Equations (2.1) and (2.2). The low viscosity of the aqueous KOH electrolyte favors higher ionic transport as well as facilitates the easy accommodation of small hydroxide ions within the microporous channels of electrodes leading to larger specific capacitance. Nonetheless, the voltage window is much restricted in aqueous KOH due to water oxidation reactions, which are successfully evaded by the usage of the polymer electrolytes. Further, addition of conducting additives like carbon black increases the overall conductivity in polymer electrolytes and hence improves capacitance compared to a simple PVA–KOH medium without any changes in the operational voltage and thus indicating promising electrochemical features for real applications (Barzegar et al. 2015).

To increase the chemical resilience of these electrolytes, doping additives such as boric acid has been added to PVA–H_2SO_4 polymer electrolytes and which optimized the ionic conductivity as well as electrolyte durability significantly

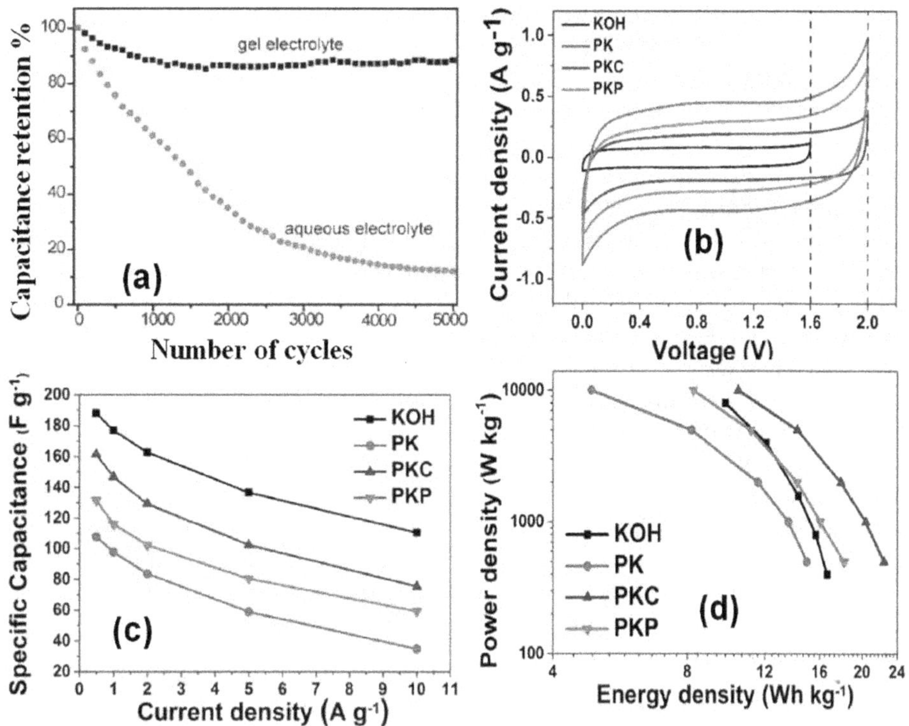

FIGURE 2.5 (a) Comparative display of cyclic stability of vanadium oxide pseudocapacitors in a 5 M LiCl aqueous electrolyte and an LiCl/PVA gel electrolyte respectively, all recorded at a voltage scan rate of 100 mV s^{-1}. (b) CV of porous activated carbon electrode based symmetric supercapacitors in various electrolytes: aqueous KOH, PVA–KOH (PK), PVA–KOH–CB (PKC), and PVA–KOH–PANI (PKP) hydrogels respectively. (c) Variation of specific capacitance at various current densities of the electrochemical cells. (d) their corresponding Ragone plots. (Wang, Lu, et al. 2012 with permission from ACS publications.Barzegar et al. 2015).

(Karaman and Bozkurt 2018). Further, to understand the role of $HClO_4$ dopant concentration on the gravimetric capacitance of black pearl carbon (EDLC) electrodes, cyclic voltammograms have been recorded that exhibit an increase in the gravimetric capacitance with rising dopant proportions, justified on the basis of the improved accessibility of the electrode surface with rising dopant ion concentrations during potential cycling (Sampath et al. 2009). Again, KOH–PVA has been reported to be the main alkaline polymer hydrogel electrolyte often used for wide cell output voltage, high capacitance, fast rate behavior, and long cyclability, although long-term retention characteristics are still not reported (Yuan et al. 2006). Nevertheless, carbon-based supercapacitors employing PVA based hydrogel electrolytes often suffer from high temperature-fluidity problems and low environmental stability due to dehydration issues (Qian 2019). To combat the problems, high cross-linked hydrogels are fabricated with several additives such as SiO_2, TiO_2, GO, and cellulose components with successful promotion results (Gao and Lian 2013; Lim, Teoh, et al. 2014). Thus, the addition of SiO_2 particles in the PVA–silicotungstic acid–H_3PO_4 hydrogel enhanced its water content retaining ability even under relatively low humid conditions, thereby demonstrating its improved environmental adaptability (Huang et al. 2014). Similarly, the introduction of redox additives, such as hydroquinone to a PVA–H_2SO_4 polymer gel electrolyte, to activated carbon electrodes derived from bio-waste improved the energy characteristics by nearly 2.25 times due to the cumulative enhancement of pseudocapacitive contributions from the electrolytes as well (Senthilkumar et al. 2012). To combat the issues related to narrow operating voltages with hydrogel electrolytes similar to that of pure aqueous electrolytes, the electrodes are so chosen to complement each other in terms of functional potential ranges that may fruitfully lead to a wider voltage window as high as 1.8 V (Lu et al. 2013; Fan et al. 2018).

Another well-known synthetically obtained polymer is polyacrylamide that is bestowed with high hydrophilicity and considerable ionic conductivity, and thus is becoming rapidly popular as an interesting GPE for flexible supercapacitors. Jin et al. (2019) proposed a highly conducting, cross-linked polyacrylamide hydrogel electrolyte-based supercapacitor that displayed high charge storing characteristics, a reduced self-discharge rate, rare leakage problems, and wide temperature tolerance. Further, a comparative study already carried out by Kim, Nam, et al. (2013) showed that the relative electrochemical performance of RuO_2-based electrodes in the presence of diverse acrylic hydrogel electrolytes, including poly(acrylic acid) (PAA), potassium polyacrylate (PAAK), and poly(2-acrylamido-2-methyl-1-propanesulfonic acid) (PAMPS), found that the charge storing efficiency followed the order: PAMPS/H_2SO_4 < PAAK/H_2SO_4 < PAA/H_2SO_4 < 1 M aqueous H_2SO_4 < PAMPS (aqueous), ascribed to high favorable proton accommodation within the side chain moieties of PAMPS. The authors claimed that the aqueous PAMPS electrolyte could house additional negatively charged oxygen sites on their side chains compared to pure PAA and PAAK and, accordingly, the former accomplishes stronger interactions to yield higher capacitance with the hydrous RuO_2 electrodes. On the other hand, PAMPS in 1 M H_2SO_4 holds back the protons on their surfaces through chemical interactions and reduces their availability, thereby resulting in the lowermost capacitance of hydrous RuO_2 electrodes. Very recently, Xia et al.(2020) reported a robust and seal-healing gel polymer electrolyte composed of Fe^{3+}-crosslinked sodium polyacrylate–LiCl hydrogel. Its 3D-interconnected spongy framework acted as an excellent stress buffer to dissipate applied physical impacts, and successfully prohibited easy cracking under repeated deformation. Choudhury et al. (2006) also explored mixed polymer blends to upgrade the water holding capacity for the fine-tuning of their ionic conductivity for supercapacitor applications. In this context, they employed acidic, alkaline, and neutral PVA/PAA blended hydrogel electrolytes containing different dopants—(a) $HClO_4$, (b) NaOH, and (c) NaCl respectively—to study the variation of charge storage in carbon-supported $RuO_x{\cdot}xH_2O$ electrodes. The capacitance of the system can be correlated with the ionic conduction of the electrolytes; they concluded that in the acidic PVA/PAA blended hydrogel electrolyte, the capacitance decreased with the increase in the fraction of PAA in the blend, while reverse variations were observed for the alkaline and neutral blended electrolytes. This contrasting feature was explained on the basis of an ultrafast proton conduction process supported via a Grotthus-type mechanism in the acidic blend electrolyte, while a much slower ionic transfer was executed by polymer chain segmental motion in alkaline and neutral blended hydrogel electrolytes. Thus, the addition of high proportions of PAA further assisted in promoting protonic conduction for the acidic blended hydrogel electrolytes, thereby improving cell capacitance, unlike the cases of neutral salt and alkali blended hydrogel electrolyte-based devices that further deteriorated the segmental motions considerably. Recently, Li et al. (2021) designed a supermolecular hydrogel composed of sodium acetate, PVA, and glycerol to record a highly stable potential window of 2 V along with a good ionic conductivity of 81.27 mS cm^{-1}. On assembling with carbon electrodes, the authors reported an observed cell gravimetric capacitance of 24 F g^{-1}, a specific energy of 12.6 W h kg^{-1} with a high capacitance retention efficacy of 110%, even on passing 5000 charging/discharging cyclic tests. Further, mixing potassium ferricyanide as a redox additive with the obtained hydrogel electrolyte elevated the cell gravimetric capacitance and specific energy to 32.7 F g^{-1} and 18.2 W h kg^{-1}, respectively. In the recent past, to address the challenge for robust electrode/electrolyte interfaces, even on subjecting them to heavy mechanical strains, a mechanically tough surface micro-structured agar/polyacrylamide/LiCl gel electrolyte was designed that indicated strong adhesion to the electrodes, thereby upgrading the capacitance, as well as its retention efficiency under repeated stretching sequences, compared to the pristine hydrogel with smooth surfaces (Fang, Cai, et al. 2019).

Thus, irrespective of some inspiring results highlighted above, most of the hydrogel-based electrolytes so far obtained for devising flexible supercapacitors often display a constricted working potential window due to the water decomposition reactions that severely restrict their commercial popularity. Also, hydrogel-based supercapacitors generally cover a much lower voltage window that leads to inferior device energy and power densities. Additionally, appropriate modifications in the chemistry of polymer hydrogels are essential to benefit them with desired qualities, such as improved mechanical tenacity, wettability and water retention capacities, along with improved tolerance and affinity, to associate high concentrations of electrolytic salt as well as strengthening the interfacial features for designing high energy, flexible, and wearable supercapacitors (Alipoori et al. 2020).

2.3.2.1.2 Natural Biopolymer-Based Hydrogel Electrolytes

Besides the synthetic polymers so far discussed, various naturally obtainable biopolymers that are widely available embrace proteins, polypeptides, DNA, polysaccharides, polyphenols, and so on, and are frequently employed as a vital component of polymer hydrogel electrolytes for smart supercapacitors (Armelin et al. 2016; Cao et al. 2017; Yang, Liu, et al. 2019). These biopolymers are gifted with unique features, such as easy availability, biocompatibility, and biodegradability, along with tunable mechanical properties. Common biopolymers, such as chitosan, gum-Arabic, and agarose, have been widely and successfully employed in gel electrolyte supercapacitors in recent times (Pérez-Madrigal et al. 2016; Khalid and Honorato 2017; Yang, Liu, et al. 2019). Chitosan has been amongst the most abundant polysaccharide frequently employed as a polymer hydrogel electrolyte material for devising supercapacitors. The precise de-acetylation of chitin, a linear N-acetylglucosamine based polysaccharide, yields chitosan having sufficient N- and O-containing functional groups and properties that are well-affected by the polymer skeletal dimensions of the precursor and the degree of deacetylation (Elieh-Ali-Komi and Hamblin 2016). For instance, Cao et al. (2017) fabricated a chitosan–lithium ion based supramolecular hydrogel electrolyte for designing a flexible asymmetric MnO_2/AC supercapacitor. Here, the right extent of complexation between Li^+ ions and the biopolymer hydrogel promoted high ionic transport and active mass transfer, along with exceptional cycling stability, even on surpassing 10,000 cycles with only the slightest capacitance loss, reporting an utmost areal capacitance of 10 mF cm^{-2}. In another report, the researchers employed a mild auto-initiated UV polymerization technique to fabricate an ionic conducting polymer water gel composed of 1-ethyl-3-methylimi dazoliumchloride, hydroxyethyl methacrylate, chitosan, and water, having non-covalent cross-linked interactions that enriched the polymer gel electrolyte with exceptional compressive durability and auto-recovering capability (Liu et al. 2014).

Other bio-polysaccharide materials, such as alginate and agarose, have also been useful in composing hydrogel electrolytes for electrochemical capacitors with appreciable outputs (Armelin et al. 2016; Elieh-Ali-Komi and Hamblin 2016; Pérez-Madrigal et al. 2016). For example, Moon et al. (2015) reported a safe, environmentally welcoming, and biodegradable gel polymer electrolyte based on NaCl-agarose for green and flexible supercapacitors that demonstrated a specific capacitance of 286 F g^{-1}, good rate capability, and a capacitance retention efficiency of 80% (Moon et al. 2015). Further, flexible PEDOT-based supercapacitors aided with various types of biopolymers, such as sodium alginate, k-carrageenan, chitosan, and gelatin, as electrolytic components, have been studied and compared (Syahidah and Majid 2013). It was observed that among them, the linear sulphated polysaccharides in k-carrageenan were the most suitable one as biopolymer hydrogel electrolytes that demonstrated an utmost gravimetric capacitance of 81 F g^{-1} and a 90% capacitance retaining efficacy, even after 3000 charging/discharging cycles with negligible leakage currents (Sudhakar and Selvakumar 2012; Sudhakar et al. 2014). Also, guar gum (GG) based on galactose- and mannose-containing polysaccharides, gum-Arabic (GA), a complex mixture of glycoproteins and polysaccharides, gelatin with a high amount of proline, hydroxyproline, and glycine moieties, have also been employed as polymer hydrogel electrolytes in energy storage applications (Wang, Zhu, et al. 2020). Recent studies infer that gelatin can be a harmless and sustainable electrolyte in different screen and stencil-printed supercapacitors for Internet of things (IoT) applications. The electrical properties were tested under severe mechanical stress for a 2 M NaCl/gelatin electrolyte that yielded appreciable results with a high degree of bendability. The leakage currents were relatively low during bending postures and no short circuits or permanent degradations were detected (Railanmaa et al. 2019). However, most gelatin-based hydrogels still suffer largely from mechanical instability and thus experience restricted commercial applications in flexible supercapacitors (Choudhury et al. 2008).

Biopolymers producing recyclable and degradable hydrogel electrolytes may be one of the solutions to the issues related to biocompatibility and high fabrication costs, although until now only limited advancements have been realized. Though exploiting biopolymers as safe, environmentally friendly electrolytes have been highly appreciated, but the so far observed capacitive and temperature operating achievements are well below the practical limits and hence sincere efforts are highly required so as to complete and replace the exploitation of synthetic polymers by designing more sustainable and ecofriendly devices in scalable quantities, as well as modifying and catering to the requirements chosen for the intended purpose.

To cope with the limitations of polymer hydrogel electrolytes due to the involvement of a water plasticizer, Morita et al. expanded the functional temperature range of the GPE and prepared non-aqueous based compositions with a poly(ethylene oxide)-modified polymethacrylate (PEO-PMA) matrix and anhydrous H_3PO_4 (Morita et al. 2004). However, employing large concentrations of phosphoric acid often restricted the working potential range to below 1 V. Thus, Łatoszyńska et al. deliberately substituted phosphoric acid with phosphate ester (aryl phosphates) that showed high acidity, polar solvent miscibility, and chemical compatibility for advanced supercapacitor applications (Łatoszyńska et al. 2015).

2.3.2.2 Polymer Organogel Electrolytes

Organogels are prepared in a similar procedure to that of hydrogels, instead replacing water by organic solvents such as acetonitrile (ACN), ethylene carbonate (EC), propylene carbonate (PC), dimethylformamide (DMF), dimethyl carbonate (DMC), tetrahydrofuran (THF), dimethylformamide (DMF), and so on, along with various salts (e.g., $LiClO_4$, $LiCF_3SO_3$, etc.) embedded in the host polymer matrix composed of poly(ethyleneoxide) (PEO), poly(vinylpyrrolidone) (PVP), poly(vinylalcohol) (PVA), poly(methylmethacrylate) (PMMA), poly(vinylidenefluoride) (PVdF), and the copolymer poly(vinylidenefluoride-*co*-hexafluoropropylene) (PVDF-HFP) and similar others, to induce high ionic conductivity, which otherwise limits the charge/discharge kinetics as well as obtains wider operating voltages between 2.5 and 3.0 V, in an attempt to highly boost the overall energy and power density (Yang et al. 2015). Some of the interesting electrochemical results for polymer organogel electrolyte-based supercapacitors are illustrated in Table 2.3.

A highly stretchy and patchable micro-supercapacitor based on a PMMA/PC/$LiClO_4$ polymer organogel electrolyte was demonstrated by Lee et al. (2015) that obtained appreciable electrochemical stability under ambient conditions, even without any encapsulation, as well as delivering a high volumetric capacitance of 8.9 F cm^{-3} that deviated slightly, even on stretching, twisting, winding, and bending deformation. In another instance, PVA was used as a polymer host for an organogel electrolyte mixed with lithium salt and organic solvent *N*-methyl-2-pyrrolidone. This gel electrolyte showed high electrochemical performance in both rechargeable batteries as well as in carbon-electrode-based electrochemical capacitors. The conductivity of this gel electrolyte was found to be as high as 2×10^{-3}–5.8×10^{-4} S cm^{-1} at room temperature (Chatterjee et al. 2010). Schroeder et al. designed a lithium doped methacrylate-based gel polymer electrolyte (high ionic conductivity of 1.8 mS cm^{-1}) based supercapacitor that could be charged to 4.0 V and attain a cell gravimetric capacitance of 24 F g^{-1}, along with a high coulombic efficiency of 100% and remarkable electrochemical stability, recording a 90% capacitance retention efficacy even on surpassing 50,000 continuous charging/discharging tests (Schroeder et al. 2013). Chiou

TABLE 2.3
Polymer Organogel Electrolytes for Supercapacitors

Electrolyte Materials	Electrode Materials	Conductivity	Specific Capacitance	Cyclic Stability	Reference
PMMA/PC/ACN/TBAPF$_6$	SWCNT	~3 mS/cm	22.2 F g^{-1}	94% capacitance retention after 500 cycle	Choudhury et al. (2008)
PMMA/PC/LiClO$_4$	MWCNT/Mn$_3$O$_4$	—	8.9 F cm^{-3} @ 0.1 A cm^{-3}	93% capacitance retention after 30,000 cycles	Łatoszyńska et al. (2015)
LiPF$_6$/PVdF-HFP	Vertically aligned CNTs	—	10 F g^{-1} @ 1 A g^{-1}	≥95% capacitance retention for 10,000 cycles	Morita et al. (2004)
PVP/PVdF-HFP/(Mg(CF$_3$SO$_3$)$_2$)	Porous carbon	2.16 × 10^{-4} S cm^{-1}	106 F g^{-1}	>95% capacitance retention after 1000 cycles	Yang et al. (2015)
Poly(ethylene glycol) blending poly(acrylonitrile)/DMF/LiClO$_4$	Carbon	9.2 × 10^{-4} S cm^{-1}	101 F g^{-1} @ 0.125 A g^{-1}	~100 % capacitance retention after 30,000 cycles	Yuksel et al. (2014)
Triblock poly(acrylonitrile)-*b*-poly(ethylene glycol)-*b*-poly(acrylonitrile)/DMF/ LiClO$_4$	Highly porous carbon derived from mesophase pitch activation	~10^{-2} S cm^{-1}	183 F g^{-1} @ 0.125 A g^{-1}	—	Lee et al. (2015)

et al. (2013) constructed LiPF$_6$/PVdF-HFP gel electrolyte supercapacitors that showed appreciable cyclic performances even at high voltages. However, PVdF-HFP copolymer electrolytes largely suffer from relatively poor mechanical stability and hence, to augment it, Yang et al. used the agent trimethylolpropane triacrylate for cross-linking PVdF-HFP film that restored almost 250% of the liquid electrolyte and revealed a higher ionic transport (Yang et al. 2015). Further, the assembled flexible supercapacitor using this cross-linked-PVdf-HFP organogel electrolyte presented an operating voltage window of 2.9 V, a volumetric capacitance of 5.2 F cm^{-3}, a volume energy density of 5.16 mW h cm^{-3}, and a specific energy of 41.4 W h kg^{-1}. Wang, Yang, and Wang (2016) further designed a highly stretchable organogel electrolyte via a covalently cross-linked fluoroelastomer, hexamethylene diamine as a host polymer, tetraethylammonium tetrafluoroborate [Et$_4$N][BF$_4$] as an electrolytic salt, and acetonitrile as a plasticizer that exhibited outstanding stretchability and an ionic conductivity of 9.9 × 10^3 S cm^{-1}, a high specific energy of 58.2 W h kg^{-1} in between 0 and 2.7 V, and an 88% capacitance retention efficiency for 10,000 consecutive GCD cycles. For more improvement in room temperature ionic conductivity, and mechanical and dimensional stability, Jain and Tripathi mixed SiO$_2$ particles with PVDF-HFP/Mg(ClO$_4$)$_2$–PC electrolytes assembled with EDLC electrodes with appreciable electrochemical responses (Jain and Tripathi 2013). Ramasamy, Palma Del Val, and Anderson (2014) prepared a PEO-based organogel polymer electrolyte mixed with sodium bis(trifluoromethanesulfonyl) imide salt, together with mixed solvents based on PC, DMC, and EC, for constructing symmetrical EDLC cells. The mixed gel electrolyte reported improved ionic conductivity (7.6 × 10^{-4} S cm^{-1}) compared to the pristine-PC gel electrolyte (5.4 × 10^{-4} S cm^{-1}). Thus, the resultant mixed organogel electrolyte-based supercapacitor achieved an improved cell gravimetric capacitance of 25 F g^{-1} and a specific energy of 20.0 W h kg^{-1}, as well as experiencing only 3% capacitance loss even after 2000 charge/discharge cycles on being charged to 2.5 V.

Nevertheless, the organogels often suffer from the disadvantages of low ionic conductivity, flammability, and electrochemical and thermal instabilities and thus further innovation in their structural and functional properties are unavoidable (Agrawal and Pandey 2008). A similar attempt was made by adding poly(ethyleneglycol)-diglycidylether-3,3,5,5-tetrabromo bisphenol A, having a polymeric network to serve as the host matrix embedded in an organic liquid electrolyte (lithium salts in propylene carbonate solvents). The polymer served as a proficient flame retardant, as well as imparting high conductivity and thermal and chemical stability to the system (Na et al. 2017). Nonetheless, such copolymer fabrication often faces low yield problems and high production expenses that restrict its abundant promotion in reality. Moreover, the organic electrolytes usually suffer from high production cost and toxicity, resulting in safety-related issues. Their hygroscopic nature and flammability necessitate the inevitable usage of well-encapsulating arrangements to ensure a dry and air-free environment, which correspondingly elevates the fabrication costs and complicates device designing procedures. Therefore, it is of major concern to design GPEs based on non-volatile, non-hygroscopic, non-toxic, inflammable solvents that display an extended working potential range and better thermal and electrochemical stabilities. Recently, a further novel class of GPEs has been proposed comprising of non-volatile ionic liquids, as highlighted in the next section.

2.3.2.3 Polymer Ionogel Electrolytes

Polymer based ionogel electrolytes involve the appropriate selection and designing of ionic liquid components and polymer hosts to cope with the difficulties faced by the previously discussed electrolyte types selected for designing flexible high-performance supercapacitors (Yan et al. 2014). Fundamentally, ionic liquids take the advantages relative to organic electrolytes in terms of improved thermal, chemical and electrochemical properties, low combustibility, good ionic conductivity, along with an expanded electrochemical stability window as high as 6 V (Armand et al. 2009; Ma et al. 2018). More significantly, the properties of these systems are vastly tailorable and strongly affected by the nature and directionalities of interactions between structural moities and inter-ionic interactions that provide a useful possibility and opportunity for planning ionic liquids with specific properties for desired applications (Nègre et al. 2016; Watanabe et al. 2017; Asbani et al. 2019). Studies reveal that the electrochemical properties of ionogels strongly depend on the designing and character of the chosen ionic liquids, polymer matrix, and their chemistries. Accordingly, Liew et al. (2014) found that the introduction of a 1-butyl-3-methylimidazolium chloride [BMIM][Cl] ionic liquid reduced the glass transition temperature of PVA polymer electrolytes well below room temperature, making it suitable for electrolytic usages in supercapacitors. Thus, a variety of host polymers such as PVA, PEO, PMMA, PEG, and PVDF-HFP has been proposed for ionogel electrolytes (Chandra et al. 2000; Pandey and Hashmi 2013; Wang et al. 2014). In a comparative study, it was reported that PVDF-HFP offered better structural stability and channel pathways for faster ionic conduction than PEO using these ionic liquids (Elamin et al. 2019). Tamilarasan and Ramaprabhu (2014) used PMMA as a polymeric host to prepare an extremely transparent and stretchable PMMA/1-butyl-3-methylimidazolium bis(trifluoromethane)sulfonimide [BMIM][TFSI] ionogel showing around a four-fold stretchability compared to a pristine polymer host. The assembled PMMA/[BMIM][TFSI]-based symmetric EDLC having graphene electrodes displayed a cell gravimetric capacitance of 83 F g^{-1}, along with a specific energy of 26.1 W h kg^{-1} @ a specific power of 5 kW kg^{-1}. Yang et al. showed that PVDF with an extra-high chemical, thermal, and mechanical tenacity can be used as an effective matrix for the [BMIM][BF_4] ionic liquid to formulate a polymeric ionogel for plastic electrochemical supercapacitors (Yang et al. 2014). Further, they combined an amorphous polyvinyl acetate with PVDF to form the resulting ionogels containing [BMIM][BF_4] that attained mechanical durability, thermal stability, as well as a high room temperature ionic conductivity. Recently, Tiruye and his team (Tiruye et al. 2015) explored a binary hybrid of polymeric ionic liquids comprising poly(diallyldimethylammonium) bis(trifluoromethanesulfonyl) imide ([pDADMA][TFSI]) and [PYR_{14}] [TFSI] (4:6 by weight ratio) as an electrolyte for supercapacitors. The resultant pyrrolidinium-based ionic liquid possessed a substantial electrochemical stability window compared to commonly obtainable imidazolium-based analogues. A similar strategy was adopted by the Muchakayala group to employ 1-methyl-1-propylpyrrolidinium bis(trifluoromethyl sulfonyl)imide entrapped in poly(vinylidenefluoride-hexafluoropropylene) for flexible solid state supercapacitor applications with good thermal, electrochemical, and morphological stability (Muchakayala et al. 2018). Further, Shirshova group (Shirshova et al. 2013) designed ionic liquid-based epoxy resin composites with high mechanical (Young's modulus of 0.2 GPa) and electrical features, displaying an elevated room temperature ionic conductivity of 0.8 mS cm^{-1}, thereby confirming its prominent importance in devising future generation flexible electronic gadgets.

As construction of highly safe and leakage-free energy storage systems necessitates the decent adherence and wettability of the ionogel electrolytes on the electrode surface, accordingly ionogel polymer electrolytes having remarkable adhesive properties were prepared using poly(1-vinyl-3-propylimidazolium bis(fluorosulfonyl)imide) and 1-ethyl-3-methyl imidazolium bis(fluorosulfonyl)imide ionic liquids that showed appreciable ionic conductivity, good adhering ability, and electrode wettability, which consequently improved the device rate capability and cyclability (Suleman et al. 2013; Hashmi 2014).

Further, polymer ionogel electrolytes have also been employed to develop flexible high quality micro-supercapacitors recently (Wang, Hsia, et al. 2014). To demonstrate this, Kim et al. designed a stretchable ionogel-based micro-supercapacitor array composed of 1-ethyl-3-methylimidazolium bis(trifluoromethanesulfonyl) imide [EMIM][NTf_2] ionic liquid and modified PMMA that recorded an appreciable specific energy of 20.7 W h kg^{-1} and a maximum specific power of 70.5 kW kg^{-1} with negligible performance degradation, even on being subjected to a 30% stretching strain (Kim, Shin, et al. 2013). Lim et al. (Lim and Yoon 2014) utilized poly(ethyleneglycol) diacrylate and 1-ethyl-3-methylimidazolium bis(trifluoromethylsulfonyl) imide [EMIM][TFSI] as an ionogel electrolyte along with CNTs as electrodes to construct biaxially stretchable arrays of micro-supercapacitors. The components and the designing of such biaxially stretchable micro-supercapacitor arrays is illustrated in Figure 2.6 along with the comparative electrochemical performances that displayed good electrochemical stability with encouraging energy storage capabilities. Table 2.4 summarizes some of the more interesting examples of the electrochemical performance of a selection of flexible supercapacitors fabricated with ionogel electrolytes to furnish readers with a better understanding.

Thus, the above studies reveal that the ionogel electrolytes, in most cases, display relatively lower capacitances primarily due to poor ionic conduction compared to hydrogel electrolytes, even though they recorded a superior working potential range and safe handling guarantees superior to the common organogels. Further, their designing involves many complex structured molecules/ionic species that make their large-scale production highly expensive and dramatically elevates the total supercapacitor fabrication costs. Additionally, deeper insights regarding the optimization of electrode/electrolyte interfacial interactions are

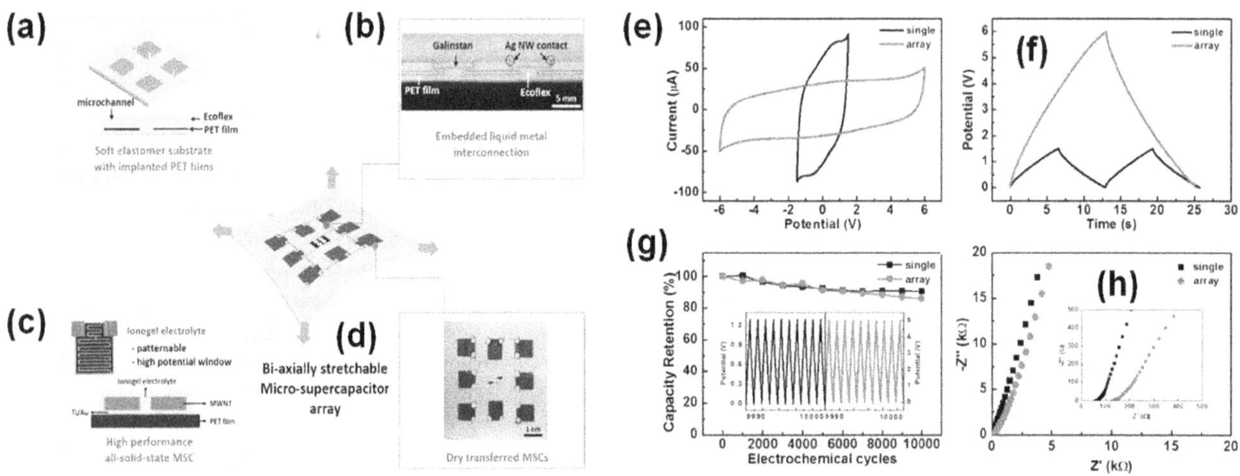

FIGURE 2.6 Schematic presentation of a biaxially stretchable micro-supercapacitor array designed as a flexible substrate. (a) Pictorial illustration of the flexible substrate, made of Ecoflex with implanted polyethylene terephthalate (PET) films entrenched in interconnected microchannels for liquid metal. (b) Cross-sectional view of liquid metal interlinks embedded within the substrate and sealed with an AgNW (silver nanowire) contact. (c) Optical image (top) and cross-sectional scheme (bottom) of the high-performance all-solid-state micro-supercapacitor containing ionogel electrolyte. (d) All-solid-state micro-supercapacitor array attached to the deformable substrate using a dry-transfer technique. (e) Comparative CVs of a single all-solid-state micro-supercapacitor (black) and an all-solid-state micro-supercapacitor array (red) obtained via layer-by-layer deposited MWNT electrodes in an iongel electrolyte at a voltage scan rate of 1 V s^{-1}. (f) Galvanostatic charging/discharging profiles for a solo-unit all-solid-state micro-supercapacitor (black) and an all-solid-state micro-supercapacitor array (red), both recorded at a current of 13.2 μA. (g) Variation of capacity retention efficiency of a solo all-solid-state micro-supercapacitor (black) and an all-solid-state micro-supercapacitor array (red) with 10,000 charging/discharging cycles at a current of 30 μA. Inset shows the observed charging/discharging characteristics between 9990 and 10,000 cycles for both the samples. (h) Nyquist plots of a solo all-solid-state micro-supercapacitor (black) and an all-solid-state micro-supercapacitor array (red) at a frequency range of 500 kHz to 0.1 Hz, respectively. (Lim, Yoon, et al. 2014 with permission of ACS publications)

TABLE 2.4
Polymer Ionogel Electrolytes for Supercapacitors

Electrolyte Materials Used	Electrode Materials Used	Specific Capacitance @ Current Density/ Voltage Scan Rate	Capacitance Retention Efficiency @ Cycle Number	Reference
PVdF-HFP/BMIBF$_4$ (1-butyl3-methylimidazolium tetrafluoroborate)	Poly(3,4-ethylenedioxythiophene) coated carbon-fiber papers	154.5 F g^{-1} (85 mF cm^{-2})	Stable 10,000 cycles	Pandey et al. (2014)
PEGDMA/PEGMA/P$_{13}$FSI	Activated carbon	36 F g^{-1} @ 2 mV s^{-1}	~ 100% for 2500 cycles	Chaudoy et al. (2017)
PVdF-HFP/SN/BMImBF$_4$	Activated carbon	176 F g^{-1} @ 0.18 A g^{-1}	80% after 10,000 cycles	Pandey et al. (2016)
PS-PEO-PS/[EMIM][NTf$_2$])	CNT	20 mF cm^{-2} @ 10 A g^{-1}	>99.5% @ 10 A g^{-1}	Kang et al. (2012)
Diol-borate ester-cross-linked polyvinyl alcohol (PVA) with 1-ethyl-3-methylimidazolium chloride ionogel electrolyte	Activated carbon	90 F g^{-1} @ 0.1 A g^{-1}	98% @ 3000 cycles	Wu, Xia, et al. (2020)

important to initiate the proper matching of the electrode porosity with the electrolyte ionic size to boost the overall energy and power characteristics of the devices. Therefore, proper tuning of the ionic conductivities of ionogels, and the fabrication of low-cost ionic liquids, are expected to be addressed in order to speed up the charge/discharge kinetics and power densities.

2.3.2.4 Proton-Conducting Gel Polymer Electrolytes

Proton-conducting gel polymer electrolytes show high ionic conductivity among all the polymer electrolytes, although they exhibit a narrower voltage window and temperature range (Gao and Lian 2014). Nafions (perfluorosulfonic acid polymers) are popular proton-conducting polymers, widely used as an electrolyte for several types of solid-state electrochemical capacitors because of its high ionic conductivity (Sekhon 2003). EDLCs assembled with a Nafion membrane have exhibited higher electrochemical stability, although the observed capacitances are often inferior to that of typical liquid cells. Thus, the introduction of a uniform inter-distributed Nafion solution on carbon electrodes has yielded superior results due to the improved adhesion of the electrolyte on the electrodes' surfaces (Lufrano and Staiti 2004). Since pure Nafion is quite expensive, several attempts have been made to modify the limitations faced in fabricating solid state supercapacitors with these electrolytes (Lufrano, Staiti and Minutoli 2004). For instance, a proton conducting gel polymer composite containing chitosan and methylcellulose was employed for ELDCs that showed an extended cell output voltage of 1.8 V (Shukur et al. 2013). Likewise, a 2-hydroxymethylmethacrylate monomer mixed with two dissimilar liquids (propylene carbonate and *N,N*-dimethylformamide) was employed as a proton conducting gel polymer electrolyte that showed a wide operating temperature range (from −40° to + 80 °C) with enhanced ionic transport features (Łatoszyńska et al. 2017). In another attempt, an ionic liquid-based proton conducting gel polymer electrolyte was designed based on PVA, ammonium acetate, and 1-butyl-3-methylimidazolium bromide. The combination of the ionic liquid increased the amorphous character of the polymer gel, extended the electrochemical voltage window, and also improved the thermal properties of the polymer electrolyte (Liew et al. 2015). The electrochemical performance of some interesting proton-conducting polymer gel electrolyte-based supercapacitors are shown in Table 2.5.

Many proton-conducting polymer electrolytes have recently emerged that are relatively low-cost and with an improved functional temperature window that promotes their commercial utilization in integrated wearable and micro-electronic gadgets. However, most of these proton-conducting gel polymer electrolytes exhibit a much inferior ionic conductivity at low operational temperatures and in low relative humidity environments, and thus necessitate further improvement for an extensive mode of applications. Additionally, they report relatively lower observed cell capacitances than the popular liquid electrolyte-based energy storage systems. They also often display restricted operating voltage ranges, analogous to hydrogels, and thus need further materialistic improvement.

2.3.3 Polyelectrolytes

Polyelectrolytes are substances that possess a polymeric skeleton with covalently bonded electrolytic functionalities (Bagchi et al. 2020). They can be categorized as salts, acids, and bases, and they generally dissociate to form ionic species in a polar medium, leading to the formation of charged polymer strands and inversely charged counter ions. Depending on the chemical character of the polyelectrolyte, the polymer chains get negatively or positively charged and are accordingly called polyanions or polycations, respectively. The bulky polymer chains are generally immobile in solid-state films while the counter ions contribute to their ionic conductivity. The studies on such indispensable polyelectrolytes are in the very nascent stage but have shown very promising results (Huang, Zhong, et al. 2015). A flexible water-deactivated polyelectrolyte hydrogel composed of carboxy methyl cellulose and methacrylamidopropyltrimethyl ammonium chloride, with good mechanical flexibility and high ionic conductivity, was assembled with activated carbon. The resultant flexible supercapacitor recorded a wide functional voltage of 2.1 V accredited to the subdued electrochemical reactivity for water within the hydrogel environment, regulated by the "molecular caging" offered by the functional groups of the polyelectrolyte to the neighboring water molecules (Wei et al. 2018). In another report, a highly conducting polyelectrolyte (ionic conductivity of 7.16 S m^{-1}) containing polyacrylic acid

TABLE 2.5
Proton Conducting Gel Electrolytes for Supercapacitors

Electrolyte Materials	Electrode Materials	Protonic Conductivity (S cm^{-1})	Gravimetric Capacitance @ Current Density/ Voltage Scan Rate	Capacitance Retention Efficiency @ Cycle Number	References
Nafion	RuO_2-Nafion	—	200 F g^{-1} @ 20 mV s^{-1}	70% @ 10,000 cycles	Park et al. (2002)
Poly (styrene sulfonic acid)	CNTs	1.5×10^{-3}	85 F g^{-1} @ 1 mA cm^{-2}	53% @ 7000 cycles	Wee et al. (2010)
Polyvinyl sulfonic acid	PANI/carbon	—	98 F g^{-1} @ 10 mV s^{-1}	>80% @ 1500 cycles	Sivaraman et al. (2006)
Nafion-115	Carbon (Vulcan XC 72)	1×10^{-2}	20 F g^{-1} @ 1 mV s^{-1}	95% @ 100 cycles	Ramya et al. (2010)
Nafion-117	PPy/PSS	3×10^{-3}	20 F g^{-1} @ 50 mV s^{-1}	—	Pickup et al. (2000)
Sulfonated poly(ether ether ketone) (S-PEEK)	Activated carbon	4.3×10^{-3}	138.4 F g^{-1} @ 2 mA cm^{-2}	~100% @ 1000 cycles	Kim, Ko, Kim, and Kim (2006)
Nafion/ polytetrafluoroethylene	Carbon (Vulcan XC 72)	1×10^{-3}	16 F g^{-1} @ 1 mV s^{-1}	95% @ 100 cycles	Subramanium et al. (2011)
Sulfonated poly(fluorenyl ether nitrile oxynaphthalate)copolymer	Activated carbon	6×10^{-3}	158 F g^{-1} @ 20 mA cm^{-2}	97% @ 1000 cycles	Seong et al. (2011)

cross-linked via methacrylate-functionalized graphene oxide was used because of its outstanding stretchability (even under a 950% strain) and self-healing efficiency (after exposure to 300% deformation), empowering its usage in self-healing and stretchable supercapacitors (Jin et al. 2018). A cationic polyelectrolyte composed of a poly(*N*-vinylpyrrolidone-*co*-diallyl dimethyl ammonium chloride) (P(NVP-*co*-DMDAAC)) copolymer and PVA was obtained by an impregnating polymer cross-linking technique. The introduction of the cationic DMDAAC monomer and polar PVA components enhanced the integral performance of the PNVP hydrogel. The P(NVP-*co*-DMDAAC)/PVA interpenetrating polymer cross-linking hydrogel showed improved ionic transportation characteristics compared to neat-PVA, PNVP, and P(NVP-*co*-DMDAAC) electrolytes. Further, increasing the PVA content, promoted the tensile strength considerably (Wang, Deng, et al. 2019). A few years ago, Guo et al. (2016b) discussed an auto-healing ferric ion interlinked supramolecular poly(acrylic acid) hydrogel, employed for supercapacitor applications with appreciable mechanical tenacity (extensibility > 700% even under stress > 400 kPa) and ionic conductivity (0.09 S cm^{-1}). The group further developed an integrated supercapacitor utilizing a physically cross-linked PVA-hydrogel electrolyte that exhibited a remarkable tensile strength of 470 kPa and a breaking strain of 380%, although it presented only a moderate mechanical healing (52.3%) efficiency (Guo et al. 2018). Further, a hydrogel electrolyte composed of poly(acrylic acid)-grafted-PVA was prepared that successfully restored its mechanical properties spontaneously and repeatedly (Wang, Tao, and Pan 2016). The same polyelectrolyte was employed in a sandwich-shaped ultracapacitor that was successful in achieving an utmost cell gravimetric capacitance of 84.8 F g^{-1} @ a current density of 1.0 A g^{-1} even after enduring cutting and healing procedures, compared to the neat values reported to be 85.4 F g^{-1} at 1.0 A g^{-1}(Liu et al. 2018).

Multifunctional polyelectrolytes motivate an intrinsic self-healing capability and high stretchability (Luo, Sun, et al. 2015). This urges high ionic conduction as well as reversible and efficient crack-activated hydrogen bonds cross-linking interactions among the polymeric strands. In another exploration, a hydrogen bonding double-cross-linked polyacrylic acid and vinyl hybrid silica nanoparticle-based polyelectrolytes exhibited a superior intrinsic self-healing capability and high stretchability (about 600% strain resistibility), undergoing a 100% capacitance retention even after 20 breaking/healing cycles (Huang, Zhong, et al. 2015). Recently, a distinctly transparent, stretchy, and auto-healing polyelectrolyte comprising of poly(2-acrylamido-2-methyl-1-propanesulfonic acid) (PAMPA)/PVA/LiCl was designed via a single-step radical polymerization technique with a significantly improved mechanical and self-healing capacity. The system demonstrated a higher stretchability of 938%, was resistible to a tension of 112.68 kPa, had an appreciable ionic conductivity of 20.6 mS cm^{-1}, and an elevated healing competence of 92.68% after a period of 1 day. The assembled supercapacitor with polypyrrole-coated single-walled carbon nanotubes as electrodes resulted in achieving a large areal capacitance of 297 mF cm^{-2} @ a current density of 0.5 mA cm^{-2} along with a decent rate capacity (218 mF cm^{-2} @ 5 mA cm^{-2}). Also, by dissecting the supercapacitor into two halves repeatedly, it underwent minimum alterations in internal resistances and successfully restored 99.2% of its original specific capacitance through an auto-healing process in 1 day under ambient conditions (Zhang, Li, Huang, et al. 2019).

A convenient strategy to successfully suppress the self-discharge of distinctive activated carbon-based supercapacitors is successfully executed by modifying their porous cellulose separators using cationic polyelectrolytes such as polyethyleneimine that minimizes the ion diffusion process guided by a concentration gradient without compromising the electrochemical performances (Wang, Zhou, et al. 2018). A highly conducting polyelectrolyte formed by combining PAMPA and silica–sulfobetaine silane zwitterion was used for devising symmetric supercapacitors that attained an appreciable gravimetric capacitance of 321 F g^{-1} @ a voltage sweep rate of 10 mV s^{-1}, almost twice as large as that of pure PAMPA, as well as a ~100% capacitance retaining efficacy even on surpassing 5000 charging/discharging cycles. The ionic channels that generated the zwitterions accelerated the ion mobility within the electrolyte (Afrifah et al. 2020). The sophisticated artificial leather garment industry is highly inspired by the introduction of wearable energy storage devices using polyurethane artificial leather supercapacitors. Here, a large sheet of electrodes embedded within the leather layers simultaneously serves as a polyelectrolyte which successfully promotes the designing of comfortable wearable electronic suits. In contrast to PVA-based acidic electrolytes, NaCl is used here for modifying the ionic conductivity of intrinsically fluorescent polyurethane leathers without any toxic effects to humans. Further, this fluorescent leather supercapacitor is readily transformable to various arbitrary substrates to acquire numerous wearable shapes, empowering multi-functionalities to a real fashion industry (Huang, Tang, et al. 2018). Similarly, polyelectrolyte-based hydrogels obtained via the single-step polymerization of acrylic acid and poly(ethylene glycol) methacrylate in the presence of oppositely charged branched polyethylenimine showed improved mechanical stretchability, a self-healing aptitude, as well as an appreciable self-recovery (Fang and Sun 2019). The synergistic influence of electrostatic and hydrogen-bonding interactions ensured a high tensile strength of ~4.7 MPa, as well as making it resistant to breaking, thereby withstanding a ~1200% strain. Further, the hydrogels recovered to their initial state within 10 min after experiencing a ~300% strain at room temperature without any external backing-up processes. Furthermore, they exhibited a rapid shape recovery speed by undergoing auto-healing from a physical cut at room temperature, thus demonstrating exceptional shape-fixing and shape-recovery ratios (Feng et al. 2020).

2.4 THE IONIC CONDUCTION MECHANISM IN VARIOUS POLYMER ELECTROLYTES

The present section highlights the fundamental understanding of the ionic conduction mechanism in the polymers employed as electrolytes for their appropriate functioning in flexible energy storage devices.

2.4.1 Ionic Conduction in Solid (Solvent-Free) Polymer Electrolytes

Ionic transport in solid polymer electrolytes is very complex and depends on a number of factors, such as salt concentration, the dielectric constant of the host polymer, the degree of salt dissociation and ion aggregation, and the mobility of polymer chains (Johansson 2001). The selection of a polymer matrix mainly depends on the presence of a polar group so that it can easily coordinate with the ions present in the electrolyte; besides, ensuring less restraint during bond rotation. The ion transportation mostly depends on polymer chain segmental motion that is favored in the amorphous region or above the glass transition temperature (T_g) of the polymer. Hence, ideally the host polymer must have a lower T_g to remain in the elastic phase in order to promote the ion conductivity under ambient conditions (Johansson 2001). Further explorations inferred that the mechanical properties (stiffness modulus), which reflect the chain dynamics at the macroscopic level, also parallel the conductivity results. In solid state polymers, the solvating sites are covalently linked together through flexible bonds (Armand 1990). Thus, although the net displacement of the ligand with the ions over macroscopic distances is forbidden, the segmental motion of the chains promotes the ionic conductivity via a solvation–desolvation process along the chains, as shown in Figure 2.7.

Thus, popular poly(ethylene oxide) (PEO) based polymer electrolytes contain both crystalline and amorphous phases, the latter playing a more contributory role in the ion conduction process (Geiculescu et al. 2014). Nonetheless, these polymer electrolytes show poor room temperature ionic conductivity (10^{-6} to 10^{-4} S m^{-1}) and hence are still far from being utilized in real-world energy storage devices. Lately, investigations with chitosan polymers and sodium triflate (NaTf) indicated that an increase in the dielectric constant at different temperatures led to a rise in DC conductivity and the ion conduction mechanism was found to follow Arrhenius behavior:

$$\sigma_{dc}(T,\varepsilon) = \sigma_0(T,\varepsilon)e^{-Ea/K_BT},$$

where σ_0 is a pre-exponential factor, E_a is the activation energy, K_B is the Boltzmann constant, and T is the temperature on the absolute scale (K) (Aziz et al. 2013). Further evidence was obtained via the exploration of the ion conduction mechanism in polyether/$Li_7La_3Zr_2O_{12}$ (LLZO) composite solid polymer electrolytes. Examination discovered that the rate of Li^+ conduction was predominantly limited by the polymer phase in such a composite system (Kato et al. 2020). Recent studies carried out on the mechanism of alkali and alkaline earth metal-ion conduction in a solid mixed-polymer matrix composed of poly(dimethyl siloxane) and poly(ethylene oxide) revealed significant improvement in the room-temperature ionic conductivity as per the proposed mechanism. *Ab initio* investigations of the interactions between the cations and the polymers showed that the oxygen sites in the polymer matrix served as the entrapment sites for these cations and that ionic conduction is likely to occur via hopping between adjacent oxygen sites. Molecular dynamics simulations confirmed that the dynamics of polymer mixes are more recurrent and that the better pronounced molecular vibrations in them are possibly triggering faster ionic activities between two consecutive oxygen entrapments, thereby accelerating the ion conduction compared to individual systems. The proposition has been experimentally validated by the enhanced observed ionic conductivity ($\sigma \approx 1.3 \times 10^{-3}$ S m^{-1} under normal conditions) and by the flexible multi-cation (Li^+, Na^+, and Mg^{2+}) conducting solid channel made up of the above mixed polymer system compared to that in the pristine polymer phases (Puthirath et al. 2020).

2.4.2 Ion Conduction in Gel Polymer Electrolytes

Gel polymer electrolytes (GPEs) are the focus of much attention because they combine the advantages of liquid electrolytes, such as high conductivity and good electrode/

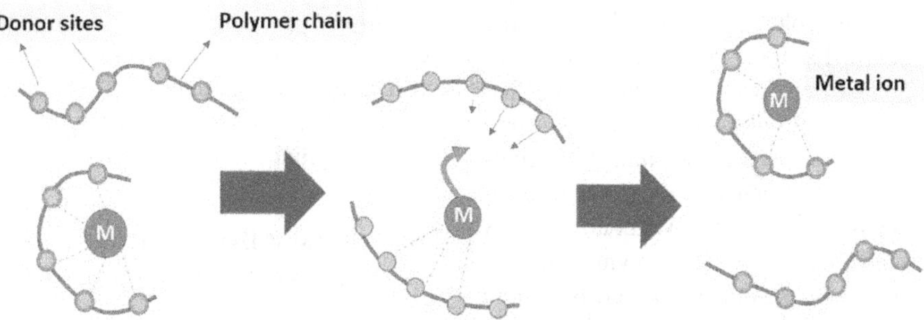

FIGURE 2.7 Schematic illustration of the movement of ions across the amorphous polymer mediated through the segmental motion of the polymer chains.

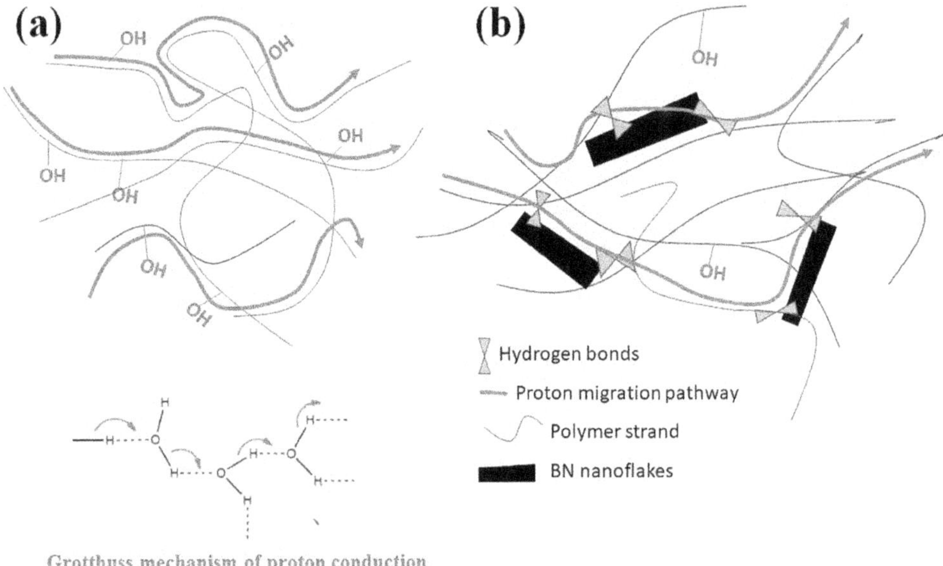

FIGURE 2.8 Illustration of the ionic (protonic) conduction mechanism in hydrogel polymer electrolytes. (a) Protonic conductivity promoted along the twisted PVA chains via intermolecular hydrogen bonds following the Grotthuss mechanism. (b) BN nanosheets/nanoflakes interconnected via intermolecular hydrogen bonds to the polar groups of the polymer, favoring faster conduction of ions (protons).

electrolyte contact, and solid electrolytes, such as safety and mechanical and thermal stability. Thus, in the popular proton conducting polymer electrolytes such as PVA–H_2SO_4, the common ion conducting mechanism is based on the Grotthuss mechanism, as illustrated in Figure 2.8(a) (Gao and Lian 2014). Here, the protons serve as the main carrier, conducting, via hopping, between the hydrogen bonds. On the addition of hexagonal boron nitride (h-BN) nanosheets to these, GPEs favor supplementary polar groups to facilitate the faster transport of H^+ ions via additional hydrogen bonds. The rigid nanosheets compared to the soft host polymer strands are more effective in shortening the transport paths, as depicted in Figure 2.8(b). Nonetheless, the high concentration of BN nanosheets introduces a steric blockade effect due to aggregation leading to suppressed ionic conductivity (Hu et al. 2017).

Likewise, a study of ionic conduction in organogel electrolytes, such as PVDF-type polymer organogel electrolytes, demonstrated that they comprised of swollen polymer chains that mainly contribute to the total conductivity of these gel electrolytes rather than the solution entrapped within the cavities (Saito et al. 2003). The ionic liquid-based GPEs are within the focus of the research owing to the necessary features of ionic liquids promoting high conductivity via free charge carriers, and also by enhancing the amorphous character of the host polymer performance of the polymer electrolyte, thereby serving as the plasticizer (Ma et al. 2016). A marked increment in the conductivity of a polymer electrolyte with an IL concentration was also reported by Gupta et al. (2018). Balo et al. (2018) described that the ionic conductivity of a polymer electrolyte (PEO + 20 wt.% LiFSI) upsurged with IL ($PYR_{13}FSI$) concentrations, but up to a critical concentration, from where the reverse effect was observed. It was proposed that the larger addition of IL into the polymer electrolyte increased the concentration of FSI^- anions that for an ion pair with Li^+ and restrict them from interacting with the etherial oxygen of PEO, thereby suppressing the overall ionic conductivity of the electrolyte.

2.5 POLYMER-BASED MULTIFUNCTIONAL FLEXIBLE SUPERCAPACITORS

To furnish steady power sources for flexible and wearable electronics, lightweight supercapacitors are being targeted based on low density smart materials without compromising energy storage performance (Sundriyal and Bhattacharya 2020). The miniaturization of electronics not only reduces the adverse consequences of environmental toxicity, but also provides the easy tagging of multi-functional properties through integrated devising for framing advanced energy harvesting cum storing gadgets (Xue et al. 2017). Thus, in recent years, enormous progress has been made on the fabrication of polymer-based flexible multifaceted supercapacitors based on special features like high stretchability, self-healing, shape memory, self-charging, and electrochromics, etc., to further augment its position in the global rankings of essential energy support systems (Guo et al. 2017).

2.5.1 Polymer-Based Stretchable or Compressible Supercapacitors

The urge to develop smart wearable electronic gadgets for supporting essential energy projects is very much associated with biomedical, defense, and astronomical sectors that are often exposed to different mechanical distortions, including stretching or elongating, twisting, bending, curving, folding, and compressing, which have necessitated the development of stretchable electronic devices (Meng et al. 2013; Huang, Zhong, et al. 2015). Such fabrication of stretchable and compressible supercapacitors needs the adequate selection of device components and meticulous configuration design. Essentially, assembling supercapacitors with all the stretchable constituents is difficult, but employing one or two major stretchable components with special conformation can well accommodate the desired stretchability. One of the most likely strategies is to employ flexible electrode materials capable of undergoing buckling and wavy conformations attached to flexible and stretchable substrates like silicone rubbers and polymer gels such as stretchable electrolytes (Huang, Zhong, et al. 2015; Huang, Zhong, et al. 2017). The adhesion between the device components such as electrolyte and electrode materials must be durable enough to withstand the strain. This requires the usage of very viscose gel electrolytes with intimate interfacial adhesion, made possible through the judicious modification of the electrode's surface to promote high affinity for the polymeric electrolyte (Yang, Deng, et al. 2013).

Typically, most of the exploration on compressible supercapacitors is based on the assembling of elastic sponges made up of composites of carbons and conducting polymers such as PANI/CNT sponges, PPy/Graphene sponges, CNT/PPy/MnO_2 sponges, and pyrolyzed-melamine foam with conducting polymer gel electrolytes (Zhao et al. 2013; Xiao et al. 2016). For instance, the complete integral device consisting of PANI/CNT sponges combined with a PVA/H_2SO_4 gel electrolyte can be arbitrarily compressed with a 60% compressible strain without any loss in electrochemical performance (Niu et al. 2015). In order to further improve functioning capability, recently a remarkably compressible, hybrid, double-cross-linked, lignin hydrogel electrolyte in combination with an H_2SO_4 solution, rigorously interconnected via physical and chemical networking (as schematically depicted in Figure 2.9a), was reported. The system attained exceptional shape regaining ability (Figure 2.9(b)) as well as high ionic conductivity of 0.08 S cm^{-1}. The corresponding compressional stress (4.74 MPa) at breaking point was found to be 40 times that of single cross-linked lignin hydrogel. Accordingly, the use of this highly flexible, double cross-linked lignin hydrogel to serve as electrolyte in the PANI/carbon cloth based flexible supercapacitor promoted appreciable specific capacitance retaining efficiency even after subjecting to wide range of deformation (Figure 2.9(c)), and compression strain (Figure 2.9(c and d)) performances respectively (Liu, Ren, Zhang, et al. 2020).

Again, a highly conducting and compressible polyacrylamide cross-linked by a methacrylated graphene oxide hydrogel polyelectrolyte was prepared that preserved admirable elasticity, even at −30 °C, along with its original conformation at 100 °C, thus outclassing the conventional PVA/H_2SO_4 gel electrolyte, as shown in Figure 2.9(e and f) respectively (Liu, Zhang, Liu, et al. 2020). Further, the device demonstrated an appreciable capacitance retention

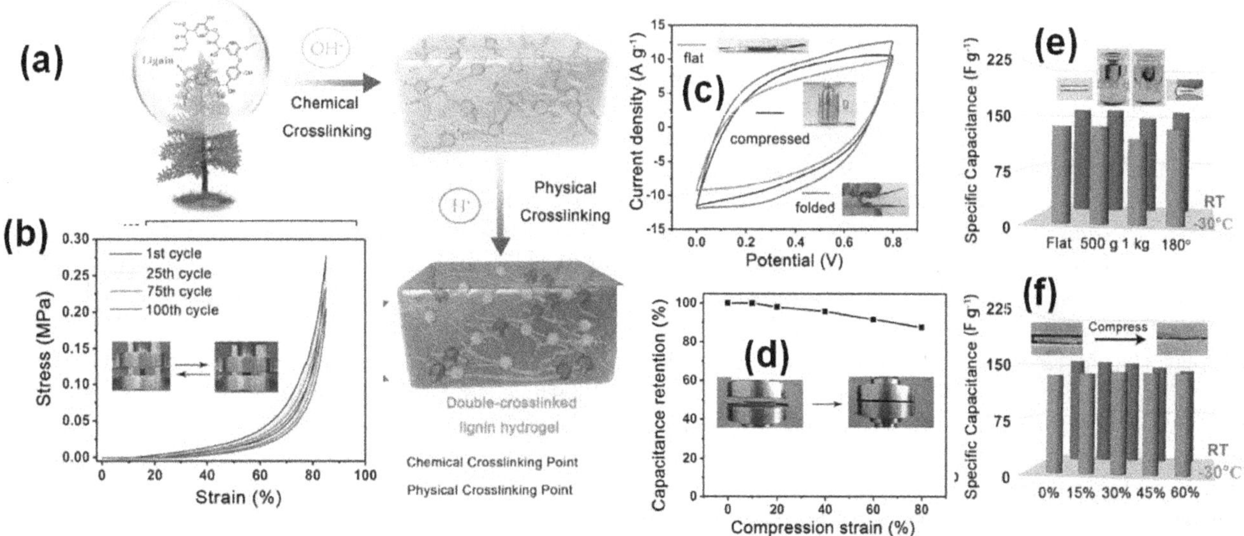

FIGURE 2.9 (a) Schematic presentation of eco-friendly, double-cross-linked, lignin hydrogel formation mediated by successive chemical and physical cross-linking approaches adopted for designing compressible and foldable supercapacitors. (b) Compressive stress versus strain profile for various loading–unloading cycles at different extents of strain application. (c) CV profiles of the fabricated supercapacitor using a double-cross-linked hydrogel electrolyte (voltage scan rate @ 75 mV s^{-1}) under different deformation positions (flat, compressed @ 500 g, folded at 180° bending) as shown by the inset images). (d) Capacitance retention efficacy of the supercapacitor under varying compression strains. Calculated specific capacitance of device based on a dual chemically cross-linked PVA double-network hydrogel mixed with an ethylene glycol-water electrolyte under. (e) varying mechanical states. (f) under different compressive strains at room temperature (ambient conditions) and −30 °C respectively. (Liu, Ren, Zhang, et al. 2020, with permission of Elsevier Publications Liu, Zhang, Liu, et al. 2020 with permission of RSC Publications)

efficacy of 93.3% even after completing 8000 cycles at temperatures as low as −30 °C, and of 76.5% after 4000 cycles under an elevated temperature of 100 °C. Even the device recorded a notable capacitance retention efficiency of 94.1% under 80% compressional forces. Polymer based electrolytes have also been successfully employed in designing highly twistable and stretchable sandwiched morphology based wearable supercapacitors that showed impressive electrochemical signatures, insensitive to severe deformations as well (Li, Yang, et al. 2014; Choi et al. 2016; Jin et al. 2019).

2.5.2 Polymer-Based Self-Healable Supercapacitors

The necessity for attaining the long-term durability of energy storage devices has mandated scheming supercapacitors based on exclusive materials with a self-healing ability so that the system can recover from damage entirely in the case of mechanical injury and resume its normal functioning immediately (Huang, Zhong, et al. 2015; Guo et al. 2017; Jin et al. 2018; Wei et al. 2018). The auto-healing property of materials or devices largely counts on the abundancy of hydrogen bonds and/or non-covalent forces existing in the interfaces interlinking the device constituents, displaying an appreciable self-healing property combined with the 100% recovery of electrochemical performance, even after being subjected several times to cutting–healing acts (Wang and Pan 2017). Thus, Wang and Pan (2017) established an omni-healing supercapacitor employing a supramolecular web-structure where hydrogen bond linkages effectively preserved the shape of the supramolecular assembly, thereby contributing to the self-healing property of the electrodes, the electrolyte, as well as their interfaces after experiencing physical damage, as well as providing strong adherence to the hydrophilic substrate. The resultant energy storage device composed of trapped active carbon materials and KCl within the integrated poly(vinyl alcohol) grafted with an *N,N,N*-trimethyl-1-(oxiran2-yl)methanaminium chloride polymer cross-linked via diol-borate ester bonds promoted more than 85.7% of the electrochemical cell capacitance retention, even on exposure to five successive fragmentations. Further, the supercapacitor showed an almost unchanged electrochemical performance even after being tailored to diverse minute geometrical forms that underwent self-healing to form new shapes. Some gel electrolytes such as PVA or PVP can autonomously undergo auto-healing owing to the presence of O–H----O hydrogen bonds in-between the polymer chains to promote high mechanical flexibility as well as exceptional electrochemical signatures (Zhong, Deng, et al. 2015; Huang et al. 2019). Self-healing in such gel electrolytes can further be promoted via cross-linking. Thus, a novel composite gel polymer electrolyte composed of hybrid vinyl-silica nanoparticles blended in polyacrylic acid was designed for self-healing supercapacitors that demonstrated an almost 100% capacitance maintenance even on being subjected to 20 cutting/healing sequences (Jin et al. 2019). The unique morphology formed on cross-linking polyacrylic acid via hydrogen bonding to the hybrid vinyl-silica nanoparticles conferred an easy healability and stretchability of more than 3700%. Similarly, a hydrogel obtained by cross-linking and employing double linkers, such as Laponite and graphene oxide (GO), to the polymer matrix was developed for self-healing supercapacitors successfully (Li, Lv, et al. 2019). The functional groups, such as –COOH and –OH in GO, and Mg^{2+} ions in the Laponite, proficiently interacted with the $–CONH_2$ moieties of the polymer strands to afford a reversible healing performance and excellent stretchability. Recently, another stretchable and instantaneous omni-healable supercapacitor was reported via sandwiching a CNT-free GCP hydrogel electrolyte in-between layers of polypyrrole-incorporated gold nanoparticle/carbon nanotube/poly(acrylamide) (GCP@PPy) hydrogel electrodes, followed by the chemical soldering of an Ag nanowire film to the hydrogel electrode to serve as the current collector. The assembled supercapacitor delivered an areal capacitance of 885 mF cm^{-2} and an areal energy density of 123 μW h cm^{-2}, which withstood an ultrahigh stretching strain of 800%, as well as having a rapid optical healing ability and a notable spontaneous healing ability (an electrical healability with an efficiency of 80% within 2 min) (Chen et al. 2019). In another recent report, an auto-healing and shape-editable supercapacitor was engineered by sandwiching poly(acrylic acid)–poly(ethylene oxide) (PAA–PEO) hydrogel electrolytes in-between two slices of CNT-coated polyurethane-poly(ε-caprolactone) (PUPCL) electrodes. The resultant device was subjected to repetitive tearing and healing, the latter made possible by heating at 75 °C for 10 min with negligible impact on the cell's capacitive performance, as indicated in Figure 2.10(a–c), demonstrating a 92.6% specific capacitance retention even after five cutting/healing cycles. A comparative Nyquist plot (Figure 2.10d) under sequential repetitive tearing and healing steps highlighted a slight variation of charge transfer resistance introduced as a result of an increase in defect concentration at the healing interfaces. Nonetheless, the devised, robust, self-healable, and shape-editable energy accumulator demonstrated an appreciable gravimetric capacitance of 37 F g^{-1} @ a current density of 0.5 A g^{-1} with a good rate capacity and a capacitance retaining efficacy of ~96.5% beyond 10,000 charging/discharging tests as depicted in Figure 2.10(e and f) (Li, Fang, et al. 2019).

Even in the recent past, an innovative conducting organo-hydrogel was designed via combinations of polyelectrolytes and glycerol that displayed exceptional elongation features as well as self-healability and self-adhesive characteristics on various material surfaces including skin (Zhang, Li, Liu, et al. 2019). Furthermore, this gel endured temperatures as low as −20 °C for several hours without freezing, with a high workability even under an AC voltage of 220 V without undergoing degradation but maintaining a decent ionic conductivity and self-healability of 96, 94, and 85% at 20, 4, and −20 °C respectively, thus proving its potentiality in a broader domain of flexible wearable electronics (Yang, Guan, et al. 2019).

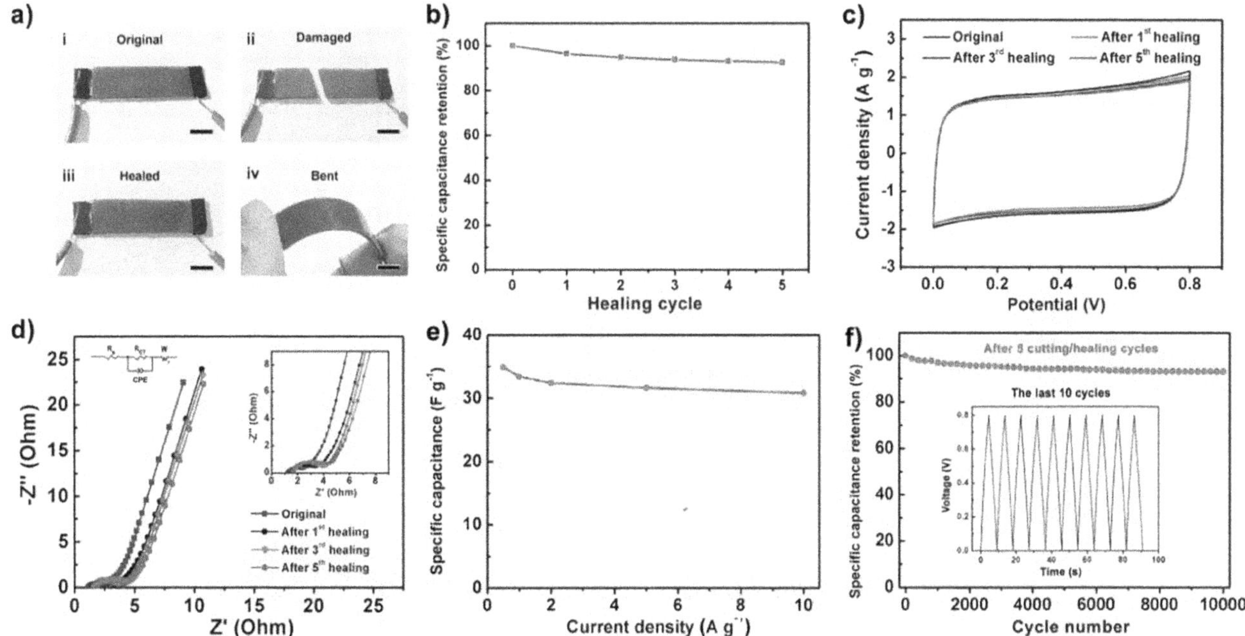

FIGURE 2.10 (a) Optical photographs showing the cutting/healing processes of the flexible supercapacitor assembled by sandwiching poly(acrylic acid)–poly(ethylene oxide) (PAA–PEO) hydrogel electrolytes in-between two slices of CNT-coated polyurethane-poly(ε-caprolactone) (PUPCL) electrodes (scale bars measure 0.5 cm). (b and c) Variation of specific capacitance versus healing cycle and corresponding CV curves at different healing cycle stages at a potential scan rate of 100 mV s^{-1}. (d) Relative Nyquist plots of the supercapacitor in the original state and after different healing states after being subjected to repetitive cutting/healing acts. Insets indicate the equivalent circuit to fit the corresponding EIS spectra and the corresponding curves at a high frequency domain. (e) Variation of gravimetric capacitances with current densities after five cutting/healing sequences. (f) Cyclic performance of the flexible supercapacitor after five cutting/healing cycles for 10,000 charging/discharging cycles @ a 5 A g^{-1} current density. Inset shows the GCD profiles for the last ten cycles between 9990 and 10,000 cyclic tests. (Li, Fang, et al. 2019 with permission of RSC Publications)

2.5.3 Polymer-Electrolyte-Based Shape Memory Supercapacitors

The shape memory property is an effective mode of restoring the electrochemical performance of flexible supercapacitors even on being subjected to substantial mechanical distortion, thus prolonging the device's cycling life (Huang, Zhu, et al. 2016). It has been realized that suitable shape memory ingredients responding to external stimuli like heat or temperature, pressure, electric forces, or a magnetic field play a key role in designing such smart supercapacitors. Usually, two strategies are projected to equip the device with the shape memory property for polymer-based supercapacitors (Heo and Sodano 2014; Deng et al. 2015; Zhong, Meng, et al. 2015). The common approaches are to employ shape memory polymers like polyurethane, polynorbornene, trans-isopolyprene, styrene-butadiene copolymers, or their appropriate derivatives, or to use shape memory alloys such as NiTi, Cu-based, Fe-based, or intermetallic systems to fabricate an electrode material for devising a shape memory fiber supercapacitor where chosen stimulating agents will assist in recovering shape deformations and restoring the pre-deformed electrochemical performance (Li, Fang, et al. 2019). Most of these systems involve the usage of quasi-solid polymer gel electrolytes as an indispensable device component to attain the desired structural properties. Thus, a smart shape memory fiber supercapacitor was constructed using a stretchable shape memory polymer nanocomposite fiber substrate, in combination with an MWCNT conductive layer and a pseudocapacitive PANI film using a layer-by-layer technique as schematically illustrated in Figure 2.11(a) (Zhong, Meng, et al. 2015). The infiltration of a PVA polymer electrolyte within a MWCNT and PANI porous layer resulted in an enhanced ionic conductivity of ~50 S cm^{-1} as well as > 400% stretchability, as well as recording an outstanding volumetric pseudocapacitance of ~427 F cm^{-3} @ a potential scan rate of 1 mV s^{-1}, along with a high rate capability, even at various stages of deformation as highlighted in Figure 2.11(b), and cyclic stability (Zhong, Meng, et al. 2015). In another work, Huang et al. introduced a shape memory alloy wire substrate for device fabrication involving a quasi-solid polymer gel (PVA/H_3PO_4) electrolyte and MnO_2/PPy composite electrodes to fabricate a shape memory supercapacitor (SMSC), as schematically shown in Figure 2.11(c). The wire supercapacitor displayed an appreciable shape memory restoring effect, induced via thermal activation without minimum compromise and with electrochemical performance, thus showing an admirable capacitance retaining capacity (85%) even after undergoing a series of deformation tests as depicted in Figure 2.11(d–e) (Huang, Zhu, et al. 2016).

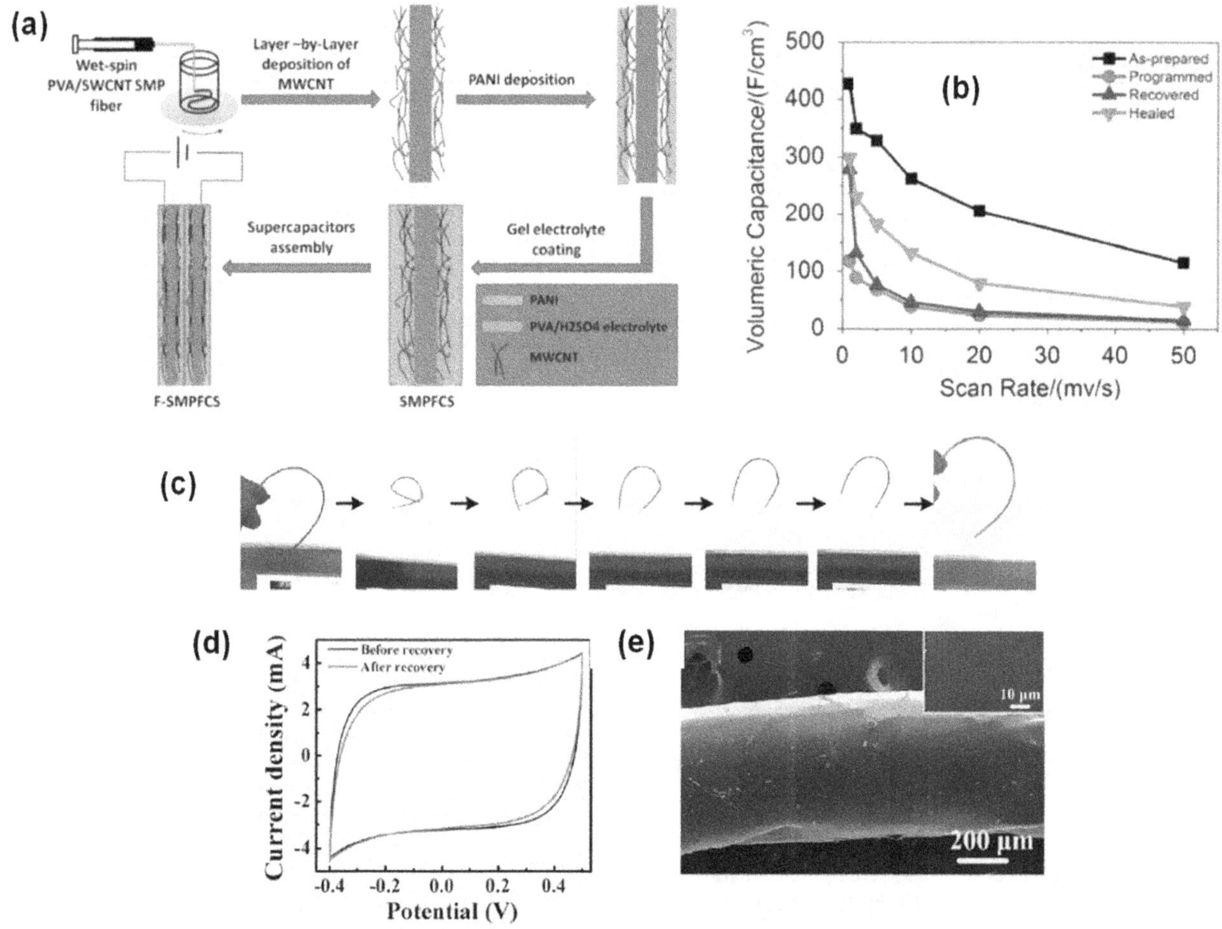

FIGURE 2.11 (a) Schematic representation of the designing of shape memory polymer fibrous supercapacitors (SMPFCS) and assembled-SMPFCS (F-SMPFCS) systems. (b) Comparative volumetric capacitance at different voltage sweep rates for varying shape memory cyclic phages. (c) Sets of instantaneous photographs of a deformed SMSC showing the steps for a shape recovery process. (d) CV curves of the SMSC before and after the shape recovery process. (e) SEM image of a single NiTi shape memory electrode coated with a solid electrolyte after shape recovery. Inset is the higher resolution image of the same. (Zhong, Meng, et al. 2015 with permission of ElsevierHuang, Zhu, et al. 2016 with permission of RSC Publications)

In another work, polyaniline coated MWCNT electrodes were assembled with a PVA based-polymer electrolyte (~50 S/cm) and showed good stretchability of more than 400% with good shape memory properties and a volumetric capacitance of about 427 F cm^{-3} (Zhong, Meng, et al. 2015). A polymer ionogel electrolyte such as poly(vinylidene fluoride-*co*-hexafluoropropylene)-[EMIM] BF_4, (PVDF-HFP)-[EMIM]BF_4 was employed to fabricate shape memory based asymmetric supercapacitors that recorded a wide voltage window of 2.5 V and demonstrated a cell gravimetric capacitance of 25.9 F g^{-1} @ a 0.5 A g^{-1} current density (Liu et al. 2016). Similarly, a shape memory fiber supercapacitor was constructed by winding well-oriented layers of CNTs on a polyurethane fiber along with a PVA gel electrolyte that yielded successful and encouraging electrochemical results, ensuring their acceptance in commercial sectors (Ma et al. 2017). Even graphene coated TiNi alloy flakes with a high shape recovering aptitude have been assembled with polymer gel electrolytes to design smart supercapacitors for bio-compatible, wearable electronics (Tai et al. 2012; Yan et al. 2019; Mackanic et al. 2020).

2.5.4 Polymer Electrolyte-Based Electrochromic Supercapacitors

Electrochromic gadgets and electrochemical energy storage systems have several features in common such as electro-active materials and their chemical and structural requirements, as well as physical and chemical working principles (Zhao et al. 2015). The charging and discharging characteristics of an electrochromic device are comparable to those of electrochemical energy storage systems (Tian et al. 2014; Cai et al. 2016). Thus, the integration of energy-storage and electrochromism functionalities within a single electrode is designed to monitor visually the degree of charge stored in these smart supercapacitors via charge insertion/de-insertion or electrochemical oxidation/

reduction processes (Chen, Lin, et al. 2014; Chen et al. 2015; Zhang, Liu, et al. 2020). A number of renowned electrochromic materials, such as WO_3, V_2O_5, MnO_x, NiO, and Prussian blue, their nanocomposites, as well as conducting polymers such as polyaniline and polypyrrole and their derivatives, have received substantial attention for constructing flexible and wearable electrochromic devices (Bi et al. 2017; Wang, Wang, et al. 2018; Guo et al. 2020; Kim, Yun, You, et al. 2020). Many successful attempts have been made in devising highly stretchable and transparent electrochromic supercapacitors with reversible color changes as an indication of electrochemical working status. Most of these devices use high ion-conducting polymer gel electrolytes that facilitate the feasibility of fabricating highly flexible and stretchable electrochromic energy storage devices (Shen et al. 2016; Zhang, Yang, et al. 2020; Wang, Guo, et al. 2020). For example, in order to achieve high-performance devices, counter-anodic materials have been directly introduced into ion gel electrolytes to avoid the need of anodic film deposition onto the opposite electrode. However, such usage of electrolyte-soluble anodic compounds led to extreme self-discharging and self-bleaching processes even under open-circuit situations due to freely mobile anodic species encountering cathodes via a concentration gradient induced diffusion phenomenon, leading to poor coulombic efficiency and device instability (Sajitha et al. 2019; Yun et al. 2019; Kim, Yun, Yu, et al. 2020). Thus, to avoid such issues, highly ion-conducing polymer gel electrolyte-insoluble film-type anodic material such as copper-doped NiO has been employed along with electrochromic WO_3 film, as depicted in Figure 2.12(a), demonstrated an areal capacitance of ~14.9 mF cm^{-2} along with appreciable transmittance contrasting characteristics, rate capacity, and extended working temperature range (0–80 °C) with consistent coulombic efficiency, as revealed from the data displayed in Figure 2.12(b–d) (Kim, Yun, Yu, et al. 2020). Similarly, H_2SO_4–PVA based gel electrolytes were employed in combination with macroporous WO_3 and PANI films as electrochromic anode and ion storage cathodes respectively, to fabricate multifunctional devices that

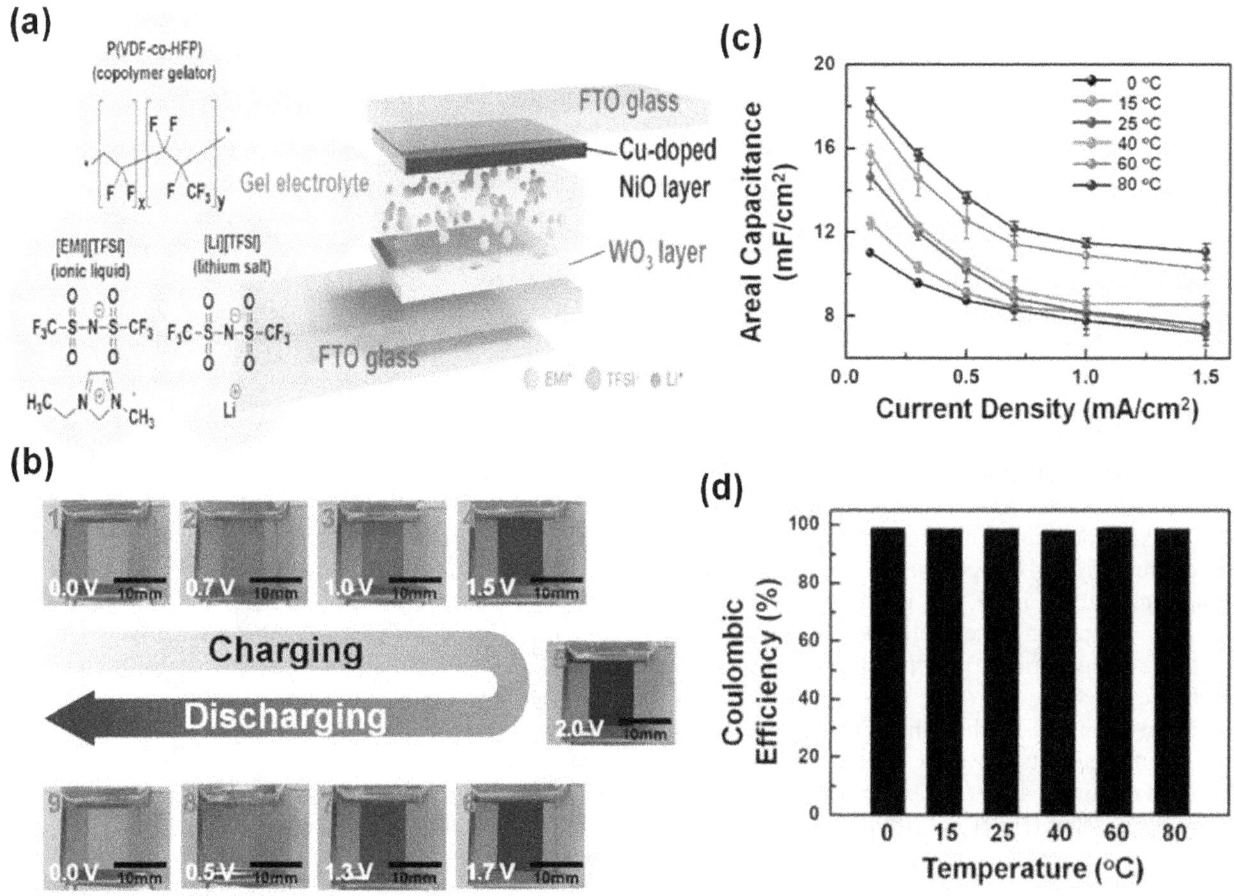

FIGURE 2.12 (a) Schematic presentation of the asymmetric cell assembled with an ion-conducing polymer gel electrolyte-using copper-doped NiO as an anodic film and an electrochromic WO_3 film as a cathode, and chemical structures of the ingredients used in the ionogel electrolyte. (b) Photographs of the device showing color change at different voltages during various stages of the charging/discharging process. (c) Variation of areal capacitance with current densities as a function of working temperatures. (d) Comparison of the coulombic efficiency of the electrochromic energy storage cell at various working temperatures. (Kim, Yun, You, et al. 2020 with permission of ACS publications)

presented an ultrafast switching speed as well as long-term electrochemical stability (Sajitha et al. 2019).

Despite having similar operating mechanisms, the criteria for optimum electrochromic and energy storage differ significantly. Generally, the coloration efficiency of electrochromic devices is given by the ratio of change in optical density to electronic charge inserted or extracted from the electrochromic material per unit area. Thus, their high coloration efficiency as well as fast switching processes require a low charge density for a distinct response in contrast to batteries or supercapacitors that always hunt for high charge storage efficiency. Thus, there lies a basic criterial contradiction in-between electrochromism and energy storage systems and, accordingly, the present challenge involves optimizing the charge density in the integrated electrochromic energy system for the simultaneous recognition of electrochromism and energy storage properties.

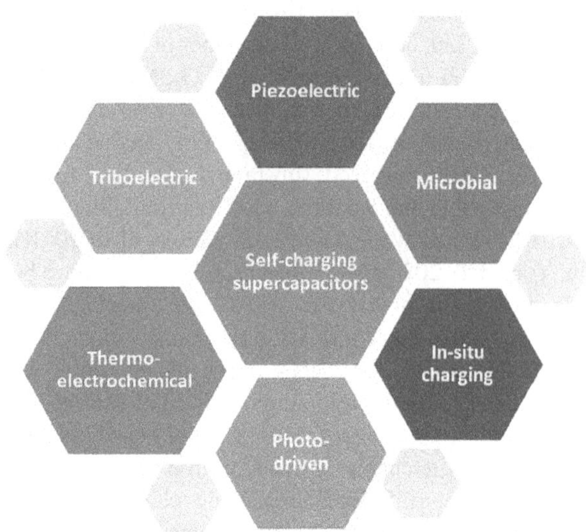

FIGURE 2.13 Some of the up-coming concepts for designing self-charging supercapacitors.

2.5.5 Polymer Electrolyte Based Self-Charging Supercapacitors

Modern day wearable and portable electronics technology are focused on designing and developing a sole device unit serving to harvest, convert, and store energy so as to efficiently meet the requirement for an uninterrupted power supply for routine applications. This has triggered the novel concept of a self-powering/charging strategy, although it will be a tough challenge to engineer flexible and light-weight self-powered systems for wearable micro-electronic gadgets (Xue et al. 2012; Xue et al. 2013; Lu, Jiang, et al. 2020). Typically, common energy harvesting units include triboelectric/piezoelectric nanogenerators (capable of converting different mechanical forces to electrical energy), photovoltaic or solar cells (capable of converting light energy to electrical energy), and thermoelectric devices (capable of converting thermal/heat energy to electrical energy), which are available to be coupled with energy storing units like supercapacitors or rechargeable batteries to form a single powering unit, although each unit has distinctive working principles, different operating mechanisms, and discrete physical components (Maitra et al. 2017; Wang, Dou, et al. 2019). Figure 2.13 illustrates some of the up-coming concepts on which self-charging supercapacitors are being designed to address the needs of self-powering electronic gadgets.

Hence, many researchers have opted for combining these segmental units via various fruitful integration strategies with the aim of achieving less expensive uninterrupted powering systems (Hu et al. 2019; Xu et al. 2019). It is observed that external integration of these energy generating units with supercapacitors frequently necessitate the addition of auxiliary power management circuits that eventually elevate the production expenses as well as complicate the device scheming procedures (Song et al. 2015; Qin et al. 2018). On the contrary, internal integration enjoys the benefits of avoiding such additional power management circuit costs; this has triggered high interest in internally integrated self-charging electronic gadgets (Yang and Urban 2018; Maitra et al. 2019; Sahoo et al. 2019). Thus, to promote smart, light-weight, and flexible self-charging supercapacitors, the proper choice of building components is crucial. Polymer based electrolytes as well as fulfilling their normal functions of ion conduction in these electrochemical cells, also contribute to mechanical flexibility and a design that can successfully inherit remarkable properties such as triboelectricity and piezoelectricity that assist in the device's self-charging process, as highlighted in the following sections.

Among the numerous techniques of transforming mechanical energy to electrical form, triboelectric nanogenerators have attracted massive attention owing to their easy operation and high output voltage (Chun et al. 2016; Zi et al. 2016). The operating principle of a typical triboelectric nanogenerator is grounded on contact and separation induced electrification and electrostatic induction (Wang 2014). They exploit different forms of substances and transform nearly all kinds of mechanical energy forms from various natural/artificial sources—such as rain drops, marine waves, air flow, bending acts, rotation, gliding, vibration energy, and body movement—into electrical energy (Wang, Chen, and Lin 2015). A characteristic triboelectric driven multi-responsive power cell comprises three components: an energy harvesting unit, a rectifier, and an energy storing unit, as shown in Figure 2.14(a) (Shi et al. 2019). The corresponding circuit diagram of the integrated self-powering system is illustrated in Figure 2.14(b). The role of the rectifier is to filter AC components from the output voltage of the nanogenerator so that it can be effectively stored as direct current (DC) components in the supercapacitor (Xue et al. 2014). The performance of the triboelectric unit depends on the selection of the contact materials (Kim, Lee, Kim, et al. 2020). The relative performance of a typical triboelectric nanogenerator driven self-powering

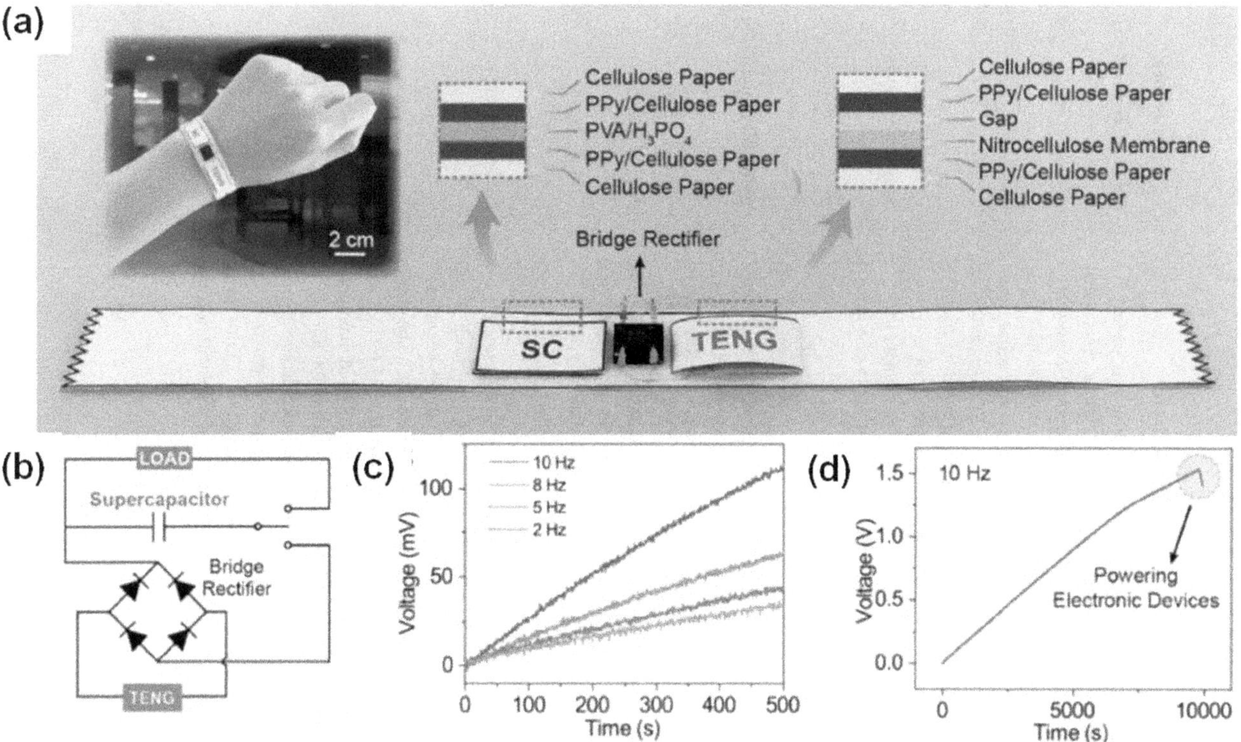

FIGURE 2.14 (a) Schematic illustration of the device components and photo image of the paper-based self-charging power wrist-band integrating the P-TENG (paper-triboelectric generator) and P-SC (paper-supercapacitor). (b) Circuit diagram showing that the P-SC charged by the P-TENG was used to power the electronics. (c) Charging profiles of the P-SC charged by the P-TENG at different frequencies. (d) Voltage–time curve of the P-SC charged by the P-TENG and then used to power an LCD. (Shi et al. 2019 with permission of ACS publications)

supercapacitor is shown in Figure 2.14(c and d). Figure 2.14(c) displays the impressive charging signatures of the paper-based SC by the paper-based TENG under a compressive stress with varying frequencies (2 to 10 Hz), indicating the uniform enhancement of the charging rate with the press frequency, reaching a voltage of 113.5 mV under stress (10 Hz) in 500 s. Figure 2.14(d) shows the voltage–time curve of a periodic compressive stress of 10 Hz to power an electronic device, showing a steady increase in voltage with the charging time up to 1.54 V in just 9793 s (~2.7 h) (Shi et al. 2019).

Recently, an ultra-thin, lightweight, transparent, and flexible touch sensor integrated with a self-charging powering electronic system for wearable flexible displays was proposed using a single-layer graphene film as a supercapacitor electrode, a touch sensor, as well as a triboelectric nanogenerator (TENG), assembled with a PVA–LiCl-soaked PAN electro-spun mat separator cum polymer electrolyte placed between two symmetric graphene film electrodes. The resultant all-in-one device showed an appreciable charge storing capability, thereby recording an areal capacitance of 3.83 μF cm^{-2} (102 F g^{-1}) with a high capacitance retaining efficacy of 99.8% even on surpassing thousand times 160° bending deformation sequences. In addition, being a highly transparent one (77.4%), the device also harvested high electric power (1.45 mW cm^{-2} with 17 kPa) in the pressure range for human physical perception, thus confirming its high degree of multifunctionality (Chun et al. 2019).

Thus, the above scheme of self-charging powering units suggests the proper integral designing of the sub-units, which essentially requires the use of a flexible electrolyte as a vital constituent part for compact packaging, to ensure smooth and effective contact, and to avoid leakage and short-circuit related issues. Further, it has been shown that triboelectric nanogenerators (TENGs) based on solid polymer electrolytes (SPEs) have produced significantly higher power output than typical metal–polymer analogues (Ryu et al. 2017). Accordingly, several researchers have proposed devising triboelectric based self-charging supercapacitors using various types of solid/gel polymer electrolytes as summarized in Table 2.6.

Rigorous studies have indicated that often the integration of energy harvesting and storage components via a rectifier considerably affects device output efficiency due to substantial energy loss across the rectifier and, thus, necessitates intensive focused research on promoting the simplistic and well-organized integration of device components (Pan et al. 2016).

Detailed investigations have proposed that the direct conversion of mechanical forces to electrochemical form with high overall efficiency can be achieved by combining

TABLE 2.6
Polymer Electrolyte-Based Self-Charging Supercapacitors

Self-Charging Approach	Electrodes Used	Electrolyte Used	Benefits	Performance	Reference
Triboelectric nanogenerator based self-charging supercapacitor	Laser-induced graphene symmetrical electrodes on polyimide substrate	PVA/H_2SO_4 gel electrolyte	High mechanical durability and electrochemical stability under several bending and pressing conditions	Power density of 0.8 W m^{-2} @ loading resistance of 20 MΩ, surface capacitance of 10.29 mF cm^{-2} @ current density of 0.01 mA cm^{-2}, can be charged to 3 V	Luo, Fan, et al. (2015)
Triboelectric nanogenerator based self-charging supercapacitor	PPy symmetrical electrodes	PVA/H_3PO_4 gel electrolyte	All-plastic based self-charging power system showing fast charging capacity and excellent stability	Power density of the supercapacitor reached 3.13 W g^{-1} @ load of 200 μA, energy density of 98 J g^{-1} and enduring 10,000 cycling tests, charging to 2.5 V	Wang, Wen, et al. (2015)
Polydimethylsiloxane based triboelectric nanogenerator for self-charging supercapacitor	CNT/paper symmetrical electrodes	PVA/H_3PO_4 gel electrolyte	Three serially connected units can drive commercial calculator working uninterruptedly as well as power electrochromic devices like smart windows	Charged to 900 mV in 3 hrs under the compressive stress frequency of 8 Hz	Song et al. (2016)
Fiber-based triboelectric nanogenerator based self-charging supercapacitor	$RuO_2 \cdot xH_2O$ on carbon fiber electrodes	PVA/H_3PO_4 gel electrolyte	Can readily harvest mechanical energy from human motion	Volume specific capacitance of 83.5 F cm^{-3} with capacitance retention of 94% @ 5000 charging/discharging cycles. Can be charged from 0.5 to 2.5 V in 873 s	Wang, Li, et al. (2015)
Fabric triboelectric nanogenerator (TENG) based self-charging flexible yarn supercapacitor	Polyester yarns successively coated with Ni layer and rGO film electrodes	PVA/H_3PO_4 gel electrolyte	High mechanical flexibility and weave-ability of both the yarn supercapacitors and the fabric TENG, to be incorporated into intricate and fashionable clothing	Charged to 2.1 V in 2009 s with a vibration motor at about 5 Hz, showing capacitance retention of 96% for 10,000 cycles	Pu et al. (2015)
Piezoelectric nanogenerator based self-charging supercapacitor	3D Au@MnO_2 symmetrical electrodes	LiCl/PVA polymer gel electrolyte	The unit can be charged simply by human finger motions, such as by tapping and sliding in all directions	Areal capacitance of 1.30 mF cm^{-2} @ current density of 2 μA cm^{-2}, with capacitance retention of 92% retention after 2000 cycles achieved. Can be charged from 0 to 2.5 V in 6102 s (~1.69 h).	Luo et al. (2016)
Piezo-polymer separator based self-charging supercapacitor	Graphene symmetrical electrodes	Porous PVDF-ionic liquid electrolyte	All-in-one energy conversion cum storage device with high flexibility and stability	The integrated device delivered maximum cell specific capacitance of 28.46 F g^{-1} (31.63 mF cm^{-2}) and specific energy of 35.58 W h kg^{-1} and specific power of 7500 W kg^{-1}. Can be charged to 112 mV under a compressive force of 20 N within 250 s	Sahoo et al. (2019)

Piezoelectric PVDF film as both separator and energy harvester	Carbon cloth symmetrical electrodes	PVA–H_2SO_4 electrolyte	External mechanical effects exert piezoelectric potential across the PVDF films that drive electrolyte ions to migrate towards the interface of the supercapacitor electrode thus assisting in storing charge	Reported power density ~32.5 mW cm^{-3} and energy storage capability ~2000 F m^{-2}	Song et al. (2015)
Piezoelectric based self-charging supercapacitor	Co-Fe_2O_3 @ACC as symmetric electrodes	PVA-KCl-$BaTiO_3$ as piezo-electrolyte	Appreciable mechanical flexibility and considerable electrochemical and self-charging capability	Can be self-charged to 120 mV via simple impacting, i.e., 7 min repetitive bendings at a frequency of 1.0 Hz	Zhou et al. (2021)
Piezoelectric based self-charging supercapacitor	2D-molybdenum di-selenide ($MoSe_2$) symmetrical electrodes	Polyvinylidene fluoride-*co*-hexafluoropropylene/ tetraethyl-ammonium tetrafluoroborate (PVDF-*co*-HFP/$TEABF_4$) ion gelled polyvinylidene fluoride/ sodium niobate (PVDF/ $NaNbO_3$) as the piezopolymer electrolyte	Appreciable mechanical flexibility and considerable electrochemical and self-charging capability	The unit delivers areal specific capacitance of 18.93 mF cm^{-2} and areal energy density of 37.90 mJ cm^{-2} @ areal power density of 268.91 μW cm^{-2} @ constant discharge current of 0.5 mA. Can be charged up to 708 mV in 100 s under compressive force of 30 N	Pazhamalai et al. (2018)
Piezoelectric based self-charging supercapacitor	MnO_2–rGO electrodes	PVA-H_3PO_4	All-in-one energy conversion cum storage device with high flexibility and stability	Maximum output voltage of 44 V, current density of 1000 nA cm^{-2}, and areal power density 193.6 μW cm^{-2} under applied mechanical force of 10 N, can store charge of 1.5×10^{-3} mC within 100 s without need of rectifier	Ramadoss et al. (2015)
Piezoelectric based self-charging supercapacitor	Graphene coated elastic styrene-ethylene-butylene-styrene electrodes	Potassium sodium niobate/ PVA/H_3PO_4 piezoelectric electrolyte	Maximum biaxial stretchability of 300%, ability to harness and transform ambient mechanical energy into electricity	Can be finally charged to about 1.0 V via simple palm patting for 300 s (2 Hz) or 0.8 V by repeated stretching for 40 s (1 Hz)	Zhou et al. (2020)
Bio-piezoelectric separator (electrolyte-soaked perforated fish swim bladder) based self-charging supercapacitor	NiCoOH-CuO@Cu foil) as a binder free positive electrode and reduced graphene oxide coated copper foil (RGO@ Cu foil) as negative electrode	PVA–KOH gel electrolyte	Bio-compatible, rectification-free all-in-one energy conversion cum storage device with high flexibility and stability	Can be charged up to 281.3 mV from its initial open circuit potential (~130.1 mV) in ~80 s under continuous human finger imparting at a frequency of 1.65 Hz	Maitra et al. (2017)
Network-like (PVDF-TrFE) film as piezoelectric separator based self-charging supercapacitor	PDMS-rGO/C film as symmetric electrodes	PVA/H_3PO_4 as gel electrolyte	Appreciable mechanical flexibility, and considerable electrochemical and self-charging capability	Bending of finger to 90° can charge the integrated device to 0.45 V in just 17.0 s with a discharge time of about 18.0 s	Lu and Jiang (2020)

(Continued)

TABLE 2.6 (Continued)

Self-Charging Approach	Electrodes Used	Electrolyte Used	Benefits	Performance	Reference
Solid-state integrated photo-supercapacitor	Ag_2S quantum dot decorated ZnO nanorod arrays as photoanode and PEDOT charge storage counter-electrode	PVP/[HEMIm][BF_4] solid-state ion-gel electrolyte	The photo-supercapacitor showed high stability, appreciable mechanical flexibility, and considerable electrochemical and self-charging capability	Recorded an areal specific capacitance of 0.667 mF cm^{-2}, with storage efficiency of 6.83%, photo-charge and discharge cycles achieving a maximum voltage of 0.33 V in 40 s	Solís-Cortés et al. (2020)
Photo assisted self-charging supercapacitor	$NiCo_2O_4$/ZnO nanorod based symmetric electrodes	KOH/PVA gel electrolyte	Appreciable mechanical flexibility, and considerable electrochemical and self-charging capability	Device displays admirable capacitance retention efficiency of 98.5 and 97% after 2000 charging/discharging cycles in the absence and presence of UV irradiation respectively	Boruah and Misra (2017)
Thermo-electrochemical cell using gel electrolytes	Au/Cr electrodes	PVA gel mixed with ferric/ferrous chloride or potassium ferricyanide/ferrocyanide couples	Charging feasible by utilizing body heat	The integrated wearable device can generate an open-circuit voltage of about 0.7 V and a short-circuit current of 2 mA by utilizing body heat, achieving an utmost output power of ~0.3 mW	Wang, Huang, et al. (2019)
Thermoelectrochemical cells (thermo-cells) designed for harvesting human body heat	3D porous poly(3,4-ethylenedioxythiophene)/polystyrene sulfonate and graphene/CNT electrodes	PVA-$FeCl_{2/3}$ as n-type electrolyte and carboxymethylcellulose-$K_{3/4}Fe(CN)_6$ electrolyte as p-type electrolyte	Flexible watch-strap shaped thermo-cell that could harvest body heat, charge supercapacitors, and illuminate a green LED light	Can show an output voltage of 0.34 V	Lechene et al. (2017)
Photothermal supercapacitor via enhanced solar energy harvest	N-doped mesoporous carbon nanosphere-intercalated 3D graphene hydrogel	PVA/H_2SO_4 gel electrolyte	Appreciable mechanical flexibility, and considerable electrochemical and self-charging capability	Achieved volumetric specific capacitance of 8.1 F cm^{-3} and volumetric energy density of 1.12 mW h cm^{-3} @ volumetric power density of 13.30 mW cm^{-3}	Zhao et al. (2020)

piezoelectric material with electrochemical cells, thus avoiding the need for rectifiers (Xue et al. 2012). Here, the energy generated by piezoelectric nanogenerators can be directly stored in electrochemical energy storage units as the output voltage is unidirectional and does not reverse, even on switching off the applied force. In this strategy, external strains or deformations on the piezoelectric nanogenerator introduce an interfacial potential difference that drives the ion migration between the electrodes after the alternative current is rectified. This means that mechanical energy can be harnessed and converted into chemical energy and charges the electrochemical capacitor directly (Ramadoss et al. 2015). Contrary to the other mechanical energy harvesters, such as triboelectric nanogenerators that necessitate structural partitions for proper functioning, piezoelectric material can be introduced in the system simply as a separator or as a matrix that is placed in-between electrodes to fit in with the electrochemical energy storage unit (Krishnamoorthy et al. 2020). Such resultant integrated units employ polymer-based electrolytes to increase mechanical durability as well as avoid constructional strains caused by liquid electrolytes in electrochemical energy storage systems as well as combating unnecessary energy loss and sluggish self-charging processes. It is worth mentioning that the gel polymer electrolytes have several merits, especially when taking into consideration safety operative issues, elevated device efficacy, and manufacturing costs. Table 2.6 also shows some of the remarkably successful results of piezoelectric based self-powering electrochemical cells using polymer electrolytes. To experimentally address the "piezo-electrochemical effect" concerned with the mechanism of energy conversion and storage in self-charging supercapacitor power cells (SCSPC), Krishnamoorthy et al. assembled siloxene sheets as symmetrical electrodes and electro-spun siloxene-incorporated PVDF piezofibers as separators coated with ionogel electrolytes. They demonstrated the self-charging mechanism as shown in Figure 2.15 (Krishnamoorthy et al. 2020). The fabricated device with no applied compressive force was (a) initially subjected to an applied compressive force that (b) generates a siloxene/PVDF interfacial piezoelectric potential that drives the electrolyte ions towards the electrode surface. Subsequently, an equilibrium state (c) is attained between the generated piezoelectric potential and the electrochemical processes within the device. On removing the compressive force (d), the piezoelectric potential tends to disappear and the equilibrium state is reattained, thereby indicating the completion of one self-charging cycle (e).

Lately, in order to further enhance the degree of stretchability as well as self-charging efficiency, piezo-electrolyte film was sandwiched between two stretchable electrodes. Zhou et al. (2020) designed graphene coated styrene-ethylene-butylene-styrene electrodes and a potassium sodium niobate mixed PVA/H_3PO_4 assembly that aided high biaxial stretchability (~300%) with simultaneous mechanical energy arrestation and storage via a persistent self-charging mode in the absence of a rectifier. Similarly, a piezo-electrolyte film comprising a solid proton conducting electrolyte (phosphotungstic acid embedded in the piezoelectric PVDF matrix) was employed as a free-standing film separator cum electrolyte assembled with graphene based-electrodes showed self-charging to 110 mV on being subjected to an external force of 2 N (Manoharan et al. 2021). Thus, the exciting results obtained so far from the above studies have definitely paved way for a new paradigm in the field of development of next-generation energy conversion cum storage devices.

In another technological approach, the light assisted automated charging of supercapacitors has been proposed. The flexible photo-charging powering devices hold the benefits of high power density, appreciable mechanical flexibility, and portability to meet the critical necessity of developing soft electronics (Liu, Takakuwa, Li, et al. 2020). Flexible photo-charging supercapacitors, fabricated using monolithically integrated photovoltaic solar cells and supercapacitors, can be effective in harvesting and storing energy simultaneously. In this context, an innovative all solid gel electrolyte-based transparent supercapacitor containing a photoactive layer (PANI) assembled with a capacitive unit containing an ultra-thin layer of gel polymer electrolyte and PANI/CNT electrodes was integrated into a single device. The resultant integrated powering cell

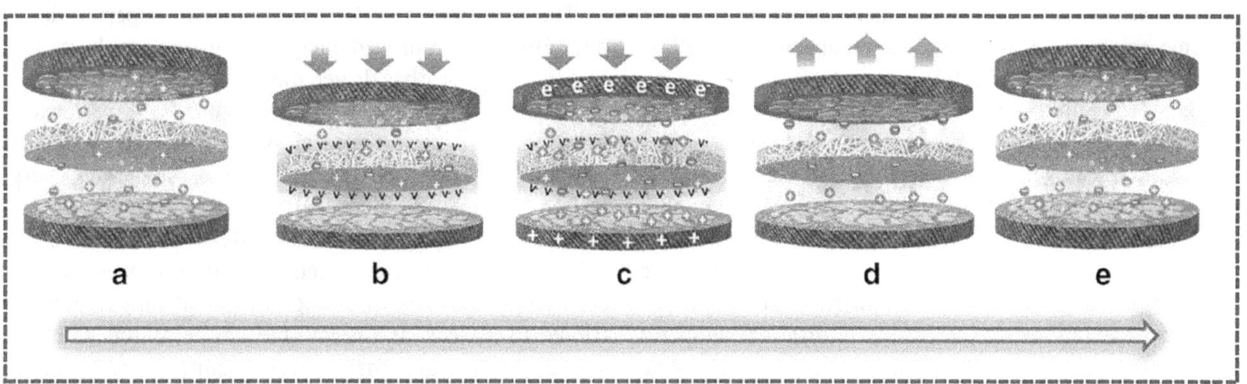

FIGURE 2.15 Schematic illustration of the cycle of a self-charging mechanism in siloxene based self-charging supercapacitor power cells. (Krishnamoorthy et al. 2020)

displayed self-charging under illumination, without the influence of any external bias, as well as demonstrating appreciable auto-storage signatures of the photogenerated electrical signal (photocurrent density of 2 mA g^{-1}). The employment of a gel electrolyte, in this case, permits the easy assembling of the components and an ability to form an ultra-thin layer which facilitates light penetration without compromising on ionic conduction to facilitate the easy migration of ions to generate a photo-current. Some additional performance results related to light-driven self-charging supercapacitors containing polymer electrolytes are shown in Table 2.6 (Zhao et al. 2020). The results clearly state that the technology is rapidly flourishing with upgraded and judicious material engineering, ultra-thin device configuration, and in situ power management measures (Yin et al. 2015; Dong et al. 2017; Shin et al. 2021). Thus, recently, a cost-effective TiO_2 nanotube-based dye sensitized solar cell in combination with a graphene-based electrical double layer capacitor was unified in a flexible architecture using photopolymer electrolytes with a high energy output efficiency of 1.02% under a 1 Sun (equivalent to 100 $mWcm^{-2}$ of irradiance) irradiation condition that further improved to 1.42 % under a reduced irradiation of 0.3 Sun. The polyelectrolyte displayed high ionic mobility under ambient temperature conditions, improving the lifetime of the dye sensitized solar cell by preventing electrolyte loss through evaporation as well as meriting structural flexibility and easy handling. The swollen polymer electrolyte membrane also served as an electrolyte separator and guaranteed the easy sealing of the flexible EDLC component as well as abating the complexity of the fabrication process, thereby certainly improving the future scaling up of the device (Scalia et al. 2017). To have better understanding of the subject, the time-dependent dynamics of the charging and discharging processes of photo-supercapacitors, obtained by combining a supercapacitor with a solar cell, was explored by means of semi-analytical modeling. It was concluded that, for a specified system, the power conversion efficiency and fill factor of the solar cell play a vital role in maximizing the energy conversion and storage efficiency (ECSE) of the photo-supercapacitor (Solís-Cortés et al. 2020). Further, the capacitance, super-capacitor series, and shunt resistances strongly influence the time constants of the photo-supercapacitor as well as the ECSE (Lechene et al. 2017).

Some interesting research in recent years unfolds the significant use of the concepts of thermo-electrochemistry to directly transform thermal energy into electricity, followed by its storage in the integrated supercapacitor unit (Al-Zubaidi et al. 2017). Here, the change in the chemical potential of a reversible redox reaction with temperature was utilized to generate a potential gradient across the device by keeping the two electrodes at different temperatures (thermal conditions), separated by an electrolyte (Wang, Huang, et al. 2019). Table 2.6 also highlights some interesting outputs of fabricated thermoelectrochemical cells available in the scientific database. Here, use of polymer electrolytes are advantageous as they successfully eliminate electrolyte leakage, enable flexible device fabrication, eliminate convection currents, and thus sustain long term thermal gradients (Wang, Feng, et al. 2015; Yang et al. 2016; Liu, Zhang, Liu, et al. 2020; Pu et al. 2020). Thus, Jin et al. (2016) fabricated a self-powering carbon electrode-based thermo-electrochemical cell containing a quasi-solid-state redox active electrolyte in a cellulose matrix that accomplished an optimum balance of mechanical stability, thermoelectric coefficients, and power outputs. Similarly, a cobalt bipyridyl complex redox couple dissolved in a 3-methoxypropionitrile solvent mixed with polyvinylidene difluoride (PVDF) and poly(vinylidene fluoride-*co*-hexafluoropropene) (PVDF-HFP) was used to investigate the gelation effect on the performance of thermo-electrochemical cells. It was revealed that a more thermally stable PVDF gel permitted the faster diffusion of the redox species through it and supported superior power performances (Taheri and MacFarlane 2018a, 2018b). The same group further explored the influence of a non-volatile redox active quasi-solid state ionic liquid on thermo-electrochemical cell performances using polyvinylidene difluoride (PVDF), mixed with a redox-active ionogel electrolyte, consisting of a $[Co(bpy)_3]^{2+/3+}$ redox couple in 1-ethyl-3-methylimidazolium bis(trifluoromethylsulfonyl) imide, to produce either soft gels or free-standing films depending on the polymer content. This quasi-solid-state electrolyte recorded an appreciable Seebeck coefficient comparable to liquid electrolytes (Taheri and MacFarlane 2018a, 2018b).

Apart from the above modes, other convenient and sustainable means of self-charging have also been explored in recent years. For example, self-charging in a symmetrical supercapacitor, fabricated with polyaniline/carbon nanotube composite (PANI/CNT) electrodes and a PVA/H_2SO_4 gel electrolyte, was induced by gravity and ascribed to the streaming potential effect in a porous medium. Here, the pressure difference initiates fluid flow to result in generating electric potential. In the absence of any external electric power supply, the resultant integrated device recorded a maximal output voltage of 0.15 V under ambient conditions (24 °C, relative humidity of about 60%) (Gao et al. 2019).

The non-stop trial for cost-effective, superior device durability, and improved energy harvesting capability with sustainable power has driven the development of innovative proposals for converting ambient energy into electricity. This has prompted the exercise of bio-electrochemical routes to harvest solar energy, the most abundant energy source, to generate electrical power even under extreme conditions (Bombelli et al. 2015; McCormick et al. 2015). Microbial photosynthesis and respiration processes occur ceaselessly in cycles all through the day and night, and contribute to a clean and renewable power source with self-sustaining potential that can be channeled into practical applications for creating a novel supercapacitor integrated with a microliter-scale photosynthetic microbial fuel cell device. The resultant device reported reasonably stable high power (38 μW cm^{-2}) and current densities (120 μA cm^{-2}) for practicability in varied environmental sensing fields. Here, Nafion based membranes have been employed with a 3D double-functional bio-anode for concurrent bio-electrocatalytic energy harvesting and charge storing processes (Liu and Choi 2019).

Last but not the least, an in-situ charging strategy has been much looked at owing to its simplistic operation and lack of further need of additional energy harvesting cum charging units. Such an approach can be applied to diverse types of supercapacitors without any complications and obviously reducing the integrated device volume significantly. Thus, such a simplistic formula will be utilized in engineering as highly flexible, portable, and wearable microelectronic gadgets for everyday usages. It has been shown that, by the simple addition of a piece of Zn foil to one electrode, a supercapacitor can stimulate the electrodes to serve dually as both supercapacitor and metal-air cell. The supercapacitor can be successfully charged to 1.0 V via the integrated metal-air cell while coming in contact with a drop of common solution or tap water or even air moisture to the Zn anode and CNT cathode, recording a highest surface energy density of 3.92 μW h cm^{-2}. The charging process can be repeated innumerable times, with higher efficiency under moist conditions, showing the capability to attain a higher output voltage on compact assembling (Luo et al. 2019).

Thus, systematic studies have inferred that the performance and popularity of these integrated energy harvesting cum storage cells can be further improved by adapting the following strategies which necessitate: (i) the determination of the appropriate proportions of components for the capacitive and charging units for adequate energy storing and driving the same for maximum utility; (ii) the correct choice of highly stable but efficient naturally occurring bio-elements that can reduce the production cost as well as functioning with natural energy sources; (iii) ensuring intimate contacts and attachment with each device segments so as to mitigate unwanted leakage currents, power loss, etc. that turn down the device's powering efficiency.

2.6 INTEGRATED SENSING DEVICES POWERED BY POLYMER ELECTROLYTE-BASED SUPERCAPACITORS

This section discusses implementing the above-mentioned polymer electrolyte based multifunctional supercapacitors into integrated self-powering sensing devices for wearable devices, biomedical and health-care instruments, human–machine interfaces (HMI), chemical and environmental monitoring, robotics, smart trafficking, smart city set-ups, fiber and fabric sensors, and so on to fulfill every desire of the current civilization. Advancements in the Internet of Things (IoT) and the smart metropolis ideology necessitate the usage of innumerable sensors in all possible locations, even in inaccessible regions, that may well include oil pipelines and gas chamber interiors, ultra-long cables, long tunnels and bridges, gigantic buildings, high-speed railway set-ups, forests, oceans, and aerospace (Wu, Cheng, and Wang 2020). Such innovative schemes mandate continuous self-powering mechanisms for a smooth operation and reliability. Additionally, self-powering sensors are very essential to implanted biomedical and therapeutic gadgets that need to function uninterruptedly as well as to be auto-maintained for a long period of time. In recent years, high demand for wearable electronics have urged the fabrication of easy body-attached, wireless, portable, and self-powered fabric clothing accessories integrated with high energy and maximum power output technology with exceptional mechanical stability and sensitivity that can promote long-lasting, light-weight, breathable, washable, deformable, human-oriented, and self-powered sensors and next-generation wearable energy harvesters (Ryu et al. 2019). The integration of energy conversion units with energy harvesting and sensing units must be properly set in suitable modules for sustaining mechanically harsh conditions. This knowledge has motivated supercapacitive sensing strategies for practical execution (Li et al. 2017). Thus, as far as the energy storage cum conversion units are concerned, such integration can be successfully made possible through the usage of not only flexible electrodes but also smart electrolyte materials such as polymer electrolytes which not only promote smooth and faster ionic conduction between electrodes but also endorse multifunctional properties such as light/thermal energy harvesting and essential sensing capabilities, as well as conferring high mechanical flexibility, thereby avoiding leakage issues and being effectually employed for designing deformable electronic textiles (Zheng et al. 2018; Lu and Chen 2020). Thus, these self-powering textiles in the form of electronic gloves, belts, wristbands, knee-caps, clothes, and so on, containing display panes/panels, sensing components, "epidermal" health monitoring systems, bio-integrated therapeutics, and so on, boosted by light-weight, flexible energy conversion devices, are fast emerging and advancing rapidly with broader application prospects (Meng et al. 2017).

Fundamentally, self-powered integrated sensing systems are being built with components capable of sensing, communicating, monitoring, and responding integrally and cooperatively with energy-harvesting cum conversion components. Common approaches to fabricating, such as integrated devices, include fitting in all building components in a single-step process continuously with significantly reduced device assembling time, and schemes of common electrode modules for promoting production efficiency (Meng et al. 2017). One such important application of the above integrated devices include capacitive force sensors that are often employed for controlling, regulating, and diagnosing various industrial processes with minimum power consumption, high sensitivity, and low temperature effects as a substitute for the popular piezoresistive force/pressure sensors that display inferior total capacitance, thereby complicating the instantaneous sensing applications (Kang et al.

2017). Thus, integrated solid-state supercapacitors have been proposed through innovative and prospective pressure sensing technologies based on the electrostatic mechanical energy harvesting principle. Consequently, to study the pressure sensing effect, a flexible solid-state supercapacitor using a carbon black/polyisoprene composite electrode assembled with a PVA/H_3PO_4 electrolyte was employed that reported an increase in capacitance with compressing pressure application. This fascinating response was attributed to the enhancement of the interfacial surface area between the electrode and the electrolyte under uniaxial pressure, thus signifying the role of deformation forces on the performance of flexible solid-state supercapacitors (Lapcinskis et al. 2019). Flexible integrated devices with sensing and energy storage functions were designed via a simple and cost-effective leaf templating procedure, involving wrinkled carbon nanotubes as the conductive layer, and patterned polydimethylsiloxane (PDMS) with a bio-inspired microstructure as a soft substrate. The assembled form acted well as a strain sensor to realize sensitive resistance responses to various distortions. Furthermore, the electro-deposition of PANI on CNT film and the sandwiching of PVA–H_3PO_4 gel electrolyte between them yielded a flexible supercapacitor that recorded a cell capacitance of 176 F g^{-1} @ a current density of 1 A g^{-1} with 88% capacitance retention even after 10,000 charging/discharging cycles along with a super-mechanical tenacity, thus recording capacitance retention of 98, 95, and 98% after 180° bending (100 cycles), 45° twisting (100 cycles), and after 100 % stretching extent (400 cycles) respectively (Zhang, Li, Huang, et al. 2019). To bring about further improvement of wearable electronics requires the designing of matched functional units for integrated energy storage devices to accommodate extensive mechanical deformations and complex environments. Taking the benefits of highly conducting CNTs and porous polydimethylsiloxane sponge scaffolds endowed with high mechanical robustness, the highly sensitive composite of CNT–polydimethylsiloxane sponge was employed as a piezoresistance sensor as well as a supercapacitor electrode combined with a PVA/H_3PO_4 polymer gel electrolyte to perform as a compressible supercapacitor that can supply power and modulate the low-power electronic devices steadily. The most significant point is when the system, when attached to the epidermal skin or clothes, can readily sense human motions, starting from speech recognition to breathing records, thus proving their usage in real-time health monitoring devices and human–machine interfaces (HMI) (Song et al. 2017). Rigorous exploration shows that broad pressure sensing operating ranges with exceptional sensitivity are very much wanted for real-world wearable electronics. Thus, a novel polymer electrolyte was composed of a room-temperature ionic liquid, 1-ethyl-3-methylimidazolium bis(trifluoromethylsulfonyl)imide with poly(vinylidene fluoride-*co*-hexafluoropropylene), promising a high-capacitance dielectric layer for interfacial capacitive pressure sensing applications, especially for global Morse code recognition, artery pulse detection, and eye blinking. Its ultra-high sensitivity offered a high potentiality for wearable devices to be utilized for health monitoring, motion detection, and e-skin electronics (Chhetry et al. 2019). Similarly, another vertical integration of the micro-supercapacitor, a strain sensor on strong adhesive film to detect bio-signals such as an arterial pulse, swallowing, and frowning of the brow, have been successfully fabricated. A high-performing flexible micro-supercapacitor for driving the vertically integrated skin-attachable strain sensor on a gecko-inspired micro-structured adhesive was composed of manganese/vanadium oxide-coated MWCNT electrodes and a sulfone-based polymer gel electrolyte that simultaneously improved the capacitance as well as functional voltage to 2 V, as well as enduring a high mechanical tenacity over 1000 repetitive bends (Du et al. 2020; Pan et al. 2020; Park et al. 2021). Wearable fiber-shaped electronics are rapidly becoming popular in scientific research fields for the development of various scalable, cost-effective, and flexible gadgets. Consequently, a high-performance fiber-shaped integrated humidity sensor was composed of flexible fiber-shaped supercapacitors with a 3D MnO_2 nanoballs@Ni cone/CNT film on MoS_2 nanosheet arrays/carbon nanotube fibers and a PVA–KOH gel electrolyte. The assembled device offered a stable power supply to the flexible humidity sensing system, displaying high sensitivity with a fast response (Zhao, Li, et al. 2018). In another study, a facile, less-expensive 3D printing procedure was adopted for designing a scalable fiber-shaped integrated device based on polymer electrolytes with high temperature responsivity (Zhao, Zhang, et al. 2018).

Electronic skin (e-skin) having inherent natural stimuli is a crucial feature in engineering prosthetics and wearable electronics (Wang, Lou, et al. 2020). Hence, stretchable energy storage e-skin supercapacitors with superior power density and long durability are integrated with stretchable sensors for fruitfully satisfying the above purpose. Accordingly, to obtain such a self-powering sensing unit, a bi-sub-layered silver nanowire/MnO_2 nanowire hybrid which conducts immobile channels into a PDMS matrix was sandwiched in-between silver nanowire/MnO_2 nanowire film electrodes and a PVA–KOH gel electrolyte. The resultant e-skin device demonstrated an exceptional areal specific capacitance, energy density, admirable capacity preservation, and coulombic efficiency for 2000 cycles with superior sensitivity during stretching and bending by hand (Wang, Lou, et al. 2020). Similarly, an integrated single unit strain sensor was designed to detect both externally applied strain and the arterial pulse on being attached to the wrist. The sensor was driven by means of the energy stored in the flexible polymer electrolyte based micro-supercapacitor harvested by the integrated solar cell, all assembled together in the same unit (Yun et al. 2018).

Highly integrated supercapacitor-sensor systems (HISS) have been proposed to gift a sensing ability to the supercapacitor for multifunctional applications. The authors proposed a geometrical designing strategy to enable ultra-knitted conducting fabrics embracing PPy@ SWCNT-coated cellulose yarns to serve as electrodes. The PPy served dually as an electroactive component as well as a photo-detecting system in the self-powering unit. When the integrated device was strained, a decrease in resistance resulted, as more and closer contact between the conducting yarns occurred, which results in the improved discharging current of the textile supercapacitor. Therefore, the technology promoted the detection of light or strain stimuli without an external power supply by using the multifunctional PPy materials (Huang, Kershaw, et al. 2016).

The target of promoting inexpensive, light-weight, environmentally friendly, and easy to use substrates has encouraged the use of paper as a substrate material for force sensors with ultra-high sensitivity and extraordinary configurability. The high sensitivity comes from the supercapacitor's response on the interaction area between the deformable electrolyte and the electrodes. Thus, highly deformable electrolytes were prepared by coating ionic gel on paper substrates, which could be tailored to different 3D geometrical shapes and sizes. The paper solubilizes in the ionic gel after being fitted into the shape of the electrolytes, leaving behind the clear electrolyte with the micro-structured fractures that originated in their high deformability. Hence, exploiting this facile paper-based flexible sensing system fabrication process, for designing different configuration processes to quantify not only force but also various normal and shear components, shows a higher sensitivity than traditional MEMS capacitive sensors, despite their being straightforwardly invented on a low costing paper substrate and avoiding the use of cleanroom facilities (Zhang et al. 2018). Further, a foldable array of a patterned graphene/ZnO nanoparticle-based UV sensor, integrated with asymmetric micro-supercapacitors on a paper substrate decorated with liquid metal interconnections, has been reported. The micro-supercapacitor was fabricated with asymmetric electrodes containing MnO_2 nanoballs deposited on MWNTs and V_2O_5-wrapped MWNTs, respectively, together with propylene carbonate-poly(methyl methacrylate)-$LiClO_4$ and an organogel electrolyte arrangement (Yun et al. 2017). In another work, a highly stretchable well-patterned graphene gas sensor with high sensitivity was self-powered by an integrated micro-supercapacitor array on a soft Ecoflex deformable substrate. The integrated micro-supercapacitor was constituted of PANI-wrapped MWCNT electrodes and a leakage-free ionogel electrolyte composed of poly(ethylene glycol)diacrylate and 1-ethyl-3-methylimidazoliumbis (trifluoromethylsulfonyl) imide to cope with the deformations it was subject to. The merit of employing a deformable substrate depends on adequately suppressing the applied strain on the gas sensor and micro-supercapacitor via the insertion of stiff platforms of epoxy-based material SU-8 underneath and twisted electrical interconnections of polymer-encapsulated long and narrow gold thin films (Yun et al., 2016). Again, Kim, Kim, Yun, et al. (2020) invented an all-in-one integrated multifunctional sensor-based solid supercapacitor using a polymer electrolyte composed of PVA/borax/agarose/$NaNO_3$ assembled with multiwalled carbon nanotubes (MWCNTs) coated on AuNS electrodes and MWCNTs with a zinc oxide nanowire composite as a multifunctional sensor for NO_2 gas and UV light. Both the supercapacitor and sensor recuperated their performance from bisectional damage via physical contact with a water supply. Very recently, a stress-free electrochemical-based sensor equipped with the IoT device for the quantification of salivary creatinine for diagnosing advanced kidney disorders was demonstrated. Such direct determination and immediate salivary creatinine response was accomplished using a supercapacitor-based sensor composed of cuprous oxide nanoparticles homogeneously distributed on a cross-linked poly(acrylic acid) gel$-Cu^{2+}$ and Nafion perfluorinated membrane on screen-printed carbon electrodes. Here, acrylic gel–Cu^{2+} ions bind irreversibly with >C=N- functional groups of creatinine (Kalasin et al. 2020).

A solar and thermal multi-sensing all-solid-state microfiber supercapacitor with a smart self-conditioned capacitance response, high mechanical stability, and consistent environmental responsivity was devised using a cellulose nanofibril-graphene-conjugated polymer simultaneously as a core-sheath microfiber electrode, PVA/H_2SO_4 gel electrolyte assembled with sensing unit. Its hierarchical synergistic design exploited entirely the exclusive characteristics and synergistic interactions of each of the multi-components to bestow maximum capacitance and mechanical robustness as well as outstanding thermal sensitivity and photothermal conversion. (Hsueh, Huang, Wu, Kuo, and Teng 2013) The integrated supercapacitor was successfully employed as photo-thermal energy conversion cum storage system to monitor body temperature and thus promoted novel future generation energy-related devices (Teng et al. 2020). Similarly, a newly conceived 3-V flexible supercapacitor was designed for AC line filtering, based on an ionic-liquid-based polymer gel electrolyte and carbon nanotube electrode material. The flexible supercapacitor exhibited an areal energy density that is more than 20 times higher than that of the previously reported systems. Further, the sensing efficiencies of the electrode and the cell established a substantial linear response to different dynamic deformation conditions (Tian et al. 2018).

Thus, it has been repeatedly observed that polymer electrolytes play an essential function in these well-designed, green, and sustainable self-powering sensing systems. However, this technological field is at a very nascent stage and much exploration regarding optimization of the components is needed to tune their output efficiency, sensitivity, durability, power management, fabrication and packaging, and production expenses to make them commercially available to all.

2.7 CONCLUSIONS

Polymer based electrolytes have been targeted to cope up with the unavoidable leakage issues, toxicity, flammability, and limited handling privileges typically associated with liquid electrolytes. The advent of this class of electrolytes has surely broadened the domain of applications of supercapacitors beyond daily usages, in nearly all technological fields, especially in the defense, biomedical, and astronomical sectors, where portable, wearable electronic devices require not only uninterrupted as well as self-powering options but also high structural and mechanical durability, thereby requiring no regular maintenance or servicing provisions.

Although this area of research is very stimulating, studies so far carried out confer that polymer-based electrolytes are still lagging in many essential areas such as ionic conductivity, stability, and manufacturing costs. Low ionic conductivity largely shrinks the capacitance as well as elevates the internal resistance of the devices, which in turn essentially worsens both the energy as well as power densities, thus erecting the main hurdles to commercialization. Since ion selectivity varies with the nature of polymer hosts' structural properties, thorough research and deep understanding of the chemistry of their inter-relative choices is necessary. Further, the use of strong acids/alkalis as dopants and mobile phases often retards the structural stability of the host polymers. Moreover, the liquid retention capability of the polymer matrix is another important issue for this class of electrolytes as they depend upon the incorporated amount of liquid, the conductivity, as well as the mechanical robustness of the polymers, which varies; and this again is of prime significance when devising flexible supercapacitors. Thus, suitable modifications of the side chains or functionalities of the polymer hosts as well as the necessary additives need to be developed and added to optimize the robustness and anti-corrosion features.

Nonetheless, the multifunctional behavior of the assembled polymeric electrolyte-based supercapacitors offers not only remarkable energy storage aptitudes but also supplementary roles in essential technological sectors operating in extreme working environments. Stretchy and compressible multifaceted properties bestowed on the device equips it with the ability to withstand large mechanical deformation without compromising the device's original electrochemical energy storage efficiency. In addition, self-healing and shape memory supercapacitors may well assist in the recovery of their initial configuration or original shape from mechanical distortion, while electrochromic supercapacitors may well be used as smart electrochemical voltage indicators due to their electrochemical color transformation features. Modern integrated devices are suitably aimed at avoiding complex structures and increasing volumes, keeping in view their portability and safe-handling criteria for vivid technological applications. Accordingly, their charging procedures must be made very simple so that the supercapacitor can be charged anytime and anywhere without requiring bulky energy sources that complicate the whole device unit.

Even though tremendous progress has been made in this stimulating field of research, the hurdles arising can hardly be ignored. For instance, these smart macromolecular based materials themselves do hold exceptional multifunctional performances, but the overall performance often gets seriously impaired when assembled into supercapacitors. Taking the stretchability criteria for instance, the entire assembled supercapacitor can often exhibit limited stretchability even after employing highly flexible electrode and electrolyte components that support many fold stretching acts. Such problems often arise due to the co-existence of stretchable and non-stretchable device components within one and the same gadget and the nature of the device's configuration often restricts the desired responses. Thus, to maximize performance, proper selection of the components and the configurational design must be meticulously considered and optimized. Apart from this, to promote the practical pertinency of these multifunctional supercapacitors, both the charge storage efficiency as well as the additional functional feature should be cumulatively upgraded. Thus, the present self-healable supercapacitors can only heal very limited numbers of cutting/healing cycles, while maintenance of practical energy storage systems obviously demands much higher cycling to at least few hundreds of cycles with unspoiled electrochemical performance. Hence, such applied efficacy urges combinational improvements in both individual material characteristics as well as effective integration at the device level.

Hence, it is quite clear that even though the current challenges stand mountain-high, the optimistic approach of worldwide scientists and technologists, all endlessly devoting their brains and hard labour in resolving the problems, certainly convince that the situation is not far-off when polymer-based flexible supercapacitors may well successfully emerge as an efficient and inevitable energy storage and conversion system, taking care of almost every step of life.

ACKNOWLEDGMENT

Dipanwita Majumdar acknowledges Chandernagore College, Chandannagar, Hooghly, West Bengal, Pin-712136, India, for providing permission to do honorary research.

DECLARATION OF COMPETING INTEREST

The authors declare no conflict of interest and that no funding assistance was received from any sources or any organization for the present project.

REFERENCES

Afrifah, V. A., Phiri, I., Hamenu, L., Madzvamuse, A., Lee, K. S., Ko, J. M., 2020, Electrochemical properties of poly(2-acrylamido-2-methylpropane sulfonic acid) polyelectrolyte containing zwitterionic silica sulfobetaine for supercapacitors, *J. Power Sources*, 479: 228657. https://doi.org/10.1016/j.jpowsour.2020.228657

Agrawal, R. C., Pandey, G. P., 2008, Solid polymer electrolytes: materials designing and all-solid-state battery applications: an overview, *J. Phys. D: Appl. Phys.*, 41: 223001. https://doi.org/10.1088/0022-3727/41/22/223001

Akinwolemiwa, B., Chen, G. Z., 2018, Fundamental consideration for electrochemical engineering of supercapattery, *J. Braz. Chem. Soc.*, 29: 960. http://dx.doi.org/10.21577/0103-5053.20180010

Akinwolemiwa, B., Peng, C., Chen, G. Z., 2015, Redox electrolytes in supercapacitors, *J. Electrochem. Soc.*, 162(5): A5054. https://doi.org/10.1149/2.0111505jes

Alipoori, S., Mazinani, S., Aboutalebi, S. H., Sharif, F., 2020, Review of PVA-based gel polymer electrolytes in flexible solid-state supercapacitors: opportunities and challenges, *J. Energy Storage*, 27: 101072. https://doi.org/10.1016/j.est.2019.101072

Al-Zubaidi, A., Ji, X., Yu, J., 2017, Thermal charging of supercapacitors: a perspective, *Sustain. Energy Fuels*, 1: 1457. https://doi.org/10.1039/C7SE00239D

Armand, M., 1990, Polymers with ionic conductivity, *Adv. Mater.*, 2(6–7): 278–286. https://doi.org/10.1002/adma.19900020603

Armand, M., Endres, F., MacFarlane, D. R., Ohno, H., Scrosati, B., 2009, Ionic-liquid materials for the electrochemical challenges of the future, *Nat. Mater.*, 8: 621–629. https://doi.org/10.1038/nmat2448

Armelin, E., Pérez-Madrigal, M. M., Alemán, C., Díaz, D. D., 2016, Current status and challenges of biohydrogels for applications as supercapacitors and secondary batteries, *J. Mater. Chem. A*, 4: 8952–8968. https://doi.org/10.1039/C6TA01846G

Asbani, B., Douard, C., Brousse, T., Bideau, J. L., 2019, High temperature solid-state supercapacitor designed with ionogel electrolyte, *Energy Storage Mater.*, 21: 439–445. https://doi.org/10.1016/j.ensm.2019.06.004

Augustyn, V., Simon, P., Dunn, B., 2014, Pseudocapacitive oxide materials for high-rate electrochemical energy storage, *Energy Environ. Sci.*, 7: 1597–1614. https://doi.org/10.1039/C3EE44164D

Aziz, S. B., Hamsan, M. H., Abdullah, R. M., Kadir, M. F. Z. 2019, A promising polymer blend electrolytes based on chitosan: methyl cellulose for EDLC application with high specific capacitance and energy density, *Molecules*, 24(13): 2503. https://doi.org/10.3390/molecules24132503

Aziz, S. B., Hazrin, Z., Abidin, Z., 2013, Electrical conduction mechanism in solid polymer electrolytes: new concepts to Arrhenius equation, *J. Soft Matter*, Article ID 323868, 8 pages. https://doi.org/10.1155/2013/323868

Aziz, S. B., Dannoun, E. M. A., Hamsan, M. H., Abdulwahid, R. T., Mishra, K., Nofal, M. M., Kadir, M. F. Z., 2021, Improving EDLC device performance constructed from plasticized magnesium ion conducting chitosan based polymer electrolytes via metal complex dispersion, *Membranes*, 11, 289 (1–20). https://doi.org/10.3390/membranes11040289

Bagchi, D., Nguyen, T. D., de la Cruz, M. O., 2020, Surface polarization effects in confined polyelectrolyte solutions, *Proc. Natl. Acad. Sci.*, 117(33): 19677–19684. https://doi.org/10.1073/pnas.2007545117

Balo, L., Gupta, H., Singh, S. K., Singh, V. K., Kataria, S., Singh, R. K., 2018, Performance of EMIMFSI ionic liquid based gel polymer electrolyte in rechargeable lithium metal batteries, *J. Ind. Eng. Chem.*, 65: 137–145. https://doi.org/10.1016/j.jiec.2018.04.022

Baños, R., Manzano-Agugliaro, F., Montoya, F. G., Gil, C., Alcayde, A., Gómez, J., 2011, Optimization methods applied to renewable and sustainable energy: a review, *Renew. Sust. Energ. Rev.*, 15(4): 1753. https://doi.org/10.1016/j.rser.2010.12.008

Barzegar, F., Dangbegnon, J. K., Bello, A., Momodu, D. Y., Johnson, A. T. C., Manyala, N., 2015, Effect of conductive additives to gel electrolytes on activated carbon-based supercapacitors, *AIP Adv.*, 5: 097171. https://doi.org/10.1063/1.4931956

Béguin, F., Presser, V., Balducci, A., Frackowiak, E., 2014, Carbons and electrolytes for advanced supercapacitors, *Adv. Mater.*, 26: 2219. https://doi.org/10.1002/adma.201304137

Bi, Z., Li, X., Chen, Y., Xu, X., Zhang, S., Zhu, Q., 2017, Bi-functional flexible electrodes based on tungsten trioxide/zinc oxide nanocomposites for electrochromic and energy storage applications, *Electrochim. Acta*, 227: 61–68. http://dx.doi.org/10.1016/j.electacta.2017.01.003

Bombelli, P., Muller, T., Herling, T. W., Howe, C. J., Knowles, T. R. J., 2015, A high power density, mediator-free, microfluidic biophotovoltaic device for cyanobacterial cells, *Adv. Energy Mater.*, 5: 1401299.

Boruah, B. D., Misra, A., 2017, Photocharge-enhanced capacitive response of a supercapacitor, *Energy Technol.*, 5: 1356–1363. https://doi.org/10.1002/ente.201600661

Brandon, E., West, W., Smart, M., Whitcanack, L., Plett, G., 2007, Extending the low temperature operational limit of double-layer capacitors, *J. Power Sources*, 170: 225–232. https://doi.org/10.1016/j.jpowsour.2007.04.001

Cai, G., Darmawan, P., Cui, M., Wang, J., Chen, J., Magdassi, S., Lee, P. S., 2016, Highly stable transparent conductive silver grid/PEDOT: PSS electrodes for integrated bifunctional flexible electrochromic supercapacitors, *Adv. Energy Mater.*, 6: 1501882. https://doi.org/10.1002/aenm.201501882

Cao, L., Yang, M., Wu, D., Lyu, F., Sun, Z., Zhong, X., Pan, H., Liu, H., Lu, Z., 2017, Biopolymer-chitosan based supramolecular hydrogels as solid state electrolytes for electrochemical energy storage, *Chem. Commun.*, 53: 1615–1618. https://doi.org/10.1039/C6CC08658F

Chakrabarti, A., Filler, R., Mandal, B. K., 2008, Borate ester plasticizer for PEO-based solid polymer electrolytes, *J. Solid State Electrochem*, 12: 269–272. https://doi.org/10.1007/s10008-007-0388-z

Chandra, S., Sekhon, S. S., Arora, N., 2000, PMMA based protonic polymer gel electrolytes, *Ionics*, 6: 112–118. https://doi.org/10.1007/BF02375554

Chatterjee, J., Liu, T., Wang, B., Zheng, J. P., 2010, Highly conductive PVA organogel electrolytes for applications of lithium batteries and electrochemical capacitors, *Solid State Ion.*, 181(11–12): 531–535. https://doi.org/10.1016/j.ssi.2010.02.020

Chaudoy, V., Van, F. T., Deschamps, M., Ghamouss, F., 2017, Ionic liquids in a poly ethylene oxide cross-linked gel polymer as an electrolyte for electrical double layer capacitor, *J. Power Sources*, 342: 872–878. https://doi.org/10.1016/j.jpowsour.2016.12.097

Chee, W. K., Lim, H. N., Zainal, Z., Huang, N. M., Harrison, I., Andou, Y., 2016, Flexible graphene-based supercapacitors: a review, *J. Phys. Chem. C*, 120(8): 4153–4172. https://doi.org/10.1021/acs.jpcc.5b10187

Chen, G. Z., 2017, Supercapacitor and supercapattery as emerging electrochemical energy stores, *Int. Mater. Rev.*, 62(4): 173. https://doi.org/10.1080/09506608.2016.1240914

Chen, M., Feng, G., Qiao, R., 2020, Water-in-salt electrolytes: an interfacial perspective, *Curr. Opin. Colloid Interface Sci.*, 47: 99–110. https://doi.org/10.1016/j.cocis.2019.12.011

Chen, Q., Li, X., Zang, X., Cao, Y., He, Y., Li, P., Wang, K., Wei, J., Wu, D., Zhu, H., 2014, Effect of different gel electrolytes on graphene-based solid-state supercapacitors, *RSC Adv.*, 4: 36253–36256. https://doi.org/10.1039/C4RA05553E

Chen, X. L., Lin, H. J., Deng, J., Zhang, Y., Sun, X. M., Chen, P. N., Fang, X., Zhang, Z. T., Guan, G. Z., Peng, H. S., 2014, Electrochromic fiber-shaped supercapacitors, *Adv. Mater.*, 26: 8126–8132. https://doi.org/10.1002/adma.201403243

Chen, C.-R., Qin, H., Cong, H.-P., Yu, S.-H., 2019, A highly stretchable and real-time healable supercapacitor, *Adv Mater.*, 31: 1900573. https://doi.org/10.1002/adma.201900573

Chen, Y., Wang, Y., Sun, P., Yang, P., Du, L., Mai, W., 2015, Nickel oxide nanoflake-based bifunctional glass electrodes with superior cyclic stability for energy storage and electrochromic applications, *J. Mater. Chem. A*, 3: 20614–20618. https://doi.org/10.1039/C5TA04011F

Chen, Y., Zhang, X. O., Zhang, D. C., Yu, P., Ma, Y. W., 2011, High performance supercapacitors based on reduced graphene oxide in aqueous and ionic liquid electrolytes, *Carbon*, 49: 573–580. https://doi.org/10.1016/j.carbon.2010.09.060

Chhetry, A., Kim, J., Yoon, H., Park, J. Y., 2019, Ultrasensitive interfacial capacitive pressure sensor based on a randomly distributed microstructured iontronic film for wearable applications, *ACS Appl. Mater. Interfaces*, 11(3): 3438–3449. https://doi.org/10.1021/acsami.8b17765

Chiou, Y.-D., Tsai, D.-S., Lam, H. H., Chang, C.-H., Lee, K.-Y., Huang, Y.-S., 2013, Cycle stability of the electrochemical capacitors patterned with vertically aligned carbon nanotubes in an LiPF6-based electrolyte, *Nanoscale*, 5: 8122–8129. https://doi.org/10.1039/C3NR01980B

Choi, C., Lee, J. M., Kim, S. H., Kim, S. J., 2016, Twistable and stretchable sandwich structured fiber for wearable sensors and supercapacitors, *Nano Lett.*, 16(12): 7677–7684. https://doi.org/10.1021/acs.nanolett.6b03739

Choudhary, N., Li, C., Moore, J., Nagaiah, N., Zhai, L., Jung, Y., Thomas, J., 2017, Asymmetric supercapacitor eand devices, *Adv. Mater.*, 29: 1605336. https://doi.org/10.1002/adma.201605336

Choudhury, N. A., Sampath, S., Shukla, A. K., 2008, Gelatin hydrogel electrolytes and their application to electrochemical supercapacitors, *J. Electrochem. Soc.*, 155(1): A74–A81. https://doi.org/10.1149/1.2803501

Choudhury, N. A., Sampath, S., Shukla, A. K., 2009, Hydrogel-polymer electrolytes for electrochemical capacitors: an overview, *Energy Environ. Sci.*, 2: 55. https://doi.org/10.1039/B811217G

Choudhury, N. A., Shukla, A. K., Sampath, S., Pitchumani, S., 2006, Cross-linked polymer hydrogel electrolytes for electrochemical capacitors, *J. Electrochem. Soc.*, 153: A614. https://doi.org/10.1149/1.2164810

Chun, S., Son, W., Lee, G., Kim, S. H., Park, J. W., Kim, S. J., Pang, C., Cho, C., 2019, Single-layer graphene-based transparent and flexible multifunctional electronics for self-charging power and touch sensing systems, *ACS Appl. Mater. Interfaces*, 11: 9301–9308. https://doi.org/10.1021/acsami.8b20143

Chun, J., Ye, B. U., Lee, J. W., Choi, D., Kang, C.-Y., Kim, S. W., Wang, Z. L., Baik, J. M., 2016, Boosted output performance of triboelectric nanogenerator via electric double layer effect, *Nat. Commun.*, 7: 12985. https://doi.org/10.1038/ncomms12985

Conway, B. E., 1991, Transition from "supercapacitor" to "battery" behavior in electrochemical energy storage, *J. Electrochem. Soc.*, 138: 1539. https://doi.org/10.1149/1.2085829

Deng, J., Zhang, Y., Zhao, Y., Chen, P., Cheng, X., Peng, H., 2015, A shape-memory supercapacitor fiber, *Angew. Chemie Int. Ed.*, 54: 15419–15423. https://doi.org/10.1002/anie.201508293

Deng, M.-J., Yeh, L-H., Lin, Y-H., Chen, J., Chou, T-H., 2019, 3D Network V_2O_5 Electrodes in a Gel Electrolyte for High-Voltage Wearable Symmetric Pseudocapacitors. *ACS Applied Materials & Interfaces*, 11(33): 29838–29848. https://doi.org/10.1021/acsami.9b07845

Dincer, I., 2000, Renewable energy and sustainable development: a crucial review, *Renew. Sust. Energ. Rev.*, 4(2): 157. https://doi.org/10.1016/S1364-0321(99)00011-8

Ding, Y., Yun, J., Liu, H., Wan, Z., Shen, M., Zhang, L., Qu, Q., Zheng, H. 2014, A safe and superior propylene carbonate-based electrolyte with high-concentration Li salt, *Pure Appl. Chem.*, 86(5): 585–591. https://doi.org/10.1515/pac-2013-1120

Dominici, F., Puglia, D., Luzi, F., Sarasini, F., Rallini, M., Torre, L. A, 2019, Novel class of cost effective and high performance composites based on terephthalate salts reinforced polyether ether ketone, *Polymers*, 11: 2097. https://doi.org/10.3390/polym11122097

Dong, P., Rodrigues, M.-T. F., Zhang, J., Borges, R. S., Kalaga, K., Reddy, A. L. M., Silva, G. G., Ajayan, P. M., Lou, J., 2017, A flexible solar cell/supercapacitor integrated energy device, *Nano Energy*, 42: 181–186. https://doi.org/10.1016/j.nanoen.2017.10.035

Dong, L., Xu, C., Li, Y., Huang, Z.-H., Kang, F., Yang, Q.-H., Zhao, X, 2016, Flexible electrodes and supercapacitors for wearable energy storage: a review by category, *J. Mater. Chem. A*, 4: 4659–4685. https://doi.org/10.1039/C5TA10582J

Du, X., Tian, M., Sun, G., Li, Z., Qi, X., Zhao, H., Zhu, S., Qu, L., 2020, Self-powered and self-sensing energy textile system for flexible wearable applications, *ACS Appl. Mater. Interfaces*, 12: 55876–55883. https://doi.org/10.1021/acsami.0c16305

Duay, J., Gillette, E., Liu, R., Lee, S. B., 2012, Highly flexible pseudocapacitor based on freestanding heterogeneous MnO_2/conductive polymer nanowire arrays, *Phys. Chem. Chem. Phys.*, 14: 3329–3337. https://doi.org/10.1039/C2CP00019A

Elamin, K., Shojaatalhosseini, M., Danyliv, O., Martinelli, A., Swenson, J., 2019, Conduction mechanism in polymeric membranes based on PEO or PVdF-HFP and containing a piperidinium ionic liquid, *Electrochim. Acta*, 299: 979–986. https://doi.org/10.1016/j.electacta.2018.12.154

Elieh-Ali-Komi, D., Hamblin, M. R., 2016, Chitin and chitosan: production and application of versatile biomedical nanomaterials, *Int. J. Adv. Res.*, 4(3): 411–427.

Ellabban, O., Abu-Rub, H., Blaabjerg, F., 2014, Renewable energy resources: current status, future prospects and their enabling technology, *Renew. Sust. Energ. Rev.*, 39: 748. http://dx.doi.org/10.1016/j.rser.2014.07.113

Fan, L., Wang, M., Zhang, Z., Qin, G., Hu, X., Chen, Q., 2018, Preparation and characterization of PVA alkaline solid polymer electrolyte with addition of bamboo charcoal, *Materials (Basel)*, 11(5): 679. https://doi.org/10.3390/ma11050679

Fang, L., Cai, Z., Ding, Z., Chen, T., Zhang, J., Chen, F., Shen, J., Chen, F., Li, R., Zhou, X., Xie, Z., 2019, Skin-inspired surface-microstructured tough hydrogel electrolytes for stretchable supercapacitors, *ACS Appl. Mater. Interfaces*, 11: 21895–21903. https://doi.org/10.1021/acsami.9b03410

Fang, X., Sun, J., 2019, One-step synthesis of healable weak-polyelectrolyte-based hydrogels with high mechanical strength, toughness, and excellent self-recovery, *ACS Macro Letters*, 8(5): 500–505. https://doi.org/10.1021/acsmacrolett.9b00189

Fei, H. J., Yang, C. Y., Bao, H., Wang, G. C., 2014, Flexible all-solid-state supercapacitors based on graphene/carbon black nanoparticle film electrodes and cross-linked poly(vinyl alcohol)–H_2SO_4 porous gel electrolytes, *J. Power Sources*, 266: 488–495. https://doi.org/10.1016/j.jpowsour.2014.05.059

Feng, E., Gao, W., Yan, Z., Li, J., Li, Z., Ma, X., Ma, L., Yang, Z., 2020, A multifunctional hydrogel polyelectrolyte based flexible and wearable supercapacitor, *J. Power Sources*, 479: 229100. https://doi.org/10.1016/j.jpowsour.2020.229100.

Fic, K., Lota, G., Meller, M., Frackowiak, E., 2012, Novel insight into neutral medium as electrolyte for high-voltage supercapacitors, *Energy Environ. Sci.*, 5: 5842–5850. https://doi.org/10.1039/C1EE02262H

Frackowiak, E., Abbas, Q., Béguin, F., 2013, Carbon/carbon supercapacitors, *J. Energy Chem.*, 22(2): 226–240. https://doi.org/10.1016/S2095-4956(13)60028-5

Gao, H., Lian, K., 2013, Effect of SiO_2 on silicotungstic acid-H_3PO_4-poly(vinyl alcohol) electrolyte for electrochemical supercapacitors, *J. Electrochem. Soc.*, 160: A505–A510. https://doi.org/10.1149/2.053303jes

Gao, H., Lian, K., 2014, Proton-conducting polymer electrolytes and their applications in solid supercapacitors: a review, *RSC Adv.*, 4: 33091–33113. https://doi.org/10.1039/C4RA05151C

Gao, D., Liu, R., Yu, W., Luo, Z., Liu, C., Fan, S., 2019, Gravity-induced self-charging in cnt/polymer supercapacitors, *J. Phys. Chem. C*, 123(9): 5249–5254 https://doi.org/10.1021/acs.jpcc.8b11644

Gao, H., Virya, A., Lian, K. 2014, Monovalent silicotungstate salts as electrolytes for electrochemical supercapacitors, *Electrochim. Acta*, 138: 240–246. https://doi.org/10.1016/j.electacta.2014.06.127

Geiculescu, O. E., Hallac, B. B., Rajagopal, R. V., Creager, S. E., Desmarteau, D. D., Borodin, O., et al., 2014, The effect of low-molecular-weight poly(ethylene glycol) (PEG) plasticizers on the transport properties of lithium fluorosulfonimide ionic melt electrolytes, *J. Physical Chemistry B*, 118: 5135–5143. https://doi.org/10.1021/jp500826c

Ghosh, A., Lee, Y. 2012, Carbon-based electrochemical capacitors, *ChemSusChem*, 5: 480–499. https://doi.org/10.1002/cssc.201100645

Gidwani, M., Bhagwani, A., Rohra, N., 2014, Supercapacitors: the near future of batteries, *Int. J. Eng. Invent.*, 4: 22.

Golodnitsky, D., Ardel, G., Peled, E., 2002, Ion-transport phenomena in concentrated PEO-based composite polymer electrolytes, *Solid State Ion.*, 147: 141–155. https://doi.org/10.1016/S0167-2738(01)01036-0

Guo, K., Yu, N., Hou, Z., Hu, L., Ma, Y., Li, H., Zhai, T., 2017, Smart supercapacitors with deformable and healable functions, *J. Mater. Chem. A*, 5: 16–30. https://doi.org/10.1039/C6TA08458C

Guo, Q., Zhao, X., Li, Z., Wang, D., Nie, G., 2020, A novel solid-state electrochromic supercapacitor with high energy storage capacity and cycle stability based on poly(5-formylindole)/WO_3 honeycombed porous nanocomposites, *Chem. Eng. J.*, 384: 123370. https://doi.org/10.1016/j.cej.2019.123370

Guo, Y., Zheng, K., Wan, P., 2018, A flexible stretchable hydrogel electrolyte for healable all-in-one configured supercapacitors, *Small*, 14: 1704497. https://doi.org/10.1002/smll.201704497

Guo, Y., Zhou, X., Tang, Q., Bao, H., Wang, G., Saha, P., 2016a, A self-healable and easily recyclable supramolecular hydrogel electrolyte for flexible supercapacitors, *J. Mater. Chem. A*, 4: 8769–8776. https://doi.org/10.1039/C6TA01441K8776.

Guo, Y., Zhou, X., Tang, Q., Bao, H., Wang, G., Saha, P., 2016b, A self-healable and easily recyclable supramolecular hydrogel electrolyte for flexible supercapacitors, *J. Mater. Chem. A*, 4: 8769. https://doi.org/10.1039/C6TA01441K

Gupta, H., Kataria, S., Balo, L., Singh, V. K., Singh, S. K., Tripathi, A. K., Verma, Y. L., Singh, R. K., 2018, Electrochemical study of ionic liquid based polymer electrolyte with graphene oxide coated $LiFePO_4$ cathode for Li battery, *Solid State Ion.*, 320: 186–192. https://doi.org/10.1016/j.ssi.2018.03.008

Gür, T. M., 2018, Review of electrical energy storage technologies, materials and systems: challenges and prospects for large-scale grid storage, *Energy Environ. Sci.*, 11: 2696–2767. https://doi.org/10.1039/C8EE01419A

Hamsan, M. H., Aziz, S. B., Kadir, M. F. Z., Brza, M. A., Karim, W. O., 2020, The study of EDLC device fabricated from plasticized magnesium ion conducting chitosan based polymer electrolyte, *Polym. Test.*, 90: 106714. https://doi.org/10.1016/j.polymertesting.2020.106714

Han, Y., Dai, L., 2019, Conducting polymers for flexible supercapacitors, *Macromol. Chem. Phys.*, 1800355. https://doi.org/10.1002/macp.201800355

Han, X., Xiao, G., Wang, Y., Chen, X., Duan, G., Wu, Y., et al., 2020, Design and fabrication of conductive polymer hydrogels and their applications in flexible supercapacitors, *J. Mater. Chem. A*, 8: 23059–23095. https://doi.org/10.1039/D0TA07468C

Hashmi, S. S. A., 2014, Quasi-solid-state pseudocapacitors using proton-conducting gel polymer electrolyte and poly(3-methyl thiophene)–ruthenium oxide composite electrodes, *J. Solid State Electrochem*, 18: 465–475. https://doi.org/10.1007/s10008-013-2276-z

He, M., Fic, K., Frackowiak, E., Novak, P., Berg, E. J., 2016, Ageing phenomena in high-voltage aqueous supercapacitors investigated by in situ gas analysis, *Energy Environ. Sci.*, 9: 623–633. https://doi.org/10.1039/C5EE02875B

Heo, Y., Sodano, H. A., 2014, Self-healing polyurethanes with shape recovery, *Adv. Funct. Mater.*, 24:5261–5268. https://doi.org/10.1002/adfm.201400299

Hong, M. S., Lee, S. H., Kim, S. W., 2002, Use of KCl aqueous electrolyte for 2 V manganese oxide/activated carbon hybrid capacitor, *Electrochem. Solid-State Lett.*, 5: A227. https://doi.org/10.1149/1.1506463

Hsueh, M.-F., Huang, C.-W., Wu, C.-A., Kuo, P.-L., Teng, H., 2013, The synergistic effect of nitrile and ether functionalities for gel electrolytes used in supercapacitors, *J. Phys. Chem. C*, 117: 16751–16758. https://doi.org/10.1021/jp4031128

Hu, J., Xie, K., Liu, X., Guo, S., Shen, C., Liu, X., Li, X., Wang, J., Wei, B., 2017, Dramatically enhanced ion conductivity of gel polymer electrolyte for supercapacitor via h-bn nanosheets doping, *Electrochim. Acta*, 227: 455–461. http://dx.doi.org/10.1016/j.electacta.2017.01.045

Hu, X., Yan, X., Gong, L., Wang, F., Xu, Y., Feng L., Zhang D., Jiang Y., 2019, Improved piezoelectric sensing performance of P(VDF–TrFE) nanofibers by utilizing BTO nanoparticles and penetrated electrodes, *ACS Appl. Mater. Interfaces*, 11: 7379–7386. https://doi.org/10.1021/acsami.8b19824

Huang, Y., Kershaw, S. V., Wang, Z., Pei, Z., Liu, J., Huang, Y., Li, H., Zhu, M., Rogach, A. L., Zhi, C., 2016, Highly integrated supercapacitor-sensor systems via material and geometry design, *Small*, 12: 3393–3399. https://doi.org/10.1002/smll.201601041

Huang, P. L., Luo, X. F., Peng, Y. Y., Pu, N. W., Ger, M. D., Yang, C. H., Wu, T. Y., Chang, J. K., 2015, Ionic liquid electrolytes with various constituent ions for graphene-based supercapacitors, *Electrochim. Acta*, 161: 371–377. https://doi.org/10.1016/j.electacta.2015.02.115.

Huang, Y., Tang, Z., Liu, Z., Wei, J., Hu, H., Zhi, C., 2018, Toward enhancing wearability and fashion of wearable supercapacitor with modified polyurethane artificial leather electrolyte, *Nanomicro. Lett.*, 10(3): 38. https://doi.org/10.1007/s40820-018-0191-7

Huang, S., Wan, F., Bi, S., Zhu, J., Niu, Z., Chen, J., 2019, A self-healing integrated all-in-one zinc-ion battery, *Angew Chem. Int. Edit.*, 58: 4313–4317 https://doi.org/10.1002/anie.201814653

Huang, C. W., Wu, C. A., Hou, S. S., Kuo, P. L., Hsieh, C. T., Teng, H., 2012, Gel electrolyte derived from poly(ethylene glycol) blending poly(acrylonitrile) applicable to roll-to-roll assembly of electric double layer capacitors, *Adv. Funct. Mater.*, 22: 4677–4685. https://doi.org/10.1002/adfm.201201342

Huang, Y. F., Wu, P. F., Zhang, M. Q., Ruan, W. H., Giannelis, E. P., 2014, Boron cross-linked graphene oxide/polyvinyl alcohol nanocomposite gel electrolyte for flexible solid-state electric double layer capacitor with high performance, *Electrochim. Acta*, 132: 103–111. https://doi.org/10.1016/j.electacta.2014.03.151

Huang, C., Zhang, J., Young, N. et al., 2016, Solid-state supercapacitors with rationally designed heterogeneous electrodes fabricated by large area spray processing for wearable energy storage applications, *Sci. Rep.*, 6: 25684. https://doi.org/10.1038/srep25684

Huang, Y., Zhong, M., Huang, Y., et al., 2015, A self-healable and highly stretchable supercapacitor based on a dual cross-linked polyelectrolyte, *Nat. Commun.* 6, 10310. https://doi.org/10.1038/ncomms10310

Huang, Y., Zhong, M., Shi, F., Liu, X., Tang, Z., Wang, Y., Huang, Y., Hou, H., Xie X., Zhi, C., 2017, An intrinsically stretchable and compressible supercapacitor containing a polyacrylamide hydrogel electrolyte, *Angew. Chemie.* 129: 9269–9273. https://doi.org/10.1002/ange.201705212

Huang, Y., Zhu, M., Pei, Z., Xue, Q., Huang, Y., Zhi, C., 2016, A shape memory supercapacitor and its application in smart energy storage textiles, *J. Mater. Chem. A*, 4: 1290–1297. https://doi.org/10.1039/C5TA09473A

Hui, C.-Y., Kan, C.-W., Mak, C.-L., Chau, K.-H., 2019, Flexible energy storage system—an introductory review of textile-based flexible supercapacitors, *Processes*, 7: 922. https://doi.org/10.3390/pr7120922

Ibrahim, Z., Gultekin, B., Singh, V., Singh, P. K., 2020, Electrochemical double-layer supercapacitor using poly(methyl methacrylate) solid polymer electrolyte, *High Perform. Polym.*, 32: 201–207. https://doi.org/10.1177/0954008319895556.

Ishimoto, S., Asakawa, Y., Shinya, M., Naoi, K., 2009, Degradation responses of activated-carbon-based EDLCs for higher voltage operation and their factors, *J. Electrochem. Soc.* 156: A563. https://doi.org/10.1149/1.3126423

Jain, A., Tripathi, S. K., 2013, Experimental studies on high-performance supercapacitor based on nanogel polymer electrolyte with treated activated charcoal, *Ionics*, 19: 549–557. https://doi.org/10.1007/s11581-012-0782-0

Jin, L., Greene, G. W., MacFarlane, D. R., Pringle, J. M., 2016, Redox-active quasi-solid-state electrolytes for thermal energy harvesting, *ACS Energy Lett.*, 1: 654–658.

Jin, X., Sun, G., Yang, H., Zhang, G., Xiao, Y., Gao, J., Zhang, Z., Qu, L., 2018, Graphene oxide-mediated polyelectrolyte with high ion-conductivity for highly stretchable and self-healing all-solid-state supercapacitors, *J. Mater. Chem. A*, 6: 19463. https://doi.org/10.1039/c8ta07373b

Jin, X., Sun, G., Zhang, G., Yang, H., Xiao, Y., Gao, J., Qu, L., 2019, A cross-linked polyacrylamide electrolyte with high ionic conductivity for compressible supercapacitors with wide temperature tolerance, *Nano Res.*, 12: 1199–1206. https://doi.org/10.1007/s12274-019-2382

Johansson, P., 2001, First principles modelling of amorphous polymer electrolytes: Li^+-PEO, Li^+-PEI, and Li^+-PES complexes, *Polymer.*, 42:4367–4373. https://doi.org/10.1016/S0032-3861(00)00731-X

Kaempgen, M., Chan, C. K., Ma, J., Cui, Y., Gruner, G., 2009, Printable thin film supercapacitors using single-walled carbon nanotubes, *Nano Lett.*, 9: 1872–1876. https://doi.org/10.1021/nl8038579

Kalasin, S., Sangnuang, P., Khownarumit, P., Tang, I. M., Surareungchai, W., 2020, Salivary creatinine detection using a Cu(I)/Cu(II) catalyst layer of a supercapacitive hybrid sensor: a wireless iot device to monitor kidney diseases for remote medical mobility, *ACS Biomater. Sci. Eng.*, 6: 5895–5910. https://doi.org/10.1021/acsbiomaterials.0c00864

Kang, Y. J., Chun, S.-J., Lee, S.-S., Kim, B.-Y., Kim, J. H., Chung, H., Lee, S.-Y., Kim, W., 2012, All-solid-state flexible supercapacitors fabricated with bacterial nanocellulose papers, carbon nanotubes, and triblock-copolymer ion gels, *ACS Nano*, 6: 6400–6406. https://doi.org/10.1021/nn301971r

Kang, M. C., Rim, C.S., Pak, Y. T., Kim, W. M., 2017, A simple analysis to improve linearity of touch mode capacitive pressure sensor by modifying shape of fixed electrode, *Sens. Actuators, A Phys.*, 263: 300–304. https://doi.org/10.1016/j.sna.2017.06.024

Karaman, B., Bozkurt, A., 2018, Enhanced performance of supercapacitor based on boric acid doped PVA-H_2SO_4 gel polymer electrolyte system, *Int. J. Hydrogen Energy*, 43: 6229–6237. https://doi.org/10.1016/j.ijhydene.2018.02.032

Karaman, B., Çevik, E., Bozkurt, A., 2019, Novel flexible Li-doped PEO/copolymer electrolytes for supercapacitor application, *Ionics*, 25: 1773–1781. https://doi.org/10.1007/s11581-019-02854-4

Kato, M., Hiraoka, K., Sek, S., 2020, Investigation of the ionic conduction mechanism of polyether/$Li_7La_3Zr_2O_{12}$ composite solid electrolytes by electrochemical impedance spectroscopy, *J. Electrochem. Soc.*, 167: 070559

Khalid, M., Honorato, A. M., 2017, Ionically conducting and environmentally safe gum Arabic as a high-performance gel-like electrolyte for solid-state supercapacitors, *J. Solid State Electrochem.*, 21: 2443–2447. https://doi.org/10.1007/s10008-017-3585-4.

Khomenko, V., Raymundo-Piñero, E., Frackowiak, E., Béguin, F., 2006, High-voltage asymmetric supercapacitors operating in aqueous electrolyte, *Appl. Phys. A*, 82: 567. https://doi.org/10.1007/s00339-005-3397-8

Kim, M. S., Kim, J. W., Yun, J., Jeong, Y. R., Jin, S. W., Lee, G., Lee, H., Kim, D. S., Keum, K., Ha, J. S., 2020, A rationally designed flexible self-healing system with a high performance supercapacitor for powering an integrated multi-functional sensor, *Appl. Surf. Sci.*, 515: 146018. https://doi.org/10.1016/j.apsusc.2020.146018

Kim, D.-W., Ko, J. M., Kim, W. J., Kim, J. H., 2006, Study on the electrochemical characteristics of quasi-solid-state electric double layer capacitors assembled with sulfonated poly(ether ether ketone), *J. Power Sources*, 163: 300–303. https://doi.org/10.1016/j.jpowsour.2005.12.059

Kim, D. W., Lee, J. H., Kim, J. K., et al., 2020, Material aspects of triboelectric energy generation and sensors, *NPG Asia Mater.*, 12: 6. https://doi.org/10.1038/s41427-019-0176-0

Kim, K. M., Nam, J. H., Lee, Y. G., Cho, W. I., Ko, J. M., 2013, Supercapacitive properties of electrodeposited RuO_2 electrode in acrylic gel polymer electrolytes, *Curr. Appl. Phys.*, 13: 1702–1706. https://doi.org/10.1016/j.cap.2013.06.016

Kim, M., Oh, I., Kim, J., 2015, Effects of different electrolytes on the electrochemical and dynamic behavior of electric double layer capacitors based on porous silicon carbide electrode, *Phys. Chem. Chem. Phys.*, 17: 16367–16374. https://doi.org/10.1039/C5CP01728A

Kim, D., Shin, G., Kang, Y. J., Kim, W., Ha, J. S., 2013, Fabrication of a stretchable solid-state micro-supercapacitor array, *ACS Nano*, 7: 7975–7982. https://doi.org/10.1021/nn403068d

Kim, B. K., Sy, S., Yu, A., Zhang, J., 2015, Electrochemical Supercapacitors for Energy Storage and Conversion. In *Handbook of Clean Energy Systems*. Edited by J. Yan, 1–25. Online © 2015 John Wiley & Sons, Ltd. https://doi.org/10.1002/9781118991978.hces112

Kim, K. W., Yun, T. Y., You, S. H., et al., 2020, Extremely fast electrochromic supercapacitors based on mesoporous WO3 prepared by an evaporation-induced self-assembly, *NPG Asia Mater.*, 12: 84. https://doi.org/10.1038/s41427-020-00257-w

Kim, S. Y., Yun, T. Y., Yu, K. S., Moon, H. C., 2020, Reliable, high-performance electrochromic supercapacitors based on metal-doped nickel oxide, *ACS Appl. Mater. Interfaces*, 12(46): 51978–51986. https://doi.org/10.1021/acsami.0c15424

Kondrat S., Pérez C. R., Presser V., Gogotsi Y., Kornyshev A. A., 2012, Effect of pore size and its dispersity on the energy storage in nanoporous supercapacitors, *Energy Environ. Sci.*, 5: 6474. https://doi.org/10.1039/C2EE03092F

Kotobuki, M., 2020, Polymer Electrolytes. In *Polymer Electrolytes*. Edited by T. Winie, A. K. Arof, S. Thomas, 1–21. https://doi.org/10.1002/9783527805457.ch1

Krishnamoorthy, K., Pazhamalai, P., Mariappan, V. K. et al., 2020, Probing the energy conversion process in piezoelectric-driven electrochemical self-charging supercapacitor power cell using piezoelectrochemical spectroscopy, *Nat. Commun.*, 11: 2351. https://doi.org/10.1038/s41467-020-15808-6

Lapcinskis, L., Linarts, A., Knite, M., Gornevs, I., Blums, J., Šutka, A., 2019, Solid-state supercapacitor application for pressure sensing, *Appl. Surf. Sci.*, 474: 91–96. https://doi.org/10.1016/j.apsusc.2018.05.036

Łatoszyńska, A. A., Taberna, P.-L., Simon, P., Wieczorek, W., 2017, Proton conducting gel polymer electrolytes for supercapacitor applications, *Electrochim. Acta*, 242: 31–37. https://doi.org/10.1016/j.electacta.2017.04.122

Łatoszyńska, A. A., Zukowska, G. Z., Rutkowska, I. A., Taberna, P.-L., Simon, P., Kulesza, P. J., Wieczorek, W., 2015, Non-aqueous gel polymer electrolyte with phosphoric acid ester and its application for quasi solid-state supercapacitors, *J. Power Sources*, 274: 1147–1154. https://doi.org/10.1016/j.jpowsour.2014.10.094

Lechene, B. P., Clerc, R., Arias, A. C., 2017, Theoretical analysis and characterization of the energy conversion and storage efficiency of photo-supercapacitors, *Sol. Energy Mater Sol. Cells*, 172: 202–212. https://doi.org/10.1016/j.solmat.2017.07.034

Lee, G., Kim, D., Kim, D., Oh, S., Yun, J., Kim, J., Lee, S.-S., Ha, J. S., 2015, Fabrication of a stretchable and patchable array of high performance micro-supercapacitors using a non-aqueous solvent based gel electrolyte, *Energy Environ. Sci.*, 8: 1764–1774. https://doi.org/10.1039/C5EE00670H

Lee, S., Lee, Y., Cho, M.-S., Nam, J.-D. 2008, New strategy and easy fabrication of solid-state supercapacitor based on polypyrrole and nitrile rubber, *J. Nanosci Nanotechnol.*, 8(9): 4722–4725. https://doi.org/10.1166/jnn.2008.ic43

Lee, J., Srimuk, P., Fleischmann, S., Su, X., Alan, H. T., Presser, V. 2019, Redox-electrolytes for non-flow electrochemical energy storage: a critical review and best practice, *Prog. Mater. Sci.*, 101: 46–89. https://doi.org/10.1016/j.pmatsci.2018.10.005

Lee, K. T., Wu, N.-L., 2008, Manganese oxide electrochemical capacitor with potassium poly(acrylate) hydrogel electrolyte, *J, Power Sources*, 179: 430–434. https://doi.org/10.1016/j.jpowsour.2007.12.057

Lewandowski, A., Olejniczak, A., Galinski, M., Stepniak, I., 2010, Performance of carbon–carbon supercapacitors based on organic, aqueous and ionic liquid electrolytes, *J. Power Sources*, 195(17): 5814–5819. https://doi.org/10.1016/j.jpowsour.2010.03.082.

Li, T., Fang, X., Pang, Q., Huang, W., Sun, J., 2019, Healable and shape editable supercapacitors based on shape memory polyurethanes, *J. Mater. Chem. A*, 7: 17456–17465. https://doi.org/10.1039/C9TA04673A

Li, H., Lv, T., Sun, H., et al., 2019, Ultrastretchable and superior healable supercapacitors based on a double cross-linked hydrogel electrolyte. *Nat. Commun.*, 10: 536. https://doi.org/10.1038/s41467-019-08320-z

Li, R., Si, Y., Zhu, Z., Guo, Y., Zhang, Y., Pan, N., Sun, G., Pan, T., 2017, Supercapacitive iontronic nanofabric sensing, *Adv. Mater.*, 29: 1700253. https://doi.org/10.1002/adma.201700253

Li, P., Yang, Y., Shi, E., Shen, Q., Shang, Y., Wu, S., Wei, J., Wang, K., Zhu, H., Yuan, Q., 2014, Core-double-shell, carbon nanotube@polypyrrole@MnO_2 sponge as freestanding, compressible supercapacitor electrode, *ACS Appl. Mater. Interfaces*, 6: 5228–5234. https://doi.org/10.1021/am500579c

Li, X., Zang, X., Cao, Y., He, Y., Li, P., Wang, K., Wei, J., Wu, D., Zhu, H., Chen, Q., 2014, Effect of different gel electrolytes on graphene-based solid-state supercapacitors, *RSC Adv.*, 4: 36253. https://doi.org/10.1039/C4RA05553E

Li, G., Zhang, X., Sang, M., Wang, X., Zuo, D., Xu, J., Zhang, H., 2021, A supramolecular hydrogel electrolyte for high-performance supercapacitors, *J. Energy Storage*, 33: 101931. https://doi.org/10.1016/j.est.2020.101931

Liew, C. W., Ramesh, S., Arof, A. K., 2014, Good prospect of ionic liquid based-poly(vinyl alcohol) polymer electrolytes for supercapacitors with excellent electrical, electrochemical and thermal properties, *Int. J. Hydrogen Energy*, 39: 2953–2963. https://doi.org/10.1016/j.ijhydene.2013.06.061

Liew, C.-W., Ramesh, S., Arof, A.K., 2015, Characterization of ionic liquid added poly(vinyl alcohol)-based proton conducting polymer electrolytes and electrochemical studies on the supercapacitors, *Int. J. Hydrogen Energy*, 40: 852–862. http://dx.doi.org/10.1016/j.ijhydene.2014.09.160

Lim, C. S., Teoh, K. H., Liew, C. W., Ramesh, S., 2014, Capacitive behavior studies on electrical double layer capacitor using poly (vinyl alcohol)–lithium perchlorate based polymer electrolyte incorporated with TiO_2, *Mater. Chem. Phys.*, 143: 661–667. https://doi.org/10.1016/j.matchemphys.2013.09.051

Lim, Y., Yoon, J., Yun, J., Kim, D., Hong, S. Y., Lee, S.-J., Zi, G., Ha, J. S., 2014, Biaxially stretchable, integrated array of high performance microsupercapacitors, *ACS Nano*, 8: 11639–11650. https://doi.org/10.1021/nn504925s

Lin, R., Taberna, P. L., Chmiola, J., Guay, D., Gogotsi, Y., Simon, P., 2009, Microelectrode study of pore size, ion size, and solvent effects on the charge/discharge behavior of microporous carbons for electrical double-layer capacitors, *J. Electrochem. Soc.*, 156(1): A7. https://doi.org/10.1149/1.3002376

Liu, L., Choi, S., 2019, A self-charging cyanobacterial supercapacitor, *Biosens. Bioelectron.*, 140: 111354. https://doi.org/10.1016/j.bios.2019.111354

Liu, T., Ren, X., Zhang, J., Liu, J., Ou, R., Guo, C., Yu, X., Wang, Q., Liu, Z., 2020, Highly compressible lignin hydrogel electrolytes via double-crosslinked strategy for superior foldable supercapacitors, *J. Power Sources*, 449: 227532. https://doi.org/10.1016/j.jpowsour.2019.227532

Liu, L., Shen, B., Jiang, D., Guo, R., Kong, L., Yan, X., 2016, Watchband-like supercapacitors with body temperature inducible shape memory ability, *Adv. Energy Mater.*, 6: 1600763. https://doi.org/10.1002/aenm.201600763

Liu, R., Takakuwa, M., Li, A., Inoue, D., Hashizume, D., Yu, K., Umezu, S., Fukuda, K., Someya, T., 2020, An efficient ultra-flexible photo-charging system integrating organic photovoltaics and supercapacitors, *Adv. Energy Mater.*, 10: 2000523. https://doi.org/10.1002/aenm.202000523

Liu F., Wang J., Pan Q., 2018, An all-in-one self-healable capacitor with superior performance, *J. Mater. Chem. A*, 6: 2500–2506. https://doi.org/10.1039/C7TA10323A

Liu, X., Wu, D., Wang, H., Wang, Q., 2014, Self-recovering tough gel electrolyte with adjustable supercapacitor performance, *Adv. Mater.*, 26: 4370–4375. https://doi.org/10.1002/adma.201400240

Liu, Z., Zhang, J., Liu, J., Long, Y., Fang, L., Wang, Q., Liu, T., 2020, Highly compressible and superior low temperature tolerant supercapacitors based on dual chemically cross-linked PVA hydrogel electrolytes, *J. Mater. Chem. A*, 8: 6219. https://doi.org/10.1039/C9TA12424A

Liu, Y., Zhang, S., Zhou, Y., Buckingham, M. A., Aldous, L., Sherrell, P. C., Wallace, G. G., Ryder, G., Faisal, S., Officer, D. L., Beirne, S., Chen, J., 2020, Wearable thermocells for body heat harvesting, *Adv. Energy Mater.*, 10: 48. https://doi.org/10.1002/aenm.202002539

Lu, C., Chen, X., 2020, Latest advances in flexible symmetric supercapacitors: from material engineering to wearable applications, *Acc. Chem. Res.*, 53: 1468–1477. https://doi.org/10.1021/acs.accounts.0c00205

Lu, Y., Jiang, Y., Lou, Z., Shi, R., Chen, D., Shen, G., 2020, Wearable supercapacitor self-charged by P(VDF-TrFE) piezoelectric separator, *Prog. Nat. Sci.: Mater. Int.* 30(2): 174–179. https://doi.org/10.1016/j.pnsc.2020.01.023

Lu, X., Yu, M., Wang, G., Tong, Y., Li, Y., 2014, Flexible solid-state supercapacitors: design, fabrication and applications, *Energy Environ. Sci.*, 7: 2160–2181. https://doi.org/10.1039/C4EE00960F

Lu, X., Yu, M., Zhai, T., Wang, G. M., Xie, S. L., Liu, T. Y., Liang, C. L., Tong, Y. X., Li, Y., 2013, High energy density asymmetric quasi-solid-state supercapacitor based on porous vanadium nitride nanowire anode, *Nano Lett.*, 13: 2628–2633. https://doi.org/10.1021/nl400760a

Lufrano, F., Staiti, P., 2004, Performance improvement of Nafion based solid state electrochemical supercapacitor, *Electrochim. Acta*, 49: 2683–2689. https://doi.org/10.1016/j.electacta.2004.02.021

Lufrano, F., Staiti, P., Minutoli, M., 2004, Influence of nafion content in electrodes on performance of carbon supercapacitors, *J. Electrochem. Soc.*, 151: A64. https://doi.org/10.1149/1.1626670

Luo, J., Fan, F. R., Jiang, T., Wang, Z., Tang, W., Zhang, C., et. al., 2015, Integration of micro-supercapacitors with triboelectric nanogenerators for a flexible selfcharging power unit, *Nano Res.*, 8: 3934–3943. https://doi.org/10.1007/s12274-015-0894-8

Luo, Z., Liu, C., Fan, S., Liu, E., 2019, A universal in situ strategy for charging supercapacitors, *J. Mater. Chem. A*, 7: 15131. https://doi.org/10.1039/C9TA04105B

Luo, F., Sun, T.L., Nakajima, T., Kurokawa, T., Zhao, Y., Sato, K., Ihsan, A. B., Li, X., Guo, H., Gong, J. P., 2015, Oppositely charged polyelectrolytes form tough, self-healing, and rebuildable hydrogels, *Adv. Mater.*, 27: 2722–2727. https://doi.org/10.1002/adma.201500140

Luo, J., Tang, W., Fan, F. R., Liu, C., Pang, Y., Cao, G., Wang, Z. L., 2016, Transparent and flexible self-charging power film and its application in a sliding unlock system in touchpad technology, *ACS Nano*, 10: 8078–8086. https://doi.org/10.1021/acsnano.6b04201

Luo, B., Ye, D., Wang, L., 2017, Recent progress on integrated energy conversion and storage systems, *Adv. Sci.*, 4: 1700104. https://doi.org/10.1002/advs.341

Ma, L., Chen, S., Li, H., Ruan, Z., Tang, Z., Liu, Z., Wang, Z., Huang, Y., Pei, Z., Zapien, J. A., Zhi, C., 2018, Initiating a mild aqueous electrolyte Co_3O_4/Zn battery with 2.2 V-high voltage and 5000-cycle lifespan by a Co(III) rich-electrode, *Energy Environ. Sci.* 11: 2521–2530. https://doi.org/10.1039/C8EE01415A

Ma, Y., Li, L. B., Gao, G. X., Yang, X. Y., You, J., Yang, P. X., 2016, Ionic conductivity enhancement in gel polymer electrolyte membrane with N-methyl-N-butyl-piperidine-bis(trifluoromethylsulfonyl) imide ionic liquid for lithium ion battery, *Colloids Surf. A Physicochem. Eng.*, 502: 130–138. https://doi.org/10.1016/j.colsurfa.2016.05.011

Ma, G., Li, J., Sun, K., Peng, H., Mu, J., Lei, Z., 2014, High performance solid-state supercapacitor with PVA–KOH–$K_3[Fe(CN)_6]$ gel polymer as electrolyte and separator, *J. Power Sources*, 256: 281–287. https://doi.org/10.1016/j.jpowsour.2014.01.062

Ma, Y.-Y., Yi, G.-B., Wang, J.-C., Wang, H., Luo, H.-S., Zu, X.-H., 2017, Shape-controllable and -tailorable multi-walled carbon nanotube/MnO_2/shape-memory polyurethane composite film for supercapacitor, *Synth. Met.*, 223: 67–72. https://doi.org/10.1016/j.synthmet.2016.12.007

Mackanic, D. G., Chang, T., Huang, Z., Cui, Y., Bao, Z., 2020, Stretchable electrochemical energy storage devices, *Chem. Soc. Rev.*, 49: 4466. https://doi.org/10.1039/D0CS00035C

Maitra, A., Karan, S. K., Paria, S., Das, A. K., Bera, R., Halder, L., Si, S. K., Bera, A., Khatua B. B., 2017, Fast charging self-powered wearable and flexible asymmetric supercapacitor power cell with fish swim bladder as an efficient natural bio-piezoelectric separator, *Nano Energy*, 40: 633–645. https://doi.org/10.1016/j.nanoen.2017.08.057

Maitra, A., Paria, S., Karan, S. K., Bera, R., Bera, A., Das, A. K., Si, S. K., Halder, L., De, A., Khatua, B. B., 2019, Triboelectric nanogenerator driven self-charging and self-healing flexible asymmetric supercapacitor power cell for direct power generation, *ACS Appl. Mater. Interfaces*, 11: 5022–5036. https://doi.org/10.1021/acsami.8b19044

Majumdar, D., 2016, Functionalized-graphene/polyaniline nanocomposites as proficient energy storage material: an overview, *Innov Ener Res.*, 5(2): 145.

Majumdar, D., 2019a, "Polyaniline as Proficient Electrode Material for Supercapacitor Applications: PANI Nanocomposites for Supercapacitor Applications." In *Polymer Nanocomposites for Advanced Engineering and Military Applications.* Edited by Noureddine Ramdani, 190–219. Hershey, PA: IGI Global. http://doi:10.4018/978-1-5225-7838-3.ch007

Majumdar, D., 2019b, "Polyaniline Nanocomposites: Innovative Materials for Supercapacitor Applications – PANI Nanocomposites for Supercapacitor Applications." In *Polymer Nanocomposites for Advanced Engineering and Military Applications.* Edited by Noureddine Ramdani, 220–253. Hershey, PA: IGI Global. http://doi:10.4018/978-1-5225-7838-3.ch008

Majumdar, D., 2021a, Recent progress in copper sulfide based nanomaterials for high energy supercapacitor applications, *J. Electroanal. Chem.*, 880: 114825. https://doi.org/10.1016/j.jelechem.2020

Majumdar, D., 2021b, Review on current progress of MnO_2-based ternary nanocomposites for supercapacitor applications, *ChemElectroChem*, 8(2): 291–336. https://doi.org/10.1002/celc.202001371

Majumdar, D., Baugh, N., Bhattacharya, S. K., 2017, Ultrasound assisted formation of reduced graphene oxide-copper (II) oxide nanocomposite for energy storage applications, *Colloids Surf. A: Physicochem. Eng. Asp.*, 512: 158–170. https://doi.org/10.1016/j.colsurfa.2016.10.010

Majumdar, D., Bhattacharya, S. K., 2017, Sonochemically synthesized hydroxy-functionalized graphene–MnO_2 nanocomposite for supercapacitor applications, *J. Appl. Electrochem.*, 47(7): 789. https://doi.org/10.1007/s10800-017-1080-3

Majumdar, D., Ghosh, S., 2021, Recent advancements of copper oxide based nanomaterials for supercapacitor applications, *J. Energy Storage*, 34: 101995. https://doi.org/10.1016/j.est.2020.101995

Majumdar, D., Maiyalagan, T., Jiang, Z., 2019, Recent progress in ruthenium oxide-based composites for supercapacitor applications, *ChemElectroChem*, 6: 4343. https://doi.org/10.1002/celc.201900668

Majumdar, D., Mandal, M., Bhattacharya, S. K., 2019, V_2O_5 and its' carbon-based nanocomposites for supercapacitor applications, *ChemElectroChem*, 6: 1623. https://doi.org/10.1002/celc.201801761

Majumdar, D., Mandal, M., Bhattacharya, S. K., 2020, Journey from supercapacitors to supercapatteries: recent advancements in electrochemical energy storage systems, *Emerg. Mater.*, 3: 347–367. https://doi.org/10.1007/s42247-020-00090-5

Majumdar, D., Saha, S. K., 2010, Poly (3-hexylthiophene) nanotubes with superior electronic and optical properties, *Chem. Phys. Lett.*, 489(4–6): 219–224. https://doi.org/10.1016/j.cplett.2010.03.017

Majumdar, D., Saha, S. K., 2015, Charge transport in polypyrrole nanotubes, *J. Nanosci. Nanotechnol.*, 15(12), 9975–9981. https://doi.org/10.1166/jnn.2015.11708

Manoharan, S., Pazhamalai, P., Mariappan, V. K., Murugesan, K., Subramanian, S., Krishnamoorthy, K., Kim, S.-J., 2021, Proton conducting solid electrolyte-piezoelectric PVDF hybrids: novel bifunctional separator for self-charging supercapacitor power cell, *Nano Energy*, 83: 105753. https://doi.org/10.1016/j.nanoen.2021.105753

McCormick, A. J., Bombelli, P., Bradley, R. W., Thorne, R., Wenzel, T., Howe, C. J., 2015, Biophotovoltaics: oxygenic photosynthetic organisms in the world of bioelectrochemical systems, *Energy Environ. Sci.*, 8: 1092–1109. https://doi.org/10.1039/C4EE03875D

Meng, F., Li, Q., Zheng, L., 2017, Flexible fiber-shaped supercapacitors: design, fabrication, and multi-functionalities, *Energy Storage Mater.*, 8: 85–109. https://doi.org/10.1016/j.ensm.2017.05.002

Meng, C., Liu, C., Chen, L., Hu, C., Fan, S., 2010, Highly flexible and all-solid-state paperlike polymer supercapacitors, *Nano Lett.*, 10(10): 4025–4031. https://doi.org/10.1021/nl1019672

Meng, Y., Zhao, Y., Hu, C., Cheng, H., Hu, Y., Zhang, Z., Shi, G., Qu, L., 2013, All-graphene core-sheath microfibers for all-solid-state, stretchable fibriform supercapacitors and wearable electronic textiles, *Adv. Mater.*, 25: 2326–2331. https://doi.org/10.1002/adma.201300132

Merrill, M. D., Montalvo, E., Campbell, P. G., Wang, Y. M., Stadermann, M., Baumann, T. F., Biener, J., Worsley, M. A., 2014, Optimizing supercapacitor electrode density: achieving the energy of organic electrolytes with the power of aqueous electrolytes, *RSC Adv.*, 4: 42942–42946. https://doi.org/10.1039/C4RA08114E

Mondal, S., Das, S., Nandi, A. K., 2020, A review on recent advances in polymer and peptide hydrogels, *Soft Matter*, 16: 1404–1454. https://doi.org/10.1039/C9SM02127B

Mondal, R., Sarkar, K., Dey, S., Majumdar, D., Bhattacharya, S. K., Sen, P., Kumar, S., 2019, Magnetic, pseudocapacitive, and H_2O_2-electrosensing properties of self-assembled superparamagnetic $Co_{0.3}Zn_{0.7}Fe_2O_4$ with enhanced saturation, *ACS Omega*, 4(7): 12632–12646. https://doi.org/10.1021/acsomega.9b01362

Moon, W. G., Kim, G.-P., Lee, M., Song, H. D., Yi, J., 2015, A biodegradable gel electrolyte for use in high-performance flexible supercapacitors, *ACS Appl. Mater. Interfaces*, 7: 3503–3511. https://doi.org/10.1021/am5070987

Morita, M., Qiao, J.-L., Yoshimoto, N., Ishikawa, M., 2004, Application of proton conducting polymeric electrolytes to electrochemical capacitors, *Electrochim. Acta*, 50: 837–841. https://doi.org/10.1016/j.electacta.2004.02.053

Muchakayala, R., Song, S., Wang, J., Fan, Y., Bengeppagari, M., Chen, J., Tan, M., 2018, Development and supercapacitor application of ionic liquid-incorporated gel polymer electrolyte films, *J. Ind. Eng. Chem.*, 59: 79–89. https://doi.org/10.1016/j.jiec.2017.10.009

Na, R., Huo, P., Zhang, X., Zhang, S., Du, Y., Zhu, K., Wang, G. A., 2016, Flexible solid-state supercapacitor based on a poly(aryl ether ketone)–poly(ethylene glycol) copolymer solid polymer electrolyte for high temperature applications, *RSC Adv.*, 6(69): 65186–65195. https://doi.org/10.1039/c6ra11202a

Na, R., Lu, N., Zhang, S., Huo, G., Yang, Y., Zhang, C., Mu, Y., Luo, Y., Wang, G., 2018, Facile synthesis of a high-performance, fire-retardant organic gel polymer electrolyte for flexible solid-state supercapacitors, *Electrochim. Acta*, 290: 262–272. https://doi.org/10.1016/j.electacta.2018.09.074

Na, R., Su, C., Su, Y., Chen, Y., Chen, Y., Wang, G., Teng, H., 2017, Solvent-free synthesis of an ionic liquid integrated ether-abundant polymer as a solid electrolyte for flexible electric double-layer capacitors, *J. Mater. Chem. A*, 5: 19703. https://doi.org/10.1039/C7TA05358D

Nègre, L., Daffos, B., Turq, V., Taberna, P.-L., Simon, P., 2016, Ionogelbased solid-state supercapacitor operating over a wide range of temperature, *Electrochim. Acta*, 206: 490–495. http://dx.doi.org/10.1016/j.electacta.2016.02.013

Ngai, K. S., Ramesh, S., Ramesh, K. et al., 2016, A review of polymer electrolytes: fundamental, approaches and applications, *Ionics*, 22: 1259–1279. https://doi.org/10.1007/s11581-016-1756-4

Niu, Z., Zhou, W., Chen, X., Chen, J., Xie, S., 2015, Highly compressible and all-solid-state supercapacitors based on nanostructured composite sponge, *Adv. Mater.*, 27: 6002–6008. https://doi.org/10.1002/adma.201502263

Nohara, S., Asahina, T., Wada, H., Furukawa, N., Inoue, H., Sugoh, N., Iwasaki, H., Iwakura, C., 2006, Hybrid capacitor with activated carbon electrode, $Ni(OH)_2$ electrode and polymer hydrogel electrolyte, *J. Power Sources*, 157: 605–609. https://doi.org/10.1016/j.jpowsour.2005.07.024

Ohtaki, H., 2002, "Ionic Solvation in Aqueous and Nonaqueous Solutions." In *Highlights in Solute-Solvent Interactions*. Edited by W. Linert, Vienna: Springer. https://doi.org/10.1007/978-3-7091-6151-7_1

Orita, A., Kamijima, K., Yoshida, M., 2010, Allyl-functionalized ionic liquids as electrolytes for electric double-layer capacitors, *J. Power Sources*, 195: 7471. https://doi.org/10.1016/j.jpowsour.2010.05.066

Orita, A., Kamijima, K., Yoshida, M., et al., 2010, Application of sulfonium-, thiophenium-, and thioxonium-based salts as electric double-layer capacitor electrolytes, *J. Power Sources*, 195(19): 6970–6976. https://doi.org/10.1016/j.jpowsour.2010.04.028

Pal, B., Yang, S., Ramesh, S., Thangadura, V., Jose, R., 2019, Electrolyte selection for supercapacitive devices: a critical review, *Nanoscale Adv.*, 1: 3807–3835. https://doi.org/10.1039/C9NA00374F

Pan, S., Ren, J., Fang, X., Peng, H., 2016, Integration: an effective strategy to develop multifunctional energy storage devices, *Adv. Energy. Mater.*, 6: 1501867. https://doi.org/10.1002/aenm.201501867

Pan, Z., Yang, J., Li, L., Gao, X., Kang, L., Zhang, Y., Zhang, Q., et. al., 2020, All-in-one stretchable coaxial-fiber strain sensor integrated with high-performing supercapacitor, *Energy Storage Materials*, 25: 124–130. https://doi.org/10.1016/j.ensm.2019.10.023

Panda, P. K., Grigoriev, A., Mishra, Y. K., Ahuja, R., 2020, Progress in supercapacitors: roles of two dimensional nanotubular materials, *Nanoscale Adv.*, 2: 70–108. https://doi.org/10.1039/C9NA00307J

Pandey, G. P., Hashmi, S. A., 2013, Ionic liquid 1-ethyl-3-methylimidazolium tetracyanoborate-based gel polymer electrolyte for electrochemical capacitors, *J. Mater. Chem. A*, 1: 3372–3378. https://doi.org/10.1039/C2TA01347A

Pandey, G. P., Liu, T., Hancock, C., Li, Y., Sun, X. S., Li, J., 2016, Thermostable gel polymer electrolyte based on succinonitrile and ionic liquid for high-performance solid-state supercapacitors, *J. Power Sources*, 328: 510–519. https://doi.org/10.1016/j.jpowsour.2016.08.032

Pandey, G., Rastogi, A., Westgate, C. R., 2014, All-solid-state supercapacitors with poly(3,4-ethylenedioxythiophene)-coated carbon fiber paper electrodes and ionic liquid gel polymer electrolyte, *J. Power Sources*, 245: 857–865. https://doi.org/10.1016/j.jpowsour.2013.07.017

Park, K.-W., Ahn, H.-J., Sung, Y.-E., 2002, All-solid-state supercapacitor using a Nafion® polymer membrane and its hybridization with a direct methanol fuel cell, *J. Power Sources*, 109: 500–506. https://doi.org/10.1016/S0378-7753(02)00165-9

Park, H., Song, C., Jin, S. W., Lee, H., Keum, K., Lee, Y. H., Lee, G., Jeong, Y. R., Ha, J. S., 2021, High performance flexible micro supercapacitor for powering a vertically integrated skin-attachable strain sensor on a bio-inspired adhesive, *Nano Energy*, 83: 105837. https://doi.org/10.1016/j.nanoen.2021.105837

Pazhamalai, P., Krishnamoorthy, K., Mariappan, V. K., Sahoo, S., Manoharan, S., Kim, S.-J., 2018, A high efficacy self-charging $MoSe_2$ solid-state supercapacitor using electrospun nanofibrous piezoelectric separator with ionogel electrolyte, *Adv. Mater. Interfaces*, 5: 1800055. https://doi.org/10.1002/admi.201800055

Peng, X., Liu, H., Yin, Q., Wu, J., Chen, P., Zhang, G., Liu, G., Wu, C., Xie, Y., 2016, A zwitterionic gel electrolyte for efficient solid-state supercapacitors, *Nat. Commun.*, 7: 11782. https://doi.org/10.1038/ncomms11782

Peng, H., Lv, Y., Wei, G., Zhou, J., Gao, X., Sun, K., Ma, G., Lei, Z., 2019, A flexible and self-healing hydrogel electrolyte for smart supercapacitor, *J. Power Sources*, 431: 210–219. https://doi.org/10.1016/j.jpowsour.2019.05.058

Pérez-Madrigal, M. M., Estrany, F., Armelin, E., Díaz, D. D., Alemán, C., 2016, Towards sustainable solid-state supercapacitors: electroactive conducting polymers combined with biohydrogels, *J. Mater. Chem. A*, 4: 1792–1805. https://doi.org/10.1039/C5TA08680A

Pickup, P. G., Kean, C. L., Lefebvre, M. C., Li, G., Qi, Z., Shan, J., 2000, Electronically conducting cation-exchange polymer powders: synthesis, characterization and applications in pem fuel cells and supercapacitors, *J. New Mater. Electrochem. Syst.*, 3: 21–26.

Poonam, Sharma K., Arora, A., Tripathi, S. K., 2019, Review of supercapacitors: materials and devices, *Journal of Energy Storage*, 21: 801. https://doi.org/10.1016/j.est.2019.01.010

Pu, X., Li, L., Liu, M., Jiang, C., Du, C., Zhao, Z., Hu, W., Wang, Z. L., 2015, Wearable self- charging power textile based on flexible yarn supercapacitors and fabric nanogenerators, *Adv. Mater.*, 28: 98–105. https://doi.org/10.1002/adma.201504403

Pu, S., Liao, Y., Chen, K., Fu, J., Zhang, S., Ge, L., Conta, G., Bouzarif, S., Cheng, T., Hu, X., Liu, K., Chen, J., 2020, Thermogalvanic hydrogel for synchronous evaporative cooling and low-grade heat energy harvesting, *Nano Lett.*, 5: 3791–3797. https://doi.org/10.1021/acs.nanolett.0c00800

Puthirath, A. B., Tsafack, T., Patra, S., Thakur, P., Chakingal, N., Saju, S. K., Baburaj, A., Kato, K., Ganguli, B., Narayanan, T. N., Ajayan, P. M., 2020, Lithium, sodium and magnesium ion conduction in solid state mixed polymer electrolytes, *Phys. Chem. Chem. Phys.*, 22: 19108–19119. https://doi.org/10.1039/D0CP02609C

Qian, L., 2019, "Cellulose-Based Composite Hydrogels: Preparation, Structures, and Applications." In *Cellulose-Based Superabsorbent Hydrogels. Polymers and Polymeric Composites: A Reference Series*. Edited by M. Mondal, 655–704. Cham: Springer. https://doi.org/10.1007/978-3-319-77830-3_23

Qin, S., Zhang, Q., Yang, X., Liu, M., Sun, Q., Wang, Z. L., 2018, Hybrid piezo/triboelectric-driven self-charging electrochromic supercapacitor power package, *Adv. Energy Mater.*, 8: 1800069. https://doi.org/10.1002/aenm.201800069

Railanmaa, A., Kujala, M., Keskinen, J. et al., 2019, Highly flexible and non-toxic natural polymer gel electrolyte for printed supercapacitors for IoT, *Appl. Phys. A*, 125: 168. https://doi.org/10.1007/s00339-019-2461-8

Ramadoss, A., Saravanakumar, B., Lee, S.W., Kim, Y.-S., Kim, S.J., Wang, Z.L., 2015, Piezoelectric driven self-charging supercapacitor power cell, *ACS Nano*, 9(4): 4337–4345. https://doi.org/10.1021/acsnano.5b00759

Ramasamy, C., del Vel, J. P., Anderson, M., 2014, An activated carbon supercapacitor analysis by using a gel electrolyte of sodium salt-polyethylene oxide in an organic mixture solvent, *J. Solid State Electrochem.*, 18: 2217–2223. https://doi.org/10.1007/s10008-014-2466-3

Ramasamy, C., Palma Del Val, J., Anderson, M., 2014, A 3-V electrochemical capacitor study based on a magnesium polymer gel electrolyte by three different carbon materials, *J. Solid State Electrochem.*, 18(10): 2903–2911. https://doi.org/10.1007/s10008-014-2557-1

Ramya, C. S., Subramaniam, C. K., Dhathathreyan, K. S., 2010, Perfluorosulfonic acid based electrochemical double-layer capacitor, *J. Electrochem. Soc.*, 157: A600–A605. https://doi.org/10.1149/1.3328180

Raza, W., Ali, F., Raza, N., Luo, Y., Kim, K.-H., Yang, J., Kumar, S., Mehmood, A., Kwon, E. E., 2018, Recent advancements in supercapacitor technology, *Nano Energy*, 52: 441. https://doi.org/10.1016/j.nanoen.2018.08.013

Roy, P., Srivastava, S. K., Kandula, S., Kim, N. H., Lee, J. H., Lee, J. H., 2019, Polymer-Based Flexible Electrodes for Supercapacitor Applications. In *Nanomaterials for Electrochemical Energy Storage Devices*. Edited by P. Roy, S.K. Srivastava. https://doi.org/10.1002/9781119510000.ch11

Ruch, P. W., Hahn, M., Rosciano, F., Holzapfel, M., Kaiser, H., Scheifele, W., Schmitt, B., Novák, P., Kötz, R., Wokaun, A., 2007, In situ X-ray diffraction of the intercalation of (C2H5)4N+ and BF4− into graphite from acetonitrile and propylene carbonate based supercapacitor electrolytes, *Electrochim. Acta*, 53(3): 1074–1082. https://doi.org/10.1016/j.electacta.2007.01.069

Ryu, H., Lee, J.-H., Kim, T.-Y., Khan, U., Lee, J. H., Kwak, S. S., Yoon, H.-J., Kim, S.-W., 2017, High-performance triboelectric nanogenerators based on solid polymer electrolytes with asymmetric pairing of ions, *Adv. Energy Mater.*, 7: 1700289. https://doi.org/10.1002/aenm.201700289

Ryu, H., Yoon, H. J., Kim, S. W., 2019, Hybrid energy harvesters: toward sustainable energy harvesting, *Adv Mater.*, 31(34): e1802898. https://doi.org/10.1002/adma.201802898

Sahoo, S., Krishnamoorthy, K., Pazhamalai, P., Mariappan, V. K., Manoharan, S., Kim, S.-J., 2019, High performance self-charging supercapacitors using a porous PVDF-ionic liquid electrolyte sandwiched between two-dimensional graphene electrodes, *Journal of Materials Chemistry A*, 7: 21693–21703. https://doi.org/10.1039/C9TA06245A

Saito, Y., Stephan, A. M., Kataoka, H., 2003, Ionic conduction mechanisms of lithium gel polymer electrolytes investigated by the conductivity and diffusion coefficient, *Solid State Ionics*, 160(1–2): 149–153. https://doi.org/10.1016/s0167-2738(02)00685-9

Sajitha, S., Aparna, U., Deb, B., 2019, Ultra-thin manganese dioxide-encrusted vanadium pentoxide nanowire mats for electrochromic energy storage applications, *Adv. Mater. Interfaces*, 6: 1901038. https://doi.org/10.1002/admi.201901038

Salanne, M., Rotenberg, B., Naoi, K., Kaneko, K., Taberna, P.-L., Grey, C.P., Dunn B., Simon, P., 2016, Efficient storage mechanisms for building better supercapacitors, *Nat. Energy*, 1: 16070. https://doi.org/10.1038/nenergy.2016.70

Sampath, S., Choudhury, N. A., Shukla, A. K., 2009, Hydrogel membrane electrolyte for electrochemical capacitors, *J. Chem. Sci.*, 121: 727. https://doi.org/10.1007/s12039-009-0087-7

Scalia, A., Bella, F., Lamberti, A., Bianco, S., Gerbaldi, C., Tresso, Elena M., Pirri, C. F., 2017, A flexible and portable powerpack by solid-state supercapacitor and dye-sensitized solar cell integration, *J. Power Sources*, 359: 311–321. https://doi.org/10.1016/j.jpowsour.2017.05.07

Schroeder, M., Isken, P., Winter, M., Passerini, S., Lex-Balducci, A., Balducci, A., 2013, An investigation on the use of a methacrylate-based gel polymer electrolyte in high power devices, *J. Electrochem. Soc.*, 160: A1753–A1758. https://doi.org/10.1149/2.067310jes

Sekhon, S. S., 2003, Conductivity behaviour of polymer gel electrolytes: role of polymer, *Bull. Mater. Sci.*, 26: 321–328. https://doi.org/10.1007/BF02707454

Senthilkumar, S. T., Selvan, R. K., Ponpandian, N., Melo, J. S., 2012, Redox additive aqueous polymer gel electrolyte for an electric double layer capacitor, *RSC Adv.*, 2: 8937–8990. https://doi.org/10.1039/C2RA21387

Seong, Y.-H., Choi, N.-S., Kim, D.-W., 2011, Quasi-solid-state electric double layer capacitors assembled with sulfonated poly(fluorenyl ether nitrile oxynaphthalate) membranes, *Electrochim. Acta*, 58: 285–289. https://doi.org/10.1016/j.electacta.2011.09.061

Shao, Y., El-Kady, M. F., Sun, J., Li, Y., Zhang, Q., Zhu, M., Wang, H., Dunn, B., Kaner, R. B., 2018, Design and mechanisms of asymmetric supercapacitors, *Chem. Rev.*, 118: 9233. https://doi.org/10.1021/acs.chemrev.8b00252

Shen, L., Du, L., Tan, S., et al., 2016, Flexible electrochromic supercapacitor hybrid electrodes based on tungsten oxide films and silver nanowires, *Chem. Commun.*, 52(37): 6296–6299. https://doi.org/10.1039/c6cc01139j

Shi, X., Chen, S., Zhang, H., Jiang, J., Ma, Z., Gong, S., 2019, Portable self-charging power system via integration of a flexible paper-based triboelectric nanogenerator and supercapacitor, *ACS Sustain. Chem. Eng.*, 7(22): 18657–18666. https://doi.org/10.1021/acssuschemeng.9b05129

Shin, J., Tran, V.-H., Nguyen, D. C. T., Kim, S.-K., Lee, S.-H., 2021, Integrated photo-rechargeable supercapacitors formed via electrode sharing, *Org. Electron.*, 89: 106050. https://doi.org/10.1016/j.orgel.2020.106050

Shirshova, N., Bismarck, A., Carreyette, S., Fontana, Q. P. V., Greenhalgh, E. S., Jacobsson, P., Johansson, P., Marczewski, M. J., Kalinka, G., Kucernak, A. R. J., Scheers, J., Shaffer, M. S. P., Steinke, J. H. G., Wienrich, M., 2013, Structural supercapacitor electrolytes based on bicontinuous ionic liquid–epoxy resin systems, *J. Mater. Chem. A*, 1: 15300–15309. https://doi.org/10.1039/C3TA13163G

Shukur, M. F., Ithnin, R., Illias, H. A., Kadir, M. F. Z., 2013, Proton conducting polymer electrolyte based on plasticized chitosan–PEO blend and application in electrochemical devices, *Opt. Mater.*, 35: 1834–1841. https://doi.org/10.1016/j.optmat.2013.03.004

Shukur, M. F., Ithnin, R., Kadir, M. F. Z., 2014, Electrical properties of proton conducting solid biopolymer electrolytes based on starch–chitosan blend, *Ionics*, 20: 977–999. https://doi.org/10.1007/s11581-013-1033-8

Sivaraman, P., Rath, S. K., Hande, V. R., Thakur, A. P., Patri, M., Samui, A. B., 2006, All-solid-supercapacitor based on polyaniline and sulfonated polymers, *Synth. Met.*, 156: 1057–1064. https://doi.org/10.1016/j.synthmet.2006.06.017

Solís-Cortés, D., Navarrete-Astorga, E., Schrebler, R., Peinado-Pérez, J. J., Martín, F., Ramos-Barrado, J. R., Dalchiele, E. A., 2020, A solid-state integrated photo-supercapacitor based on ZnO nanorod arrays decorated with Ag_2S quantum dots as the photoanode and a PEDOT charge storage counter-electrode, *RSC Adv.*, 10: 5712. https://doi.org/10.1039/C9RA10635A

Song, Y., Chen, H., Su, Z., Chen, X., Miao, L., Zhang, J., Cheng, X., Zhang, H., 2017, Highly compressible integrated supercapacitor–piezoresistance-sensor system with CNT–PDMS sponge for health monitoring, *Small*, 13: 1702091. https://doi.org/10.1002/smll.201702091

Song, Y., Cheng, X., Chen, H., Huang, J., Chen, X., Han, M., Su, Z., Meng, B., Song, Z., Zhang, H., 2016, Integrated self-charging power unit with flexible supercapacitor and triboelectric nanogenerator, *J. Mater. Chem. A*, 4: 14298–14306. https://doi.org/10.1039/C6TA05816G

Song, R., Jin, H., Li, X., Fei, L., Zhao, Y., et. al., 2015, A rectification-free piezo-supercapacitor with a polyvinylidene fluoride separator and functionalized carbon cloth electrodes, *J. Mater. Chem. A*, 3: 14963–14970. https://doi.org/10.1039/C5TA03349G

Sorrell, S., 2015, Reducing energy demand: a review of issues, challenges and approaches, *Renew. Sust. Energ. Rev.*, 47: 74. https://doi.org/10.1016/j.rser.2015.03.0025

Staffell, I., Scamman, D., Abad, A. V., Balcombe, P., Dodds, P. E., Ekins, P., Shahd, N., Ward, K. R., 2019, The role of hydrogen and fuel cells in the global energy system, *Energy Environ. Sci.*, 12: 463. https://doi.org/10.1039/C8EE01157E

Subramani, K., Sudhan, N., Divya, R., Sathish, M., 2017, All-solid-state asymmetric supercapacitors based on cobalt hexacyanoferrate-derived CoS and activated carbon, *RSC Adv.*, 7: 6648. https://doi.org/10.1039/C6RA27331A

Subramaniam, C. K., Ramya, C. S., Ramya, K., 2011, Performance of EDLCs using Nafion and Nafion composites as electrolyte, *J. Appl. Electrochem.*, 41: 197–206. https://doi.org/10.1007/s10800-010-0224-5

Sudhakar, Y. N., Selvakumar, M., 2012, Lithium perchlorate doped plasticized chitosan and starch blend as biodegradable polymer electrolyte for supercapacitors, *Electrochim. Acta*, 78: 398–405. https://doi.org/10.1016/j.electacta.2012.06.032

Sudhakar, Y., Selvakumar, M., Bhat, D. K., 2014, Tubular array, dielectric, conductivity and electrochemical properties of biodegradable gel polymer electrolyte, *Mater. Sci. Eng. B*, 180: 12–19. https://doi.org/10.1016/j.mseb.2013.10.013

Suleman, M., Kumar, Y., Hashmi, S. A., 2013, Structural and electrochemical properties of succinonitrile-based gel polymer electrolytes: role of ionic liquid addition, *J. Phys. Chem. B*, 117: 7436–7443. https://doi.org/10.1021/jp312358x

Sundaram, M. M., Appadoo, D., 2020, Traditional salt-in-water electrolyte vs. water-in- salt electrolyte with binary metal oxide for symmetric supercapacitors: capacitive vs. faradaic, *Dalton Trans.*, 49: 11743–11755. https://doi.org/10.1039/d0dt01871f

Sundriyal, P., Bhattacharya, S., 2020, Textile-based supercapacitors for flexible and wearable electronic applications, *Sci. Rep.*, 10: 13259. https://doi.org/10.1038/s41598-020-70182-z

Syahidah, S. N., Majid, S., 2013, Super-capacitive electrochemical performance of polymer blend gel polymer electrolyte (GPE) in carbon-based electrical double-layer capacitors, *Electrochim. Acta*, 112: 678–685. https://doi.org/10.1016/j.electacta.2013.09.008

Taheri, A., MacFarlane, D. R., Pozo-Gonzalo, C., Pringle, J. M., 2018a, Flexible and non-volatile redox active quasi-solid state ionic liquid based electrolytes for thermal energy harvesting, *Sustain. Energy Fuels*, 2: 1806. https://doi.org/10.1039/C8SE00224J

Taheri, A., MacFarlane, D., Pozo-Gonzalo, C., Pringle, J. M., 2018b, Quasi-solid state electrolytes for low-grade thermal energy harvesting using a cobalt redox couple, *ChemSusChem*, 11: 2788. https://doi.org/10.1002/cssc.201800794

Tai, Z., Yan, X., Xue, Q., 2012, Shape-alterable and -recoverable graphene/polyurethane bi-layered composite film for supercapacitor electrode, *J Power Sources*. 213: 350–357. https://doi.org/10.1016/j.jpowsour.2012.03.086

Tamilarasan, P., Ramaprabhu, S., 2014, Stretchable supercapacitors based on highly stretchable ionic liquid incorporated polymer electrolyte, *Mater. Chem. Phys.*, 148: 48–56. https://doi.org/10.1016/j.matchemphys.2014.07.010

Teng, Y., Wei, J., Du, H., Mojtaba, M., Li, D., 2020, A solar and thermal multisensing microfibersupercapacitor with intelligent self-conditioned capacitance and body temperature monitoring, *J. Mater. Chem. A*, 8: 11695. https://doi.org/10.1039/D0TA02894K

Tian, Y., Cong, S., Su, W., Chen, H., Li, Q., Geng, F., Zhao, Z., 2014, Synergy of $W_{18}O_{49}$ and polyaniline for smart supercapacitor electrode integrated with energy level indicating functionality, *Nano. Lett.*, 14: 2150–2156. https://doi.org/10.1021/nl5004448

Tian, H., Shi, H., Khalili, N., Morrison, T., Naguib, H. E., 2018, Self-assembled nanorod structures on nanofibers for textile electrochemical capacitor electrodes with intrinsic tactile sensing capabilities, *ACS Appl. Mater. Interfaces*, 10: 19037–19046. https://doi.org/10.1021/acsami.8b03779

Tiruye, G. A., Muñoz-Torrero, D., Palma, J., Anderson, M., Marcilla, R., 2015, All-solid state supercapacitors operating at 3.5 V by using ionic liquid based polymer electrolytesx, *New J. Power Sources*, 279: 472–480. https://doi.org/10.1016/j.jpowsour.2015.01.039

Vraneš, M., Zec, N., Tot, A., Papović, S., Dožić, S., Gadžurić, S., 2014, Density, electrical conductivity, viscosity and excess properties of 1-butyl-3-methylimidazolium bis(trifluoromethylsulfonyl)imide+propylene carbonate binary mixtures, *J. Chem. Thermodyn.*, 68: 98–108. https://doi.org/10.1016/j.jct.2013.08.034.

Wang, Z. L., 2014, Triboelectric nanogenerators as new energy technology and self-powered sensors–principles, problems and perspectives, *Faraday Discuss*, 176: 447–458. https://doi.org/10.1039/C4FD00159A

Wang, Z. L., Chen, J., Lin, L., 2015, Progress in triboelectric nanogenerators as a new energy technology and self-powered sensors, *Energy Environ. Sci.*, 8: 2250–2282. https://doi.org/10.1039/C5EE01532D

Wang, J., Deng, M., Xiao, Y., Hao, W., Zhu, C., 2019, Dielectric and transport properties of cationic polyelectrolyte membrane P(NVP-co-DMDAAC)/PVA for solid-state supercapacitors, *New J. Chem.*, 43: 4815–4822. https://doi.org/10.1039/C9NJ00468H

Wang, J., Dong, S., Ya, B. D., Xiaodong, W., Hui, H., Yongyao, D., Zhang, X. X., 2017, Pseudocapacitive materials for electrochemical capacitors: from rational synthesis to capacitance optimization, *Natl. Sci. Rev.*, 4(1): 71. https://doi.org/10.1093/nsr/nww072

Wang, N., Dou, W., Hao, S., Cheng, Y., Zhou, D., Huang, X., Jiang, C., Cao, X., 2019, Tactile sensor from self-chargeable piezoelectric supercapacitor, *Nano energy*, 56: 868–874. https://doi.org/10.1016/j.nanoen.2018.11.065

Wang, J., Feng, S.-P., Yang, Y., Hau, N. Y., Munro, M., Ferreira-Yang, E., Chen, G., 2015, Thermal charging phenomenon in electrical double layer capacitors, *Nano Lett.*, 15: 5784–5790. https://doi.org/10.1021/acs.nanolett.5b01761

Wang, Y., Guo, J., Wang, T., Shao, J., Wang, D., Yang, Y. W., 2015, Mesoporous transition metal oxides for supercapacitors, *Nanomaterials*, 5: 1667. https://doi.org/10.3390/nano5041667

Wang, L., Guo, M., Zhan, J., Jiao, X., Chen, D., Wang, T., 2020, A new design of an electrochromic energy storage device with high capacity, long cycle lifetime and multicolor display, *J. Mater. Chem. A*, 8: 17098. https://doi.org/10.1039/D0TA04824K

Wang, S., Hsia, B., Carraro, C., Maboudian, R., 2014, High-performance all solid-state micro-supercapacitor based on patterned photoresist-derived porous carbon electrodes and an ionogel electrolyte, *J. Mater.Chem. A*, 2: 7997–8002. https://doi.org/10.1039/C4TA00570H

Wang, X., Huang, Y. T., Liu, C., et al., 2019, Direct thermal charging cell for converting low-grade heat to electricity, *Nat. Commun.*, 10: 4151. https://doi.org/10.1038/s41467-019-12144-2

Wang, J., Li, X., Zi, Y., Wang, S., Li, Z., Zheng, L., et al., 2015, A flexible fiberbased supercapacitor–triboelectric-nanogenerator power system for wearable electronics, *Adv. Mater.*, 27: 4830–4836. https://doi.org/10.1002/adma.201501934

Wang, J., Lou, H., Meng, J., Peng, Z., Wang, B., Wan, J., 2020, Stretchable energy storage E-skin supercapacitors and body movement sensors, *Sens. Actuators B: Chem.*, 305: 127529. https://doi.org/10.1016/j.snb.2019.127529

Wang, G. M., Lu, X. H., Ling, Y. C., Zhai, T., Wang, H. Y., Tong, Y. X., Li, Y., 2012, LiCl/PVA gel electrolyte stabilizes vanadium oxide nanowire electrodes for pseudocapacitors, *ACS Nano*, 6: 10296–10302. https://doi.org/10.1021/nn304178b

Wang, Z, Pan, Q. 2017, An omni-healable supercapacitor integrated in dynamically cross-linked polymer networks, *Adv. Funct. Mater.*, 27: 1700690. https://doi.org/10.1002/adfm.201700690

Wang, Y., Song, Y., Xia, Y., 2016, Electrochemical capacitors: mechanism, materials, systems, characterization and applications, *Chem. Soc. Rev.*, 45: 5925–5950. https://doi.org/10.1039/C5CS00580A

Wang, Z., Tao, F., Pan, Q., 2016, A self-healable polyvinyl alcohol-based hydrogel electrolyte for smart electrochemical capacitors, *J. Mater. Chem. A*, 4: 17732–17739. https://doi.org/10.1039/C6TA08018A

Wang, W., Wang, X., Xia, X., Yao, Z., Zhong, Y., Tu, J., 2018, Enhanced electrochromic and energy storage performance in mesoporous WO_3 film and its application in a bi-functional smart window, *Nanoscale*, 10: 8162–8169. https://doi.org/10.1039/C8NR00790J

Wang, J., Wen, Z., Zi, Y., Zhou, P., Lin, J., Guo, H., Xu, Y., Wang, Z. L., 2015, All-plastic-materials based self-charging power system composed of triboelectric nanogenerators and supercapacitors, *Adv. Funct. Mater.*, 26: 1070–1076. https://doi.org/10.1002/adfm.201504675

Wang, F., Xiao, S., Hou, Y., Hu, C., Liu, L., Wu, Y., 2013, Electrode materials for aqueous asymmetric supercapacitors, *RSC Adv.*, 3: 13059–13084. https://doi.org/10.1039/C3RA23466E

Wang, X., Yang, C., Wang, G., 2016, Stretchable fluoroelastomer quasi-solid-state organic electrolyte for high-performance asymmetric flexible supercapacitors, *J. Mater. Chem. A*, 4: 14839–14848. https://doi.org/10.1039/C6TA05299A

Wang, G., Zhang, L., Zhang, J., 2012, A review of electrode materials for electrochemical supercapacitors, *Chem. Soc. Rev.*, 41: 797. https://doi.org/10.1039/C1CS15060J

Wang, Y., Zhong, W. H., 2015, Development of electrolytes towards achieving safe and high-performance energy-storage devices: a review, *ChemElectroChem*, 2: 22–36. https://doi.org/10.1002/celc.201402277

Wang, H. Y., Zhou, Q. Q., Yao, B. W., Ma, H. Y., Zhang, M., Li, C., Shi, G. Q., 2018, Suppressing the self-discharge of supercapacitors by modifying separators with an ionic polyelectrolyte, *Adv. Mater. Interfaces*, 5: 1701547. https://doi.org/10.1002/admi.201701547

Wang, Z., Zhu, M., Pei, Z., Xue, Q., Li, H., Huang, Y., Zhi, C., 2020, Polymers for supercapacitors: boosting the development of the flexible and wearable energy storage, *Mater. Sci. Eng. R Rep.*, 139: 100520. https://doi.org/10.1016/j.mser.2019.100520

Watanabe, M., Thomas, M. L., Zhang, S., Ueno, K., Yasuda, T., Dokko, K., 2017, Application of ionic liquids to energy storage and conversion materials and devices, *Chem. Rev.*, 117(10): 7190–7239. https://doi.org/10.1021/acs.chemrev.6b00504

Wee, G., Larsson, O., Srinivasan, M., Berggren, M., Crispin, X., Mhaisalkar, S., 2010, Effect of the ionic conductivity on the performance of polyelectrolyte-based supercapacitors, *Adv. Funct. Mater.*, 20: 4344–4350. https://doi.org/10.1002/adfm.201001096

Wei, J., Zhou, J., Su, S., Jiang, J., Feng, J., Wang, Q., 2018, Water-deactivated polyelectrolyte hydrogel electrolytes for flexible high-voltage supercapacitors, *ChemSusChem*, 11: 3410. https://doi.org/10.1002/cssc.201801277

Winter, M., Brodd, R. J., 2004, What are batteries, fuel cells, and supercapacitors? *Chem. Rev.*, 104: 4245. https://doi.org/10.1021/cr020730k

Wu, Z., Cheng, T., Wang, Z. L., 2020, Self-powered sensors and systems based on nanogenerators, *Sensors (Basel, Switzerland)*, 20(10): 2925. https://doi.org/10.3390/s20102925

Wu, Z. S., Parvez, K., Feng, X., Mullen, K., 2013, Graphene-based in-plane micro-supercapacitors with high power and energy densities, *Nat. Commun.*, 4: 1–8. https://doi.org/10.1038/ncomms3487

Wu, J., Xia, G., Li, S., Wang, L., Ma, J., 2020, A flexible and self-healable gelled polymer electrolyte based on a dynamically cross-linked PVA ionogel for high-performance supercapacitors, *Ind. Eng. Chem. Res.*, 59(52): 22509–22519. https://doi.org/10.1021/acs.iecr.0c04741

Xia, L., Huang, L., Qing, Y., Zhang, X., Wu, Y., Jiang, W., Lu, X., 2020, In situ filling of a robust carbon sponge with hydrogel electrolyte: a type of omni-healable electrode for flexible supercapacitors, *J. Mater. Chem. A*, 8: 7746–7755. https://doi.org/10.1039/D0TA01764G

Xia, H., Meng, Y. S., Yuan, G., Cui, C., Lu, L., 2012, A symmetric RuO_2/RuO_2 supercapacitor operating at 1.6 V by using a neutral aqueous electrolyte electrochem, *Solid-State Lett.*, 15: A60–A63. http://dx.doi.org/10.1149/2.023204esl

Xiao, K., Ding, L.X., Liu, G., Chen, H., Wang, S., Wang, H., 2016, Freestanding, hydrophilic nitrogen-doped carbon foams for highly compressible all solid-state supercapacitors, *Adv. Mater.*, 28: 5997–6002. https://doi.org/10.1002/adma.201601125

Xu, M., Zhao, T., Wang, C., Zhang, S. L., Li, Z., Pan, X., Wang, Z. L., 2019, High power density tower-like triboelectric nanogenerator for harvesting arbitrary directional water wave energy, *ACS Nano*, 13: 1932–1939. https://doi.org/10.1021/acsnano.8b08274

Xue, X., Deng, P., He, B. Nie, Y., Xing, L., Zhang, Y., Wang, Z. L., 2014, Flexible self-charging power cell for one-step energy conversion and storage, *Adv. Energy Mater.*, 4: 1–5. https://doi.org/10.1002/aenm.201301329

Xue, J. X., Deng, P., Yuan, S., Nie, Y., He, B., Xing, L., Zhang, Y., 2013, CuO/PVDF nanocomposite anode for a piezo-driven self-charging lithium battery, *Energy Environ. Sci.*, 6: 2615. https://doi.org/10.1039/C3EE41648H

Xue, Q., Sun, J., Huang, Y., Zhu, M., Pei, Z., Li, H., Wang, Y., Li, N., Zhang, H., Zhi, C., 2017, Recent progress on flexible and wearable supercapacitors, *Small*, 13: 1701827. https://doi.org/10.1002/smll.201701827

Xue, X., Wang, S., Guo, W., Zhang, Y., Wang, Z. L., 2012, Hybridizing energy conversion and storage in a mechanical-to-electrochemical process for self-charging power cell, *Nano Lett.*, 12: 5048–5054. https://doi.org/10.1021/nl302879t

Yadav, N., Mishra, K., Hashmi, S. A., 2017, Optimization of porous polymer electrolyte for quasi-solid-state electrical double layer supercapacitor, *Electrochim. Acta*, 235: 570–582. https://doi.org/10.1016/j.electacta.2017.03.101

Yan, J., Wang, Q., Wei, T., Fan, Z., 2014, Supercapacitors: recent advances in design and fabrication of electrochemical supercapacitors with high energy densities, *Adv. Energy Mater.*, 4: 1300816. https://doi.org/10.1002/aenm.201470017

Yan, Y., Xia, H., Qiu, Y., Xu, Z., Ni, Q.-Q., 2019, Multi-layer graphene oxide coated shape memory polyurethane for adjustable smart switches, *Compos. Sci. Technol.*, 172: 108–116. https://doi.org/10.1016/j.compscitech.2019.01.013.

Yang, Z., Deng, J., Chen, X., Ren, J., Peng, H., 2013, A highly stretchable, fiber-shaped supercapacitor, *Angew Chem. Int. Edit.*, 52: 13453–13457. https://doi.org/10.1002/anie.201307619

Yang, Y., Guan, L., Li, X., Gao, Z., Ren, X., Gao, G., 2019, Conductive organohydrogels with ultrastretchability, anti-freezing, self-healing, and adhesive properties for motion detection and signal transmission, *ACS Appl. Mater. Interfaces*, 11: 3428–3437. https://doi.org/10.1021/acsami.8b17440

Yang, L., Hu, J., Lei, G., Liu, H., 2014, Ionic liquid-gelled polyvinylidene fluoride/polyvinyl acetate polymer electrolyte for solid supercapacitor, *Chem. Eng. J.*, 258: 320–326. https://doi.org/10.1016/j.cej.2014.05.149

Yang, P., Liu, K., Chen, Q., Mo, X., Zhou, Y., Li, S., Feng, G., Zhou, J., 2016, Wearable thermocells based on gel electrolytes for the utilization of body heat, *Angew. Chem. Int. Ed.*, 55: 12050. https://doi.org/10.1002/ange.201606314

Yang, H., Liu, Y., Kong, L., Kang, L., Ran, F., 2019, Biopolymer-based carboxylated chitosan hydrogel film crosslinked by HCl as gel polymer electrolyte for all-solid-sate supercapacitors, *J. Power Sources*, 426: 47–54. https://doi.org/10.1016/j.jpowsour.2019.04.023.

Yang, C., Sun, M., Wang, X., Wang, G., 2015, A novel flexible supercapacitor based on cross-linked PVDF-HFP porous organogel electrolyte and carbon nanotube paper@π--conjugated polymer film electrodes, *ACS Sustain. Chem. Eng.*, 3(9): 2067–2076. https://doi.org/10.1021/acssuschemeng.5b00334

Yang, Y., Urban, M. W., 2018, Self-healing of polymers via supramolecular chemistry, *Adv. Mater. Interfaces*, 5: 1800384. https://doi.org/10.1002/admi.201800384

Yang, X., Zhang, F., Zhang, L., Zhang, T., Huang, Y., Chen, Y. 2013, A high-performance graphene oxide-doped ion gel as gel polymer electrolyte for all-solid-state supercapacitor applications, *Adv. Funct. Mater.*, 23: 3353–3360. https://doi.org/10.1002/adfm.201203556

Yassine, M., Fabri, D., 2017, Performance of commercially available supercapacitors, *Energies*, 10: 1340. https://doi.org/10.3390/en10091340

Yi, J., Huo, Z., Asiri, A. M., Alamry, K. A., Li, J. 2018, Development and application of electrolytes in supercapacitors, *Prog. Chem.*, 30(11): 1624. https://doi.org/10.7536/PC180314

Yin, Y., Feng, K., Liu, C., Fan, S., 2015, A polymer supercapacitor capable of self-charging under light illumination, *J. Phys. Chem. C*, 119(16): 8488–8491. https://doi.org/10.1021/acs.jpcc.5b00655

Yu, D., Li, X., Xu, J., 2019, Safety regulation of gel electrolytes in electrochemical energy storage devices, *Sci. China Mater.*, 62(11): 1556–1573. https://doi.org/10.1007/S40843-019-9475-4 https://engine.scichina.com/doi/10.1007/S40843-019-9475 4

Yuan, C., Zhang, X., Wu, Q., Gao, B., 2006, Effect of temperature on the hybrid supercapacitor based on NiO and activated carbon with alkaline polymer gel electrolyte, *Solid State Ion.*, 177: 1237–1242. https://doi.org/10.1016/j.ssi.2006.04.052

Yuksel, R., Sarioba, Z., Cirpan, A., Hiralal, P., Unalan, H. E., 2014, Transparent and flexible supercapacitors with single walled carbon nanotube thin film electrodes, *ACS Appl. Mater. Interfaces*, 6: 15434–15439. https://doi.org/10.1021/am504021u

Yun, T. Y., Li, X., Bae, J., Kim, S. H., Moon, H. C., 2019, Non-volatile, Li-doped ion gel electrolytes for flexible WO_3-based electrochromic devices, *Mater. Des.*, 162: 45–51. https://doi.org/10.1016/j.matdes.2018.11.016

Yun, J., Lim, Y., Jang, G. N., Kim, D., Lee, S. J., Park, H., Hong, S. Y., Lee, G., Zi, G., Ha, S. J., 2016, Stretchable patterned graphene gas sensor driven by integrated micro-supercapacitor array, *Nano Energy*, 19: 401–414. https://doi.org/10.1016/j.nanoen.2015.11.023

Yun, J., Lim, Y., Lee, H., Lee, G., Park, H., Hong, S. Y., Jin, S. W., Lee, Y. H., Lee, S.-S., Sook, J., 2017, A patterned graphene/ZnO UV sensor driven by integrated asymmetric micro-supercapacitors on a liquid metal patterned foldable paper, *Adv. Funct. Mater.*, 27: 1700135. https://doi.org/10.1002/adfm.201700135

Yun, J., Song, C., Lee, H., Park, H., Jeong, Y. R., Kim, J. W., Jin, S. W., Oh, S. Y., Sun, L., Zi, G., Ha, J. S., 2018, Stretchable array of high-performance micro-supercapacitors charged with solar cells for wireless powering of an integrated strain sensor, *Nano Energy*, 49: 644–654. https://doi.org/10.1016/j.nanoen.2018.05.017

Zhang, C., Li, H., Huang, A., Zhang, Q., Rui, K., Lin, H., Sun, G., Zhu, J., Peng, H., Huang, W., 2019, Rational design of a flexible CNTs@PDMS film patterned by bio-inspired templates as a strain sensor and supercapacitor, *Small*, 15: 1805493. https://doi.org/10.1002/smll.201805493

Zhang, B., Li, J., Liu, F., Wang, T., Wang, Y., Xuan, R., Zhang, G., Sun, R., Wong, C.-P., 2019, Self-healable polyelectrolytes with mechanical enhancement for flexible and durable supercapacitors, *Chem. Eur. J.*, 25: 11715. https://doi.org/10.1002/chem.201902043

Zhang, L., Liu, D., Wu, Z.-S., Lei, W., 2020, Micro-supercapacitors powered integrated system for flexible electronics, *Energy Storage Mater.*, 32: 402–417. https://doi.org/10.1016/j.ensm.2020.05.025

Zhang, Y., Sezen, S., Ahmadi, M., Cheng, X., Rajamani, R., 2018, Paper-based supercapacitive mechanical sensors, *Sci. Rep.*, 8: 16284. https://doi.org/10.1038/s41598-018-34606-1

Zhang, J., Yang, J., Leftheriotis, G., Huang, H., Xia, Y., Liang, C., Gan, Y., Zhang, W., 2020, Integrated photo-chargeable electrochromic energy-storage devices, *Electrochim. Acta*, 345: 136235. https://doi.org/10.1016/j.electacta.2020.136235

Zhao, Y., Ding, Y., Li, Y., Peng, L., Byon, H. R., Goodenough, J. B., Yu, G., 2015, A chemistry and material perspective on lithium redox flow batteries towards high-density electrical energy storage, *Chem. Soc. Rev.*, 44: 7968. https://doi.org/10.1039/C5CS00289C

Zhao, M., Li, Y., Lin, F., Xu, Y., Chen, L., Jiang, W., Jiang, T., Yang, S., Wang, Y., 2020, A quasi-solid-state photothermal supercapacitor via enhanced solar energy harvest, *J. Mater. Chem. A*, 8: 1829. https://doi.org/10.1039/C9TA11793H

Zhao, J., Li, L., Zhang, Y., Li, C., Zhang, Q., Peng, J., Zhao, X., Li, Q., Wang, X., Xie, J., Sun, J., He, B., Lu, C., Lu, W., Zhang, T., Yao, Y., 2018, Novel coaxial fiber-shaped sensing system integrated with an asymmetric supercapacitor and a humidity sensor, *Energy Storage Mater.*, 15: 315–323. https://doi.org/10.1016/j.ensm.2018.06.007

Zhao, Y., Liu, J., Hu, Y., Cheng, H., Hu, C., Jiang, C., Jiang, L., Cao, A., Qu, L., 2013, Highly compression-tolerant supercapacitor based on polypyrrole-mediated graphene foam electrodes, *Adv.Mater.*, 25:591–595. https://doi.org/10.1002/adma.201203578

Zhao, H., Liu, L., Vellacheri, R., Lei, Y., 2017, Recent advances in designing and fabricating self-supported nanoelectrodes for supercapacitors, *Adv. sci.*, 4(10): 1700188. https://doi.org/10.1002/advs.201700188

Zhao, J., Zhang, Y., Huang, Y., Xie, J., Zhao, X., Li, C., Qu, J., Zhang, Q., Sun, J., He, B., Li, Q., Lu, C., Xu, X., Lu, W., Li, L., Yao, Y., 2018, 3D printing fiber electrodes for an all fiber integrated electronic device via hybridization of an asymmetric supercapacitor and a temperature sensor, *Adv. Sci.*, 5: 1801114. https://doi.org/10.1002/advs.201801114

Zheng, H., Zi, Y., He, X., Guo, H., Lai, Y. C., Wang, J., Zhang, S. L., Wu, C., Cheng, G., Wang, Z. L., 2018, Concurrent harvesting of ambient energy by hybrid nanogenerators for wearable self-powered systems and active remote sensing, *ACS Appl. Mater. Interfaces*, 10(17): 14708–14715. https://doi.org/10.1021/acsami.8b01635

Zhong, C., Deng, Y., Hu, W., Qiao, J., Zhang, L., Zhang, J., 2015, A review of electrolyte materials and compositions for electrochemical supercapacitors, *Chem. Soc. Rev.*, 44: 7484–7539. https://doi.org/10.1039/C5CS00303B

Zhong, J., Meng, J., Yang, Z., Poulin, P., Koratkar, N., 2015, Shape memory fiber supercapacitors, *Nano Energy*, 17: 330–338. https://doi.org/10.1016/j.nanoen.2015.08.024

Zhou, D., Wang, F., Yang, J., Fan, L., 2021, Flexible solid-state self-charging supercapacitor based on symmetric electrodes and piezo-electrolyte, *Chem. Eng. J.*, 406: 126825. https://doi.org/10.1016/j.cej.2020.126825

Zhou, D., Wang, N., Yang, T., Wang, L., Cao, X., Wang, Z., 2020, A piezoelectric nanogenerator promotes highly stretchable and self-chargeable supercapacitors, *Mater. Horiz.*, 7: 2158–2167. https://doi.org/10.1039/D0MH00610F

Zhu, Q., Song, Y., Zhu, X., Wang, X., 2007, Ionic liquid-based electrolytes for capacitor applications, *J. Electrochem. Soc.*, 601(1–2): 229–236. https://doi.org/10.1016/j.jelechem.2006.11.016

Zi, Y., Wang, J., Wang, S., Li, S., Wen, Z., Guo, H., Wang, Z. L., 2016, Effective energy storage from a triboelectric nanogenerator, *Nat. Commun.*, 7: 10987 https://doi.org/10.1038/ncomms10987

Zuo, W., Li, R., Zhou, C., Li, Y., Xia, J., Liu, J., 2017, Battery-supercapacitor hybrid devices: Recent progress and future prospects, *Adv. Sci.*, 4: 1600539. https://doi.org/10.1002/advs.201600539

3 Polyaniline-Based Ternary Composites for Energy Accumulation in Electrochemical Capacitors

Mohammad Faraz Ahmer and Adnan
Mewat Engineering College, Nuh, Haryana, India

CONTENTS

3.1 INTRODUCTION

In order to address the concerns related to the over-utilization of fossil fuel resources and environmental pollution, numerous attempts have been made to create sustainable and clean energy alternative systems [1, 2]. Accordingly, wind, tidal, hydro, and biomass, as renewable energy sources, have been utilized after being converted into electric energy [3]. These energy sources have the limitation of heavily depending on natural climatic conditions. Thus, in order to quench the thirst for developing novel green energy storage devices, electrochemical supercapacitors (ESCs), rechargeable batteries, and fuel cell related research areas have been aggressively pursued over recent years. Among these, supercapacitors (also called electrochemical capacitors) have been preferred as green energy storing devices because of their several fascinating characteristics that include high output power density (about 10 kW/kg), reasonable cyclic sustainability (>10^5 cycles), expeditious charging–discharging capability (a few seconds), low maintenance cost, and a non-polluting nature [4, 5]. Furthermore, it is possible to utilize electrochemical supercapacitors in diverse energy harvesting applications after hybridizing with fuel cells and/or rechargeable batteries. Basically, both ESCs and batteries are electrochemical energy storage systems, but in the former case, energy is stored directly as charge whereas in the letter case the energy generated due to chemical reactions is stored as charge. On the basis of charge storing mechanisms as well as the nature of the active electrode materials used for ESCs, supercapacitors can be categorized into three broad groups [6]: (i) electrical double layer capacitors (EDLCs), (ii) pseudo-capacitors (PSCs) or redox supercapacitors, and (iii) hybrid capacitors (HBCs) as illustrated in Figure 3.1.

In EDLCs, charge is electrostatically accumulated in the double layer via charge separation onto the interface between the electrode surface and the liquid electrolyte through reversible adsorbing/desorbing of charged ions from the electrolyte without any chemical reaction or electron transfer between the electrodes. The capacitance, thus, depends upon the electrolyte concentration as well as on the electrode material properties. The electrode materials commonly used for EDLCs include activated carbon (AC), graphene (G), carbon nanofibers (CNFs), and carbon nanotubes (CNTs). These materials create favorable electrochemical situations to provide wider surface area, reasonable power density, good conductivity, and fantastic cyclic performance. However, the lower specific capacitance values of these carbonaceous materials have restricted their widespread usage in EDLCs [7–11].

The other type of ESC is a pseudo-supercapacitor in which the charge storing mechanism involves the fast reversible redox reactions of electroactive species onto the surface of the electrode. In this case, electron transfer affects charge accumulation and the charge migration is Faradaic. The specific capacitance of pseudo-capacitors is usually better than EDLCs because of the use of metallic oxides/hydroxides, chalcogenides, and organic conductive polymers as electrode materials. However, the low degree of power density and the inferior cyclability have been the main limitations of pseudo-capacitors [12, 13].

DOI: 10.1201/9781003169727-3

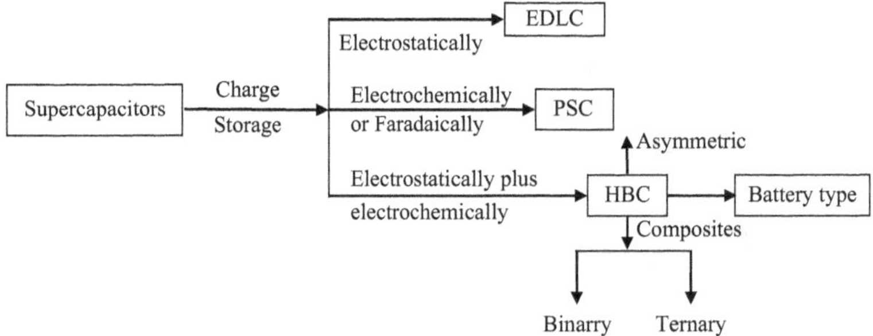

FIGURE 3.1 Classification of supercapacitors.

The third category of ESCs belongs to hybrid supercapacitors which have been developed by combining both EDLCs and pseudo-supercapacitors [6, 14]. To fabricate these capacitors, the composite materials either of two or more different components are used, which show higher specific capacitance, enhanced energy density, and a better life cycle than either EDLCs or pseudo-capacitors [15–18]. In the case of hybrid supercapacitors, both Faradaic and non-Faradaic charge transfer mechanisms are operative. Basically, there are three types of hybrid supercapacitors: (i) composites, (ii) asynometric, and (iii) battery-type. These are currently in use. These hybrid capacitors have removed the shortcomings of EDLCs and pseudo-capacitors by coupling the advantageous features of EDLCs (e.g., excellent cyclic stability and superb electrical conductivity) and pseudo-capacitors (e.g., high specific capacitance and significant energy density).

Amongst three essential components (electrode material, separator, and electrolyte) of supercapacitors, the electrode material plays a pivotal role in deciding the efficiency of ESCs in terms of specific capacitance, cyclic lifespan, power density, charging–discharging frequency, and environmental stability. During the last few years, three different types of active materials as mentioned below have received extensive attention for use in supercapacitor technology.

1. Conducting organic polymers;
2. Carbon-based materials;
3. Metallic oxides/hydroxides.

The merits and demerits of these electrode materials are listed briefly in Table 3.1 and their comparative electrochemical performance trends are presented in Table 3.2.

TABLE 3.1
Merits and Demerits of Individual Components of Conductive Composite Electrode Materials [4, 19–22]

Component	Merits	Demerits
Conducting polymers	• High specific capacitance • Affordable cost • Excellent morphological flexibility • Ease of fabrication • Reasonable energy density and power density • Rapid charging/discharging capability • Quick response time • Usability at ambient temperature	• Short-cycle life span • Lack of long-term durability due to structural damage during ion doping/dedoping process due to volumetric swelling and shrinking • Poor solubility and hydrophobicity
Carbon-based material	• Large accessible surface area • Tunable unique porosity • High electrochemical stability • Efficient non-scattered electronic transportation • High power density • Good corrosion resistance • Low cost	• Limited energy density • Low specific capacitance • Reduction in effective surface area due to agglomeration and stacking of graphene layers in graphite • Restricted graphene dispersibility
Metal oxides/ hydroxides	• Availability of multiple oxidation states with low activation energy • Wide charge/discharge potential range	• Low specific capacitance • Poor cyclic stability • Inferior environmental/chemical stability • Limited surface area • High cost • Toxicity • Poor conductivity

TABLE 3.2
Comparative Trends of the Electrochemical Performance of Conductive Polymers, Carbon-Based Materials, and Metallic Oxides Individually as Supercapacitor Materials [22]

Electrochemical Properties	Trends[a]
Chemical stability	CBM > CP > MO
Conductivity	CBM > CP > MO
Energy density	MO > CP > CBM
Power density	CBM > CP > MO
Cycle life time	CBM > CP ≈ MO
Faradaic capacitance	CP ≈ MO > CBM
Non-Faradaic capacitance	CBM > MO ≈ CP
Flexibility	CP > CBM > MO
Fabrication processibility	CP > CBM > MO
Cost	MO > CBM ≈ CP

a CP = conducting polymer, CBM = carbon-based material, MO = metal oxide

Note: None of these materials is ideal as an electrode for supercapacitors but their ternary composites through various combinations of CP, CBM, and MO into a single electrode are expected to show superior capacitance because of synergistic effects.

Among the different types of supercapacitor materials, CP-based composites have received enormous attention in the drive to satisfy the requirement of high performance supercapacitors. In this chapter, the CP-based ternary composites that have been reported over the past few years for supercapacitors will be presented to highlight their improved supercapacitive performance compared to single component energy storage capacitors.

3.2 COMPOSITE MATERIALS FOR SUPERCAPACITORS

To cope with the limitation of the lower available specific energy of supercapacitors compared to batteries as identified from Ragone plots (plots of specific power, (W/kg) vs. specific energy W h/kg), strategies have been formulated to produce multi-component composites for supercapacitors to maintain their performance in optimum working regions [22]. As such, many composite supercapacitor materials like graphene-based ternary composites [23], ternary metal oxide nanostructured materials [5], mixed transition metal oxide composites [6], carbonaceous quantum dot composites [24], and poly(*p*-phenylenediamine) based ternary composites containing rGO and ZnO nanorods [25] were developed in addition to conducting polymer based ternary composites [26]. Mixed transition metal oxides ($NiMn_2O_4$, $CoMn_2O_4$, $MnCo_2O_4$, $NiCo_2O_4$, $MnFe_2O_4$, and $CoFe_2O_4$) produced in the single phase have generated a new category of novel electrode materials for supercapacitor applications [27–30]. Though fewer reports are available on the utilization of intermixed single-phased transition metal oxide nano-composite materials for designing supercapacitors, an interesting new multistep synthetic route was developed by Sahoo et al. [6] for synthesizing novel porous ternary nanocomposites using reduced graphene, $NiMn_2O_4$, and PANI. The typical complex synthetic process involving the *in situ* polymerization of aniline monomers along with graphene oxide and metal cations is illustrated in Figure 3.2.

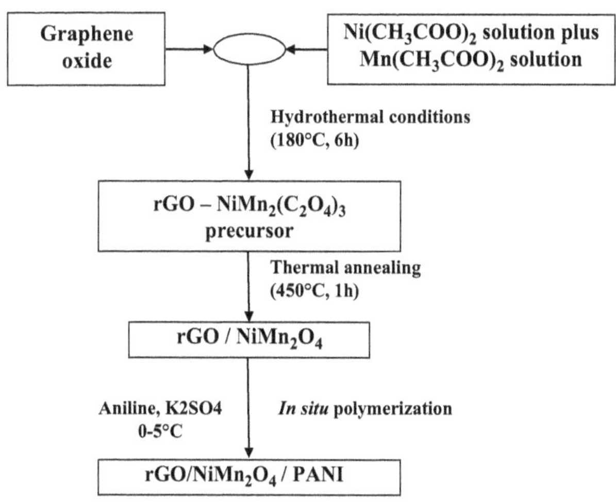

FIGURE 3.2 Illustration of multistep synthetic route for the preparation of ternary composites consisting of PANI, reduced graphene, and $NiMn_2O_4$.

3.3 CONDUCTING ORGANIC POLYMER (COP) BASED TERNARY COMPOSITES FOR SUPERCAPACITORS

In the recent past, ternary composites comprising of carbonaceous matter, metal oxides, and conductive organic polymers have been widely investigated as electrode materials utilizing both Faradaic and non-Faradaic charge storage mechanisms to achieve improved cell voltage, higher specific capacitance, longer life cycles, and enhanced energy density [4, 6, 31–34]. The better electrochemical functioning of ternary-composite electrodes in comparison to their individual components may be due to two factors: (i) the synergistic property of the constituents and (ii) the decrease in the diffusive path length due to the increase in the specific surface area.

Among several COPs (fluorinated, non-fluorinated, polyimides, polyarylenes, polyvinylpyrrolidone, and carbonhydrate polymers), the most frequently investigated polymers for supercapacitor applications have been PANI (polyaniline), PPy (polypyrol), and PEDOT (poly-3,4-ethylene-dioxythiophene). These polymers have the ability to store energy via doping processes (n-and/or p-doping). While p-doping involves the release of electrons (oxidative doping), n-doping creates a net negative charge (reductive doping). PEDOT has the ability to undergo both n as well as p-doping, but PANI and PPy go through only p-doping [35–37]. The specific capacitance of these polymers is in the order: PANI > PPy > PEDOT.

TABLE 3.3
Beneficial Features and Shortcomings of PANI for Use in Supercapacitors as an Electrode Material [19]

Favorable Properties	Shortcomings
• Easy synthesis and processability • High theoretical specific capacitance • Thermally stable intrinsically conductive polymer • Unique nanostructures with enhanced surface area and improved conductivity • Useful conductivity in doped state (doped emeraldine base with conductivity range of about 01–100 S cm^{-1}) • Adequate stability and high electro-activity • Low cost	• Requirement of protic ionic liquid (or protic solvent) for being completely charged/discharged • Slower ion transfer during reversible redox reactions • Short cyclic life • Moderate capacitance retention especially at high charging/discharging rates

In this chapter, the basic objective is to focus on the use of PANI as the most extensively investigated one of the three components (conductive polymers, metallic oxides, and carbon materials) of ternary electrode materials for supercapacitor applications.

3.3.1 Polyaniline

Polyaniline, synthesized initially in 1886 by the chemical oxidative polymerization of aniline, is an intrinsically conductive and thermally stable p-polymer. Due to its excellent pseudo-capacitance, high doping level, and environmental stability, it has now become a promising candidate for application in supercapacitors. The merits and demerits of PANI as an electrode material are listed in Table 3.3.

In fact, PANI-based materials have demonstrated a wider specific capacitance range (i.e., 30–3000 F/g) depending upon several factors such as the polymerization route, structural morphological characteristics, type/concentration of dopant, and ionic diffusion range [19, 21, 38]. PANI can exist in multiple redox states to acquire a large surface change potential while transferring from one oxidation state to another state. The three PANI states [39] are known as leucoemeraldine (fully reduced form), pernigraniline (fully oxidized state), and emeraldine base (partially oxidized, known as emeraldine salt).

The doped emeraldine salt is the most conductive form of PANI. However, the structural damage to PANI as a result of swelling/shrinking during the processes of charging/discharging is the major drawback that has been overcome by converting PANI into nanostructures (nanorods and nanoparticles). The ordered nanostructures of PANI have shown better cyclic stability, enhanced specific surface area, excellent energy-storing characteristics, and superior rate performance in comparison to randomly connected geometries of PANI [40]. Furthermore, a variety of metallic oxides (or hydroxides) and functional carbon-based materials have been incorporated into PANI nanostructures to produce ternary nanocomposite hybrids for the development of well-functioning supercapacitors [41–45].

3.3.2 Carbon-Based Materials

Various carbonaceous materials (graphene, carbon nanofibers, activated C, and multiwall carbon nanotubes) have been extensively used as supercapacitor material components owing to their high levels of chemical resistance and thermal stability, extended accessible active surface sites, superior ionic conductivity, unique porosity, and reasonable mechanical strength. Graphene with its unique one-atom-thick 2D structure has been the most favorable since its discovery in 2004, but its low dispersibility has been the main drawback. In the case of CNTs, the occurrence of a defective thin amorphous film on the outer surface facilitates better charge accumulation, leading to improved specific capacitance. The lower electrochemical capacitance compared to metal oxides and intrinsic high resistance have been the limitations of CNTs. Graphene oxide (GO) bearing oxygen containing functional groups (> CO, –OH, –COOH, etc.) is the chemically modified version of graphene and is a nonelectrical conductive material. However, reduced graphene oxide (rGO) obtained by deoxygenating GO partially via electrochemical or chemical processes is a highly conductive material retaining the graphitic structure of C-atoms. The various carbon materials normally show a specific capacitance in the decreasing order of: carbon nanotubes > graphene oxide > graphene > carbon nanofibers > activated carbon.

3.3.3 Metal Oxides

Among the metal oxides in use, one of the components of polymer-based ternary composites, RuO_2 and MnO_2, have been the most widely used active materials. In fact RuO_2 exhibits a high capacitance value (≈1475 F g^{-1}) but its toxicity and high cost have put restrictions on its large-scale use for supercapacitors. Therefore, the focus was shifted to using MnO_2 as an inexpensive and highly redox metal oxide exhibiting a theoretical capacitance of about 1400 F g^{-1}. MnO_2 is naturally abundant and environmentally friendly. However, in view of its brittleness and dense morphology, thin layers of MnO_2 on textile fibers have been used in a composite form with conducting polymers. According to their respective

TABLE 3.4
Electrochemical Characteristics of PANI-Based Ternary Composite Electrode Materials Consisting of a Carbon Material/Metal Compound and Two Other Components

Composition	Electrochemical Performance	Remark	Reference
PANI/activated carbon/MnO_2	Gravimetric capacitance (1292) at current density (4.0 mA cm^{-2}) in 1 M $LiCO_4$/acetonitrile with 82% capacity retention after 1500 cycles	PANI/MnO_2/activated carbon displayed specific energy (61 W h/kg) associated with specific power of 172 W/kg	[53]
PANI/MW CNT/MnO_2	Gravimetric (330 F g^{-1}) and volumetric (296 F cm^{-3}) capacitances at scane rate of 5 mV s^{-1} in 0.5 M Na_2SO_4 with 77% capacity retention over 1000 cycles	Excellent energy storage performance of this ternary composite material was attributed to unhindered reversible electronic charge transfer from the MnO_2 nanoflakes	[54]
PANI/porous carbon spheres/ RuO_2	Specific capacitance (531 F g–^{1}) in 1 M hydrochloric acid	Use of nano-RuO_2 in combination with PANI and porous C-spheres in fabrication of well-functioning capacitor electrodes	[55]
PANI/carbon/Ni	Specific capacitance (725 F g^{-1}) at 0.5 A g^{-1} in 1 M Na_2SO_4	Development of free-standing three-dimensional PANI/CNT/nickel-fiber nanocomposite electrodes for supercapacitor applications	[56]
PANI/GO/TiO_2	Specific capacitance (430 F g^{-1}) at 1 A g^{-1} in 1 M H_2SO_4	Facile synthesis of PANI/graphene/TiO_2 nanocomposite for high profile supercapacitors	[57]
PANI/rGO/MoO_3	Specific capacitance (363 F g^{-1}) at 1 mV s^{-1} scane rate	Excellent ternary composite for developing high performance supercapacitors	[58]
PANI/graphene/Fe_2O_3	Specific capacitance (638 F g^{-1}) with 92% capacity retention at 2 A g^{-1} for 5000 cycles	Remarkable power densities displayed by this ternary nanocomposite material were 351 W kg^{-1} at 107 W h kg^{-1} and 4407 W kg^{-1} at 17 W h kg^{-1}	[59]
PANI/mesoporous carbon/MnO_2	Gravimetric capacitance (695 F g^{-1}) at current density (0.3 A g^{-1}) in 1 M H_2SO_4 with 88% electrical capacity retention over 1000 GCD cycles.	Improved electrochemical stability along with excellent capacitive efficiency of ternary composite electrodes due to stronger interaction between PANI and mesoporous carbon through MnO_2	[60]
H/Ni Co-doped PANI/CNT	Specific capacitance (781 F g^{-1} at 0.5 A g^{-1}) with retention of 92% for 700 cycles in 1 M H_2SO_4	Evaluation of supercapacitive performance of electrode material based on Hb and Ni 2p co-doped polyaniline–MWCNTs nanocomposite	[61]
PANI/sulphonated graphene/ MnO_2	Specific capacitance (276 F g^{-1}) at 1 A g^{-1} in 1 M Na_2SO_4 with 88% original capacitance retention after 3000 cycles	Very good electrochemical capacitance retention between 0.2 and 20 A g^{-1}. The utilization of about 70% MnO_2 in ternary composite provided higher structural stability.	[62]
PANI-coupled Co_3O_4/rGO	Specific capacitance (1063 F/g) at 1 A g^{-1} with 95% retention of original capacitance for 2500 cycles in 6.0 M aqueous KOH solution	Use of PANI as linker in development of PANI-based multifunctional two-dimensional metal oxide/hydroxide graphene nanohybrids for supercapacitor applications	[63]
$MnCl_2$-doped PANI/SWCNTs	Specific capacitance (546 F g^{-1}) at current density of 0.5 A g^{-1}	Better specific capacitance, homogeneous covering of Mn-doped polyaniline on the surface of single-walled carbon nanotube (SWCNTs) surface and strong affinity between conjugated structure of PANI and π-bonded SWCNT surface	[64]
Vertically aligned tunable PANI/ graphene/ZrO_2	Specific capacitance (1360 F g^{-1}) at 1 mV s^{-1} with capacitance retention of 93% for 1000 cycles in 1 M H_2SO_4	Vertically aligned tunable polyaniline growth on graphene/ZrO_2 composites resulted in the formation of ternary nanocomposite for application to supercapacitive energy-storage devices	[65]
PANI/GO/MnO_2 nanorods	Retention of 97% original gravimetric capacitance (512 F g^{-1}) at current density of 0.25 A g^{-1} in 1 M Na_2SO_4 for 5000 cycles	The ternary nanocomposite exhibited optimum capacitance and superior charge transport at 70% MnO_2 loading. The polyaniline coated manganese dioxide nanorods inserted between graphene oxide sheets facilitated better charge transfer.	[66]
PANI/graphene/MnO_2	Specific capacitance (755 F g at 0.5 A g^{-1} current density with 87% ultimate retention for 1000 cycles	PANI/graphene/MnO_2 composite obtained by growing MnO_2/polyaniline intersecting nanorod arrays via coating of graphene on vitreous carbon surface exhibited good electrical conductivity	

(*Continued*)

TABLE 3.4 (Continued)

Composition	Electrochemical Performance	Remark	Reference
MW CNT/MnO_2	The mixed and coaxial nanocomposites showed 365 and 270 F g^{-1} gravimetric capacitances respectively at 1 A g^{-1} current density in 0.5 M aqueous Na_2SO_4 solution with 88.2% retention after 2000 galvanostatic charge–discharge (GCD) cycles	MnO_2 on the surface of PANI/MW CNT nanofibers resulted in the formation of two types (mixed and co-axial) of ternary nanocomposites of different morphologies. Better dispersion and more loading of MnO_2 were observed in mixed ternary nanocomposite than that of coaxial nanocomposite.	[68]
PANI/rGO/MnO_2	Specific capacitance (395 F g^{-1}) at current density of 10 mA cm^{-2} in 1 M H_2SO_4	About 92% of initial specific capacitance up to 1200 cycles was retained.	[69]
PANI/graphene/MnO_x	Power density = 4500 W kg^{-1} and energy density = 61.2 W h kg^{-1} along with excellent cycle stability	The 3-D hydrogel composite was synthesized with controlled morphology of MnO_x by hydrothermal method for use as binder free supercapacitor electrode	[70]
PANI/graphene/MnO_2 (use of Ni foam as current collector)	Gravimetric capacitance value of 1081 F g^{-1} at 1 mV s^{-1} in 6 M KOH with 99% capacitance retention after 2000 cycles was observed	Use of Ni foam as current collector. Improved interfacial conduction due to alternate arrangement of PANI and MnO_2 nanoparticles on Ni-foam coated graphene surface.	[71]
PANI/graphene/SnO_2	Specific capacitances (1012 F g^{-1}) at 4 A g^{-1} and (1291 F g^{-1}) at 1 A g^{-1} (in 1 M Na_2SO_4) with 91% retention after 1500 cycles.	The ternary composite synthesized by following two-step approach using ionic liquid and water as solvents demonstrated excellent reversibility during charge/discharge process in 1 M Na_2SO_4	[33]
PANI/graphene nanoplates (GNP)/MoO_3	Specific capacitance (593 F g^{-1}) at 1 A g^{-1} with 92.4% capacitance retention over 1000 cycles.	Ternary hybrid prepared via *in situ* polymerization of aniline monomer in combination with MoO_3 and graphene nanoplates showed extraordinary electrochemical performance as supercapacitor	[72]
PANI/graphene/MoS_2	Specific capacitance of 618 F g^{-1} at 1 A g^{-1} with the retention of 78% for 2000 cycles in Na_2SO_4 (1 M)	The 3D ternary nanocomposites made of molybdenum disulfide polyaniline reduced graphene oxide aerogel displayed expanded performance for supercapacitor applications	[73]
PANI/MWCNT/MnO_2 nanoflake coaxial ternary nanocomposite	Gravimetric capacitance (348.5 at 1 A g^{-1} in 1 M KOH) with 88.2% retention after 2000 cycles	Improved capacitance and cyclic stability compared to PANI/MW CNT binary composite. *In situ* growth of MnO_2 layer improved the structural stability.	[74]
Graphene/MnO_2 nanoflakes coated on PANI nanowires	Gravimetric capacitance (695 F g^{-1}) at current density of 4 A g^{-1} in 1 M Na_2SO_4 with 90% capacity retention for 1000 cycles	The hierarchical ternary nanocomposites composed of PANI/MnO_2/graphene were obtained through coupling *in situ* polymerization and reductive chemical processes. The resultant ternary composite exhibited strong interfacial connectivity, minimal charge transfer resistance, and better cyclability.	[75]
PANI/rGO/CuO	Specific capacitance (634.4 F g^{-1}), power density (114 W h/kg), and energy density (126.8 W h/kg) at 1 A g^{-1} with retention of 97.4% original capacitance after 10,000 cycles	Self-assembled ternary hydrogel composite with negligible IR drop and porous microstructure morphology was obtained through *in situ* polymerization in combination with hydrothermal treatment	[76]
PANI/graphene nanosheet/NiO	Specific capacitance (1409 F g^{-1}) at 1 A g^{-1} with retention of 92% capacity for 2500 cycles	Graphene/PANI composite thin film coated with nano-nickel oxide was developed for use as flexible supercapacitor	[77]
PANI/graphene/Co_3O_4	Specific capacitance (1247 F g^{-1} at 1 A g^{-1}) with almost 100% cyclic stability up to 3500 cycles along with energy (190 W h/kg) and power (190 W/kg) densities	Ecofriendly as well as cost effective ternary composite electrode material with excellent energy performance for devising supercapacitors	[78]
PANI/rGO/MnO_2	Specific capacitance (1090.2 F g^{-1}) at 0.5 A g^{-1} in 1 M KOH with 82.3% retention of original capacitance for 5000 cycles	The ternary composite was produced using *in situ* polymerization method where rGO/PANI composite sheets were treated with PANI coated rGO nanomaterial in the presence of $KMnO_4$ as oxidizing agent	[79]
PANI/MWCNT/transition metal ions (Cu^{2+} or Ni^{2+})	Specific capacitance (1337 F g^{-1}) at 5 mV s^{-1}	The metal doped PANI/carbon nanotube composites demonstrated better electrochemical performance because of the effective coordination between transition metal cations and nitrogen atoms of conducting polymers	[80]

PANI/graphene nanoribbons/ MnO_2	Gravimetric capacitance (472 F g^{-1}) in 1 M Na_2SO_4 aqueous solution at 1 A g^{-1} with retention of 79.7% after 5000 cycles	Better ion diffusion through 3D porous structure created by MnO_2 and graphene within polyaniline matrix. The resultant ternary composite showed superior capacitive retention and higher charging/discharging efficiency compared to MnO_2 and MnO_2/graphene.	[81]
PANI/rGO/Fe_3O_4	Specific capacitance (283.4 F g^{-1}) at 1 A g^{-1} with energy (47.7 W h/kg) and power (550 W/kg) densities. The retention capacitance of 78% was displayed by ternary composite for 5000 cycles	The ternary composite was synthesized with the use of scalable soft-template method. The obtained composite illustrated low interfacial charge transfer resistance.	[82]
PANI/3D-graphene/Fe_3O_4	Specific capacitance (486.5 F g^{-1}) with higher retention ratio of 52.1% compared to that of PANI (38.9%)	An excellent ternary composite consisting of 3D-graphene, Fe_3O_4, and polyaniline with excellent electrochemical performance for flexible supercapacitors has been developed	[83]
PANI/Ni/activated carbon	Specific capacitance (1661 F/g) at current density of 1 A g^{-1} in 1 M aqueous Na_2SO_4 with 93% retention after 2000 cycles	Development of PANI-based ternary nanocomposites using charcoal and nickel compound as other components for use as electrode material in electrochemical systems	[84]
PANI/graphene/MnO_2 ultra-thin nanosheets	Specific capacitance (245 F g^{-1}) at 0.5 A g^{-1} in 1 M Na_2SO_4 with 91.5% capacitive retention after 10,000 cycles, energy density = 30.6 W h/kg and specific power = 11,804 W/kg	Superior electrochemical performance and better charge transport characteristics were due to uniform anchoring of pseudo-capacitive MnO_2 nanosheets on graphene surface in the presence of polyaniline	[85]
PANI/rGO/V_3O_7	Specific capacitance (579 F/g) at 0.2 A g^{-1} with 94% retention of original capacitance after 2500 cycles in 1 M H_2SO_4	Brush like polyaniline in combination with vanadium oxide trapped reduced graphene oxide acted as effective electrode material for supercapacitors	[86]
PANI/rGO/MnO_2 treated carbon nanofiber sewed on carbon sponge derived from melamine	Specific capacitance of 481 F g^{-1} at 1 mA cm^{-2} in 1 M Li_2SO_4 with 89% cyclic stability for 8900 cycles	The composite material was developed using carbonization, wet precipitation, and chemical polymerization processes in combination. Carbon nanofibers facilitated strong connectivity among components leading to enhanced electronic conductivity whereas rGO improved effective surface area.	[87]
PANI/graphene/MnO_2	Specific capacitance (3.5 F cm^{-2}) at 5 mA cm^{-2} in 1 M H_2SO_4 with 90% retention after 1000 bending cycles along with energy density (5.2 mW h cm^{-3}) and power density (8.4 mW cm^{-3})	Application of *in situ* layer-by-layer growth technique in association with vacuum filtration method for developing unique composite of porous structure involving polyaniline/graphene paper wrapped by manganese dioxide nanoflowers	[88]
PANI/sulphonated graphene oxide/NiO	Specific capacitance (1350 F g^{-1}) at 1 A g^{-1} with excellent capacitance retention of 92.23% over 5000 cycles	This ternary electrode material synthesized by following hydrothermal and chemical oxidation processes exhibited unique microstructure with excellent electrochemical performance	[89]
PANI/porous carbon nanofiber/ MnO_2	Specific capacitance (289 F g^{-1}) in 1 M H_2SO_4 at current density of 1 A g^{-1} associated with specific energy (119 W h/kg) and power density (322 W/kg)	Electro-deposition of nanoparticles of MnO_2 onto porous fibrous carbon thin sheet followed by chemical polymerization of PANI. Better rate capacity than binary composite and pure PANI.	[90]
PANI/graphene oxide/MnS	Specific capacitance (773 F g^{-1}) at current density of 1 A g^{-1} as measured by galvonostatic charge–discharge measurements	Nanoparticles of manganese sulfide prepared by using template-free method were used in combination with PANI and graphene oxide to obtain PANI/graphene oxide/MnS nanocomposite. Graphene oxide acted as the acidic dopant during polymerization process of aniline.	[91]

TABLE 3.5
Electrochemical Characteristics of PANI-Based Ternary Composite Electrode Materials Consisting of Carbon Material (or Metal Compound) and Other Polymers

Composition	Electrochemical Performance	Remark	Reference
PANI/cellulose/CNT	Specific capacitance 495.2 F g^{-1} with 81% retaining initial value after 1000 cycles	Development of novel ternary electrode material through *in situ* generation of PANI within the matrix of cellulose	[92]
Binder free PANI/ polyethylene oxide (PEO)/MWCNTs	Specific capacitance (385 F g^{-1}) at 0.5 A g^{-1} current density	The ternary nanocomposite prepared by using electrospinning technique exhibited superior electrochemical activities	[93]
PANI/PPy/MoS2	Specific capacitive value (1273 F/g) at 0.5 A/g with superior cyclic stability retaining about 83% of original capacitance over 3000 charging/discharging cycles	The synthesized Pizza-type ternary composite for supercapacitors exhibited superior cyclic stability and high energy density	[94]
PANI/PPy/graphite	Areal capacitance (55.35 mF cm^{-2}), power density (87.1 μW cm^{-2}), and energy density (4.92 W h cm^{-2}) at 0.3 mA cm^{-2}	Development of inexpensive binder-free flexible supercapacitor based on PANI/polypyrrole/graphite on gold coated sand paper ternary composite	[95]
PANI/3-D bacterial cellulose/graphene	Specific capacitance (645 F g^{-1}) with 82.2% retention of initial capacitance for 1000 cycles	Use of novel ternary nanocomposite for constructing highly flexible freestanding electrodes for supercapacitors	[96]

theoretical specific capacitances (F/g), different metal oxides can be arranged in the following order [21]:

$$RuO_2 > MnO_2 > NiO > Co_3O_4 > V_2O_5 > Fe_2O_3 > SnO_2$$

3.3.4 Polyaniline-Based Ternary Composites

In the last decade, supercapacitors with three broad types of electrode materials (conducting polymers, carbonaceous materials, and metallic oxides/hydroxides) in combination have been extensively investigated with the aim of utilizing their positive synergistic effects, such as the high conductivity/specific energy of polymers, the high mechanical strength and large surface area of carbonaceous materials, and the adequate specific capacitance of metallic oxides. PANI-based ternary composite materials reported for use in supercapacitors over the last decade can be classified into following groups:

1. PANI/carbonaceous material/metal (oxide/hydroxide)
2. PANI/carbonaceous material/other metallic compounds
3. PANI/other polymer/metal (oxide/hydroxide)
4. PANI/other polymer/carbon material
5. PANI/two other carbon-based materials
6. PANI/two other metal compounds such as (i) PANI/two different single metals and (ii) PANI/two metals in a single bimetallic form.

Some recent interesting reports on PANI containing electrode materials, not covered under the above heads, include: PANI nanofiber-wrapped graphene/PANI nanorods [26], PANI coated graphene paper [46], PANI coated on a nonporous gold support [47], PANI/stretchable isotropic buckled carbon nanotubes [48], PANI cotton-shaped fiber/carbon nanotubes [49], 3-D PANI on pillared reduced graphene oxide [50], PANI nanofiber embedded reduced graphene oxide hydrogel film [51], and polylactic acid/carbon nanotubes/PANI [52].

A summary of the most important developments in the research area of PANI-based ternary composites (PANI/carbon matter/metal compounds) during the last decade is presented in Tables 3.4–3.7. From the tables, it is evident that ternary nanocomposites have tremendous scope for their application in high performance energy storage systems. The ternary PANI-based electrode materials presented in Tables 3.3–3.7 were synthesized using diversified synthetic methods [4, 19, 23, 45, 90, 126] such as: electrodeposition, chemical oxidative polymerization, multistep chemical synthesis, filtration and electrodeposition, sonochemical coprecipitation and a chemical functionalization process, *in situ* polymerization coupled with chemical synthesis processes, *in situ* hydrothermal polymerization, electrodeposition electrospun and chemical polymerization, *in situ* oxidative polymerization in combination with hydrothermal processes, and their supercapacitive characteristics were evaluated chemical following electrochemical techniques:

1. Cyclic voltammetry (CV).
2. Galvanostatic charging–discharging (GCD) cycle stability.
3. Electrochemical impedance spectroscopy (EIS).

From these measurements, various electrochemical parameters such as specific capacitance, energy density, power density, and cyclic stability of developed electrode materials were determined.

TABLE 3.6
Electrochemical Characteristics of PANI-Based Ternary Composite Electrode Materials Consisting of Two Different Carbon Materials as Other Components

Composition	Electrochemical Performance	Remark	Reference
PANI–graphene nanosheets/CNT/CB	Specific capacitance (450 F g^{-1}) with retention of 84% for 1000 cycles	Study on relationship between electrochemical performance and the morphology of carbon nanoparticles of ternary electrode material	[97]
PANI–graphene/ CNTs	Excellent supercapacitive energy (188.4 W h/kg) and power (200.5 kW/kg) densities	The carbon nanotubes served as binder and the resulted ternary composite electrode material was used for fabricating asymmetric high energy capacitors	[98]
PANI/rGO/nitrogen-doped carbon fiber cloth	Specific capacitance (1145 F g^{-1}) at 1 A g^{-1} in 1 M H_2SO_4 with energy (25.4 W h/kg) and power (92.2 kW/kg) densities with addition of 94% capacity retention over 5000 cycles	Flexible hybrid supercapacitor electrode material consisting of N-doped carbon fabric and graphene-wrapped PANI nanowire arrays has shown outstanding supercapacitive performance	[99]
PANI/rGO/carbon fibers (CFs)	Specific capacitance (257 mF/cm^2) with poor retention (70%) after 1000 cycles. The ternary composite exhibited reasonably favorable energy (19.6 W h/kg) and r power (2.2 kW/kg) densities.	The fabrication process of ternary composite involved electropolymerization of GO onto carbon fibers, reduction of GO into rGO–CFs, and the electropolymerization of PANI on rGO–CFs	[100]
PANI/rGO/vulcan carbon	Specific capacitance (347 F/g^{-1}) at 5 mV s^{-1} with loss of 13% of original capacitance after 2000 cycles	The use of inexpensive vulcan carbon to develop porous nanocomposite displaying maximum specific capacitance at 10% concentration level of vulcan carbon in the composite	[101]
PANI/GO/MWCNTs	Excellent specific capacitance (696 F g^{-1} at 20 mV s^{-1}) with good cycle stability (89% after 3000 cycles)	Synthesis of coherent ternary composites for development of asymmetric supercapacitors	[102]
PANI/graphene/CNTs	Specific capacitance (0.79 F cm^{-2} at 1.5 mA cm^{-2}) with maintenance of 76% original capacity after 3000 cycles	High-performance textile supercapacitor electrode material with 3D CNT/graphene conductive network	[103]
PANI/carbon nanofiber CNT	Specific capacitance (5.6 F m^{-2}) at 2 mA cm^{-2} with 85% retention over 5000 cycles	Use of PANI/CNT-decorated activated carbon fiber felt as flexible supercapacitor electrodes for energy storage systems	[104]
PANI/rGO/graphene carbon spheres	Specific capacitance (446 F g^{-1}) at 5 mV s^{-1} with 88.7% retention over 5000 cycles	An outstanding hierarchical PANI-based electrode material for supercapacitors	[105]
PANI/carbon nanofibers/CNTs	Specific capacitance (794 F g^{-1})	The composite was synthesized using polyacrylonitrile as precursor to which PANI was added through *in situ* polymerization of aniline monomer. The carbon nanotubes increased the diameter of ternary composite nanofibrous network which can be brought in the range 279–433 nm after carbonization.	[106]
PANI/CNT/rGO	Specific capacitance (312 F g^{-1})	The rGO islands present in the ternary composite were well connected through conductive pathways created by CNTs	[107]
PANI/rGO/CNTs	Specific capacitance (359 F g^{-1}) at 1 A g^{-1} with retaining of 80.5% original capacity for 2000 cycles	Three-dimensional PANI-based ternary composite system consisting of reduced graphene and unzipped carbon nanotubes as two other components showed excellent cyclability and electrochemical performance	[108]
PANI/carbon nanotube fiber/CNTs	Specific capacitance (0.067 F cm^{-2}) at 0.5 mA cm^{-2} along with 90% preservation of original capacity for 5000 cycles	PANI-based double core-sheathed structured ternary composite consisting of carbon nanotube fiber and carbon nanotubes as other components for designing flexible solid-state supercapacitors	[109]
PANI/carbon fiber/ CNTs	Specific capacitance (660 F g^{-1}) at 1 A g^{-1} with excellent cycle stability and 90.4% capacitance retention over 1000 cycles	Carbon fiber surface covered with helical carbon nanotubes in combination with PANI as ternary electrode material for use in the form of flexible supercapacitor	[110]
PANI/rGO/ functionalized carbon cloth	Specific capacitance (471 mF cm^{-2}) at 0.5 mA cm^{-2} with 75.5% retention up to 10,000 cycles	Graphene-wrapped PANI nanowire array modified functionalized carbon cloth composite for development of flexible energy storage devices	[111]
rGO @ carbonized PANI/CNTs stainless steel	Areal capacitance (465 mF cm^{-2}) at 1 mA cm^{-2} with maintenance of 84% original capacitance up to 1000 galvanostatic charge–discharge (GCD) cycles	Investigation of electrochemical performances of carbonized PANI/Go/CNT composite as binder-free electrode material for supercapacitors	[112]
Spongy PANI/ graphene on carbon paper	Mass capacitance (55.1 F g^{-1}) at 0.5 A g^{-1} current density	Development of porous PANI conducting fibrous network via *in situ* electropolymerization for use as solid-state supercapacitor	[113]

TABLE 3.7
Electrochemical Characteristics of PANI-Based Ternary Composite Electrode Materials Consisting of Two Different Metallic Components

(I)
PANI/Two Metals in Different Metallic States

Composition	Electrochemical Characteristics	Remark	Reference
PANI/Ag/MnO_2	Specific capacitance (800 F g^{-1}) at 1 mA g^{-1} in 0.5 M $LiClO_4$	Studies on electrochemical supercapacitive performance of PANI/Ag/MnO_2 nanocomposite material for supercapacitor application	[114]
PANI/TiN/MnO_2	Specific capacitance (674 F g^{-1}) at 1 A g^{-1} in 1 M Na_2SO_4	Application of ternary nanostructured composite consisting of polyaniline, manganese dioxide, and titanium nitride nanowire array for designing supercapacitor electrodes	[115]
PANI-grafted MWCNT/ TiO_2 nanotubes/Ti	Specific capacitance (708 F g^{-1}) with retention of 88% original capacitive value for 1000 cycles	Development of PANI grafted MWCNT–TiO_2 nanotubes–Ti ternary composite for supercapacitor applications	[116]
PANI/TiO_2/fluorine-doped SnO_2	Specific capacitance (78 F g^{-1}) under current density of 1 A g^{-1}	Development of PANI-based ternary composite consisting of TiO_2 tetragonal prism array grown on F-doped SnO_2 for fabricating efficient high energy nanocomposite electrodes	[117]

(II)
PANI/Two Metals in Single Bimetallic State

Composition	Electrochemical Characteristics	Remark	Reference
PANI/$CoMoO_4$	Specific capacitance (380 F g^{-1}) at 1 mA g^{-1} with 90% capacitance retention after 1000 cycles	PANI-wrapped 1D heterostructured $CoMoO_4$ electrode material for fabrication of high energy storage devices	[118]
PANI/graphene/ $CoFe_2O_4$	Specific capacitance (1133 F g^{-1}) at 1 mV s^{-1} scane rate for three-electrode system with retention of 96% over 5000 cycles	The ternary nanocomposite was synthesized through dispersion of $CoFe_2O_4$ nanoparticles on graphene sheet by facile hydrothermal method with subsequent coating of PANI via *in situ* polymerization	[119]
PANI/graphene/Mn ferrite	Specific capacitance (454.8 F g^{-1}) under current density of 0.2 A g^{-1}	Ternary PANI/graphene/manganese ferrite nanostructured material with increased electrochemical capacitance performance for supercapacitors	[120]
PANI/graphene/M-type hexaferrite	Specific capacitance (342 F g^{-1}) at 10 mV s^{-1} scane frequency	Graphene decorated hexagonal-shaped M-type ferrite with PANI wrapper ternary material for energy storage systems of excellent electrochemical characteristics	[121]
PANI/graphene/ $MnFe_2O4$	Specific capacitance (241 F g^{-1}) at 0.5 mA cm^{-2}	The ternary mixed-metal oxide composite material was synthesized via combination of $MnFe_2O_4$ with graphene and PANI for use as negative electrode in supercapacitor	[122]
PANI/$NiCo_2O_4$	Specific capacitance (901 F g^{-1}) at 1 A g^{-1} in 1 M H_2SO_4 with capacitive retention of 91% over 3000 cycles	Development of core–shell nanostructured $NiCo_2O_4$@ PANI nanocomposite using $NiCo_2O_4$ porous nanorod arrays as support for use as promising electrode material with desirable energy accumulating activities	[123]
PANI/rGO/$NiMn_2O_4$	Specific capacitive value (757 F/g^{-1}) at 1 A g^{-1}, energy density (70 W h/kg), and capacity retention (93%) for 2000 cycles	This ternary composite material synthesized following hydrothermal-supported annealing procedure with subsequent wrapping of PANI through *in situ* polymerization. Exhibited outstanding electrochemical performance.	[6]
PANI/rGO/$CoFe_2O_4$	Length capacitance (1.41 F m^{-1}) and specific energy (270 × 10^{-8} W h cm^{-1})	The sandwiched covalently grafted $CoFe_2O_4$/rGO/PANI nanocomposite was used to fabricate flexible fiber supercapacitor. The fabricated device lighted LED when connected.	[124]
PANI/2-D host nanosheets from MgFe layered double hydroxide (MgFe – LDH)	Specific capacitance (592.5 F g^{-1}) at current level of 2 A g^{-1} with retention of 87% original capacitance for 500 cycles	Facile fabrication of PANI/LDH composite as cost-effective electrode material for supercapacitors	[125]

3.4 CONCLUSIONS

This chapter has highlighted the latest developments in the research field related to PANI-based ternary composite electrode materials along with their applications in supercapacitors. PANI-based asymmetric supercapacitors have demonstrated reasonable cyclic stability, excellent electrochemical capacitance, a wider range of working potential voltage, and improved energy storage performance. The positive synergistic effect of transition metals (oxides/hydroxides/sulphides), carbonaceous materials, and PANI in ternary composites has contributed to superb electrical conductivity as well as higher power and energy densities. PANI in nano-morphology has been pioneering conducting polymers, paving the way to designing versatile supercapacitors with superb capacitive potentiality in the area of energy accumulation devices.

In recent years, PANI-based ternary composites have made tremendous improvements in the research area of supercapacitors. However, certain issues related to (i) diminution and augmentation of PANI, (ii) poor mutual interfacial linking between carbon and PANI, (iii) optimization of microstructure of composites, and (iv) the development of binder-free nanostructured composites are still to be solved.

In the future, extensive research is therefore expected on the development of binder-free nanostructured composites to maximize electrochemical performance. Furthermore, PANI in conjunction with other conducting polymers is hoped to be tested for its application in supercapacitors. It is interesting to note that stretchable PANI-based composites are gaining immense acceptability as proficient supercapacitor materials for usage in flexible energy accumulative systems.

REFERENCES

[1] S. Chu, Y. Cui and N. Liu, *Nat Mater*, 2016, 16, 16–22.

[2] S. Chu and A. Majumdar, *Nature*, 2012, 488, 294–303.

[3] W. Raza, F. Ali, N. Raza, Y. Luo, K.-H. Kim, J. Yang, S. Kumar, A. Mehmood and E. E. Kwon, *Nano Energy*, 2018, 52, 441–473.

[4] A. Ehsani, A. A. Heidari and H. M. Shiri, *Chem Rec*, 2019, 19, 908–926.

[5] D. Chen, Q. Wang, R. Wang and G. Shen, *J Mater Chem A*, 2015, 3, 10158–10173.

[6] S. Sahoo, S. Zhang and J. -J. Shim, *Electrochem Acta*, 2016, 216, 386–396.

[7] S. Bose, T. Kuila, A. K. Mishra, R. Rajasekar, N. H. Kim and J. H. Lee, *J Mater Chem*, 2012, 22(3), 767–784.

[8] E. Frackowiak, *Phys Chem Chem Phys*, 2007, 9(15), 1774–1785.

[9] H. Pan, J. Li and Y. P. Feng, *Nanoscale Res Lett*, 2010, 5, 654–668.

[10] Y. Chen, M. Han, Y. Tang, J. Bao, S. Li, Y. Lan and Z. Dai, *Chem Commun*, 2015, 51, 12377–12380.

[11] A. Ehsani, E. Kowsari, M. D. Najfi, R. Safari and H. M. Shiri, *J Colloid Interface Sci*, 2017, 500, 315–320.

[12] N. A. Kumar and J. B. Back, *Chem Commun*, 2014, 50, 6298–6308.

[13] P. Asen and S. Shahrokhian, *Int J Hydrogen Energy*, 2017, 42, 21073–21085.

[14] E. Kowsari, A. Ehsani, M. D. Najafi, M. Bigdeloo and J. Colloid. *Interface Sci*, 2018, 512, 346–352.

[15] A. Moyscowicz, A. Sliwaka, E. Miniach and G. Gryglewicz, *Compos Part B – Eng*, 2017, 109, 23–29.

[16] S. Giri, D. Ghosh, A. Mondal and C. K. Das, *Macromol Symp*, 2013, 327(1), 54–63.

[17] P. Li, Y. Yang, E. Shi, Q. Shen, Y. Shang, S. Wu, J. Wei, K. Wang, H. Zhu and Q. Yuan, *ACS Appl Mater Interfaces*, 2014, 6, 5228–5234.

[18] D.-W. Kim, K.-S. Kim and S.-J. Park, *Carbon Lett*, 2014, 127, 53–55.

[19] J. Banerjee, K. Dutta, M. A. Kader and S. K. Nayak, *Polym Adv Technol*, 2019, 1–20. https://doi.org/10.1002/Pat.4624

[20] H. Jiang, P. S. Lee and C. Li, *Energ Environ Sci*, 2013, 6, 41–53.

[21] I. Shown, A. Ganguly, L.-C. Chen and K.-H. Chen, *Energy Sci Eng*, 2015, 3(1), 2–26.

[22] P. Simon and Y. Gogotsi, *Nat Mater*, 2008, 7(11), 845–854.

[23] N. H. N. Azman, M. S. Mamat @ Mat Nazir, L. H. Ngee and Y. Sulaiman, *Int J Energy Res*, 2018, 42, 2104–2116. https://doi.org/10.1002/er.4001

[24] S. Y. Lim, W. Shen and Z. Gao, *Chem Soc Rev*, 2015, 44, 362–381.

[25] J. Li, Y. Sun, D. Li, H. Yang, X. Zhang and B. Lin, *J Alloys Compd*, 2017, 708, 787–795.

[26] J. Li, D. Xiao, Y. Ren, H. Liu, Z. Chen and J. Xiao, *Electrochem Acta*, 2019, 300, 193–201.

[27] D. H. Deng, H. Pang, J. M. Du, J. W. Deng, S. J. Li, J. Chem and J. S. Zhang, *Cryst Res Technol*, 2012, 47, 1032–1038.

[28] S.-L. Kuo and N.-L. Wu, *Electrochem Solid State Lett*, 2005, 8, A495–A499.

[29] V. H. Nguyen and J.-J. Shin, *Mater Chem Phys*, 2016, 176, 6–11.

[30] M. Zhang, S. Guo, L. Zheng, G. Zhang, Z. Hao, L. Kang and Z.-H. Lu, *Electrochim Acta*, 2013, 87, 546–553.

[31] W. Wang, W. Lei, T. Yao, X. Xia, W. Huang, Q. Hao and X. Wang, *Electrochim Acta*, 2013, 108, 118–126.

[32] Q. Wu, M. Chen, S. Wang, X. Zhang, L. Huan and G. Diao, *Chem Engg J*, 2016, 304, 29–38.

[33] J.-J. Shim, *Synth Met*, 2015, 207, 110–115.

[34] R. Yuksel and H. E. Unalan, *Int J Energy Res*, 2015, 39, 2042–2052.

[35] R. Ramya, R. Sivasubramanian and M. V. Sangaranarayanan, *Electrochim Acta*, 2012, 101, 109–129.

[36] G. Wang, L. Zhang and J. Zhang, *Chem Soc Rev*, 2012, 4(2), 797–828

[37] O. Gorduk, S. Gorduk, M. Gencten, M. Sahin and Y. Sahin, *Int J Energy Res*, 2020, 44, 9093–9111.

[38] S. K. Simotwo and V. Kalra, *Curr Optinion Chem Engg*, 2016, 13, 150–160.

[39] K. Dutta, P. Kumar, S. Das and P. P. Kundu, *Polym Rev*, 2014, 54(1), 1–32.

[40] Y. Li, X. Zhao, P. Yu and Q. Zhang, *Langmuir*, 2013, 29, 493–500.

[41] J. Wang, H. Xian, T. Peng, H. Sun and F. Zheng, *RSC Adv*, 2015, 5(18), 13607–13612.

[42] M. Yu, Y. Ma, J. Liu and S. Li, *Carbon*, 2015, 87, 98–105.,

[43] K. Zhou, Y. He, Q. Xu, et al., *ACS Nano*, 2018, 12(6), 5888–5894.

[44] B. C. Kim, J. S. Kwon, J. M. Ko, J. H. Park, C. O. Too and G. G. Wallace, *Synth Met*, 2010, 160, 94–98.

[45] X. Wang, D. Wu, X. Song, W. Du, X. Zhao and D. Zhang, *Molecules*, 2019, 24, 2263. https://doi.org/10.3390/molecules24122263

[46] K. Li, X. Liu, S. Chen, W. Pan and J. Zhang, *J Energy Chem*, 2019, 32, 166–173.
[47] K.-U. Lee, J. Y. Byun, H. J. Shin and S. H. Kim, *J Alloys Compd*, 2019, 779, 74–80.
[48] J. Yu, W. Lu, S. Pei, et al., *ACS Nano*, 2016, 10(5), 5204–5211.
[49] D. Xie, Q. Jiang, G. Fu, Y. Deing, X. Kang, W. Cao and Y. Zhao, *Rare Metals*, 2011, 30(S1), 94–97.
[50] X. Zhang, X. Li, M. Zhu, X. Li, Z. Zhen, Y. He, K. Wang, J. Wei, F. Kang and H. Zhu, *Nanoscale*, 2015, 7(16), 7318–7322.
[51] N. Hu, L. Zhang, Y. Yang, et al., *Sci Rep*, 2016, 6(1), 19777.
[52] Q. Wang, H. Wang, P. Du, J. Liu, D. Liu and P. Liu, *Electrochim Acta*, 2019, 294, 312–324.
[53] W. Zou, W. Wang, B. He, M. Sun and Y. Yin, *J Power Sources*, 2010, 195, 7489–7493.
[54] Q. Li, J. Liu, J. Zou, A. Chunder, Y. Chen and L. Zhai, *J Power Sources*, 2011, 196, 565–572.
[55] D. Zhao, X. Guo, Y. Gao and F. Gao, *ACS Appl Mater Interfaces*, 2012, 4(10), 5583–5589.
[56] Y. Li, Y. Fang, H. Liu, X. Wu and Y. Lu, *Nanoscale*, 2012, 4(9), 2867–2869.
[57] H. Su, T. Wang and S. Zhang, J. Song, C. Mao, H. Niu, B. Jin, J. Wu and Y. Tian, *Solid State Sci*, 2012, 14(6), 677–681.
[58] X. Xia, Q. Hao, W. Lei, W. Wang, H. Wang and X. Wang, *J Mater Chem*, 2012, 22, 8314–8320.
[59] X. Xia, Q. Hao, W. Lei, W. Wang, D. Sun and X. Wang, *J Mater Chem*, 2012, 22, 16844–16850.
[60] Y. Yan, Q. Cheng, V. Pavlinek, P. Saha and C. Li, *Electrochim Acta*, 2012, 71, 27–32.
[61] D. Ghosh, S. Giri, A. Mandal and C. K. Das, *RSC Adv*, 2013, 3(29), 11676–11685.
[62] G. Wang, Q. Tang, H. Bao, X. Li and G. Wang, *J Power Sources*, 2013, 241, 231–238.
[63] S. Li, D. Wu and C. Cheng, J. Wang, F. Zhang, Y. Su and X. Fang. *Angew Chem Int Ed*, 2013, 52(46), 12105–12109.
[64] S. Dhibar, P. Bhattacharya, G. Hatui, S. Sahoo, C. K. Das and ACS Sustain. *Chem Eng*, 2014, 2(5), 1114–1127.
[65] S. Giri, D. Ghosh and C. K. Das, *Adv Funct Mater*, 2014, 24(9), 1312–1324.
[66] G. Han, Y. Liu, L. Zhang, E. Kan, S. Zhang, J. Tang and W. Tang, *Sci Rep*, 2014, 4(1–7), 4824.
[67] L. Yu, M. Gan, L. Ma, H. Huang, H. Hu, Y. Li, Y. Tu, C. Ge, F. Yang and J. Yan, *Synth Met*, 2014, 198, 167–174.
[68] S. Grover, S. Shekhar, R. K. Sharma and G. Singh, *Electrochim Acta*, 2014, 116, 137–145.
[69] B. Mu, W. Zhang, S. Shao and A. Wang, *Phys Chem Chem Phys*, 2014, 16, 7872–7880.
[70] A. Jayakumar, Y.-J. Yoon, R. Wang and J.-M. Lee, *RSC Adv*, 2015, 5(114), 94388–94396.
[71] M. Usman, L. Pan, M. Asif and Z. Mahmood, *J Mater Res*, 2015, 30, 3192–3200.
[72] A. K. Das, S. K. Karan and B. Khatua, *Electrochem Acta*, 2015, 180, 1–15.
[73] C. Sha, B. Lu and H. Mao, J. Cheng, X. Pan, J. Lu and Z. Ye, *Carbon*, 2016, 99, 26–34.
[74] W. Liu, S. Wang, Q. Wu, L. Huan, X. Zhang, C. Yao and M. Chen, *Chem Eng Sci*, 2016, 156, 178–185.
[75] C. Pan, H. Gu and L. Dong, *J Power Sources*, 2016, 303, 175–181.
[76] S. Zhu, M. Wu, M.-H. Ge, H. Zhang, S.-K. Li and C.-H. Li, *J Power Sources*, 2016, 306, 593–601.
[77] X. Wu, Q. Wang, W. Zhang, Y. Wang and W. Chen, *Electrochim Acta*, 2016, 211, 1066–1075.
[78] H. Lin, Q. Huang, J. Wang, J. Jiang, F. Liu, Y. Chin, C. Wang, D. Lu and S. Han, *Electrochim Acta*, 2016, 191, 444–451.
[79] Q. Wu, M. Chen, S. Wang, X. Zhang, L. Huan and G. Diao, *Chem Eng J*, 2016, 304, 29–38.
[80] A. K. Sharma, G. Chaudhary, P. Bhardwaj, I. Kaushal and S. Duhan, *Curr Anal Chem*, 2017, 13, 277–284.
[81] T. Wu, C. Wang, Y. Mo, X. Wang, J. Fan, Q. Xu and Y. Min, *RSC Adv*, 2017, 7, 33591–33599.
[82] S. Mondal, U. Rana and S. Malik, *J Phys Chem C*, 2017, 121, 7573–7583.
[83] M. M. Mezgebe, Z. Yan, G. Wei, S. Gaung and H. Xu, *Mater Today Energy*, 2017, 5, 164–172.
[84] E. Elanthamilan, A. Sathiyan, S. Rajkumar, E. J. Sheryl and J. P. Merlin, *Sustain Energy and Fuels*, 2018, 2(4), 811–819.
[85] L. Wang, Y. Ouyang, X. Jiao, X. Xia, W. Lei and Q. Hao, *Chem Eng J*, 2018, 334, 1–9.
[86] K. Y. Yasoda, A. A. Mikhaylov A. G. Medvedev, et al., *J Energy Storage*, 2019, 22, 188–193.
[87] S. Li, Y. Chang, G. Han, H. Song, Y. Chang and Y. Xiao, *J Phys Chem Solid*, 2019, 130, 100–110.
[88] N. Song, Y. Wu, W. Wang, D. Xiao, H. Tan and Y. Zhao, *Mater Res Bull*, 2019, 111, 267–276.
[89] C. Huang, C. Hao, W. Zheng, S. Zhou, L. Yang, X. Wang, C. Jiang and L. Zhu, *Appl Surf Sci*, 2020, 505, 144589.
[90] M. Dirican, M. Yanilmaz, A. M. Asiri and X. Zhang, *J Electroanal Chem*, 2020, 861, 113995. https://doi.org/10.1016/j.jelechem.2020.113995.
[91] K. Y. Yasoda, S. Kumar, M. S. Kumar, K. Ghose and S. K. Batabyal, *Materials Today Chemistry*, 2021, 19, 100394 (10 pages).
[92] X. Shi, Y. Hu, M. Li, et al., *Cellulose*, 2014, 21(4), 2337–2347.
[93] S. K. Simotwo, C. Delre and V. Kalra, *ACS Appl Mater Interfaces*, 2016, 8(33), 21261–21269.
[94] K. Wang, L. Li, Y. Liu, C. Zhang and T. Liu, *Adv Mater Interface*, 2016, 3(19), 1600665 (9 pages).
[95] J. J. Alcaraz-Espinoza and H. P. de Oliveira, *Electrochim Acta*, 2018, 274, 200–207.
[96] H. Luo, J. Dong, Y. Zhang, G. Li, R. Guo, G. Zuo, M. Ye, Z. Wang, Z. Yang and Y. Wan, *Chem Eng J*, 2018, 334, 1148–1158.
[97] G.-M. Zhou, D.-W. Wang, F. Li, L.-L. Zhang, Z. Weng and H.-M. Cheng, *New Carbon Mater*, 2011, 26, 180–186.
[98] Q. Cheng, J. Tang, N. Shinya and L. C. Qin, *J Power Sources*, 2013, 241, 423–428.
[99] P. Yu, Y. Li, X. Zhao, L. Wu and Q. Zhang, *Langmuir*, 2014, 30, 5306–5313.
[100] X. Jiang, Y. Cao, P. Li, J. Wei, K. Wang, D. Wu and H. Zhu, *Mater Lett*, 2015, 140, 43–47.
[101] M. Hwang, J. Oh, J. Kang, K.-D. Seong and Y. Piao, *Electrochim Acta*, 2016, 221, 23–30.
[102] M. Hao, Y. Chen, W. Xiong, L. Zhang, L. Wu, Y. Fu, T. Mei, J. Wang, J. Li and X. Wang, *Electrochim Acta*, 2016, 191, 165–172.
[103] L.-N. Jin, F. Shao, C. Jin, J.-N. Zhang, P. Liu, M.-X. Guo and S.-W. Bian, *Electrochim Acta*, 2017, 249, 387–394.
[104] L. Yang and C. Chao, *J Mater Sci*, 2017, 52, 12348–12357.
[105] L. Liu, Y. Wang, Q. Meng and B. Cao, *J Mater Sci*, 2017, 52, 7969–7983.
[106] F. O. Agyemang, G. M. Tomboc, S. Kwofie and H. Kim, *Electrochim Acta*, 2018, 259, 1110–1119.
[107] M. S. Kumar, K. Y. Yasoda, S. K. Batabyal and N. K. Kothurkar, *Mater Res Express*, 2018, 5(4), 045505.
[108] Y. Huang, J. Zhou, N. Gao, Z. Yin, H. Zhou, X. Yang and Y. Kuang, *Electrochim Acta*, 2018, 269, 649–656.

[109] J.-H. Liu, X.-Y. Xu, W. Lu, X. Xiong, X. Ouyang, C. Zhao, F. Wang, S.-Y. Qin, J.-L. Hong, J.-N. Tang and D.-Z. Chen, *Electrochim Acta*, 2018, 283, 366–373.
[110] H. Luo, H. Lu and J. Qiu, *J Electroanal Chem*, 2018, 828, 24–32.
[111] P. Du, Y. Dong, H. Kang, X. Yang, Q. Wang, J. Niu, S. Wang and P. Liu, *ACS Sustain Chem Eng*, 2018, 6, 14723–14733.
[112] B. Vedhanarayanan, T.-H. Huang and T.-W. Lin, *Inorg Chim Acta*, 2019, 489, 217–223.
[113] Y. He, X. Wang, H. Huang, P. Zhang, B. Chen and Z. Guo, *Appl Surf Sci*, 2019, 469, 446–455.
[114] J. Kim, H. Ju, A. L. Inamdar, Y. Jo, J. Han, H. Kim and H. Im, *Energy*, 2014, 70, 473–477.
[115] C. Xia, Y. Xie, H. Du and W. Wang, *J Nanopart Res*, 2015, 17(1–12), 30.
[116] M. Faraji, P. N. Moghadam and R. Hasanzadeh, *Chem Eng J*, 2016, 304, 841–851.
[117] S. Chen, B. Liu and X. Zhang, F. Chen, H. Shi, C. Hu and J. Chen, *Electrochim Acta*, 2019, 300, 373–379.
[118] M. Mandal, D. Ghosh, S. Giri, I. Shakir and C. K. Das, *RSC Adv*, 2014, 4(58), 30832–30839.
[119] P. Xiong, H. J. Huang and X. Wang, *J Power Sources*, 2014, 245, 937–946.
[120] P. Xiong, C. Hu, Y. Fan, W. Zhang, J. Zhu and X. Wang, *J Power Sources*, 2014, 266, 384–392.
[121] P. Bhattacharya, S. Dhibar, G. Hatui, A. Mandal, T. Das and C. K. Das, *RSC Adv*, 2014, 4, 17039–17053.
[122] K. V. Sankar and R. K. Selvan, *J Power Sources*, 2015, 275, 399–407.
[123] N. Jabeen, Q. Xia, M. Yang and H. Xia, *ACS Appl Mater Interfaces*, 2016, 8(9), 6093–6100.
[124] K. V. Sankar and R. K. Selvan, *Electrochim Acta*, 2016, 213, 469–481.
[125] X. Cao, H.-Y. Zeng, S. Xu, J. Yuan, J. Han and G.-F. Xiao, *Appl Clay Sci*, 2019, 168, 175–183.
[126] D. Majumdar, *ChemElectroChem*. https://doi.org/10.1002/celc.202001371

4 Self-Healing Gel Electrolytes for Flexible Supercapacitors

A. M. Vinu Mohan, Vinoth Rajendran, and Mathiyarasu Jayaraman
CSIR-Central Electrochemical Research Institute (CECRI), Karaikudi, India

CONTENTS

4.1 INTRODUCTION

Emerging flexible electronics have attracted great interest in the development of flexible and bendable electronic displays [1], wearable sensors [2–5], artificial electronic skin [6], bendable electronic gadgets [7], and so on [8]. These flexible devices mandate the next generation energy storage or harvesting devices [9–11] as a power source to realize light, portable, and shape-conformable platforms with promising mechanical properties [12–14]. However, majority of the available energy storage devices like batteries and supercapacitors are rigid, bulky, and not suitable to be integrated with flexible systems [15]. The burgeoning area of wearable electronics necessitates soft, flexible, and stretchable energy storage devices which can conformally be mounted on the irregular, curvilinear surfaces of the epidermis [16–18]. Personalized healthcare monitoring devices [19, 20] depend on skin-interfaced flexible and stretchable patches with appreciable mechanical resiliency [21]. Such devices mandate conformal contact with the human body for continuously analyzing relevant biomarkers from body fluids like sweat [22, 23], saliva [24, 25], tears [26], and interstitial fluid [27–29]. Also, the seamless attachment is crucial for precise sample collection and transport to the transducer region. Integrating soft, secondary skin-like supercapacitor can function as a flexible power source for these biosensing systems [30].

Supercapacitors, also known as electrochemical capacitors or ultra-capacitors, are one of the major energy storage devices owing to their fast charge–discharge rate, long cycle life, high volumetric capacitance, and power density [31, 32]. The conventional sandwich type solid-state supercapacitor consists of typically electrodes (anode and cathode) and a polymeric gel (which functions as both electrolyte and separator) [33, 34]. Based on the charge storage mechanism, supercapacitors can be categorized into three types such as double layer capacitors, pseudo-capacitors, and asymmetric capacitors [35]. In double layer capacitors, the electric charge accumulates at the electrode–electrolyte interface as an electrical double layer by a non-Faradaic process [36]. In the case of pseudo-capacitors, capacitance originates from the interfacial electron transfer during the Faradaic redox reaction [37]. Both of these charge storage mechanisms apply together for developing asymmetric supercapacitors [38].

The incorporation of stretchable electrode materials realizes absolute flexible devices, which can mitigate the developed strain during intense physical activities, muscle movements, body motion, or other external stresses. However, the excessive strain imposed on these stretchable devices undergoes microscopic level cracks and minute damages [39]. Hence, in order to mimic the skin-like properties, the wearable devices should have appreciable self-healing properties. Taking a cue from nature, some damage and biological functions are intrinsically curable [40]. The use of artificial self-healing electrodes and electrolytes enables the fabrication of self-healing energy storage devices. Autonomous healing of the damage by the device itself, without involving assistance from an external trigger, instantaneously restores the electrical and ionic conductivities. The self-healing process facilitates long life, restores original device activity, and maintains stable performances. It is important to introduce smart self-healing hydrogel electrolytes into flexible supercapacitors, which would enable the restoration of physical damage or cracks spontaneously,

DOI: 10.1201/9781003169727-4

and would maintain its stable performances [41–43]. This chapter reveals the promising developments associated with the synthesis of novel supramolecular self-healing hydrogels and their advantages as an ionic conductor and separator, when integrated with flexible supercapacitors.

4.2 AN OVERVIEW OF SELF-HEALING GEL ELECTROLYTES

Researchers are focused on developing smart self-healing gels which are responsive towards various external stimuli such as magnetic fields [44], temperature [45], electrical fields [46], and light [47]. Hydrogels are promising material for electronic skin [48], wearable sensors (interfacial layer) [49], biocompatible adhesive layers [50], tissue engineering [51], water purification [52], biomedical engineering [53], drug delivery [54], and regenerative medicine [55]. Hydrogels are cross-linked 3D networks of polymeric chains which swell in the presence of water instead of dissolving. Hydrogels with superior ionic conductivity are popular semisolid electrolytes, exploited in solid state energy storage devices like batteries and supercapacitors. Self-healing ionic gels utilize polyelectrolytes and reversible cross-linking agents. Polyelectrolytes possess good ionic conductivity (10^{-4} to 10^{-3} S cm^{-1}), tunable mechanical properties, and appreciable chemical compatibility while integrating with electrodes, which makes them suitable for energy storage applications [56]. The preferable features required for self-healing electrolyte gels include high molecular weight, low glass transition temperature, rapid segmental movement along the polymer sequence, and high degradation temperature [57, 58]. The self-healing process involves the recovery of microscopic-level internal properties with a reorientation of molecular structure and geometry that leads to re-establishing of its viscoelastic properties [59].

Based on the associated mechanism, self-healing substances can be classified into extrinsic and intrinsic materials. Figure 4.1 shows schematic representation of the self-healing process by (a) extrinsic and (b) intrinsic materials. The extrinsic materials provide self-healing by releasing the preloaded healing agents into the bulk of the gel network and restoring the bonding interactions [60]. Microcapsule-based self-healing systems are widely reported which encapsulates healing agents like specific solvents that facilitate repolymerization at the damaged spot and repair the cracks [61].

The major drawback associated with this method is its incapability to restore multiple damages at the same site. The releasing of healing agents augments the polymerization reaction only one or two times and the successive healing processes are prevented due to the lack of active microcapsules with healing agents [62]. These challenges can be addressed by establishing typical covalent or non-covalent interactions within the polymeric network that facilitate multiple healing at the same location. The dynamic healing process involves the reversible formation of chemical bonds that drive the restoration of ionic conductivity in a real-time fashion [63]. Depending on the driving force, the healing chemistry can be classified into physical and chemical processes (Figure 4.2). The physical healing route deals with the creation of various non-covalent interactions such as hydrogen bonding [64], metal coordination [65], π–π stacking [66], hydrophobic interaction [67], and

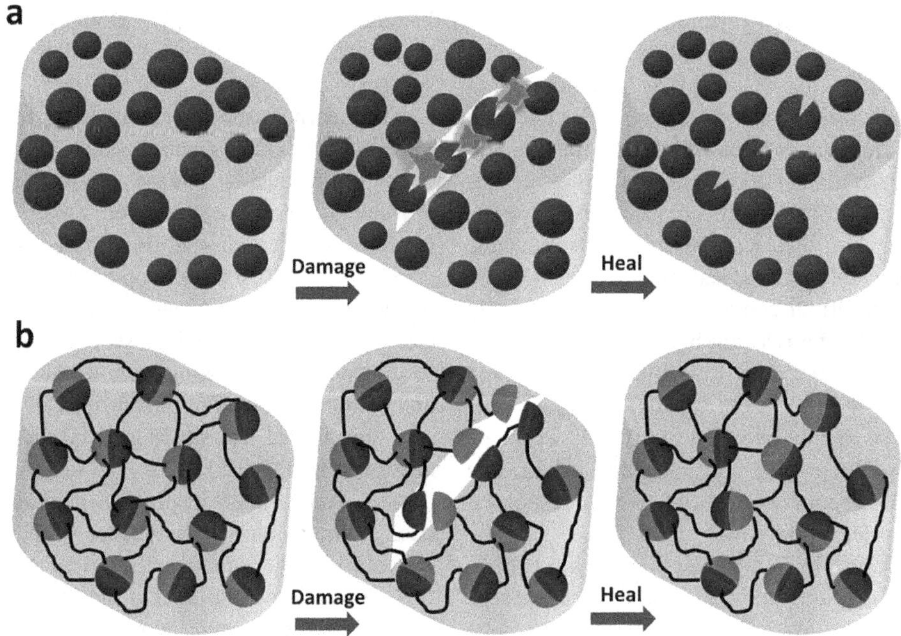

FIGURE 4.1 Schematic representation of the self-healing process by (a) the expulsion of healing agents, pre-encapsulated within the active materials which can restore the polymer network, and (b) the restoration of reversible chemical bonds through supramolecular interactions.

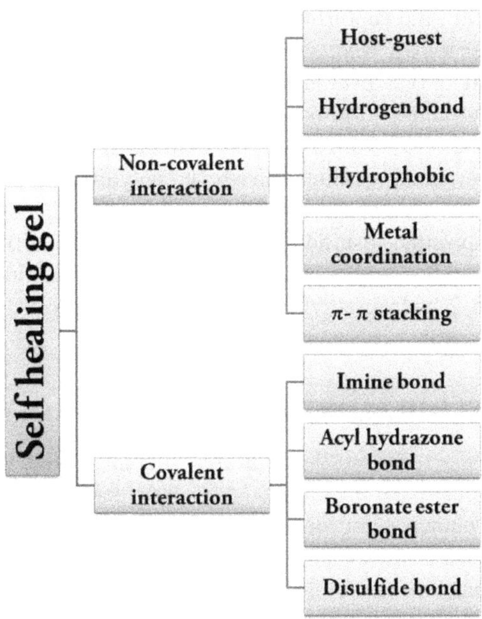

FIGURE 4.2 Illustration of different procedures employed for preparing self-healing gel electrolytes.

host–guest bonding [68], while the chemical healing depends on the restoration of reversible chemical bonds like acylhydrazone bonds [69], imine bonds [70], boronate ester bonds [63], and disulfide bonds [71]. In addition, low molecular weight ionic gels can be judiciously prepared by self-assembling tiny organic molecules with a non-covalent force of attraction [72].

The qualitative assessment of self-healing processes can be achieved by different methods such as slicing the gel into separate parts, punching a hole through it, or scratching the surfaces and then allowing them to join together. But the quantitative analysis of self-healing ability requires more precise methods. Stress–strain measurements are generally adapted for tough elastic gels, and the tensile strength value provides progress on the self-healing process. In contrast, the soft gels are evaluated by compressive stress–strain trials. The self-healing efficiency (ε_h) can be determined from the ratio between the elongation distance of cured (λ_b) and pristine gel ($\lambda_{b,0}$) [58]:

$$\varepsilon_h = \lambda_b/\lambda_{b,0} \tag{4.1}$$

4.3 SYNTHESIS OF SELF-HEALING GELS BASED ON NON-COVALENT INTERACTIONS

Hydrogen bonding interaction is one of the reversible supramolecular interactions which form small molecules based dynamic self-healing gels with appreciable visco-elastic or thixotropic behavior [73, 74]. These interactions yield intrinsically self-healing gels with good flexibility and water uptake properties [75]. A polyacrylamide–chitosan-based supramolecular tough hydrogel was developed with stretchable and self-healing properties [76]. The hybrid hydrogel was prepared by forming interchain hydrogen bonds as cross-linking networks between the short chain chitosan and polyacrylamide via free radical polymerization. The gel could withstand multiple compression–relaxation strain cycles and is stretchable up to 30 times its initial length. In another attempt, Chirila and coworkers prepared a self-healing supramolecular gel of a ureido-pyrimidinone molecule, (2-(*N′*-methacryloyloxyethylureido)-6-(1-adamantyl)-4[1*H*]-pyrimidinone), further copolymerized with *N*,*N*-dimethylacrylamide in a medium of lithium chloride in *N*,*N*-dimethylacetamide [77]. The assembly of the copolymer is based on reversible quadruple hydrogen bonds between the side chains of the monomers. The self-healing capacity is found to be ideal for developing injectable substitutes essential for the eye's vitreous humor.

The π-electron-rich polymeric gels also provide self-healing activity by π–π stacked supramolecular interactions. The π-electron-rich bis-perylene-terminated polyether, blended with electron deficient chain-folding polydiimide, forms elastomeric gels with rapid self-healing ability [78]. Rheometric analysis revealed that the blend possesses excellent retention of tensile modulus (10 MPa) over three crack–heal cycles, and good recovery of the modulus of toughness. Exploiting metal–ligand coordination is yet another building tool for constructing multiple bonded, cross-linked, self-responsive supramolecular polymers with autonomous self-healing properties [79]. The coordination chemistry of Zn^{2+} with adenosine monophosphate was utilized for synthesizing mechanically induced self-healing gels [80]. Such a low molecular weight gel undergoes large-scale gelation, triggered simply by centrifugation; the broken pieces can rejoin into a large monolithic gel.

Dynamic micelle formation chemistry can be used for preparing tough hydrogels with self-healing capabilities [81]. The incorporation of hydrophobic functionality over long alkyl side chains of hydrophilic polymers generates an energy dissipation mechanism. These hydrophobic associations establish reversible cross-linkages within the polymer matrix and facilitate elasticity, water uptake properties, and self-healing ability. The copolymerizatin of hydrophobic monomers such as stearyl methacrylate and dococyl acrylate with the hydrophilic monomer, acrylamide, in a surfactant solution of sodium dodecyl sulfate depicts one such example [67].

The reversible nature of host–guest interaction is also ideal for preparing tough, maintenance-free polymeric gels with a self-crack-healing mechanism. Cyclodextrins are a class of cyclic oligosaccharides, having a macrocyclic ring of glucose subunits connected by α-1,4-glycosidic bonds [82]. The ability of cyclodextrins to form water-soluble inclusion complexes with many lipophilic sparingly soluble compounds facilitates the formation of an ideal host–guest complex. A water-soluble polymer like poly(acrylamide) can be modified with β-cyclodextrin, and adamantane at the side chains provides the visoelastic gels with scratch curable properties [83].

4.4 SYNTHESIS OF SELF-HEALING GELS BASED ON COVALENT INTERACTIONS

Dynamic covalent chemistry offers three-dimensional cross-linked networks in polymer gel with self-healing properties, owing to its reversible, evenly distributed covalent bonds [84, 85]. Acylhydrazone bonds are predominantly covalent in nature and the polymer gel network is steady under ambient conditions. By tuning the optimum pH of the system, the gel can be molded to possess reversible sol–gel phase transitions and self-healing behavior by regenerating acylhydrazone bonds. Deng et al. reported the functionalization of acylhydrazine moieties at both ends of a poly(ethylene oxide) using aldehyde groups in tris[(4-formylphenoxy)methyl]ethane. Such a cross-linked, acylhydrazone-based gel revealed an outstanding autonomous self-healing property, which is independent of any external contributions [84].

The imine bond (Schiff base) incorporation between the aldehyde groups and the amino functionality are characteristic dynamic covalent bonds, which could be utilized to prepare biocompatible, self-healing, drug delivery systems [70]. An imine bond is a strong covalent linkage having a bond dissociation energy of 147 kcal mol^{-1}, which can facilitate rapid degenerative bond exchange devoid of any considerable side reactions. Chao and coworkers prepared such a covalent polymeric gel consisting of imine cross-linkages utilizing the condensation polymerization of poly(ethylene glycol) bis(3-aminopropyl) along with 1,3,5-triformylbenzene; the polymer networks show malleability and appreciable self-healing properties. The imine–imine bond exchange rate was analyzed by ^1H NMR spectroscopy and the results corroborated that the dynamic imine bond exchange occurs faster in polar solvents than in a non-polar medium [86].

Boronic acids can readily establish dynamic covalent bonds while reacting with diols to yield boronic or boronate esters, ideally in basic aqueous solutions or in non-aqueous organic solvents. Boronate esters are widely utilized for preparing self-healing gel networks by forming fresh boronate ester bonds at the edge of damage [87]. Chen et al. reported a versatile approach to establishing reversible boronic ester linkages with styrene-butadiene rubber (SBR). A boronic ester, 2,2′-(1,4-phenylene)-bis[4-mercaptan-1, 3,2-dioxaborolane], having two thiol groups, was synthesized and mechanically mixed with SBR [88]. The thermally activated thiol-ene "click" chemistry between thiol groups of the ester and pendent vinyl groups of SBR chains facilitates the formation of boronic ester bonds. These boronic esters cross-linked with polymeric gels are capable of self-healing the damage by reversible dynamic covalent interactions.

The mechanical cleavage of disulfide bonds is easier because their bond strength is lower than carbon–carbon bonds. The underlying chemistry behind the self-healing ability of disulfide bonds is mainly a [2+2] metathesis and a [2+1] radical-mediated mechanism. The metathesis process deals with the rupturing and formation of disulfide bonds simultaneously, whereas the latter mechanism involves the cracking of one disulfide bond which results in the creation of sulfur-centered radicals that further attack other disulfide bonds [89]. Wan and Chen showcased a novel method to prepare a self-healing water-based polyurethane gel from poly(e-caprolactone) glycol, isophorone diisocyanate, 2-hydroxyethyl disulfide, and 2-bis(hydroxymethyl) propionic acid. The chain exchange reactions of disulfide linkages in 2-hydroxyethyl disulfide activate the self-healing process [90].

4.5 SELF-HEALING IONIC GEL ELECTROLYTES

Liquid electrolytes offer high ionic conductivities (10^{-3} to 10^{-2} S cm^{-1}) and apparent contact with electrode surfaces, which makes them ideal for energy storage applications. But the organic solvent-based poisonous electrolytes possess some threats of leakage and even combustion. The solid polymer electrolytes are hazard-free, promising materials for tackling the safety problems, owing to the lack of solvents, and have attracted considerable attention as ionic conductors for various applications. But the poor ionic conductivities (10^{-8} to 10^{-5} S cm^{-1}) of such solid electrolytes limit their energy storage applications, and eventually affect their cyclic stability. Recently researchers have paid much attention to gel polymer electrolytes, owing to their good ionic conductivities (10^{-4} to 10^{-3} S cm^{-1}), mechanical strength, workability, flexibility, elasticity, and so on. These ionic conductors can coordinate with the transporting of charged species, suitable for a variety of applications such as anion and proton exchange membrane-based fuel cells, electronic skin, electrolytes in dye sensitized solar cells, batteries, and supercapacitors.

Smart electrochemical supercapacitors for wearable application necessitate flexible and stretchable electrodes, separators, and gel electrolytes. The merging of self-healing properties facilitates stable device performance by spontaneously restoring its electrical conductivity while undergoing cracking or physical cutting-off. Thus, self-healing ionically conductive gel electrolytes are considered as an efficient component for developing wearable supercapacitors. Poly(vinyl alcohol) (PVA) based hydrogels are widely exploited for supercapacitor applications as an ionic conductor. In addition, the self-repairing property PVA is useful for developing stretchable, self-healable, energy storage devices. The shortcomings of the PVA gels are lack of interfacial adhesion and recyclability, which are prerequisite properties for fabricating flexible energy storage devices. Hu et al. reported a phytic acid (PA) based PVA gel which can overcome these issues by rendering distinctive electrode contact, and can be recycled by hot pressing [91]. PA is a natural product having six phosphate groups and its typical orientation offers multiple hydrogen bonding interactions with PVA. The gel can be stretched up to 1000%, and upon damage it undergoes self-repairing at room temperature without external support. Also, the PVA/PA gel was used as an electrolyte for a carbon cloth supported poly(aniline) supercapacitor; the repeated bending tests revealed robust charge storage performance and a long cycle life with 92% capacitance retention.

The fabrication of an all-in-one flexible and stretchable supercapacitor with intrinsic self-healing features is crucial for powering wearable devices. The sandwich-type multi-layered supercapacitor with additional self-healing layers limits its performance due to the delamination or detachment of layers, which adversely affects the interfacial contact resistance and decreases the specific capacitance. Current research is focused on simplifying the complex integration procedure for developing sandwich-like self-healable supercapacitors. Guo et al. configured such all-in-one device by depositing nanocomposite electrode materials onto both faces of an ionically conductive self-healing hydrogel separator, via an *in situ* polymerization method [92]. The dynamic hydrogen bonding based PVA gel provides self-healing ability by restoring the cracked zones through non-covalent interactions. The PVA–H_2SO_4 hydrogel possessed appreciable ionic conductivity (136.4 mS cm^{-1}) and consistent mechanical resiliency up to 380%. The assembled supercapacitor displayed excellent flexibility, cycle stability, and an autonomous self-healing ability with 80% of capacitance retention. Furthermore, the seamless integration provides enhanced energy storage performance when compared with the capacitive behavior of pure electrode materials.

The ionic conductivity in the polymer electrolytes can be augmented by merging suitable materials with the polymer matrix which can assist the aggregation of polymer sequences in gels [93]. The excellent ionic conductivity (2.1 ± 0.2 S cm^{-1}) of graphene oxide (GO) is being utilized for developing high performance gel electrolytes for various energy related applications [94, 95]. The exfoliated GO layers have a high surface area and more oxygen-based functional groups, which enhance the ionic conductivity and the van der Waals force of attractions with polyelectrolytes. Peng et al. reported a flexible supercapacitor having a self-healable diol-borate ester bond based hydrogel electrolyte via doping with GO [96]. The cross-linked PVA gel exploits both the dynamic ester based self-healing ability and the intermolecular hydrogen bonding of GO with the polymer. The self-healing gel can repair the damage and restore its original arrangement within 5 min. Furthermore, the presence of GO enhances the supercapacitor's performance; the specific capacitance can be restored for seven damage–heal cycles, without the assistance of any external stimulation.

Self-healing ability can be augmented by developing hybrid double network ionic gels via a dual physical cross-linking mechanism (DPCL). The DPCL platform combines both hydrogen bonding and ionic interactions, and shows an enviable energy dissipating mechanism. Such mechanically flexible cross-links can self-heal by themselves without outside stimulus. Lin et al. synthesized a ferric ion enhancing DPCL polyelectrolyte by copolymerizing acrylic acid, a hydrophilic monomer and stearyl methacrylate, a hydrophobic monomer in the presence of ferric ions, and cetyltrimethylammonium bromide, a cationic surfactant [97]. The poly(acrylic) acid (PAA) provided non-sacrificial intermolecular hydrogen bonding and Fe^{3+} facilitated ionic and covalent cross-linkages with carboxylic acid groups. The presence of a low concentration of H_2SO_4 provided movable protons which facilitated electrode capacitance, ionic conductivity (>30 mS cm^{-1}), and exceptional self-healing properties. The DPCL gel possesses appreciable ionic conductivity (>30 mS cm^{-1}) and can be stretched above 2400%. The polyelectrolyte interfaced PANI/carbon fiber supercapacitors exhibit good capacitive behavior and a self-healing ability with 86% capacitance retention even after seven autonomous restoring cycles. Guo et al. constructed a similar ferric ion cross-linked PAA hydrogel (KCl–Fe^{3+}/PAA) with a recycling and self-healing ability [41]. The hydrogel electrolytes exploited both the ionic bonding between the Fe^{3+} and carboxylic groups and the dynamic intra- and intermolecular hydrogen bonding interactions. The gel shows good conductivity (0.09 S cm^{-1}) and exceptional stretchability (>700%). The hydrogel was integrated with graphene foam supported polypyrrole electrodes, and the flexible supercapacitor demonstrated excellent specific capacitance (87.4 F g^{-1}) at a current density of 0.5 A g^{-1}. The full device performance was stable for 5000 charging–discharging cycles with an 89% capacitance retention.

The autonomous self-healing capacity and enormous stretchability of a multifunctional polyelectrolyte mandates efficient ionic conduction, crack responsive formation of reversible hydrogen bonds, and cross-linking interactions among polymer networks. Huang et al. developed dual cross-linked PAA by non-sacrificial hydrogen bonding with vinyl hybrid silica nanoparticles [98]. The silica nanoparticles provide covalent cross-linking interactions and stress dissipation throughout the polyelectrolyte network. PAA polymer chains offer adequate intra- and intermolecular hydrogen bonding which can dynamically restore the cracks and evenly transfer the strain during stretching. Also, the presence of movable protons in the hydrogel augments ionic conductivity. The ionic gel can be stretched over 3700% without any damage which corroborates the efficacy of reversible dual cross-linking interactions. The ionic gel interfaced supercapacitor, based on polypyrrole-deposited carbon nanotube paper electrodes having a pre-stretched wavy structure, demonstrates stable performance for more than 20 repeated damage–heal cycles. The self-healing supercapacitor showed enhanced energy storage performance even at 600% strain, owing to the exposure of more electrode materials.

The performance of supercapacitors at low temperature is challenging for many applications like wearable devices and automobiles in a cold atmosphere. The diminished ion transfer at low temperatures due to the high viscosity causes device failure. The supercapacitors integrated with organic solvents or ionic liquids having low freezing points facilitate the ultra-low temperature performance. There is a vital demand for developing stretchable and self-healable supercapacitors with a stable gel electrolyte for low temperature operation. Li et al. constructed such flexible electrochemical capacitor having high specific capacitance at low temperature by amalgamating biochar-based reduced-graphene-oxide electrodes and a polyampholyte ionic gel electrolyte [99]. The tough polyampholyte gel offers exceptional mechanical resiliency and self-healing capability. The assembled supercapacitor showed enhanced performance at −30 °C due to the presence of non-freezable water molecules close to the hydrophilic polymer networks. The polyampholyte hydrogels prevent monolithic ice formation in cold environments, and possess significant tear-resistance, adaptable adhesion, and self-restoration efficiency. Wang et al. reported another low temperature self-repairing gel with regeneration capacity [42]. The renewable hydrogel is judiciously synthesized by copolymerizing vinylimidazole and hydroxypropyl acrylate. The polyelectrolyte interfaced activated carbon based supercapacitor autonomously heals and restores its energy storage capabilities over a wide temperature window, ranging from 25 to −15 °C, soon after repeated mechanical damage for five cycles.

Ionogels are distinctive composite materials in which an ionic liquid is incorporated in a three-dimensional solid-like matrix, which has been widely studied for various electrochemical applications. Ionogels are promising candidates for electrochemical energy storage and conversion systems, due to their good ionic conductivity, high temperature stability, and stable broad potential range. Trivedi et al. reported a carbanilated agarose-based ionogel prepared with a protic–aprotic mixed-ionic liquid (1-butyl-3-methylimidazolium chloride and *N*-(2-hydroxyethyl) ammonium formate) [100]. The ionogel showed remarkable self-healing behavior on account of its typical inter–intra hydrogen-bonding interactions with the functionalized agarose moieties. The solid ionogel electrolyte was assembled with an activated-carbon-based supercapacitor device, and showed a stable charge–discharge performance for 1000 cycles.

Attempts have been made to build up a self-healing solid gel electrolyte with appreciable stretchable, compressible, and recycling properties. Dai et al. reported a poly(acrylic acid-co-acrylamide) based self-healing gel, cross-linked with Co^{2+} [101]. The metal coordination with carboxylic acid groups, and the inter- and intramolecular hydrogen bonding networks facilitate a highly flexible ionic gel which can adapt well to various mechanical deformations. The solid gel electrolyte accommodates elongation of more than 1200% and 90% compressibility under an applied stress of 600 kPa. The solid electrolyte integrated activated carbon based double layer supercapacitor demonstrated stable energy storage ability over 5000 cycles, even after being subjected to repeated bending, twisting, and compressing agitations. The self-healing nature of the electrolyte provided 90% capacitance retention after five break–heal cycles.

4.6 REDOX-ACTIVE SELF-HEALING GEL ELECTROLYTES

The interfacial electrolyte layer is a major component in the fabrication of supercapacitors which play a vital role in fundamental energy storage mechanisms. A stable electrolyte system with significant ionic conductivities and high electrochemical sustainability is inevitable for a proficient supercapacitor. Various types of electrolytes have been explored for supercapacitors such as aqueous, organic, and ionic liquids. Roldan et al. reported a redox mediator incorporated hybrid electrolyte platform which can augment the specific capacitance by combining the double-layer charging characteristics of carbon-based supercapacitors and the Faradaic reaction features of batteries. Figure 4.3 is a schematic illustration of a double layer supercapacitor with gel electrolytes having redox active additives.

The addition of a small quantity of redox mediators provides additional Faradaic redox reactions which can also contribute to the overall device performance. Several redox mediators have been extensively studied to develop a reliable platform for augmenting the energy storage abilities of the assembled supercapacitors. Organic redox active species such as methylene blue [102], hydroquinone [103], indigo carmine [104], lignosulfonates [105], *p*-phenylenediamine [106], and *m*-phenylenediamine [107] showed exceptional pseudo-capacitive properties when integrated with supercapacitor electrodes. Several ionic redox active molecules such as KI [108], Na_2MoO_4 [109], $VOSO_4$ [110], and $CuCl_2$ [111] have been extensively explored for supercapacitor application.

Iodide and iodine (I^-/I_2) are a typical redox active complex, amalgamated with a poly(ethylene oxide)/lithium aluminate electrolyte to fabricate activated carbon composite based supercapacitors [112]. Interestingly, the supercapacitor possessed 27-fold enhancements in specific capacitance when compared to the system devoid of the redox couple. Ma et al. reported an indigo carmine based effective redox gel

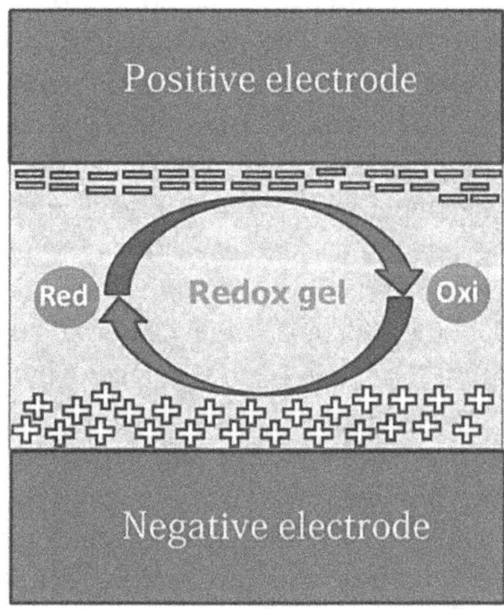

FIGURE 4.3 Schematic representation of a double layer capacitor with redox active gel electrolytes.

electrolyte by doping a PVA–H_2SO_4 system [35]. It was observed that the ionic conductivity of the gel increased by 188% and to 20.27 mS cm^{-1}. The solid state gel was assembled with activated carbon-based supercapacitor electrodes; the capacitance maintained 80% even after 3000 charge–discharge cycles. 2-Mercaptopyridine embedded with PVA–H_2SO_4 [113], 1,5-diaminoanthraquinone doped with PVA–H_3PO_4 [114], sodium molybdate with PVA–H_3PO_4 [109], and so on are some of the other high performance supercapacitor systems based on redox active gel electrolytes.

4.7 REDOX-ACTIVE SELF-HEALING ELECTROLYTES FOR SUPERCAPACITORS

Solid state self-healing gel electrolytes with reversible redox shuttling of electrons enhances the rate of Faradaic electrode reactions and augments the ionic conductivity and overall pseudo-capacitance. Even though extensive research has been focused on the capacitance boosting properties of the redox mediators, very few articles are reported for redox species based self-healing hydrogels with supercapacitor application. Recently Feng et al. developed a zwitterionic supramolecular hydrogel by the radical copolymerization of acrylic acid and 3-dimethyl (methacryloyloxyethyl) ammonium propane sulfonate in an aqueous acid medium containing bromamine acid sodium as a redox active species [115]. The as-prepared hydrogel demonstrates the self-healing of the damaged parts within 8 min and the substantial stretchability of over 2000%. The dual cross-linked gel depends on dynamical cross-linking by the hydrogen bonding interactions between the carbonyl and hydroxyl groups and the ionic bonding between sulfonic acid and quaternary ammonium moieties. The presence of a redox responsive additive enhanced the specific capacitance of the carbon based supercapacitor, and maintained its stable performance for more than 400 bending–releasing cycles and 5000 charge–discharge cycles.

The multifunctional gel electrolytes incorporated with redox active functionalities, superior self-healing ability, and stretchability provide a spectacular platform for realizing advanced flexible electrochemical supercapacitors for powering wearable devices. For developing next generation energy storage devices, self-healing redox electrolytes should seamlessly be merged with flexible and stretchable electrodes having self-healing properties. Such facile integration provides electrically and ionically conductive high performance supercapacitors. The extensive research on self-healing electrodes and electrolytes prove the substantial performance enhancements, like specific capacitance, cycling stability, and the long life of the devices. In fact, there is a lack of progress in the development of multifunctional self-healing gel electrolytes with redox active species, and in the amalgamation of self-healing electrodes and electrolytes. Hence, further fundamental and translational research should focus in the future on realizing reliable supercapacitors for wearable applications.

4.8 CONCLUSION

The present chapter has provided a brief outline of the recent advances in smart self-healing gels for energy storage application. The various synthesis protocols of self-healing gels, their mechanical and ion transport properties, and their application as an electrolyte medium in flexible and stretchable supercapacitors has been discussed. Multifunctional gel electrolytes with a self-healing ability, ionic conductivity, fatigue resistance, mechanical endurance, recyclability, and the presence of redox active compounds enhance the overall performance of the supercapacitor, including its long device life and cycling stability. The merging of stretchable supercapacitor electrodes having a self-healing ability with self-healing solid state gel electrolytes enables the facile fabrication of next generation wearable supercapacitors.

ACKNOWLEDGMENTS

Authors acknowledge the Department of Science and Technology (DST) for providing the INSPIRE Faculty award (DST/INSPIRE/04/2016/001601), and the Council of Scientific & Industrial Research (CSIR) for financial support.

REFERENCES

[1] Chen, Y., Au, J., Kazlas, P., Ritenour, A., Gates, H., McCreary, M., Flexible active-matrix electronic ink display. *Nature*, *423*, 136, 2003.

[2] Bandodkar, A.J., Jia, W., Wang, J., Tattoo-based wearable electrochemical devices: A review. *Electroanalysis*, *27*, 562, 2015.

[3] Bandodkar, A.J., Wang, J., Non-invasive wearable electrochemical sensors: A review. *Trends in Biotechnology*, *32*, 363, 2014.

[4] Jia, W., Bandodkar, A.J., Valdés-Ramírez, G., Windmiller, J.R., Yang, Z., Ramírez, J., Chan, G., Wang, J., Electrochemical tattoo biosensors for real-time noninvasive lactate monitoring in human perspiration. *Analytical Chemistry*, *85*, 6553, 2013.

[5] Windmiller, J.R., Wang, J., Wearable electrochemical sensors and biosensors: A review. *Electroanalysis*, *25*, 29, 2013.

[6] Zhao, X., Hua, Q., Yu, R., Zhang, Y., Pan, C., Flexible, stretchable and wearable multifunctional sensor array as artificial electronic skin for static and dynamic strain mapping. *Advanced Electronic Materials*, *1*, 1500142, 2015.

[7] Choi, K.-H., Kim, S.-H., Ha, H.-J., Kil, E.-H., Lee, C.K., Lee, S.B., Shim, J.K., Lee, S.-Y., Compliant polymer network-mediated fabrication of a bendable plastic crystal polymer electrolyte for flexible lithium-ion batteries. *Journal of Materials Chemistry A*, *1*, 5224, 2013.

[8] Tricoli, A., Nasiri, N., De, S., Wearable and miniaturized sensor technologies for personalized and preventive medicine. *Advanced Functional Materials*, *27*, 1605271, 2017.

[9] Bandodkar, A.J., You, J.-M., Kim, N.-H., Gu, Y., Kumar, R., Mohan, A.V., Kurniawan, J., Imani, S., Nakagawa, T., Parish, B., Soft, stretchable, high power density electronic skin-based biofuel cells for scavenging energy from human sweat. *Energy & Environmental Science*, *10*, 1581, 2017.

[10] Chen, X., Yin, L., Lv, J., Gross, A.J., Le, M., Gutierrez, N.G., Li, Y., Jeerapan, I., Giroud, F., Berezovska, A., Stretchable and flexible buckypaper-based lactate biofuel cell for wearable electronics. *Advanced Functional Materials*, *29*, 1905785, 2019.

[11] Jeerapan, I., Sempionatto, J.R., Wang, J., On-body bioelectronics: Wearable biofuel cells for bioenergy harvesting and self-powered biosensing. *Advanced Functional Materials*, 30, 1906243, 2020.

[12] Bandodkar, A.J., Jeerapan, I., You, J.-M., Nuñez-Flores, R., Wang, J., Highly stretchable fully-printed cnt-based electrochemical sensors and biofuel cells: Combining intrinsic and design-induced stretchability. *Nano Letters*, *16*, 721, 2016.

[13] Bandodkar, A.J., Nuñez-Flores, R., Jia, W., Wang, J., All-printed stretchable electrochemical devices. *Advanced Materials*, *27*, 3060, 2015.

[14] Gates, B.D., Flexible electronics. *Science*, *323*, 1566, 2009.

[15] Chen, D., Avestro, A.J., Chen, Z., Sun, J., Wang, S., Xiao, M., Erno, Z., Algaradah, M.M., Nassar, M.S., Amine, K., A rigid naphthalenediimide triangle for organic rechargeable lithium-ion batteries. *Advanced Materials*, *27*, 2907, 2015.

[16] Mohan, A.V., Kim, N., Gu, Y., Bandodkar, A.J., You, J.M., Kumar, R., Kurniawan, J.F., Xu, S., Wang, J., Merging of thin-and thick-film fabrication technologies: Toward soft stretchable "island–bridge" devices. *Advanced Materials Technologies*, *2*, 1600284, 2017.

[17] Park, S.I., Brenner, D.S., Shin, G., Morgan, C.D., Copits, B.A., Chung, H.U., Pullen, M.Y., Noh, K.N., Davidson, S., Oh, S.J., Soft, stretchable, fully implantable miniaturized optoelectronic systems for wireless optogenetics. *Nature Biotechnology*, *33*, 1280, 2015.

[18] Rajendran, V., Mohan, A.V., Jayaraman, M., Nakagawa, T., All-printed, interdigitated, freestanding serpentine interconnects based flexible solid state supercapacitor for self powered wearable electronics. *Nano Energy*, *65*, 104055, 2019.

[19] Trung, T.Q., Lee, N.E., Flexible and stretchable physical sensor integrated platforms for wearable human-activity monitoringand personal healthcare. *Advanced Materials*, *28*, 4338, 2016.

[20] Zhao, J., Guo, H., Li, J., Bandodkar, A.J., Rogers, J.A., Body-interfaced chemical sensors for noninvasive monitoring and analysis of biofluids. *Trends in Chemistry*, *1*, 2019. DOI:10.1016/j.trechm.2019.07.001

[21] Imani, S., Bandodkar, A.J., Mohan, A.V., Kumar, R., Yu, S., Wang, J., Mercier, P.P., A wearable chemical–electrophysiological hybrid biosensing system for real-time health and fitness monitoring. *Nature Communications*, *7*, 1, 2016.

[22] Bandodkar, A.J., Molinnus, D., Mirza, O., Guinovart, T., Windmiller, J.R., Valdés-Ramírez, G., Andrade, F.J., Schöning, M.J., Wang, J., Epidermal tattoo potentiometric sodium sensors with wireless signal transduction for continuous non-invasive sweat monitoring. *Biosensors and Bioelectronics*, *54*, 603, 2014.

[23] Kim, J., Jeerapan, I., Imani, S., Cho, T.N., Bandodkar, A., Cinti, S., Mercier, P.P., Wang, J., Noninvasive alcohol monitoring using a wearable tattoo-based iontophoretic-biosensing system. *Acs Sensors*, *1*, 1011, 2016.

[24] Kim, J., Imani, S., de Araujo, W.R., Warchall, J., Valdés-Ramírez, G., Paixão, T.R., Mercier, P.P., Wang, J., Wearable salivary uric acid mouthguard biosensor with integrated wireless electronics. *Biosensors and Bioelectronics*, *74*, 1061, 2015.

[25] Kim, J., Valdés-Ramírez, G., Bandodkar, A.J., Jia, W., Martinez, A.G., Ramírez, J., Mercier, P., Wang, J., Non-invasive mouthguard biosensor for continuous salivary monitoring of metabolites. *Analyst*, *139*, 1632, 2014.

[26] Kim, J., Kim, M., Lee, M.-S., Kim, K., Ji, S., Kim, Y.-T., Park, J., Na, K., Bae, K.-H., Kim, H.K., Wearable smart sensor systems integrated on soft contact lenses for wireless ocular diagnostics. *Nature Communications*, *8*, 1, 2017.

[27] Mishra, R.K., Goud, K.Y., Li, Z., Moonla, C., Mohamed, M.A., Tehrani, F., Teymourian, H., Wang, J., Continuous opioid monitoring along with nerve agents on a wearable microneedle sensor array. *Journal of the American Chemical Society*, *142*, 5991–5995, 2020.

[28] Mishra, R.K., Mohan, A.V., Soto, F., Chrostowski, R., Wang, J., A microneedle biosensor for minimally-invasive transdermal detection of nerve agents. *Analyst*, *142*, 918, 2017.

[29] Mohan, A.V., Windmiller, J.R., Mishra, R.K., Wang, J., Continuous minimally-invasive alcohol monitoring using microneedle sensor arrays. *Biosensors and Bioelectronics*, *91*, 574, 2017.

[30] Lu, X., Yu, M., Wang, G., Tong, Y., Li, Y., Flexible solid-state supercapacitors: Design, fabrication and applications. *Energy & Environmental Science*, *7*, 2160, 2014.

[31] Jeerapan, I., Poorahong, S., Flexible and stretchable electrochemical sensing systems: Materials, energy sources, and integrations. *Journal of the Electrochemical Society*, *167*, 037573, 2020.

[32] Yan, J., Wang, Q., Wei, T., Fan, Z., Recent advances in design and fabrication of electrochemical supercapacitors with high energy densities. *Advanced Energy Materials*, *4*, 1300816, 2014.

[33] Jiang, H., Zhao, T., Li, C., Ma, J., Hierarchical self-assembly of ultrathin nickel hydroxide nanoflakes for high-performance supercapacitors. *Journal of Materials Chemistry*, *21*, 3818, 2011.

[34] Zhang, L.L., Zhao, X., Carbon-based materials as supercapacitor electrodes. *Chemical Society Reviews*, *38*, 2520, 2009.

[35] Ma, G., Dong, M., Sun, K., Feng, E., Peng, H., Lei, Z., A redox mediator doped gel polymer as an electrolyte and separator for a high performance solid state supercapacitor. *Journal of Materials Chemistry A*, *3*, 4035, 2015.

[36] Shi, H., Activated carbons and double layer capacitance. *Electrochimica Acta*, *41*, 1633, 1996.

[37] Chen, K., Xue, D., Formation of electroactive colloids via in situ coprecipitation under electric field: Erbium chloride alkaline aqueous pseudocapacitor. *Journal of Colloid and Interface Science*, *430*, 265, 2014.

[38] Shao, Y., El-Kady, M.F., Sun, J., Li, Y., Zhang, Q., Zhu, M., Wang, H., Dunn, B., Kaner, R.B., Design and mechanisms of asymmetric supercapacitors. *Chemical Reviews*, *118*, 9233, 2018.

[39] Kim, J., Kumar, R., Bandodkar, A.J., Wang, J., Advanced materials for printed wearable electrochemical devices: A review. *Advanced Electronic Materials*, *3*, 1600260, 2017.

[40] Trask, R., Williams, H., Bond, I., Self-healing polymer composites: Mimicking nature to enhance performance. *Bioinspiration & Biomimetics*, *2*, P1, 2007.

[41] Guo, Y., Zhou, X., Tang, Q., Bao, H., Wang, G., Saha, P., A self-healable and easily recyclable supramolecular hydrogel electrolyte for flexible supercapacitors. *Journal of Materials Chemistry A*, *4*, 8769, 2016.

[42] Wang, J., Liu, F., Tao, F., Pan, Q., Rationally designed self-healing hydrogel electrolyte toward a smart and sustainable supercapacitor. *ACS Applied Materials & Interfaces*, *9*, 27745, 2017.

[43] Wang, Z., Tao, F., Pan, Q., A self-healable polyvinyl alcohol-based hydrogel electrolyte for smart electrochemical capacitors. *Journal of Materials Chemistry A*, *4*, 17732, 2016.

[44] Ahmed, A.S., Ramanujan, R.V., Magnetic field triggered multicycle damage sensing and self healing. *Scientific Reports*, *5*, 13773, 2015.
[45] Li, C.-H., Wang, C., Keplinger, C., Zuo, J.-L., Jin, L., Sun, Y., Zheng, P., Cao, Y., Lissel, F., Linder, C., A highly stretchable autonomous self-healing elastomer. *Nature Chemistry*, *8*, 618, 2016.
[46] Peng, R., Yu, Y., Chen, S., Yang, Y., Tang, Y., Conductive nanocomposite hydrogels with self-healing property. *RSC Advances*, *4*, 35149, 2014.
[47] Haraguchi, K., Uyama, K., Tanimoto, H., Self-healing in nanocomposite hydrogels. *Macromolecular Rapid Communications*, *32*, 1253, 2011.
[48] Zhang, J., Wan, L., Gao, Y., Fang, X., Lu, T., Pan, L., Xuan, F., Highly stretchable and self-healable mxene/polyvinyl alcohol hydrogel electrode for wearable capacitive electronic skin. *Advanced Electronic Materials*, *5*, 1900285, 2019.
[49] Liao, M., Wan, P., Wen, J., Gong, M., Wu, X., Wang, Y., Shi, R., Zhang, L., Wearable, healable, and adhesive epidermal sensors assembled from mussel-inspired conductive hybrid hydrogel framework. *Advanced Functional Materials*, *27*, 1703852, 2017.
[50] Yamanlar, S., Sant, S., Boudou, T., Picart, C., Khademhosseini, A., Surface functionalization of hyaluronic acid hydrogels by polyelectrolyte multilayer films. *Biomaterials*, *32*, 5590, 2011.
[51] Daniele, M.A., Adams, A.A., Naciri, J., North, S.H., Ligler, F.S., Interpenetrating networks based on gelatin methacrylamide and peg formed using concurrent thiol click chemistries for hydrogel tissue engineering scaffolds. *Biomaterials*, *35*, 1845, 2014.
[52] Chen, Y., Chen, L., Bai, H., Li, L., Graphene oxide–chitosan composite hydrogels as broad-spectrum adsorbents for water purification. *Journal of Materials Chemistry A*, *1*, 1992, 2013.
[53] Gaharwar, A.K., Peppas, N.A., Khademhosseini, A., Nanocomposite hydrogels for biomedical applications. *Biotechnology and Bioengineering*, *111*, 441, 2014.
[54] Hamidi, M., Azadi, A., Rafiei, P., Hydrogel nanoparticles in drug delivery. *Advanced Drug Delivery Reviews*, *60*, 1638, 2008.
[55] Slaughter, B.V., Khurshid, S.S., Fisher, O.Z., Khademhosseini, A., Peppas, N.A., Hydrogels in regenerative medicine. *Advanced Materials*, *21*, 3307, 2009.
[56] Cheng, X., Pan, J., Zhao, Y., Liao, M., Peng, H., Gel polymer electrolytes for electrochemical energy storage. *Advanced Energy Materials*, *8*, 1702184, 2018.
[57] Jeon, I., Cui, J., Illeperuma, W.R., Aizenberg, J., Vlassak, J.J., Extremely stretchable and fast self-healing hydrogels. *Advanced Materials*, *28*, 4678, 2016.
[58] Li, J., Geng, L., Wang, G., Chu, H., Wei, H., Self-healable gels for use in wearable devices. *Chemistry of Materials*, *29*, 8932, 2017.
[59] Chen, D., Wang, D., Yang, Y., Huang, Q., Zhu, S., Zheng, Z., Self-healing materials for next-generation energy harvesting and storage devices. *Advanced Energy Materials*, *7*, 1700890, 2017.
[60] Zhu, D.Y., Rong, M.Z., Zhang, M.Q., Self-healing polymeric materials based on microencapsulated healing agents: From design to preparation. *Progress in Polymer Science*, *49*, 175, 2015.
[61] Bandodkar, A.J., Mohan, V., López, C.S., Ramírez, J., Wang, J., Self-healing inks for autonomous repair of printable electrochemical devices. *Advanced Electronic Materials*, *1*, 1500289, 2015.
[62] Zhao, Y., Fickert, J., Landfester, K., Crespy, D., Encapsulation of self-healing agents in polymer nanocapsules. *Small*, *8*, 2954, 2012.
[63] Cromwell, O.R., Chung, J., Guan, Z., Malleable and self-healing covalent polymer networks through tunable dynamic boronic ester bonds. *Journal of the American Chemical Society*, *137*, 6492, 2015.
[64] Lin, Y., Li, G., An intermolecular quadruple hydrogen-bonding strategy to fabricate self-healing and highly deformable polyurethane hydrogels. *Journal of Materials Chemistry B*, *2*, 6878, 2014.
[65] Häring, M., Díaz, D.D., Supramolecular metallogels with bulk self-healing properties prepared by in situ metal complexation. *Chemical Communications*, *52*, 13068, 2016.
[66] Shen, Z., Jiang, Y., Wang, T., Liu, M., Symmetry breaking in the supramolecular gels of an achiral gelator exclusively driven by π–π stacking. *Journal of the American Chemical Society*, *137*, 16109, 2015.
[67] Tuncaboylu, D.C., Sari, M., Oppermann, W., Okay, O., Tough and self-healing hydrogels formed via hydrophobic interactions. *Macromolecules*, *44*, 4997, 2011.
[68] Kakuta, T., Takashima, Y., Nakahata, M., Otsubo, M., Yamaguchi, H., Harada, A., Preorganized hydrogel: Self-healing properties of supramolecular hydrogels formed by polymerization of host–guest-monomers that contain cyclodextrins and hydrophobic guest groups. *Advanced Materials*, *25*, 2849, 2013.
[69] Kuhl, N., Bode, S., Bose, R.K., Vitz, J., Seifert, A., Hoeppener, S., Garcia, S.J., Spange, S., van der Zwaag, S., Hager, M.D., Acylhydrazones as reversible covalent cross-linkers for self-healing polymers. *Advanced Functional Materials*, *25*, 3295, 2015.
[70] Han, X., Meng, X., Wu, Z., Wu, Z., Qi, X., Dynamic imine bond cross-linked self-healing thermosensitive hydrogels for sustained anticancer therapy via intratumoral injection. *Materials Science and Engineering: C*, *93*, 1064, 2018.
[71] Yang, W.J., Tao, X., Zhao, T., Weng, L., Kang, E.-T., Wang, L., Antifouling and antibacterial hydrogel coatings with self-healing properties based on a dynamic disulfide exchange reaction. *Polymer Chemistry*, *6*, 7027, 2015.
[72] Yu, X., Chen, L., Zhang, M., Yi, T., Low-molecular-mass gels responding to ultrasound and mechanical stress: Towards self-healing materials. *Chemical Society Reviews*, *43*, 5346, 2014.
[73] Cui, J., del Campo, A., Multivalent h-bonds for self-healing hydrogels. *Chemical Communications*, *48*, 9302, 2012.
[74] Yan, X., Chen, Q., Zhu, L., Chen, H., Wei, D., Chen, F., Tang, Z., Yang, J., Zheng, J., High strength and self-healable gelatin/polyacrylamide double network hydrogels. *Journal of Materials Chemistry B*, *5*, 7683, 2017.
[75] Han, J., Ding, Q., Mei, C., Wu, Q., Yue, Y., Xu, X., An intrinsically self-healing and biocompatible electroconductive hydrogel based on nanostructured nanocellulose-polyaniline complexes embedded in a viscoelastic polymer network towards flexible conductors and electrodes. *Electrochimica Acta*, *318*, 660, 2019.
[76] Dutta, A., Maity, S., Das, R.K., A highly stretchable, tough, self-healing, and thermoprocessable polyacrylamide–chitosan supramolecular hydrogel. *Macromolecular Materials and Engineering*, *303*, 1800322, 2018.
[77] Chirila, T.V., Lee, H.H., Oddon, M., Nieuwenhuizen, M.M., Blakey, I., Nicholson, T.M., Hydrogen-bonded supramolecular polymers as self-healing hydrogels: Effect of a bulky adamantyl substituent in the ureido-pyrimidinone monomer. *Journal of Applied Polymer Science*, *131*, 2014. DOI:10.1002/app.39932

[78] Hart, L.R., Nguyen, N.A., Harries, J.L., Mackay, M.E., Colquhoun, H.M., Hayes, W., Perylene as an electron-rich moiety in healable, complementary π–π stacked, supramolecular polymer systems. *Polymer*, *69*, 293, 2015.

[79] Wang, Z., Xie, C., Yu, C., Fei, G., Wang, Z., Xia, H., A facile strategy for self-healing polyurethanes containing multiple metal–ligand bonds. *Macromolecular Rapid Communications*, *39*, 1700678, 2018.

[80] Liang, H., Zhang, Z., Yuan, Q., Liu, J., Self-healing metal-coordinated hydrogels using nucleotide ligands. *Chemical Communications*, *51*, 15196, 2015.

[81] Okay, O. Self-healing hydrogels formed via hydrophobic interactions. In Sebastian Seiffert (Ed.) *Supramolecular polymer networks and gels*; Springer: 2015; pp 101–142.

[82] Kitamura, S. Cyclic oligosaccharides and polysaccharides. In E.R. Semlyen, J. A. Semlyen (Eds.) *Cyclic polymers*; Springer: 2000; pp 125–160.

[83] Nakahata, M., Takashima, Y., Harada, A., Highly flexible, tough, and self-healing supramolecular polymeric materials using host–guest interaction. *Macromolecular Rapid Communications*, *37*, 86, 2016.

[84] Deng, G., Tang, C., Li, F., Jiang, H., Chen, Y., Covalent cross-linked polymer gels with reversible sol–gel transition and self-healing properties. *Macromolecules*, *43*, 1191, 2010.

[85] Imato, K., Nishihara, M., Kanehara, T., Amamoto, Y., Takahara, A., Otsuka, H., Self-healing of chemical gels cross-linked by diarylbibenzofuranone-based trigger-free dynamic covalent bonds at room temperature. *Angewandte Chemie International Edition*, *51*, 1138, 2012.

[86] Chao, A., Negulescu, I., Zhang, D., Dynamic covalent polymer networks based on degenerative imine bond exchange: Tuning the malleability and self-healing properties by solvent. *Macromolecules*, *49*, 6277, 2016.

[87] Cash, J.J., Kubo, T., Bapat, A.P., Sumerlin, B.S., Room-temperature self-healing polymers based on dynamic-covalent boronic esters. *Macromolecules*, *48*, 2098, 2015.

[88] Chen, Y., Tang, Z., Zhang, X., Liu, Y., Wu, S., Guo, B., Covalently cross-linked elastomers with self-healing and malleable abilities enabled by boronic ester bonds. *ACS Applied Materials & Interfaces*, *10*, 24224, 2018.

[89] Nevejans, S., Ballard, N., Miranda, J.I., Reck, B., Asua, J.M., The underlying mechanisms for self-healing of poly (disulfide) s. *Physical Chemistry Chemical Physics*, *18*, 27577, 2016.

[90] Wan, T., Chen, D., Synthesis and properties of self-healing waterborne polyurethanes containing disulfide bonds in the main chain. *Journal of Materials Science*, *52*, 197, 2017.

[91] Hu, R., Zhao, J., Wang, Y., Li, Z., Zheng, J., A highly stretchable, self-healing, recyclable and interfacial adhesion gel: Preparation, characterization and applications. *Chemical Engineering Journal*, *360*, 334, 2019.

[92] Guo, Y., Zheng, K., Wan, P., A flexible stretchable hydrogel electrolyte for healable all-in-one configured supercapacitors. *Small*, *14*, 1704497, 2018.

[93] Adhikari, B., Nanda, J., Banerjee, A., Pyrene-containing peptide-based fluorescent organogels: Inclusion of graphene into the organogel. *Chemistry – A European Journal*, *17*, 11488, 2011.

[94] Cao, Y.-C., Xu, C., Wu, X., Wang, X., Xing, L., Scott, K., A poly (ethylene oxide)/graphene oxide electrolyte membrane for low temperature polymer fuel cells. *Journal of Power Sources*, *196*, 8377, 2011.

[95] Yang, X., Zhang, F., Zhang, L., Zhang, T., Huang, Y., Chen, Y., A high-performance graphene oxide-doped ion gel as gel polymer electrolyte for all-solid-state supercapacitor applications. *Advanced Functional Materials*, *23*, 3353, 2013.

[96] Peng, H., Lv, Y., Wei, G., Zhou, J., Gao, X., Sun, K., Ma, G., Lei, Z., A flexible and self-healing hydrogel electrolyte for smart supercapacitor. *Journal of Power Sources*, *431*, 210, 2019.

[97] Lin, Y., Zhang, H., Liao, H., Zhao, Y., Li, K., A physically crosslinked, self-healing hydrogel electrolyte for nanowire pani flexible supercapacitors. *Chemical Engineering Journal*, *367*, 139, 2019.

[98] Huang, Y., Zhong, M., Huang, Y., Zhu, M., Pei, Z., Wang, Z., Xue, Q., Xie, X., Zhi, C., A self-healable and highly stretchable supercapacitor based on a dual crosslinked polyelectrolyte. *Nature Communications*, *6*, 1, 2015.

[99] Li, X., Liu, L., Wang, X., Ok, Y.S., Elliott, J.A., Chang, S.X., Chung, H.-J., Flexible and self-healing aqueous supercapacitors for low temperature applications: Polyampholyte gel electrolytes with biochar electrodes. *Scientific Reports*, *7*, 1, 2017.

[100] Trivedi, T.J., Bhattacharjya, D., Yu, J.S., Kumar, A., Functionalized agarose self-healing ionogels suitable for supercapacitors. *ChemSusChem*, *8*, 3294, 2015.

[101] Dai, L.X., Zhang, W., Sun, L., Wang, X.H., Jiang, W., Zhu, Z.W., Zhang, H.B., Yang, C.C., Tang, J., Highly stretchable and compressible self-healing p (aa-co-aam)/$cocl_2$ hydrogel electrolyte for flexible supercapacitors. *ChemElectroChem*, *6*, 467, 2019.

[102] Zhong, J., Fan, L.-Q., Wu, X., Wu, J.-H., Liu, G.-J., Lin, J.-M., Huang, M.-L., Wei, Y.-L., Improved energy density of quasi-solid-state supercapacitors using sandwich-type redox-active gel polymer electrolytes. *Electrochimica Acta*, *166*, 150, 2015.

[103] Dubal, D.P., Suarez-Guevara, J., Tonti, D., Enciso, E., Gomez-Romero, P., A high voltage solid state symmetric supercapacitor based on graphene–polyoxometalate hybrid electrodes with a hydroquinone doped hybrid gel-electrolyte. *Journal of Materials Chemistry A*, *3*, 23483, 2015.

[104] Feng, E., Ma, G., Sun, K., Ran, F., Peng, H., Lei, Z., Superior performance of an active electrolyte enhanced supercapacitor based on a toughened porous network gel polymer. *New Journal of Chemistry*, *41*, 1986, 2017.

[105] Li, F., Wang, X., Sun, R., A metal-free and flexible supercapacitor based on redox-active lignosulfonate functionalized graphene hydrogels. *Journal of Materials Chemistry A*, *5*, 20643, 2017.

[106] Zhang, Z.J., Zhu, Y.Q., Chen, X.Y., Cao, Y., Pronounced improvement of supercapacitor capacitance by using redox active electrolyte of p-phenylenediamine. *Electrochimica Acta*, *176*, 941, 2015.

[107] Wang, H., Zhang, W., Chen, H., Zheng, W., Towards unlocking high-performance of supercapacitors: From layered transition-metal hydroxide electrode to redox electrolyte. *Science China Technological Sciences*, *58*, 1779, 2015.

[108] Yu, H., Wu, J., Fan, L., Xu, K., Zhong, X., Lin, Y., Lin, J., Improvement of the performance for quasi-solid-state supercapacitor by using pva–koh–ki polymer gel electrolyte. *Electrochimica Acta*, *56*, 6881, 2011.

[109] Veerasubramani, G.K., Krishnamoorthy, K., Pazhamalai, P., Kim, S.J., Enhanced electrochemical performances of graphene based solid-state flexible cable type supercapacitor using redox mediated polymer gel electrolyte. *Carbon*, *105*, 638, 2016.

[110] Wang, A.-Y., Chaudhary, M., Lin, T.-W., Enhancing the stability and capacitance of vanadium oxide nanoribbons/3d-graphene binder-free electrode by using voso4 as redox-active electrolyte. *Chemical Engineering Journal*, *355*, 830, 2019.
[111] Yadav, N., Yadav, N., Singh, M.K., Hashmi, S.A., Nonaqueous, redox-active gel polymer electrolyte for high-performance supercapacitor. *Energy Technology*, *7*, 1900132, 2019.
[112] Yin, Y., Zhou, J., Mansour, A.N., Zhou, X., Effect of nai/i2 mediators on properties of peo/lialo2 based all-solid-state supercapacitors. *Journal of Power Sources*, *196*, 5997, 2011.
[113] Pan, S., Deng, J., Guan, G., Zhang, Y., Chen, P., Ren, J., Peng, H., A redox-active gel electrolyte for fiber-shaped supercapacitor with high area specific capacitance. *Journal of Materials Chemistry A*, *3*, 6286, 2015.
[114] Hashmi, S., Suematsu, S., Naoi, K., All solid-state redox supercapacitors based on supramolecular 1, 5-diaminoanthraquinone oligomeric electrode and polymeric electrolytes. *Journal of Power Sources*, *137*, 145, 2004.
[115] Feng, E., Gao, W., Li, J., Wei, J., Yang, Q., Li, Z., Ma, X., Zhang, T., Yang, Z., Stretchable, healable, adhesive, and redox-active multifunctional supramolecular hydrogel-based flexible supercapacitor. *ACS Sustainable Chemistry & Engineering*, *8*, 3311, 2020.

5 Polymeric Nanogenerators

Rutuja S. Bhoje
Institute of Chemical Technology, Mumbai, India

Parag R. Nemade
Institute of Chemical Technology Matunga, Mumbai, India

CONTENTS

5.1 INTRODUCTION

Many electronic devices improve our quality of life, such as GPS devices, health monitoring devices, sensors, and cell phones (Baker et al. 2007). A conventional power supply like lithium batteries cannot fulfill the enhanced power requirements for such devices. In the energy scavenging process, energy in the form of thermal, electrostatic, and solar energy is captured and stored as electrical energy at ambient temperature. There has been increased focus on renewable energy sources on account of the adverse climate effects of fossil fuels. Micro-energy harvesting technology utilizes alternative sources of energy for power supplies. Nanogenerators are a type of micro-energy harvesting device that converts the vibrations, mechanical energy, waste thermal energy, sunlight, room light, movement of the human body, and so on into electrical energy. Nanogenerators hold the capacity to convert light/mechanical/thermal energy into electricity on the nanoscale. If nanogenerators are integrated with electronic devices, the power requirements from the battery may be reduced significantly. In certain situations, the device itself may even function as a power generating device and act as a power solution. This way of energy generation can help to resolve the issue of the energy crisis.

Polymers are attractive as a material for nano energy harvesters. Polymers with good structural flexibility and robustness, carry the potential to be an inexpensive fabrication for nanogenerators. Polymers can readily be modified at the nanoscale at room temperature. Several technologies are available in the market for the scale-up of the respective nanogenerators. Polymer materials that act as nanogenerators consist of distinct functional groups, like carboxyl (–COOH), ester (–COOR), fluorine (–F), cyano (–CN), acyl (–CON), and nitro (–NO_2) groups; as electron-withdrawing groups (EWG); as alkoxy (–OR), amide (–CONH), amine (NH_2), and hydroxy (–OH) groups; as electron-donating groups (EDG); or as an electron releasing group (ERG). These functional groups facilitate the transfer or withholding of charge due to their unique hybrid orbital configurations. It has been shown that a simple wearable fabric could

DOI: 10.1201/9781003169727-5

generate power using human activity and store it in an integrated supercapacitor (Jung et al. 2014). Polymer generators can generate electrical energy from natural things like the sun, wind, water flows, human body activity, and mechanical energy such as vibrations. Different transducing mechanisms have been investigated for mechanical energy harvestings, such as piezoelectricity, triboelectricity, electrostatic induction, and electromagnetic induction. This chapter covers the application of polymer nanomaterials in energy harvesting. Fabrication and incorporation in energy harvesting devices of polymeric nanomaterials in different energy harvesting applications are explained systematically.

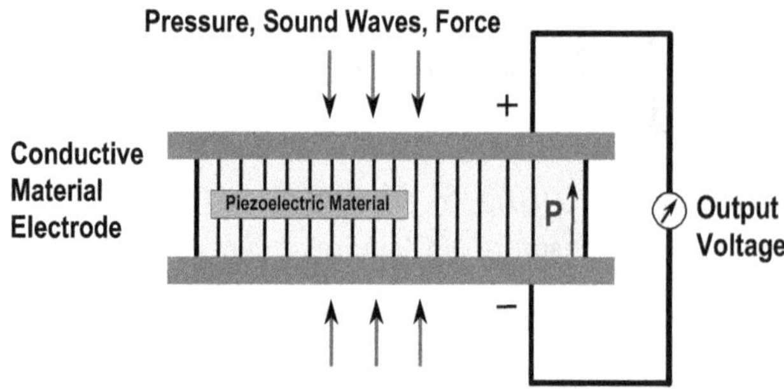

FIGURE 5.2 Schematic diagram of the piezoelectric effect.

5.2 PIEZOELECTRIC NANOGENERATORS

The word "piezoelectricity" has roots in the Greek "piezein," meaning "to press or push," and "elecktron," from "amber", a natural plastic material; in short, piezoelectricity is essentially pressure-induced electricity. The first piezoelectric nanogenerator (PENG) device was produced in 2006 (Wang and Song 2006), as shown in Figure 5.1. The device proved its potential to convert mechanical energy into electrical energy by using the piezoelectric properties of zinc oxide nanowire. An atomic force microscopy tip moves swiftly across the vertically grown ZnO nanowire. Deflection of the aligned zinc oxide nanowires caused a strain field by combining the piezoelectric and semiconducting properties of zinc oxide. After this positive attempt, many researchers developed nanogenerators with different designs and different materials. The piezoelectric phenomenon represents the generation of an electric spark or the opposite sign of an electric charge by striking material. In other words, the piezoelectric effect is the generation of electric polarization by mechanical stress, as shown in Figure 5.2. The piezoelectric materials are divided into different classes, such as ceramic, crystal, and polymer. Due to the lack of a center of symmetry in a crystalline form, the piezoelectric material is readily compressed or stressed. In the absence of short-circuited points, when the material is connected to an external load, a free-electron flow is generated in the external circuit to balance the polarization charges which forms a fresh equilibrium state. Many piezoelectric materials have been studied by researchers, such as zinc oxide (Lu et al. 2009; Wang and Song 2006), lead zirconium titanate (Chen et al. 2010; Zhu et al. 2020), barium titanate (Marino et al. 2019), gallium nitride (Gaska, Shur, and Bykhovski 1999; Ramesh et al. 2012), and sodium niobate (Ke et al. 2008).

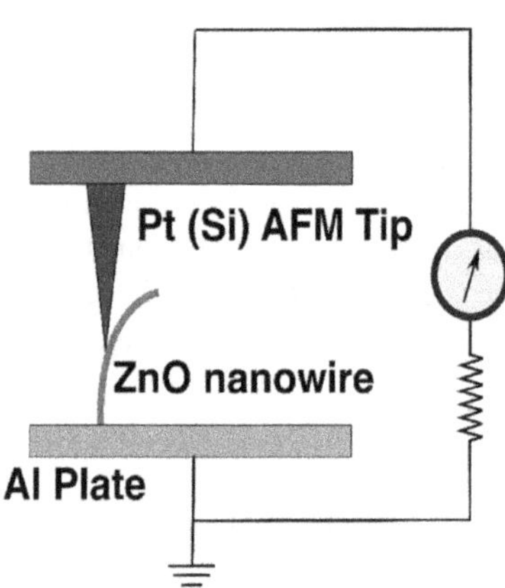

FIGURE 5.1 PENG device designed by Wang and Song 2006.

5.2.1 Polyvinylidene Fluoride and Its Co-Polymers

5.2.1.1 Polyvinylidene Fluoride

Many polymers consist of piezoelectric properties used in polymer-based nanogenerator devices where polymeric layers are placed in electrical contact at both ends and kept on a flexible and-stretchable substrate. Piezoelectric polymers such as polyvinylidene fluoride (PVDF) and its co-polymers are the most used and studied due to their beneficial piezoelectric properties such as structural flexibility, ease of processing, good chemical resistance, and mechanical robustness (Gusarov 2015; Kawai 1969). Piezoelectric materials like inorganic ceramics and metal oxides have less preference when these materials are compared with piezoelectric polymers due to their high fracture strain, which gives them an upper hand in the application of stretchable energy harvesting devices. In the PVDF structure, each carbon in the chain forms a bond with fluorine and two hydrogen atoms alternatively, inclined to opposing sides of the carbon backbone. PVDF and co-polymers show piezoelectric properties due to a significant electronegativity difference between fluorine and carbon atoms in comparison with that between carbon and hydrogen atoms. This leads to the formation of a dipole moment from fluorine to the hydrogen side (Chunga and Petchsuk 2003). Among all polymers, PVDF-based piezoelectric devices exhibit higher piezoelectricity due to their highly robust mechanical flexibility (Klimiec et al. 2011; Rocha et al. 2010). A polymer energy generator exhibits good performance in a typical piezoelectric crystalline phase (Cauda et al. 2013; García-Gutiérrez et al. 2010). The crystalline phases of PVDF are shown in Figure 5.3. The β phase of PVDF shows a self-poled effect, indicating it does not require external electric poling to demonstrate a piezoelectric effect (Lovinger 1983). Typically, poling with an electric field or mechanical

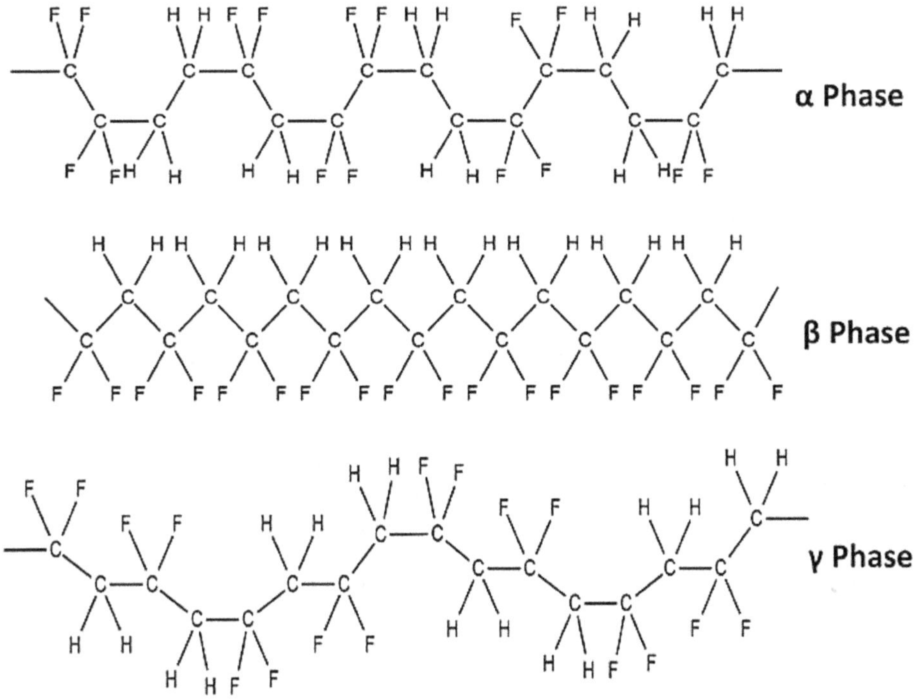

FIGURE 5.3 Crystalline phases of PVDF.

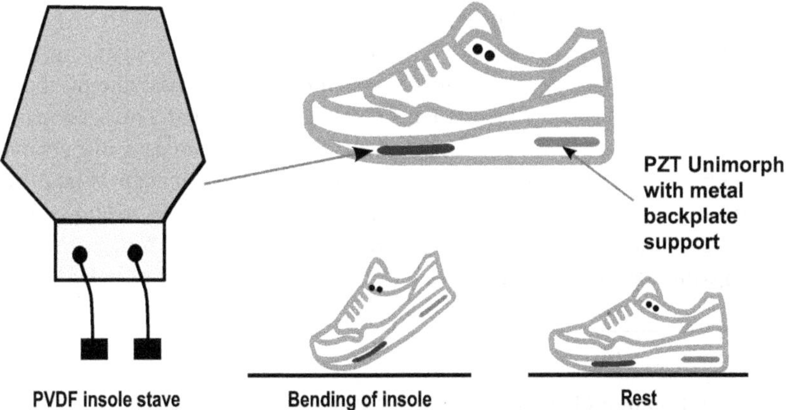

FIGURE 5.4 Schematic representation of piezoelectric shoes.

stretching may push the PVDF polymer into the ferroelectric β phase. Out of 90 to 95% of cases, the β phase shows a strong piezoelectric effect for PVDF polymers (Salimi and Yousefi 2003; Whiter, Narayan, and Kar-Narayan 2014). Some hydrogen atoms are replaced by fluorine in PVDF polymers, making them more crystalline in the polar phase due to steric factors.

PVDF-based nanogenerators have several applications. PVDF-based piezoelectric shoes fabricated by introducing sole staves made up of stacks of PVDF foils. Piezoelectric shoes generate power due to the bending stress on the shoes during walking or running. Approximately 1 mW of power was generated from PVDF insole stave nanogenerators in shoes (Kymissis et al. 1998). Figure 5.4 shows a schematic representation of piezoelectric shoes. A cantilever type PVDF-based piezoelectric energy harvester (PEH) fabricated by using a laminated PVDF polymer with an NdFeB magnet at the tip of the cantilever using coupled model analysis (Jiang et al. 2009) is shown in Figure 5.5. and operated in resonant mode. At a frequency of 14 Hz and an acceleration of 0.92 g, this PEH generated a maximum output voltage of 32 V and a power output of 16 μW at an acceleration of 1.2 g under resonant frequency. A similar PVDF-based cantilever-based PEH device also gave a maximum power output of about 200 μW within a volume of 1 cm^3 at 120 Hz, with an acceleration amplitude of 0.25 g. A cantilever-based PVDF-based PENG device also studied with uncoupled model analysis and operated on resonant mode produced 112.8 μW of power measured within a volume of 13.1 mm^3 (Song et al. 2017). Various PVDF/PVDF-copolymer and piezoelectric nanoparticle composites that have been reported as PENGs are listed in Table 5.1.

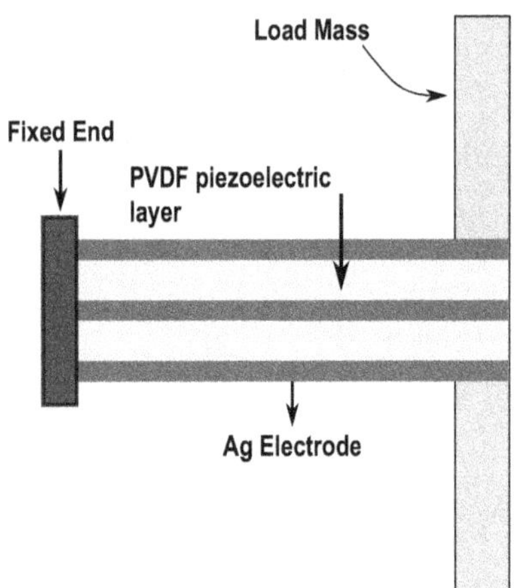

FIGURE 5.5 Cantilever type PVDF-based piezoelectric energy harvester (PEH).

5.2.1.2 Polyvinylidene Fluoride-Trifluoroethylene

Nanofibers of PVDF, blends with co-polymers or derivatives with different morphologies and aggregation, have been reported in the literature. The piezoelectricity of an electrically poled polyvinylidene fluoride-trifluoroethylene P(VDF-TrFE) nanograss structure gives a 5.2-fold higher output performance than the flat thin films of P(VDF-TrFE) under similar treatment conditions (Hong et al. 2012). The systematically arranged cylindrical cavities increase deformation thereby enhancing the performance of P(VDF-TrFE) nanofiber films upon exposure to an external force. Energy harvesting by human walking shows a maximum output voltage and short-circuit current (SCC) of 5 V and 1.2 μA, respectively (Gui et al. 2018). A solution-based PVDF-TrFE piezoelectric thin film placed with a cellulose substrate was manufactured as a portable and flexible power generator for a wearable device application (Won et al. 2015), as shown in Figure 5.6. Synthesis of the co-polymer in optimized conditions results in the formation of a β-phase, responsible for generating a maximum voltage of 1.5 V at 1 Hz in an open circuit from a paper power generator.

The piezoelectricity generated from an electrospun PVDF-TrFE nanofiber due to preferentially oriented induced dipoles was caused by a single-stage electrospinning process (Mandal, Yoon, and Kim 2011). PVDF-TrFE nanofibers have great application in flexible piezoelectric nanogenerators and nano-pressure sensors. PVDF based co-polymers were synthesized using a variety of processes, including spin coating (Suresh et al. 2010), hot pressing, electrospinning, solvent casting (Jain et al. 2013), and tape casting (Seema, Dayas, and Varghese 2007). A spin coating technique uses a green solvent like 1,3-dioxolane (DXL) to synthesize a PVDF-TrFE thin film (Nunes-Pereira et al. 2016). The dielectric constant of the PENG was about 11, while the piezoelectric coefficient was around 21 pC N^{-1} and was obtained for the green solvent-based PVDF-TrFE films which were stable up to 100 °C. These green solvent-based PVDF-TrFE films show a promising future in sensors and actuator applications. PVDF-TrFE exhibits an enhanced piezoelectric effect by combining with piezoelectric non-materials. Modified PVDF-TrFE with a ZnO composite film shows a large specific area with high crystallinity, which helps to increase the piezoelectric effect. This work shows that modified ZnO nanoparticles were synthesized with *n*-propylamine as a dispersant and a silane coupling agent perfluorooctyltriethylsilane. A PVDF-TrFE film with a modified ZnO thin film shows an increase in the crystallinity by 36.12% and a piezoelectric constant d_{33} value by 73.5% compared to normal PVDF-TrFE (Li et al. 2019).

TABLE 5.1
PVDF/PVDF-Copolymers and Piezoelectric Nanoparticles Composite-Based PENG.

Polymer	Composite Material	Voltage (V)	Current (μA)	Current Density (μA cm^{-2})	Power Density	References
PVDF	PVDF/GO nanofiber	21				Li et al. (2018)
PVDF	rGO–Ag/PVDF nanocomposite	18	1.05		28 W m^{-3}	Pusty, Sinha, and Shirage (2018)
PVDF	$BiCl_3$/PVDF nanofibers	1.1				Chen et al. (2020)
PVDF	Graphene/PVDF	7.9	4.5			Abolhasani, Shirvanimoghaddam, and Naebe (2017)
PVDF	Pt-PVDF nanofiber	30	6000		22 μW cm^{-2}	Ghosh and Mandal (2018)
P(VDF-TrFE)	P(VDF-TrFE)/$BaTiO_3$	13.2		0.33		Chen et al. (2017)
P(VDF-TrFE)	Polydopamine Dopped-$BaTiO_3$- P(VDF-TrFE)	6	1.5			Guan, Xu, and Gong (2020)
P(VDF-TrFE)	P(VDF-TrFE)-PMN-PT and PDMS-rGO	8.5	3		6.1 μW cm^{-1}	Yaqoob et al. (2019)
PVDF-HFP	PVDF-HFP/Ni-doped ZnO nano-composite	1.2	0.019–0.021		10.16 W m^{-3}	Parangusana, Ponnamma, and Al Ali AlMaadeed (2017)
PVDF-HFP	PVDF-HFP/carbon black composite films	3.68			28.3 W m^{-3}	Wu et al. (2014)

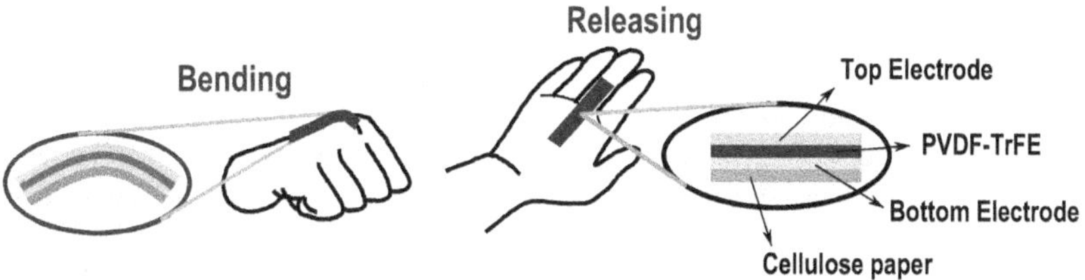

FIGURE 5.6 PVDF-TrFE piezoelectric thin film placed with cellulose paper made into a portable and flexible power generator for wearable devices.

In the energy harvesting application, a $BaTiO_3$-PVDF-TrFE composite was synthesized by a spin coating technique, where the composite acts as a filler material. The respective composite shows an energy harvesting performance of around 0.28 μW by using 20% filler material with 10 nm sized particles and a 5% filler material with 100 and 500 nm sized particles (Nunes-Pereira et al. 2015).

5.2.1.3 Polyvinylidene Fluoride Hexafluoro-Propylene

Polyvinylidene fluoride hexafluoro-propylene P(VDF-HFP) shows a high piezoelectric effect and a high electrostrictive value. The crystallinity results of the piezoelectricity of P(VDF-HFP) films were extensively explored (Hu et al. 2015). The P(VDF-HFP) shows improved and better properties than PVDF, such as higher solubility and mechanical strength (Wang et al. 2018). The composite, where the H_2O content shows a significant effect on electrical properties (Ma et al. 2018) of 15 wt.% of MMT and 1.72 wt.% of H_2O, was proven to be the optimum concentration with P(VDP-HFP). PVDF-HFP/Li-H_2O exhibited a high dielectric constant, high open-circuit voltage (OCV), and good current density due to Li^+ polarization and a β-phase nucleation effect generated by a hydrogen bond.

5.2.2 Polyamide

Polyamide(PA) has proved its capability in PENG applications. Polyamide exhibits stable dipole moments due to the presence of -C=O and -NH in the amide groups in a similar direction to generate a piezoelectric effect (Mathur, Scheinbeim, and Newman 1984). Nylon 11 shows extensive polymorphism with different crystalline phases, such as α, α', Υ, and δ'. Among these phases, the δ' phase is the only phase that shows high piezoelectric behavior due to the presence of a polar smectic pseudohexagonal chain with insignificant inter-chain hydrogen bonds (Choi et al. 2018; Nair, Ramesh, and Tashiro 2006). Generally, its electric properties are based on its crystal structure which depends on its processing conditions (Liu et al. 2007). The electrical property of polyimide 11 maximizes its electrical polarization by using its specific crystalline phase as a meta-stable pseudohexagonal phase (Scheinbeim, Lee, and Newman 1992; Zhang, Litt, and Zhu 2016). A polyamide 11/barium titanate ($BaTiO_3$)/graphene composite shows a significant increase in piezoelectric properties because of the unique disrupted graphene network with a porous structure (Jin et al. 2020). Here, graphene interfaces act as an electrode. The piezoelectric coefficient (d_{33}) of this piezoelectric composite obtained 3.8 pC N^{-1} with an OC voltage of 16.2 ± 0.4 V. The piezoelectric properties of the polyamide 11/lead-zirconate-titanate/carbon nanotube (polyamide11/PZT/CNT) tri-composite have been studied (Carponcin et al. 2014). Here, polyamide 11 was the polymer, and PZT and carbon nanotubes acted as fillers. Carbon nanotubes helped to produce a maximum piezoelectric coefficient (d_{33}), 26% higher for a poling field of 2 kV mm^{-1} than normal polyimide 11/PZT composite results. Aliphatic and aromatic hyperbranched polyamides have also been used for making piezoelectric immune sensors (Liu et al. 2007). A PENG based on a castor oil-derived, cellulose, nanocrystal composite film was fabricated using a simple casting process (Ram et al. 2019). A crystalline α-phase of neat Nylon 11, completely converted into the polar γ-phase by incorporating cellulose nanocrystals with a low concentration (2–5%), leads to an increase in the piezoelectric properties. Formation of a strong hydrogen bond between cellulose nanocrystals and amide groups in Nylon 11 with a 5 wt.% Nylon 11/CNC composite film-based PENG generates a 2.6-fold higher OCV than a regular nylon-based PENG.

5.2.3 Polyvinyl Chloride

Polyvinyl chloride (PVC) is primarily an amorphous polymer. PVC films obtained by solution casting with stretching and corona poling increases the piezoelectric coefficient (Bharti, Kaura, and Nath 1995). The piezoelectric properties of a PZT/PVC composite have been studied by adding graphite to the composite (Liu et al. 2006). A C/PZT/PVC composite has been synthesized by a hot-pressing method using nanocrystalline PVC. The piezoelectric coefficient (d_{33}) for a 0.005C/0.5PZT/0.495PVC composite was increased by 50% as compared to a control without graphitic carbon. Graphite particles form coagulation around the piezoelectric composites, which results in higher electrical conduction. A two-dimensional fabric like PVC fiber-based PENG was synthesized using a $BaTiO_3$ nanowire and a PVC fiber (Zhang et al. 2015); the structure is shown in Figure 5.7. The presence of the active $BaTiO_3$ nanowire enhances the piezoelectric properties of a PENG. The respective PENG attached to a protective garment like an elbow pad creates electricity due to the bending movement

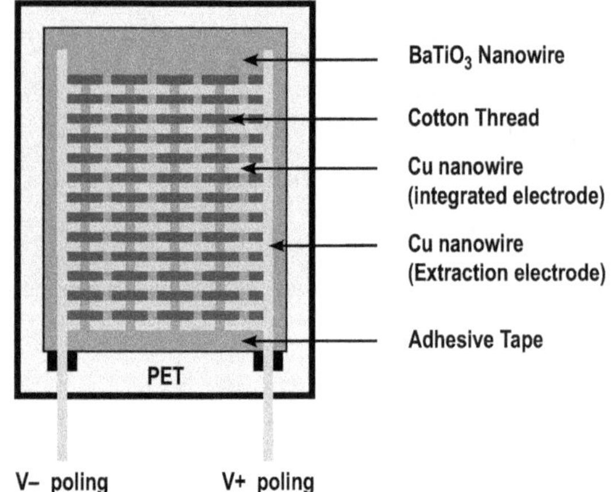

FIGURE 5.7 A $BaTiO_3$ nanowire and a PVC fiber-based PENG.

and the friction of the human arm. This device could generate 1.9 V in an open circuit. The SCC was 24 nA, and the generated energy was sufficient to power an LCD. $ZnSnO_3$ (35 wt.%) nanoparticle-PVC composites fabricated by a solution casting method generated an OCV of 40 V with an SCC of 1.4 μA at a power density of 3.7 μW cm^{-2}. These nanogenerators charged a capacitor of 2.2 μF to about 6.7 V in 129 seconds. The power generated from the respective composite-based nanogenerators were utilized to power seven different colored LED bulbs, a calculator, a mobile LCD, and a wristwatch.

5.2.4 Poly-L-Lactic Acid

Poly-L-lactic acid (PLLA) shows piezoelectric behavior due to its chiral nature. In the PLLA structure, each substantial group shows distinct optical properties from the others (Zhang et al. 2005). This property makes PLLA a more attractive polymer for a researcher to study as an alternative to conventional polymeric and inorganic piezoelectric materials (Sawano et al. 2010). PLLA nanofibers have been synthesized by the electrospinning process, which shows piezoelectric behavior with fiber direction (d_{33}) (Zhao et al. 2017). This also indicates that supercritical CO_2 treatment of a PLLA fiber increases the piezoelectricity behavior, which arises from enhanced crystallinity; similar results were also reported by other researchers (Cuong et al. 2020). A folded electrospun PLLA nanofiber web with the parallel connection of an electrode assembly used as a sensor shows increased piezoelectric signals. The energy generated from the PLLA-based PENG was utilized to power LEDs (Lee, Arun, and Kim 2015). Piezoelectric PLLA nanofibers on a comb electrode was synthesized by a direct current electric field electrospinning process, which could be used in harvesting energy from human motion as well as sensing strain (Zhu, Jia, and Huang 2017). Such PLLA piezoelectric nanofibers generate charges which flow through the comb electrode for measurement. This nanofiber-based PENG generates an OCV of 0.55 V and an SCC of 230 pA, at a strain deformation angle of 28.9°. This PLLA-based PENG produced a highest peak power of 19.5 nW from a human joint motion energy harvesting application. The piezoelectric properties of PLLA nanofibers mainly based on the β crystalline phase of the polymer increased by applying an electric field in the electrospinning process (Lee, Arun, and Kim 2015). They also concluded that shear piezoelectricity was enhanced by the orientation of the polymer web stacking and the thickness of the nanofiber web. PLLA/graphene electrospun composite nanofibers synthesized by an electrospinning process results in a 30% increase in the crystallinity and a 2048% increase in the d_{14} (piezoelectric coefficient) compared to the normal PLLA (Li et al. 2020). A piezoelectric bioelectronic skin was made by staking an electrospun PLLA/graphene nanofiber mesh over with a polyester cloth and poly-dimethyl siloxane. On touching, the composite shows an OCV of 184.6 V and an SCC of 10.8 μA. The generated power could then be converted into a digital signal for applications such as wearable sensors for monitoring body pulse.

There are a number of other materials and composites that show piezoelectric properties, such as PEO (Khanbareh, Zwaag, and Groen 2015), ionic gel (ionic liquid immobilized by a polymeric matrix) (Villa et al. 2019), polypyrrole (Ebarvia, Cabanilla, and Sevilla 2005; Vigmond et al. 1992), polyurethane (Shafeeq et al. 2020; Souri, Nam, and Lee 2015), PVA (Hashim et al. 2017), PMMA (Chelu et al. 2020), and PCL (Liu et al. 2020), which show considerable electrical output.

5.3 TRIBOELECTRIC NANOGENERATORS

A triboelectric nanogenerator (TENG) is a cost-effective and clean technology. A TENG works on the triboelectrification effect along with electrostatic induction, which transforms mechanical energy into electrical energy at ambient temperatures (Cao et al. 2018, 2016; Wang, Wang, and Yang 2016; Wang and Wang 2019). The working modes of triboelectric effect-based devices are shown in Figure 5.8. A TENG offers high output performance, ease in manufacturing, high power density, high efficiency and flexibility, lightweight with low cost, high stability, and environment-friendly properties. Polymer materials have been widely studied in TENG applications (Xu et al. 2017; Zhang et al. 2017).

Wang and his research group developed TENGs in 2012 (Fan, Tian, and Wang 2012) in which a polymer-based TENG was made by stacking two different triboelectric properties carrying materials with a deposition of a metal sheet on the top and bottom. Under mechanical deformation, surface roughness creates friction in the two films, resulting in the generation of equal and opposite charges on both sides. The triboelectric potential layer generated at the interface acts as a driving force for electron transport with the change in the capacitance. This polymer shows a 3.3 output voltage at a 10.4 mW cm^{-3} power density. The polymeric materials used in TENG-based applications are discussed below.

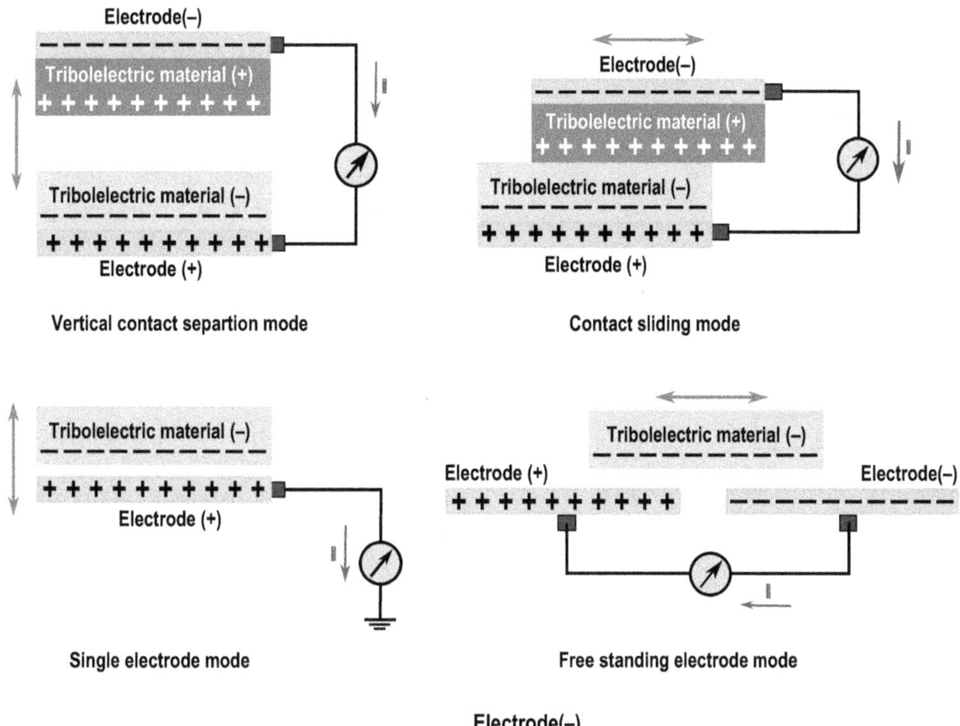

FIGURE 5.8 Working modes of triboelectric effect-based devices.

5.3.1 Polytetrafluoroethylene

Polytetrafluoroethylene (PTFE) is an electronegative polymeric widely used as a triboelectricity generator (Burgo et al. 2012; Liu et al. 2018). Triboelectric materials PTFE and Al with negative and positive tendencies, respectively, are used to synthesize TENGs by a thermal imprinting lithograph process (Zhong et al. 2019), which can easily and efficiently harvest energy from stable or unstable or fluctuating surfaces as tilting sensitive TENGs. In this process, PTFE is modified by producing micro-grooved architectures, which increase surface roughness, increasing the contact area between PTFE and Al, enhancing the performance of the respective TENG. The peak current and output voltage obtained after structural optimization were 25.59 μA and 302.87 V, respectively, with a wave amplitude of 2.5 V at a frequency of 1.2 Hz. When matched with an optimum resistance of about 20.6 MΩ, the peak and average power of the device were 2.84 mW and 0.2 mW, respectively. A high working efficiency at low-frequency conditions makes this PTFE-based TENG a promising contender for powering many of the sensor devices on a ship, which were demonstrated to power 30 LEDs and many capacitors.

A self-powered sensor for touchpad applications harvesting mechanical energy was developed and powered by an arch-type single electrode-based sliding TENG using plates of PTFE and Al (Yang et al. 2013). A PTFE plate and aluminum plates were placed parallel to each other, but the inner sides of both plates were in intimate contact. During sliding, a periodic change in the contact area was coupled with the electrostatic induction and triboelectric effect. The TENG generated an OCV of nearly 1100 V and an SCC density of around 6 mA m^{-2}. At an external loading resistance of 100 MΩ, a maximum power density of 350 mW m^{-2} was obtained. The generated energy was utilized to power 100 green light-emitting diodes.

An assembly of an electrode consisting of metal deposited on PTFE film on one side and in contact with another metal electrode (contact electrode) on the other side was made into a triboelectric nanogenerator (Zhu et al. 2014). Such a PTFE-based TENG could generate an average power output of 3 W with a nearly 50% average energy conversion with an overall area of 60 cm^2, a volume of 0.2 cm^3, a weight of 0.6 g, and a sliding velocity of 10 m s^{-1}. A polytetrafluorethylene/polyethylene oxide (PTFE/PEO) membrane was synthesized using an emulsion electrospinning process, eliminating the carrier PEO by a sintering process to obtain pure PTFE nanofibers. An electrode assembly of a porous electrospun PTFE membrane and a polyamide 6 (PA 6) layer generated by phase inversion was used to fabricate a PTFE/PA6-based vertical contact-mode TENG with a nanostructured rough surface (Zhao et al. 2018). Further, PTFE was plasma treated to generate negative charges on the surface to increase the power output. The TENG could generate an OCV of about 900 V, an SCC density of 20 mA m^{-2}, and a charge density of 149 μC m^{-2}, which was nearly 14-fold higher than the individual values before the ion injection process.

PTFE-based triboelectric-electromagnetic hybridized nanogenerators (TENG-EMG), designed to harvest wave energy, solved the power supply problem to wireless sensors

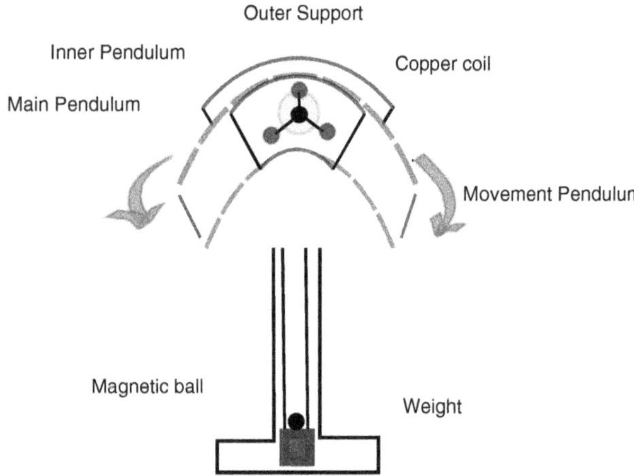

FIGURE 5.9 Schematic diagram of the hybridized (TENG-EMG) nanogenerator.

in a marine environment monitoring system (Chen et al. 2020). The schematic representation of the TENG-EMG is shown in Figure 5.9. When these hybridized nanogenerators are agitated by generating waves, the PTFE film on the inner pendulum side on the interdigitated electrodes (IDEs) generates electrical energy using oscillating mechanical energy. Three magnetic balls inside the chaotic pendulum get displaced due to their own gravity and external magnetic conditions; this leads to a changing magnetic flux in the copper coils, which are connected to the inner side of the chaotic pendulum. This shows that hybridized nanogenerators efficiently harvest the oscillating mechanical energy of waves and convert low mechanical vibration energy into electric energy, which can efficiently light up around 100 LEDs. The maximum power observed for the TENG was 15.21 μW while for the EMG it was 1.23 mW.

5.3.2 Fluorinated Ethylene Propylene

Fluorinated ethylene propylene (FEP) has strong electrophilic capabilities and was used as an electronegative material in a pair of triboelectric materials (Lin et al. 2015; Xie et al. 2014). A FEP-based spherical whirling-folded TENG (WF-TENG) with a 3D-printed substrate has been studied (An et al. 2019). The WF-TENG efficiently generated a peak power of 6.5 mW and an average power of 0.28 mW. This WF-TENG was used in a self-powered temperature sensing device and for self-monitoring the working state of the device. This spherical TENG with a spring-assisted multi-layered structure gave a maximum OCV of 560 V, an output current of 120 μA, and a peak power of 7.96 mW at a frequency of 1.0 Hz (Xiao et al. 2018). A TENG array studied for four spherical units connected in parallel was optimized and fabricated, which produced a peak output power of 15.97 mW at the matched resistance of 2.21 MΩ, and the generated electrical energy was utilized to power a dozen LEDs and an electronic thermometer.

A FEP-based U-tube TENG has been developed by considering a liquid–solid mode where 11 liquids were applied to check the impact of liquid properties on the output performance of the TENG (Pan et al. 2018). A Cu electrode, a FEP U-tube, and a liquid solution are the three parts of the device. A FEP material was selected due to its carbon–fluorine molecular structure, electronegative nature, and complete fluorination. Eleven different liquids were studied for their effect on the OCV, SCC, and peak power generated by the TENG. A pure water TENG showed the best performance. This TENG harvested mechanical energy in a shaking mode and a shifting mode, and was driven by a shaker and linear motor selectively. This TENG showed the best output performance, and generated an OCV of 93.0 V and an SCC of 0.48 μA, in the horizontal shifting mode at a frequency of 1.25 Hz. This performance for the shaking mode was 81.7 V and an SCC of 0.26 μA. It also showed promising results for harvesting water wave energy, which delivered an OCV of 350 V, an SCC of 1.75 μA, and a power density of 2.04 W m^{-3}.

Similarly, the number of FEP-based TENGs with different structural designs and assemblies was used for the generation of electric energy from water wave energy and provided a potential blue energy harvesting approach (Li et al. 2017; Wang et al. 2015; Wang, Jiang, and Xu 2017; Xu et al. 2018). At frequency 1.43 Hz, a water-wave-driven TENG lit up 10 LEDs and powered an electrical thermometer for 20 min (Wang et al. 2015). A nanowire-based TENG with a contact area around 30 cm^2 used wave energy to generate an OCV and SCC of 200 V and 10 μA, respectively (Li et al. 2017). The power generated from a nanowire-based TENG can charge a 3.3 μF capacitor to a voltage capacity of 1.1 V in just 15 s. This TENG was applied to power a variety of devices such as a stopwatch, thermometer, humidity sensor, and ethanol sensor, in addition to powering an infrared-based wireless data transmission system and used as a self-powered wireless infrared system for communication.

An oblate spheroidal TENG (OS-TENG) arranged with dual triboelectric components, the upper one which was a steel plate made up of a FEP and the lower consisting of an iron shot with a copper-coated polymer (FEP and PET) made film, was used to harvest the energy from a rough and tranquil condition of the sea (Liu et al. 2019). The two parts of the OS-TENG work in the contact and separation mode. The maximum output generated by the upper part was an OCV of 281 V and an SCC of 76 μA at a frequency of 4 Hz with an amplitude of 12.5 mm. The output energy charged 4.7 μF capacitors to a voltage of 5 V in just 2 min, while 10 μF and 2.2 μF capacitors required 150 s and 35 s, respectively.

5.3.3 Cellulose

A cellulose-based TENG leads to a more eco-friendly and green approach. Cellulose nanofibrils (CNFs) were utilized to fabricate a high-performance TENG. Cellulose was used as the raw material. Attaching the methyl and nitro groups to the cellulose structure, a composite molecule results in a change in the tribopolarities of the CNF, which causes a significant increase in the triboelectric output (Yao et al. 2017). A CNF has weak polarization, limiting its capability to produce a surface charge and generating less electrical output than synthetic fibers. Using a cellulose structure with many hydroxyl groups makes it possible for the chemical reaction approach to introduce various functionalities to the CNF, which gives it different tribopolarities. The surface charge density in nitro-CNF and methyl-CNF of −85.8 and 62.5 μC m^{-2}, respectively, is significantly greater than the pristine CNF value of −13.3 μC m^{-2}. A nitro-CNF and methyl-CNF pair-based TENG generates an average OCV of 8 V and an SCC of 9 μA.

A pair-based Teng (P-TENG) made from a nitro-cellulose membrane (NCM) and crepe cellulose paper (CCP) shows potential in self-powered human-machine interfaces and sensors (Chen et al. 2019). In the P-TENG structure, CCP and NCM are used as friction layers, and the print layer acts as a substrate. Different tribopolarities and porous structures of the CCP and NCM generate excellent triboelectric performance for the P-TENG. A dry-creeping process was used to make the CCP, as shown in Figure 5.10. The doctor used such a CCP to create many microscopic folds, which increase the surface area of the tribopositive frictional layer, resulting in an output voltage of 196.8 V, an output current of 31.5 μA, and power density of 16.1 W m^{-2}. The P-TENG showed good stability in the output performance and more than during the 10,000 press/release cycle. A vacuum-filtration process was used to fabricate a cellulose nanofiber-based TENG with the various conditions of a cellulose nanofiber and an Ag nanoweb (Kim et al. 2018). A cellulose nanofiber was used as a free-standing triboelectric nanogenerator, with an Ag nanoweb layer employed on it to form an electrode and create a triboelectric counterlayer. For the CN-TENG, maximum output performance obtained with 0.1 wt.% of Ag nanowires and 20 passes in 1000 bar of CNF paper leads to a generated peak OCV and SCC of 21 V and 2.5 μA, respectively. The maximum power density of 693 mW m^{-2} for 10 MΩ of circuit resistance was obtained.

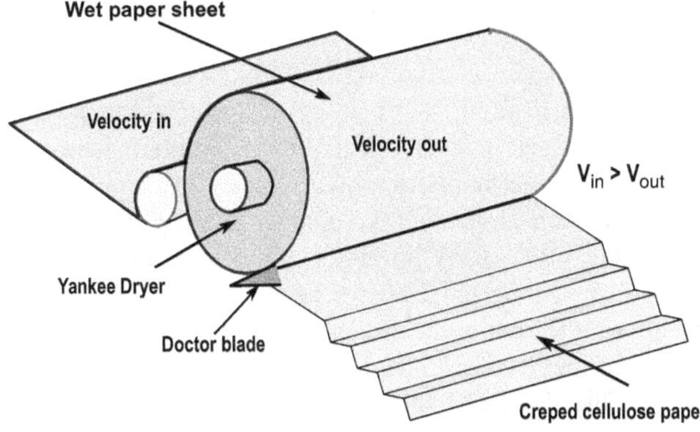

FIGURE 5.10 Dry-creeping process to synthesis crepe cellulose paper.

5.3.4 Polyvinylidene Fluoride

PVDF and its co-polymers possess strong piezoelectric properties, and due to their inert nature, satisfactory mechanical robustness, and preferable flexibility makes them an attractive polymer material for triboelectric energy generation technology (Fatma et al. 2020; Kaur et al. 2016; Lee, Lee, and Baik 2018; Xia et al. 2018). A humidity of 10% PVDF in combination with bacterial cellulose (BC) based triboelectric energy generators produces a maximum OCV of 410 V, an SCC of 14.8 μA, with a current and power density of 0.57 μA cm^{-2} and 3.5 mW, respectively (Fatma et al. 2020). The obtained maximum power density of around 0.136 mW cm^{-2} occurred when an external resistance of 33 MΩ was added in the circuit. This TENG device was used in a biomechanical energy harvesting application and lit up more than 100 LEDs and a five-bit seven-segment LCD display. The respective device also showed application for powering a motion sensor. PVDF and PTFE pair-based TENGs were used for powering temperature sensors (Xia et al. 2018). This flexible TENG was fabricated by using a triboelectric pair of PTFE and PVDF, a conductive electrode of copper, and silicone paper with a PVDF board for support. The OCV was found to be dependent on the relative humidity of the surrounding environment. At a load of 10 MΩ for a temperature of 20 °C, the output voltages of the TENG were observed as 42, 37, and 32 V for a relative humidity 70, 80, and 90%, respectively, which explained that the increase in relative humidity in the atmosphere reduced the output performance of the TENG device. In particular, the OCV of the TENG increased with temperature. Polarized PVDF thin films were utilized in TENG technology, fabricated by using two layers. The first was uniformly sized with well-distributed nanopores carrying aluminum foil; the second was composite and made up of a copper deposited PVDF thin film and acrylic sheets (Bai et al. 2014). The aligned dipoles generated by the bond charges in the polarized PVDF affected the potential energy on the surface of the polarized PVDF. At a constant contact force 50 N, the output OCV was observed in the range 72 to 215 V.

A novel TENG was fabricated using electrospun PVDF and nylon nanowires as triboelectric pairs with Ag wires. This was the first time a TENG was used to power a direct current (DC) motor (Zheng et al. 2014). The power generated by the TENG due to the vibrational energy of 5 Hz frequency with an amplitude of 20 mm showed an OCV of 1163 V, an SCC density of 11.5 mA cm^{-2}, and a peak power density of 26.6 W m^{-2}. This TENG demonstrated a self-powered ultraviolet radiation (UV-R) detection device that indicates the UV-R's level directly with no additional components and shows the potential in the different public health applications.

A PVDF-based TENG was fabricated by using modified multiwalled carbon nanotubes (PBA-MWCNTs) (Kim, Song, and Heller 2017). The surface of the MWCNTs was modified by 1-pyrenebutyric acid (PBA) using a chemical process. These modified nanoparticle PBA-MWCNTs

(PCNTs) show compatibility with PVDF polymers. A β-Phase nucleation site was provided to the fluorine in the PVDF polymer by the carboxylic acid groups of the PCNTs. The maximum content of the β-crystalline phase of PVDF polymer was observed at a concentration of 1.0 wt.% of the PCNTs. The output performance of a 1.0% PCNTs/PVDF composite-based TENG showed a maximum OCV of 16 volts. The arch-shaped TENG was fabricated by adding reduced graphene oxide nanoribbons (rGONRs) in a PVDF polymer (Kaur et al. 2016). Here the PVDF polymer was used as a binder. The presence of rGONRs in a PVDF polymer leads to an increase in the average surface roughness of the rGONRs/PVDF composite thin film. The film with functional groups and enhanced average surface area improves charge storage capacity. After comparing the redox peaks obtained from cyclic voltammetry for rGONRs/PVDF with pristine rGONR, the increase in the charge transfer capability was confirmed. The maximum output voltage was observed to be 0.35 V. The PVDF-ZnO-NWs/nylon-ZnO-NWs based TENG was fabricated by an electrospinning process, which resulted in a high proportion of the β-crystalline phase (Pu et al. 2020). This TENG showed a maximum power density of 3.0 W/m^2 under an external load of 10–20 MΩ; the generated energy was used to light up more than 100 LEDs. A Europium-doped PVDF nanofiber (Eu-PVDF nanofiber) was produced by electrospinning and used in a TENG as an active layer (Kim and Park 2018). The TENG's output performance showed a power density enhancement from 13 to 26 μW/cm^2 as the Eu increased from 0 to 2.7 wt.%. When the Eu content increases beyond 5.3 wt.%, the power density drops to 4.9 μW cm^{-2}.

The TENG can also be fabricated by a co-polymer of PVDF-TrFE with controlled high-performance piezoelectric and triboelectric properties. A PVDF-TrFE-based TENG gives around 65 and 75% improvement in output voltage and current, respectively, compared to a PVDF-based TENG (Kim et al. 2017). The TENG based on PVDF-TrFE nanofibers and PDMS/MWCNT shows a flexible nature and generates a peak–peak voltage of 25 V, power of 98.56 μW, and power density of 1.98 mW cm^{-3} under a pressure force of 5 N (Xingzhao Wang et al. 2016). A PVDF-TrFE-based TENG shows application in high output wearable devices. A PVDF-HEP and PDMS ion gel-based TENG obtained a maximum power density of about 0.9 W m^{-2} and generated energy used to light up 100 LEDs (Lin et al. 2018).

5.3.5 Polyamide

Electrical polarization in odd-numbered nylon, such as Nylon 11, enhances the surface charge density. Therefore, Nylon 11 has been widely utilized as a tribopositive material in TENGs (Lee et al. 2015; Zhou et al. 2014). A Nylon 11 nanowire-based TENG device was fabricated by using a δ'-phase-Nylon 11 nanowire (Choi et al. 2017). The quenching technique at a temperature of 210 °C and mechanical stretching or a high voltage (140 V) electrical poling process has been used for rapid crystallization required for δ'-phase Nylon 11 film. But novel one-step, gas-flow-assisted nano-template (GANT) infiltration was used to fabricate a crystalline self-poled' δ'-phase Nylon 11 nanowire structure within nonporous anodized aluminum oxide (AAO) templates with no use of mechanical stretching/poling. A preferable crystal orientation was observed in GANT-Nylon 11 nanowires due to template-induced nano-confinement as compared to the orientation of crystals in the melt-quenched Nylon 11 films. The electrical output performance for Al, melt-quenched Nylon 11 and GANT-Nylon 11 nanowire-based TENGs was obtained, with output OCVs of 40, 62, 110 V; SCC densities of 13, 21, 38 mA m^{-2}; and power densities of 1.03, 0.19, 0.099 W m^{-2}, respectively, at a 5 Hz frequency at 0.5 mm amplitude. GANT-Nylon 11 nanowire-based devices showed a nearly six and ten times increase in output power density than those of a control TENG made from films of melt-quenched Nylon 11 and aluminum, respectively. The a-phase-nylon 11 nanowire-based contact-separation mode triboelectric generator (TEG) was fabricated by the optimized thermally assisted nano-template infiltration (TANI) method (Choi et al. 2020). The performance of the Al-based and d′-phase nanowire-based TEG was compared to that of an a-phase Nylon 11 nanowire-based TEG. The generated SCC was about 38 and 74 mA m^{-2}, observed from an Al-based d′-phase nanowire-based and a-phase nanowire-based TENG, respectively. The generated output power from an a-phase nanowire-based TEG was nearly three-fold higher than the d′-phase nanowire-based TEG. The performance was about 35-fold higher than the Al-wire-based TEG. This enhancement in the performance was attributed to the high net dipole moment and charge-donating property of Nylon 11.

5.3.6 Polydimethylsiloxane

Polydimethylsiloxane (PDMS) showed an ideal behavior for flexible wearable device fabrication due to its inherent properties, like excellent elasticity and biocompatibility (Wang et al. 2014). The TENG device was mainly fabricated with two friction layers which generate electrical energy. During friction, many polymeric layers get damaged, resulting in a reduction in TENG performance. So the high-performing self-healing triboelectric nanogenerator (SH-TENG) was fabricated by polydimethylsiloxane (PDMS) and thermoplastic polyurethane (TPU) polymeric film with semi-embedded Ag nanowires (Niu et al. 2018). The result was a peak power of 2.5 W m^{-2} along with a 100 μC m^{-2} triboelectric charge density. Even after the five cycles of the cutting/healing process, the output performance of the SH-TENG device reached 99% of its original value. The SH-TENG shows great potential application in flexible power electronic devices that needed long-lasting resilience and reliability.

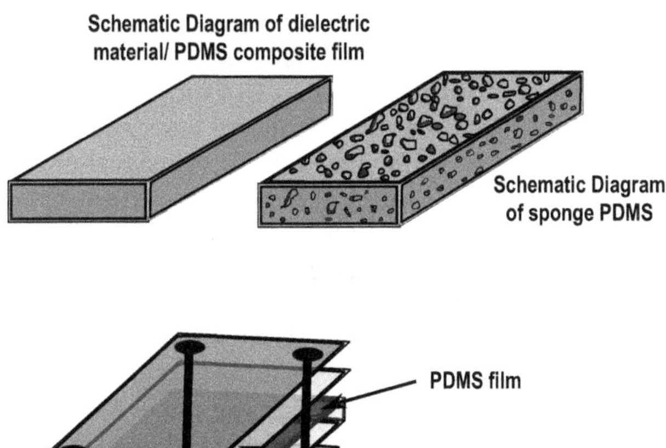

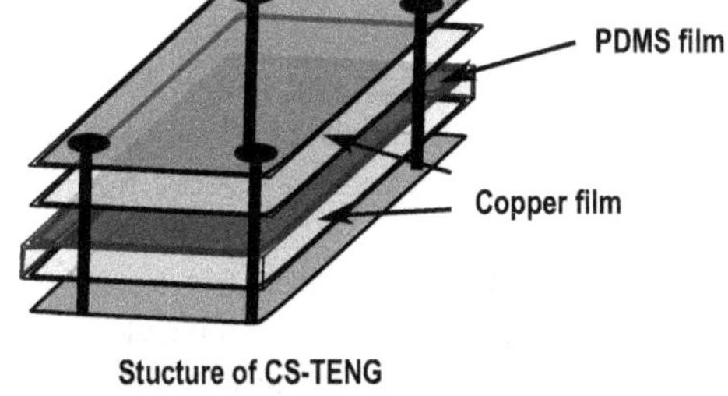

FIGURE 5.11 The design of flat and spongy PDMS films and PDMS and copper-film-based TENG.

A spongy structure-based PDMS-based TENG gives better results than solid or flat PDMS. Spongy or porous structure PDMS polymeric film was achieved by embedding micro- or nanoparticles in a PDMS polymer and removing them later. The design of PDMS films is shown in Figure 5.11 (Chen et al. 2016; Mao et al. 2014; Zheng et al. 2018). Micro- and nano-sphere particles were utilized to fabricate porous PDMS material, which shows maximum output performance of an output voltage of 130 V and the current density reaches 0.10 mA cm^{-2} (Lee et al. 2014). The power generated by a porous PDMS-based TENG was ten times higher than the flat-film-based TENG. Similarly, ZnO nanoparticles used in a PDMS-based TENG exhibited a 3.7 times higher electrical output than flat PDMS film (He et al. 2016).

A TENG was fabricated by using modified triboelectric material (PDMS) with high permittivity nanoparticles and pores (Chen et al. 2016). The study indicates the amount of charge transferred as well as the surface charge density strongly affect relative permittivity, surface area, thickness, and porosity of the tribomaterial. A composite sponge-based TENG (CS-TENG) shows the enhancement of permittivity due to an optimized amount of filler 10% $SrTiO_3$ nanoparticles (ca. 100 nm size) with a pore volume of 15%. It shows a charge density of about 19 nC cm^{-2}, an OCV of 338 V, and a power density of 6.47 W m^{-2} at a frequency of 2.5 Hz. This CS-TENG gives a five times increase in output power compared to a pure PDMS film-based TENG (P-TENG) and was used to power 44 LEDs. The enhancement in the electrical output performance of the TENG shows a linear relation with porosity for a specific range due to the increase in the ratio of the actual friction area to the thickness of the polymeric film. The PDMS-based TENG was fabricated utilizing a 3D printer, which is cylindrical shaped and completely packed, carrying a PDMS material made bumpy by balls inside its volume and an Al film linearly attached to its inner surface (Lee et al. 2017). The mold-casting technique was used to synthesize PDMS bumpy/rough balls. This newly designed PDMS-based TENG shows a significant enhancement in the output power at around 45 mW and can operate in harsh conditions like water, and generate enough DC power to charge a smartwatch battery. When a cylinder-shaped 3D printer shakes the PDMS balls, it creates contact with the electrodes, which transfer the negative charge to the PDMS balls. The linear arrangement of the Al film makes more contact with the PDMS balls with an effective moment, which results in the enhancement in transferred charge density. For this modified TENG, it was observed that the decibel level of the noise ranged from 45 to 52 dB during the working of the device, which corresponds to a normal conversation noise level. So, it is concluded that reduction of the noise level by 20 dB occurs without loss of output power.

5.3.7 Polyimide

Polyimide (PI) has been widely used in TENGs as a negative friction layer because of its high negatively charged behavior (Jing et al. 2014; Zhao et al. 2019; Zhu et al. 2015). PI shows excellent stability or solidity as a friction layer in TENGs under recurring deformation or external pressure. The polyimide aerogel film-based TENG device offers the potential for energy harvesting and sensing applications (Saadatnia et al. 2019). Polyimide-based aerogel film shows negative electron affinity in TENGs, so positive electron affinity carrying copper film was used as a contact material in the effective triboelectric material pair. The performance of polyimide aerogel film-based TENGs has been tested for different percentages of the porosity levels of aerogel films in the polyimide-based aerogel composite. The polyimide aerogel film-based TENG showed better output performance than the dense polyimide layer due to the presence of pores which increase the effective surface area, relative capacitance, and generation of charges inside the open cells of the aerogel. Polyimide aerogel film-based TENGs with a 50% porosity of aerogel film showed the maximum output performance: an OCV of 40 V, a peak SCC of 5 μA, and a power of 47 μW, and was eight-fold higher than PENGs made of solid polyimide film.

PI is a highly chemical resistant solvent, so it is pretty challenging to make PI nanofibers using electrospinning by employing PI directly. So, modification in PI makes itself a compatible material for electrospinning. PI nanofibers were electrospun from PI ink consisting of powdered polyimide suspended in fluid media (Kim, Wu, and Oh 2020). PI nanofibers synthesized with 15 wt.% and 20 wt.% PI in a doped solution showed a continuous and fibrous PI nanofiber structure which is employed as a friction layer in TENGs. A PI-TENG with 20 wt.% of polymer showed a maximum output OCV, SCC, and power density of around 753 V, 10.79 μA, and 2.61 W m^{-2}, respectively, at a load resistance of 100 MΩ. The TENG harvested the electrical energy during 10,000 cycles of the tapping experiment. The generated energy was utilized to power tens of LEDs and a few low power electronic devices; PI nanofiber-based TENGs also show application in wearable devices. A wind-driven TENG, combined with a PI/rGO foam-based pressure-sensitive elastic polyimide, offers an excellent output performance effect (Zhao et al. 2019). A wind-driven TENG was fabricated using Ag nanoparticle coated photographic paper as electrode and nylon vibration film. PI/rGO nanocomposite foam is the pressure-sensitive element in this system. A wind-driven TENG with PI/rGO foam of volume 5.88 cm^3 generated an output voltage of 130 V and a current of about 7.5 μA by using a 100 mm × 15 mm effective contacting area.

C_{60}-containing block polyimide (PI-*b*-C_{60})-based TENGs work in non-contact mode to generate an electrical output (Jae et al. 2021). PI-*b*-C_{60} was synthesized using a cycloaddition reaction in which azide groups from polyimide react with C_{60}. The performance output of the PI-*b*-C_{60} based TENG shows a 1.6-fold higher charge density and a 4.3-fold higher output power than a perfluoroalkoxy alkane (PFA) film-based TENG. A TENG based on PI-*b*-C_{60} maintained about 255 μC m^{-2}, 85% of the maximum charge density, under a 3 Hz mechanical vibration after 5 h. Under a similar condition, a PFA-based TENG showed a decrease of nearly 50% of its maximum value in charge density value. C_{60} shows excellent charge-retention characteristics due to its high negative electrostatic potential within the backbone, leading to enhanced TENG output performance.

5.4 ELECTROSTATIC NANOGENERATORS

Electrostatic nanogenerators (ESGs) were the first mechanical energy harvester that produces electrical energy based on the principle of electrostatic induction or triboelectricity. There are three types of operational mode of electret generators: gap-closing, in-plane oscillation, and in-plane oscillation. Figure 5.12 shows the principle of electrostatic induction. In 1929 Van de Graff fabricated a machine that can produce a high voltage (Furfari 2005). Electret plays a vital role in electrostatic generators. Organic polymer-based electrets like PTFE and related materials have been studied (Sessler and West 1973). The stability of thermal charge storage can be improved by the unidirectional stretching of porous PTFE films at a high temperature (Xia, Wedel, and Danz 2003). In addition, the surface potential was studied at higher temperatures, such as 200 and 300 °C, which showed that no potential surface decay was observed at 200 °C for several hours and a 10% reduction of surface potential at 300 °C over five hours (Xia et al. 1999). An electret film of porous PTFE showed practical application for promoting fast healing and the recovery of injured tissues in the human body.

The first study reported parylenes as electret materials in 1980 (Raschke and Nowlin 1980). A charged parylene HT® thin polymer film was synthesized using chemical vapor deposition, which was used to develop the electret power generator. A charged parylene HT® with a 7.2 μm thick film showed a 3.69 mC m^{-2} surface charge density, which was studied using the corona discharge method. Microfabricated parylene HT®-based ESGs consisted of a sliding door with electrets that delivered a peak output power of 17.98 μW, when the oscillation frequency was 50 Hz against an external resistance of 80 MΩ, as well as at a lower frequency such as 10 and 20 Hz, which showed a power output of 7.7 and 8.23 μW, respectively.

CYTOP is a MEMS-compatible amorphous perfluoropolymer, a polymer electret (Arakawa, Suzuki, and Kasagi 2004). A CYTOP-based ESG showed a surface density of 1.3 mC m^{-2} for films of 15 μm in thickness and displayed a peak output power and peak-to-peak voltage of 37.7 mW and 150 V, respectively (Tsutsumino et al. 2006). Using aminosilane additives in a CYTOP polymer-based ESG provided a surface charge density around 1.5 mC m^{-2} for 15 μm thick films and showed a nearly 2.5-fold increase in power generation at a 20 Hz oscillation frequency as compared to a normal CYTOP-based ESG. It is theorized that

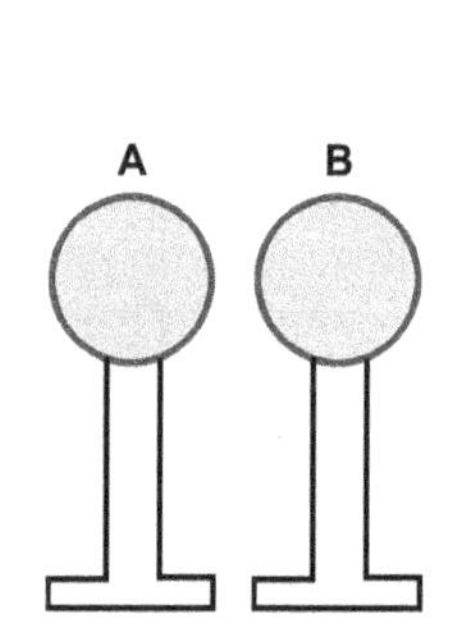

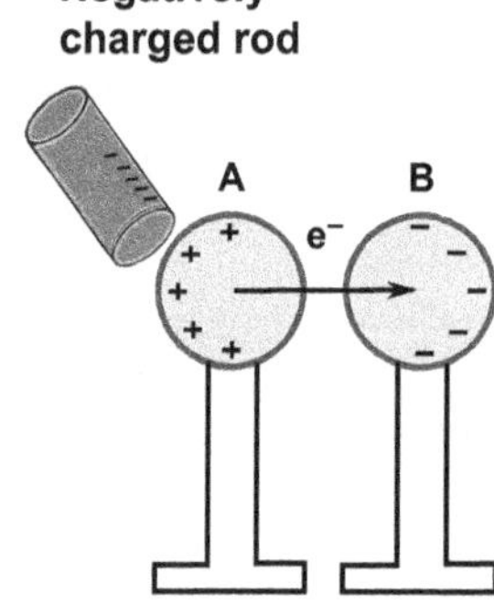

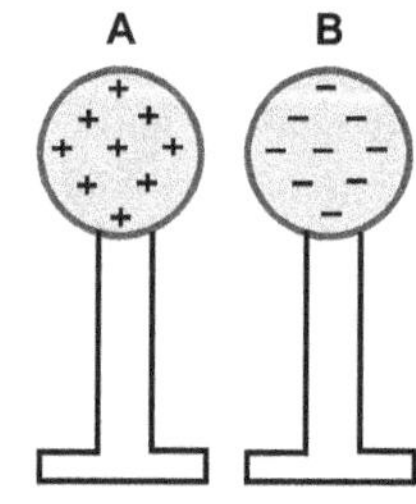

FIGURE 5.12 Schematic representation of the principle of electrostatic induction.

aminosilane traps the charge in the CYTOP electrets and makes it a more efficient electret material. Introducing aminosilane derivatives like 3-aminopropyl (diethoxy) methylsilane (3-APMS), 3-aminopropyl (triethoxy) silane (APTMS), and *N*-(2-aminoethyl)-3-aminopropyl-(triethoxy) silane (APTES) into a CYTOP solution and later dissolving that solution into a perfluorinated solvent to prepare a CYTOP-based perfluorinated polymer film shows a successful formation of nanoclusters of the organic siloxanes in it successfully (Kashiwagi et al. 2011). The study suggests that nanoclusters present in the polymeric electret serve as the charge trap, enhancing the thermal stability of the entrapped charges, thereby increasing the surface charge density. Almost 1.6 kV of surface charge potential is developed by the electret under a continuous corona discharge. CYTOP-based microelectromechanical system (MEMS) electret generators were used in various energy harvesting applications (Suzuki et al. 2010).

5.5 ELECTROMAGNETIC INDUCTION NANOGENERATORS

Many devices developed that convert mechanical energy to electrical energy using electromagnetic induction are electromagnetic induction generators (EMIGs), where a conductor is located within the magnetic field and generates electricity (Beeby, Tudor, and White 2006). In 1831, Faraday discovered electromagnetic induction, shown in Figure 5.13. He used two coils wound around the opposite sides of a soft iron ring: when the current passes through one coil, a momentary current is generated in the second coil. He also demonstrated that it induced an electric current to flow through the wire whenever a magnet moves around the wire's loop. The principle of electrotactic induction by the moving a magnet and moving a wire model is shown in Figure 5.14.

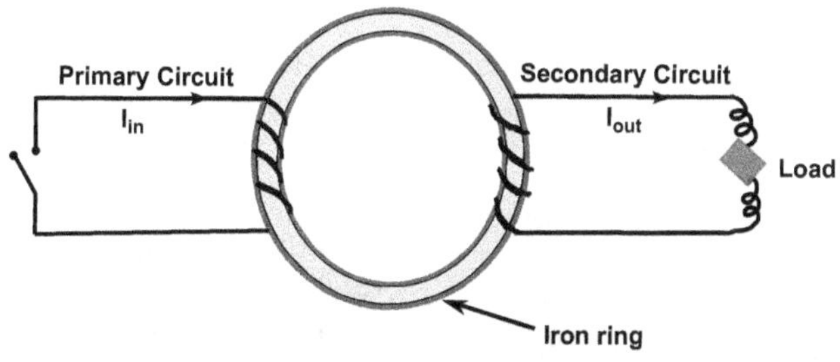

FIGURE 5.13 Schematic of Faraday's electromagnetic induction experimental set up.

In the environment, most of the vibration frequencies are lower than 10 Hz. EMIGs generate an output power based on the operational frequency; conventional energy generators barely work at a frequency below 50 Hz (Zi et al. 2016). Silicon-based electromagnetic generators (EMGs) were utilized to generate electricity from low vibration frequencies (Kulah and Najafi 2004). Polymer-like silicon has a relatively high resonant frequency, including less than 100 Hz ambient vibration frequencies. Using a silicon-based device for low/ambient environmental vibration frequency conversion leads to an open, effective option for a problem in the energy harvesting field. In the silicon-based generator assembly, an upper resonator with a resonant frequency of 25 Hz is a diaphragm with a NdFeB permanent magnet on it. They demonstrated this by fabricating a silicon generator which consists of the upper diaphragm having a resonance frequency of 25 Hz. A magnetic pull is employed to drive a lower diaphragm into resonance using an NdFeB magnet on the upper diaphragm. The resonance frequency of the lower diaphragm is roughly 11 kHz. The lower resonator is a cantilever beam with a higher resonance frequency, supporting the cantilever, which converts the mechanical frequency into electric energy; it also holds a magnetic tip that is attracted toward the NdFeB magnet. The coils used in the lower frequency vibration-based energy generated are arranged on the lower structure. The theoretical maximum power from a single cantilever was 2.5 μW for the microscale design set up, and the researchers obtained a maximum power of 4 nW for a milliscale design set up.

A high sensitivity (17 to 33 Ω cm) silicon substrate-based micromachine generator was fabricated by a chemical polishing process (Koukharenko, Tudor, and Beeby 2008). An HNA (hydrofluoric, fuming nitric, and acetic acid) solution with a 27:43:30 ratio was used in the chemical polishing process. The deep reactive ion etching (DRIE) process was used to carve a silicon wafer to fabricate the silicon paddle

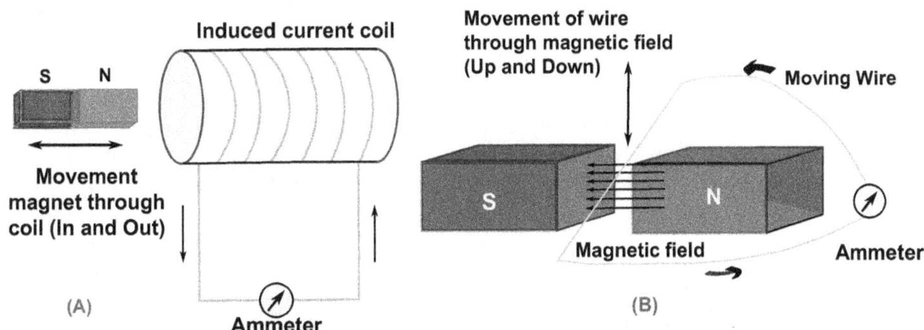

FIGURE 5.14 Schematic representation of the principle of the electromagnetic induction model by (A) a moving magnet and (B) a moving wire.

and to reduce its sidewall roughness. This microgenerator uses electromagnetic transduction to convert the ambient vibrations into electricity, whose output performance is based on the maximum amplitude of the movement, which is limited by the fracture strength of the silicon paddle. A multi-frequency energy harvester based on electromagnetic induction was fabricated using three pairs of bi-layer copper coils, three permanent magnets, and an acrylic beam support (Yang et al. 2009). The vibration energy under the three environments, which were excited at the frequencies of 369, 938, and 1184 Hz, was harvested: the first vibration mode gave an OCV and power of 1.38 mV and 0.6 μW, respectively, while those from the second mode of vibration were found to be 3.2 mV and 3.2 μW, respectively.

When the direct current EMIG output performance was compared with DC-TENGs, especially when lighter material was used in the TENG, the DC-TENG showed a higher maximum output power than the DC-EMIG. TENGs are self-powered, and flexible for any design of the device. They offer a wide range of materials for selection and for making TENG a potential option for future applications compared to EMIGs.

5.6 CONCLUSION

This chapter has discussed polymer-based piezoelectric, triboelectric, and electrostatic/electromagnetic nanogenerators and their application with operating mechanisms, device design, and performance. Polymer generators have excellent properties like flexibility, an active mechanical mechanism, and sensing ability. Polymeric nanogenerator application gives a broader aspect to converting wasted energy into electricity. Polymeric nanogenerators successfully scavenge energy from sources like ambient vibrations, wind and water flow, human motion, and sea wave vibrations. Nanogenerator devices show significant application in fields like wearable and stretchable devices, sensors devices, photovoltaic textiles, wind-driven nanogenerators, and blue energy harvesting. Though nanogenerators are very well developed, there are still several challenges and much scope for improvement, as the mechanism behind triboelectrification has many theories but which are not justified yet. A commercial viewpoint on polymeric nanogenerators has also yet to be found; however, there is a need to explore areas like the selection and optimization of materials for the system, the design of the NG, enhancement in the output performance of NG, reproducibility of results, and optimal units in energy harvesting devices like power management and signal processing energy storage elements. PNG devices also need mechanical stability and long durability.

REFERENCES

Abolhasani, Mohammad Mahdi, Kamyar Shirvanimoghaddam, and Minoo Naebe. 2017. "PVDF/Graphene Composite Nanofibers with Enhanced Piezoelectric Performance for Development of Robust Nanogenerators." *Composites Science and Technology* 138 (January): 49–56. https://doi.org/10.1016/j.compscitech.2016.11.017

An, Jie, Zi Ming Wang, Tao Jiang, Xi Liang, and Zhong Lin Wang. 2019. "Whirling-Folded Triboelectric Nanogenerator with High Average Power for Water Wave Energy Harvesting." *Advanced Functional Materials* 29 (39): 1904867. https://doi.org/10.1002/adfm.201904867

Arakawa, Yasuhiro, Yuji Suzuki, and Nobuhide Kasagi. 2004. "Micro Seismic Power Generator Using Electret Polymer Film." *The Proceedings of the National Symposium on Power and Energy Systems* 2004. 9 (June). https://doi.org/10.1299/jsmepes.2004.9.37

Bai, Peng, Guang Zhu, Yu Sheng Zhou, Sihong Wang, Jusheng Ma, Gong Zhang, and Zhong Lin Wang. 2014. "Dipole-Moment-Induced Effect on Contact Electrification for Triboelectric Nanogenerators." *Nano Research* 7 (7): 990–97. https://doi.org/10.1007/s12274-014-0461-8

Baker, Chris R., Kenneth Armijo, Simon Belka, Merwan Benhabib, Vikas Bhargava, Nathan Burkhart, Artin Der Minassians, and Gunes Dervisoglu. 2007. "Wireless Sensor Networks for Home Health Care." In Niagara Falls, Ont., Canada: *IEEE Conference Publication*. https://ieeexplore.ieee.org/abstract/document/4224209/.

Beeby, S. P., M. J. Tudor, and N. M. White. 2006. "Energy Harvesting Vibration Sources for Microsystems Applications." *Measurement Science and Technology* 17 (12): R175. https://doi.org/10.1088/0957-0233/17/12/R01

Bharti, V., T. Kaura, and R. Nath. 1995. "Improved Piezoelectricity in Solvent-Cast PVC Films." *IEEE Transactions on Dielectrics and Electrical Insulation* 2 (6): 1106–10. https://doi.org/10.1109/TDEI.1995.8881940

Burgo, Thiago A. L., Telma R. D. Ducati, Kelly R. Francisco, Karl J. Clinckspoor, Fernando Galembeck, and Sergio E. Galembeck. 2012. "Triboelectricity: Macroscopic Charge Patterns Formed by Self-Arraying Ions on Polymer Surfaces." *Langmuir* 28 (19): 7407–16. https://doi.org/10.1021/la301228j

Cao, Xia, Yang Jie, Ning Wang, and Zhong Lin Wang. 2016. "Triboelectric Nanogenerators Driven Self-Powered Electrochemical Processes for Energy and Environmental Science – Cao – 2016 – Advanced Energy Materials – Wiley Online Library." *Advanced Energy Materials* 6 (23). https://doi.org/10.1002/aenm.201600665

Cao, Xia, Meng Zhang, Jinrong Huang, Tao Jiang, Jingdian Zou, Ning Wang, and Zhong Lin Wang. 2018. "Inductor-Free Wireless Energy Delivery via Maxwell's Displacement Current from an Electrodeless Triboelectric Nanogenerator." *Advanced Materials* 30 (6): 1704077. https://doi.org/10.1002/adma.201704077

Carponcin, Delphine, Eric Dantras, Jany Dandurand, Gwenaëlle Aridon, Franck Levallois, Laurent Cadiergues, and Colette Lacabanne. 2014. "Electrical and Piezoelectric Behavior of Polyamide/PZT/CNT Multifunctional Nanocomposites." *Advanced Engineering Materials* 16 (8): 1018–25. https://doi.org/10.1002/adem.201300519

Cauda, Valentina, Stefano Stassi, Katarzyna Bejtka, and Giancarlo Canavese. 2013. "Nanoconfinement: An Effective Way to Enhance PVDF Piezoelectric Properties." *ACS Applied Materials & Interfaces* 5 (13): 6430–37. https://doi.org/10.1021/am4016878

Chelu, M., H. Stroescu, M. Anastasescu, J. M. Calderon-Moreno, S. Preda, M. Stoica, Z. Fogarassy, et al. 2020. "High-Quality PMMA/ZnO NWs Piezoelectric Coating on Rigid and Flexible Metallic Substrates." *Applied Surface Science* 529 (November): 147135. https://doi.org/10.1016/j.apsusc.2020.147135.

Chen, Chong, Zikui Bai, Yunzheng Cao, Mingchao Dong, Kankan Jiang, Yingshan Zhou, Yongzhen Tao, et al. 2020. "Enhanced Piezoelectric Performance of $BiCl_3$/PVDF Nanofibers-Based Nanogenerators." *Composites Science and Technology* 192 (May): 108100. https://doi.org/10.1016/j.compscitech.2020.108100

Chen, Xin, Lingxiao Gao, Junfei Chen, Shan Lu, Hong Zhou, Tingting Wang, Aobo Wang, et al. 2020. "A Chaotic Pendulum Triboelectric-Electromagnetic Hybridized Nanogenerator for Wave Energy Scavenging and Self-Powered Wireless Sensing System." *Nano Energy* 69 (March): 104440. https://doi.org/10.1016/j.nanoen.2019.104440

Chen, J., H. Guo, X. He, G. Liu, Y. Xi, H. Shi, and C. Hu. 2016. "Enhancing Performance of Triboelectric Nanogenerator by Filling High Dielectric Nanoparticles into Sponge PDMS Film." *ACS Applied Materials & Interfaces* 8(1):736–744. https://doi.org/10.1021/acsami.5b09907

Chen, Sheng, Jingxian Jiang, Feng Xu, and Shaoqin Gong. 2019. "Crepe Cellulose Paper and Nitrocellulose Membrane-Based Triboelectric Nanogenerators for Energy Harvesting and Self-Powered Human-Machine Interaction." *Nano Energy* 61 (July): 69–77. https://doi.org/10.1016/j.nanoen.2019.04.043

Chen, Xiaoliang, Xiangming Li, Jinyou Shao, Ningli An, Hongmiao Tian, Chao Wang, Tianyi Han, Li Wang, and Bingheng Lu. 2017. "High-Performance Piezoelectric Nanogenerators with Imprinted P(VDF-TrFE)/$BaTiO_3$ Nanocomposite Micropillars for Self-Powered Flexible Sensors." *Small* 13(23) April: 1604245. https://onlinelibrary.wiley.com/doi/full/10.1002/smll.201604245

Chen, Xi, Shiyou Xu, Nan Yao, and Yong Shi. 2010. "1.6 V Nanogenerator for Mechanical Energy Harvesting Using PZT Nanofibers | Nano Letters." *Nano Letters* 10 (6): 2133–37. https://doi.org/10.1021/nl100812k

Choi, Yeon Sik, Qingshen Jing, Anuja Datta, Chess Boughey, and Sohini Kar-Narayan. 2017. "A Triboelectric Generator Based on Self-Poled Nylon-11 Nanowires Fabricated by Gas-Flow Assisted Template Wetting." *Energy & Environmental Science* 10 (10): 2180–89. https://doi.org/10.1039/C7EE01292F

Choi, Yeon Sik, Sung Kyun Kim, Michael Smith, Findlay Williams, Mary E. Vickers, James A. Elliott, and Sohini Kar-Narayan. 2020. "Unprecedented Dipole Alignment in α-Phase Nylon-11 Nanowires for High-Performance Energy-Harvesting Applications." *Science Advances* 6 (24): eaay5065. https://doi.org/10.1126/sciadv.aay5065

Choi, Yeon Sik, Sung Kyun Kim, Findlay Williams, Yonatan Calahorra, James A. Elliott, and Sohini Kar-Narayan. 2018. "The Effect of Crystal Structure on the Electromechanical Properties of Piezoelectric Nylon-11 Nanowires." *Chemical Communications (Cambridge, England)* 54 (50): 6863–66. https://doi.org/10.1039/c8cc02530d

Chunga, T.C.Mike, and A. Petchsuk. 2003. *Polymers, Ferroelectric – ScienceDirect.* Encyclopedia of Physical Science and Technology (Third Edition). https://www.sciencedirect.com/science/article/pii/B0122274105005949.

Cuong, Nguyen Thai, Sophie Barrau, Malo Dufay, Nicolas Tabary, Antonio Da Costa, Anthony Ferri, Roberto Lazzaroni, Jean-Marie Raquez, and Philippe Leclère. 2020. "On the Nanoscale Mapping of the Mechanical and Piezoelectric Properties of Poly (L-Lactic Acid) Electrospun Nanofibers." *Applied Sciences* 10 (2): 652. https://doi.org/10.3390/app10020652

Ebarvia, Benilda S., Sharlene Cabanilla, and Fortunato Sevilla. 2005. "Biomimetic Properties and Surface Studies of a Piezoelectric Caffeine Sensor Based on Electrosynthesized Polypyrrole." *Talanta* 66 (1): 145–52. https://doi.org/10.1016/j.talanta.2004.10.009

Fan, Feng-Ru, Zhong-Qun Tian, and Zhong Lin Wang. 2012. "Flexible Triboelectric Generator." *Nano Energy* 1 (2): 328–34. https://doi.org/10.1016/j.nanoen.2012.01.004

Fatma, Bushara, Shashikant Gupta, Chandrachur Chatterjee, Ritamay Bhunia, Vivek Verma, and Ashish Garg. 2020. "Triboelectric Generators Made of Mechanically Robust PVDF Films as Self-Powered Autonomous Sensors for Wireless Transmission Based Remote Security Systems." *Journal of Materials Chemistry A* 8 (30): 15023–33. https://doi.org/10.1039/D0TA04716C

Furfari, F. A. 2005. "A History of the Van de Graaff Generator." *IEEE Industry Applications Magazine* 11 (1): 10–14. https://doi.org/10.1109/MIA.2005.1380320

García-Gutiérrez, Mari-Cruz, Amelia Linares, Daniel R. Rueda, and Jaime J. Hernández. 2010. "Confinement-Induced One-Dimensional Ferroelectric Polymer Arrays | Nano Letters." *Nano Letters* 10 (4): 1472–76.

Gaska, R., M. S. Shur, and A. D. Bykhovski. 1999. "Pyroelectric and Piezoelectric Properties of GaN-Based Materials." *Materials Research Society Internet Journal of Nitride Semiconductor Research* 4 (S1): 57–68. https://doi.org/10.1557/S1092578300002246

Ghosh, Sujoy Kumar, and Dipankar Mandal. 2018. "Synergistically Enhanced Piezoelectric Output in Highly Aligned 1D Polymer Nanofibers Integrated All-Fiber Nanogenerator for Wearable Nano-Tactile Sensor." *Nano Energy* 53 (November): 245–57. https://doi.org/10.1016/j.nanoen.2018.08.036

Guan, Xiaoyang, Bingang Xu, and Jianliang Gong. 2020. "Hierarchically Architected Polydopamine Modified $BaTiO_3$@P(VDF-TrFE) Nanocomposite Fiber Mats for Flexible Piezoelectric Nanogenerators and Self-Powered Sensors." *Nano Energy* 70 (April): 104516. https://doi.org/10.1016/j.nanoen.2020.104516

Gui, Jinzheng, Yezi Zhu, Lingling Zhang, Xi Shu, Wei Liu, Shishang Guo, and Xingzhong Zhao. 2018. "Enhanced Output-Performance of Piezoelectric Poly(Vinylidene Fluoride Trifluoroethylene) Fibers-Based Nanogenerator with Interdigital Electrodes and Well-Ordered Cylindrical Cavities." *Applied Physics Letters* 112 (7): 072902. https://doi.org/10.1063/1.5019319

Gusarov, Boris. 2015. "PVDF Piezoelectric Polymers: Characterization and Application to Thermal Energy Harvesting." -

Hashim, Ahmed, Majeed Ali Habeeb, Aseel Hadi, Qayssar M. Jebur, and Waled Hadi. 2017. "Fabrication of Novel (PVA-PEG-CMC-Fe_3O_4) Magnetic Nanocomposites for Piezoelectric Applications." *Sensor Letters* 15 (12): 998–1002. https://doi.org/10.1166/sl.2017.3935

He, Xianming, Xiaojing Mu, Quan Wen, Zhiyu Wen, Jun Yang, Chenguo Hu, and Haofei Shi. 2016. "Flexible and Transparent Triboelectric Nanogenerator Based on High Performance Well-Ordered Porous PDMS Dielectric Film." *Nano Research* 9 (12): 3714–24. https://doi.org/10.1007/s12274-016-1242-3

Hong, Chien-Chong, Sheng-Yuan Huang, Jiann Shieh, and Szu-Hung Chen. 2012. "Enhanced Piezoelectricity of Nanoimprinted Sub-20 Nm Poly(Vinylidene Fluoride–Trifluoroethylene) Copolymer Nanograss." *Macromolecules* 45 (3): 1580–86. https://doi.org/10.1021/ma202481t.

Hu, Bin, Ning Hu, Liangke Wu, Feng Liu, Yaolu Liu, Ning Huiming, Satoshi Atobe, and Hisao Fukunaga. 2015.

"Effects of Initial Crystallization Process on Piezoelectricity of PVDF-HFP Films | Request PDF." *Journal of Polymer Engineering* 35 (5): 451–61. https://doi.org/10.1515/polyeng-2014-0239

Jae, Won Lee, Sungwoo Jung, Jinhyeong Jo, Gi Hyeon Han, Dong-Min Lee, Jiyeon Oh, Hee Jae Hwang, et al. 2021. "Sustainable Highly Charged C 60 – Functionalized Polyimide in a Non-Contact Mode Triboelectric Nanogenerator." *Energy & Environmental Science*. https://doi.org/10.1039/D0EE03057K

Jain, Anjana, S. Jayanth Kumar, M. Ramesh Kumar, A. Sri Ganesh, and S. Srikanth. 2013. "PVDF-PZT Composite Films for Transducer Applications." *Mechanics of Advanced Materials and Structures* 21 (3): 181–86. https://doi.org/10.1080/15376494.2013.834094

Jiang, Yonggang, Syohei Shiono, Hiroyuki Hamada, Takayuki Fujita, Kohei Higuchi, and Kazusuke Maenaka. 2009. "Low-Frequency Energy Harvesting Using a Laminated PVDF Cantilever with a Magnetic Mas." *Power MEMS* 2010.

Jin, Yipu, Ning Chen, Yijun Li, and Qi Wang. 2020. "The Selective Laser Sintering of a Polyamide 11/BaTiO$_3$/Graphene Ternary Piezoelectric Nanocomposite." *RSC Advances* 10 (35): 20405–13. https://doi.org/10.1039/D0RA01042A

Jing, Qingshen, Guang Zhu, Peng Bai, Yannan Xie, Jun Chen, Ray P. S. Han, and Zhong Lin Wang. 2014. "Case-Encapsulated Triboelectric Nanogenerator for Harvesting Energy from Reciprocating Sliding Motion | ACS Nano." *ACS Nano* 8 (4): 3836–42. https://doi.org/10.1021/nn500694y

Jung, Sungmook, Jongsu Lee, Taeghwan Hyeon, Minbaek Lee, and Dae-Hyeong Kim. 2014. "Fabric-Based Integrated Energy Devices for Wearable Activity Monitors." *Advanced Materials* 26 (36): 6329–34. https://doi.org/10.1002/adma.201402439

Kashiwagi, Kimiaki, Kuniko Okano, Tatsuya Miyajima, Yoichi Sera, Noriko Tanabe, Yoshitomi Morizawa, and Yuji Suzuki. 2011. "Nano-Cluster-Enhanced High-Performance Perfluoro-Polymer Electrets for Energy Harvesting." *Journal of Micromechanics and Microengineering* 21 (12): 125016. https://doi.org/10.1088/0960-1317/21/12/125016

Kaur, Navjot, Jitendra Bahadur, Vinay Panwar, Pushpendra Singh, Keerti Rathi, and Kaushik Pal. 2016. "Effective Energy Harvesting from a Single Electrode Based Triboelectric Nanogenerator." *Scientific Reports* 6 (1): 38835. https://doi.org/10.1038/srep38835

Kawai, Heiji. 1969. "The Piezoelectricity of Poly (Vinylidene Fluoride)." *Japanese Journal of Applied Physics* 8 (7). https://iopscience.iop.org/article/10.1143/JJAP.8.975/meta

Ke, Tsung-Ying, Hsiang-An Chen, Hwo-Shuenn Sheu, Jien-Wei Yeh, Heh-Nan Lin, Chi-Young Lee, and Hsin-Tien Chiu. 2008. "Sodium Niobate Nanowire and its Piezoelectricity | The Journal of Physical Chemistry C." *The Journal of Physical Chemistry C* 112 (24): 8827–31.

Khanbareh, H., S. van der Zwaag, and W. A. Groen. 2015. "Piezoelectric and Pyroelectric Properties of Conductive Polyethylene Oxide-Lead Titanate Composites." *Smart Materials and Structures* 24 (4): 045020. https://doi.org/10.1088/0964-1726/24/4/045020

Kim, Inkyum, Hyejin Jeon, Dabum Kim, Jungmok You, and Daewon Kim. 2018. "All-in-One Cellulose Based Triboelectric Nanogenerator for Electronic Paper Using Simple Filtration Process." *Nano Energy* 53 (November): 975–81. https://doi.org/10.1016/j.nanoen.2018.09.060

Kim, Jihye, Jeong Hwan Lee, Hanjun Ryu, Ju-Hyuck Lee, Usman Khan, Hang Kim, Sung Soo Kwak, and Sang-Woo Kim. 2017. "High-Performance Piezoelectric, Pyroelectric, and Triboelectric Nanogenerators Based on P(VDF-TrFE) with Controlled Crystallinity and Dipole Alignment – Kim – 2017 – Advanced Functional Materials – Wiley Online Library." *Advanced Functional Materials* 27 (22). https://onlinelibrary.wiley.com/doi/abs/10.1002/adfm.201700702

Kim, Hong-Seok, and Il-Kyu Park. 2018. "Enhanced Output Power from Triboelectric Nanogenerators Based on Electrospun Eu-Doped Polyvinylidene Fluoride Nanofibers." *Journal of Physics and Chemistry of Solids* 117 (June): 188–93. https://doi.org/10.1016/j.jpcs.2018.02.045

Kim, Sejung, Youngjun Song, and Michael J. Heller. 2017. "Influence of MWCNTs on *β*-Phase PVDF and Triboelectric Properties." Research Article. *Journal of Nanomaterials* Hindawi. July 16, 2017. https://doi.org/10.1155/2017/2697382.

Kim, Yeongjun, Xinwei Wu, and Je Hoon Oh. 2020. "Fabrication of Triboelectric Nanogenerators Based on Electrospun Polyimide Nanofibers Membrane." *Scientific Reports* 10 (1): 2742. https://doi.org/10.1038/s41598-020-59546-7

Klimiec, Ewa, Krzysztof Zaraska, Wiesław Zaraska, and Szymon Kuczyński. 2011. "Micropower Generators and Sensors Based on Piezoelectric Polypropylene PP and Polyvinylidene Fluoride PVDF Films – Energy Harvesting from Walking | Scientific.Net." *Applied Mechanics and Materials* 110–116 (October): 1245–51. https://doi.org/10.4028/www.scientific.net/AMM.110-116.1245

Koukharenko, E., M. J. Tudor, and S. P. Beeby. 2008. "Performance Improvement of a Vibration-Powered Electromagnetic Generator by Reduced Silicon Surface Roughness." *Materials Letters* 62 (4): 651–54. https://doi.org/10.1016/j.matlet.2007.06.050

Kulah, H., and K. Najafi. 2004. "An Electromagnetic Micro Power Generator for Low-Frequency Environmental Vibrations." In *17th IEEE International Conference on Micro Electro Mechanical Systems. Maastricht MEMS 2004 Technical Digest*, 237–40. https://doi.org/10.1109/MEMS.2004.1290566

Kymissis, J., C. Kendall, J. Paradiso, and N. Gershenfeld. 1998. "Parasitic Power Harvesting in Shoes." In *Digest of Papers. Second International Symposium on Wearable Computers (Cat. No.98EX215)*, 132–39. https://doi.org/10.1109/ISWC.1998.729539

Lee, Sol Jee, Anand Prabu Arun, and Kap Jin Kim. 2015. "Piezoelectric Properties of Electrospun Poly(l-Lactic Acid) Nanofiber Web." *Materials Letters* 148 (June): 58–62. https://doi.org/10.1016/j.matlet.2015.02.038

Lee, Keun Young, Jinsung Chun, Ju-Hyuck Lee, Kyeong Nam Kim, Na-Ri Kang, Ju-Young Kim, Myung Hwa Kim, et al. 2014. "Hydrophobic Sponge Structure-Based Triboelectric Nanogenerator." *Advanced Materials* 26 (29): 5037–42. https://doi.org/10.1002/adma.201401184

Lee, Ju-Hyuck, Ronan Hinchet, Tae Yun Kim, Hanjun Ryu, Wanchul Seung, Hong-Joon Yoon, and Sang-Woo Kim. 2015. "Control of Skin Potential by Triboelectrification with Ferroelectric Polymers – Lee – 2015 – Advanced Materials – Wiley Online Library." *Advanced Materials* 27 (37): 5553–58. https://doi.org/10.1002/adma.201502463

Lee, Jin Pyo, Jae Won Lee, and Jeong Min Baik. 2018. "The Progress of PVDF as a Functional Material for Triboelectric Nanogenerators and Self-Powered Sensors." *Micromachines* 9 (10): 532. https://doi.org/10.3390/mi9100532

Lee, Jin Pyo, Byeong Uk Ye, Kyeong Nam Kim, Jae Won Lee, Won Jun Choi, and Jeong Min Baik. 2017. "3D Printed Noise-Cancelling Triboelectric Nanogenerator." *Nano Energy* 38 (August): 377–84. https://doi.org/10.1016/j.nanoen.2017.05.054

Li, Xuan, Sai Chen, Xianye Zhang, Jinhao Li, Haihui Liu, Na Han, and Xingxiang Zhang. 2020. "Poly-l-Lactic Acid/Graphene Electrospun Composite Nanofibers for Wearable Sensors." *Energy Technology* 8 (5): 1901252. https://doi.org/10.1002/ente.201901252

Li, Kaidi, Xia Liu, Yifeng Liu, and Xiaohong Wang. 2018. "A Piezoelectric Generator Based on PVDF/GO Nanofiber Membrane." *Journal of Physics: Conference Series* 1052 (July): 012110. https://doi.org/10.1088/1742-6596/1052/1/012110

Li, Xiaoyi, Juan Tao, Jing Zhu, and Caofeng Pan. 2017. "A Nanowire Based Triboelectric Nanogenerator for Harvesting Water Wave Energy and Its Applications." *APL Materials* 5 (7): 074104. https://doi.org/10.1063/1.4977216

Li, Jie, Chunmao Zhao, Kai Xia, Xi Liu, Dong Li, and Jing Han. 2019. "Enhanced Piezoelectric Output of the PVDF-TrFE/ZnO Flexible Piezoelectric Nanogenerator by Surface Modification." *Applied Surface Science* 463 (January): 626–34. https://doi.org/10.1016/j.apsusc.2018.08.266

Lin, Long, Yannan Xie, Simiao Niu, Sihong Wang, Po-Kang Yang, and Zhong Lin Wang. 2015. "Robust Triboelectric Nanogenerator Based on Rolling Electrification and Electrostatic Induction at an Instantaneous Energy Conversion Efficiency of ∼55% | ACS Nano." *ACS Nano* 9 (1): 922–30. https://doi.org/10.1021/nn506673x

Lin, Meng-Fang, Jiaqing Xiong, Jiangxin Wang, Kaushik Parida, and Pooi See Lee. 2018. "Core-Shell Nanofiber Mats for Tactile Pressure Sensor and Nanogenerator Applications." *Nano Energy* 44 (February): 248–55. https://doi.org/10.1016/j.nanoen.2017.12.004

Liu, Guanlin, Hengyu Guo, Sixing Xu, Chenguo Hu, and Zhong Lin Wang. 2019. "Oblate Spheroidal Triboelectric Nanogenerator for All-Weather Blue Energy Harvesting." *Advanced Energy Materials* 9 (26): 1900801. https://doi.org/10.1002/aenm.201900801

Liu, JinYan, Xiao Yue Hu, Hu Min Dai, Zhi San, Fu Ke Wang, Li Ren, and GuoYuan Li. 2020. "Polycaprolactone/Calcium Sulfate Whisker/Barium Titanate Piezoelectric Ternary Composites for Tissue Reconstruction." *Advanced Composites Letters* 29 (January). https://doi.org/10.1177/2633366X19897923

Liu, Meihua, Guangyu Shen, Weijian Xu, and Guoli Shen. 2007. "Synthesis of a Novel Aromatic–Aliphatic Hyperbranched Polyamide and Its Application in Piezoelectric Immunosensors." *Polymer International* 56 (11): 1432–39. https://doi.org/10.1002/pi.2299

Liu, Xiao-fang, Chuan-xi Xiong, Hua-jun Sun, Lijie Dong, Ri Li, and Yang Liu. 2006. "Piezoelectric and Dielectric Properties of PZT/PVC and Graphite Doped with PZT/PVC Composites." *Materials Science and Engineering B-Advanced Functional Solid-State Materials* 127 (February): 261–66. https://doi.org/10.1016/j.mseb.2005.10.022

Liu, Shirui, Wei Zheng, Bao Yang, and Xiaoming Tao. 2018. "Triboelectric Charge Density of Porous and Deformable Fabrics Made from Polymer Fibers." *Nano Energy* 53 (November): 383–90. https://doi.org/10.1016/j.nanoen.2018.08.071

Lovinger, Andrew J. 1983. "Ferroelectric Polymers | Science." *Science* 220 (4602): 1115–21. https://doi.org/10.1126/science.220.4602.1115

Lu, Ming-Pei, Jinhui Song, Ming-Yen Lu, Min-Teng Chen, Yifan Gao, Lih-Juann Chen, and Zhong Lin Wang. 2009. "Piezoelectric Nanogenerator Using P-Type ZnO Nanowire Arrays." *Nano Letters* 9 (3): 1223–27. https://doi.org/10.1021/nl900115y

Ma, Yuan, Wangshu Tong, Wenjiang Wang, Qi An, and Yihe Zhang. 2018. "Montmorillonite/PVDF-HFP-Based Energy Conversion and Storage Films with Enhanced Piezoelectric and Dielectric Properties." *Composites Science and Technology* 168 (November): 397–403. https://doi.org/10.1016/j.compscitech.2018.10.009

Mandal, Dipankar, Sun Yoon, and Kap Jin Kim. 2011. "Origin of Piezoelectricity in an Electrospun Poly(Vinylidene Fluoride-Trifluoroethylene) Nanofiber Web-Based Nanogenerator and Nano-Pressure Sensor." *Macromolecular Rapid Communications* 32 (11): 831–37. https://doi.org/10.1002/marc.201100040

Mao, Yanchao, Ping Zhao, Geoffrey McConohy, Hao Yang, Yexiang Tong, and Xudong Wang. 2014. "Sponge-Like Piezoelectric Polymer Films for Scalable and Integratable Nanogenerators and Self-Powered Electronic Systems." *Advanced Energy Materials* 4 (7): 1301624. https://doi.org/10.1002/aenm.201301624

Marino, Attilio, Enrico Almici, Simone Migliorin, Christos Tapeinos, Matteo Battaglini, Valentina Cappello, Marco Marchetti, et al. 2019. "Piezoelectric Barium Titanate Nanostimulators for the Treatment of Glioblastoma Multiforme." *Journal of Colloid and Interface Science* 538 (March): 449–61. https://doi.org/10.1016/j.jcis.2018.12.014

Mathur, S. C., J. I. Scheinbeim, and B. A. Newman. 1984. "Piezoelectric Properties and Ferroelectric Hysteresis Effects in Uniaxially Stretched Nylon-11 Films." *Journal of Applied Physics* 56 (9): 2419–25. https://doi.org/10.1063/1.334294

Nair, Smitha S., C. Ramesh, and K. Tashiro. 2006. "Crystalline Phases in Nylon-11: Studies Using HTWAXS and HTFTIR | Macromolecules." *Macromolecules* 39 (8): 2841–48. https://doi.org/10.1021/ma052597e

Niu, Huidan, Xinyu Du, Shuyu Zhao, Zuqing Yuan, Xiuling Zhang, Ran Cao, Yingying Yin, Chi Zhang, Tao Zhou, and Congju Li. 2018. "Polymer Nanocomposite-Enabled High-Performance Triboelectric Nanogenerator with Self-Healing Capability." *RSC Advances* 8 (54): 30661–68. https://doi.org/10.1039/C8RA05305G

Nunes-Pereira, J., P. Martins, V. F. Cardoso, C. M. Costa, and S. Lanceros-Méndez. 2016. "A Green Solvent Strategy for the Development of Piezoelectric Poly(Vinylidene Fluoride–Trifluoroethylene) Films for Sensors and Actuators Applications." *Materials & Design* 104 (August): 183–89. https://doi.org/10.1016/j.matdes.2016.05.023

Nunes-Pereira, J., V. Sencadas, V. Correia, V. F. Cardoso, Weihua Han, J. G. Rocha, and S. Lanceros-Méndez. 2015. "Energy Harvesting Performance of $BaTiO_3$/Poly(Vinylidene Fluoride–Trifluoroethylene) Spin Coated Nanocomposites." *Composites Part B: Engineering* 72 (April): 130–36. https://doi.org/10.1016/j.compositesb.2014.12.001

Pan, Lun, Jiyu Wang, Peihong Wang, Ruijie Gao, Yi-Cheng Wang, Xiangwen Zhang, Ji-Jun Zou, and Zhong Lin Wang. 2018. "Liquid-FEP-Based U-Tube Triboelectric Nanogenerator for Harvesting Water-Wave Energy." *Nano Research* 11 (8): 4062–73. https://doi.org/10.1007/s12274-018-1989-9

Parangusana, Hemalatha, Deepalekshmi Ponnamma, and Mariam Al Ali AlMaadeed. 2017. "Flexible Tri-Layer Piezoelectric Nanogenerator Based on PVDF-HFP/Ni-Doped ZnO Nanocomposites – RSC Advances (RSC Publishing)." *RSC Advances* 7 (October): 50156–65. https://doi.org/10.1039/C7RA10223B

Pu, Xue, Jun-Wei Zha, Chun-Lin Zhao, Shao-Bo Gong, Jie-Feng Gao, and Robert K. Y. Li. 2020. "Flexible PVDF/Nylon-11 Electrospun Fibrous Membranes with Aligned ZnO Nanowires as Potential Triboelectric Nanogenerators."

Chemical Engineering Journal 398 (October): 125526. https://doi.org/10.1016/j.cej.2020.125526

Pusty, Manojit, Lichchhavi Sinha, and Parasharam M. Shirage. 2018. "A Flexible Self-Poled Piezoelectric Nanogenerator Based on a RGO–Ag/PVDF Nanocomposite." *New Journal of Chemistry* 43 (1): 284–94. https://doi.org/10.1039/C8NJ04751K

Ram, Farsa, Sithara Radhakrishnan, Tushar Ambone, and Kadhiravan Shanmuganathan. 2019. "Highly Flexible Mechanical Energy Harvester Based on Nylon 11 Ferroelectric Nanocomposites." *ACS Applied Polymer Materials* 1 (8): 1998–2005. https://doi.org/10.1021/acsapm.9b00246

Ramesh, Prashanth, Sriram Krishnamoorthy, Siddharth Rajan, and Gregory N. Washington. 2012. "Fabrication and Characterization of a Piezoelectric Gallium Nitride Switch for Optical MEMS Applications." *Smart Materials and Structures* 21 (9): 094003. https://doi.org/10.1088/0964-1726/21/9/094003

Raschke, C. R., and T. E. Nowlin. 1980. "Polyparaxylylene Electrets Usable at High Temperatures." *Journal of Applied Polymer Science* 25 (8): 1639–44. https://doi.org/10.1002/app.1980.070250811

Rocha, J. G., L. M. Goncalves, P. F. Rocha, M. P. Silva, and S. Lanceros-Mendez. 2010. "Energy Harvesting From Piezoelectric Materials Fully Integrated in Footwear." *IEEE Transactions on Industrial Electronics* 57 (3): 813–19. https://doi.org/10.1109/TIE.2009.2028360

Saadatnia, Zia, Shahriar Ghaffari Mosanenzadeh, Ebrahim Esmailzadeh, and Hani E. Naguib. 2019. "A High Performance Triboelectric Nanogenerator Using Porous Polyimide Aerogel Film." *Scientific Reports* 9 (1): 1370. https://doi.org/10.1038/s41598-018-38121-1

Salimi, Ali, and Ali Akbar Yousefi. 2003. "FTIR Studies of Beta-Phase Crystal Formation in Stretched PVDF Films." *Polymer Testing* 22 (6): 699–704. https://doi.org/10.1002/smll.201702268

Sawano, Michiya, Komei Tahara, Yoshihiro Orita, Mesatoshi Nakayama, and Yoshiro Tajitsu. 2010. "New Design of Actuator Using Shear Piezoelectricity of a Chiral Polymer, and Prototype Device." *Polymer International* 59 (3): 365–70. https://doi.org/10.1002/pi.2758

Scheinbeim, J. I., J. W. Lee, and B. A. Newman. 1992. "Ferroelectric Polarization Mechanisms in Nylon 11 | Macromolecules." *Macromolecules* 25 (14): 3729–32. https://doi.org/10.1021/ma00040a019

Seema, A., K. R. Dayas, and Justin M. Varghese. 2007. "PVDF-PZT-5H Composites Prepared by Hot Press and Tape Casting Techniques – Seema – 2007 – Journal of Applied Polymer Science – Wiley Online Library." *Journal of Applied Polymer Science* 106 (1): 146–51. https://doi.org/10.1002/app.26673

Sessler, G. M., and J. E. West. 1973. "Electret Transducers: A Review." *The Journal of the Acoustical Society of America* 53: 1589. https://doi.org/10.1121/1.1913507

Shafeeq, Valiyaveetil Haneefa, Cherumannil Karumuthil Subash, Soney Varghese, and Gopalakrishna Panicker Unnikrishnan. 2020. "Nanohydroxyapatite Embedded Blends of Ethylene-co-vinyl Acetate and Millable Polyurethane as Piezoelectric Materials: Dielectric, Viscoelastic and Mechanical Features." *Polymer International* 69 (12): 1256–66. https://doi.org/10.1002/pi.6070

Song, Jundong, Guanxing Zhao, Bo Li, and Jin Wang. 2017. "Design Optimization of PVDF-Based Piezoelectric Energy Harvesters." *Heliyon* 3 (9): e00377. https://doi.org/10.1016/j.heliyon.2017.e00377

Souri, H., I. W. Nam, and H. K. Lee. 2015. "A Zinc Oxide/Polyurethane-Based Generator Composite as a Self-Powered Sensor for Traffic Flow Monitoring." *Composite Structures* 134 (December): 579–86. https://doi.org/10.1016/j.compstruct.2015.08.112

Suresh, M. B., Tsung-Her Yeh, Chih-Chieh Yu, and Chen-Chia Chou. 2010. "Dielectric and Ferroelectric Properties of Polyvinylidene Fluoride (PVDF)-$Pb_{0.52}Zr_{0.48}TiO_3$ (PZT) Nano Composite Films." *Ferroelectrics* 381 (1): 80–86. https://doi.org/10.1080/00150190902869699

Suzuki, Yuji, Daigo Miki, Masato Edamoto, and Makoto Honzumi. 2010. "A MEMS Electret Generator with Electrostatic Levitation for Vibration-Driven Energy-Harvesting Applications." *Journal of Micromechanics and Microengineering* 20 (10): 104002. https://doi.org/10.1088/0960-1317/20/10/104002

Tsutsumino, Takumi, Yuji Suzuki, Nobuhide Kasagi, and Yoshihiko Sakane. 2006. Seismic Power Generator Using High-Performance Polymer Electret. *Proceedings of the IEEE International Conference on Micro Electro Mechanical Systems (MEMS).* 98–101. https://doi.org/10.1109/MEMSYS.2006.1627745

Vigmond, Stephen J., Krishna M. R. Kallury, Vida Ghaemmaghami, and Michael Thompson. 1992. "Characterization of the Polypyrrole Film-Piezoelectric Sensor Combination." *Talanta* 39 (4): 449–56. https://doi.org/10.1016/0039-9140(92)80161-6

Villa, Sara Moon, Vittorio Massimo Mazzola, Tommaso Santaniello, Erica Locatelli, Mirko Maturi, Lorenzo Migliorini, Ilaria Monaco, Cristina Lenardi, Mauro Comes Franchini, and Paolo Milani. 2019. "Soft Piezoionic/Piezoelectric Nanocomposites Based on Ionogel/$BaTiO_3$ Nanoparticles for Low Frequency and Directional Discriminative Pressure Sensing." *ACS Macro Letters* 8 (4): 414–20. https://doi.org/10.1021/acsmacrolett.8b01011

Wang, Zhong Lin, Tao Jiang, and Liang Xu. 2017. "Toward the Blue Energy Dream by Triboelectric Nanogenerator Networks." *Nano Energy* 39 (September): 9–23. https://doi.org/10.1016/j.nanoen.2017.06.035

Wang, Xiaofeng, Simiao Niu, Yajiang Yin, Fang Yi, Zheng You, and Zhong Lin Wang. 2015. "Triboelectric Nanogenerator Based on Fully Enclosed Rolling Spherical Structure for Harvesting Low-Frequency Water Wave Energy." *Advanced Energy Materials* 5 (24): 1501467. https://doi.org/10.1002/aenm.201501467

Wang, Zhong Lin, and Jinhui Song. 2006. "Piezoelectric Nanogenerators Based on Zinc Oxide Nanowire Arrays." *Science* 312 (5771): 242–46. https://doi.org/10.1126/science.1124005

Wang, Zhong Lin, and Aurelia Chi Wang. 2019. "On the Origin of Contact-Electrification." *Materials Today* 30 (November): 34–51. https://doi.org/10.1016/j.mattod.2019.05.016

Wang, Shuhua, Zhong Lin Wang, and Ya Yang. 2016. "A One-Structure-Based Hybridized Nanogenerator for Scavenging Mechanical and Thermal Energies by Triboelectric–Piezoelectric–Pyroelectric Effects." *Advanced Materials* 28 (15): 2881–87. https://doi.org/10.1002/adma.201505684

Wang, Xinya, Changfa Xiao, Hailiang Liu, Qinglin Huang, and Hao Fu. 2018. "Fabrication and Properties of PVDF and PVDF-HFP Microfiltration Membranes." *Journal of Applied Polymer Science* 135 (40): 46711. https://doi.org/10.1002/app.46711

Wang, Xingzhao, Bin Yang, Jingquan Liu, Yanbo Zhu, Chunsheng Yang, and Qing He. 2016. "A Flexible Triboelectric-Piezoelectric Hybrid Nanogenerator Based on P(VDF-TrFE)

Nanofibers and PDMS/MWCNT for Wearable Devices | Scientific Reports." *Scientific Reports* 6 (November). https://www.nature.com/articles/srep36409

Wang, Xuewen, Yang Gu, Zuoping Xiong, Zheng Cui, and Ting Zhang. 2014. "Silk-Molded Flexible, Ultrasensitive, and Highly Stable Electronic Skin for Monitoring Human Physiological Signals." *Advanced Materials* 26 (9): 1336–42. https://doi.org/10.1002/adma.201304248

Whiter, Richard A., Vijay Narayan, and Sohini Kar-Narayan. 2014. "A Scalable Nanogenerator Based on Self-Poled Piezoelectric Polymer Nanowires with High Energy Conversion Efficiency." *Advanced Energy Materials4* 4 (18). https://doi.org/10.1002/aenm.201400519

Won, Sung Sik, Mackenzie Sheldon, Nicholas Mostovych, Jiyeon Kwak, Bong-Suk Chang, Chang Won Ahn, Angus I. Kingon, Ill Won Kim, and Seung-Hyun Kim. 2015. "Piezoelectric Poly(Vinylidene Fluoride Trifluoroethylene) Thin Film-Based Power Generators Using Paper Substrates for Wearable Device Applications." *Applied Physics Letters* 107 (20): 202901. https://doi.org/10.1063/1.4935557

Wu, Liangke, Weifeng Yuan, Ning Hu, Zhongchang Wang, Chunlin Chen, Jianhui Qiu, Ji Ying, and Yuan Li. 2014. "Improved Piezoelectricity of PVDF-HFP/Carbon Black Composite Films." *Journal of Physics D: Applied Physics* 47 (13): 135302. https://doi.org/10.1088/0022-3727/47/13/135302

Xia, Zhongfu, Reimund Gerhard-Multhaupt, Wolfgang Künstler, Armin Wedel, and Rudi Danz. 1999. "High Surface-Charge Stability of Porous Polytetrafluoroethylene Electret Films at Room and Elevated Temperatures." *Journal of Physics D: Applied Physics* 32 (17): L83–85. https://doi.org/10.1088/0022-3727/32/17/102

Xia, Zhongfu, A. Wedel, and R. Danz. 2003. "Charge Storage and Its Dynamics in Porous Polytetrafluoroethylene (PTFE) Film Electrets." *IEEE Transactions on Dielectrics and Electrical Insulation* 10 (1): 102–8. https://doi.org/10.1109/TDEI.2003.1176568

Xia, Kequan, Zhiyuan Zhu, Hongze Zhang, and Zhiwei Xu. 2018. "A Triboelectric Nanogenerator as Self-Powered Temperature Sensor Based on PVDF and PTFE." *Applied Physics A* 124 (8): 520. https://doi.org/10.1007/s00339-018-1942-5

Xiao, Tian, Xi Liang, Tao Jiang, Liang Xu, Jia Jia Shao, Jin Hui Nie, Yu Bai, Wei Zhong, and Zhong Lin Wang. 2018. "Spherical Triboelectric Nanogenerators Based on Spring-Assisted Multilayered Structure for Efficient Water Wave Energy Harvesting." *Advanced Functional Materials* 28 (35): 1802634. https://doi.org/10.1002/adfm.201802634

Xie, Yannan, Sihong Wang, Simiao Niu, Long Lin, Qingshen Jing, Jin Yang, Zhengyun Wu, and Zhong Lin Wang. 2014. "Grating-Structured Freestanding Triboelectric-Layer Nanogenerator for Harvesting Mechanical Energy at 85% Total Conversion Efficiency." *Advanced Materials* 26 (38): 6599–6607. https://doi.org/10.1002/adma.201402428

Xu, Wei, Long-Biao Huang, Man-Chung Wong, Li Chen, Gongxun Bai, and Jianhua Hao. 2017. "Environmentally Friendly Hydrogel-Based Triboelectric Nanogenerators for Versatile Energy Harvesting and Self-Powered Sensors." *Advanced Energy Materials* 7 (1): 1601529. https://doi.org/10.1002/aenm.201601529

Xu, Liang, Tao Jiang, Pei Lin, Jia Jia Shao, Chuan He, Wei Zhong, Xiang Yu Chen, and Zhong Lin Wang. 2018. "Coupled Triboelectric Nanogenerator Networks for Efficient Water Wave Energy Harvesting." *ACS Nano* 12 (2): 1849–58. https://doi.org/10.1021/acsnano.7b08674

Yang, Bin, Chengkuo Lee, Wenfeng Xiang, Jin Xie, Johnny Han He, Rama Krishna Kotlanka, Siew Ping Low, and Hanhua Feng. 2009. "Electromagnetic Energy Harvesting from Vibrations of Multiple Frequencies." *Journal of Micromechanics and Microengineering* 19 (3): 035001. https://doi.org/10.1088/0960-1317/19/3/035001

Yang, Ya, Hulin Zhang, Jun Chen, Qingshen Jing, Yu Sheng Zhou, Xiaonan Wen, and Zhong Lin Wang. 2013. "Single-Electrode-Based Sliding Triboelectric Nanogenerator for Self-Powered Displacement Vector Sensor System." *ACS Nano* 7 (8): 7342–51. https://doi.org/10.1021/nn403021m

Yao, Chunhua, Xin Yin, Yanhao Yu, Zhiyong Cai, and Xudong Wang. 2017. "Chemically Functionalized Natural Cellulose Materials for Effective Triboelectric Nanogenerator Development." *Advanced Functional Materials* 27 (30): 1700794. https://doi.org/10.1002/adfm.201700794

Yaqoob, Usman, Rahaman Md Habibur, Muhammad Sheeraz, and Hyeon Cheol Kim. 2019. "Realization of Self-Poled, High Performance, Flexible Piezoelectric Energy Harvester by Employing PDMS-RGO as Sandwich Layer between P(VDF-TrFE)-PMN-PT Composite Sheets." *Composites Part B: Engineering* 159 (February): 259–68. https://doi.org/10.1016/j.compositesb.2018.09.102

Zhang, Jianming, Yongxin Duan, Harumi Sato, Hideto Tsuji, Isao Noda, Shouke Yan, and Yukihiro Ozaki. 2005. "Crystal Modifications and Thermal Behavior of Poly(l-Lactic Acid) Revealed by Infrared Spectroscopy | Macromolecules." *Macromolecules* 38 (19): 8012–21. https://doi.org/10.1021/ma051232r

Zhang, Min, Tao Gao, Jianshu Wang, Jianjun Liao, Yingqiang Qiu, Quan Yang, Hao Xue, et al. 2015. "A Hybrid Fibers Based Wearable Fabric Piezoelectric Nanogenerator for Energy Harvesting Application." *Nano Energy* 13 (April): 298–305. https://doi.org/10.1016/j.nanoen.2015.02.034

Zhang, Zhongbo, Morton H. Litt, and Lei Zhu. 2016. "Unified Understanding of Ferroelectricity in N-Nylons: Is the Polar Crystalline Structure a Prerequisite? | Macromolecules." *Macromolecules* 49 (8): 3070–82. https://doi.org/10.1021/acs.macromol.5b02739

Zhang, Binbin, Lei Zhang, Weili Deng, Long Jin, Fengjun Chun, Hong Pan, Bingni Gu, et al. 2017. "Self-Powered Acceleration Sensor Based on Liquid Metal Triboelectric Nanogenerator for Vibration Monitoring." *ACS Nano* 11 (7): 7440–46. https://doi.org/10.1021/acsnano.7b03818

Zhao, Xue, Bo Chen, Guodong Wei, Jyh Ming Wu, Wei Han, and Ya Yang. 2019. "Polyimide/Graphene Nanocomposite Foam-Based Wind-Driven Triboelectric Nanogenerator for Self-Powered Pressure Sensor." *Advanced Materials Technologies* 4 (5): 1800723. https://doi.org/10.1002/admt.201800723

Zhao, Gengrui, Baisheng Huang, Jinxi Zhang, Aochen Wang, Kailiang Ren, and Zhong Lin Wang. 2017. "Electrospun Poly(l-Lactic Acid) Nanofibers for Nanogenerator and Diagnostic Sensor Applications." *Macromolecular Materials and Engineering* 302 (5): 1600476. https://doi.org/10.1002/mame.201600476

Zhao, Pengfei, Navneet Soin, Kovur Prashanthi, Jinkai Chen, Shurong Dong, Erping Zhou, Zhigang Zhu, et al. 2018. "Emulsion Electrospinning of Polytetrafluoroethylene (PTFE) Nanofibrous Membranes for High-Performance Triboelectric Nanogenerators." *ACS Applied Materials & Interfaces* 10 (6): 5880–91. https://doi.org/10.1021/acsami.7b18442

Zheng, Youbin, Li Cheng, Miaomiao Yuan, Zhe Wang, Lu Zhang, Yong Qin, and Tao Jing. 2014. "An Electrospun Nanowire-Based Triboelectric Nanogenerator and Its Application in a Fully Self-Powered UV Detector." *Nanoscale* 6 (14): 7842–46. https://doi.org/10.1039/C4NR01934B

Zheng, Qifeng, Liming Fang, Haiquan Guo, Kefang Yang, Zhiyong Cai, Mary Ann B. Meador, and Shaoqin Gong. 2018. "Highly Porous Polymer Aerogel Film-Based Triboelectric Nanogenerators." *Advanced Functional Materials* 28 (13): 1706365. https://doi.org/10.1002/adfm.201706365

Zhong, Wei, Liang Xu, Haiming Wang, Jie An, and Zhong Lin Wang. 2019. "Tilting-Sensitive Triboelectric Nanogenerators for Energy Harvesting from Unstable/Fluctuating Surfaces." *Advanced Functional Materials* 29 (45): 1905319. https://doi.org/10.1002/adfm.201905319

Zhou, Tao, Zhang Chi, Chang Bao Han, Feng-Ru Fan, Wei Tang, and Lin Zhong. 2014. "Woven Structured Triboelectric Nanogenerator for Wearable Devices." *ACS Applied Materials & Interfaces* 6 (16): 14695–701.

Zhu, Chen, Dongliang Guo, Dong Ye, Shan Jiang, and YongAn Huang. 2020. "Flexible PZT-Integrated, Bilateral Sensors via Transfer-Free Laser Lift-Off for Multimodal Measurements | ACS Applied Materials & Interfaces." *ACS Applied Materials & Interfaces* 12 (33): 37354–62.

Zhu, Jianxiong, Luyu Jia, and Run Huang. 2017. "Electrospinning Poly(l-Lactic Acid) Piezoelectric Ordered Porous Nanofibers for Strain Sensing and Energy Harvesting." *Journal of Materials Science: Materials in Electronics* 28 (16): 12080–85. https://doi.org/10.1007/s10854-017-7020-5

Zhu, Guang, Bai Peng, Jun Chen, Qingshen Jing, and Zhong Lin Wang. 2015. "Triboelectric Nanogenerators as a New Energy Technology: From Fundamentals, Devices, to Applications." *Nano Energy*, Special issue on the 2nd International Conference on Nanogenerators and Piezotronics (NGPT 2014) 14 (May): 126–38. https://doi.org/10.1016/j.nanoen.2014.11.050.

Zhu, Guang, Yu Sheng Zhou, Peng Bai, Xian Song Meng, Qingshen Jing, Jun Chen, and Zhong Lin Wang. 2014. "A Shape-Adaptive Thin-Film-Based Approach for 50% High-Efficiency Energy Generation Through Micro-Grating Sliding Electrification." *Advanced Materials* 26 (23): 3788–96. https://doi.org/10.1002/adma.201400021

Zi, Yunlong, Hengyu Guo, Zhen Wen, Min-Hsin Yeh, Chenguo Hu, and Zhong Lin Wang. 2016. "Harvesting Low-Frequency (<5 Hz) Irregular Mechanical Energy: A Possible Killer Application of Triboelectric Nanogenerator." *ACS Nano* 10 (4): 4797–4805. https://doi.org/10.1021/acsnano.6b01569

6 Pyroelectric and Piezoelectric Polymers

Pragati Kumar
Central University of Jammu, Jammu, J&K, India

Lakshmi Unnikrishnan
Central Institute of Petrochemicals Engineering & Technology (CIPET), Bengaluru, India

CONTENTS

6.1 INTRODUCTION

Piezoelectric materials are subclasses of dielectric materials, whereas pyroelectric materials are subclasses of piezoelectrics, that is all pyroelectrics exhibit piezoelectricity and all piezoelectric materials are dielectric; however, the converse is not true, as illustrated in Figure 6.1. Piezolectricity is the phenomenon of the generation of electricity by the application of force, while the phenomenon of the generation of electricity due to a temperature gradient in the material is known as pyroelectricity.

Piezoelectricity is a term coined in Greek, "piezo," meaning "to squeeze or press." The phenomenon was first observed by two French Scientist brothers, Jacques and Pierre Curie in 1880. They observed that application of pressure to quartz or certain crystals creates an electrical charge in them [1, 2]. This scientific phenomenon was later named the "piezoelectric effect." Gabriel Lippmann predicted and the Curies further verified the "inverse piezoelectric effect" wherein application of an electric field onto the crystals leads to the deformation or displacement of the crystals. "Piezoelectricity" can be defined as the mechanical strain generated by an electric field in a material. The mechanical strain on a material produces polarity in it while the applied electric field develops a strain within the material in the reverse case. The mechanical stress leads to strain developed in the material which consequently produces a dipole and net polarization in the direction of the applied mechanical stress.

For a material to act as a piezoelectric, it should not possess a centre of symmetry at all. Usually in a centrosymmetric material, the centre of the masses of the constituent +ve and ve charges coincides with the centre of symmetry and hence such materials have zero net polarization, when applied with or without mechanical pressure. Conversely, in piezoelectric materials which are non-centrosymmetric systems, the application of stress changes the centre of the masses of the constituent +ve and ve ions according to the direction of stress. This results in a net polarization leading to the creation of a potential difference between the two surfaces of the material on which stress is applied.

From Figure 6.2, it is observed that when a mechanical stress is applied to materials with a centre of symmetry, the centre of mass for the constituent +ve and ve ions lies at the same point 'C' (Figure 6.2(a)), thereby rendering the material non-piezoelectric. However, in the case of materials without a centre of symmetry, the centres of mass for the ingredient +ve and ve ions overlap at point 'O' as shown in Figure 6.2(b). When applied by a vertical force, the centres of mass for the +ve and ve ions changes, thereby developing a net non-zero polarization (Figure 6.2(c)). These materials show a piezoelectric effect and are termed "piezoelectric materials." However, when subjected to a horizontal force as shown in Figure 6.2(d), the centres of mass for the ingredient +ve and ve ions coincide with each other. Hence, depending on the crystal structure, no net polarization is created. Also, Poisson's ratio can be a key factor in the

DOI: 10.1201/9781003169727-6

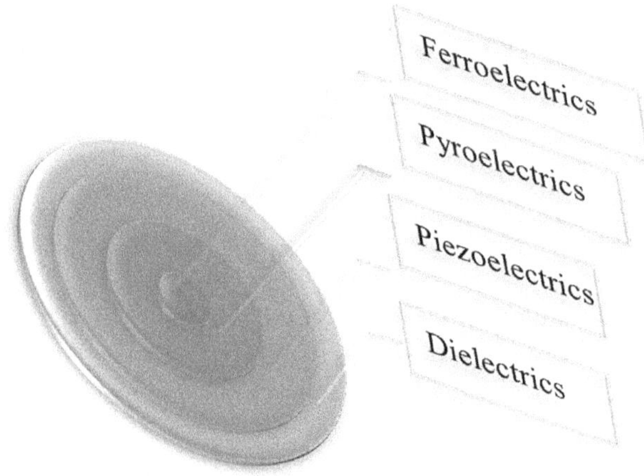

FIGURE 6.1 Classification of dielectric materials.

piezoelectric effect of a material. Here the material expands in the other directions of application of force and the ratio of change in strain in the transverse to axial direction (two directions that were not compressed) determines Poisson's ratio, which for an ideal piezoelectric material is 0, and its effect on polarization is eliminated. Thus, the net polarization induced depends on the crystal structure, the direction of application of force, and the material's Poisson's ratio.

On the other hand, pyroelectric materials are the class of dielectric materials that possess the ability to produce electrical output when they are subjected to a temperature gradient, that is a change in temperature. This category of materials shows a coupling between thermal and electrical parameters which gives rise to the phenomenon called the pyroelectric effect.

This was perhaps first observed in tourmaline crystals in approximately 400 BC in ancient Greece;, which can attract small objects due to the generation of electrostatic charges by temperature changes. Though, the first explanation based on electrical phenomena was given by Linnaeus and Aepinus in the 18th century. The term "pyroelectric" (the Greek word "pyro" means "fire") was first used tentatively 200 years ago by D. Brewster, while W. Thomson and W. Voigt were the first who proposed the theory at the end of the 19th century. The verity of materials and their promising applications regarding the pyroelectric effect began to be explored from the 1930s [4–6]. Nevertheless, serious attention was delayed until the sixth decade of the 19th century when its possible application in thermal detectors became apparent [5, 7–14] In addition, the effect was also utilized in waste heat harvesters [13, 15–18], pyroelectro-catalytic disinfection [19] and hydrogen production [20]

There exists a wide range of pyroelectric materials, ranging from single crystals to thin films, ceramics to polymers, and organics to inorganics. Akin to ferroelectric material, pyroelectric materials also exhibit spontaneous polarization without applying an electric field; polarization can, however, be switched when subjected to an electric field. Indeed, ferroelectric materials are a sub-group of pyroelectric materials, that is all the ferroelectric materials are pyroelectric materials but the converse is not true. Usually all the crystals belonging to a polar class exhibit a pyroelectric property. Out of a total of 32 classes of crystal symmetry, 20 classes (C_1, C_4, S_4, D_4, C_4v, D_{2d}, C_6, C_{3h}, D_6, C_6v, D_{3h}, C_2, C_5, D_2, C_2v, C_3, D_3, C_3v, T, T_d) are piezoelectric, whereas materials of classes C_1, C_4, C_4v, C_6, C_6v, C_2, C_5, C_2v, C_3, and C_3v only, possess a unique polar axis which may or may not be switched by an external electric field, show pyroelectricity [14]. Every unit cell of the pyroelectric material has its own dipole moment and thereby shows spontaneous polarization ($\vec{P}_S$) [21], that is the dipole moment per unit volume of the material can be changed under varying temperature conditions. Thus, a change in $\vec{P}_S$ due to a thermal fluctuation in such (polar or anisotropic) materials is referred to as a pyroelectric effect [22] Under the influence of a temporal temperature gradient (*dT/dt*) an electric current/potential is generated in materials as a consequence of a change in $\vec{P}_S$ (Figure 6.1) and a differential change in polarization ($\vec{P}$) due to varying temperature represents the pyroelectric coefficient ($\vec{p}$) [23], that is

$$\vec{p} = \frac{d\vec{D}}{dT} = \frac{d\vec{P}}{dT} \tag{6.1}$$

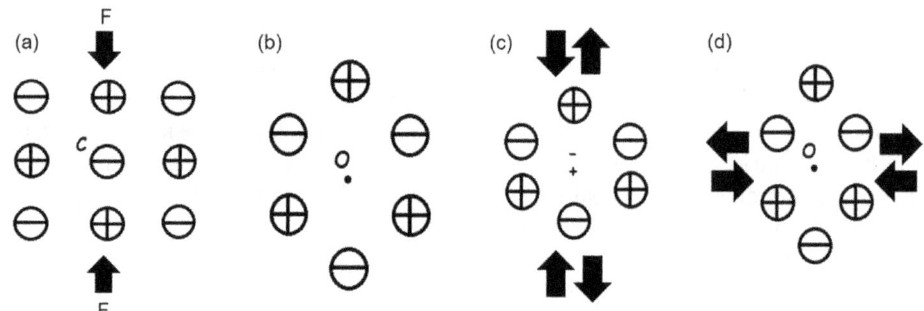

FIGURE 6.2 Schematic diagram illustrating (a) the effect of the application of mechanical stress on materials with a centre of symmetry, (b) materials without a centre of symmetry, (c) the effect of the application of vertical force on materials without a centre of symmetry, and (d) the effect of the application of horizontal force on materials without a centre of symmetry. Adopted from [3].

where dielectric displacement $\vec{D}$ is related to $\vec{P}$ by the well-known relation $\vec{D} = \varepsilon_0\vec{E} + \vec{P}$. Here, ε_0 and $\vec{E}$ are the permittivity of vacuum and the electric field vector, respectively. However, in the absence of the external field $\vec{E}$, a spontaneous polarization $\vec{P_S}$ is permitted by many of the point groups due to symmetry reasons. Thus, the total polarization is the sum of the induced and spontaneous polarization, that is $\vec{P} = \vec{P_{\text{ind}}} + \vec{P_S}$ [24]. In the absence of an electric field, $\vec{P_{\text{ind}}} = 0$ and hence $\vec{D} = \vec{P_S}$.

The origin of $\vec{P_S}$ stems from the crystal structure symmetry that allows the cataloguing of dielectric materials as illustrated in Figure 6.1. The non-existence of a centre of inversion symmetry alone allows piezoelectricity in a crystal, whereas the company of at least one unique polar direction is required for pyroelectricity. However, no further crystallographic restrictions are imposed for ferroelectricity. The pyroelectric vector is usually tackled as a scalar quantity $p = |\vec{p}|$ due to the existence of only one polar direction in most of the pyroelectric materials [23]. Moreover, pyroelectric materials also possess an inverse effect known as the electrocaloric effect, like the piezoelectric effect, which can result in a temperature fluctuation due to varying polarization [25].

6.1.1 The Concept of Piezoelectricity and the Figure of Merits

Piezoelectricity is associated with the capability of a material, when subjected to mechanical stresses, to produce electric surface charges because of macroscopic polarization within it. For a material to act as piezoelectric, it should not possess a centre of symmetry. The centre of the masses of the +ve and ve ions of the material overlaps to the centre of symmetry of such a material when applied with or without mechanical pressure, thereby nulling the net polarization. On the other hand, the application of stress in piezoelectric materials results in the adjustment of the centre of masses of the +ve and ve charges depending on the direction of the applied stress. This results in a net polarization leading to the creation of a potential difference between the two surfaces of the material on which the stress acts (Figure 6.3) [26].

When applied by a tensile or compressive force (Figure 6.3), the dipoles in the piezoelectric material cancel out each other, while they are not cancelled on the surface. This produces a polarity given by the expression:

$$P = dT = d\frac{F}{A} \tag{6.2}$$

where T, P, and d are the mechanical stress applied, the polarization induced, and the piezoelectric coefficient, respectively.

The polarization charge generated is given by:

$$P = \frac{Q}{A} \tag{6.3}$$

where Q is the charge and A is the cross-sectional area of the piezoelectric material where the stress acts.

The voltage induced as a result of polarization is given by:

$$V = \frac{Q}{C} = \frac{AP}{\frac{\varepsilon_0\varepsilon_r A}{r}} = \frac{LP}{\varepsilon_0\varepsilon_r} = \frac{L\left(d\frac{F}{A}\right)}{\varepsilon_0\varepsilon_r} = \frac{dLF}{\varepsilon_0\varepsilon_r A} \tag{6.4}$$

where ε_0, ε_r, L, and C are the vacuum permittivity, relative permittivity, the piezoelectric material's length, and capacitance, respectively.

Contrariwise, the strain S produced upon action of an electric field E on a piezoelectric material is represented by:

$$S = dE \tag{6.5}$$

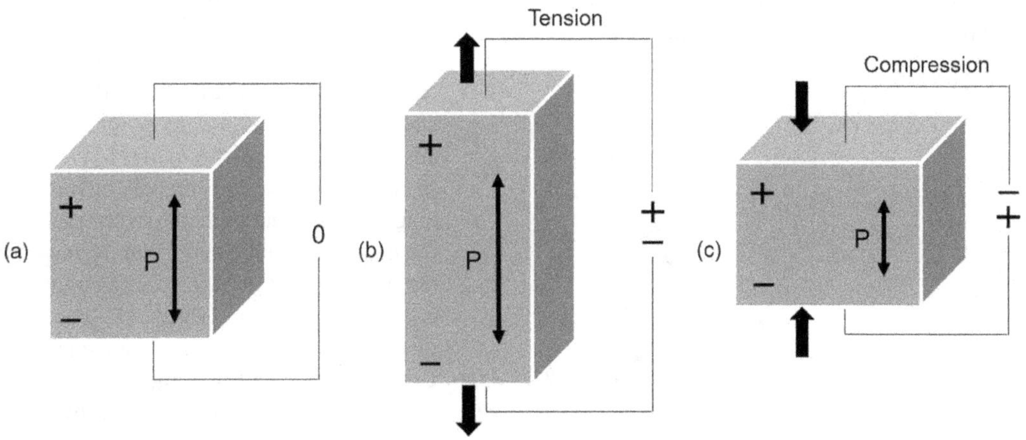

FIGURE 6.3 Piezoelectric effect inducing a current depending on the type of stress and direction. [26] adopted with permission © 2016 Woodhead Publishing.

This is known as an inverse piezoelectric effect, where ve ions tend to move to the +ve end of the electric field, while the +ve ions move towards the ve end. This shift triggers a change in the dimensions of the material, creating a strain within. It is worth mentioning here that the piezoelectric coefficient remains the same for both the direct and inverse piezoelectric effect on application of the same magnitude of either mechanical stress and induced polarization or an electric field and induced strain.

The piezoelectric effect can be induced by the following methods:

- Forcing a structural change in non-polar dielectrics to remove central symmetry by applying a strong electrical field. Materials like rutile, calcium titanate, and strontium titanate etc. show electrically induced piezoelectricity.
- Inducing a change in the sound velocity with the help of an external electric field to control the coefficient of elasticity in materials like quartz, langasite, potassium dihydrogen phosphate, niobium lithium, and silicosillenite.
- Influencing a change in the domain orientation and, hence, polarization in the case of ferroelectric materials. The electric field modifies the attenuation as well as velocity of sound waves, causing domain reorientation.
- Application of an electric field changing the frequency of the soft transverse optical mode of lattice oscillation by the application of an electric field.

6.1.1.1 Piezoelectric Coefficients

The interplay between the electrical and mechanical behaviour of a material defines the piezoelectric effect. The direction of polarization is commonly named as the z-axis of the crystallographic materials, as shown in Figure 6.4.

The directions 1, 2, and 3 represent the x, y, and z axes respectively while 4, 5, and 6 represent the shear along the respective axes. The final piezoelectric material properties depend on the electric field applied, the corresponding displacement, and the stress as well as the strain.

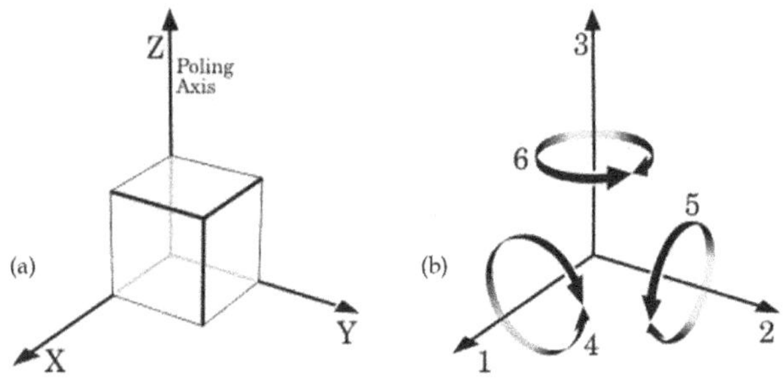

FIGURE 6.4 Schematic representation of (a) the polarization direction and (b) the axes, directions, and shear in the axes. Adopted from [27]

The major piezoelectric coefficients are designated where the superscripts and subscripts are used to indicate direction. Planar modes are expressed with a subscript p. The electrical field direction is related to the voltage applied or charge developed and gives the direction of mechanical stress or strain direction which are represented by the first and second subscripts respectively, while the consistent boundary conditions of the mechanical or electrical force are indicated by superscripts. The detailed scheme is given in Figure 6.5.

The piezoelectricity of piezo-polymers can be increased by modifying the β phase content so as to enhance the energy harvesting. Some of the commonly used methods are uniaxial and biaxial stretching, electrical and corona poling, and the addition of various fillers to the polymer matrix.

6.1.1.1.1 Stretching

Polymers, which are semi-crystalline in nature, require crystalline transformation, which can be effectively achieved for generating piezoelectricity by thermal annealing, mechanical orientation, or applying a high voltage [28]. Mechanical stretching of polymers causes molecular orientation of amorphous strands in the planar direction. The characteristics of piezo-polymers depend on the mode of stretching, that is uniaxial or biaxial, as well as the temperature at which they are stretched [28–30].

6.1.1.1.2 Poling

This is the procedure where the crystallites (molecular dipoles) are reoriented to enhance the piezoelectricity in polymers [31]. Applying a high electric field (E) at a moderate temperature produces the permanent alignment of the molecular dipoles of the polymer along the electric field direction. The alignment of the dipole in a single direction results in polymer materials due to the generation of spontaneous polarization on the application of an electric field and the majority of the dipoles are sealed in an arrangement of similar orientation, even after the removal of the field [32].

- Direct contact method: A high voltage (DC or AC in the form of sinusoidal or triangular low-frequency signals) is applied across the polymer surfaces through the deposition of conductive electrodes. The electric field strength applied across the polymer surface will usually be in the range of 5 to 100 MV/m. The piezoelectric coefficient depends broadly on three factors: (i) the magnitude and duration of the electric field applied; (ii) the magnitude and extent of uniformity of the applied temperature; and (iii) the extent of impurity and uniformity of the polymer surface [33]. The electrical poling method is widely used for commercial films.
- Corona poling: In this method, the charging is done on one surface of the polymer either by using electrodes or without them. An extremely sharp and conducting needle is firstly kept in a high voltage (~8–20 kV). Then it is placed over a grid which is kept over the

FIGURE 6.5 Representation of various piezoelectric coefficients. Adopted from [27]

piezoelectric polymer. The polymer is kept in an inert ambient atmosphere like dry air or argon at a sufficiently lower DC voltage of 0.2–3 kV. Inert gas is ionized around the sharp tip and moved towards the polymer and charges the surface. The charging of the polymer surface is controlled by optimizing the grid position and magnitude of the voltage applied [34].

The third way of enhancing the piezoelectricity of polymers is reported in the literature by the incorporation of fillers—using a variety of materials like ceramics, carbon allotropes and compounds, metal oxides, and composites of these materials—into the piezoelectric polymer matrix [35–38]. This will be discussed later in detail.

6.1.2 The Concept of Pyroelectricity and the Figures of Merit

The concept of the pyroelectric effect lies in the temporal temperature gradient induced by spontaneous polarization. This can be understood easily by considering the material as a thin parallel-sided one which is cut in such a way that its crystallographic symmetry axis always remains normal to the flat surfaces [22]. In the same direction, the components of the dipole moment existing in each unit cell sum up. This presence of the dipole moment per unit volume without applying any electric field can be referred to a space charge region of bound charges on every flat surface of the pyroelectric material and hence attracts neighbouring free charge carriers like ions or electrons. If the external circuit is designed via the fabrication of the conductive electrodes on the surfaces, and if these electrodes are connected using an ammeter of low internal resistance, the circuital current remains zero until the temperature of the sample is steady. Though a rise in the temperature causes a net non-zero dipole moment in most of the pyroelectric materials, which turns into a decrease of $\vec{P}_S$ and thereby the number of the bound charge. Subsequently, to balance this change in the bound charge, free charges are redistributed and results in a flow of current. Conversely, the sign of the current is reversed if the temperature of the sample is lowered instead of raised. The pyroelectric current lasts until there exists a temporal temperature gradient. Figure 6.6 illustrates the mechanism of the pyroelectric effect as discussed above.

Since, the thermodynamically reversible processes can take place between the thermal, mechanical, and electrical phenomena of a crystal, there are many conceivable ways to determine the total p as illustrated by the triangular diagram in Figure 6.7. Here, the magnetic properties are overlooked for simplicity. The reversible small changes in the two pair of variables are represented by the lines connecting the pairs of circles. The physical properties of the variables—like the thermal, elastic, and electric—are defined by heat capacity, elasticity, and electrical permittivity respectively and represented by the three short bold lines.

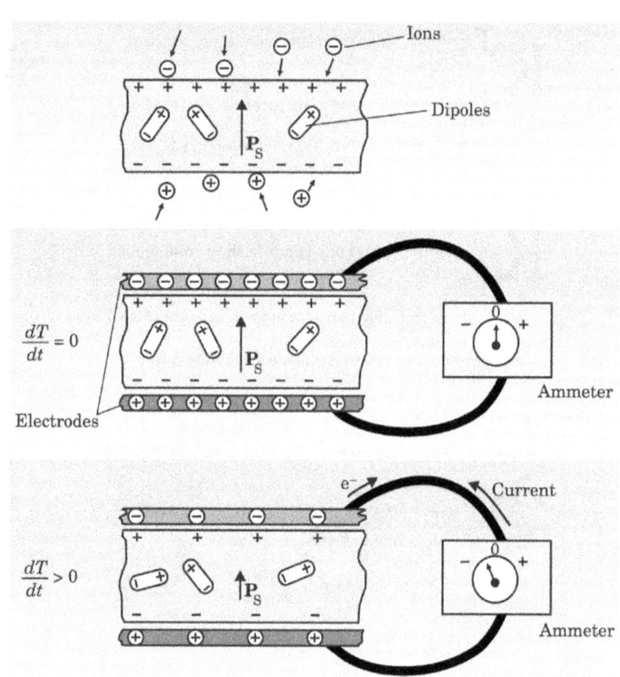

FIGURE 6.6 A pyroelectric crystal with an inherent dipole moment with attraction of free charges from the surroundings (top) and compensation of free charges with applied electric field (middle); a rise in temperature T results in charge distribution and a flow of current (bottom). [22, 39] adopted with permission from ©2005 American Institute of Physics

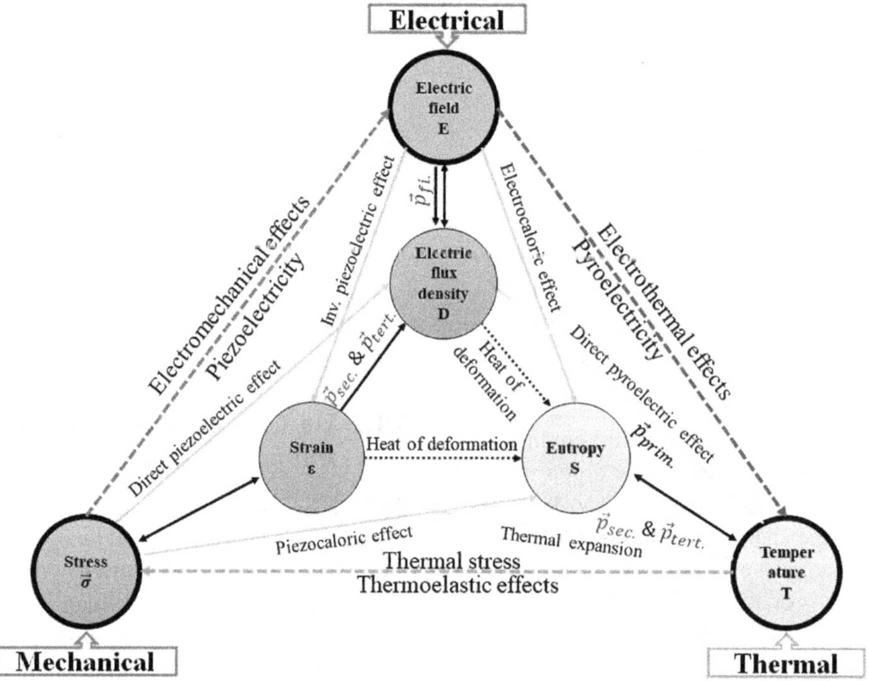

FIGURE 6.7 The Heckmann diagram, illustrating direct and cross-coupling phenomena between thermal, electrical, and mechanical state variables. Here, the intensive and extensive state variables are represented by bold and unbold circles respectively. To simplify, the magnetic properties are omitted. [22, 23] adopted and reproduced with permission from ©2005 American Institute of Physics

The pyroelectric effect may be produced in diverse materials in various ways [40] and can be categorised as:

1. Primary pyroelectric effect (p_{prim}): this is generated in polar materials as a consequence of an actual change in $\overrightarrow{P}_S$ due to temperature fluctuations [23, 24, 40].
2. Secondary pyroelectric effect (p_{sec}): this arises due to a change in mechanical strain e induced by thermal expansion and the piezoelectric effect [23, 24, 40].
3. Tertiary pyroelectric effect (p_{tert}): this is produced in non-piezoelectric materials due to a change in polarization resulting from the existence of spatial strain gradients $\partial e/\partial r$, which can also be produced by deformation caused by non-uniform rises or falls in the temperature of the material [40–44].
4. Field-induced pyroelectric effect (p_{fi}): this is caused by a change in the dielectric permittivity of materials as a function of temperature in an external electric field. In particular, this effect is dominant in materials in which the dielectric constant strongly depends on the temperature [23, 40].

The total magnitude of the pyroelectric coefficient p is estimated by four significant parts in the case when all the involved state variables are unfixed [23] and is given as:

$$p_n = \frac{\partial P_{S,n}}{\partial T} + \sum_{i,j} \frac{\partial d_{n,i,j}\sigma_{ij}}{\partial T} + \sum_{i,j,k} \frac{\partial \mu_{n,i,j,k} \frac{\partial e_{i,j}}{\partial r_k}}{\partial T} + \sum_{i,\varepsilon_0} E_i \frac{\partial \varepsilon_{n,i}}{\partial T} \tag{6.6}$$

Here the first, second, third, and fourth terms of Eq. 6.6 represent primary, secondary, tertiary, and field-induced pyroelectric contributions. The piezoelectric tensor, mechanical stress, the flexoelectric tensor, and the strain tensor are denoted by d, σ, μ, and e respectively, with indices n, i, j, and k which refer to the tensor components taken over x, y, and z, and r is the position vector. Notice that n is excluded from the above equation for materials having one polar direction ($n = z$) and $p = |p_z|$.

It is worth mentioning here that pyroelectric behaviour particularly of polymers can be confused with the behaviour of an electret [40] as they may show finite polarization without applying an electric field, which may not be initiated because of the crystal structure. The finite polarization may result from quasi-permanent electric charges (for which τ_d up to years) acquired due to the positional separation of frozen-in +ve and ve charge carriers stemming from an earlier action [45] Therefore, electrets (a piece of dielectric substance possessing a quasi-permanent electrical charge) also display a hysteresis, which is a characteristic of a ferroelectric.

Jachalke and co-workers [23] discussed a variety of techniques in a review article to precisely quantify p, which is essential for the identification of pyroelectric substances and the fabrication of devices based on them. They discuss the pros and cons of the each techniques for specific samples and were also focused on the opportunity to isolate diverse involvements to the pyroelectric coefficient, to omit thermally stimulated currents, the ability to localize the measurement of p, and the necessity of metal electrodes. They summarize all these considerations for every characteristic technique in a tabulated form (Table 6.1) and also represented them in a schematic diagram (Figure 6.8).

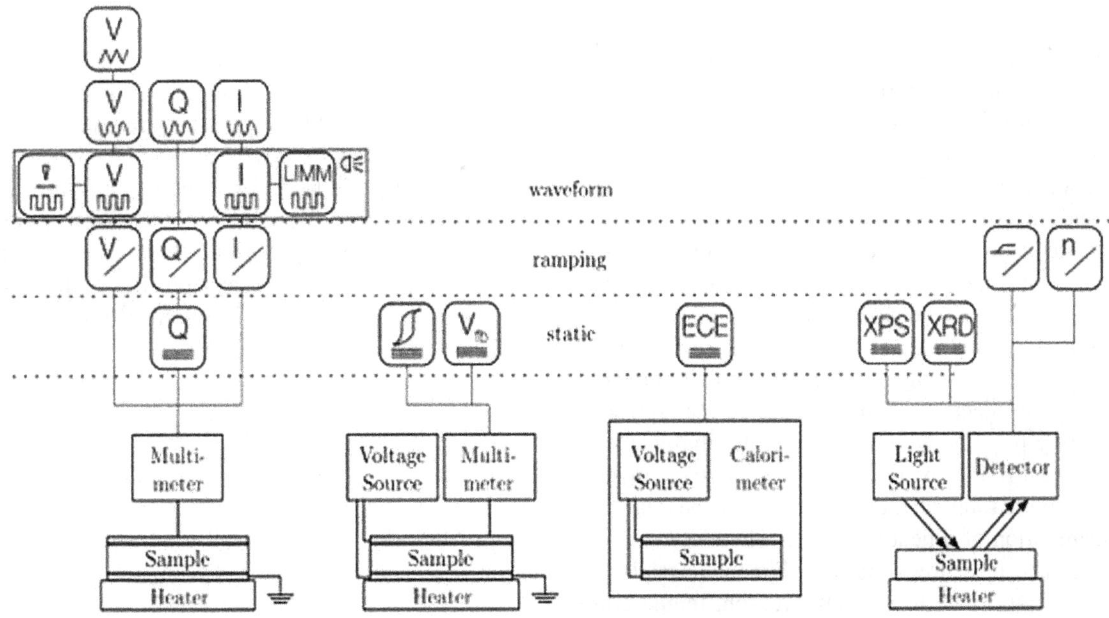

FIGURE 6.8 Systematization and pictorial representation of the four different pyroelectric measurement methods described in Table 6.1. [23] adopted with permission from ©2017 American Institute of Physics

TABLE 6.1
Classification and Evaluation Overview

Type of Mode/ Excitation	Name	Measures Signals	Separation of					Metallic Electrodes	Reference
			p_{sec}	p_{tert}	p_{fi}	TSC	Mapping		
Static/constant temperature	Charge compensation	Q (with V_{ref}) T	(✓)	(✓)	✓	×	(✓≧ A_{crit})	✓	[265]
	Sawyer-Tower hysteresis	P_s (with V_{in}(t), V_{out} (t)), T	(✓)	(✓)	×	×	(✓≈ 0.1 lm)	✓	[266]
	Electrocaloric measurement	dT, T	×	(✓)	×	(✓)	(✓≈ 10 lm)	✓	[267, 268]
	Flatband voltage shift	C,T	(✓)	(✓)	×	×	(✓)	✓	[269]
	Photoelectron spectroscopy	V_{sf}, T	×	(✓)	✓	×	(✓)	×	[270]
	X-ray diffraction/density function theory	T, $I_{x\text{-ray}}$	×	(✓)	✓	×	×	×	[271]
Dynamic/ temperature ramping	Lang-Steckel method	T (t), V(t)	(✓)	(✓)	✓	×	✓	✓	[272]
	Glass method	T (t), V(t)	(✓)	(✓)	✓	×	(✓≧ A_{crit})	✓	[273]
	Byer-Roundy method	T (t), I(t)	(✓)	(✓)	✓	(✓)	(✓≧ A_{crit})	✓	[274]
	Birefringence change	T (t), n(t)	(✓)	(✓)	✓	(✓)	×	×	[275]
	Refractive index change	T (t), n(t)	(✓)	(✓)	✓	(✓)	×	×	[276]
Step-wise temperature	Thermal pulse	F_0(t), I(f)	(✓)	(✓)	✓	(✓)	(✓≧ A_{crit})	✓	[277]
	Schein method	F_0(t), V(t)	✓	✓	✓	×	✓ ≈ 0.1 mm	×	[278, 279]
	PyroSPM	F_0(t), V_{tip} (t)	×	(✓)	✓	×	✓ ≈ 0.1 lm	×	[280]
	LIMM	F_0 (f), I (f), f	×	×	✓	✓	✓ ≈ 1 lm	✓	[281]
Waveform temperature	Daglish method	T(t), I(t)	(✓)	(✓)	✓	×	(✓≧ A_{crit})	✓	[282]
	Sussner method	Q (with V(t)), T(t)	(✓)	(✓)	✓	(✓)	(✓≧ A_{crit})	✓	[283]
	Hartley method	T(t), V(t)	(✓)	(✓)	✓	(✓)	✓	✓	[284]
	Sharp-Garn method	T(t), I(t)	(✓)	(✓)	✓	✓	(✓≧ A_{crit})	✓	[285, 286]

Source: [23] adopted with permission from ©2017 American Institute of Physics.

Notes: A check mark indicates the unrestricted capability of the method. A check in parentheses indicates a, Whereas the method, which is viable capable in principle, but which the original work did not consider. ed or might be Complicated to Realize, Is Illustrated by Check Mark in Parentheses. A Cross Indicates the Impossibility of an Evaluation Criterion. Capacity (C), heat flux (F_0), frequency (f), current (I), X-ray Intensity ($I_{x\text{-ray}}$), charge (Q), spontaneous polarization (P_S), temperature (T), time (t), voltage (V), refractive index (n), critical area (A_{crit}).

The scope of this chapter is polymeric materials only, therefore we present here an overview of approaches used for the precise measurement of p and the separation of various contributions of p in them. In the case of these materials, significant tertiary signals are a major issue which are produced due to thermal properties like low thermal conductivity and a high thermal expansion coefficient, and mechanical properties like small Young's modulus and growth induced strain gradients [40, 42]. This issue can be addressed by lateral homogeneous heating. Further, to avoid the poling problem with metallic contact, corona poling [46, 47] is often used to pole polymeric samples. Over a long time, a large number of charge carriers may be injected and trapped in the material during the poling process, which results in thermally stimulated currents (TSCs). A TSC (non-pyroelectric current) contributes to the total current, consequently the quantitative determination of p becomes false. Therefore, current and capacitance measurements of polymer materials are not advisable [40]. Due to the same reason, temperature ramping methods are one type of dynamic method that is also not recommended. On the other hand, the capability of the Sharp–Garn and Hartley waveform methods not only to discriminate TSCs from output signals but also to satisfy almost all relevant aspects, make them suitable for each type of pyroelectric material. Also, the reduced instrumentation and modelling effort, and the very few effortlessly manageable parameters are additional advantages of these methods [23].

Further, p is used to derive three major figures of merit (FOMs), namely potential current responsivity performance (F_i), voltage responsivity (F_v), and detection capability (F_D), to compare the responsivity and sensing performances of the various pyroelectric materials [48]. Mathematically, the FOMs are given by:

$$F_i = \frac{p}{C_V} \tag{6.7}$$

$$F_v = \frac{p}{C_V \varepsilon_0 \varepsilon_r} \tag{6.8}$$

$$F_D = \frac{p}{C_V \sqrt{\varepsilon_0 \varepsilon_r \tan\delta}} \tag{6.9}$$

where C_v, ε_0, ε_r, and $\tan\delta$ are the volume specific heat, free space permittivity in a vacuum, dielectric constant, and dielectric loss respectively [22, 49, 50].

6.1.3 Piezoelectric and Pyroelectric Materials

Pyroelectric materials are the subclass of piezoelectric materials without the necessity of all piezoelectrics exhibiting pyroelectric behaviour. Usually, dielectric materials without centrosymmetry behave as piezoelectrics, whereas pyroelectric materials are indeed polar dielectrics. Both types of material are categorized as functional materials which produce an electrical signal as a consequence of fluctuation in temperature or application of stress. Indeed, this class of materials are also ferroelectric and are usually very sensitive to ambient fluctuations. Thus mostly commercial piezo/pyroelectric devices employ ferroelectric materials [51].

6.1.3.1 Types of Piezoelectric and Pyroelectric Materials

There is a variety of materials showing piezoelectricity and/or pyroelectricity. Both piezo- and pyroelectricity exist in some materials offered by nature, such as tourmaline, polyvinyl fluorides and its copolymers [52–55], phenyl pyridine [56] and its derivatives, potassium sodium tartrate tetrahydrate (Rochelle salt) [57, 58], topaz [59], quartz [59], gallium nitride [60], lithium tantalite [61], protein amino acids [62], cesium nitrate [63], cobalt phthalocyanine, and sucrose (table sugar) [64, 65]. On the other hand, naturally available materials like berlinite show a piezoelectric effect only [59]. Piezo- and pyroelectric materials can be categorized broadly in two: lead based and lead free. Among them, lead compounds and composites have been mostly used as conventional piezo/pyroelectric material, but they may shortly be barred owing to their highly toxic nature. Therefore, a serious concern is paying for search of high performance lead-free piezo- and pyroelectric materials over the last decade. Many significant advancements, including applications ranging from a typical sensor to energy harvesting and saving devices, have been carried out experimentally for lead-free piezo- and pyroelectrics in different forms. The classification of lead-free piezo- and pyroelectrics is illustrated in Figure 6.9.

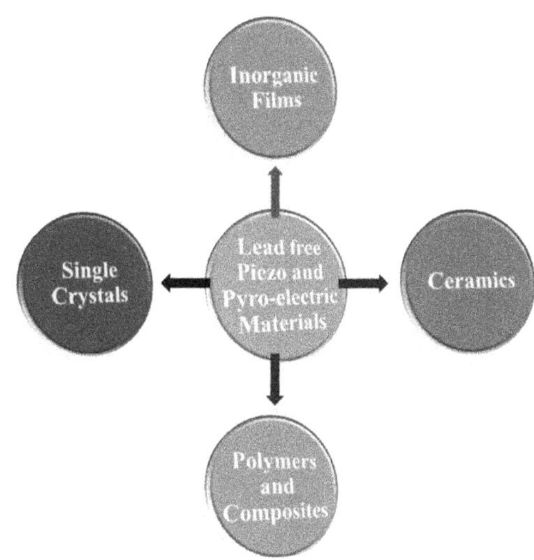

FIGURE 6.9 Classification of lead-free piezo- and pyroelectric materials.

6.1.3.1.1 Single Crystals

Single crystals are extensively studied for piezo- and pyroelectric application and show excellent detectivity and stability depending on the control of the crystal orientation and domain structures [12, 48, 66–71]. One of the major families of single crystal pyroelectrics since the 1960s is the triglycine sulphate (TGS) family [71–79]. In spite of showing a high p and FOMs due to their single crystal materials, their uses are greatly limited due to the expensive and complex growth methods and practical difficulty in getting real single crystals in reasonable sizes.

6.1.3.1.2 Ceramics

This particular class of materials plays a pretty significant role in piezo- and pyroelectric applications because of the manifold benefits over other materials like reliable transport properties, economics, easy production of large samples, and impressive mechanical properties. Using various easy approaches like incorporating suitable dopants and additives, phase boundary optimization, and modified synthesis methods, etc., one can tune the electrical properties of ceramics. There is a variety of lead-less ceramics which have been extensively explored for their piezo- and pyroelectric properties, including perovskite [80–88], bismuth layer-structured ferroelectrics (BLSF) [89–95], sodium niobate [96–99], and tungsten bronze [100–104] etc.

6.1.3.1.3 Inorganic Films

Significant attention was focused on piezo- and pyroelectric thin and thick films of inorganic materials because they can deliver a lower heat capacity and cost effective fabrication as compared to the above-mentioned categories, which are valuable for piezo/pyroelectric applications. It is observed that the piezo/pyroelectric performances of thin films mainly depend on various factors, such as composition of material, choice of substrate, synthesis route, buffer layer used, film thickness, microstructure present in the film, grain alignment, boundaries, and residual stress etc. [105–112].

6.1.3.1.4 Polymers

Polymers are extensively used as electrical insulators due to their non-conductive nature at low temperature. Despite this, over the last few decades, not only conducting polymers, but also piezoelectric, pyroelectric, and ferroelectric polymers have been well explored. Piezo/pyroelectric polymers comprise of carbon-based materials with long macromolecular chains with excellent flexibility and can endure a large amount of bending and twisting. The spontaneous polarization of the crystalline phase of these semi-crystalline polymers results in their piezo/pyroelectricity. These materials also possess high mechanical strength, high impact resistance, a low ε, low elastic stiffness, and low mass density. Besides, piezo/pyroelectric polymers possess certain advantages over inorganic piezo/pyroelectric materials, like low dielectric permittivity and low acoustic impedance, light weight, flexibility, and easy and cost-effective large area fabrication. Polyvinylidene fluoride (PVDF) [52, 54, 113, 114] and their copolymers [38, 54, 56, 115–119], polyamides (PA) [120–126], polyvinyl fluoride (PVF) [127–134], polyviny chloride (PVC) [128, 129, 134, 135], polyacrylonitrile (PAN) [134, 136, 137], and polyacry lamide (PAA) [134] are some well-known piezo/pyroelectric polymers. They find distinct applications in energy-related devices, particularly transducers like ultrasonic, audio, and medical displays, shock and pressure sensors, and vibrometers. Also, polylactic acids (PLA) [138–140] and cellulose and their derivatives [141–143] are polymers used for piezoelectricity only. However, these polymers in composite or doped form can generate pyroelectricity too [144, 145].

6.1.3.1.4.1 *Poly(Vinylidene Fluoride) (PVDF)*

PVDF and copolymers of this group are undoubtedly the best ever known class of high-performing and most sought polymers for piezoelectric and ferroelectric applications. Through three decades of research and investigation, the piezo/pyroelectricity and electromechanical properties of this family of polymers have been improved remarkably. Even today, PVDF and its copolymers still exhibit the maximum electromechanical responses over a wide range of temperatures. PVDF, owing to repeated units of ($-CH_2-CF_2-$), stands unique for its very high piezoelectric coefficient (d_{33} = 13–28 pC/N [146], d_{31} = 12–23 pC/N [147]) and pyroelectric coefficient (p = 25, μC m^{-2} K^{-1}), reduced dielectric constant (ε_r = 9), and thermally stable up to a sufficiently high temperature of ~120 °C, as compared to other polymeric materials. These extraordinary properties of PVDF enable us to fabricate piezo/pyroelectric devices with high FOMs [54, 113, 114]. Among the five crystallized forms α, β, γ, δ, and ε of PVDF[51, 52] (Figure 6.10a), only one form, that is β with the TTTT chain conformation, exhibits a net dipole moment with an orthorhombic *Amm*2 symmetry, which results in the co-exhibition of pyroelectricity along with piezoelectricity [51, 146, 148–150]. Figure 6.10b illustrates the alignment of hydrogen and fluorine atoms after poling. It was Heiji Kawaï [151] who discovered the piezoelectric property in PVDF, first in 1969; the very next year pyroelectricity was observed by Bergman and coworkers [152] in the same polymer material. Indeed, under mechanical stretching, the α form having a TGTG` chain and a $\gamma$ form with a TTTGTTTG` chain of PVDF can be transformed into the β form [113, 153].

The PVDF sheets or films are commonly stretched and poled to generate piezo/pyroelectric properties. Upon poling, the hydrogen atoms with a net positive charge and the fluorine atoms with a net negative charge align along opposite sides of the sheet, thereby creating a pole direction, which can exist either at the top or bottom of the sheet. Depending on the direction of the applied electric field across the sheets, either contraction in thickness and expansion along the stretched direction, or expansion in thickness and contraction along the stretched direction, results. This happens because the positively charged H atoms are attracted towards the –ve electrode and repelled by the +ve electrode. In the same way, the fluorine atoms carrying a –ve charge drifted towards the +ve electrode and are repelled by the –ve side of the applied electric field.

Although the piezo/pyroelectric coefficients d_{33}/p of polymers are lower (–24 pCN^{-1}/200 μC m^{-2} K^{-1}) than ceramics (300–1000 pCN^{-1}/550 μC m^{-2} K^{-1}), the low permittivity values, design flexibility, low density, light weight, low refractive index, and low cost make them important for energy harvesting applications. Mostly, piezo/pyroelectric polymers possess a semi-crystalline nature where crystalline domains are immersed within an amorphous matrix. These polymers contain non-polarizable phases that are created during synthesis when they are cooled after the molten state, that is polymerization. For these polymers to exhibit piezo/pyroelectric properties, either uniaxial or biaxial stretching is required to convert the non-polarizable phases into polarizable ones, or else inducing polarity via a solution casting and alignment of the dipoles with a corona poling treatment [113].

Further, the copolymers of PVDF (Figure 6.10b) like polyvinylidene fluoride-trifluoroethylene (P(VDF-TrFE)) [55, 115–117, 154], polyvinylidene fluoride-hexafluoropropylene (P(VDF-HFP)) [53, 117–119], polyvinylidene fluoride-chlorotrifluoroethylene (P(VDF-CTFE)) [117, 155], and (P(VDF-TrFE-CTFE)) [115, 117] are also found to be piezo/pyroelectrics. The electrical properties of these polymers can be tuned by varying the VDF/copolymer ratio. The structure of copolymers is illustrated in Figure 6.10c.

6.1.3.1.4.2 *Polyvinylidene Fluoride-Trifluoroethylene (P(VDF-TrFE))*

The introduction of the 20–50% content of the TrFE comonomer into VDF results in the crystallization of the product into the ferroelectric β form directly without the thermo-mechanical processes which are generally applied to PVDF samples[157]. In contrast to PVDF, P(VDF-TrFE) exhibits a larger remnant polarization due to better crystallinity and the favoured alignment of well-grown crystallites [158] The transition temperature of the copolymer decreases to ≈130 °C with an increase in the molar ratio of TrFE between 55 and 88 mol% [113]. In spite of showing mechanical and dielectric properties similar to PVDF, P(VDF-TrFE) shows larger piezo/pyroelectric coefficients d_{33}, k_{33}, and p compared with pure PVDF [159].

6.1.3.1.4.3 *Polyvinylidene Fluoride-Hexafluoropropylene (P(VDF–HFP))*

This copolymer has poor crystallinity and is chemically non-reactive among PVDFs and their copolymers [113]. This copolymer shows processing conditions dependent on

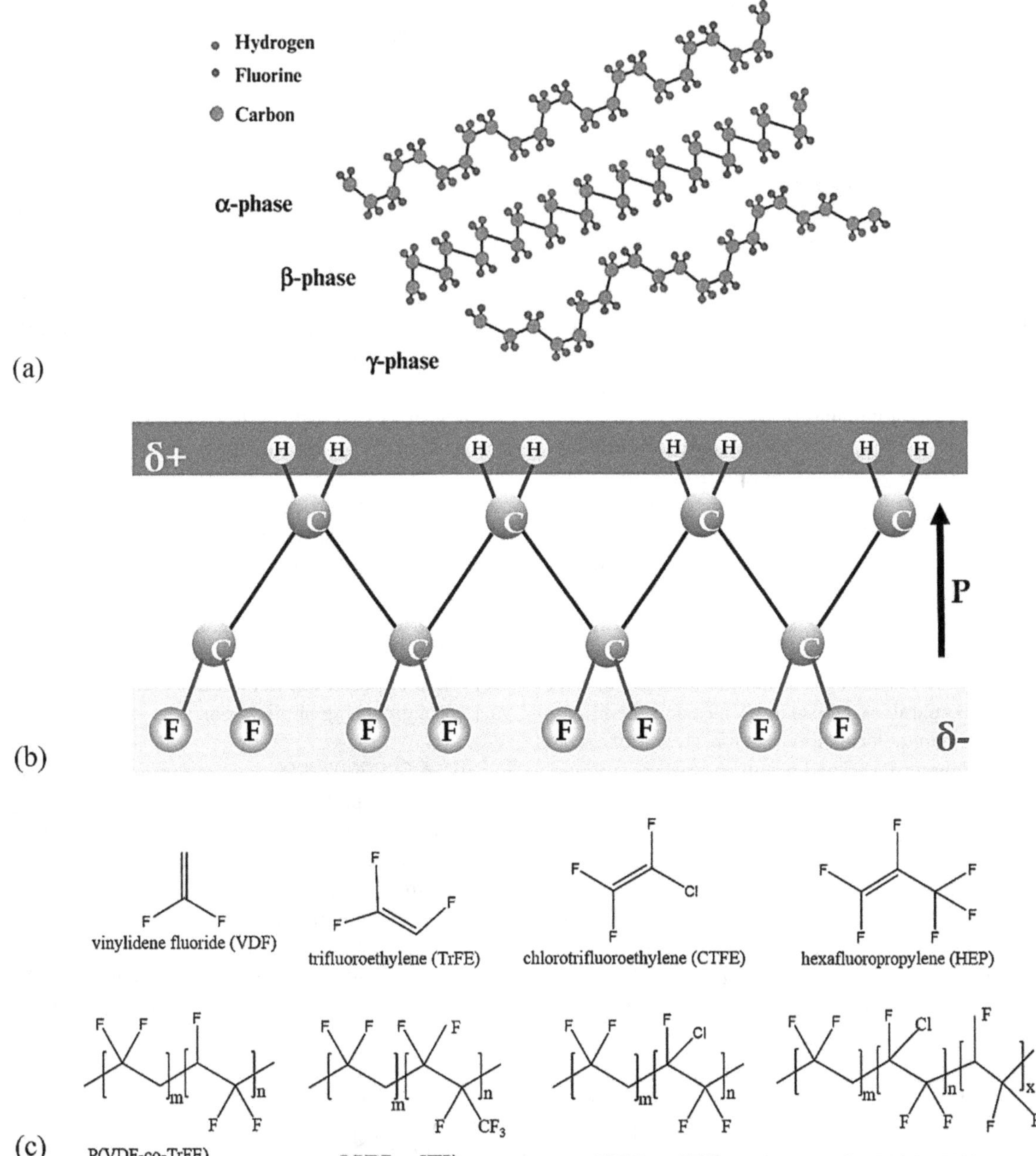

FIGURE 6.10 (a) Schematic representation of the chain conformation for the α, β, and γ phase of PVDF. (b) Alignment of hydrogen and fluorine atoms after poling. (c) Co-polymers of PVDF. (a) [113] adopted with the permission of ©2014 Elsevier. (b) Adopted from [156]

ferroelectric properties and a maximum remnant polarization of 80 mC m^{-2} for low HFP contents ~5% which was obtained after processing by solvent casting[113]. The remnant polarization and hence piezo/pyroelectricity decreases with increasing HFP concentration.

6.1.3.1.4.4 Polyvinylidene Fluoride-Chlorotrifluoroethylene (P(VDF–CTFE))

This copolymer exists in two states: a semicrystalline state with a CTFE concentration below 16 mol% and an amorphous state for higher CTFE concentrations [160]. P(VDF-CTFE) shows advanced piezo/pyroelectric properties, a larger electrostrictive strain response, and a higher dielectric constant compared to PVDF. An easier orientation of dipoles under an applied electric field is possible in this copolymer due to reduction in the packing fraction of structures by the introduction of a bulky CTFE.

6.1.3.1.4.5 Poly(Vinylidene Fluoride-Tri Fluoroethylene-Chlorotri Fluoroethylene) (P(VDF-TrFE-CTFE))

The dielectric constant along with the electromechanical properties of polymers can further be significantly improved through terpolymers. The conjugation of CTFE in P(VDF-TrFE) results in (P(VDF-TrFE-CTFE)). This terpolymer possess not only high polarization and a higher dielectric constant relative to other phases and copolymers of PVDF polymers but also a high electromechanical response and elastic energy density. Therefore, it is considered a very promising piezo/pyro/ferroelectric material [161].

6.1.3.1.5 Polyamides (PA)

Polyamides are commercially known as nylons and possess amide (–CO–NH–) group as linkages in the repeating units of hydrocarbon in the polymer chain [162], as shown in Figure 6.11. Nylons can broadly be categorized as even or odd, where even or odd represents the number of C atoms attached to the repeating unit [162]. Even numbers of methylene groups and an amide group with a dipole moment forms the monomer unit of odd nylons (nylon-5, nylon-7, or nylon-11) [161]. The alignment of dipoles in one direction in every unit cell of these odd nylons (crystalline phase) enables them to show good piezo/pyro/ferroelectric properties as a large dipole moment and spontaneous polarization results from all-transconformation [162].

A polyamide is a semi-crystalline polymer with high temperature stability with hydrogen bonds. Due to the presence of extensive hydrogen bonds, nylon 11 exhibits a better packing fraction and a stable molecular configuration. Although nylon 11 displays good thermal stability and ferroelectricity comparable to those of PVDFs and their copolymers, the remnant polarization of nylon 11 cannot exceed that of fluoropolymers [121] and results in lower piezo/pyroelectric coefficients (d_{31} = 14 pCN^{-1}at 180°C/p = 100 pC m^{-2} K^{-1}) [162]. Among all the crystalline phases of nylon 11, the metastable δ'-phase, with relatively rare chain packing and randomly distributed hydrogen bonds, has shown ferroelectric properties [122, 124]. In the case of the thermodynamically stable α-phase, the dense packing of hydrogen bonds restrains the rotation of dipoles up to the point of electrical breakdown. Hence this phase is polar but non-ferroelectric [121, 125] The application of an electric field triggers the molecular alignment and remnant polarization can be generated; the resulting dipole alignment provides piezo/pyroelectricity [120].

6.1.3.1.6 Polyureas

Polyureas are made by the fast reaction occurring between isocyanates and amine. They are hard and exhibit outstanding resistance to chemicals, temperature, and organic solvents.[163] Their hard domain with a high T_g and soft domain with a low T_g display microphase separation. Polyureas possess more excellent ferroelectric properties than the other piezoelectric polymers and show a large dipole moment of 4.9 D for the urea bond, which is nearly twice the dipole moment of P(VDF-TrFE) [164]. Furthermore, the remnant polarization (P_r) of aliphatic polyureas is 440 mC m^{-2} [164]; they can be synthesized at lower temperatures but their piezoelectric properties are less than aromatic polyureas. The first polyurea structures that identified as piezoelectric were aromatic polyureas which show almost constant pyro and piezoelectric properties, even at an elevated temperature of 200 °C with maximum piezo/pyroelectric coefficients d_{31} = 10 pCN^{-1}/p_3 = 18 μC m^{-2} K^{-1} [163]. The overall properties of polyureas make them a suitable candidate for various engineering applications.

6.1.3.1.7 Biopolymers

Biomedical applications like tissue engineering [165] and biodevices demand biocompatible materials like biopolymers. Biopolymers can be categorized in two types: natural and synthetic. Collagen, polypeptides, that is poly(γ-methylglutamate) and poly(γ-benzyl-l-glutamate), oriented films of DNA, the poly(α-hydroxyacid) family, and chitin are the well know natural biopolymers [166] Usually the poor mechanical, electrical, and electro-mechanical

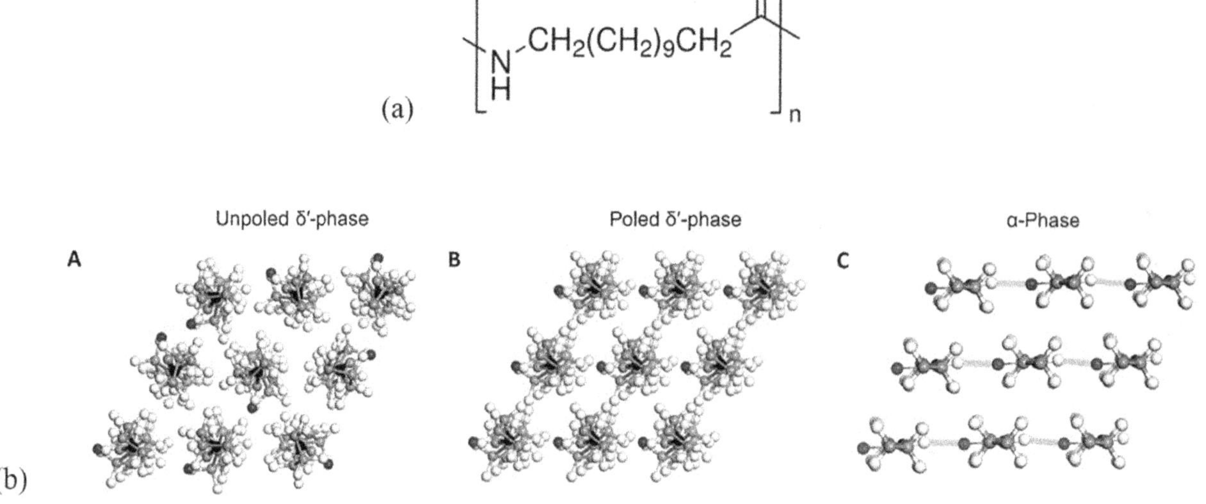

FIGURE 6.11 (a) Structure of nylon 11. (b) Molecular structure and chain packing of nylon 11: A unpoled δ'-phase; B poled δ'-phase; C crystal structure of α-phase nylon 11. Adopted from [121].

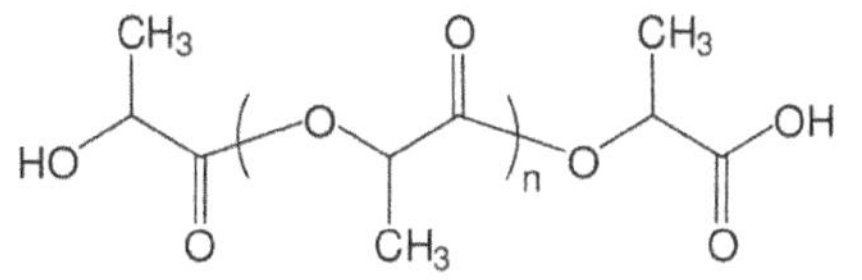

FIGURE 6.12 Structure of PLA.

properties of natural polymers restrict them to being processed in different forms [165]. On the other hand, synthetic biopolymers are bio-erodible materials because of their controllable biodegradability and relatively easy processability. Polyglycolide, polylactide (PLA), and its copolymers, including poly(lactide-co-glycolide) (PLGA) and poly-l-lactic acid (PLLA) are commonly synthetic biopolymers. The structure of PLA is shown in Figure 6.12.

PLA is prepared via the ring-opening polymerization of lactide, a cyclic dimer of lactic acid. The presence of asymmetric carbon lactic acid possess chirality. Therefore, L-form and D-form are present in PLA, and the corresponding polymers are known as poly L-lactic acid (PLLA) and poly D-lactic acid (PDLA), respectively. The left-handed and right-handed helix structure is the principal chain of PLLA and PDLA respectively. PLLA, in its commercially available form, is a helical polymer having dipoles of large value in the helix axis direction (C-axis direction).

In the case of PLLA, the crystal structure is packed with dipoles along the C-axis, facing opposite directions alternately, the dipoles cancelling with each other to be zero. When a PLLA sheet is stretched along the "3" axis, they become uniaxially oriented and show piezoelectricity. The piezoelectric constant of PLLA is $d_{14} = -10$ pC N^{-1}, with T_g at 64 °C, T_C at 107 °C, and T_m at 171 °C [163].

The structure and properties of major piezoelectric polymers are given in Table 6.2 and 6.3 illustrates the properties of the pyroelectric PVDF polymer and its family.

6.1.3.2 Polymer Nanocomposites for Piezo/Pyroelectricity

In spite of the higher piezo/pyroelectric coefficients of a single crystal and ceramic than the polymers, their FOMs are mainly limited by their high dielectric permittivity and thermal conductivity. Therefore, piezo/pyroelectric polymer composites could be a solution to advance FOMs by combining the advantages from both sides, that is the piezo/pyroelectric efficacy of the filler material and mechanical flexibility of the polymeric matrix. Diverse modified properties, like the inferior dielectric constant as compared to ceramics, the lower specific heat, the light weight, the enriched mechanical flexibility, and the low fabrication cost may all be expected in polymer composites. The properties of composites depend largely on the structural arrangement of the reinforcing materials and the connectivity of the components. Various configurations have been studied to ensure proper connectivity between the composite phases, that is 0–3, 1–3, 2–2, and 3–3, wherein the first number corresponds to the arrangement of the electromechanically active ceramic phase while the second number refers to the connectivity of the polymeric phase, which is electromechanically passive.

0–3 connectivity: The polymer phase loaded with a powdered piezo/pyro phase. This pattern facilitates easy fabrication and the mass production of flexible thin sheets, extruded bars, fibres, and other moulded shapes, even curved surfaces. The particle size and relative density affect the piezoelectric coefficient; larger particles show improved efficiency and moderate pressure dependency. On the other hand, the smaller particles show no or negligible pressure dependency, despite showing some piezoelectric coefficients. It is also noted that an increase in the volume of the ceramic phase improves the pressure dependency, indicating improved connectivity.

1–3 connectivity: The piezo/pyroelectric phase is uninterruptedly connected in one dimension (1D) while the polymer phase is connected in all three dimensions. In such composites, it is observed that the stress amplification reduced because the internal stress created opposed the applied stress, hence limiting the piezoelectric coefficient. The high Poisson ratio of polymers was accounted for by this behaviour. There was research to decrease Poisson's ratio by introducing a third factor like porosity and transverse reinforcement [167].

2–2 connectivity: An arrangement of alternating layers of the two phases make up a 2–2 pattern, which may be produced by the tape-casting of multilayer capacitors with alternating layers of metal and ceramic. Here, both the phases are self-connected in lateral X and Y directions but not connected perpendicular to the layers along Z [168].

3–0 connectivity: These composites are formed when the piezo/pyro phase is self-connected in three dimensions while the other phase is not at all connected in any dimension. The hot pressing method used commonly for such composites yields an irregular array of polymer chains surrounded by the piezo-ceramic phase.

3–1 and 3–2 connectivity: This pattern constitutes a piezo/pyro phase connected in all three dimensions, while the polymeric matrix is connected in one or two dimensions. Honeycomb patterns and perforated composites fall under this category. The stress distribution reduced the d_{31} coefficient in both cases [167, 169].

3–3 connectivity: Such composites display a pattern where both component phases are in three dimension and are self-connected. They are also intimately in contact with each other. Many such composites have been studied [170–172].

TABLE 6.2
Piezoelectric Properties of Piezo-Polymers

Polymers	Structure	Piezoelectric Coefficient (pCN^{-1})		Piezoelectric Voltage Co-Efficient (10^{-3} VmN^{-1})	Relative Permittivity	Electro-Mechanical Coupling Coefficient	
		d_{33}	d_{31}	g_{33}	ε_r	k_{33}	k_{31}
PVDF	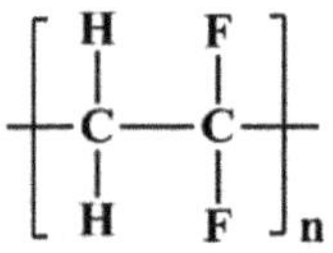	−24 to −34	8 to 22	−330	6–12	0.20	0.12
PVDF-TrFE	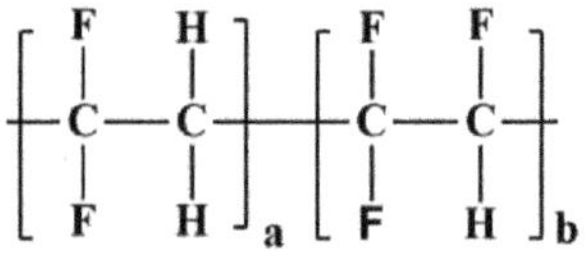	−25 to −40	12 to 25	380	18	0.29	0.16
PVDF-HFP	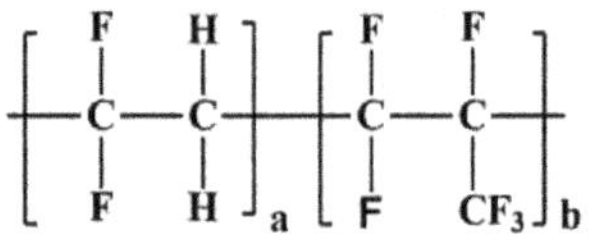	−24	30	—	11	0.36	0.187
PLA	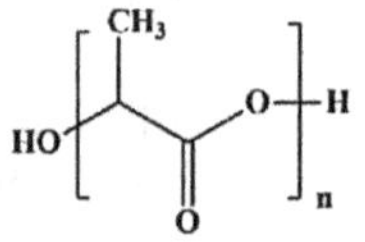	—	1.58	—	3–4	—	—C
Nylon 11	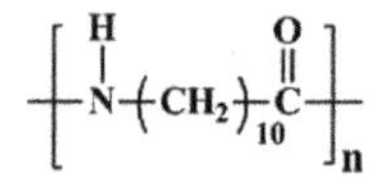	4	14 (at 100–200 °C)	—	5	—	0.049
Cellulose	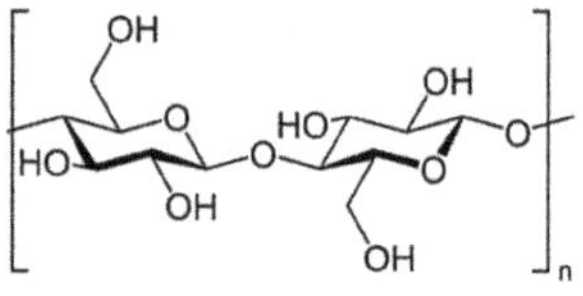	5.7 ± 1.2 (for CNF)	1.88–30.6 (at 0 °C, 45 °C & 90 °C)	—	—	—	—
Polyurethane	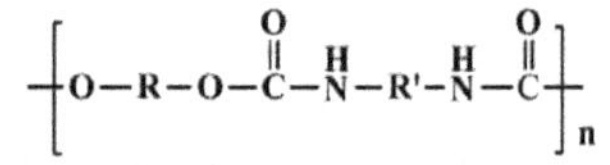	—	27.2	—	4.8–6.8	—	—
Polyurea	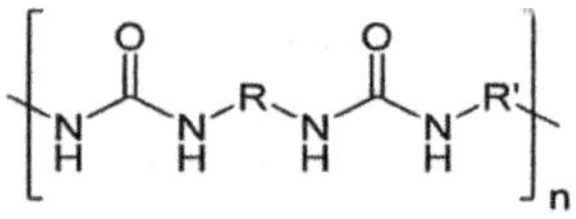	19 at 60 °C & 21 at 180 °C	10	—	—		0.08
Polyimide	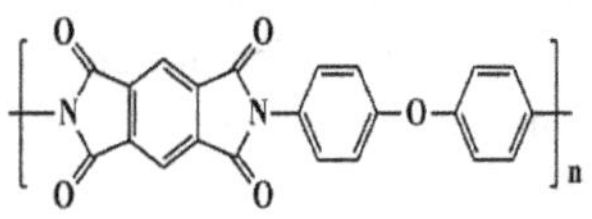	2.5–16.5	—	—	3.34	—	—
PAN	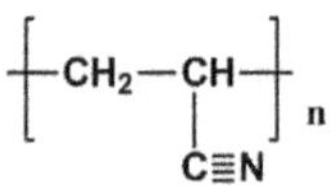	—	2	—	—	—	—

Source: [257] adapted with permission from ©2019 John Wiley and Sons.

A schematic diagram of piezoelectric ceramic-polymer composites in diverse connectivity is shown in Figure 6.13 and the advantages and applications of these are summarized in Table 6.4.

However, pyroelectric composites are frequently synthesized in 0–3 and 1–3 connectivity by embedding ceramic particles and fibres respectively in the polymer matrix. Commonly, high piezo/pyroelectric performance polymers like PVDF, P(VDF/TrFE), cellulose and its derivatives, polyamides, and polylactic acids were used as the organic matrix. Polymeric materials like PVC and HA have also been studied. A few polymer composites with various fillers are described as follows.

TABLE 6.3
Pyroelectric Properties of Representative Polymers at Room Temperature

Material	ε_r	tanδ	*Polarization* μCm^{-2}	*P* μCm^{-2}K^{-1}	F_i 10^{-10} mV^{-1}	F_v m^{2}C^{-1}	F_D 10^{-5}Pa$^{-1/2}$	Reference
PVDF	12	0.015		27	0.11	0.105	0.88	[114, 287]
P(VDF/TrFE) 80/20	7	0.015		31	0.13	0.217	1.40	[288, 289]
P(VDF/TrFE) 60/40	19.4	0.041		51	0.225	0.13	0.838	[290]
P(VDF/TrFE) 70/30	7.4	0.017		33	0.14	0.220	1.36	[289, 291]
P(VDF/TrFE) 60/40	29			45	0.20	0.076		[292]
P(VDF/TrFE) 56/44	18	0.05		52	0.24	0.148	0.81	[293]
P(VDF/TrFE) 50/50	18	0.05		40	0.17	0.109		[49, 293]
(P(VDF–HFP))			24	49				[118]
(P(VDF–HFP))				49				[53]
(P(VDF–HFP))	13.1		24					[119]
(P(VDF–CTFE))			17.1					[155]
Polyviny fluoride (PVF)	5	0.05		18	0.08	0.177	0.53	[134, 294, 295]
Polyviny chloride (PVC)	5	0.01		1				[134, 135, 294]
Polyacrylonitrile (PAN)	7.7	0.2		1				[134, 136, 294]

Source: [25] reproduced with permission from ©2019 The Royal Society of Chemistry.

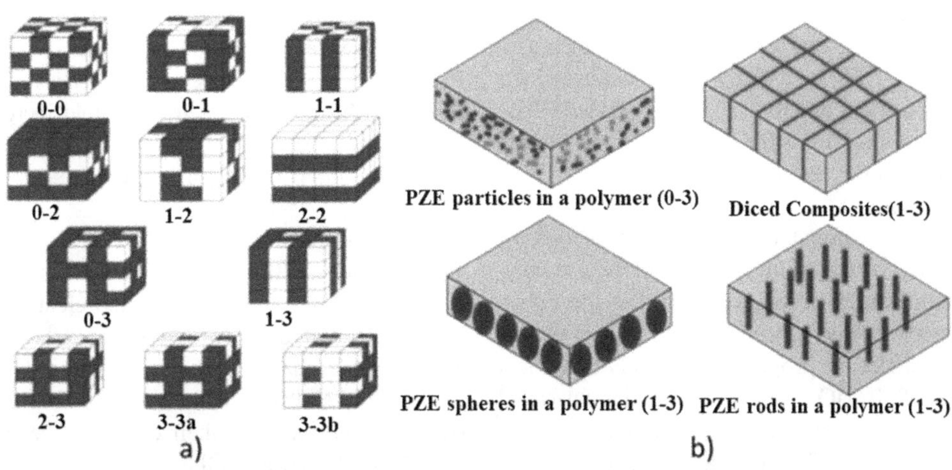

FIGURE 6.13 (a) Connectivity patterns for piezoelectric ceramic and polymer composites. (b) Connectivity of piezo composites [173]. [174] adopted with the permission of ©2012 Elsevier.

6.1.3.2.1 PVDF and Its Copolymers

Amongst flexible piezo/pyroelectric materials, PVDF has a high piezo/pyroelectric coefficient, decent flexibility, and outstanding biocompatibility; hence, it is commonly used to develop flexible electromechanical/electrothermal devices. PVDF is the widely used piezo/pyro-polymer, majorly in combination with a piezo/pyro-filler. The piezo/pyroelectric properties greatly depend on the types of fillers used. Several categories of fillers, such as ceramic fillers, carbon-based fillers, hybrid fillers, and bio-based fillers, have been studied and are detailed below.

6.1.3.2.1.1 Ceramic Fillers

Distinguished ceramic fillers include lead-zirconate-titanate (PZT), barium titanate (BaTiO3), sodium potassium niobate (KNN), and bismuth sodium titanate (BNT) which display high piezo- or piezo/pyroelectric coefficients. The synergism obtained by blending piezo-ceramics with polymeric matrices has prompted extensive research in this area. The incorporation of PZT as a filler showed improved electrical and structural properties [175] of PVDF. With an increase in the PZT particle size, the energy loss was observed to decrease. However, the piezoelectric coefficient

TABLE 6.4
Advantages and Applications of Different Piezo-Composite Connectivities

Connectivity	Advantages	Applications
0–3	Good flexibility High piezoelectric voltage Complex shapes	Potential use as piezoelectric paint for "built-in" vibration modal sensors Sensors on wind turbine blades
3–3	Higher piezoelectric coefficients compared to 0–3 composites	Acoustic transducers Medical imaging Non-destructive testing (NDT)
2–2	High acoustic impedance backing	Multilayer actuators Bimorph actuators or sensors Piezoelectric stack actuators Medical linear array transducers
1–3	Highest electromechanical coupling factors	Medical imaging NDT Underwater acoustic transducers
Piezoelectric composites	Lightweight Robust construction Increased actuation strain energy density	Structural control applications Structural health monitoring Active damping of high-frequency vibration

Source: [296] adapted with permission.

also decreased with the particle size of PZT. Calcined PZT showed improved pyroelectric properties when reinforced in a polymer matrix, due to the enhanced coupling interactions between the composite constituents [176] Researchers have also found that the dielectric permittivity also increases in the case of PVDF-TrFE/PZT composites, because of the increased ß-phase content due to the ceramic phase [177].

Further, incorporation of $BaTiO_3$ showed high energy output, remarkably higher than the polymer matrix [178]. When the PVDF/$BaTiO_3$ composites were cross-linked with dopamine, the hydroxyl (–OH) and amine (–NH_2) groups of dopamine reacted with $BaTiO_3$, forming a polydopamine structure which becomes attached to the fluorinated polymer matrix. This induced more all trans ß-phases and improved the dielectric properties. A steady increment in the loading of $BaTiO_3$ nanoparticles demonstrated increased breakdown strength and energy storage density of the piezo-composite [179]. The calcination of $BaTiO_3$ led to a decrease in dielectric loss (tan δ) when subjected to low frequency (<104 Hz) and an increase in tan δ at high frequency (>104 Hz), which is primarily related to an interfacial polarization effect. Recently, another ceramic filler, gallium ferrite nanoparticles (GFO), was also studied for its contribution towards the piezoelectric properties of PVDF and its copolymers. The output voltage of the composite was found to be considerably high at a loading of 30 wt.% of the ceramic phase [180].

Lam et al. [181] investigated the piezo- and pyroelectric properties of $(Bi_{0.5}Na_{0.5})_{0.94}Ba_{0.06}TiO_3$ or BNBT/P(VDF/TrFE) 0–3 connected composites. They varied the BNBT volume fraction from 5 to 30% in the polyvinylidene fluoride-trifluoroethylene [P(VDF-TrFE) 70/30 mol%] copolymer. They observed that the 0–3 BNBT/P(VDF-TrFE) composites had comparable piezoelectricity (d_{33} = –21pCN^{-1}) but a higher pyroelectric figure of merit (p = 47.3 μC m^{-2} K^{-1} and F_p = 2.29 μC m^{-2} K^{-1}) compared with PZT/P(VDF-TrFE) 0–3 composites. The piezoelectric properties of P(VDF-TrFE) (70/30% mol)/(60 vol%) $BaTiO_3$ nanoparticle (BTNPs) composite film were investigated by Genchi and co-workers [182]. In their study the piezoelectric properties of the composite films significantly improved (a 4.5-fold increase in d_{31}) and attained d_{31} = 53.5 pmV^{-1} and g_{31} = 0.24 mVN^{-1} from d_{31} = 11.8 pmV^{-1} and g_{31} = 0.11 mVN^{-1} of bare P(VDF-TrFE).

6.1.3.2.1.2 Carbon-Based Fillers

Their innate multifunctional features, such as high stiffness, low density, and thermal and electrical conductivity [183] have created a lot of interest in carbon-based fillers as a reinforcing phase. Carbon nanotubes (CNTs), graphene oxide (GO), and reduced graphene oxide (rGO) have been studied extensively for their effect on the improvement of piezo/pyroelectric as well as dielectric characteristics.

It was found that even low poling voltages can bring in a permanent piezoelectric effect in the PVDF/CNT composite owing to the reaction between the electronegative fluorine atom with the π-electron rich surface of MWCNT, triggering the all conformation [184]. In the case of a single walled CNT, it was seen that the piezoelectric response decreased with concentration due to cluster formation [185]. When doped with PVDF/PZT, multi-walled CNTs showed an increase in the dielectric constant at lower frequencies [178]. Batra et al. [186] measured the dielectric, pyro/pyroelectric properties of MWCNT/PVDF composite films and calculated the FOMs for IR detectors and thermal-vidicons. They found a huge improvement of the FOMs of composites compared to pristine PVDF. Similar, improved FOMs of a PVDF/MWCNT nanocomposite compared to pristine PVDF were observed by Edward and co-workers [187] during studies of the pyroelectric and dielectric properties of film for uncooled infrared detectors.

The incorporation of graphene oxide nano-sheets (GOns) increased the output voltage by 293% of that of PVDF [188] while the piezoelectric constant and remnant polarization were found to increase by 78.6% and 69.3%, respectively [189]. GO bi-layer films with a superior piezoelectric energy harvesting capability further increased the cycling stability of the composite under compression pressure, where the output voltage remained constant even after 1000 repeating cycles [190]. Carbon black was also studied as a reinforcing agent in the fluorinated polymer where the harvested electrical power density of the polymer matrix increased by 464% for an AC test and 561% for a DC test[191]. Polyaniline-grafted-graphene oxide (G-graft-PANI) led to the conversion of the non-polar α-phase to 91% of the polar β-phase of PVDF and an increase of the dielectric permittivity to 264 and a dielectric loss to 1.125 [192]. A new approach using a material combination like graphene ink/PVDF for developing smart materials and structures, wireless technologies, and Internet of Thing (IoT) devices was presented by Zabek et al [193] Computational calculations focused on the pyroelectric properties of the graphene (G) or graphene oxide (GO)/PVDF composites were carried out by Bystrov and co-workers [194]. Numerous molecular dynamic simulation-based models employing a quantum-chemical semi-empirical PM3 method from a HyperChem tool were used to model the pyroelectric effect and to calculate pyroelectric coefficients. The results presented new prospects for further studies of such nanocomposite materials and their applications. Recently, Li et al. [195] investigated the pyroelectric efficiencies and solar harvesting properties of polyethyleneimine chemically modified reduced graphene oxide (rGO-PEI) and a pyroelectric layer based on a polarized PVDF film. The author and his team demonstrated its practical viability by integrating the S-PENG into a wearable outdoor bracelet.

6.1.3.2.1.3 Metal-Based Fillers

Metallic filler particles like zinc oxide (ZnO), titanium dioxide (TiO2), ferrites, and iron oxide (Fe3O4) can significantly augment the piezo- and/or pyro-response of polymers to a major extent. The PVDF-TrFE/ZnO composites showed enhanced piezoelectricity and remnant polarization while being mechanically flexible. While poled, they showed almost ten times increase in the output voltage and reported a higher open-circuit AC voltage, piezoelectric charge constant (d_{33}), and piezoelectric voltage constant (g_{33}) for higher ZnO content. When doped with another metal like aluminium (Al), PVDF/ZnO composites displayed dielectric constant almost twice that of the undoped composite [196, 197]. From the studies, it was observed that a lower loading of ZnO nanoparticles (0.25 wt.%) with a 0–3 connectivity of ZnO/PVDF nanocomposites considerably improved the pyroelectric coefficient of PVDF by 25%. Additionally, a lower poling field was required by the nanocomposite films compared to virgin PVDF thin films to form a polar-δ phase [198]. It was also established that the ZnO nanoparticles promoted the conversion of the α phase to the β phase of PVDF in ZnO/PVDF nanocomposites during the process of electrospinning and finally improved the piezo/pyroelectric properties of electrospun composite fibers [199].

PVDF/Fe_3O_4 composites were found to have a better storage modulus and thermal stability [200] The thermomechanical behaviour and dielectric response were also studied for Fe_3O_4 nanocomposites. The incorporation of TiO_2 nanoparticles into PVDF improved the pyroelectric properties and led to an easier poling process [201]. Singh et al. [202] have demonstrated a lead-free, solution processed flexible piezoelectric energy generator based on an MgO nanoparticle/P(VDF-TrFE) nanocomposite film. A dramatic improvement was recorded in the piezoelectric, ferroelectric, and leakage current with varying concentrations of MgO nanoparticles; a piezoelectric response at 2 wt.% MgO was increased by nearly 50% compared to the pure P(VDF-TrFE). Besides, the metal oxides, metal (silver) nanoparticles, and nanowires incorporated into the PVDF-TrFE [203, 204] and nanowire based composite exhibited an improved output voltage.

6.1.3.2.1.4 Hybrid Fillers

Hybrid fillers are constituted of two components where one is organic and the other is inorganic. They not only provide synergism, but also help in lowering the filler loading and hence the processing difficulties as well as cost. Qiu et al. [205] and Yang et al. [206] have studied the effect of TiO_2-MWCNT composite films and found that the PVDF/TiO_2@MWCNT composite showed an enhanced dielectric constant, breakdown strength, and a high piezoelectric coefficient at low filler concentrations. Iron-doped RGO (Fe-RGO)/PVDF films—because of the electrostatic interaction between the oxygen functionality as well as the delocalized π-electrons of Fe-doped RGO and the dipoles of PVDF (CH_2 and CF_2) through ion-dipole and/or hydrogen bonding interaction—displayed a high output voltage [207]. RGO/TiO_2 hybrids were also studied [208], which showed an increase in the mechanical properties, dielectric permittivity, and pressure sensitivity of the PVDF composite. Further trials by Abdelhamid et al. [209], where RGO, ZnO, and Fe_3O_4 were hybridized, showed a poor response in terms of dielectric behaviour and ß-phase content. This revealed that excess hybridization would lead to a loss of properties.

In addition to the above discussed categories, recently Karan and co-workers [210] designed a bio-inspired vitamin assisted single-structured self-powered piezoelectric energy harvester with high energy conversion efficiency. They gained a high out-of-plane piezoelectric coefficient of ~ −50.3pC/N from bio-inspired vitamin B_2 assisted PVDF. Shehzad and Wang [115] investigated the effect of blending a small amount of crystalline P(VDF-TrFE) into P(VDF-TrFE-CTFE) on pyroelectric properties. The pyroelectric coefficient and FOMs of TGS/PVDF composites were investigated by Batra et al.[211]. The composite with 80 vol% TGS showed a high p = 90 μCm^{-2}K^{-1} (much higher than that of the pure PVDF), a low dielectric constant of ε_r = 12, and considerably high FOMs.

6.1.3.2.2 Polylactic Acid (PLA)

This biodegradable polymer also shows piezo/pyroelectric behaviour and hence has been used for developing biocompatible fibre composites. Solvent blending and electrospinning techniques have been reported for the development of composite films. Electrospun composite mats reported a higher mechanical integrity and piezo/pyroelectric coefficient, indicating its viability for commercial applications [212]. The PLA/$BaTiO_3$ composite could light a small LCD and was also utilized in harvesting vibrational energy [213]. This could be obtained by the enhanced crystallinity of PLLA when annealed at 140 °C for 24 hrs, which led to an increased output voltage (9.4 V) and power (≈14.45 μW). Attempts were also made to prepare yarns from PLLA in 2017 [214] and it was observed that the extension and contraction of twisted yarns leads to the generation of an electric field. This results in an antibacterial effect of the as-prepared fabric, thereby opening up new opportunities for piezo-textiles. Thermal diffusivity, thermal effusivity, specific heat capacity, and the thermal conductivity of a PLA/PPy (polypyrrole) composite have been investigated using photoacoustic calorimetry and photopyroelectric techniques [215].

6.1.3.2.3 Polyurethanes (PU)

Another class of polymers showing piezo/pyroelectricity is polyurethane (PU). PU displays transverse piezoelectricity depending on the electrostriction and bias DC field. Under optimum conditions, the material exhibits a piezoelectric constant double the value of poled PVDF film and an electromechanical coupling factor similar to PVDF [216]. PU/ceramic (PZT) composites were studied and approaches made to facilitate a shorter and easier poling process by increasing electrical conductivity [217, 218] with the addition of a carbon filler as well as polyaniline. Flexible PU foams were also investigated for their piezoelectric constant after incorporating the composite with polar dopant molecules [219]. It was found that even when the human body movement presented a very low applied force, the doped foam composite showed a high piezo-responsiveness; sensitivity was directly proportional to the concentration and dipole moment of the dopant molecule. The stored capacitance and maximum stored energy for temperature fluctuations over a period of time in the order of 140 s were 1 μF and 14 μW for a PZT/PU composite loaded with 40% PZT [220]. The ambient pyroelectric coefficients of the PZT/PU composite thin films were found to have a linear increase with the PZT volume fraction and reached 90 μCm^{-2}K^{-1} for 30% PZT, which was more than tenfold that of a PZT/PVDF composite of the same ceramic volume fraction [221]. The pyroelectric coefficients of poled strontium barium niobate/PU composites increased from 81 to 385 μCm^{-2}K^{-1} and the FOMs F_i, F_v, and F_D increased respectively from 48 to 283, 6.83 to 20.99, and 191to 792, all in units of 10^{-3} μCmJ^{-1} as the volume fraction of SBN30 increases from 0 to 0.25 [222]. Recently, Sakamoto et al. [223] observed the pyroelectric coefficient of 70 μCm^{-2}K^{-1} was higher than that of β-PVDF (10 μCm^{-2}K^{-1}) for PZT/PU composites at 343 K.

6.1.3.2.3.1 Polyamides (PA)

Polyamides (PA), semi-crystalline polymers commonly known as nylons, show a low level of piezo/pyroelectricity. However, in the case of odd polyamides such as nylon 5 or nylon 11, the interaction between an even-numbered methylene group and a one amide group led to aligned dipoles and hence a net dipole moment, whereas the amide dipoles cancel each other out in the case of even-numbered polyamides [224]. Although the stretching and poling of nylons has been studied for analysing piezoelectric properties, their low piezoelectric values have triggered researchers to focus on filled nylons [225, 226]. A variety of fillers such as PZT [227], sodium niobate ($NaNbO_3$) [228], CNTs [229], layered silicates [230], and $BaTiO_3$ [231] etc. have been incorporated into nylons for improving the piezo or piezo/pyroelectric behaviour. The temperature and time of poling greatly affects the performance of the composite. A comparative analysis by David et al. [232] revealed the improved efficacy of $NaNbO_3$ over $BaTiO_3$ as a piezo-filler for PA. However, the addition of $BaTiO_3$ influenced the crystallinity behaviour of PA and improved the thermal stability. Another hybrid system studied comprised PA, PZT, and CNT where it was observed that the incorporation of carbon-based fillers increased the dielectric permittivity of piezo-polymer/piezo-ceramic composites [233]. The three-phase multifunctional system was found to have huge potential for energy harvesting applications. Leveque et al. [230] found that embedding layered silicates with polar functionalities induces more polarity to the PA composite system thereby providing better piezo-responsiveness. BT/PA-11 composites showed interesting pyro-piezoelectric activity and pyroelectric FOMs were increased linearly and attained a limiting value of 0.3 for a volume fraction ϕ = 0.1 [234].

6.1.3.2.3.2 Cellulose and Its Derivatives

Cellulose, a major constituent of plant cell walls, is a widely available biomaterial in nature. The dipolar alignment and confined charges of the polymeric structure make them a choice of interest for energy harvesting and related applications [190]. Another major benefit is the shear piezoelectricity of the cellulose-based biopolymers such as wood, amylase, chitin, and starch. Cellulose nanocrystals (CNC) showed increased piezoelectric properties when aligned [142]. Wood cellulosic fibres were also studied and piezoelectric paper was fabricated, which was functionalized and doped with $BaTiO_3$. The composite paper displayed a high d_{33} value, but at a higher loading of the piezo-ceramic phase [235]. Similar electroactive papers based on cellulose are being developed for biomimetic actuators exhibiting large output with low power consumption, low actuating voltage, low cost, lighter weight, and biodegradability. The researchers [200, 236] found that it showed a high piezoelectric response under ambient conditions and that the properties are sensitive to the fabrication process and material orientation. Chitosan/cellulose nanofibril (CNF) composites showed piezoelectric values very close to PVDF, while self-standing CNF films also showed improved piezo-performance, thereby ascertaining its suitability for energy harvesting and storage application [141, 237]. Metal (ZnO) filled cellulose films, when investigated, showed properties similar to electroactive paper with a remarkable piezoelectric coefficient and mechanical properties [238].

6.2 OTHER POLYMER COMPOSITE SYSTEMS

Polymeric materials such as polypropylene (PP) [239], polyethylene oxides (PEO) [240], polyureas [241], polyvinylidene cyanide [242], polyacrylo nitrile (PAN) [243], and polyimides (PI) [244] could also be promising materials, when modified with appropriate fillers. Health monitoring applications have opted for cellular PP based generators where the d_{33} values remained stable for weeks, even at elevated temperatures. A similar investigation of a PEO/ceramic composite and composites made from a copolymer of vinylidene cyanide and vinyl acetate showed their potential for use as flexible piezo-sensors while a PAN/$BaTiO_3$ composite system showed remarkable triboelectric, piezoresistive, and piezoelectric effects. Further, polyureas also showed piezoelectric properties upon poling; these materials are highly stable even up to 200 °C, thus widening their application arena. Depending on their structure (aliphatic or aromatic) and poling temperature, polyureas show variable ferroelectric/piezoelectric properties. Polyimides (PI), due to their high glass transition, are one of the favourite high temperature amorphous piezoelectric polymers. However, the residual polarization and piezoelectric coefficients strongly depends on the quantity of nitrile dipoles present on the polymer backbone, the imidization process, and the poling parameters as well [243]. The electrical properties of $BaTiO_3$ (BTO)/PVC with different BTO volume percentages were studied by Olszowy et al. [245]. They illustrated that the value of the pyroelectric coefficient increased with an increasing ceramic volume fraction. A composite with 40 vol% ceramic showed small dielectric dispersion and loss with a relatively large $p = 106\ \mu Cm^{-2}K^{-1}$.

6.2.1 Piezo/PyroElectric Polymers for Energy Harvesting

6.2.1.1 Energy Harvesting: Principles and Methods

The increasing demand for self-powered autonomous or low power consumption electronics has been stressing the energy sector worldwide; the scientific community has been extensively studying highly efficient energy harvesters. Solid-state batteries, which require periodic recharging/replacement, are currently widely used as an energy source in electronic devices. Hence, a number of strategies have been proposed to harvest energy from unexploited energy sources and to enable provision for a sustainable energy supply for small electronic devices [246, 247]. The piezoelectric energy harvesters of mechanical energy[247] and the thermal energy harvesters using temperature gradients through thermoelectricity or pyroelectricity [248] are among them. Generally, there are three parts for an energy harvesting system: [249]

- *Energy source*: The source from which energy will be harvested for conversion; this may be sunlight, heat, wind, lightning, human heat, or vibrations [250].
- *Harvesting mechanism*: An assembly for converting the energy source into electrical energy.
- *Load*: The sink storing electrical energy output.

Most common energy sources are sunlight, electromagnetic radiation, human body heat, and environmental and human body mechanical energy. Energy harvesters working on human body factors can be assimilated into daily human activities to power a variety of devices, unlike environment-dependent energy sources such as solar energy, electromagnetic radiation, and environmental mechanical energy [251]. The primary principle of the energy harvesting technique is the utilization of the piezo/pyroelectric conversion effect of the piezo/pyro-materials. The piezoelectric material deforms when subjected to an external pressure and generates a voltage from the conversion of mechanical energy to electrical energy. On the other hand, pyroelectric materials produce a temporal temperature change which results in a change in the internal spontaneous polarization of the material and generates an electric current.

Human body and environmental mechanical energy are an impending source of kinetic energy and are extensively exploited because of their easy availability in daily life [252]. Power produced by scavenging these mechanical energies can ensure long-term autonomy for self-powered systems [253]. For example, upper limb movement can harvest around 10 mW while typing motion can harvest 1 mW; breathing can generate around 100 mW and walking up to 1 W [254] The harvested power density (P_{res}) for mechanical energy depends on the frequency and magnitude of motion: [255]

$$P_{res} = 4\pi^3 m f_{res}^3 y Z_{max}$$

where m represents the inertial mass, Z_{max} is the maximum displacement, f_{res} is the resonance frequency, and y is the amplitude of the vibration of the housing.

The approximate frequency of different mechanical energy sources is shown in Figure 6.14.

A piezoelectric energy harvester basically has two parts: [256]

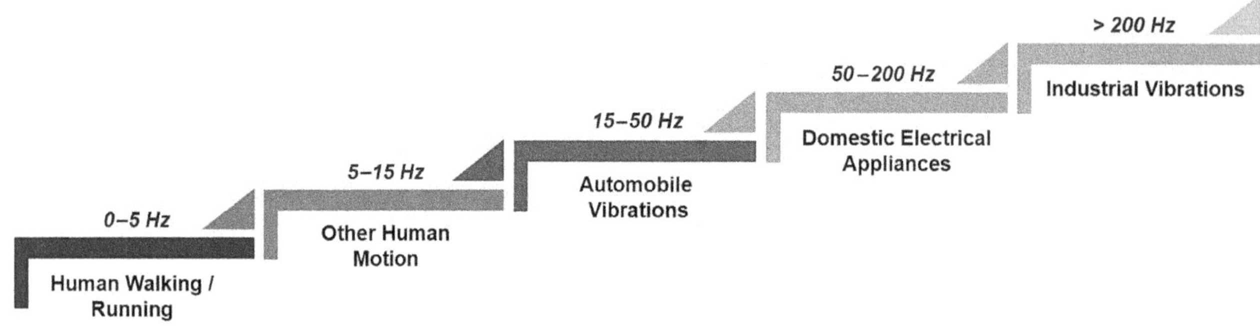

FIGURE 6.14 Frequencies of various mechanical energy sources. [252]

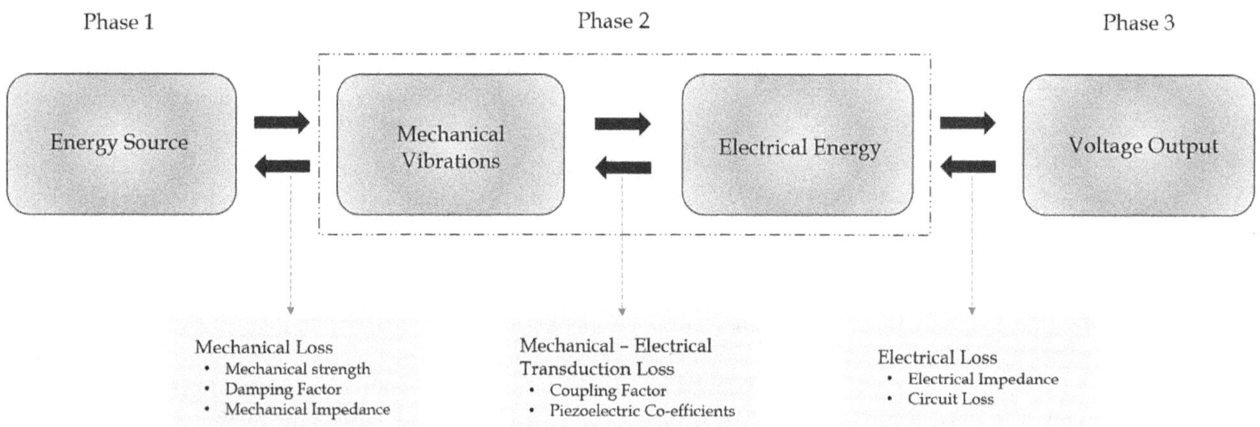

FIGURE 6.15 Phases of piezoelectric energy harvesting. [257] with permission ©2019 John Wiley and Sons

- A mechanical module for generating electrical energy;
- An electrical module comprising an electrical circuit for converting and modifying the generated voltage.

As a result, the effectiveness of an energy harvester is dependent on both the piezoelectric transducers and its integration with the electrical circuit. In general, the piezoelectric energy harvesting system works towards the following three phases of energy conversion: [257, 258]

- *Mechanical energy to mechanical energy*: Linked with the compliance of mechanical impedance of piezoelectric energy harvester and its mechanical strength under high stresses.
- *Mechanical energy into electrical energy*: Correlated with the electromechanical coupling factor and its coefficients of the piezoelectric energy harvester.
- *Electrical energy into electrical energy*: Encompassing the compliance of electrical impedance.

The three phases associated with the piezoelectric energy harvester are illustrated in Figure 6.15.

The thermoelectric or seeback effect and the pyroelectric effect are the two major phenomena used to harvest thermal energy. The former requires a spatial temperature gradient (*dT/dx*) whereas a temporal temperature gradient (*dT/dt*) is required for the latter. Pyroelectric materials having a number of advantages over thermoelectric materials, particularly in the application of thermal harvesting, are: (i) pyroelectric materials can be used in those environments where the temperature is uniformly distributed and only certain temperature fluctuations occur, whereas the seeback effect fails due to the need of a temperature difference between the two coupling ends; (ii) the stability of many pyroelectric materials even under extremely high temperatures (~1200 °C), unlike thermoelectrics, enables effective energy harvesting from high-temperature sources. Therefore, the pyroelectric effect can be extremely beneficial in solar energy harvesting and utilizing wasted thermal energy from different sources [259, 260]. Furthermore, methods to convert stationary spatial gradients to transient temperature gradients have been introduced [261], which enable the utility of both thermoelectric and pyroelectric effects for the development of hybrid energy harvesters [262].

Generally, the electro-thermal coupling factor (k) which indicates the effectiveness, the pyroelectric energy harvesting figure of merit (F_E) that provides a comparison ability of the power generation, and the modified figure of merit (F_E`) that considers specific heat are used to characterize the energy harvesting capability of a pyroelectric material and are mathematically expressed as: [25]

$$k^2 = \frac{p^2 T_h}{C_v \varepsilon_r} \quad (6.10)$$

TABLE 6.5
Performances of Representative Lead-Free Materials for Energy Harvesting

Material	Type	p (mC m^{-2} K^{-1})	e_r	C_v (J K^{-1} cm^{-3})	k^2 (T_{hot} = 300 K)	F_E (J m^{-3} K^{-2})	FE^0 (10^{-12} m^3 J^{-1})
TGS	Single crystal	350	40	2.6	6.272	345.89	51.17
$LiTaO_3$	Single crystal	190	47	3.2	0.813	86.75	8.47
$LiNbO_3$	Single crystal	96	31	2.7	0.373	33.58	4.61
Mn:94.6BNT–5.4BT <111>	Single crystal	588	279	2.9 2	1.448	139.96	16.64
SBN-50	Single crystal	550	400	2.2	1.165	85.41	17.65
ZnO	Single crystal	9.4	11	3.4	0.008	0.91	0.08
$BaTiO_3$	Ceramic	200	1200	2.5	0.045	3.76	0.60
$Ba_{0.85}Ca_{0.1}Sr_{0.05}Zr_{0.1}Ti_{0.9}O_3$	Ceramic	900	2667	2.3	0.447	34.30	6.48
$0.93(Bi_{0.5}Na_{0.5})TiO_3–0.07Ba(Ti_{0.945}Zr_{0.055})O_3$	Ceramic	570	1052	2.8	0.374	34.88	4.45
$0.94(Bi_{0.5}Na_{0.5})TiO_3–0.06BaTiO_3$	Ceramic	315	400	2.8	0.300	28.02	3.57
$[(K_{0.5}Na_{0.5})_{0.96}Li_{0.04}](Nb_{0.84}Ta_{0.1}Sb_{0.06})O_3$	Ceramic	190	1520	2.0	0.040	2.68	0.67
$0.95AgNbO_3–0.05LiTaO_3$	Ceramic	368	252	2.7	0.674	60.70	8.33
0.1 wt% $Mn:(Na_{0.5}Bi_{0.5})_{0.95}Ca_{0.05}Bi_4Ti_4O_{15}$	Ceramic	100	134	2.9	0.087	8.43	1.00
0.2 wt% $Mn:CaBi_4Ti_{3.95}Nb_{0.05}O_{15}$	Ceramic	84	99	2.4	0.101	8.05	1.40
0.5 mol% $Cu:CaBi_4Ti_4O_{15}$	Ceramic	86	145	2.6	0.066	5.76	0.85
$Sr_{0.5}Ba_{0.5}Nb_2O_6$	Ceramic	205	971	2.1	0.070	4.89	1.11
$Ca_{0.15}Sr_{0.425}Ba_{0.425}Nb_2O_6$	Ceramic	361	933	2.1	0.225	15.78	3.58
$Ca_{0.3}Ba_{0.7}Nb_2O_6$	Ceramic	130	355	2.1	0.077	5.38	1.22
$Bi_{1.5}Zn_{1.0}Nb_{1.5}O_7$-$BaTi_{0.85}Sn_{0.15}O_3$	Film	2470	600	2.2	15.660	1148.42	237.28
$LiNbO_3$	Film	71	30	3.2	0.178	18.98	1.85
$BaZr_{0.25}Ti_{0.75}O_3$	Film	74	635	2.6	0.011	0.97	0.14
1 mol% $MnO:0.82Bi_{0.5}Na_{0.5}TiO_3–0.18Bi_{0.5}K_{0.5}TiO_3$	Film	380	620	2.8	0.282	26.30	3.56
PVDF	Polymer	27	12	2.4	0.086	6.86	1.19
P(VDF/TrFE) 60/40	Polymer	45	29	2.3	0.103	7.89	1.49
P(VDF–TrFE) 56/44	Polymer	52	18	2.2	0.231	16.97	3.51
15 vol% $K_{0.5}Na_{0.5}NbO_3$–P(VDF/TrFE) 70/30	Composite	63	30	2.8	0.160	14.94	1.91

Source: [25] adapted with permission ©2019 The Royal Society of Chemistry.

$$F_E = \frac{p^2}{\varepsilon_r} \tag{6.11}$$

$$F_E' = \frac{p^2}{\varepsilon_r C_v^2} \tag{6.12}$$

where T_h is the maximum working temperature.

A low conversion efficiency is yielded from materials possessing low electrothermal coupling factors, in particular, for PVDF, $k^2 = 0.001194$. However, conversion efficiency can be increased by stacking up several nonlinear pyroelectric elements; [263] for ferroelectric polymers, when the Olsen cycle (aka Ericsson cycle) is used [16, 148, 264], up to 70% of Carnot's efficiency can be achieved. The Olsen cycle consists of a hysteresis loop of two isothermal and two isoelectric processes. To perform the Olsen cycle [13, 15, 16], a pyroelectric material is sandwiched between two electrodes in contact with a cold (T_c) and hot (T_h) source under the application of electric fields. The Olsen cycle has been reported to be successfully implemented in P(VDF-TrFE) thin films. For high yield energy harvesting performances, a pyroelectric material with a high pyroelectric coefficient is needed with the assumption of nearly constant volume specific heat. The energy harvesting FOMs for lead-free pyroelectric materials are listed in Table 6.5.

6.3 CONCLUSION

In summary, this chapter has described the highly sought pyro/piezo-polymers and the mechanism for obtaining the pyro/piezoelectricity in them. Pyro/piezoelectric polymers are the thrust area for future foldable and wearable devices that can harvest energy from the environment. The best method of measurement particularly for pyroelectricity with the separation of pyroelectric coefficients induced by various phenomena is discussed. The advantages of polymer composite piezo/pyroelectric materials over both the polymer as well as other materials have been described, and these composite materials were proven to advance the various FOMs utilizing the synergetic effect of both constituents. The utility and progress in particular harvesting applications using piezo/pyroelectric polymers and their composites have also been described.

REFERENCES

1. Carter, R.; Kensley, R.; Introduction to piezoelectric transducers, 2017. https://piezo.com/pages/history-of-piezoelectricity. (Accessed 14 Feb. 2021).
2. Manbachi, A.; Cobbold, R.S.C.; Development and application of piezoelectric materials for ultrasound generation and detection, *Ultrasound* 19(4) (2011) 187–196.
3. https://eng.libretexts.org/Bookshelves/Materials_Science/Supplemental_Modules_(Materials_Science)/Electronic_Properties/Piezoelectricity. (Accessed 20 Feb. 2021)
4. Chynoweth, A.G.; Dynamic method for measuring the pyroelectric effect with special reference to barium titanate, *Journal of Applied Physics* 27(1) (1956) 78–84.
5. Cooper, J.; A fast response total-radiation detector, *Nature* 194 (1962) 269–271.
6. Chynoweth, A.G.; Pyroelectricity, internal domains, and interface charges in triglycine sulfate, *Physical Review* 117(5) (1960) 1235–1243.
7. Beerman, H.P.; Improvement in the pyroelectric infrared radiation detector, *Ferroelectrics* 2(1) (1971) 123–128.
8. Cooper, J.; Minimum detectable power of a pyroelectric thermal receiver, *Review of Scientific Instruments* 33(1) (1962) 92–95.
9. Porter, S.G.; A brief guide to pyroelectric detectors, *Ferroelectrics* 33(1) (1981) 193–206.
10. Watton, R.; Smith, C.; Jones, G.R.; Pyroelectric materials: Operation and performance in the pyroelectric camera tube, *Ferroelectrics* 14(1) (1976) 719–721.
11. Whatmore, R.W.; Pyroelectric ceramics and devices for thermal infra-red detection and imaging, *Ferroelectrics* 118(1) (1991) 241–259.
12. Whatmore, R.W.; Osbond, P.C.; Shorrocks, N.M.; Ferroelectric materials for thermal IR detectors, *Ferroelectrics* 76(1) (1987) 351–367.
13. Olsen, R.B.; Briscoe, J.M.; Bruno, D.A.; Butler, W.F.; A pyroelectric energy converter which employs regeneration, *Ferroelectrics* 38(1) (1981) 975–978.
14. Das-Gupta, D.K.; Pyroelectricity in polymers, *Ferroelectrics* 118(1) (1991) 165–189.
15. Nguyen, H.; Navid, A.; Pilon, L.; Pyroelectric energy converter using co-polymer P(VDF-TrFE) and Olsen cycle for waste heat energy harvesting, *Applied Thermal Engineering* 30(14–15) (2010) 2127–2137.
16. Navid, A.; Pilon, L.; Pyroelectric energy harvesting using Olsen cycles in purified and porous poly(vinylidene fluoride-trifluoroethylene) [P(VDF-TrFE)] thin films, *Smart Materials and Structures* 20(2) (2011) 025012.
17. McKinley, I.M.; Lee, F.Y.; Pilon, L.; A novel thermomechanical energy conversion cycle, *Applied Energy* 126 (2014) 78–89.
18. Kouchachvili, L.; Ikura, M.; Improving the efficiency of pyroelectric conversion, *International Journal of Energy Research* 32(4) (2008) 328–335.
19. Gutmann, E.; Benke, A.; Gerth, K.; Böttcher, H.; Mehner, E.; Klein, C.; Krause-Buchholz, U.; Bergmann, U.; Pompe, W.; Meyer, D.C.; Pyroelectrocatalytic disinfection using the pyroelectric effect of nano- and microcrystalline $LiNbO_3$ and $LiTaO_3$ particles, *The Journal of Physical Chemistry C* 116(9) (2012) 5383–5393.
20. Kakekhani, A.; Ismail-Beigi, S.; Ferroelectric oxide surface chemistry: Water splitting via pyroelectricity, *Journal of Materials Chemistry A* 4(14) (2016) 5235–5246.
21. Lingam, D.; Parikh, A.R.; Huang, J.; Jain, A.; Minary-Jolandan, M.; Nano/microscale pyroelectric energy harvesting: Challenges and opportunities, *International Journal of Smart and Nano Materials* 4(4) (2013) 229–245.
22. Lang, S.B.; Pyroelectricity: From ancient curiosity to modern imaging tool, *Physics Today* 58(8) (2005) 31–36.
23. Jachalke, S.; Mehner, E.; Stöcker, H.; Hanzig, J.; Sonntag, M.; Weigel, T.; Leisegang, T.; Meyer, D.D.; How to measure the pyroelectric coefficient, *Applied Physics Reviews* 4(2) (2017) 021303.
24. Damjanovic, D.; Ferroelectric, dielectric and piezoelectric properties of ferroelectric thin films and ceramics, *Reports on Progress in Physics* 61 (1998) 1267–1324.

25. He, H.; Lu, X.; Hanc, E.; Chen, C.; Zhang, H.; Lu, L.; Advances in lead-free pyroelectric materials: A comprehensive review, *Journal of Materials Chemistry C* 8(5) (2020) 1494–1516.
26. Soin, N.; Anand, S. C.; Shah, T. H.; Energy harvesting and storage textiles, in Anand, S. C.; Horrocks, A. R. (Eds.) *Handbook of Technical Textiles*, Woodhead Publishing, Cambridge, 2016, pp. 357–396.
27. https://sensortechcanada.com/technical-notes/piezoelectric-fundamentals. (Accessed 20 Feb. 2021)
28. Feng, Y.; Peng, C.; Deng, Q.; Li, Y.; Hu, J.; Wu, Q.; Annealing and stretching induced high energy storage properties in all-organic composite dielectric films, *Materials (Basel)* 11(11) (2018) 2279.
29. Li, L.; Zhang, M.; Rong, M.; Ruan, W.; Studies on the transformation process of PVDF from α to β phase by stretching, *RSC Advances* 4(8) (2014) 3938–3943.
30. Sencadas, V.; Moreira, M.V.; Lanceros-Méndez, S.; Pouzada, A.S.; Gregório Filho, R.; α- to β transformation on PVDF films obtained by uniaxial stretch, *Materials Science Forum* 514–516 (2006) 872–876.
31. Damjanovic, D.; The science of hysteresis, in I. a. B. Mayergoyz, G.(Ed.) *Hysteresis in Materials*, Elsevier, Amsterdam, 2005, p. 744.
32. Dineva, P.; Gross, D.; Müller, R.; Rangelov, T.; *Solid Mechanics and Its Applications*, Springer; Dordrecht, New York; London, 2014.
33. Dargaville, T.R.; Celina, T.R.; Elliott, J.M.; Chaplya, P.M.; Jones, G.D.; Mowery, D.M.; Assink, R.A.; Clough, R.L.; Martin, J.W.; Characterization, Performance and Optimization of PVDF as a Piezoelectric Film for Advanced Space Mirror Concepts, Sandia National Laboratories Albuquerque, New Mexico and Livermore, California, 2005, p. 49.
34. Mahadeva, S.K.; Berring, J.; Walus, K.; Stoeber, B.; Effect of poling time and grid voltage on phase transition and piezoelectricity of poly(vinyledene fluoride) thin films using corona poling, *Journal of Physics D: Applied Physics* 46(28) (2013) 285305.
35. Bae, S.H.; Kahya, O.; Sharma, B.K.; Kwon, J.; Cho, H.J.; Ozyilmaz, B.; Ahn, J.H.; Graphene-P(VDF-TrFE) multilayer film for flexible applications, *ACS Nano* 7(4) (2013) 3130–3138.
36. Ding, R.; Zhang, X.; Chen, G.; Wang, H.; Kishor, R.; Xiao, J.; Gao, F.; Zeng, K.; Chen, X.; Sun, X.W.; Zheng, Y.; High-performance piezoelectric nanogenerators composed of formamidinium lead halide perovskite nanoparticles and poly(vinylidene fluoride), *Nano Energy* 37 (2017) 126–135.
37. Harstad, S.; D'Souza, N.; Soin, N.; El-Gendy, A.A.; Gupta, S.; Pecharsky, V.K.; Shah, T.; Siores, E.; Hadimani, R.L.; Enhancement of β-phase in PVDF films embedded with ferromagnetic Gd_5Si_4 nanoparticles for piezoelectric energy harvesting, AIP Advances 7(5) (2017) 056411.
38. Shi, K.; Sun, B.; Huang, X.; Jiang, P.; Synergistic effect of graphene nanosheet and $BaTiO_3$ nanoparticles on performance enhancement of electrospun PVDF nanofiber mat for flexible piezoelectric nanogenerators, *Nano Energy* 52 (2018) 153–162.
39. Lang, S.B.; *Modern Bioelectricity* A. A. Marino (Ed.) Marcel Dekker, New York, 1988.
40. Lubomirsky, I.; Stafsudd, O.; Invited review article: Practical guide for pyroelectric measurements, *Review of Scientific Instruments* 83(5) (2012) 051101.
41. Cross, L.E.; Flexoelectric effects: Charge separation in insulating solids subjected to elastic strain gradients, *Journal of Materials Science* 41(1) (2006) 53–63.
42. Lu, J.; Lv, J.; Liang, X.; Xu, M.; Shen, S.; Improved approach to measure the direct flexoelectric coefficient of bulk polyvinylidene fluoride, *Journal of Applied Physics* 119(9) (2016) 094104.
43. Ma, W.; Cross, L.E.; Strain-gradient-induced electric polarization in lead zirconate titanate ceramics, *Applied Physics Letters* 82(19) (2003) 3293–3295.
44. Tagantsev, A.K.; Piezoelectricity and flexoelectricity in crystalline dielectrics, *Physical Review B: Condensed Matter* 34(8) (1986) 5883–5889.
45. Sessler, G.M.; *Topics in Applied Physics* Second ed., Springer, Verlag Berlin Heidelberg, 1987.
46. Bharti, V.; Kaura, T.; Nath, R.; Ferroelectric hysteresis in simultaneously stretched and corona-poled PVDF films, *IEEE Transactions on Dielectrics and Electrical Insulation* 4 (1997) 738–741.
47. Giacometti, J.A.; Oliveira Jr., O.N.; Corona charging of polymers, *IEEE Transactions on Electrical Insulation* 27 (1992) 924–943.
48. Whatmore, R.W.; Pyroelectric devices and materials, *Reports on Progress in Physics* 49(12) (1986) 1335–1386.
49. Lang, S.B.; Das-Gupta, D.K.; *Handbook of Advanced Electronic and Photonic Materials and Devices* First ed., Elsevier, 2001.
50. Muralt, P.; Micromachined infrared detectors based on pyroelectric thin films, *Reports on Progress in Physics* 64(10) (2001) 1339–1388.
51. Costa, P.; Nunes-Pereira, J.; Pereira, N.; Castro, N.; Gonçalves, S.; Lanceros-Mendez, S.; Recent progress on piezoelectric, pyroelectric, and magnetoelectric polymer-based energy-harvesting devices, *Energy Technology* 7(7) (2019) 1800852. DOI:10.1002/ente.201800852
52. Abbasipour, M.; Khajavi, R.; Yousefi, A.A.; Yazdanshenas, M.E.; Razaghian, F.; Akbarzadeh, A.; Improving piezoelectric and pyroelectric properties of electrospun PVDF nanofibers using nanofillers for energy harvesting application, *Polymers for Advanced Technologies* 30(2) (2019) 279–291.
53. Künstler, W.; Wegener, M.; Seiß, M.; Gerhard-Multhaupt, R.; Preparation and assessment of piezo- and pyroelectric poly (vinylidene fluoride-hexafluoropropylene) copolymer films, *Applied Physics A: Materials Science & Processing* 73(5) (2001) 641–645.
54. Lang, S.B.; Muensit, S.; Review of some lesser-known applications of piezoelectric and pyroelectric polymers, *Applied Physics A* 85(2) (2006) 125–134.
55. Ng, C.Y.B.; Gan, W.C.; Velayutham, T.S.; Goh, B.T.; Hashim, R.; Structural control of the dielectric, pyroelectric and ferroelectric properties of poly(vinylidene fluoride-co-trifluoroethylene) thin films, *Physical Chemistry Chemical Physics* 22(4) (2020) 2414–2423.
56. Quek, S.Y.; Kamenetska, M.; Steigerwald, M.L.; Choi, H.J.; Louie, S.G.; Hybertsen, M.S.; Neaton, J.B.; Venkataraman, L.; Mechanically controlled binary conductance switching of a single-molecule junction, *Nature Nanotechnologys* 4(4) (2009) 230–234.
57. Andrusyk, A.; Ferroelectrics: Physical effects, piezoelectric effect in Rochelle salt, in M. Lallart (Ed). *Ferroelectrics: Physical Effects*, Intech Open, Croatia, 2011, 195–220.
58. Mueller, H.; Properties of Rochelle salt, *Physics Review* 47 (1935) 175–191.
59. Helman, D.S.; Symmetry-based electricity in minerals and rocks: A review with examples of centrosymmetric minerals that exhibit pyro-and piezoelectricity, *Periodico di Mineralogia* 85 (2016) 201–248.

60. Gaska, R.; Shur, M.S.; Bykhovski, A.D.; Pyroelectric and piezoelectric properties of GaN-based materials, *MRS Internet Journal of Nitride Semiconductor Research* 4(S1) (2014) 57–68.
61. Jacob, M.V.; Hartnettb, J.G.; Mazierska, J.; Krupkad, J.; Tobar, M.E.; Lithium tantalate – A high permittivity dielectric material for microwave communication systems, *TENCON 2003. Conference on Convergent Technologies for Asia-Pacific Region*, IEEE, Bangalore, India, 2003, pp. 1362–1366.
62. De Matos Gomes, E.; Viseu, T.; Belsley, M.; Almeida, B.; Costa, M.M.R.; Rodrigues, V.H.; Isakov, D.; Piezoelectric and pyroelectric properties of DL-alanine and L-lysine amino-acid polymer nanofibres, *Materials Research Express* 5(4) (2018) 045049.
63. Bury, P.C.; McLare, A.C.; Pyroelectric properties of rubidium, cesium, and thallium nitrates, *Physica Status Solidi* 31 (1969) 799–806.
64. Muratsugu, M.; Kurosawa, S.; Kamo, N.; Detection of antistreptolysin 0 antibody: Application of an initial rate method of latex piezoelectric immunoassay, *Analytical Chemistry* 64 (1992) 2403–2407
65. Najdoski, M.; Pejov, L.; Petruševski, V.M.; Snetsinger, P.; Juergens, F.; Pyroelectric effect of a sucrose monocrystal, *Journal of Chemical Education* 76(3) (1999) 360–361.
66. Hossain, A.; Rashid, M.H.; Pyroelectric detectors and their applications, *IEEE Transactions on Industry Applications* 27(5) (1991) 824–829.
67. Luo, W.; Luo, J.; Shuai, Y.; Zhang, K.; Wang, T.; Wu, C.; Zhang, W.; Infrared detector based on crystal ion sliced $LiNbO_3$ single-crystal film with BCB bonding and thermal insulating layer, *Microelectronic Engineering* 213 (2019) 1–5.
68. Xu, Q.; Zhao, X.; Li, X.; Li, L.; Yang, L.; Di, W.; Jiao, J.; Luo, H.; Novel electrode layout for relaxor single crystal pyroelectric detectors with enhanced responsivity and specific detectivity, *Sensors and Actuators A: Physical* 234 (2015) 82–86.
69. Wang, Y.; Luo, C.; Wang, S.; Chen, C.; Yuan, G.; Luo, H.; Viehland, D.; Large piezoelectricity in ternary lead-free single crystals, *Advanced Electronic Materials* 6(1) (2019) 1900949.
70. Quan, N.D.; Huu Bac, L.; Thiet, D.V., Hung, V.N.; Dung, D.D.; Current development in lead-free$Bi_{0.5}(Na,K)_{0.5}TiO_3$-based piezoelectric materials, *Advances in Materials Science and Engineering* 2014 (2014) 1–13.
71. Drozhdin, S.N.; Golitsyna, O.M.; Nikishina, A.I.; Kostsov, A.M.; Pyroelectric and dielectric properties of triglycine sulphate with an impurity of phosphorus (TGSP), *Ferroelectrics* 373(1) (2010) 93–98.
72. Hussain, A., Sinha, N.; Joseph, A.J.; Goel, S.; Singh, B.; Bdikin, I.; Kumar, B.; Mechanical investigations on piezo-/ferrolectric maleic acid-doped triglycine sulphate single crystal using nanoindentation technique, *Arabian Journal of Chemistry* 13(1) (2020) 1874–1889.
73. Banan, M.; Lal, R.B.; Batra, A.; banan1992.pdf, *Journal of Materials Science* 27 (1992) 2291–2297.
74. Bdikin, I.K.; Wojtaś, M.; Kiselev, D.; Isakov, D.; Kholkin, A.L.; Ferroelectric-paraelectric phase transition in triglycine sulphate via piezoresponse force microscopy, *Ferroelectrics* 426(1) (2012) 215–222.
75. Ikeda, T.; Tanaka, Y.; Toyoda, H.; Piezoelectric properties of triglycine-sulphate, *Japanese Journal of Applied Physics* 1(1) (1962) 1–21.
76. Lal, R.B.; Batra, A.K.; Growth and properties of triglycine sulfate (TGS) crystals: Review, *Ferroelectrics* 142(1) (1993) 51–82.
77. Neumann, N.; Modified triglycine sulphate for pyroelectric infrared detectors, *Ferroelectrics* 142(1) (1993) 83–92.
78. Plyushch, A.; Macutkevic, J.; Samulionis, V.; Banys, J.; Bychanok, D.; Kuzhir, P.; Mathieu, S.; Fierro, V.; Celzard, A.; Synergetic effect of triglycine sulfate and graphite nanoplatelets on dielectric and piezoelectric properties of epoxy resin composites, *Polymer Composites* 40(S2) (2018) E1181–E1188.
79. Varikash, V.M.; Shuvalov, L.A.; Tarasevich, E.V.; Lagutina, J.P.; Piezoelectric properties of triglycine sulphate (TGS) type crystals with imperfections, *Ferroelectrics* 42(1) (2011) 47–56.
80. Patel, S.; Srikanth, K.S.; Steiner, S.; Vaish, R.; Frömling, T.; Pyroelectric and impedance studies of the $0.5Ba(Zr_{0.2}Ti_{0.8})O_3$-$0.5(Ba_{0.7}Sr_{0.3})TiO_3$ ceramics, *Ceramics International* 44(17) (2018) 21976–21981.
81. Srikanth, K.; Patel, S.; Vaish, R.; Pyroelectric performance of $BaTi_1$-$xSnxO_3$ ceramics, *International Journal of Applied Ceramic Technology* 15(2) (2018) 546–553.
82. Guo, F.; Yang, B.; Zhang, S.; Wu, F.; Liu, D.; Hu, P.; Sun, Y.; Wang, D.; Cao, W.; Enhanced pyroelectric property in (1−x)$(Bi_{0.5}Na_{0.5})TiO_3$-$xBa(Zr_{0.055}Ti_{0.945})O_3$: Role of morphotropic phase boundary and ferroelectric-antiferroelectric phase transition, *Applied Physics Letters* 103(18) (2013) 182906.
83. Halim, N.A.; Velayutham, T.S.; Majid, W.H.A.; Pyroelectric, ferroelectric, piezoelectric and dielectric properties of $Na_{0.5}Bi_{0.5}TiO_3$ ceramic prepared by sol-gel method, *Ceramics International* 42(14) (2016) 15664–15670.
84. Shen, M.; Li, W.; Li, W.H.A.; Liu, H.; Xu, J.; Qiu, S.; Zhang, G.; Lu, Z.; Li, H.; Jiang, S.; High room-temperature pyroelectric property in lead-free BNT-BZT ferroelectric ceramics for thermal energy harvesting, *Journal of the European Ceramic Society* 39(5) (2019) 1810–1818.
85. Li, S.; Nie, H.; Wang, G.; Liu, N.; Zhou, M.; Cao, F.; Dong, X.; Novel $AgNbO_3$-based lead-free ceramics featuring excellent pyroelectric properties for infrared detecting and energy-harvesting applications via antiferroelectric/ferroelectric phase-boundary design, *Journal of Materials Chemistry C* 7(15) (2019) 4403–4414.
86. Wang, Q.; Bowen, C.R.; Lei, W.; Zhang, H.; Xie, B.; Qiu, S.; Li, M.-Y.; Jiang, S.; Improved heat transfer for pyroelectric energy harvesting applications using a thermal conductive network of aluminum nitride in PMN–PMS–PZT ceramics, *Journal of Materials Chemistry A* 6(12) (2018) 5040–5051.
87. Li, S.; Zhao, Z.; Zhao, J.; Zhang, Z.; Li, X.; Zhang, J.; Recent advances of ferro-, piezo-, and pyroelectric nanomaterials for catalytic applications, *ACS Applied Nano Materials* 3(2) (2020) 1063–1079.
88. Bai, Y.; Siponkoski, T.; Peräntie, J.; Jantunen, H.; Juuti, J.; Ferroelectric, pyroelectric, and piezoelectric properties of a photovoltaic perovskite oxide, *Applied Physics Letters* 110(6) (2017) 063903.
89. Takenaka, T.; Sakata, K.; Piezoelectric and pyroelectric properties of calcium-modified and grain-oriented (NaBi)1/$2Bi_4Ti_4O_{15}$ ceramics, *Ferroelectrics* 94(1) (1989) 175–181.
90. Takenaka, T.; Sakata, K.; Pyroelectric properties of bismuth layer-structured ferroelectric ceramics, *Ferroelectrics* 118(1) (1991) 123–133.
91. Tang, Y.; Shen, Z.-Y.; Zhang, S.; Shrout, T.R.; Feteira, A.; Improved pyroelectric properties of $CaBi_4Ti_4O_{15}$ ferroelectrics ceramics by Nb/Mn co-doping for pyrosensors, *Journal of the American Ceramic Society* 99(4) (2016) 1294–1298.

92. Tang, Y.; Shen, Z.-Y.; Zhao, X.; Wang, F.; Shi, W.; Sun, D.; Zhou, Z.; Zhang, S.; Pyroelectric properties of Mn-doped aurivillius ceramics with different pseudo-perovskite layers, *Journal of the American Ceramic Society* 101(4) (2018) 1592–1597.
93. Takenaka, T.; Piezoelectric and acoustic materials for transducer applications, in A.S.K. Akdoğan (Ed.) *Piezoelectric and Acoustic Materials for Transducer Applications*, Springer, New York, 2008.
94. Takenaka, T.; Sakata, K.; Toda, K.; Piezoelectric properties of bismuth layer-structured ferroelectric $Na_{0.5}Bi_{4.5}Ti_4O_{15}$ ceramic, *Japanese Journal of Applied Physics* 24 (1985) 730–732.
95. Wu, J.; *Advances in Lead-Free Piezoelectric Materials*, Springer, Singapore, 2018.
96. Morshed, T.; Haq, E.U.; Silien, C.; Tofail, S.A.M.; Zubair, M.A.; Islam, M.F.; Piezo and pyroelectricity in spark plasma sintered potassium sodium niobate (KNN) ceramics, *IEEE Transactions on Dielectrics and Electrical Insulation* 27(5) (2020) 1428–1432.
97. Kawada, S.; Hayashi, H.; Ishii, H.; Kimura, M.; Ando, A.; Omiya, S.; Kubodera, N.; Potassium sodium niobate-based lead-free piezoelectric multilayer ceramics co-fired with nickel electrodes, *Materials (Basel)* 8(11) (2015) 7423–7438.
98. Wu, J.; Xiao, D.; Zhu, J.; Potassium-sodium niobate lead-free piezoelectric materials: Past, present, and future of phase boundaries, *Chemical Reviews* 115(7) (2015) 2559–2595.
99. Bowen, C.R.; Taylor, J.; LeBoulbar, E.; Zabek, D.; Chauhan, A.; Vaish, R.; Pyroelectric materials and devices for energy harvesting applications, *Energy & Environmental Science* 7(12) (2014) 3836–3856.
100. Venet, M.; Santos, I.; Eiras, J.; Garcia, D.; Potentiality of SBN textured ceramics for pyroelectric applications, *Solid State Ionics* 177(5–6) (2006) 589–593.
101. Zhang, J.; Dong, X.; Cao, F.; Guo, S.; Wang, G.; Enhanced pyroelectric properties of $Cax(Sr_{0.5}Ba_{0.5})_1{-}_xNb_2O_6$ lead-free ceramics, *Applied Physics Letters* 102(10) (2013) 102908.
102. Chen, H.; Guo, S.; Dong, X.; Cao, F.; Mao, C.; Wang, G.; Ca $Sr_{0.3}{-}Ba_{0.7}Nb_2O_6$ lead-free pyroelectric ceramics with high depoling temperature, *Journal of Alloys and Compounds* 695 (2017) 2723–2729.
103. Jiang, W.; Cao, W.; Yi, X.; Chen, H.; The elastic and piezoelectric properties of tungsten bronze ferroelectric crystals $(Sr_{0.7}Ba_{0.3})_2NaNb_5O_{15}$ and $(Sr_{0.3}Ba_{0.7})_2NaNb_5O_{15}$, *Journal of Applied Physics* 97(9) (2005) 094106.
104. Spinola, D.U.P.; Moreira, E.N.; Bhsora, E.N.; Eiras, I.A.; Garcia, D.; Pyroelectric and piezoelectric properties of SBN ceramics, J.S. Schoenwald (Ed.) *IEEE Ultrasonics Symposium IEEE*, San Antonio, Texas, U.S, 1996, pp. 523–526.
105. Lee, J.S.; Park, J.S.; Kim, J.S.; Lee, J.H.; Lee, Y.H.; Hahn, S.R.; Preparation of (Ba, Sr)TiO_3 thin films with high pyroelectric coefficients at ambient temperatures, *Japanese Journal of Applied Physics* 38(5B) (1999) L574–L576.
106. Jin, F.; Auner, G.W.; Naik, R.; Schubring, N.W.; Mantese, J.V.; Catalan, A.B.; Micheli, A.L.; Giant effective pyroelectric coefficients from graded ferroelectric devices, *Applied Physics Letters* 73(19) (1998) 2838–2840.
107. Sengupta, S.; Sengupta, L.C.; Synowczynski, J.; Rees, D.A.; Novel pyroelectric sensor materials, *IEEE Transactions on Ultrasonics Engineering* 45(6) (1998) 1444–1452.
108. Wang, S.J.; Lu, L.; Lai, M.O.; Fuh, J.Y.H.; The role of oxygen pressure and thickness on structure and pyroelectric properties of $Ba(Ti_{0.85}Sn_{0.15})O_3$ thin films grown by pulsed laser deposition, *Journal of Applied Physics* 105(8) (2009) 084102.
109. Zhang, T.; Ni, H.; Pyroelectric and dielectric properties of sol–gel derived barium–strontium–titanate (Ba 0.64Sr 0.36TiO_3) thin films, *Sensors and Actuators A: Physical* 100(2) (2002) 252–256.
110. Kim, D.B.; Park, K.H.; Cho, Y.S.; Origin of high piezoelectricity of inorganic halide perovskite thin films and their electromechanical energy-harvesting and physiological current-sensing characteristics, *Energy & Environmental Science* 13(7) (2020) 2077–2086.
111. Li, L.; Miao, L.; Zhang, Z.; Pu, X.; Feng, Q.; Yanagisawa, K.; Fan, Y.; Fan, M.; Wen, P.; Hu, D.; Recent progress in piezoelectric thin film fabrication via the solvothermal process, *Journal of Materials Chemistry A* 7(27) (2019) 16046–16067.
112. Muralt, P.; Polcawich, R.G.; Trolier-McKinstry, S.; Piezoelectric thin films for sensors, actuators, and energy harvesting, *MRS Bulletin* 34 (2009) 658–664.
113. Martins, P.; Lopes, A.C.; Lanceros-Mendez, S.; Electroactive phases of poly(vinylidene fluoride): Determination, processing and applications, *Progress in Polymer Science* 39(4) (2014) 683–706.
114. Ueberschlag, P.; PVDF piezoelectric polymer, *Sensor Review* 21(2) (2001) 118–126.
115. Shehzad, M.; Wang, Y.; Structural tailing and pyroelectric energy harvesting of P(VDF-TrFE) and P(VDF-TrFE-CTFE) ferroelectric polymer blends, *ACS Omega* 5(23) (2020) 13712–13718.
116. Wang, S.; Li, Q.; Design, synthesis and processing of PVDF-based dielectric polymers, *IET Nanodielectrics* 1(2) (2018) 80–91.
117. Wang, Y.; Zhou, X.; Chen, Q.; Chu, B.; Zhang, Q.; Recent development of high energy density polymers for dielectric capacitors, *IEEE Transactions on Dielectrics and Electrical Insulation* 17(4) (2010) 1036–1042.
118. Wegener, M.; Kunstler, W.; Gerhard-Multhaupt, R.; Piezo-, pyro- and ferroelectricity in poly(Vinylidene Fluoride-Hexafluoropropylene) copolymer films, *Integrated Ferroelectrics* 60(1) (2004) 111–116.
119. Wegener, M.; Künstler, W.; Richter, K.; Gerhard-Multhaupt, R.; Ferroelectric polarization in stretched piezo- and pyroelectric poly(vinylidene fluoride-hexafluoropropylene) copolymer films, *Journal of Applied Physics* 92(12) (2002) 7442–7447.
120. Frubing, P.; Kremmer, A.; Neumann, W.; Gerhard-Multhaupt, R.; Guy, I.L.; Dielectric relaxation in piezo-, pyro- and ferroelectric polyamide 11, *IEEE Transactions on Dielectrics and Electrical Insulation* 11(2) (2004) 271–279.
121. Choi, Y.S.; Kim, S.K.; Smith, M.; Williams, F.; Vickers, M.E.; Elliott, J.A.; Kar-Narayan, S.; Unprecedented dipole alignment in α-phase nylon-11 nanowires for high-performance energy-harvesting applications, *Science Advances* 6 (24) (2020) eaay5065.
122. Lee, J.W.; Takase, Y.; Newman, B.A.; Scheinbeim, J.I.; Ferroelectric polarization switching in nylon-11, *Journal of Polymer Science: Part B: Polymer Physics* 229 (1991) 273–277.
123. Newman, B.A.; Chen, P.; Pae, K.D.; Scheinbeim, J.I.; Piezoelectricity in nylon 11, *Journal of Applied Physics* 51(10) (1980) 5161.
124. Scheinbeim, J.I.; Lee, J.W.; Newman, B.A.; Ferroelectric polarization mechanisms in nylon 11, *Macromolecules* 25 (1992) 3729–3732.
125. Wu, G.; Yano, O.; Soen, T.; Dielectric and piezoelectric properties of nylon 9 and nylon 11, *Polymer Journal* 18(1) (1986) 51–61.

126. Ibos, L.; Maraval, C.; Berneas, A.; Teysseadre, G.; Lacabanne, C.; Wu, S.L.; Scheinbeim, J.I.; Thermal behavior of ferroelectric polyamide 11 in relation to pyroelectric properties, *Journal of Polymer Science: Part B: Polymer Physics* 37 (1999) 715–723.
127. Alaaeddin, M.H.; Sapuan, S.M.; Zuhri, M.Y.M.; Zainudin, E.S.; Al-Oqla, F.M.; Properties and common industrial applications of polyvinyl fluoride (PVF) and polyvinylidene fluoride (PVDF), *IOP Conference Series: Materials Science and Engineering* 409 (2018) 012021.
128. Alaaeddin, M.H.; Sapuan, S.M.; Zuhri, M.Y.M.; Zainudin, E.S.; Al-Oqla, F.M.; Polyvinyl fluoride (PVF); Its properties, applications, and manufacturing prospects, *IOP Conference Series: Materials Science and Engineering* 538 (2019) 012010.
129. Broadhurst, M.G.; Davis, G.T.; *Piezo- and Pyroelectric Properties of Electrets*, Institute for Materials Research National Bureau of Standards Washington, Washington, U.S., 1975, pp. 1–62.
130. Cohen, J.; Edelman, S.; Vezzetti, C.F.; Pyroelectric effect in polyvinyl fluoride, *Nature Physical Science* 233 (1971) 12.
131. Lang, S.B.; DeReggi, A.S.; Broadhurst, M.G.; Thomas Davis, G.; Effects of poling field and time on pyroelectric coefficient and polarization uniformity in polyvinyl fluoride, *Ferroelectrics* 33(1) (2011) 119–125.
132. Li, Z.; Wang, J.; Wang, X.; Yang, Q.; Zhang, Z.; Ferro- and piezo-electric properties of a poly(vinyl fluoride) film with high ferro- to para-electric phase transition temperature, *RSC Advances* 5(99) (2015) 80950–80955.
133. Reddy, P.J.; Sirajuddin, M.; Pyroelectric polymer films for infrared detection, *Bulletin of Materials Science* 8(3) (1986) 365–371.
134. Srinivasan, M.; Pyroelectric materials, *Bulletin of Materials Science* 6(2) (1984) 317–325.
135. Bharti, V.; Nath, R.; Piezo-, pyro- and ferroelectric properties of simultaneously stretched and corona poled extruded poly(vinyl chloride) films, *Journal of Physics D: Applied Physics* 34 (2001) 667–672.
136. Likhovidov, V.S.; Golovai, V.V.; Vannikov, A.V.; Pyroelectric properties and the nature of the polarization of polyacrylonitrile, *Polymer Science U.S.S.R* 20 (1978) 80–87.
137. Street, R.M.; Minagawa, M.; Vengrenyuk, A.; Schauer, C.L.; Piezoelectric electrospun polyacrylonitrile with various tacticities, *Journal of Applied Polymer Science* 136(1) (2019) 47530.
138. Ando, M.; Kawamura, H.; Kitada, H.; Sekimoto, Y.; Tajitsu, Y.; New human machine interface devices using a piezoelectric poly(L-lactic acid) film, *2013 Joint UFFC, EFTF and PFM Symposium*, IEEE, Prague, Czech Republic, 2013, pp. 236–239.
139. Ando, M.; Tamakura, D.; Inoue, T.; Takumi, K.; Yamanaga, T.; Todo, R.; Hosoya, K.; Onishi, O.; Electric antibacterial effect of piezoelectric poly(lactic acid) fabric, *Japanese Journal of Applied Physics* 58 (2019) SLLD09. DOI: 10.7567/1347-4065/ab3b1b
140. Lee, S.J.; Arun, A.P.; Kim, K.J.; Piezoelectric properties of electrospun poly(l-lactic acid) nanofiber web, *Materials Letters* 148 (2015) 58–62.
141. Rajala, S.; Siponkoski, T.; Sarlin, E.; Mettänen, M.; Vuoriluoto, M.; Pammo, A.; Juuti, J.; Rojas, O.J.; Franssila, S.; Tuukkanen, S.; Cellulose nanofibril film as a piezoelectric sensor material, *ACS Applied Materials & Interfaces* 8(24) (2016) 15607–15614.
142. Csoka, L.; Hoeger, I.C.; Rojas, O.J.; Peszlen, I.; Pawlak, J.J.; Peralta, P.N.; Piezoelectric effect of cellulose nanocrystals thin films, *ACS Macro Letters* 1(7) (2012) 867–870.
143. Tuukkanen, S.; Rajala, S.; Nanocellulose as a piezoelectric material, In Dimitroula, Matsouka; Savvas, G. Vassiliadis (Eds.) *Piezoelectricity – Organic and Inorganic Materials and Applications*, IntechOpen, Croatia, 2018.
144. Zhang, S.; Liang, Y.; Qian, X.; Hui, D.; Sheng, K.; Pyrolysis kinetics and mechanical properties of poly(lactic acid)/bamboo particle biocomposites: Effect of particle size distribution, *Nanotechnology Reviews* 9(1) (2020) 524–533.
145. Sharma, A.K.; Ramu, C.; Pyroelectricity in Fe-doped cellulose acetate films, *Materials Letters* 10(11,12) (1991) 517–520.
146. Kaczmarek, H.; Królikowski, B.; Klimiec, E.; Chylińska, M.; Bajer, D.; Advances in the study of piezoelectric polymers, *Russian Chemical Reviews* 88(7) (2019) 749–774.
147. Li, H.; Tian, C.; Deng, Z.D.; Energy harvesting from low frequency applications using piezoelectric materials, *Applied Physics Reviews* 1(4) (2014).
148. Li, Q.; Wang, Q.; Ferroelectric polymers and their energy-related applications, *Macromolecular Chemistry and Physics* 217(11) (2016) 1228–1244.
149. Dillon, D.R.; Tenneti, K.K.; Li, C.Y.; Ko, F.K.; Sics, I.; Hsiao, B.S.; On the structure and morphology of polyvinylidene fluoride–nanoclay nanocomposites, *Polymer* 47(5) (2006) 1678–1688.
150. Cauda, V.; Canavese, G.; Stassi, S.; Nanostructured piezoelectric polymers, *Journal of Applied Polymer Science* 132(13) (2015) 41667.
151. Kawai, H.; The piezoelectricity of poly (vinylidene Fluoride), *Japanese Journal of Applied Physics* 8 (1969) 975–976.
152. Bergman, J.G.; McFee, J.H.; Crane, G.R.; Pyroelectricity and optical second harmonic generation in polyvinylidene fluoride films, *Applied Physics Letters* 18(5) (1971) 203–205.
153. Marcus, M.A.; Ferroelectric polymers and their applications, *Ferroelectrics* 40(1) (2011) 29–41.
154. Shin, Y.-E.; Sohn, S.-D.; Han, H.; Park, Y.; Shin, H.-J.; Ko, H.; Self-powered triboelectric/pyroelectric multimodal sensors with enhanced performances and decoupled multiple stimuli, *Nano Energy* 72 (2020) 104671.
155. Chen, X.Z.; Li, Z.W.; Cheng, Z.X.; Zhang, J.Z.; Shen, Q.D.; Ge, H.X.; Li, H.T.; Greatly enhanced energy density and patterned films induced by photo cross-linking of poly(vinylidene fluoride-chlorotrifluoroethylene), *Macromolecular Rapid Communications* 32(1) (2011) 94–99.
156. https://physics.montana.edu/eam/polymers/piezopoly.html. (Accessed 20 Feb. 2021)
157. Oliveira, F.; Leterrier, Y.; Månson, J.-A.; Sereda, O.; A. Neels; A. Dommann; D. Damjanovic, Process influences on the structure, piezoelectric, and gas-barrier properties of PVDF-TrFE copolymer, *Journal of Polymer Science Part B: Polymer Physics* 52(7) (2014) 496–506.
158. Ruan, L.; Yao, X.; Chang, Y.; Zhou, L.; Qin, G.; Zhang, X.; Properties and applications of the beta phase poly(vinylidene fluoride), *Polymers (Basel)* 10(3) (2018) 228.
159. Ramadan, K.S.; Sameoto, D.; Evoy, S.; A review of piezoelectric polymers as functional materials for electromechanical transducers, *Smart Materials and Structures* 23(3) (2014). DOI: 10.1088/0964-1726/23/3/033001.

160. Ameduri, B.; From vinylidene fluoride (VDF) to the applications of VDF-containing polymers and copolymers: Recent developments and future trends, *Chemical Reviews* 109 (2009) 6632–6686.
161. Kamberi, M.; Pinson, D.; Pacetti, S.; Perkins, L.E.L.; Hossainy, S.; Mori, H.; Rapoza, R.J.; Kolodgie, F.; Virmani, R.; Evaluation of chemical stability of polymers of XIENCE everolimus-eluting coronary stents in vivo by pyrolysis-gas chromatography/mass spectrometry, *Journal of Biomedical Materials Research. Part B, Applied Biomaterials* 106(5) (2018) 1721–1729.
162. Nalwa, H.S.; *Ferroelectric Polymers: Chemistry, Physics, and* Applications, Marcel Dekker, New York, USA, 1995.
163. Fukada, E.; Recent developments of polar piezoelectric polymers, *IEEE Transactions on Dielectrics and Electrical Insulation* 13(5) (2006) 1110–1119.
164. Cho, Y.; Ahn, D.; Park, J.B.; Pak, S.; Lee, S.; Jun, B.O.; Hong, J.; Lee, S.Y.; Jang, J.E.; Hong, J.; Morris, S.M.; Sohn, J.I.; Cha, S.N.; Kim, J.M.; Enhanced ferroelectric property of P(VDF-TrFE-CTFE) film using room-temperature crystallization for high-performance ferroelectric device applications, *Advanced Electronic Materials* 2(10) (2016) 1600225. DOI: 10.1002/aelm.201600225
165. Ribeiro, C.; Sencadas, V.; Correia, D.M.; Lanceros-Mendez, S.; Piezoelectric polymers as biomaterials for tissue engineering applications, *Colloids and Surfaces. B, Biointerfaces* 136 (2015) 46–55.
166. Yadav, P.; Yadav, H.; Shah, V.G.; Shah, G.; Dhaka, G.; Biomedical biopolymers, their origin and evolution in biomedical sciences: A systematic review, *Journal of Clinical and Diagnostic Research* 9(9) (2015) ZE21–ZE25.
167. Shorrocks, N.M.; Brown, M.E.; Whatmore, R.W.; Ainger, F.W.; Piezoelectric composites for underwater transducers, *Ferroelectrics* 54(1) (2011) 215–218.
168. Newnham, R.E.; Skinner, D.P.; Cross, L.E.; Connectivity and piezoelectric-pyroelectric composites, *Materials Research Bulletin* 13 (1978) 525–536.
169. Safari, A.; Halliyal, A.; Newnham, R.E.; Lachman, I.M.; Transvers honeycomb composite transducers, *Materials Research Bulletin* 17 (3) (1982) 301–308.
170. Rittenmyer, K.; Shrout, T.; Schulze, W.A.; Newnham, R.E.; Piezoelectric 3–3 composites, *Ferroelectrics* 41(1) (2011) 189–195.
171. Shrout, T.R.; Schulze, W.A.; Biggers, J.V.; Simplified fabrication of PZT/polymer composites, *Materials Research Bulletin* 14 (1979) 1553–1559.
172. Skinner, D.P.; Newnham, R.E.; Cross, L.E.; Flexible composite transducers, *Materials Research Bulletin* 13 (1978) 599–607.
173. Sappati, K.K.; Bhadra, S.; Piezoelectric polymer and paper substrates: A review, *Sensors (Basel)* 18(11) (2018) 3605.
174. Dang, Z.-M.; Yuan, J.-K.; Zha, J.-W.; Zhou, T.; Li, S.-T.; Hu, G.-H.; Fundamentals, processes and applications of high-permittivity polymer–matrix composites, *Progress in Materials Science* 57(4) (2012) 660–723.
175. Greeshma, T.; Balaji, R.; Jayakumar, S.; PVDF phase formation and its influence on electrical and structural properties of PZT-PVDF composites, *Ferroelectric Letters Section* 40(1–3) (2013) 41–55.
176. Wu, C.-G.; Cai, G.-Q.; Luo, W.-B.; Peng, Q.X.; Sun, X.Y.; Zhang, W.L.; Enhanced pyroelectric properties of PZT/PVDF-TrFE composites using calcined PZT ceramic powders, *Journal of Advanced Dielectrics* 3(01) (2013) 1350004.
177. Tiwari, V., Srivastava, G.; Structural, dielectric and piezoelectric properties of 0–3 PZT/PVDF composites, *Ceramics International* 41(6) (2015) 8008–8013.
178. Batra, A.K.; Edwards, M.E.; Alomari, A.; Elkhaldy. A.; Dielectric behavior of P(VDF-TrFE)/PZT nanocomposites films doped with multi-walled carbon nanotubes (MWCNT), *American Journal of Materials Science* 5(3A) (2015) 55–61.
179. Kakimoto, K. I.; Fukata, K.; Ogawa, H.; Fabrication of fibrous $BaTiO_3$-reinforced PVDF composite sheet for transducer application, *Sensors and Actuators, A: Physical* 200 (2013) 21–25.
180. Mishra, M.; Roy, A.; Dash, S.; Mukherjee, S.; Flexible nano-GFO/PVDF piezoelectric-polymer nano-composite films for mechanical energy harvesting, *IOP Conference Series: Materials Science and Engineering* 338(1) (2018) 012026.
181. Lam, K.H.; Wang, X.; Chan, H.L.W.; Piezoelectric and pyroelectric properties of $(Bi_{0.5}Na_{0.5})_{0.94}Ba_{0.06}TiO_3$/P(VDF-TrFE) 0–3 composites, *Composites Part A: Applied Science and Manufacturing* 36(11) (2005) 1595–1599.
182. Genchi, G.G.; Ceseracciu, L.; Marino, A.; Labardi, M.; Marras, S.; Pignatelli, F.; Bruschini, L.; Mattoli, V.; Ciofani, G.; P(VDF-TrFE)/$BaTiO_3$ nanoparticle composite films mediate piezoelectric stimulation and promote differentiation of SH-SY5Y neuroblastoma cells, *Advanced Healthcare Materials* 5(14) (2016) 1808–1820.
183. Turku, I.; Kärki, T.; The influence of carbon-based fillers on the flammability of polypropylene-based co-extruded wood-plastic composite, *Fire and Materials* 40(3) (2016) 498–506.
184. Kim, J.; Loh, K.J.; Lynch, J.P., Piezoelectric polymeric thin films tuned by carbon nanotube fillers, In Sensors and Smart Structures Technologies for Civil, Mechanical, and Aerospace Systems. *International Society for Optics and Photonics*, 6932 (2008) 693232.
185. Batth, A.; Mueller, A.; Rakesh, L.; Mellinger, A.; Electrical properties of poly(vinylidene fluoride-hexafluoropropylene) (PVDF-HFP) blended with carbon nanotubes, presented at the *2012 Annual Report Conference on Electrical Insulation and Dielectric Phenomena(CEIDP)*, Montreal, QC, Canada, October 2012, pp. 28–31
186. Batra, A.K.; Edwards, M.E.; Guggilla, P.; Aggarwal, M.D.; Lal, R.B.; Pyroelectric properties of PVDF: MWCNT nanocomposite film for uncooled infrared detectors and medical applications, *Integrated Ferroelectrics* 158 (1) (2014) 98–107.
187. Edwards, M.E.; Batra, A.K.; Chilvery, A.K.; Guggilla, P.; Curley, M.; Aggarwal, M.D.; Pyroelectric properties of PVDF:MWCNT nanocomposite film for uncooled infrared detectors, *Materials Sciences and Applications* 3 (2012) 851–855.
188. El Achaby, M.; Arrakhiz, F.Z.; Vaudreuil, S.; Essassi, E.M.; Qaiss. A.; Piezoelectric β-polymorph formation and properties enhancement in graphene oxide–PVDF nanocomposite films, *Applied Surface Science* 258(19) (2012) 7668–7677.
189. Huang, L.; Lu, C.; Wang, F.; Wang, L.; Preparation of PVDF/graphene ferroelectric composite films by in situ reduction with hydrobromic acids and their properties, *RSC Advances* 4(85) (2014) 45220–45229.
190. Khan, A.; Abas, Z.; Kim, H. S.; Kim, J.; Recent progress on cellulose-based electro-active paper, its hybrid nanocomposites and applications, *Sensors* 16(8) (2016) 1172.
191. Wu, L.; Yuan, W.; Hu, N.; Wang, Z.; Chen, C.; Qiu, J.; Ying, J.; Li, Y.; Improved piezoelectricity of PVDF-HFP/carbon black composite films, *Journal of Physics D: Applied Physics* 47(13) (2014) 135302.

192. Maity, N.; Mandal, A.; Nandi, A.K.; Hierarchical nanostructured polyaniline functionalized graphene/poly (vinylidene fluoride) composites for improved dielectric performances, *Polymer* 103 (2016) 83–97.
193. Zabek, D.; Seunarine, K.; Spacie, C.; Bowen, C.; Graphene ink laminate structures on poly(vinylidene di fluoride) (PVDF) for pyroelectric thermal energy harvesting and waste heat recovery, *ACS Applied Materials & Interfaces* 9 (2017) 9161–9167.
194. Bystrov, V.S.; Bdikin, I.K.; Silibin, M.V.; Meng, X.J.; Lin, T.; Wang, J.L.; Karpinsky, D.V.; Bystrova, A.V.; Paramonova, E.V.; Pyroelectric properties of ferroelectric composites based on polyvinylidene fluoride (PVDF) with graphene and graphene oxide, *Ferroelectrics* 541 (2019) 17–24.
195. Li, H.; Koh, C.S.L.; Lee, Y.H.; Zhang, Y.; Phan-Quang, G.C.; Zhu, C.; Liu, Z.; Chen, Z.; Sim, H.Y.F.; Lay, C.L.; An, Q.; Ling, X.Y.; A wearable solar-thermal-pyroelectric harvester: Achieving high power output using modified rGO-PEI and polarized PVDF, *Nano Energy* 73 (2020) 104723.
196. Peihai, J.; Ling, W.; Lizhu, L.; Xiaorui, Z.; Preparation and characterisation of Al-doped ZnO and PVDF composites, *High Voltage* 1(4) (2016) 166–170.
197. Choi, K.; Choi, W.; Yu, C.; Park, W.T.; Enhanced piezoelectric behavior of PVDF nanocomposite by AC dielectrophoresis alignment of ZnO nanowires. *Journal of Nanomaterials* 2017 (2017) 6590121.
198. Tan, K.S.; Gan, W.C.; Velayutham, T.S.; Majid, W.H.A.; Pyroelectricity enhancement of PVDF nanocomposite thin films doped with ZnO nanoparticles, *Smart Materials and Structures* 23 (2014) 125006.
199. Li, G.Y.; Zhang, H.D.; Guo, K.; Ma, X.S.; Long, Y.Z.; Fabrication and piezoelectric-pyroelectric properties of electrospun PVDF/ZnO composite fibers, *Materials Research Express* 7 (2020) 095502.
200. Zhao, L.J.; Tang, C.P.; Gong, P.; Correlation of direct piezoelectric effect on EAPap under ambient factors, *International Journal of Automation and Computing* 7(3) (2010) 324–329.
201. Gan, W. C.; Abd Majid, W.H.; Effect of TiO_2 on enhanced pyroelectric activity of PVDF composite. *Smart Materials and Structures* 23(4) (2014) 045026.
202. Singh, D.; Choudhary, A.; Garg, A.; Flexible and robust piezoelectric polymer nanocomposites based energy harvesters, *ACS Applied Materials & Interfaces* 10(3) (2018) 2793–2800.
203. Chen, H.J.; Han, S.; Liu, C.; Luo, Z.; Shieh, H.P.D.; Hsiao, R.S.; Yang, B.R.; Investigation of PVDF-TrFE composite with nanofillers for sensitivity improvement, *Sensors and Actuators A: Physical* 245 (2016) 135–139.
204. AIssa, A.A.; Al-Maadeed, M.A.; Luyt, A.S.; Ponnamma, D.; Hassan, M.K.; Physico-mechanical, dielectric, and piezoelectric properties of PVDF electrospun mats containing silver nanoparticles, *Journal of Carbon Research* 3(4) (2017) 30.
205. Yang, L.; Qiu, J.; Ji, H.; Zhu, K.; Wang, J.; Enhanced dielectric and ferroelectric properties induced by TiO_2@MWCNTs nanoparticles in flexible poly (vinylidene fluoride) composites, *Composites Part A: Applied Science and Manufacturing* 65(2014) 125–134.
206. Yang, L.; Ji, H.; Zhu, K.; Wang, J.; Qiu, J.; Dramatically improved piezoelectric properties of poly (vinylidene fluoride) composites by incorporating aligned TiO_2@MWCNTs, *Composites Science and Technology* 123 (2016) 259–267.
207. Karan, S.K.; Mandal, D.; Khatua, B.B.; Self-powered flexible Fe-doped RGO/PVDF nanocomposite: An excellent material for a piezoelectric energy harvester, *Nanoscale* 7(24) (2015) 10655–10666.
208. David, C.; Capsal, J.-F.; Laffont, L.; Dantras, E.; Lacabanne, C.; Piezoelectric properties of polyamide 11/$NaNbO_3$ nanowire composites, *Journal of Physics D: Applied Physics* 45(41) (2012) 415305.
209. Abdelhamid, E.H.; Jayakumar, O.D.; Kotari, V.; Mandal, B.P.; Rao, R.; Naik, V.M.; Naik, R.; Tyagi, A.K.; Multiferroic PVDF–Fe_3O_4 hybrid films with reduced graphene oxide and ZnO nanofillers, *RSC Advances* 6(24) (2016) 20089–20094.
210. Karan, S.K.; Maiti, S.; Agrawal, A.K.; Das, A.K.; Maitra, A.; Paria, S.; Bera, A.; Bera, R.; Halder, L.; Mishra, A.K.; Kim, J.K.; Khatua, B.B.; Designing high energy conversion efficient bio-inspired vitamin assisted single-structured based self-powered piezoelectric/wind/acoustic multi-energy harvester with remarkable power density, *Nano Energy* 59 (2019) 169–183.
211. Batra, A.K.; Aggarwal, M.D.; Edwards, M.E.; Bhalla, A.; Present Status of Polymer: Ceramic Composites for Pyroelectric Infrared Detectors, *Ferroelectrics* 366(1) (2010) 84–121.
212. Varga, M.; Morvan, J.; Diorio, N.; Buyuktanir, E.; Harden, J.; West, J.L.; Jákli, A.; Direct piezoelectric responses of soft composite fiber mats, *Applied Physics Letters* 102(15) (2013) 153903.
213. Zhao, C.; Zhang, J.; Wang, Z.L.; Ren, K.; A Poly (l-Lactic Acid) Polymer-based thermally stable cantilever for vibration energy harvesting applications, *Advanced Sustainable Systems* 1(9) (2017) 1700068.
214. Ando, A.; Takeda, K.; Ohta, T.; Ito, M.; Hiramatsu, M.; Ishikawa, K.; Kondo H.; Sekine, M.; Suzuki, T.; Inoue, S.; Ando, Y.; Hori, M.; Characteristics of optical emissions of arc plasma processing for high-rate synthesis of highly crystalline single-walled carbon nanotubes, *Japanese Journal of Applied Physics* 56(3) (2017) 035101.
215. Rodríguez-Roldán, G.; Cruz-Orea, A.; Suaste-Gómez, E.; Thermal characterization of a PPy/PLA composite by photoacoustic calorimetry and photopyroelectric techniques, *International Journal of Thermophysics* 40 (2019) 16.
216. Fukada, E.; Date, M.; Ochiai, T.; Piezoelectricity in polyurethane films induced by electrostriction and a bias electric field, Presented at the *10th IEEE International Symposium on Electrets*, Greece, 1999, p. 655.
217. Sakamoto, W.K.; De Souza, E.; Das-Gupta, D.K.; Electroactive properties of flexible piezoelectric composites, *Materials Research* 4(3) (2001) 201–204.
218. Sakamoto, W.K.; Higuti, R.T.; Crivelini, E.B.; Nagashima, H.N.; *IEEE International Symposium on Applications of Ferroelectric and Workshop on the Piezoresponse Force Microscopy (ISAF/PFM)*, Prague, Czech Republic, July 2013, p. 295.
219. Moody, M. J.; Marvin, C.W.; Hutchison, G.R.; Molecularly-doped polyurethane foams with massive piezoelectric response, *Journal of Materials Chemistry C* 4(20) (2016) 4387–4392.
220. Tabbai, Y.; Belhora, F.; Moznine, R.El; Hajjaji, A.; El Ballouti, A.; Pyroelectric effect in lead zirconate titanate/polyurethane composite for thermal energy harvesting, *European Physical Journal Applied Physics* 86 (2019) 10902.
221. Lam, K.S.; Wong, Y.W.; Tai, L.S.; Poon, Y.M.; Shin, F.G.; Dielectric and pyroelectric properties of lead zirconate titanate/polyurethane composites, *Journal of Applied Physics* 96 (7) (2004) 3896.

222. Jayalakshmy, M.S.; Philip, J.; Pyroelectricity in strontium barium niobate/polyurethane nanocomposites for thermal/infrared detection, *Composites Science and Technology* 109 (2015) 6–11.
223. Sakamoto, W.K.; Kanda, D.H.F.; Gupta, D.K.D.; Dielectric and pyroelectric properties of a composite of ferroelectric ceramic and polyurethane, *Materials Research Innovations* 5 (2002) 257–260.
224. Ounaies, Z.; Harrison, J.S.; Piezoelectric Polymers, NASA/CR-2001-211422, ICASE Report No. 2001–43, USA 2001, 1, https://ntrs.nasa.gov/search.jsp?R=20020044745.
225. Kawai, H.; The piezoelectricity of poly (vinylidene fluoride). *Japanese Journal of Applied Physics* 8(7) (1969) 975.
226. Takase, Y.; Lee, J.W.; Scheinbeim, J.I.; Newman, B.A.; High-temperature characteristics of nylon-11 and nylon-7 piezoelectrics. *Macromolecules* 24(25) (1991) 6644–6652.
227. Babu, I.; Van Den Ende, D.A.; Processing and characterization of piezoelectric 0-3 PZT/LCT/PA composites. *Journal of Physics D: Applied Physics* 43(42) (2010) 425402.
228. David, C.; Capsal, J.-F.; Laffont, L.; Dantras, E.; Lacabanne, C.; Piezoelectric properties of polyamide 11/$NaNbO_3$ nanowire composites, *Journal of Physics D: Applied Physics* 45(41) (2012) 415305.
229. Carponcin, D.; Dantras, E.; Michon, G.; Dandurand, J.; Aridon, G.; Levallois, F.; Cadiergues, L.; Lacabanne, C.; New hybrid polymer nanocomposites for passive vibration damping by incorporation of carbon nanotubes and lead zirconatetitanate particles, *Journal of Non-Crystalline Solids* 409 (2015) 20–26.
230. Leveque, M. C.; Rguiti, D.M.; Prashantha, K.; Courtois, C.; Lacrampe, M.-F.; Krawczak, P.; Vibrational energy-harvesting performance of bio-sourced flexible polyamide 11/layered silicate nanocomposite films, *International Journal of Polymer Analysis and Characterization* 22(1) (2017) 72–82.
231. Hua, Z.; Shi, X.; Chen, Y.; Preparation, structure, and property of highly filled polyamide 11/$BaTiO_3$ piezoelectric composites prepared through solid-state mechanochemical method, *Polymer Composites* 40(S1) (2019) E177–E185.
232. David, Charlotte; Capsal, Jean-Fabien; Laffont, Lydia; Dantras, Eric; Lacabanne, Colette; Piezoelectric properties of polyamide 11/$NaNbO_3$ nanowire composites, *Journal of Physics D: Applied Physics* 45(41) (2012) 415305.
233. Carponcin, Delphine; Dantras, Eric; Dandurand, Jany; Aridon, Gwenaëlle; Levallois, Franck; Cadiergues, Laurent; Lacabanne, Colette; Electrical and piezoelectric behavior of polyamide/PZT/CNT multifunctional nanocomposites, *Advanced Engineering Materials* 16(8) (2014) 1018–1025.
234. Capsal, J.F.; Dantras, E.; Dandurand, J.; Lacabanne, C.; Electroactive influence of ferroelectric nanofillers on polyamide 11 matrix properties, *Journal of Non-Crystalline Solids*, 353 (47–51) (2007) 4437–4442.
235. Mahadeva, S.K.; Walus, K.; Stoeber, B.; Piezoelectric paper fabricated via nanostructured barium titanate functionalization of wood cellulose fibers, *ACS Applied Materials & Interfaces* 6(10) (2014) 7547–7553.
236. Kim, H.S.; Yun, S.; Kim, J.H.; Yun, G.Y.; Kim, J.; Fabrication and characterization of piezo-paper made with cellulose, *Behavior and Mechanics of Multifunctional and Composite Materials. International Society for Optics and Photonics, 2008*, 6929, 2008, p. 692900.
237. Hänninen, A.; Rajala, S.; Salpavaara, T.; Kellomäki, M.; Tuukkanen, S.; Piezoelectric sensitivity of a layered film of chitosan and cellulose nanocrystals, *Procedia Engineering* 168 (2016) 1176–1179.
238. Kim, J.H.; Ko, H.U.; Zinc Oxide-Cellulose Nanocomposite and Preparation There of.US Patent US 20140272397A1, 2014.
239. Wu, N.; Cheng, X.; Zhong, Q.; Zhong, J.; Li, W.; Wang, B.; Hu, B.; Zhou, J.; Cellular polypropylene piezoelectret for human body energy harvesting and health monitoring, *Advanced Functional Materials* 25(30) (2015) 4788–4794.
240. Khanbareh, H.; Zwaag, S.V.D.; Groen, W.A.; Piezoelectric and pyroelectric properties of conductive polyethylene oxide-lead titanate composites, *Smart Materials and Structures* 24(4) (2015) 045020.
241. Zhang, F.; Jiang, X.; Zhu, X.; Chen, Z.; Kong, X.Z.; Preparation of uniform and porous polyurea microspheres of large size through interfacial polymerization of toluene diisocyanate in water solution of ethylene diamine, *Chemical Engineering Journal* 303 (2016) 48–55.
242. Miyata, S.; Yoshikawa, M.; Tasaka, S.; Ko, M.; Piezoelectricity revealed in the copolymer of vinylidene cyanide and vinyl acetate, *Polymer Journal* 12(12) (1980) 857–860.
243. Gonzalo, B.; Breczewski, T.; Vilas, J.L.; Jubindo, M.A.P.; Fuente, M.R. De La; Dios, J.R.; León, L.M.; Dielectric properties of piezoelectric polyimides, *Ferroelectrics* 370(1) (2008) 3–10.
244. Ounaies, Z.; Park, C.; Harrison, J.S.; Smith, J.G.; Hinkley, J.A.; Structure property study of piezoelectricity in polyimides, *Smart Structures and Materials 1999: Electroactive Polymer Actuators and Devices, International Society for Optics and Photonics*, 3669, 1999, pp. 171–178.
245. Olszowy, M.; Pawlaczyk, C.; Markiewicz, E.; Kułek, J.; Dielectric and pyroelectric properties of $BaTiO_3$-PVC composites, *Physica Status Solidi A* 202(9) (2005) 1848–1853.
246. Annapureddy, V.; Kim, M.; Palneedi, H.; Lee, H.-Y.; Choi, S.-Y.; Yoon, W.-H.; Park, D.-S.; Choi, J.-J.; Hahn, B.-D.; Ahn, C.-W.; Kim, J.-W.; Jeong, D.-Y.; Ryu, J.; Low-loss piezoelectric single-crystal fibers for enhanced magnetic energy harvesting with magnetoelectric composite, *Advanced Energy Materials* 6(24) (2016) 1601244. DOI:10.1002/aenm.201601244
247. Priya, S.; Song, H.-C.; Zhou, Y.; Varghese, R.; Chopra, A.; Kim, S.-G.; Kanno, I.; Wu, L.; Ha, D.S.; Ryu, J.; Polcawich, R.G.; A review on piezoelectric energy harvesting: Materials, methods, and circuits, *Energy Harvesting and Systems* 4(1) (2019) 3–39.
248. Kishore, R.A.; Priya, S.; A review on low-grade thermal energy harvesting: Materials, methods and devices, *Materials (Basel)* 11(8) (2018) 1433.
249. Covaci, C.; Gontean, A.; Piezoelectric energy harvesting solutions: A review, *Sensors (Basel)* 20(12) (2020) 3512.
250. Shaikh, F.K.; Zeadally, S.; Energy harvesting in wireless sensor networks: A comprehensive review, *Renewable and Sustainable Energy Reviews* 55 (2016) 1041–1054.
251. Kumar, S.S.; Narayanan, L.A.J.; Kaviyaraj, R.; Saleekha; Energy harvesting by piezoelectric sensor array in road using internet of things, *5th International Conference on Advanced Computing & Communication Systems (ICACCS)*, IEEE, India, 2019, pp. 482–484.
252. Maamer, B.; Boughamoura, A.; Fath El-Bab, A.M.R.; Francis, L.A.; Tounsi, F.; A review on design improvements and techniques for mechanical energy harvesting using piezoelectric and electromagnetic schemes, *Energy Conversion and Management* 199 (2019) 111973.

253. Rjafallah, A.; Hajjaji, A.; Guyomar, D.; Kandoussi, K.; Belhora, F.; Boughaleb Y.; Modeling of polyurethane/lead zirconate titanate composites for vibration energy harvesting, *Journal of Composite Materials* 53(5) (2018) 613–623.
254. Calio, R.; Rongala, U.B.; Camboni, D.; Milazzo, M.; Stefanini, C.; De Petris, G.; Oddo, C.M.; Piezoelectric energy harvesting solutions, *Sensors (Basel)* 14(3) (2014) 4755–4790.
255. Mitcheson, P.D.; Green, T.C.; Yeatman, E.M.; Holmes, A.S.; Architectures for vibration-driven micropower generators, *Journal of Microelectromechanical Systems* 13(3) (2004) 429–440.
256. Gareh, S.; Kok, B.C.; Yee, M.H.; Borhana, A.A.; Alswed, S.K.; Optimization of the compression-based piezoelectric traffic model (CPTM) for road energy harvesting application, *International Journal of Renewable Energy Research* 9(3) (2019) 1272–1282.
257. Mishra, S.; Unnikrishnan, L.; Nayak, S.K.; Mohanty, S.; Advances in piezoelectric polymer composites for energy harvesting applications: A systematic review. *Macromolecular Materials and Engineering* 304(1) (2019) 1800463.
258. Uchino, K.; Ishii, T.; Energy Flow Analysis in Piezoelectric Energy Harvesting Systems, *Ferroelectrics* 400(1) (2010) 305–320.
259. Sharma, M.; Chauhan, A.; Vaish, R.; Chauhan, V.S.; Pyroelectric materials for solar energy harvesting: A comparative study, *Smart Materials and Structures* 24(10) (2015) 105013.
260. Yang, Y.; Guo, W.; Pradel, K.C.; Zhu, G.; Zhou, Y.; Zhang, Y.; Hu, Y.; Lin, L.; Wang, Z.L.; Pyroelectric nanogenerators for harvesting thermoelectric energy, *Nano Letters* 12(6) (2012) 2833–2838.
261. Ravindran, S.K.T.; Huesgen, T.; Kroener, M.; Woias, P.; A self-sustaining micro thermomechanic-pyroelectric generator, *Applied Physics Letters* 99(10) (2011) 104102.
262. Fang, J.; Frederich, H.; Pilon, L.; Harvesting nanoscale thermal radiation using pyroelectric materials, *Journal of Heat Transfer* 132(9) (2010) 092701.
263. Sebald, G.; Guyomar, D; Agbossou, A.; On thermoelectric and pyroelectric energyharvesting, *Smart Materials and Structures* 18 (2009) 125006.
264. Lee, F.Y.; Navid, A.; Pilon, L.; Pyroelectric waste heat energy harvesting using heat conduction, *Applied Thermal Engineering* 37 (2012) 30–37.
265. Ackermann, W.; Beobachtungen uber pyroelektrizitat in ihrer abhangigkeit von der temperatur, *Annals of Physics* 351 (1915) 197–220.
266. Sawyer, C. B.; Tower, C. H.; Rochelle salt as a dielectric, *Physical Review* 35 (1930) 269–273.
267. Kutnjak, Z.; Rozic, B.; Pirc, R.; *Electrocaloric Effect: Theory, Measurements, and Applications*, John Wiley & Sons, Inc., Hoboken, NJ, USA, 2015.
268. Wiseman, G.G.; Kuebler, J. K.; Electrocaloric effectin ferroelectric Rochelle salt, *Physical Review* 131 (1963) 2023–2027.
269. Matocha, K.; Chow, T. P.; Gutmann, R. J.; Positive flatband voltage shift in MOS capacitors on n-type GaN, *IEEE Electron Device Letters* 23 (2002) 79–81.
270. Ehre, D.; Cohen, H.; Contact-free pyroelectric measurements using x-ray photoelectron spectroscopy, *Applied Physics Letters* 103 (2013) 052901.
271. Weigel, T.; Elektronendichteanalyse von pyroelektrischem $LiMO_3$-Rontgeneinkristalldiffraktometrie und ab-initio modellierung Master's thesis (Institut of Experimental Physics, TU Bergakademie, Freiberg, 2017).
272. Lang, S.B.; Steckel, F.; Method for the measurement of the pyroelectric coefficient, dc dielectric constant, and volume resistivity of a polar material, *The Review of Scientific Instruments* 36 (1965) 929–932.
273. Glass, A. M.; Investigation of the electrical properties of $Sr_{1-x}Ba_xNb_2O_6$ with special reference to pyroelectric detection, *Journal of Applied Physics* 40 (1969) 4699.
274. Byer, R. L.; Roundy, C. B.; Pyroelecric coefficient direct measurement technique and application to a nsec response time detector Ferroelectrics, *IEEE Transactions on Sonics and Ultrasonics* 3 (1972) 333–338.
275. Parravicini, J.; Safioui, J.; Degiorgio, V.; Minzioni, P.; Chauvet, M.; All-optical technique to measure the pyroelectric coefficient in electrooptic crystals, *Journal of Applied Physics* 109 (2011) 033106.
276. Popescu, S. T.; Petris, A.; Vlad, I.; Interferometric measurement of the pyroelectric coefficient in lithium niobate, *Journal of Applied Physics* 113 (2013) 043101.
277. Chynoweth, A. G.; Dynamic method for measuring the pyroelectric effect with special reference to barium titanate, *Journal of Applied Physics* 27 (1956) 78–84.
278. Schein, L. B.; Cressman, P. J.; Cross, L. E.; Electrostatic measurements of unusually large secondary pyroelectricity in partially clamped $LiNbO_3$, *Ferroelectrics* 22 (1978) 937–943.
279. Schein, L. B.; Cressman, P. J.; Cross, L. E.; Electrostatic measurements of tertiary pyroelectricity in partially clamped $LiNbO_3$, *Ferroelectrics* 22 (1978) 945–948.
280. Groten, J.; Zirkl, M.; Jakopic, G.; Leitner, A.; Stadlober, B.; Pyroelectric scanning probe microscopy: A method for local measurement of the pyroelectric effect in ferroelectric thin films, *Physics Review* 82 (2010) 054112.
281. Bauer, S.; Ploss, B.; A heat wave method for the measurement of thermal and pyroelectric properties of pyroelectric films, *Ferroelectrics* 106 (1990) 393–398.
282. Daglish, M.; A dynamic method for determining the pyroelectric response of thin films, *Integrated Ferroelectrics* 22 (1998) 473–488.
283. Sussner, H.; Horne, D. E.; Yoon, D. Y.; A new method for determining the pyroelectric coefficient of thin polymer films using dielectric heating, *Applied Physics Letters* 32 (1978) 137–139.
284. Hartley, N.P.; Squire, P. T.; Putley, E. H.; A new method of measuring pyroelectric coefficients, *Journal of Physics E: Scientific Instruments* 5 (1972) 787–789.
285. Garn, L. E.; Sharp, E. J.; Use of low-frequency sinusoidal temperature waves to separate pyroelectric currents from nonpyroelectric currents Part I. Theory, *Journal of Applied Physics* 53 (1982) 8974.
286. Sharp, E. J.; Garn, L. E.; Use of low-frequency sinusoidal temperature waves to separate pyroelectric currents from nonpyroelectric currents, Part II. Experiment, *Journal of Applied Physics* 53 (1982) 8980.
287. Ploss, B.; Domig, A.; Static and dynamic pyroelectric properties of PVDF, *Ferroelectrics* 159(1) (1994) 263–268.
288. Bauer, S.; Lang, S.B.; Pvroelectric polvmer electrets, *IEEE Transactions on Dielectrics and Electdcal Insulation* 3(5) (1996) 647–676.
289. Neumann, N.; Köhler, R.; Hofmann, G.; Pyroelectric thin film sensors and arrays based on P(VDF/TrFE), *Integrated Ferroelectrics* 6(1–4) (2006) 213–230.
290. Navid, A.; Lynch, C.S.; Pilon, L.; Purified and porous poly(vinylidene fluoride-trifluoroethylene) thin films for

pyroelectric infrared sensing and energy harvesting, *Smart Materials and Structures* 19(5) (2010) 055006(13).
291. Whatmore, R.W.; Watton, R.; Pyroelectric ceramics and thin films for uncooled thermal imaging, *Ferroelectrics* 236(1) (2000) 259–279.
292. Alpay, S.P.; Mantese, J.; Trolier-McKinstry, S.; Zhang, Q.; Whatmore, R.W.; Next-generation electrocaloric and pyroelectric materials for solid-state electrothermal energy interconversion, *MRS Bulletin* 39(12) (2014) 1099–1111.
293. Dietze, M.; Krause, J.; Solterbeck, C.H.; Es-Souni, M.; Thick film polymer-ceramic composites for pyroelectric applications, *Journal of Applied Physics* 101(5) (2007).
294. Stephens, A.; Levine, A.; Fech Jr, J., Zrebiec, T.; Cafiero, A.; Garofalo, A.; Pyroelectric polymer films, *Thin Solid Films* 24 (2) (1974) 361–379.
295. Lang, S. B.; DeReggi, A. S.; Broadhurst, M. G.; Davis, G. T.; Effects of poling field and time on pyroelectric coefficient and polarization uniformity in polyvinyl fluoride, *Ferroelectrics* 33 (1981) 119–125.
296. Hamdi, Q.; Mighri, F.; Rodrigue, D.; Piezoelectric polymer films: Synthesis, applications, and modeling, In Bouhfid, R.; Qaiss, A. el K.; Jawaid, M. (Ed.) *Polymer Nanocomposite-Based Smart Materials: From Synthesis to Application*, Woodhead Publishing, Cambridge, 2020, pp. 79–101.

7 Polymers and Their Composites for Solar Cell Applications

Abdelaal S. A. Ahmed
Al-Azhar University, Assiut, Egypt

Abdullah Jan
Wuhan University of Technology, Wuhan, P. R. China

Gomaa A. M. Ali
Al-Azhar University, Assiut, Egypt

CONTENTS

LIST OF ABBREVIATIONS

1D One-dimensional
CB Conduction band
CE Counter electrode
CPs Conductive polymers
CVD Chemical vapor deposition
DSSCs Dye-sensitized solar cells
MPa Megapascal
PANI Polyaniline
PCE Power conversion efficiency
PSCs Perovskite solar cells
PVP Polyvinylpyrrolidone
PVs Photovoltaics

7.1 INTRODUCTION

The rise of the Industrial Revolution in the mid-18th century led to an increase in energy demand, which in turn led to clear impacts on the environment and the future of humanity [1]. As energy needs increase, fossil fuel production and use create serious environmental concerns such as greenhouse gas emissions like CO_2 which are the main reason for global climate change. Thus, finding alternative clean energy resources is highly required. Solar energy has received much attention as the most promising sustainable energy for future generations [2]. Many techniques have been utilized to make use of solar energy, such as solar cells or photovoltaics (PVs), which convert sunlight into

DOI: 10.1201/9781003169727-7

electrical energy via the photovoltaic effect [3]. However, the high production cost is the main reason for restricting the wide application of silicon-based PVs [4]. During recent decades, emergency solar cells such as dye-sensitized solar cells (DSSCs) [5] and perovskite solar cells [6] have shown advantages such as being lightweight and of low cost. Despite the ability of DSSCs being available on the market at a low price, the ease of leakage of liquid electrolyte and solvent evaporation is the main challenge which damages their long-term stability and environmental safety [1, 7]. Although the solid-state perovskite solar cell overcomes the limitations of DSSC, their maximum power conversion efficiency (PCE) still does not exceed 23.3%.

Moreover, device stability is still a problem that limits their wide application [8–10]. Each component has a significant task in the assembled devices; thus, it is important to optimize all components in the device. Polymers have been extensively studied in various electrochemical applications [11, 12]. According to their three-dimensional structures, polymers are usually utilized as templates for the production of mesoporous materials and as a polymeric matrix in solid electrolytes [13–15]. On the other hand, polymers' high catalytic activity makes them a promising candidate as counter-electrodes for DSSCs [16, 17]. In perovskite solar cells, the polymers are usually utilized to adjust the perovskite morphologies, mainly attributed to their various functional groups.

The high transporter mobilities make polymers efficient electron and hole transporter materials [18–20]. The ease of coating polymers via conventional techniques such as spin-coating, spray coating, and drop coating allows them to be inserted into micro/nanostructure devices [21–23]. Here, we mainly emphasize the recent progress of the application of polymer-based materials for DSSCs and PSC devices.

7.2 POLYMER COMPOSITES FOR DSSC APPLICATIONS

Although the idea of sensitizing semiconductors was started in 1968 by Gerischer [24] and in 1976 by Tsubomura [25], where ZnO was sensitized with different photosensitizers, later Kalyan et al. in 1987 sensitized TiO_2 by zinc porphyrins; the related assembled PV devices showed relatively low PCE values [26]. In 1991, Grätzel achieved a breakthrough by sensitizing a porous TiO_2 layer via a ruthenium dye; the achieved PCE was higher than 7% [5]. During recent decades, DSSCs received meaningful interest as an alternative to silicon-based cells due to their low cost and easy production process [2, 27]. Typically DSSCs consist of a dye-sensitized porous nanocrystalline TiO_2 anode, a cathode, and an electrolyte with a redox couple as shown in Figure 7.1 [2]. The work function of DSSCs has been described in the literature [2, 28, 29]. When the incident sunlight hits the working electrode's surface, the dye molecules are excited, and the generated electrons are inserted into the conduction band (CB) of the semiconductor and then to the CE by an outdoor circuit. The oxidized form of

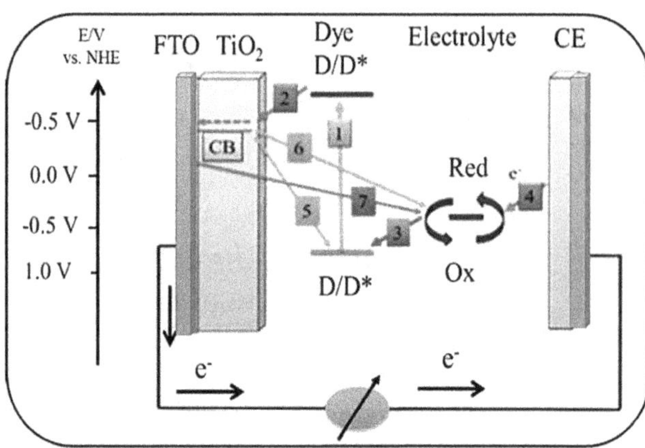

FIGURE 7.1 Operating principle of DSSCs. [2] with permission of the American Chemical Society.

the sensitizer captures an electron from the redox couple in the electrolyte [30]. Thus, the wise choosing and fabricating of the constituent materials for the three parts displayed a crucial role in the total performance of the DSSC device. Due to their various properties, polymers are usually used to design flexible substrates, porous structures in the anode, gel electrolytes, and cathodes [31]. The appliances of polymers in DSSCs are demonstrated in the next sections.

7.2.1 Polymer Composites for Flexible Substrates in DSSCs

According to the work function of DSSC, the conductive substrate is required for collecting and transferring the photogenerated electrons from the photoanode to the counter-electrode. Currently, indium-doped tin oxide (ITO), fluorine-doped tin oxide (FTO), and both ITO/polyethylene terephthalate (PET) and ITO/polyethylene naphthalate (PEN) are widely used conductive substrates [32–35]. Based on their low cost, significant transparency, and quiet resistance, both ITO/PEN and ITO/PET are widely utilized as substrates in DSSCs [33]. In fabricating photoanodes, sintering at high temperature (≈400 °C(is required to ensure the high crystallinity of TiO_2 and the high contact of TiO_2 film with the conductive glass substrate [36]. Also, a high temperature (≈350 °C(is applied to fabricate platinum counter-electrodes by pyrolysis [37]. However, ITO/PEN and ITO/PET are able to deform at 150 °C, and even melt at 235 °C, restricting their uses in high-temperature processes [33, 38]. Therefore, intensive efforts have been devoted to developing lower temperature deposition techniques to be workable with these substrates.

7.2.2 Polymer Composites for Mesoporous TiO_2 Photoanodes in DSSCs

The photoanode displays a multifunctional role, where the incident light is harvested, then the generated electrons are injected and transported, and support dye adsorption [39]. Since the first report on DSSCs by the Gratzel group, TiO_2

has been known as the photoanode art material state due to its attractive performance [39–42].

Typically a TiO_2 photoanode is fabricated by coating the precursor on a conductive substrate to be a porous film with a large surface area which is desirable to enhance the dye loading [41, 43, 44]. In a photoanode, the porosity can be regulated by adding 50–60% of polymers as a pore-forming agent, leaving the pores after the sintering process [45, 46]. However, polymers' higher contents reduce the interconnections between particles, which directly reduces the charge collection [47]. Elayappan et al. used polyvinylpyrrolidone (PVP) in different portions (5, 8, and 10 wt.%) as a template and pore-forming agent to prepare porous TiO_2 nanofibers (NFs) at 475 °C via an electrospinning technique [48]. In this report, the TiO_2 NFs obtained by 5 wt.% of PVP showed a higher surface area. Thus, the assembled DSSC device with a 5 wt.% PVP-T2 photoanode displayed the best performance with a PCE of 4.81%. Furthermore, Hou et al. utilized a higher content (15–35 wt.%) of PVP to fabricate mesoporous TiO_2 photoanodes [49]. The morphological studies showed that the film with 20 wt.% PVP gives the maximum surface area and larger pores with a thickness of 5.5 μm. Moreover, the assembled DSSCs displayed a superior performance with a PCE of 8.39%. This is mainly assigned to the higher specific surface area, boosting dye adsorption and electron transfer and electrolyte diffusion ability. However, the poor interfacial charge transfer or slow electron mobility is the main challenge in TiO_2 nanoparticle systems, thus increasing the possibility of the recombination process [42]. Intensive efforts have been made to modify titania materials to overcome this problem. Utilizing different morphological patterns such as nanofibers and nanotubes is a promising technique. For instance, Kokubo et al. used a polyvinyl acetate (PVAc)/titania composite to fabricate multi-core cable-like TiO_2 nanofibrous membranes by an electrospinning technique and a hot-pressing pre-treatment process [50]. The assembled DSSCs with a TiO_2 nanofibrous membrane photoanode of 9.21 micrometers exhibited a PCE of 5.77%. The polymer can be utilized to make a scattering layer to reduce light loss, as reported by Mustafa et al. [51]. In their report, polyvinyl alcohol/titanium dioxide (PVA/TiO_2) nanofibers were prepared to be a light scattering layer over the photoanode. The photovoltaics study, utilizing the PVA/TiO_2 nanofiber scattering layer, remarkably enhanced the assembled device's total performance with a PCE value of 4.06%. This reduces the radiation loss, increases the electron's excitation, and increases the electron's lifetime and charge collection efficiency. Surface modification of TiO_2 photoanodes is an encouraging technique for enhancing the total performance of the DSSCs. For example, Roy et al successfully wrapped the surface of a TiO_2 rutile nanorod by polyaniline (PANI) [52]. The assembled DSSC device with TiO_2-PANI composite films as photoanodes displayed a promising photovoltaic performance with a PCE of 6.23%, which is superior to that achieved by the device with TiO_2 nanorods (4.28%) and with Degussa P25 (3.95%). This is attributed to enhancing the rate of electron conduction of TiO_2 on illumination due to the PANI layer's presence. From the above example, one can conclude that the electron transfer at the TiO_2 photoanode surface can be enhanced by utilizing polymers as pore agents and regulating the morphologies of photoanodes.

7.2.3 Polymer Composites as Counter-Electrodes for DSSCs

In DSSCs, CEs act as electron transporters to the redox couple in the electrolyte, hence regenerating the excited dye molecules to their original states [29, 35, 53]. Therefore the utilized materials should have a low resistance and elevated catalytic activity toward the used redox couple [29]. Usually, thermally deposited platinum on the conductive glass superstrate is utilized as a CE of DSSCs. However, Pt is a highly expensive material, as well as displaying corrosion ability in the presence of a I_3^- / I^- redox shuttle [54–56]. Thus, great attention has been given to developing alternative catalytic materials [57]. Thus, many materials such as carbons [58], transitional metals [59], conductive polymers (CPs) [16], and their corresponding composites [60], have been widely utilized in DSSCs. During the last decade, various polymers such as polypyrrole (PPy), polyaniline (PANI), and poly(3,4-ethylenedioxythiophene) (PEDOT) have been widely investigated as CEs for DSSC applications due to their low cost, high catalytic activity, facile synthesis, and remarkable stability. [61] The chemical structures of these polymers are shown in Figure 7.2. Recent progress will be discuses in the following sections.

7.2.3.1 Polypyrrole (PPy)-Based CEs for DSSCs

Polypyrrole (PPy) has shown itself to be an attractive Pt-free CE, due to its outstanding catalytic ability and chemical stability. It can be easily synthesized by polymerizing the pyrrole monomer via chemical or electrochemical processes and its viability for large-scale device processing [62]. Wu et al. prepared PPy nanoparticles by iodine oxidation of a pyrrole monomer; the prepared PPy showed a particle size of 40–60 nm, and the assembled PPy film on

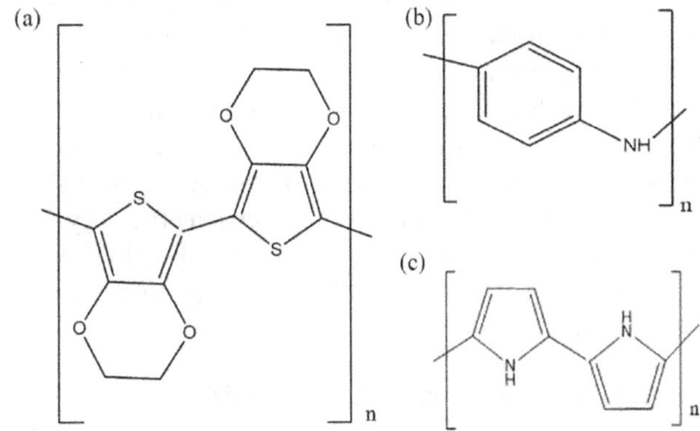

FIGURE 7.2 The chemical structures of (a) PEDOT (b) PANI (c) PPy.

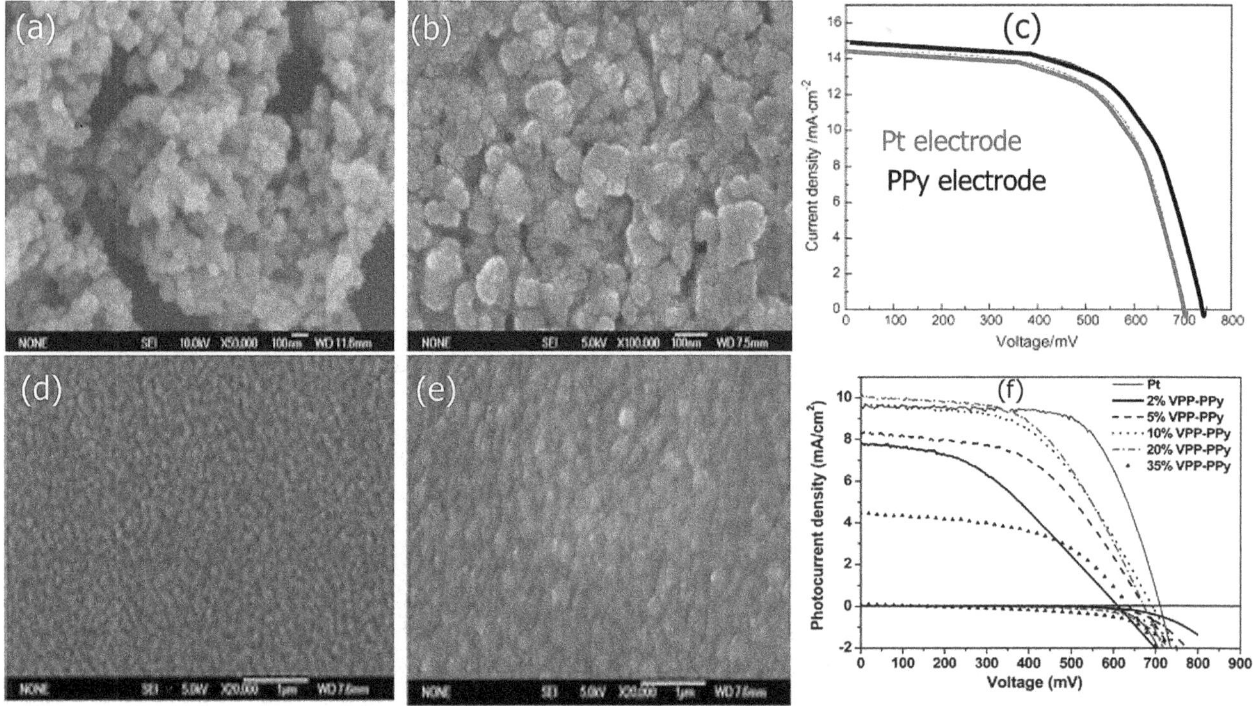

FIGURE 7.3 (a) SEM images of the PPy nanoparticle (b) PPy/FTO CE (c) *J–V* curve of DSSCs (d) SEM images of 20% VPP-PPy CE (e) EP-PPy Ces (f) *J–V* curves of DSSCs with different counter-electrodes. [63] with permission of Elsevier, 2008. [62] with permission of the Royal Society of Chemistry.

FTO glass showed a porous network as in Figure 7.3 [63]. The PPy CE revealed lower charge-transfer resistance (R_{CT}) toward I_3^- reduction compared with the Pt CE. Moreover, the device assembled with PPy CE showed a PCE of 7.66%. Both vapor phase polymerization (VPP) and electro-polymerization (EP) techniques have been utilized by Xia et al. to prepare VPP-PPy and EP-PPy CEs in DSSCs [62]. In preparing VPP-PPy CEs, Fe-TsO in various concentrations (2–35 wt.%) was used as an oxidant. The electrochemical analysis showed that 20% of VPP-PPy CEs showed catalytic activity like that obtained by Pt CE simultaneously, which is much better than that achieved by EP CE. The assembled devices with Pt, 20% VPP-PPy, and EP-PPy CEs showed PCE values of 4.4, 3.4, and 3.2%, respectively.

To further enhance the total performance of the devices with PPy CEs, Lu et al., doped the electropolymerized PPy film with both iodine and *p*-toluenesulfonate (TsO^-) [64]. The morphological studies showed that PPy film doped with I is much rougher than the pristine PPy film, hence a lower charge transfer resistance. The *J–V* of the fabricated DSSCs with iodine doped PPy CEs showed superior short current density and higher power conversion efficiency. The PCE of the device with PPy CE doped with LiI (10 mM) was 5.42%, which is about 20% greater than that obtained by the device without doping.

7.2.3.2 Polyaniline-Based CEs for DSSCs

Polyaniline (PANI) is another polymer that is among the widely used CEs in DSSCs due to its promising catalytic activity, reasonably low cost, and various oxidation states with various colors. The first work utilized PANI, where an aqueous oxidative polymerization reaction was used to prepare porous PANI nanoparticles with a diameter size of 100 nm [65]. In comparison to Pt CE, the PANI CE showed superior catalytic activity toward tri-iodide reduction, and the PCE of the assembled DSSCs with PANI was 7.15%, while the value obtained by the Pt-based device was 6.90%. The PANI can also be fabricated as a transparent CE to fabricate bi-facial DSSCs to irradiate from the front and the rear sides. Bu et al successfully prepared PANI CEs via *in situ* polymerizations of pyrrole on FTO glass [65]. The prepared CEs showed elevated transparency in the visible region with high catalytic activity comparable with the standard Pt CE. The PANI was inserted in bi-facial DSSCs, and the transparent DSSCs showed PEC values of 5.74 and 3.06% under front and rear illumination, respectively. Moreover, the fabricated DSSCs showed long-term stability due to the high adhesion of PPy/FTO. Further study by the same technique has been reported by Hou et al where a flexible polyaniline nanoribbon (PANI NR) CE with a serrated and ultrathin nanostructure was prepared by the *in situ* polymerization of aniline on V_2O_5 as a template and oxidant at the same time as shown in Figure 7.4a [66]. Due to its high catalytic activity, the DSSCs assembled with PANI NR CE showed promising performance with a PCE of 7.23%, which is very close to that achieved by Pt-CE (7.42%). The oxidative chemical polymerization is a recommended method that has been used to prepare prickly polyaniline nanorods (PPNR) and microgranules (PPMGs) by Wang et al., [67]. The diameter

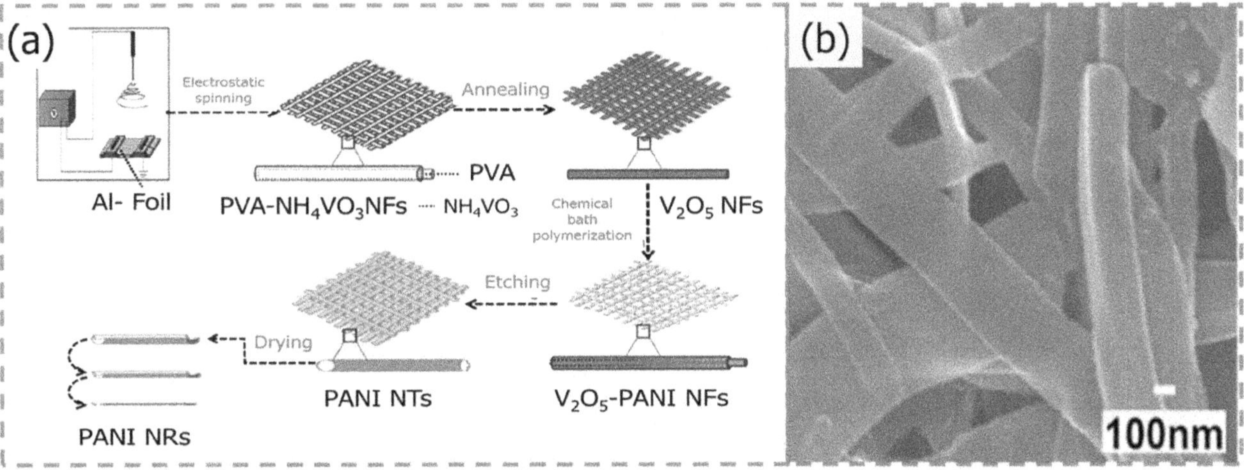

FIGURE 7.4 (a) Schematic diagram of the fabrication processes of the PANI NRs (b) FESEM image of PANI NR. [66] with permission of Elsevier, 2016

TABLE 7.1
Photovoltaic Factors of DSSCs Fabricated by PANI-Carbon Based Material Counter-Electrodes

CEs	Electrolyte	V_{OC} (mV)	J_{SC} (mA cm^{-2})	PCE (%)	Reference
PANI/MWCNT	I_3^- / I^-	721	13.53	6.24	[69]
Transparent-PANI/MWCNT*	I_3^- / I^-	691	22.25	9.24	[70]
Axle-sleeve MWCNT/PANI	I_3^- / I^-	780	16.08	7.21	[71]
GN/PANI	I_3^- / I^-	650	10.3	4.31	[72]
Water-dispersible PANI/GO oxide	I_3^- / I^-	710	12.91	6.12	[73]
PANi–SWCNT/GO**	I_3^- / I^-	—	—	6.88	[74]
PANi–SWCNT	I_3^- / I^-	734	14.88	7.81	[75]
PANI/carbon nanodots	I_3^- / I^-	770	13.8	7.37	[76]
PPy/graphene-coated Al_2O_3 (GCA)	I_3^- / I^-	717	6.54	7.33	[77]

* Irradiation from both front and rear sides together; ** Authors did not mention these data in the main text.

of the PPNR is ~80 nm, while the size of the PPMGs is ~ 400 nm. The as-prepared PPANR CEs for DSSCs showed greater electrocatalytic activity for the I_3^- reduction, this mainly assigned to the unique structure that offers abundant electrocatalytic active sites and fast charge transport. Subsequently, the fabricated DSSCs with PPNR CE provided a PCE value of 6.86%, close to the efficiency of a Pt CE (7.21%).

The fabricated CEs not only possess an elevated catalytic activity toward the I_3^- / I^- redox couple, but also showed a boosting catalytic ability toward the $[Co(bpy)_3]^{2+/3+}$ redox couple as reported by Wang et al, where an oriented PANI nanowire array was grown *in situ* on FTO [68]. The prepared array exhibited much higher electrocatalytic activity compared with PANI film with a random network, and even than Pt CE. Subsequently, the PCE of the assembled DSSCs with the oriented nanowire array was 8.42%, while the platinized device showed a 6.78% value. To promote enrichment of the catalytic activity and thus the total performance of the assembled devices, PANI was mixed with numerous carbon materials as shown in Table 7.1.

7.2.3.3 Poly(3,4-ethylenedioxythiophene) (PEDOT)-Based CEs for DSSCs

PEDOT is a famous polymer with high catalytic activity toward the I_3^- reduction with a conductivity in the range of 300–500 S cm^{-1}, which is much better than those of PANI (0.1–5 S cm^{-1}) and PPy (10–50 S cm^{-1}) [78]. The conductivity of PEDOT was remarkably enhanced to about 4600 S cm^{-1} by doping with poly(styrene sulfonate) (PSS) [79]. Therefore, PEDOT shows the potential ability to replace the expensive Pt-based materials. This is mainly attributed to the elevated conductivity and catalytic activity, but also to the high thermal and chemical stability, as well as extensive electrochemical reversibility [79]. Therefore, many research groups worldwide are performing more research studies on utilizing PEDOT or PEDOT: PSS as a CE for DSSCs. For instance, Pringle et al utilized the electrochemical

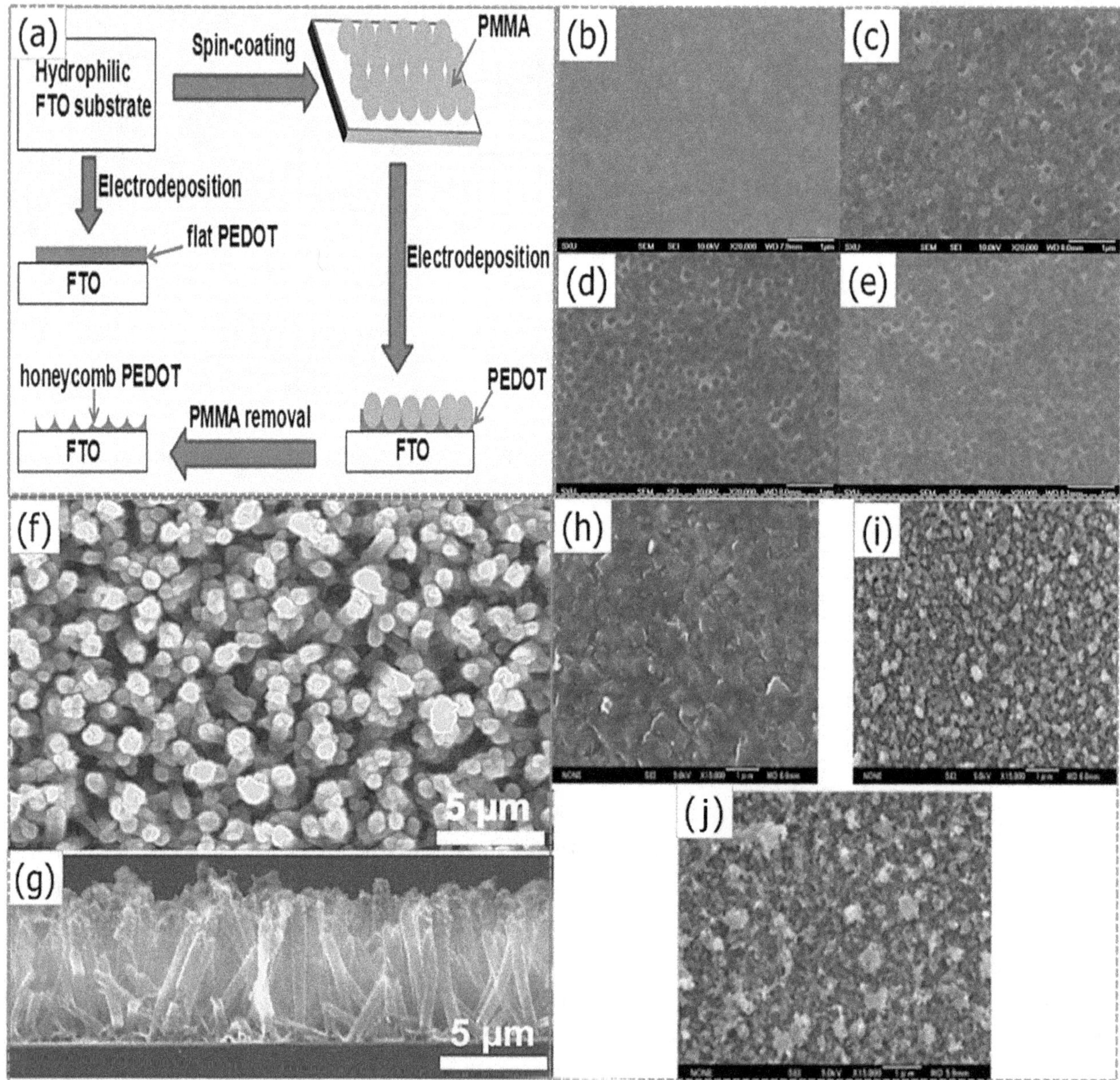

FIGURE 7.5 (a) Schematic diagram of the fabrication of the flat and honeycomb PEDOT CEs. SEM images of (b) flat-PEDOT (c) honey-like-PEDOT-1 (d) PEDOT-2 (e) h-PEDOT-3 CEs (f)SEM images of a PEDOT NT counter electrode (g) top and cross-section views (h) SEM images of electropolymerized PEDOT–ClO_4 (i) PEDOT–PSS (j) PEDOT–TsO. [81] with permission from Elsevier, 2017. [82] with permission from Willy, 2011. [83] with permission from the Royal Society of Chemistry, 2007

polymerization technique to fabricate PEDOT/plastic CEs for DSSCs [80]. The maximum efficiency of the assembled device was 8%. To regulate the morphology of the PEDOT films, more than one technique can be utilized as described by Li et al, where both electrodeposition and poly(methyl methacrylate) (PMMA) as a sacrificial template were combined to prepare honeycomb-like PEDOT CEs as shown in Figure 7.5a [81]. The as-prepared CEs showed good transparency in the visible range and significant catalytic performance comparable to the flat PEDOT CEs. Subsequently, the fabricated DSSCs with the honeycomb-like PEDOT CE exhibit an astonishing power conversion efficiency with PCE values of 9.12 and 5.75%, under the front and rear irradiations, respectively. Furthermore, Trevisan et al utilized electrodeposition and template techniques to fabricate PEDOT nanotubes, as shown in Figure 7.5f and g. The assembled device's PCE with PEDOT NT CE was 8.3%, which is quite similar to that obtained by a Pt-based device (8.5%) [82]. As the doping is an effective way for enhancing the catalytic activity of PEDOT, Xia et al reviewed the impacts of doping various ions on the morphology of PEDOT (Figure 7.5h–j) and corresponding device performances [83]. Electrochemical and photoelectrochemical measurements reveal that the interaction between polymers and FTO substrates directly influences the fill factor (FF) in *J*–*V* curves. Moreover, PEDOT showed high catalytic

TABLE 7.2
Photovoltaic Parameters of DSSC Devices Fabricated by PEDOT-Composite Counter-Electrodes

CEs	Electrolyte		V_{OC} (mV)	J_{SC} (mA cm^{-2})	PCE (%)	Reference
Ni-NPs/PEDOT: PSS	I_3^-/I^-		740	15.56	7.81	[84]
Nitrogen-doped graphene (NGr)/PEDOT	I_3^-/I^-		739	15.60	8.30	[85]
TiS_2/PEDOT: PSS	I_3^-/I^-		681	15.78	7.04	[86]
Electropolymerized PEDOT (Transparent)	$[Co(bpy)_3]^{2+/3+}$	Front	872	14.5	8.65	[31]
		Rear	876	11.9	7.48	
MoS_2/PEDOT (Transparent)	I_3^-/I^-	Front	743	13.73	7.00	[87]
		Rear	730	8.71	4.82	
ZnO-NC/PEDOT: PSS	I_3^-/I^-		740	15.20	8.05	[88]
	$[Co(bpy)_3]^{2+/3+}$		836	13.58	8.12	
ZIF-8/PEDOT: PSS	I_3^-/I^-		797	14.67	7.02	[36]
	$[Co(bpy)_3]^{2+/3+}$		856	8.52	5.03	
(Zn_3N_2, ZnO, ZnS, and ZnSe)/PEDOT: PSS	I_3^-/I^-		740–810	14.99–15.77	7.40–8.73	[89]
Si_3N_4, SiO_2, SiS_2, and $SiSe_2$/PEDOT: PSS	I_3^-/I^-		720–760	15.05–16.98	5.98–8.20	[90]

activity toward $[Co(bpy)_3]^{2+/3+}$ redox couples. For instance, Kang et al successfully prepared transparent PEDOT via an electrochemical deposition technique [31].

The data analysis showed that the fabricated DSSC devices display a remarkable total performance with a maximum PCE higher than those achieved by devices with Pt CE under irradiation from the front and rear sides. Combined PEDOT with nanomaterials is another promising technique due to the synergetic effect between the polymer matrices' high conductivity and the roughness of the nanomaterials [36]. Various materials showed remarkable catalytic performance, such as carbon-based materials, transition metals, and 2D materials. The total performance of these composite materials as CEs in DSSCs is listed in Table 7.2.

7.3 POLYMER-BASED ELECTROLYTES OF DSSCs

Electrolytes act as charge carrier transporters in DSSCs, which play a significant role in total performance. The electrolyte typically consists of a redox couple (e.g., I_3^-/I^-) dissolved in an organic solvent, as well as some additives [1, 91]. Generally, the electrolyte has a significant effect on all photovoltaic parameters [2, 47]. The redox couple transfer rate directly influences the short current density (J_{SC}) [2, 47]. Open-circuit voltage (V_{OC}) mainly alters the Fermi energy level of the photoanode and the redox potential of the redox couples [47]. At the electrolyte/electrode interfaces, the charge transfer impedance has a meaningful influence on the FF [2, 47]. The liquid electrolyte systems are widely utilized in DSSCs with the highest photoelectric conversion efficiency [5, 92, 93]. However, the low stability of the CE materials, the desorption of dye, the outflow of liquid electrolytes, and the evaporation of the organic solvent are potential problems that have impeded their commercial uses [30, 53]. Thus, both solid-state and quasi-solid-state electrolyte systems have been proposed as promising alternatives [94, 95]. However, the photovoltaic conversion efficiency of the solid-state DSSCs is still quite low compared to that achieved by liquid-based devices, which mainly results from the poor contact with the TiO_2 anode and the low charge transport. On the other hand, the quasi-solid-state electrolyte shows high ionic conductivity, high wettability, charge transfer, and prevents liquid leakage [96, 97]. Therefore the quasi-solid-state electrolyte is an excellent option for DSSCs [93, 98] and other applications such as fuel cells [99] and batteries [100]. The polymer-based electrolyte is often made by mixing the liquid electrolyte with organic polymers [101]. The polymer electrolyte can be defined as thermoplastic, thermosetting, or a composite polymer electrolyte depending on the physical state. In the following sections, a brief description of each kind will be given.

7.3.1 Thermoplastic Polymer Electrolytes

Generally, a polymer-based electrolyte involves a blended polymer with the organic solvent, inorganic salts, and the additives. The polymer is the framework to adsorb, gel, and interact with the electrolyte components. In this system, the polymer is called a gelator, and the solvent is a plasticizer [93]. By incorporation of the polymer into the electrolyte, the system regularly adapts to a gel state, where the solvent weakens the interaction between polymer chains by creating hydrogen bonds, weak van der Waals forces, and an electrostatic interface [93]. The physical crosslinking mainly varies with the temperature; therefore, this electrolyte is called a thermoplastic polymer electrolyte (TPPE). Up to the present, various kinds of polymers have been utilized as TPPEs, such as poly(acrylonitrile) (PAN), poly(ethylene oxide) (PEO), poly(ethylene glycol) (PEG), PVP, poly(vinylidene ester) (PVE), poly(vinyl chloride)

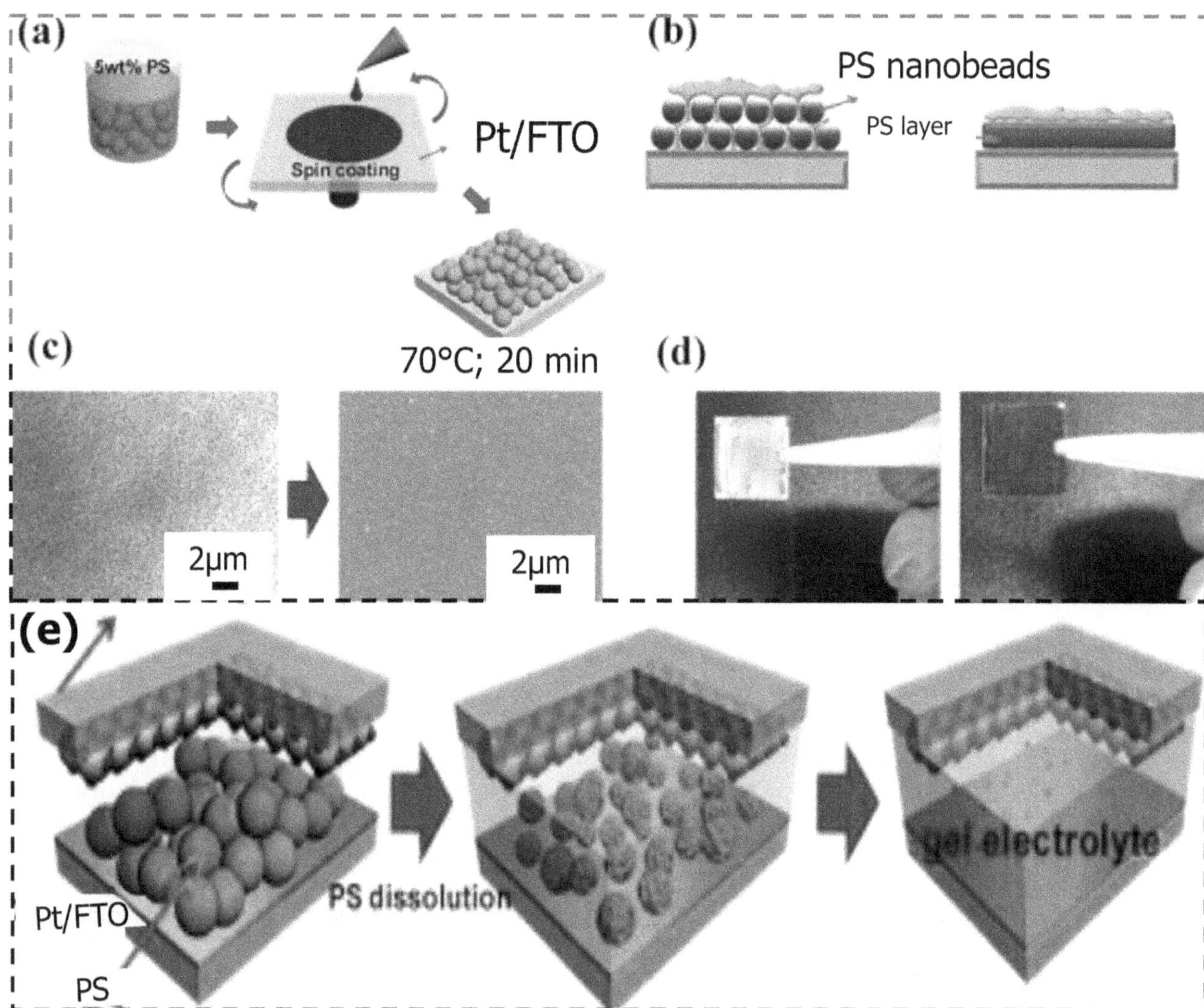

FIGURE 7.6 (a) Scheme for PS nanobead deposition on Pt CE (b) scheme for pore-filling before (left) and after (right) dissolving the PS nanobeads (c and d) SEM and photographic images of the PS nanobeads on the Pt CE before (left) and after (right) the dissolution process (e) scheme of transfer of the liquid electrolyte to the gel form. [105] with permission of the American Chemical Society

(PVC), and poly(vinylidene fluoride) (PVDF). Cao and co-workers in 1995 firstly reported utilizing solid-state DSSCs by blending PAN with a liquid electrolyte based on an I_2/NaI redox couple [102]. The fabricated devices showed a significant total conversion efficiency with a maximum PCE of 5% with remarkable stability. Later Wu et al fabricated a polymer-based electrolyte by mixing poly(acrylonitrile-*co*-styrene) with a mixed organic solvent (ethylene carbonate and propylene carbonate) and *N*-methyl pyridine iodide and I_2 [103]. The PCE of the assembled device under full irradiation was 3.10%. Wu et al. prepared a polymer-based electrolyte from PC/PEG/KI/I_2 components [104]. The related assembled devices showed a total performance with a PCE of 7.22%, which is very close to that achieved by a liquid electrolyte (PCE = 7.60%). Although gel-based devices have a relatively high efficiency, the high viscosity is a challenge that wastes most of the prepared materials. To overcome this matter, Lee et al. injected a liquid electrolyte into the DSSCs fabricated by Pt CE and covered with a layer of PS as shown in Figure 7.6 [105]. After injecting the liquid, the PS was dissolved and became viscous, and the device achieved a maximum efficiency of 7.54%, with long-term stability. Naturally extracted dyes combined with a polymer gel electrolyte in DSSCs were recently reported by Prima et al. [97]. In this study, the anthocyanin of cyanidin-3-glucoside extracted from Indonesian black rice equally combined with a ruthenium photosensitizer in quasi-solid-state DSSCs based on PEO with an I_3^-/I^- electrolyte; the total conversion efficiency of the device was 3.51%.

PVDF and its copolymer poly(vinylidenefluoride-*co*-hexafluoropropylene) (PVDF-HFP) show a promising ability for preparing polymer electrolytes. For instance, Noor et al. fabricated a 40 wt.%PVDF-HFP-10 wt.%KI-50 wt.% (EC + PC) + 10 wt.%I_2 gel electrolyte. The estimated conductivity at room temperature was 1.10×10^{-3} S cm^{-1} [106]. The DSSCs with several dyes such as Ruthenizer 535 (N3), anthocyanin, chlorophyll, and a mixture of anthocyanin and

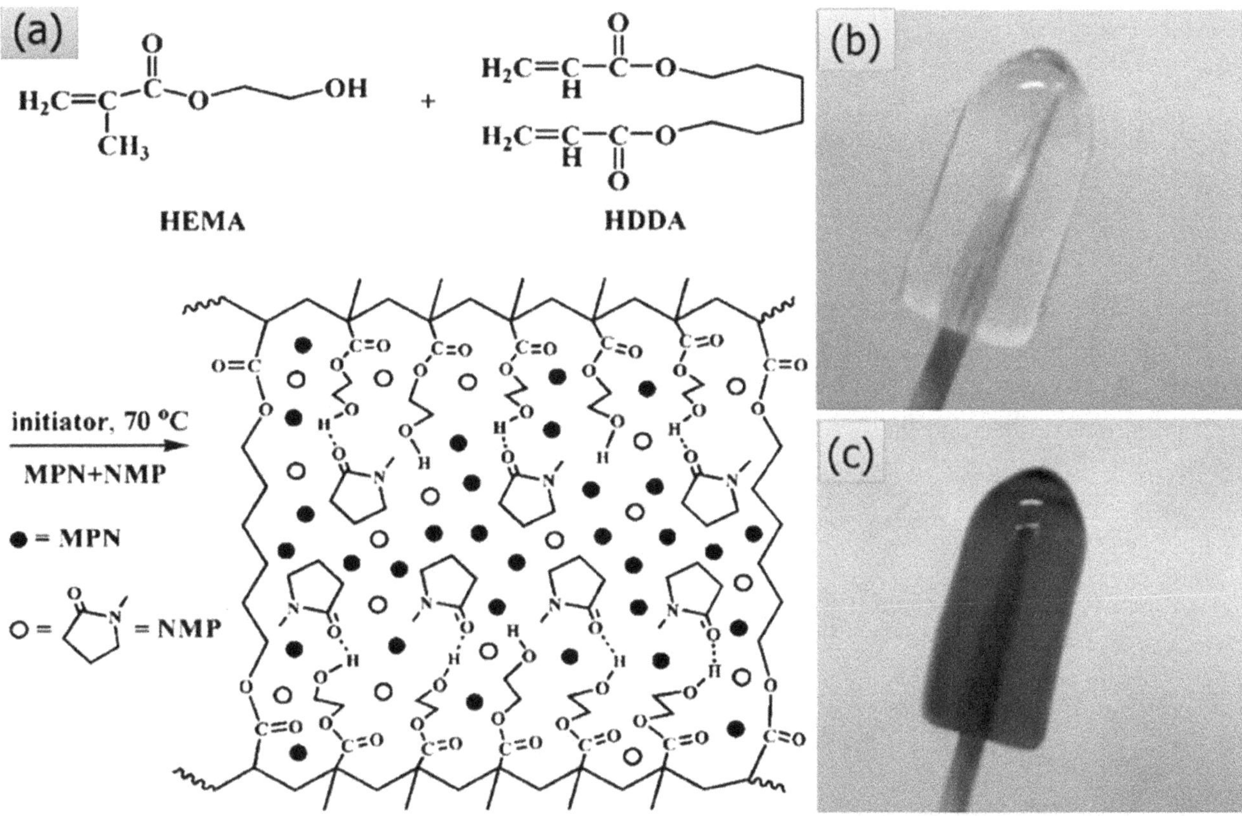

FIGURE 7.7 (a) Synthesis of the crosslinked PHEMA-based organo gel (b) the formed organo gel before soaking in liquid electrolyte (c) the formed organo gel after soaking in liquid electrolyte. [110] with permission from the Royal Society of Chemistry, 2011

chlorophyll (v/v = 1) have been prepared. The device with Ruthenizer 535 (N3) dye produced the greatest performance among them all. Further study has been reported by Wang et al., where the polymer electrolyte was formed by blending 5 wt.% PVDF-HFP with an MPN-based liquid electrolyte. The assembled device achieved a total conversion efficiency of more than 6% with excellent stability for a long time at 80 °C. Recently, Li et al successfully prepared quasi-solid-state cobalt-complex electrolytes for indoor DSSCs using PVDF-HFP and PMMA polymers [107]. By regulating the PVDF-HFP/PMMA ratio and the electrolyte composition, the resultant quasi-solid-state cell showed a PCE value of 25% under 1000-lux fluorescent lighting and possessed long-term stability for 2000 h.

7.3.2 Thermosetting Polymer Electrolytes

A thermosetting polymer electrolyte (TSPE) is prepared by crosslinking the organic molecules in the mixed solution via chemical or covalent interactions [93]. The status of the polymer cannot change with temperature, thus it is called a TSPE. Typically, a TSPE can be obtained by *in situ* light, *in situ* heat polymerization, or liquid electrolyte adsorption [93]. Matsumoto et al utilized *in situ* photopolymerization to polymerize α-methacryloyl-ω-methoxyocta (oxyethylene) in porous TiO_2 film and an absorbing liquid electrolyte [108]. The assembled device with an iodide redox couple showed a PCE of 2.62%. Parvez et al. firstly inserted both PEG and PEGDA [poly(ethylene glycol) diacrylate] monomers in the TiO_2 film, followed by UV light treatment [109]. Both PEG and PEGDA formed a crosslinked gel structure. The assembled DSSCs achieved a PCE of 4.18%. Yu et al prepared poly(*b*-hydroxyethyl methacrylate), a PHEMA-based organogel electrolyte as shown in Figure 7.7 [110]. The QS-DSSCs with a PHEMA gel electrolyte offered high energy conversion efficiency (PCE = 7.5%) with good long-term stability after 1000 h. Xu et al utilized a free radical polymerization method to prepare a PMMA-based electrolyte with a different viscosity [111]. In this work, the content of MMA was regulated in acetone with a 50, 40, 30, and 20 volume percentage. The obtained results showed that PMMA (40 vol%) and PMMA (50 vol%) have a higher viscosity, while PMMA (20 vol%) has a low viscosity, and a PMMA (30 vol%) gel sample showed moderate liquidity and dimensional stability. Moreover, the ionic conductivity decreased with the increase of the PMMA viscosity and concentration. The related assembled QS-DSSCs showed outstanding stability with a high conversion efficiency.

7.3.3 Composite Polymer Electrolytes for DSSCs

Usually, the composite polymer-based electrolytes are made by introducing inorganic nanoparticles into the polymer-based electrolyte to enlarge the electrolyte's amorphous

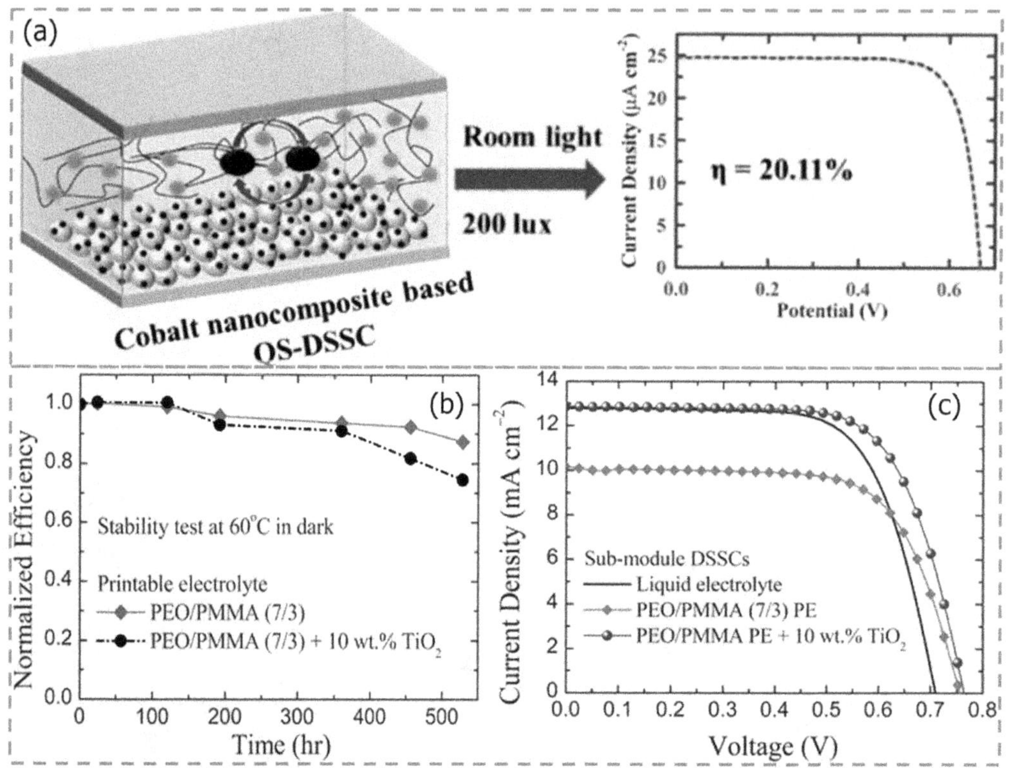

FIGURE 7.8 (a) Graphical abstract of the QS-DSSCs using PVDF-HFP PGEs with ZnO nanofillers (b) Stability of the QS-DSSCs (c) *J–V* curves of the liquid cell and QS-DSSCs using a PEO/PMMA (7/3) polymer electrolyte. [94] with permission from the American Chemical Society, 2019. [112] with permission from the Royal Society of Chemistry, 2018

phase [112]. Introducing inorganic fillers can enrich the conductivity and long-term stability of the assembled devices [94]. Venkatesan et al blended 7 wt.% PVDF-HFP + 4 wt.% ZnO to the $[Co(bpy)_3]^{2+/3+}$ liquid via a simple mechanical mixing technique [94]. The assembled device achieved a power conversion efficiency of 20.11% under 200 lx irradiation with an excellent stability for a long time at 35 °C (Figure 7.8a). Recently Liu et al designed a DSSC device with a double layer of polymer electrolyte PEO/PVDF in the ratio of 8/2 [113]. The results show that the DSSCs with a 9 wt.% PEO/PVDF double-layered electrolyte exhibits an efficiency of 7.92%. Further enhancement to 7.97% was achieved by adding 0.6 wt.% ZnO nanoparticles. Venkatesan et al blended PEO and PMMA in the ratio of 7/3 with the iodine liquid electrolyte [112]. Although the device assembled with a liquid-based electrolyte showed conversion efficiency with a PCE value of 8.32%, the value achieved with a polymer electrolyte system was higher (8.48%). Moreover, when inserting a small portion of TiO_2 nanoparticles (10 wt.%) into the polymer electrolyte, the PCE increased to 9.12%, with excellent stability (Figure 7.8b). This is mainly attributed to increasing the ionic conductivity of the electrolyte. Further enhancements were achieved by Chen et al, in which poly(acrylonitrile-*co*-vinyl acetate) (PAN-VA) and TiO_2 fillers were utilized as a gelation agent to prepare gel-state electrolytes for DSSC applications [101]. The related assembled DSSCs achieved a conversion efficiency of 10.58%, which is higher than that based on liquid electrolyte (10.23%). This is mainly assigned to the *in situ* gelation of the gel electrolyte. In this electrolyte system, the PAN-VA is the charge transfer, and TiO_2 fillers play a significant role in enhancing the charge transfer at the Pt/electrolyte interface. Furthermore, Liu et al made a printable electrolyte system based on an I_3^- / I^- redox couple with a 9 wt.% PEO/PVDF (8/2) + 4 wt.% TiO_2 composition [114]. The corresponding device showed a conversion efficiency of 8.32%, close to that achieved by the liquid cell. Carbon nanomaterials can be utilized as nanofillers to upgrade the total performance of DSSCs. Recently Sakali et al added small portions of carbon nanotube (CNT) to the PAN-based electrolyte to evaluate the effect of CNT on the total performance of fabricated DSSC devices [115]. The maximum ionic conductivity (4.45 mS cm^{-1}) was achieved by incorporating a small portion of CNT (11 wt.%); and the maximum PCE value achieved was 8.87%.

7.4 APPLICATION OF POLYMERS IN PEROVSKITE SOLAR CELLS

Generally, the perovskite is a compound with a structure of ABX_3, where A is a cation such as formamidinium (FA^+), methylammonium (MA^+), or cesium (Cs^+), while B is a metal cation such as Pb^{2+} or Sn^{2+}, and X is a halogen ion such as F^-, Cl^-, or Br^- [116, 117]. The organic–inorganic hybrid PSC is considered a mark in the photovoltaics due to its superior performance [117]. The typical structure of a

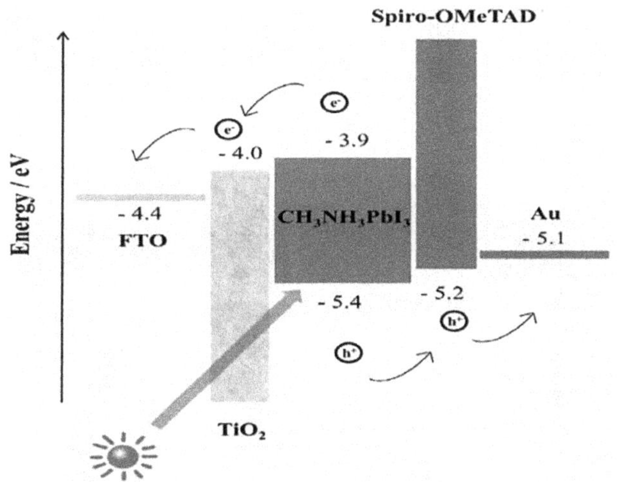

FIGURE 7.9 Energy diagram of a PSC device with the energy levels of materials used in the various layers. [118] adapted with permission from Elsevier, 2018

PSC is very similar to DSC and composed of six layers: the FTO glass substrate, a compact semiconductor layer, a mesoporous semiconductor film, a perovskite absorbing layer, a hole transfer material (HTM), and an electrically conductive back contact. The working function of a PSC is shown in Figure 7.9 [118]. Under illumination, the perovskite layer produces electrons and holes, the excited electrons are inserted into the CB of the perovskite layer, and the holes are left to the valence band (VB). Then the electrons transfer to the FTO, flow through the external circuit, in parallel with the holes in the perovskite/HTL interface, then transferred to the cathode layer through the HTL. Therefore, all components, the crystal morphology of the perovskite layer, as well as the interfacial properties all influence the total performance and stability of the assembled device.

7.4.1 Polymers for Regulating the Morphology of the Perovskite Layer

The rapidly increasing PSC efficiency is mainly due to the encouraging properties of the organic–inorganic hybrid perovskite, such as its elevated absorption facility, direct and tunable bandgap, and extreme carrier movement with long lifetimes. The morphology of the perovskite layer can significantly influence the light absorption and carrier transportation dynamics, thus playing a key role in the total performance of the assembled device. However, the rapid reaction between precursors accelerates the carrier recombination, which is the core cause for harming the total performance. Thus, much intensive effort to diminish the deserts and pinholes and obtain the perovskite compressed layer has been made. The additives of fullerene [119], metal halide salts [120], nanoparticles [121], and polymers present themselves as efficient for controlling crystal growing [122]. Based on its ionic nature, perovskite can chemically interact many times with various functional groups. Moreover, the presence of the empty orbitals in Pb ions results in a high ability to coordinate with molecules containing heteroatoms; the H atoms in MA^+ or FA^+ can create hydrogen bonds with a small radius and electronegativity atoms (F, O, and N). It is well known that polymers are organic macromolecules with various functional groups and heteroatoms, which can be utilized as additives in a PSC. This is mainly utilized to boost the contact between granules in perovskite films, improving the device stability and reducing the contact angle resulting from high solubility in polar solvents [123, 124]. Recently Zheng et al. utilized PANI as an additive due to its promising electrical and optical properties [124]. The data analysis showed enhancement in the perovskite layer properties in terms of structure, light absorption, and surface morphology, leading to faster charge transfer. Comparable to the pristine device with a PCE of 16.96%, the device with PANI exhibited a higher value (19.09%) with lower hysteresis behavior, as shown in Figure 7.10f. Moreover, the PANI-based devices were exposed to high humidity and heat resistance and a bigger water contact angle (Figure 7.10e). A further study by Wu et al showed that the device fabricated with methylammonium (MA)/formamidinium (FA) mixed-cation perovskite as the active layer and a partially substituted Pb^{2+} with Ba^{2+} device showed the best performance [123]. Moreover, the device with a 10.0 mol% Ba-doped MA/FA mixed-cation showed an efficiency of 16.1%.

7.4.2 Polymers as Hole Transport Layers

As described in the PSC work function, the time needed to inject electrons into the CB of the perovskite layer to the CB of the TiO_2 film is about 0.2 picoseconds (ps), while the time to inject the holes from the VB of the perovskite layer into the HTL is about 0.75 ps. This means a higher chance of hole recombination at the perovskite/HTL interface, thus it is urgently necessary to develop an efficient hole transfer material to enhance the electron–hole separation. Currently, three kinds of materials, including (a) organic molecules, (b) inorganic materials, and (c) polymeric-based materials are usually utilized as HTL. For the first kind, Spiro-OMeTAD HTL is widely used; however, the total performance of the assembled devices are quite poor due to the low hole-mobility, while in the second category there are various materials including NiMgLiO, CuSCN, and CuI which have been widely investigated and mainly attributed to their being simply prepared with high hole mobility and excellent stability. However, the solvent used for the deposition of inorganic HTLs can destroy the perovskite film structure, and thus damage the device performance. Despite the limitations of the first two categories, the polymer HTLs possess high stability, which is important for PSCs. Triarylamine (TAA)-based polymers, conductive polymers, and poly-3-hexylthiophene (P3HT) are widely used as HTLs. Recently *Kim* et al successfully prepared methoxy-PTAA as HTLs in PSC [125]. The PCE of the device with CH_3O-PTAA was over 20%, with thermal stability under darkness at 85 °C, and 85% relative humidity (RH) for more than 1000 h with only a 3% reduction presented.

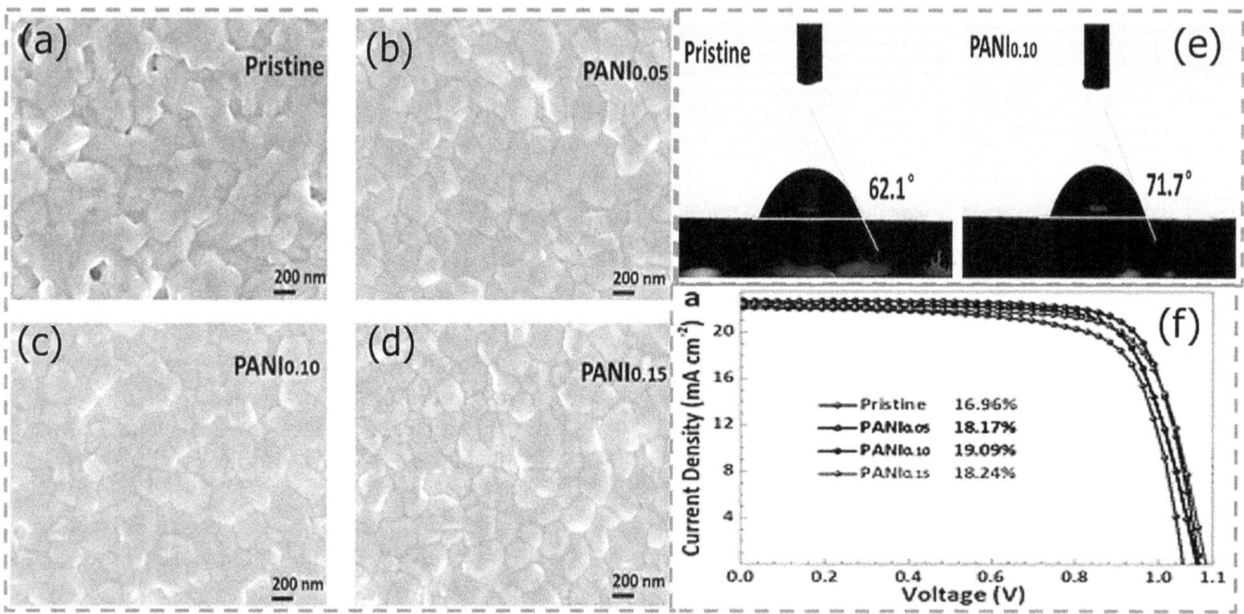

FIGURE 7.10 SEM images of the (a) pristine and (b–d) perovskite films with different amounts of PANI (e) the water contact angles of pristine and PANI0.10 perovskite films (f) *J–V* curves of the perovskite solar cells with various amounts of PANI. [124] with permission of the Royal Society of Chemistry, 2019

A famous group of organic materials called conductive polymers are usually utilized as the hole transfer materials in PSCs, due to their outstanding properties such as easy fabrication process, low price, and semiconducting nature (p-type). Kanwat et al [126] reported the influence of ethylene glycol (EG) on the conductivity of both PEDOT: PSS and PEDOT: PSS: WOx films. The EG treatments not only remarkably increased the conductivity of PEDOT: PSS: WOx thin film, but also exhibited high transmittance (< 95%) in the visible range with high thermal stability. The PCE of the device with PEDOT: PSS: WOx was 12.69%. Further enhancement by Wang et al. [127] successfully treated GO and PEDOT: PSS with ammonia. The PCE of the devices assembled with ammonia-treated GO was over 14%, while the PCE of the devices with ammonia treated PEDOT: PSS was 13.38%. Recently, Reza et al. [128] treated the PEDOT: PSS layer with the solvent-engineered approach to remove the predominant PSS in PEDOT: PSS. The treated PEDOT: PSS-based devices showed greater charge carrier lifetime, reduced charge transport time, and poorer transfer impedance. Moreover, the assembled PSCs displayed high stability with a PCE of 18.18%, which is much higher than that achieved by devices with the untreated PEDOT: PSS. Furthermore, PANI has been utilized for HTLs; for instance Mabrouk et al. prepared a new composite HTL by doping PEDOT: PSS with PANI and GO dopants [129]. Adding GO to a PEDOT: PSS/PANI composite significantly increased the V_{OC} to 1.05 V for the related PSC device. The PANI doped films showed superior electrical conductivity compared with those without PANI. Thus, the PSC device showed a higher efficiency of 18.12%, demonstrating better stability than the pristine PEDOT: PSS cell.

7.4.3 Polymers as Electron Transport Layers

According to the PSC work function, the ETL gathers electrons from the perovskite layer to the external circuit. Hence, the ETL material should display high electron mobility with an energy level well-matched with perovskite. Typically, TiO_2 and ZnO are commonly used materials. Mainly due to their low price, tunable bandgap, and high electron mobility, polymers are widely used as electron transfer materials. Chen et al. [130] utilized two n-type polymers poly(2,2′-bithiophene-3,3′-dicarboxyimide) (PBTI) and 2,2′-bithiazolothienyl-4,4′,10,10′-tetracarboxydiimide (PDTzTI) as ETL in inverted PSCs. The devices fabricated with PDTzTI ETL demonstrated the best power conversion efficiency of 20.86% due to their high electron mobility and well-matched energy level alignment with the passivation of interfacial traps/defects. Owning to the highly hydrophobic properties and the mobile ion blocking capability of polymers, the PDTzTI ETL-based device also exhibited excellent long-term stability. Further study has been done by Yuang et al. [131] using mixed n-type polymeric materials with phenyl C61 butyric acid methyl ester (PCBM) to form a homogeneous bulk-mixed (HBM) continuous film with high electron mobility and suitable energy level. HBM film significantly enhances the electron extraction and reduces the critical electron capture radius, resulting in reduced nonradiative recombination at the perovskite/HBM interface. The PCE of the PSC device is over 20.6%. Moreover, the device exhibits long-term stability under ambient air conditions, with just a 20% loss of efficiency after 45 days.

7.4.4 Polymers as the Interlayer

As the PSCs are composed of multilayer structures (the inverted p-i-n, the FTO/HTL, the HTL/perovskite, and the perovskite/ETL, ETL/Au), the charge separation and extraction occur at the layer interfaces. Therefore, interface regulation is a key factor for improving the total performance of the device. The polymer is one of the promising materials for overcoming these limitations. At the HTL/perovskite interface, Malinkiewicz et al introduced a poly(*N*,*N*′-bis(4-butylphenyl)-*N*,*N*′-bis(phenyl)benzidine) (polyTPD) layer, while Wen et al. [132] utilized PS in the perovskite films in order to form a tunneling junction at the perovskite/HTL interface. The PCE of the assembled device enhanced from 15.90 to 17.80%. Furthermore, Chaudhary et al. [133] modified the surface of perovskite film with a PVP solution to passivate the under-coordinated lead atoms via the pyridine Lewis base side chains and simultaneously act as an electron barrier at the perovskite/HTL to reduce charge recombination. The PCE increased by 3% to 5% more than the untreated. To control the inserted polymer layer, Byranvand et al. [134] reported a chemical vapor deposition (CVD) polymerization of poly(*p*-xylylene) (PPX) layers for the surface passivation of perovskite films. The obtained PCE of the device enhanced to 20.4% compared to 19.4% for the non-passivated one. At the ETL/metal interface, Zhang et al. [135] inserted a polyelectrolyte interlayer of polyethyleneimine (PEIE) and poly[3-(6-trimethylammoniumhexyl) thiophene] (P3TMAHT). The PCE increased significantly from 8.53 to 12.01% (PEIE) and 11.28% (P3TMAHT). Using the polymers at the different PSC interfaces is an effective technique for improving the total performance, according to the literature.

7.5 SUMMARY AND FUTURE PERSPECTIVES

The conjugated polymers display a promising potential in third-generation photovoltaic devices. In DSSCs, the polymers can be incorporated in all components such as substrates and as a pore-forming agent in mesoporous photoanode preparation. Also, the conductive polymers and their corresponding composites can be utilized as Pt-free CEs due to their high catalytic performance toward the commonly used redox couples. Various kinds of polymers have been utilized to form quasi-solid-state electrolytes to overcome the leakage of liquid electrolytes. Although the DSSCs with a polymer-based electrolyte have long-term stability, the low conductivity of these electrolytes is still a challenge that needs to be dealt with. In PSCs, the polymers are usually utilized to adjust the crystallization of perovskite films and boost device stability. Although the polymers successfully work as hole transfer materials due to their excessive hole mobility, designing polymers with a higher hole mobility is a big challenge. Moreover, the polymers show they significantly enhance the PSC devices' long-term stability, which opens the way for their practical application.

REFERENCES

1. Gong, J., Sumathy, K., Qiao, Q., and Zhou, Z. 2017. Review on dye-sensitized solar cells (DSSCs): advanced techniques and research trends. *Renewable and Sustainable Energy Reviews* 68: 234–246.
2. Anders Hagfeldt, G.B., Sun, Licheng, Kloo, Lars, and Pettersson, Henrik 2010. Dye-Sensitized Solar Cells. *Chemical Reviews* 110: 6595–6663.
3. Grätzel, M. 2001. Photoelectrochemical cells. *Nature* 414: 338–344.
4. Chapin, D.M., Fuller, C.S., and Pearson, G.L. 1954. A new silicon p-n junction photocell for converting solar radiation into electrical power. *Journal of Applied Physics* 25(5): 676–677.
5. O'Regan, B. and Gratzel, M. 1991. A low-cost, high-effeciency solar cell based on dye-sensitized colloidal TiO_2 films. *Nature* 353: 737–740
6. Yang, S., Fu, W., Zhang, Z., Chen, H., and Li, C.-Z. 2017. Recent advances in perovskite solar cells: efficiency, stability and lead-free perovskite. *Journal of Materials Chemistry A* 5(23): 11462–11482.
7. Gao, J., Yang, Y., Zhang, Z., Yan, J., Lin, Z., and Guo, X. 2016. Bifacial quasi-solid-state dye-sensitized solar cells with poly(vinyl pyrrolidone)/polyaniline transparent counter electrode. *Nano Energy* 26: 123–130.
8. Futscher, M.H. and Ehrler, B. 2017. Modeling the performance limitations and prospects of perovskite/Si tandem solar cells under realistic operating conditions. *ACS Energy Lett* 2(9): 2089–2095.
9. Pazos-Outon, L.M., Xiao, T.P., and Yablonovitch, E. 2018. Fundamental efficiency limit of lead iodide perovskite solar cells. *The Journal of Physical Chemistry Letters* 9(7): 1703–1711.
10. Zohar, A., Kulbak, M., Levine, I., Hodes, G., Kahn, A., and Cahen, D. 2018. What limits the open-circuit voltage of bromide perovskite-based solar cells? *ACS Energy Letters* 4(1): 1–7.
11. Tajik, S., Beitollahi, H., Nejad, F.G., Shoaie, I.S., Khalilzadeh, M.A., Asl, M.S., Van Le, Q., Zhang, K., Jang, H.W., and Shokouhimehr, M. 2020. Recent developments in conducting polymers: applications for electrochemistry. *RSC Advances* 10(62): 37834–37856.
12. Kondratenko, M.S., Elmanovich, I.V., and Gallyamov, M.O. 2017. Polymer materials for electrochemical applications: processing in supercritical fluids. *The Journal of Supercritical Fluids* 127: 229–246.
13. Zuin, V.G., Budarin, V.L., De Bruyn, M., Shuttleworth, P.S., Hunt, A.J., Pluciennik, C., Borisova, A., Dodson, J., Parker, H.L., and Clark, J.H. 2017. Polysaccharide-derived mesoporous materials (Starbon(R)) for sustainable separation of complex mixtures. *Faraday Discuss* 202: 451–464.
14. Kierys, A., Zaleski, R., Grochowicz, M., Gorgol, M., and Sienkiewicz, A. 2020. Polymer–mesoporous silica composites for drug release systems. *Microporous and Mesoporous Materials* 294: 109881.
15. Sang, X., Peng, L., Zhang, J., Han, B., Xue, Z., Liu, C., and Yang, G. 2014. Template-free synthesis of mesoporous polymers. *Chem Commun (Camb)* 50(60): 8128–8130.
16. Alemu, D., Wei, H.-Y., Ho, K.-C., and Chu, C.-W. 2012. Highly conductive PEDOT: PSS electrode by simple film treatment with methanol for ITO-free polymer solar cells. *Energy & Environmental Science* 5(11): 9662.

17. Gentian Yue, X.M., Jiang, Qiwe, Tan, Furui, Jihuai, Wu, Chen, Chong, Li, Fumin, Li, Qinghua 2014. PEDOT: PSS and glucose assisted preparation of molybdenum disulfide/ single-wall carbon nanotubes counter electrode and served in dye-sensitized solar cells. *Electrochimica Acta* 142: 68–75.
18. Wang, L., Zhang, F., Liu, T., Zhang, W., Li, Y., Cai, B., He, L., Guo, Y., Yang, X., Xu, B., Gardner, J.M., Kloo, L., and Sun, L. 2021. A crosslinked polymer as dopant-free hole-transport material for efficient n-i-p type perovskite solar cells. *Journal of Energy Chemistry* 55: 211–218.
19. Lim, J., Kong, S.Y., and Yun, Y.J. 2018. Hole transport behaviour of various polymers and their application to perovskite-sensitized solid-state solar cells. *Journal of Nanomaterials* 2018: 1–6.
20. You, G., Zhuang, Q., Wang, L., Lin, X., Zou, D., Lin, Z., Zhen, H., Zhuang, W., and Ling, Q. 2019. Dopant-free, donor–acceptor-type polymeric hole-transporting materials for the perovskite solar cells with power conversion efficiencies over 20%. *Advanced Energy Materials* 10(5): 1903146.
21. Limongi, T., Tirinato, L., Pagliari, F., Giugni, A., Allione, M., Perozziello, G., Candeloro, P., and Di Fabrizio, E. 2017. Fabrication and applications of micro/nanostructured devices for tissue engineering. *Nanomicro Letters* 9(1): 1.
22. Zhang, F., Xia, Y., Liu, Y., and Leng, J. 2020. Nano/micro-structures of shape memory polymers: from materials to applications. *Nanoscale Horiz* 5(8): 1155–1173.
23. Wu, Z.L., Qi, Y.N., Yin, X.J., Yang, X., Chen, C.M., Yu, J.Y., Yu, J.C., Lin, Y.M., Hui, F., Liu, P.L., Liang, Y.X., Zhang, Y., and Zhao, M.S. 2019. Polymer-based device fabrication and applications using direct laser writing technology. *Polymers (Basel)* 11(3): 553.
24. Gerischer, H., Michel-Beyerle, M.E., Rebentrost, F., and Tributsch, H. 1968. Sensitization of charge injection into semiconductors with large band gap. *Electrochimica Acta* 13(6): 1509–1515.
25. Tsubomura, H., Matsumura, M., Nomura, Y., and Amamiya, T. 1976. Dye sensitised zinc oxide: aqueous electrolyte: platinum photocell. *Nature* 261(5559): 402–403.
26. Kalyanasundaram, K., Vlachopoulos, N., Krishnan, V., Monnier, A., and Graetzel, M. 1987. Sensitization of titanium dioxide in the visible light region using zinc porphyrins. *The Journal of Physical Chemistry* 91(9): 2342–2347.
27. Roy-Mayhew, Joseph D., and Aksay, Ilhan A. 2014. Graphene materials and their use in dye-sensitized solar cells. *Chemical Reviews* 114: 6323–6348.
28. Richhariya, G., Kumar, A., Tekasakul, P., and Gupta, B. 2017. Natural dyes for dye sensitized solar cell: a review. *Renewable and Sustainable Energy Reviews* 69: 705–718
29. Thomas, S., Deepak, T.G., Anjusree, G.S., Arun, T.A., Nair, S.V., and Nair, A.S. 2014. A review on counter electrode materials in dye-sensitized solar cells. *Journal of Materials Chemistry A* 2(13): 4474–4490.
30. Sharma, K., Sharma, V., and Sharma, S.S. 2018. Dye-sensitized solar cells: fundamentals and current status. *Nanoscale Res Lett* 13(1): 381.
31. Kang, J.S., Kim, J., Kim, J.Y., Lee, M.J., Kang, J., Son, Y.J., Jeong, J., Park, S.H., Ko, M.J., and Sung, Y.E. 2018. Highly efficient bifacial dye-sensitized solar cells employing polymeric counter electrodes. *ACS Appl Mater Interfaces* 10(10): 8611–8620
32. Chen, L.L., Liu, J., Zhang, J.B., Zhou, X.W., Zhang, X.L., and Lin, Y. 2010. Low temperature fabrication of flexible carbon counter electrode on ITO-PEN for dye-sensitized solar cells. *Chinese Chemical Letters* 21(9): 1137–1140.
33. Zardetto, V., Brown, T.M., Reale, A., and Di Carlo, A. 2011. Substrates for flexible electronics: a practical investigation on the electrical, film flexibility, optical, temperature, and solvent resistance properties. *Journal of Polymer Science Part B: Polymer Physics* 49(9): 638–648.
34. Subbiah, V., Landi, G., Wu, J.J., and Anandan, S. 2019. MoS_2 coated CoS_2 nanocomposites as counter electrodes in Pt-free dye-sensitized solar cells. *Physical Chemistry Chemical Physics* 21(45): 25474–25483.
35. Theerthagiri, J., Senthil, A.R., Madhavan, J., and Maiyalagan, T. 2015. Recent progress in non-platinum counter electrode materials for dye-sensitized solar cells. *ChemElectroChem* 2(7): 928–945.
36. Ahmed, A.S.A., Xiang, W., Saana Amiinu, I., and Zhao, X. 2018. Zeolitic-imidazolate-framework (ZIF-8)/PEDOT:PSS composite counter electrode for low cost and efficient dye-sensitized solar cells. *New Journal of Chemistry* 42: 17303–17310.
37. Zhou, Z., Sigdel, S., Gong, J., Vaagensmith, B., Elbohy, H., Yang, H., Krishnan, S., Wu, X.-F., and Qiao, Q. 2016. Graphene-beaded carbon nanofibers with incorporated Ni nanoparticles as efficient counter-electrode for dye-sensitized solar cells. *Nano Energy* 22: 558–563.
38. Yun, S., Freitas, J.N., Nogueira, A.F., Wang, Y., Ahmad, S., and Wang, Z.-S. 2016. Dye-sensitized solar cells employing polymers. *Progress in Polymer Science* 59: 1–40.
39. Son, Y.J., Kang, J.S., Yoon, J., Kim, J., Jeong, J., Kang, J., Lee, M.J., Park, H.S., and Sung, Y.-E. 2018. Influence of TiO_2 particle size on dye-sensitized solar cells employing an organic sensitizer and a cobalt(III/II) redox electrolyte. *The Journal of Physical Chemistry C* 122(13): 7051–7060.
40. Sun, K.C., Qadir, M.B., and Jeong, S.H. 2014. Hydrothermal synthesis of TiO_2 nanotubes and their application as an over-layer for dye-sensitized solar cells. *RSC Advances* 4(44): 23223.
41. Jalali, M., Siavash Moakhar, R., Kushwaha, A., Goh, G.K.L., Riahi-Noori, N., and Sadrnezhaad, S.K. 2015. Enhanced dye loading-light harvesting TiO_2 photoanode with screen printed nanorod-nanoparticles assembly for highly efficient solar cell. *Electrochimica Acta* 169: 395–401.
42. Pandikumar, A., Lim, S.-P., Jayabal, S., Huang, N.M., Lim, H.N., and Ramaraj, R. 2016. Titania@gold plasmonic nanoarchitectures: an ideal photoanode for dye-sensitized solar cells. *Renewable and Sustainable Energy Reviews* 60: 408–420.
43. Jung, H.-G., Kang, Y.S., and Sun, Y.-K. 2010. Anatase TiO_2 spheres with high surface area and mesoporous structure via a hydrothermal process for dye-sensitized solar cells. *Electrochimica Acta* 55(15): 4637–4641.
44. Mali, S.S., Betty, C.A., Bhosale, P.N., Shinde, P.S., Pramod, P.S., Jadkar, S.R., and Patil, P.S. 2012. Efficient dye-sensitized solar cells based on hierarchical rutile TiO_2 microspheres. *CrystEngComm* 14(23): 8156.
45. Arla, S.K., Sana, S.S., Badineni, V., and Boya, V.K.N. 2020. Effect of nature of polymer as a binder in making photo-anode layer and its influence on the efficiency of dye-sensitized solar cells. *Bulletin of Materials Science* 43: 210.
46. Charbonneau, C., Tanner, T., Davies, M.L., Watson, T.M., and Worsley, D.A. 2016. Effect of TiO_2 photoanode porosity on dye diffusion kinetics and performance of standard dye-sensitized solar cells. *Journal of Nanomaterials* 2016: 1–10.
47. Benkstein, K.D., Kopidakis, N., van de Lagemaat, J., and Frank, A.J. 2003. Influence of the percolation network geometry on electron transport in dye-sensitized titanium dioxide solar cells. *The Journal of Physical Chemistry, B* 107: 7759–7767.

48. Elayappan, V., Panneerselvam, P., Nemala, S., Nallathambi, K.S., and Angaiah, S. 2015. Influence of PVP template on the formation of porous TiO_2 nanofibers by electrospinning technique for dye-sensitized solar cell. *Applied Physics A* 120(3): 1211–1218.
49. Hou, W., Xiao, Y., Han, G., Zhou, H., Chang, Y., and Zhang, Y. 2016. Preparation of mesoporous titanium dioxide anode by a film- and pore-forming agent for the dye-sensitized solar cell. *Materials Research Bulletin* 76: 140–146.
50. Kokubo, H., Ding, B., Naka, T., Tsuchihira, H., and Shiratori, S. 2007. Multi-core cable-like TiO_2 nanofibrous membranes for dye-sensitized solar cells. *Nanotechnology* 18(16): 165604.
51. Mustafa, M.N., Shafie, S., Wahid, M.H., and Sulaiman, Y. 2019. Light scattering effect of polyvinyl-alcohol/titanium dioxide nanofibers in the dye-sensitized solar cell. *Scientific Reports* 9(1): 14952.
52. Roy, A., Mukhopadhyay, S., Devi, P.S., and Sundaram, S. 2019. Polyaniline-layered rutile TiO_2 nanorods as alternative photoanode in dye-sensitized solar cells. *ACS Omega* 4(1): 1130–1138.
53. Wu, M., Lin, X., Wang, Y., Wang, L., Guo, W., Qi, D., Peng, X., Hagfeldt, A., Grätzel, M., and Ma, T. 2012. Economical Pt-free catalysts for counter electrodes of dye-sensitized solar cells. *Journal of the American Chemical Society* 134(7): 3419–3428.
54. Wang, W., Yao, J., Zuo, X., and Li, G. 2018. High efficiency nitrogen-doped core-shell carbon spheres as counter electrodes for dye-sensitized solar cells. *Materials Letters* 227: 172–175.
55. Wu, C.-S., Chang, T.-W., Teng, H., and Lee, Y.-L. 2016. High performance carbon black counter electrodes for dye-sensitized solar cells. *Energy* 115: 513–518.
56. Espen Olsen, G.H., and Lindquist, Sten Eric 2000. Dissolution of platinum in methoxy propionitrile containing LiI/I_2. *Solar Energy Materials & Solar Cells* 63: 267–273.
57. Ahmad, S., Yum, J.H., Butt, H.J., Nazeeruddin, M.K., and Gratzel, M. 2010. Efficient platinum-free counter electrodes for dye-sensitized solar cell applications. *Chemphyschem* 11(13): 2814–2819.
58. Ahmed, A.S.A., Xiang, W., Li, Z., Amiinu, I.S., and Zhao, X. 2018. Yolk-shell m-SiO_2@ Nitrogen doped carbon derived zeolitic imidazolate framework high efficient counter electrode for dye-sensitized solar cells. *Electrochimica Acta* 292: 276–284.
59. Bakr, Z.H., Wali, Q., Yang, S., Yousefsadeh, M., Padmasree, K.P., Ismail, J., Ab Rahim, M.H., Yusoff, M.M., and Jose, R. 2019. Characteristics of ZnO–SnO_2 composite nanofibers as a photoanode in dye-sensitized solar cells. *Industrial & Engineering Chemistry Research* 58(2): 643–653.
60. Bagavathi, M., Ramar, A., and Saraswathi, R. 2016. Fe_3O_4–carbon black nanocomposite as a highly efficient counter electrode material for dye-sensitized solar cell. *Ceramics International* 42(11): 13190–13198.
61. Gentian Yue, J.W., Xiao, Yaoming, Lin, Jianming, Huang, Miaoliang, and Lan, Zhang 2012. Application of poly(3,4-ethylenedioxythiophene): polystyrenesulfonate/polypyrrole counter electrode for dye-sensitized solar cells. *The Journal of Physical Chemistry, C* 116(34): 18057–18063.
62. Xia, J., Chen, L., and Yanagida, S. 2011. Application of polypyrrole as a counter electrode for a dye-sensitized solar cell. *Journal of Materials Chemistry* 21(12): 4644.
63. Wu, J., Li, Q., Fan, L., Lan, Z., Li, P., Lin, J., and Hao, S. 2008. High-performance polypyrrole nanoparticles counter electrode for dye-sensitized solar cells. *Journal of Power Sources* 181(1): 172–176.
64. Lu, S., Zhang, X., Feng, T., Han, R., Liu, D., and He, T. 2015. Preparation of polypyrrole thin film counter electrode with pre-stored iodine and resultant influence on its performance. *Journal of Power Sources* 274: 1076–1084.
65. Li, Q., Wu, J., Tang, Q., Lan, Z., Li, P., Lin, J., and Fan, L. 2008. Application of microporous polyaniline counter electrode for dye-sensitized solar cells. *Electrochemistry Communications* 10(9): 1299–1302.
66. Hou, W., Xiao, Y., Han, G., Fu, D., and Wu, R. 2016. Serrated, flexible and ultrathin polyaniline nanoribbons: an efficient counter electrode for the dye-sensitized solar cell. *Journal of Power Sources* 322: 155–162.
67. Wang, G., Yan, C., and Zhang, W. 2017. Prickly polyaniline nano/microstructures as the efficient counter electrode materials for dye-sensitized solar cells. *Journal of Nanoparticle Research* 19(12): 395.
68. Wang, H., Feng, Q., Gong, F., Li, Y., Zhou, G., and Wang, Z.-S. 2013. In situ growth of oriented polyaniline nanowires array for efficient cathode of Co(III)/Co(II) mediated dye-sensitized solar cell. *Journal of Materials Chemistry A* 1(1): 97–104.
69. Xiao, Y., Lin, J.-Y., Wu, J., Tai, S.-Y., Yue, G., and Lin, T.-W. 2013. Dye-sensitized solar cells with high-performance polyaniline/multi-wall carbon nanotube counter electrodes electropolymerized by a pulse potentiostatic technique. *Journal of Power Sources* 233: 320–325.
70. Zhang, H., He, B., Tang, Q., and Yu, L. 2015. Bifacial dye-sensitized solar cells from covalent-bonded polyaniline–multiwalled carbon nanotube complex counter electrodes. *Journal of Power Sources* 275: 489–497.
71. Niu, H., Qin, S., Mao, X., Zhang, S., Wang, R., Wan, L., Xu, J., and Miao, S. 2014. Axle-sleeve structured MWCNTs/polyaniline composite film as cost-effective counter-electrodes for high efficient dye-sensitized solar cells. *Electrochimica Acta* 121: 285–293.
72. Qin, Q., He, F., and Zhang, W. 2016. One-step electrochemical polymerization of polyaniline flexible counter electrode doped by graphene. *Journal of Nanomaterials* 2016: 1–7.
73. Lemos, H.G., Barba, D., Selopal, G.S., Wang, C., Wang, Z.M., Duong, A., Rosei, F., Santos, S.F., and Venancio, E.C. 2020. Water-dispersible polyaniline/graphene oxide counter electrodes for dye-sensitized solar cells: influence of synthesis route on the device performance. *Solar Energy* 207: 1202–1213.
74. Wang, M., Tang, Q., Chen, H., and He, B. 2014. Counter electrodes from polyaniline–carbon nanotube complex/graphene oxide multilayers for dye-sensitized solar cell application. *Electrochimica Acta* 125: 510–515.
75. He, B., Tang, Q., Liang, T., and Li, Q. 2014. Efficient dye-sensitized solar cells from polyaniline–single wall carbon nanotube complex counter electrodes. *Journal of Materials Chemistry A* 2(9): 3119.
76. Lee, K., Cho, S., Kim, M., Kim, J., Ryu, J., Shin, K.-Y., and Jang, J. 2015. Highly porous nanostructured polyaniline/carbon nanodots as efficient counter electrodes for Pt-free dye-sensitized solar cells. *Journal of Materials Chemistry A* 3(37): 19018–19026.
77. Thuy, C.T.T., Jung, J.H., Thogiti, S., Jung, W.-S., Ahn, K.-S., and Kim, J.H. 2016. Graphene coated alumina-modified polypyrrole composite films as an efficient Pt-free counter electrode for dye-sensitized solar cells. *Electrochimica Acta* 205: 170–177.
78. Wei, W., Wang, H., and Hu, Y.H. 2014. A review on PEDOT-based counter electrodes for dye-sensitized solar cells. *International Journal of Energy Research* 38(9): 1099–1111.

79. Worfolk, B.J., Andrews, S.C., Park, S., Reinspach, J., Liu, N., Toney, M.F., Mannsfeld, S.C.B., and Bao, Z. 2015. Ultrahigh electrical conductivity in solution-sheared polymeric transparent films. *Proceedings of the National Academy of Sciences* 112(46): 14138–14143.
80. Pringle, J.M., Armel, V., and MacFarlane, D.R. 2010. Electrodeposited PEDOT-on-plastic cathodes for dye-sensitized solar cells. *Chem Commun (Camb)* 46(29): 5367–5369.
81. Li, H., Xiao, Y., Han, G., and Hou, W. 2017. Honeycomb-like poly(3,4-ethylenedioxythiophene) as an effective and transparent counter electrode in bifacial dye-sensitized solar cells. *Journal of Power Sources* 342: 709–716.
82. Trevisan, R., Döbbelin, M., Boix, P.P., Barea, E.M., Tena-Zaera, R., Mora-Seró, I., and Bisquert, J. 2011. PEDOT Nanotube arrays as high performing counter electrodes for dye sensitized solar cells. Study of the interactions among electrolytes and counter electrodes. *Advanced Energy Materials* 1(5): 781–784.
83. Xia, J., Masaki, N., Jiang, K., and Yanagida, S. 2007. The influence of doping ions on poly(3,4-ethylenedioxythiophene) as a counter electrode of a dye-sensitized solar cell. *Journal of Materials Chemistry* 17(27): 2845.
84. Chang, L.-Y., Li, Y.-Y., Li, C.-T., Lee, C.-P., Fan, M.-S., Vittal, R., Ho, K.-C., and Lin, J.-J. 2014. A composite catalytic film of Ni-NPs/PEDOT: PSS for the counter electrodes in dye–sensitized solar cells. *Electrochimica Acta* 146: 697–705.
85. Chen, P.-Y., Li, C.-T., Lee, C.-P., Vittal, R., and Ho, K.-C. 2015. PEDOT-decorated nitrogen-doped graphene as the transparent composite film for the counter electrode of a dye-sensitized solar cell. *Nano Energy* 12: 374–385.
86. Chun-Ting Li, C.-P.L., Li, Yu-Yan, Yeh, Min-Hsin, and Ho, Kuo-Chuan 2013. A composite film of TiS_2/PEDOT: PSS as the electrocatalyst for the counter electrode in dye-sensitized solar cells. *Journal of Materials Chemistry A* 1(47): 14888–14896.
87. Xu, T., Kong, D., Tang, H., Qin, X., Li, X., Gurung, A., Kou, K., Chen, L., Qiao, Q., and Huang, W. 2020. Transparent MoS_2/PEDOT composite counter electrodes for bifacial dye-sensitized solar cells. *ACS Omega* 5(15): 8687–8696.
88. Ahmed, A.S.A., Xiang, W., Amiinu, I.S., Li, Z., Yu, R., and Zhao, X. 2019. ZnO-nitrogen doped carbon derived from a zeolitic imidazolate framework as an efficient counter electrode in dye-sensitized solar cells. *Sustainable Energy & Fuels* 3(8): 1976–1987.
89. Chun-Ting Li, H.-Y.C., Li, Yu-Yan, Huang, Yi-June, Yu-Lin Tsai, R., Vittal, Yu-Jane Sheng, and Ho, Kuo-Chuan 2015. Electrocatalytic zinc composites as the efficient counter electrodes of dye-sensitized solar cells: study on the electrochemical performances and density functional theory calculations. *ACS Appl. Mater. Interfaces* 7(51): 28254–28263.
90. Chun-Ting Li, Y.-L.T., and Ho, Kuo-Chuan 2016. Earth abundant silicon composites as the electrocatalytic counter electrodes for dye-sensitized solar cells. *ACS Appl. Mater. Interfaces* 8(11): 7037–7046.
91. Gong, J., Liang, J., and Sumathy, K. 2012. Review on dye-sensitized solar cells (DSSCs): fundamental concepts and novel materials. *Renewable and Sustainable Energy Reviews* 16(8): 5848–5860
92. Yum, J.H., Baranoff, E., Kessler, F., Moehl, T., Ahmad, S., Bessho, T., Marchioro, A., Ghadiri, E., Moser, J.E., Yi, C., and Nazeeruddin, M.K. 2012. A cobalt complex redox shuttle for dye-sensitized solar cells with high open-circuit potentials. *Nature Communications* 3: 631.
93. Wu, J., Lan, Z., Lin, J., Huang, M., Huang, Y., Fan, L., and Luo, G. 2015. Electrolytes in dye-sensitized solar cells. *Chemical Reviews* 115(5): 2136–2173.
94. Venkatesan, S., Liu, I.P., Li, C.-W., Tseng-Shan, C.-M., and Lee, Y.-L. 2019. Quasi-solid-state dye-sensitized solar cells for efficient and stable power generation under room light conditions. *ACS Sustainable Chemistry & Engineering* 7(7): 7403–7411.
95. Jang, Y., Thogiti, S., Lee, K.-y., and Kim, J. 2019. Long-term stable solid-state dye-sensitized solar cells assembled with solid-state polymerized hole-transporting material. *Crystals* 9(9): 452.
96. Su'ait, M.S., Rahman, M.Y.A., and Ahmad, A. 2015. Review on polymer electrolyte in dye-sensitized solar cells (DSSCs). *Solar Energy* 115: 452–470.
97. Prima, E.C., Nugroho, H.S., Nugraha, Refantero, G., Panatarani, C., and Yuliarto, B. 2020. Performance of the dye-sensitized quasi-solid state solar cell with combined anthocyanin-ruthenium photosensitizer. *RSC Advances* 10(60): 36873–36886.
98. Michio Suzuka, N.H., Takashi, Sekiguchi, Kouichi, Sumioka, Masakazu, Takata, Noriko, Hayo, Hiroki, Ikeda, Kenichi, Oyaizu and Hiroyuki, Nishide 2016. A quasi-solid state DSSC with 10.1% efficiency through molecular design of the charge-separation and -transport. *Scientific Reports* 6: 28022.
99. Aili, D., Jensen, J.O., and Li, Q., Polymers for Fuel Cells, in S. Kobayashi and K. Müllen (eds), *Encyclopedia of Polymeric Nanomaterials*. Berlin: Springer, 2014, pp. 1–13.
100. Lopez, J., Mackanic, D.G., Cui, Y., and Bao, Z. 2019. Designing polymers for advanced battery chemistries. *Nature Reviews Materials* 4(5): 312–330.
101. Chen, C.L., Chang, T.W., Teng, H., Wu, C.G., Chen, C.Y., Yang, Y.M., and Lee, Y.L. 2013. Highly efficient gel-state dye-sensitized solar cells prepared using poly(acrylonitrile-co-vinyl acetate) based polymer electrolytes. *Physical Chemistry Chemical Physics* 15(10): 3640–3645.
102. Fei Cao, G.O., and Searson, Peter C. 1995 A solid state, dye sensitized photoelectrochemical cell. *The Journal of Physical Chemistry* 99: 17071–17073.
103. Wu, J., Lan, Z., Wang, D., Hao, S., Lin, J., Huang, Y., Yin, S., and Sato, T. 2006. Gel polymer electrolyte based on poly(acrylonitrile-co-styrene) and a novel organic iodide salt for quasi-solid state dye-sensitized solar cell. *Electrochimica Acta* 51(20): 4243–4249.
104. Wu, J.H., Hao, S.C., Lan, Z., Lin, J.M., Huang, M.L., Huang, Y.F., Fang, L.Q., Yin, S., and Sato, T. 2007. A thermoplastic gel electrolyte for stable quasi-solid-state dye-sensitized solar cells. *Advanced Functional Materials* 17(15): 2645–2652.
105. Lee, K.S., Jun, Y., and Park, J.H. 2012. Controlled dissolution of polystyrene nanobeads: transition from liquid electrolyte to gel electrolyte. *Nano Letters* 12(5): 2233–2237.
106. Noor, M.M., Buraidah, M.H., Yusuf, S.N.F., Careem, M.A., Majid, S.R., and Arof, A.K. 2011. Performance of dye-sensitized solar cells with (PVDF-HFP)-KI-EC-PC electrolyte and different dye materials. *International Journal of Photoenergy* 2011: 1–5.
107. Liu, I.P., Cho, Y.-S., Teng, H., and Lee, Y.-L. 2020. Quasi-solid-state dye-sensitized indoor photovoltaics with efficiencies exceeding 25%. *Journal of Materials Chemistry A* 8(42): 22423–22433.
108. Masamitsu Matsumoto, Y.W., Kitamura, Takayuki, Shigaki, Kouichiro, Inoue, Teruhisa, Ikeda, Masaaki, and Yanagida, Shozo 2001. Fabrication of solid-state dye-sensitized TiO_2 electrolyte. *Bulletin of the Chemical Society of Japan* 74: 387–393.

109. Parvez, M.K.I., In, I., Park, J.M., Lee, S.H., Kim, S.R. 2011. Long-term stable dye-sensitized solar cells based on UV photo-crosslinkable poly(ethylene glycol) and poly(ethylene glycol) diacrylate based electrolytes. *Solar Energy Materials and Solar Cells* 95(1): 318–322.
110. Yu, Z., Qin, D., Zhang, Y., Sun, H., Luo, Y., Meng, Q., and Li, D. 2011. Quasi-solid-state dye-sensitized solar cell fabricated with poly(β-hydroxyethyl methacrylate) based organogel electrolyte. *Energy & Environmental Science* 4(4): 1298.
111. Xu, T., Li, J., Gong, R., Xi, Z., Huang, T., Chen, L., and Ma, T. 2017. Environmental effects on the ionic conductivity of poly(methyl methacrylate) (PMMA)-based quasi-solid-state electrolyte. *Ionics* 24(9): 2621–2629.
112. Venkatesan, S., Liu, I.P., Lin, J.-C., Tsai, M.-H., Teng, H., and Lee, Y.-L. 2018. Highly efficient quasi-solid-state dye-sensitized solar cells using polyethylene oxide (PEO) and poly(methyl methacrylate) (PMMA)-based printable electrolytes. *Journal of Materials Chemistry A* 6(21): 10085–10094.
113. Liu, I.P., Chen, Y.-Y., Cho, Y.-S., Wang, L.-W., Chien, C.-Y., and Lee, Y.-L. 2021. Double-layered printable electrolytes for highly efficient dye-sensitized solar cells. *Journal of Power Sources* 482: 228962.
114. Liu, I.P., Hung, W.-N., Teng, H., Venkatesan, S., Lin, J.-C., and Lee, Y.-L. 2017. High-performance printable electrolytes for dye-sensitized solar cells. *Journal of Materials Chemistry A* 5(19): 9190–9197.
115. Sakali, S.M., Khanmirzaei, M.H., Lu, S.C., Ramesh, S., and Ramesh, K. 2018. Investigation on gel polymer electrolyte-based dye-sensitized solar cells using carbon nanotube. *Ionics* 25(1): 319–325.
116. Kim, J.Y., Lee, J.W., Jung, H.S., Shin, H., and Park, N.G. 2020. High-efficiency perovskite solar cells. *Chemical Reviews* 120(15): 7867–7918.
117. Roy, P., Kumar Sinha, N., Tiwari, S., and Khare, A. 2020. A review on perovskite solar cells: evolution of architecture, fabrication techniques, commercialization issues and status. *Solar Energy* 198: 665–688.
118. Mesquita, I., Andrade, L., and Mendes, A. 2018. Perovskite solar cells: materials, configurations and stability. *Renewable and Sustainable Energy Reviews* 82: 2471–2489.
119. Zhen, J., Zhou, W., Chen, M., Li, B., Jia, L., Wang, M., and Yang, S. 2019. Pyridine-functionalized fullerene additive enabling coordination interactions with $CH_3NH_3PbI_3$ perovskite towards highly efficient bulk heterojunction solar cells. *Journal of Materials Chemistry A* 7(6): 2754–2763.
120. Wang, L., Moghe, D., Hafezian, S., Chen, P., Young, M., Elinski, M., Martinu, L., Kena-Cohen, S., and Lunt, R.R. 2016. Alkali metal halide salts as interface additives to fabricate hysteresis-free hybrid perovskite-based photovoltaic devices. *ACS Appl Mater Interfaces* 8(35): 23086–23094.
121. Gao, Y., Wu, Y., Lu, H., Chen, C., Liu, Y., Bai, X., Yang, L., Yu, W.W., Dai, Q., and Zhang, Y. 2019. $CsPbBr_3$ perovskite nanoparticles as additive for environmentally stable perovskite solar cells with 20.46% efficiency. *Nano Energy* 59: 517–526.
122. Furukawa, Y., Ikawa, S., Kiyohara, H., Sendai, Y., and Bahtiar, A. 2020. Inorganic-organic hybrid perovskite solar cells fabricated with additives. *Key Engineering Materials* 860: 3–8.
123. Wu, M.-C., Li, Y.-Y., Chan, S.-H., Lee, K.-M., and Su, W.-F. 2020. Polymer additives for morphology control in high-performance lead-reduced perovskite solar cells. *Solar RRL* 4(6): 2000093.
124. Zheng, H., Xu, X., Xu, S., Liu, G., Chen, S., Zhang, X., Chen, T., and Pan, X. 2019. The multiple effects of polyaniline additive to improve the efficiency and stability of perovskite solar cells. *Journal of Materials Chemistry C* 7(15): 4441–4448.
125. Kim, Y., Kim, G., Jeon, N.J., Lim, C., Seo, J., and Kim, B.J. 2020. Methoxy-functionalized triarylamine-based hole-transporting polymers for highly efficient and stable perovskite solar cells. *ACS Energy Letters* 5(10): 3304–3313.
126. Kanwat, A., Rani, V.S., and Jang, J. 2018. Improved power conversion efficiency of perovskite solar cells using highly conductive WOx doped PEDOT:PSS. *New Journal of Chemistry* 42(19): 16075–16082.
127. Wang, Y., Hu, Y., Han, D., Yuan, Q., Cao, T., Chen, N., Zhou, D., Cong, H., and Feng, L. 2019. Ammonia-treated graphene oxide and PEDOT:PSS as hole transport layer for high-performance perovskite solar cells with enhanced stability. *Organic Electronics* 70: 63–70.
128. Reza, K.M., Gurung, A., Bahrami, B., Mabrouk, S., Elbohy, H., Pathak, R., Chen, K., Chowdhury, A.H., Rahman, M.T., Letourneau, S., Yang, H.-C., Saianand, G., Elam, J.W., Darling, S.B., and Qiao, Q. 2020. Tailored PEDOT:PSS hole transport layer for higher performance in perovskite solar cells: enhancement of electrical and optical properties with improved morphology. *Journal of Energy Chemistry* 44: 41–50.
129. Mabrouk, S., Bahrami, B., Elbohy, H., Reza, K.M., Gurung, A., Liang, M., Wu, F., Wang, M., Yang, S., and Qiao, Q. 2019. Synergistic engineering of hole transport materials in perovskite solar cells. *InfoMat* 2(5): 928–941.
130. Chen, W., Shi, Y., Wang, Y., Feng, X., Djurišić, A.B., Woo, H.Y., Guo, X., and He, Z. 2020. N-type conjugated polymer as efficient electron transport layer for planar inverted perovskite solar cells with power conversion efficiency of 20.86%. *Nano Energy* 68: 104363.
131. Yang, D., Zhang, X., Wang, K., Wu, C., Yang, R., Hou, Y., Jiang, Y., Liu, S., and Priya, S. 2019. Stable efficiency exceeding 20.6% for inverted perovskite solar cells through polymer-optimized pcbm electron-transport layers. *Nano Letters* 19(5): 3313–3320.
132. Wen, X., Wu, J., Ye, M., Gao, D., and Lin, C. 2016. Interface engineering via an insulating polymer for highly efficient and environmentally stable perovskite solar cells. *Chemical Communications* 52(76): 11355–11358.
133. Chaudhary, B., Kulkarni, A., Jena, A.K., Ikegami, M., Udagawa, Y., Kunugita, H., Ema, K., and Miyasaka, T. 2017. Poly(4-Vinylpyridine)-based interfacial passivation to enhance voltage and moisture stability of lead halide perovskite solar cells. *ChemSusChem* 10(11): 2473–2479.
134. Malekshahi Byranvand, M., Behboodi-Sadabad, F., Alrhman Eliwi, A., Trouillet, V., Welle, A., Ternes, S., Hossain, I.M., Khan, M.R., Schwenzer, J.A., Farooq, A., Richards, B.S., Lahann, J., and Paetzold, U.W. 2020. Chemical vapor deposited polymer layer for efficient passivation of planar perovskite solar cells. *Journal of Materials Chemistry A* 8(38): 20122–20132.
135. Zhang, H., Azimi, H., Hou, Y., Ameri, T., Przybilla, T., Spiecker, E., Kraft, M., Scherf, U., and Brabec, C.J. 2014. Improved high-efficiency perovskite planar heterojunction solar cells via incorporation of a polyelectrolyte interlayer. *Chemistry of Materials* 26(18): 5190–5193.

8 Polymers and Composites for Fuel Cell Applications

Anita Samage and S.K. Nataraj
Jain University, Bangalore, India

CONTENTS

8.1 INTRODUCTION

The over-exploitation of fossil fuels has caused an exhaustive decrease in conventional energy resources. On the other hand, extensive fossil fuel usage has increased CO_2 concentration in the atmosphere and a deterioration of the ecological environment. Now due to the continuous depletion of fossil-based fuels, it has become an unavoidable necessity to explore and use abundant, renewable, and clean energy sources [1]. Concerning this question, the development of renewable sources and effective procedures for energy storage and conversion devices are in high interest, among which fuel cells (FC), solar cells (SC), supercapacitors, and batteries are the most favorable options [2]. Fuel cells are motionless energy conversion devices but they can be carried on mobile vehicles or carriers that convert chemical reactions between hydrogen-containing fuel and oxygen from the air into electrical energy and release a by-product as water [3]. Although the old-style heat engines produce electrical energy from chemical energy with the help of mechanical energy as an intermediate, the conversion efficiency is reduced compared to that of fuel cells. However, the fuel cell follows the best topographies of both engines and batteries to work longer without any intermediate, while having the similar characteristics of loaded batteries [2]. For economic considerations, the performance of various energy conversion and storage devices is compared in terms of their energy and power density, using a Ragone plot [4].

Therefore, the values of energy density are plotted against power density that allows us to compare the performance of various devices as shown in Figure 8.1a. Comparing all other devices, fuel cells (FCs) offer more advantages, like zero-emission, scalability, modularity, reliability, high energy conversion efficiency, and rapid set up, which provides better opportunities for cogeneration strategies. By this time FC technology has achieved the highest electrical efficiencies from mid- to lower power conversion ranges and has the potential to be advantageous regarding vital power production in the future [5, 6].

FCs have been known for over 200 years since Humphry Davy, a chemist, in 1801, first illustrated the conception of a fuel cell [7]. In 1838, Christian Friedrich Schönbein published an innovative paper explaining the hydrogen and oxygen reaction producing a by-product as water in *The Philosophical Magazine*. A significant breakthrough occurred, in 1839, when William Grove, a British amateur chemist, invented the earliest FC based on reversing the electrolysis of water to create electrical energy using hydrogen and oxygen gases [3]. This system was initially known as the "gas voltaic battery," but was later renamed the "Grove battery." The Grove battery is made up of two electrodes of a noble metal (platinum) mounted on two small

DOI: 10.1201/9781003169727-8

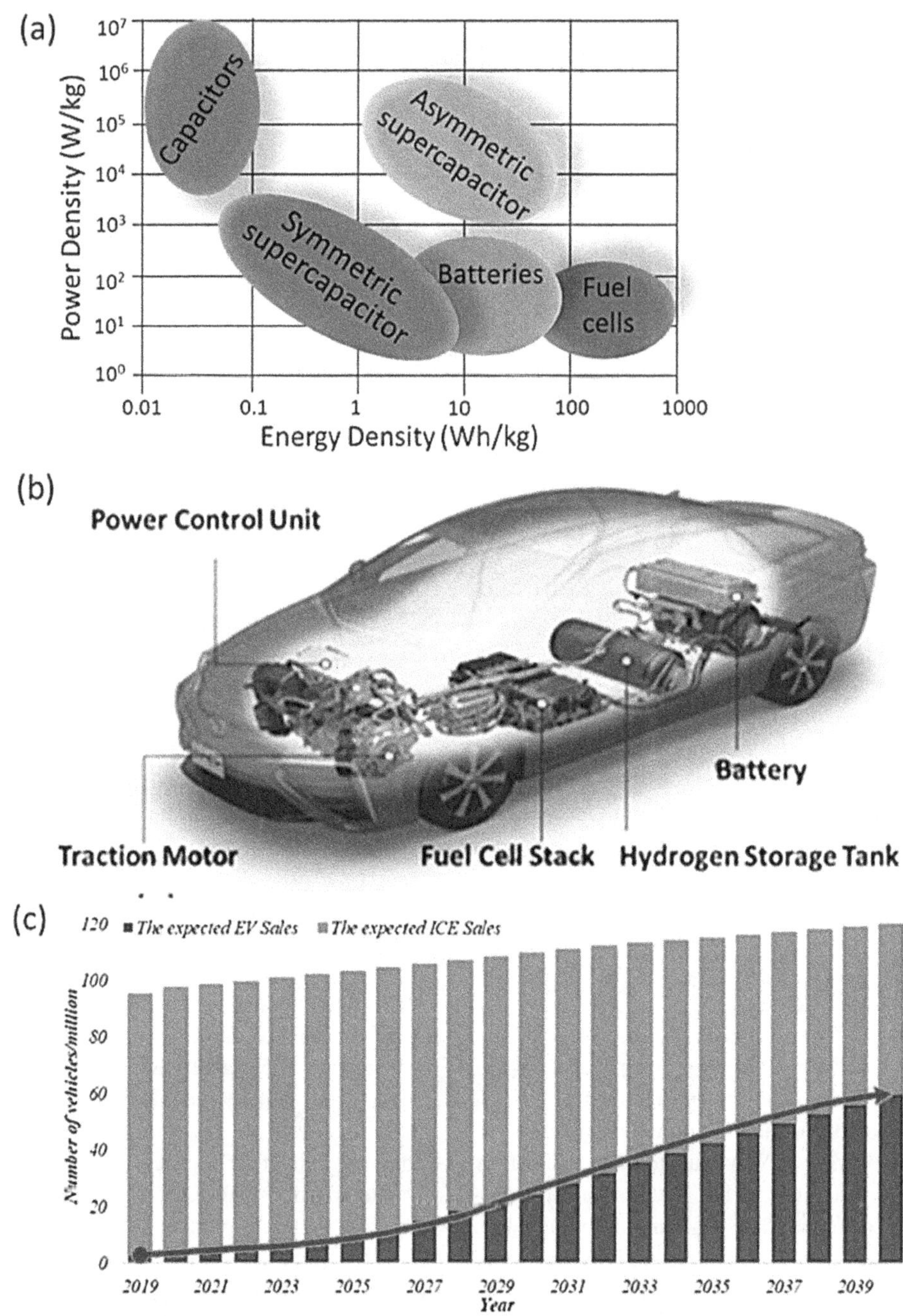

FIGURE 8.1 A general survey of the energy storage and conversion performance of devices in the form of (a) a Ragone plot, (b) the application of a fuel cell in a car, (c) the projected growth of fuel cell application in the market.

upward-tilted vessels, one with hydrogen gas and the other with oxygen (air), and these vessels are submerged in an electrolyte container with sulphuric acid. However, Grove's battery has a problem with corroding electrodes which makes the device unstable [3]. Nonetheless, Charles Langer and Ludwig Mond tried to construct an FC by using air and coal gas, and coined the word "fuel cell" in 1889. During the Industrial Revolution, for nearly a half-century, the invention of the traditional diesel/petrol engine forced the FC invention out of the spotlight. However, there were continuous efforts to develop alternative energy conversion and storage devices. Francis Bacon fruitfully returns to the clue of the fuel cell in the 1930s, thanks to a strategy by Langer and Mond, and created an archetype alkaline fuel cell (AFC) in 1932, using potassium hydroxide as the electrolyte and nickel foam as the electrodes, which substituted for sulfuric acid [8]. Bacon continued to refine the design over the next 30 years, presenting the first usable operational alkaline fuel cell in 1959, capable of generating 5 kW of electric energy. In the same year, Allis-Chalmers and his team created an AFC to power a tractor, which was extensively displayed across the country. Simultaneously, Willard Thomas Grubb and Lee Niedrach developed their prototypes of proton exchange membrane fuel cells (PEMFCs), which were later used in many lunar missions, including Gemini V, the Apollo Program, Apollo–Soyuz, Skylab, and the Space Shuttle. Beginning in the mid-1960s, FCs have widely been used for stationary power, portable devices, and transportation purposes with the help of funding agencies like the governments of the USA, Canada, and Japan [9].

Further, in 1966, General Motors applied FCs to ground transportation in the name of Electrovan, a concept shown schematically in Figure 8.1(b), which was powered by hydrogen fuel. This popularized the commercialization of FCs and their extensive research among academia and manufacturers. Fuel cells were successfully commissioned in submarines by the US Navy in the 1980s, where their massively powerful, zero-emission, and near-silent operational qualities offered significant advantages. During the 1990s, fuel cell technology became popular for small stationary applications such as residential micro-combined heat and power.

This led to the huge funding flow into fuel cell research which has resulted in the modernization of different cell types such as solid oxide fuel cells (SOFCs), molten carbonate fuel cells (MCFCs), and direct methanol fuel cells (DMFCs). With this, the commercialization process of fuel cells accelerated in the early 2000s. As a result of their high performance, zero emissions, and ease of refueling, FCs have gained widespread acceptance in public transportation buses. Smart energy, the first "neutral emission vehicle" (NEV), created a motorcycle model entirely made of fuel cells in 2005. By using DMFCs as a power station, the company was able to overcome miniaturization barriers. Between 2006 and 2009, several FC-powered buses were on road across European countries, Australia, and China. This also encouraged all vehicle manufacturers to develop personal and cargo vehicles, such as cars, trucks, buses, forklifts, motorcycles, scooters, submarines, ferries, aircraft, and electric boats, all using FC technology. The Mirai was the first fuel-cell-powered car in 2014 rolled out in Japan, meaning the start of an innovative episode in the history of automobiles. This began the revolution in electric vehicles which forced all major auto manufacturers including Ford, Honda, GM, Mitsubishi, Mercedes, and other companies to invest in FC-based car manufacturing. The global fuel cell market is currently projected to be worth USD15 billion in 2022, and it is estimated to increase exponentially as demand for alternative energy sources grows, as shown in Figure 8.1(c).

8.2 THE WORKING PRINCIPLE OF THE FUEL CELL

As discussed earlier, a fuel cell is an *electro-chemical structure in which the chemical energy of the fuel is continuously converted into electric energy* with water and heat as derivatives. Thermodynamical conditions play a very crucial role in fuel cell performance, and in fuel cells energy transfer takes place at constant pressure and temperature. The basic structure of a fuel cell consists of a cathode, anode, and electrolytic system as shown in Figure 8.2

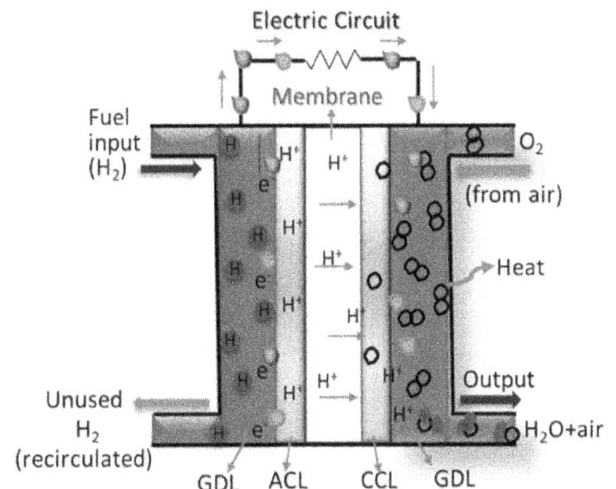

FIGURE 8.2 Working principle of the fuel cell.

[10]. The fuel cell is made of an electrolytic chamber or polyelectrolytic membrane that is interacted by two electrodes on each face. The hydrogen-composed fuel (methanol in DMFCs) is continuously supplied to the anode, while the oxygen (air) (oxidant) is continuously introduced to the cathode. At the anode electrode, hydrogen-composed fuel oxidizes into hydronium ions (protons) and electrons with the help of an electrocatalyst on the anode side. The intermediate electrolyte or polyelectrolytic casing allows only the hydronium ions (protons) to move from anode to cathode and behaves as an insulator for electrons that are moving through the external circuit and will reunite on the further side of the membrane until the structure becomes stable. The association of protons and electrons with the oxidizer takes place at the cathode to produce pure water as shown in the reaction mechanism below (Equations 8.1–8.3) [11].

Reaction at the anode:

$$H_2 \rightarrow 2H^+ + 2e^- \quad (8.1)$$

Reaction at the cathode:

$$1/2O_2 + 2H^+ + 2e^- \rightarrow H_2O \quad (8.2)$$

Overall reaction:

$$H_2 + 1/2O \rightarrow H_2O \quad (8.3)$$

Now, there are various types and forms of FCs; the most individual unit produces between 0.5 and 1.23 volts of DC electricity. In all cases, elementary fuel cells have essentially the same fundamental component, namely a membrane electrode assembly (MEA) which is the core and brain of the fuel cell. This MEA is made up of an anode, cathode, and electrolyte (or ion exchange membrane). An electron conductor is an anode collector plate, which is in touch with the anode gas diffusion layer which purifies fuel particulates, if any, but generally is coated with a hydrophobic polytetrafluorethylene (PTFE) polymer. Through oscillation and diffusion, the gas diffusion layer of the anode introduces the fuel cell by a reactant mixture containing hydrogen or hydrogen-enriched gases. The diffusion layer attached at both ends to the anode and cathode distributes reactants/products to and from the catalyst layers on the cathode and anode and conducts electrons to the anode collector plates. Momentum, transport, concentration, and pressure gradients are the main factors that affect the feeding of reactant mixture constituents through the pores of the anode diffusion layer. Meanwhile, the gas diffuser pores are filled with liquid water flowing from the MEA. The electrons are conducted between the current collectors and coatings of the catalysts of the MEA and through the graphite matrix of the gas diffusers. The reactions (electrochemical) happen in the anode catalyst layers; the diffused gas will oxidize according to the reaction into protons and electrons; liquid water that permeates the polyelectrolyte will dissolve protons. Water and protons will be able to pass through the polymer membrane while the reactants H_2 and O_2 will be separated. Due to pressure gradients, convection, dispersion, diffusion, and electro-osmosis, forces pulled by the traveling protons and water in the polyelectrolyte are transferred.

The oxygen reduction takes place on the cathode side. The oxygen from the cathode gas diffusion channels dissolves in the runny water and reacts with both hydronium ions (protons) released through the membrane electrolyte in this section. The electrons, on the other hand, pass from the anode through the outside circuit towards the cathode [12]. Meanwhile, diffused water from the MEA fills the gas diffusion layer pores and must be removed, which is essential to protect the MEA from flooding. The graphite matrix in the gas diffusion layer serves as a platform for electrons to travel between collector plates, allowing the catalyst layers to operate continuously and prevent electron recombination in the MEA. The air mixture in the cathode diffusion channels allows partial transportation towards the cathode electrocatalyst layer via the cathode gas diffusion method. The electron conductor is the cathode collector plate, which is made of porous carbon coated with PTFE and is in contact with the cathode gas diffusion sheet [13]. If the free electrons or any substances are left alone in the MEA then they will interrupt the chemical reaction either by combining or recombining at the anode or cathode, together with hydrogen and oxygen as water, which drains from the cell. The polyelectrolyte, which is a proton exchange membrane, plays an important role in preventing this by allowing selectively the necessary ions to move between the negative and positive electrode. As a result, a fuel cell can yield electricity as long as it is supplied with hydrogen fuel and oxygen (air) [14].

Due to rapid growth in technology and the fundamental understanding of the materials and their properties, there are many types of currently available fuel cells, most of which are on the market. They are conventionally categorized according to their fuel, electrolyte material, and working principle, which depends on the materials used for the anode, cathode, and electrolytes. This also influences their power outputs, operating temperatures, electrical efficiencies, and typical applications.

Fuel cells are generally categorized by the polyelectrolyte employed in the cell; a list of them is provided and schematically displayed in Figure 8.3. The direct methanol fuel cell (DMFC), which is a proton exchange membrane fuel cell in which methanol and water mixture is electrochemically oxidized directly inside an anode in various concentrations, is an exception to this classification. A second classification is based on each fuel cell's operating temperature. Therefore, based on the range of low to high temperature, fuel cells are categorized in the following order of increasing operating temperature: (a) alkaline fuel cell [15], (b) proton exchange membrane fuel cell, (c) direct methanol fuel cell [16], (d) phosphoric acid [17], (e) molten carbonate [17], and (f) solid oxide [18]. Further, an overview of all fuel cell operating temperatures, reactions, charge carriers, and their advantages and disadvantages is listed in Table 8.1.

8.3 POLYMERS IN FUEL CELLS

Jöns Jakob Berzelius coined the word "polymer" for the first time in 1833. The principle of the polymer was introduced, and the speedy advancement of polymer materials was initiated in the 1920s [19, 20]. But, subsequently in the 1930s, polymers entered a golden age, when new forms were exposed and soon found commercial applications. A polymer is a large molecule or macromolecule that is made up of many subunits ("poly" means many + "mers" means units or parts). Polymers can now be used in thousands of ways and are present everywhere, ranging from the strands of our DNA, which is a naturally occurring biopolymer, to polypropylene, which is used as a plastic all over the world. Many of the polymers which are used commonly, for example those found in consumer products, are referred to by a common or trivial name. Polymers such as PE (polyethylene), PP (polypropylene), PVA (polyvinyl alcohol), and PMMA (poly(methyl methacrylate)) are commonly utilized in our day-to-day lives because of their potential benefits, for example their high machinability, negligible weight, and being more economical when compared to metal and ceramic materials. It is possible to precisely regulate the characterization of polymers by modifying the functional groups and shapes of the backbones and branched chains to meet extensive requirements. Prior to the invention of electrically conductive polymers, polymers with non-conducting behaviors were used typically as insulators or separators and packing materials in electronic devices. Polymers were previously so uncommon that they were not considered for electrodes or active substances, though both can be electrically conducting materials. But when polyacetylene was discovered in 1977, polymers came to fascinate and interest both academic and commercial enterprises [21] Electrically conducting polymeric materials, for example, polyacetylene,

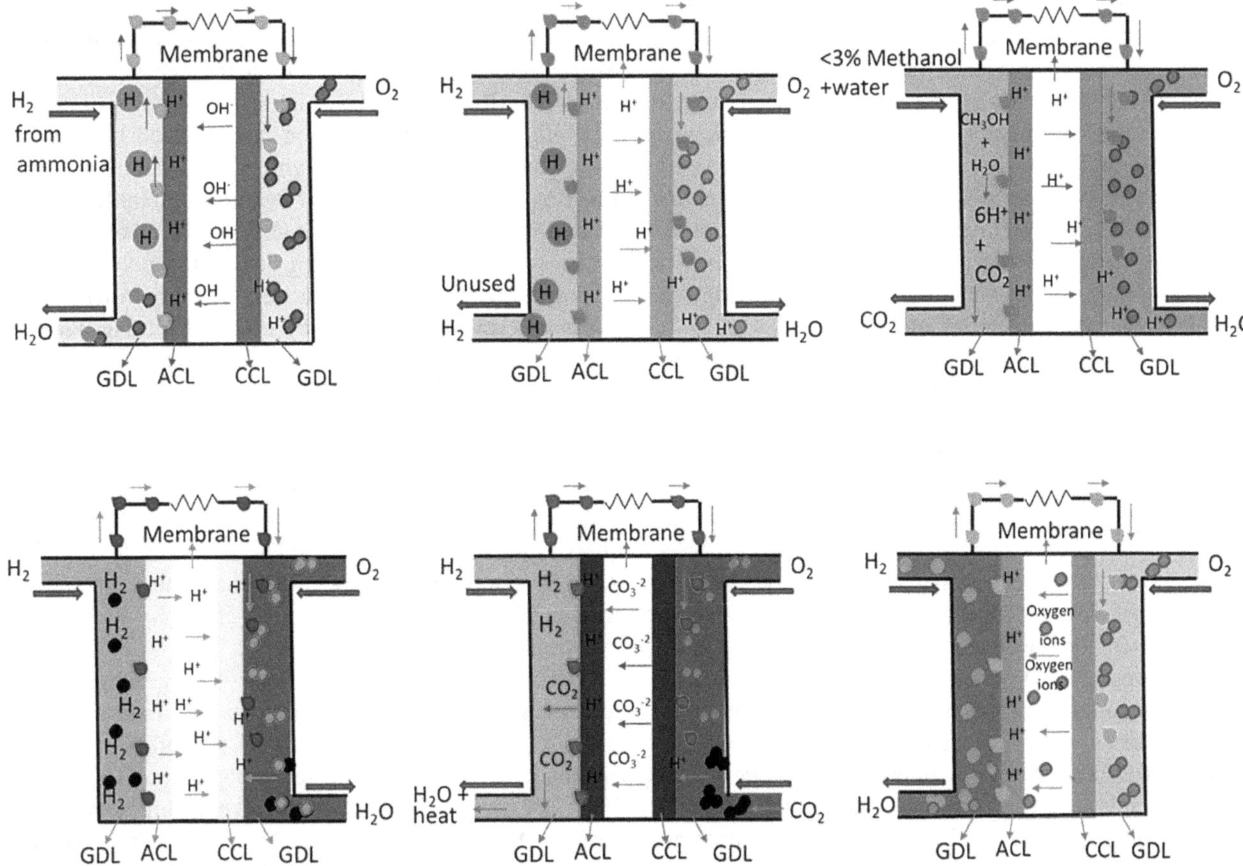

FIGURE 8.3 Working principle of different types of fuel cells. (a) alkaline fuel cell [15], (b) proton exchange membrane fuel cell, (c) direct methanol fuel cell [16], (d) phosphoric acid [17], (e) molten carbonate [17], (f) solid oxide [18].

TABLE 8.1
Details of Different Fuel Cells and Their Constituent Specifications

	Alkaline Fuel Cell (AFC)	Proton Exchange (PEMFC)	Direct Methanol (DMFC)	Phosphoric Acid (PAFC)	Molten Carbonate (MCFC)	Solid Oxide (SOFC)
Temperature (°C)	60–90 °C	80–100 °C	80–100 °C	160–200 °C	600–800 °C	800–1000 °C
Anode Feed	Pure H_2	Pure H_2	Methanol	Pure H_2	H_2/natural gas	Gasoline/natural gas
Cathode Feed	O_2/air	O_2/air	O_2/air	O_2/air	O_2/air	O_2/air
Common Electrolyte	Aq. KOH	Acidic polymer	Membrane	H_3PO_4 in SiC matrix	Molten $LiCO_3$ in $LiAlO_2^-$	Yttria-stabilized support
Charge Carrier	OH^-	H^+	H^+	H^+	CO_3^{2-}	O^{2-}
Advantages	Simple system design	Faster startup, no leakage of electrolyte	Methanol is easier to store	Higher operating temperature reduces CO problem	High efficiencies especially when combined with a gas turbine, internal reforming possible, CO and CO_2 tolerant	No electrolyte creeping, high efficiency possible
Disadvantages	Not CO_2 tolerant	Not CO_2 tolerant, water management problem in membrane	Not CO tolerant, water management problems, methanol cross-over	Liquid electrolyte leaks, lower phosphoric acid conductivity	Longer startup, electrolyte creep possible	Longer startup

polyaniline, and polypyrrole, exhibit high potential as electrode materials and active substances for different energy and electronic systems because of their large and adjustable electrical movability in their matrix. Electrical and ionic conducting polymers became the subject of research in the early 2000s because of their intriguing characters, for example an extensive range of conductivity, ease of fabrication, mechanical solidity, featherlight weight, economy, and effortlessness with which they can be nanostructured or used to fabricate devices with unique uses such as supercapacitors, rechargeable batteries, and fuel cells.

Graphitic materials were previously used for anode preparation in energy storing systems such as metal-ion batteries, but graphitic resources had only a limited capacity and could not completely utilize the capacity of lithium [22]. To address this major issue, silicon as an anode and sulfur as a cathode material were created by merging carbon precipitate and adhesive polymers to generate an active material fixed in a conductive medium that moderately addressed the energy storage device's capacity problem [23]. K. Shamara et al. reported for the first time more porous azo group linked polymers (ALPs) which have been used since as the modern oxidation-reduction active electrode material for Na-ion batteries. ALPs are hugely cross-linked polymers that remove the solubility issue of organic electrode materials in popular electrolytes, which is particularly prevalent in small organic molecules and brings about a rapid decrease in power. Furthermore, the high specific surface area of ALPs, collectively with their microporous form of conjugated-π systems, aids electrolyte adsorption in the pores, as well as rapid ionic transport and charge transfer. The polymer leads to electronic and ionic properties which maintain better rate capability, higher cyclic stability, and remove the complications of electrolyte dissolution.

Flexible cross-linked polymer architecture creates a new approach for developing effective and eco-friendly electrode materials for rechargeable Na-ion batteries and Li-ion batteries [24]. Because of their high capacitance and inherent flexibility, structural fluctuation, and reduced long-lasting cycle stability, conjugated polymers, for example polyaniline (PANI) and polypyrrole (PPy), are commonly adopted in flexible supercapacitor electrodes. On the other hand, the cyclic stability of freestanding conjugated polymer films during the charge–discharge process seriously restricts their use in appliances, as shown in Figure 8.4. Xinming Wu et al. combined a decentralized conjugated polymer (PDT) string with layered MXene ($Ti_3C_2T_x$) to produce a free-standing composite film. In terms of capacitance and cycling stability, conjugated polymer-based film electrodes, such as a decentralized polymer chain (PDT) and fixed $Ti_3C_2T_x$ material, showed cyclic stability over 10,000 charge and discharge cycles in an acidic electrolyte solution, and this PDT/$Ti_3C_2T_x$ composite electrode exhibited the high performance of 284 mF cm^{-2} at 50 mA cm^{-2} and long-lasting cyclic stability [25].

When compared to other energy storage devices, FC technology is superior and competitive since it can be used for the plug-in-use kind of flexibility. [27, 28]. However, even after several technological advancements, FCs face many obstacles that make commercialization difficult, including inadequate longevity, reliability, and the high cost of the materials used. The efficiency and cost of FCs are primarily determined by three major components: catalysts, bipolar plates, and electrolytes (membranes) [29].

Catalysts: In FCs, electrocatalyst and support materials play a critical role in performance and durability. The FC electrocatalyst and support should have properties

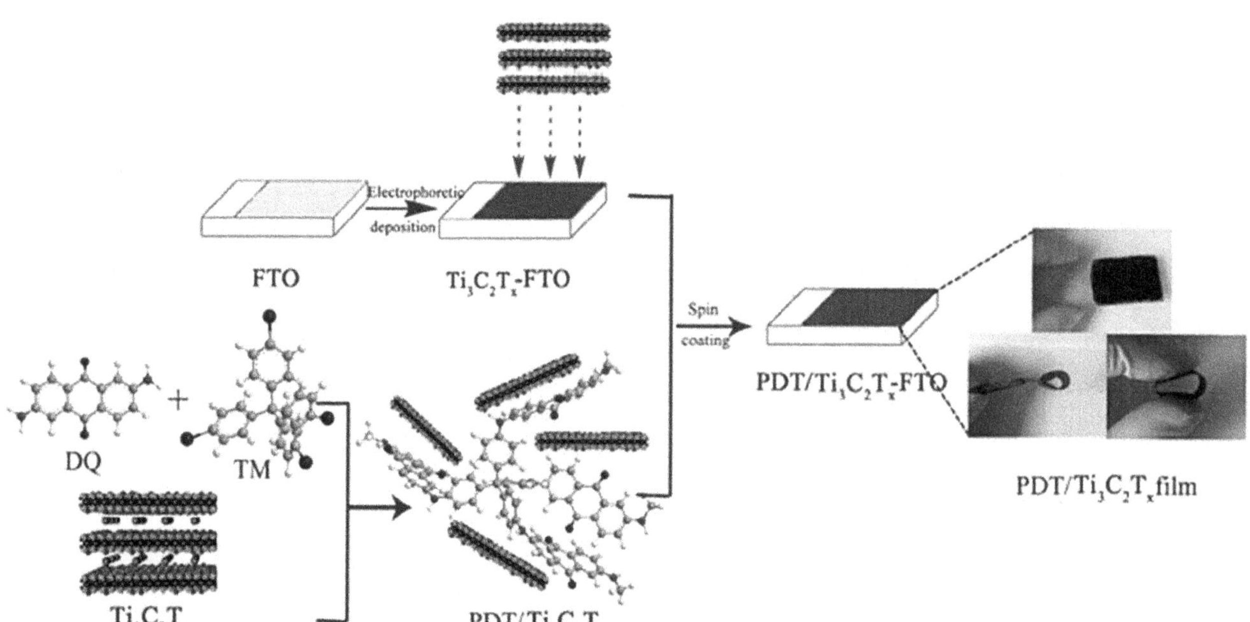

FIGURE 8.4 Synthesis mechanism of a polymer and mxene composite [26].

including excellent electronic conductance, reduced corrosion, uniform particle size distribution, high specific surface area, and easy formation of the uniform dispersion of the electrocatalyst on a conductive support material with strong cohesive forces to achieve high efficiency. In FCs, the most used electrocatalysts are either platinum or platinum-based alloys on carbon materials [30]. When it comes to the small strength of PEMFC catalyst material, the weakening of carbonaceous support materials has been reported as the primary cause of catalyst failure. This is due to platinum-based catalysts undergoing dissolution which then leads to the agglomeration of the platinum catalyst. Even though carbon-based supports have been widely used, they are not robust enough for fuel cell longevity when in use [31]. During FC activity, the cathode catalyst is exposed to the higher potential of an electrode in the presence of oxygen when cycling at low to negligible loads of the catalyst. When unstable carbon support is oxidized, CO_2 is produced, which causes Pt particles to separate from the carbon support, resulting in a loss of FC output.

For the anode electrocatalyst, the carbon backing can also be reacted in the condition of fuel starvation [32]. To overcome the problem of traditional carbon-based backing materials in PEMFCs, the non-carbon matrix is used to reduce the corrosion of the traditional material, for example nitride [33–35], carbides, silica, metal oxides [36], as well as conducting polymers, which are non-carbon supports [37]. These also provide a high surface area and a favored dispersion of the electrocatalyst, electrochemical stability under FC operating conditions of low solubility in alkaline and acidic media, as well as electrical conductivity. Because of their electrical conductance over wide specific surface areas, and long-lasting stability, polymers with conductance have been used as an electrocatalyst backing for FCs for many years. By combining metal-based nanostructures with a polymer pattern, nanocomposites can efficiently enhance the specific surface area of these polymeric materials, improving their catalytic performance. Heet et al. used a polymer with inherent microporosity for catalyst support to resolve the substantial loss of the working power of the cell caused by commercial noble-metal-based (Pt-based) catalysts, which prevents FC technologies from being popular. In accelerated catalyst corrosion experiments, a highly rigid and porous polymer with an inherent microporosity (PIM-EA-TB) and with a specific surface area of 1027 $m^2\ g^{-1}$ was used. Thus, composites of porous sheets of PIM-EA-TB provide effective support to avoid corrosion of the Pt/C catalyst without hindering the flux of reagents. With these newly developed

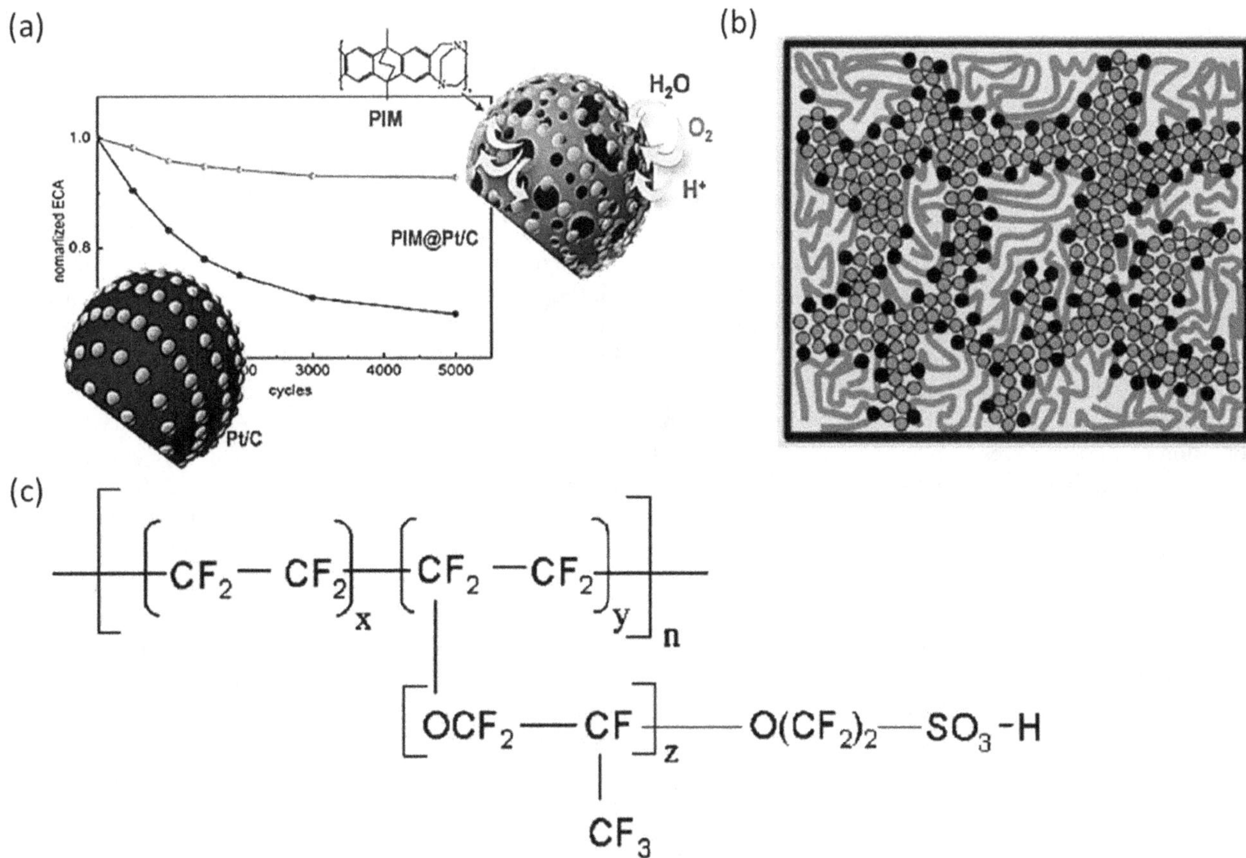

FIGURE 8.5 (a) Performance of PIM-EA-TB as a catalyst in fuel cells [26], (b) Structure of a Nafion membrane showing water channels which help in proton transfer, (c) Chemical formula of Nafion.

materials, electrochemical cycling stability tests show that the PIM-EA-TB composed Pt/C shows a significantly developed durability as compared to a conventional Pt/C catalyst as shown in Figure 8.5(a) [26]. Further, Zhanget et al. reported an Au-nanoparticle—prepared from an $HAuCl_4$ precursor loaded conductive polymer poly(3,4-ethylenedioxythiophene) polystyrene sulfonate (PEDOT: PSS) nanocomposite—was obtained by a one-step synthesis approach based on a direct current plasma liquid interaction. The resulting AuNPs/PEDOT: PSS nanocomposites performed better in alkaline direct ethanol fuel cells [38].

Bipolar plates: Bipolar plates represent nearly 80% of the weight of the FC stack and cost approximately 30 to 40% of the cost [39, 40]. Bipolar plates perform very crucial functions in the overall performance of FCs, such as (i) isolating the fuel and oxidant (air) gases and avoiding leakage of gases, (ii) collecting the current created by electrochemical reactions, (iii) evenly distributing gases throughout the cell, (iv) ensuring the stack's mechanical power, and (v) discharging the water emitted. In the fuel cell, bipolar plate application must exist in a balance of various electrochemical and physical properties [41]. The electric and mechanic properties, in addition to its stability, are the most important factors, as they influence the FC stack productivity [32]. Materials for bipolar plates include stainless steel, titanium, and aluminum, all of which have strong electrical conductivity, alongside stainless steel being the supreme expectant for marketing [42]. Furthermore, metallic bipolar plates typically have a large amount of electrical and thermal conduction, outstanding motorized properties, and low gas penetrability. Innovative methods, such as continuous rolling, have been established to enable the mass manufacture of metallic bipolar plates to enhance corrosion resistance [41], minimize membrane contamination, and increase the system's durability [43]. Furthermore, to reduce the corrosion of the metals, significant efforts have been made by using a noble metal (platinum and ruthenium), steel (stainless), and different types of doped materials, for example nitrides, carbides, alloys of carbides, and also conductive polymers. Despite these advancements, the ability of resistance to the corrosion of these sources remains less than that needed by BPPs; handling BPPs with these materials is not economical [33].

Research has been continued to adopt an alternative solution to replace conventional metal-based plates using conductive polymers and polymer composites, which also helps in processability with improved flexible and mechanical properties [44]. Further, researchers are also focusing on carbon composites with conductive polymers as bipolar plates. In this direction, Afiqahet et al. carried out research to see whether milled carbon fiber could be used as a conductive filler in a polymer. Incorporating carbon nanotube (CNT) and carbon black (CB) as secondary fillers on carbon fiber (CF) reinforced epoxy composites has helped to improve the composites' electrical conductivity. The addition of secondary fillers such as CNT and CB to hybrid systems increased the strength of the polymer composite as a bipolar layer, according to the report [45]. Further, the same research group has reported on a polypropylene reinforced carbon material with high electrical conductivity which was used for a bipolar plate [17].

Membranes (electrolytes): The polyelectrolyte or ion-exchange membrane, which is a semi-porous polymer casing acting as an electrolyte ionic conductor between the anodic and cathodic sides while also acting as a separator for two raw reactant gases, plays a significant role in fuel cells. Grubb and Niedrach at General Electric (GE) invented polyelectrolyte membranes for fuel cells in the 1960s [46]. The first membrane was made of poly(phenol-formaldehyde sulfonic acid) resin, which was made by the condensation of phenosulphonic acid and formaldehyde and which was brittle and hydrolyzed quickly. Further, a poly(styrenesulfonic acid) (PSSA) polyelectrolyte was prepared by cross-linking sulfonated styrene divinylbenzene into an inert fluorocarbon matrix, which showed a lifetime of 200 hours in AFC at 60 °C [47]. Later, the breakthrough was achieved in the 1980s with the development of fluorinated polymers which were then cast into ion-exchange membranes for fuel cells.

After the rediscovery of pioneering research, DuPont introduced poly(perfluorosulfonic acid) (PFSA) membranes under the brand name Nafion, see Figure 8.5(b), which revolutionized FC technology. However, the high-cost of Nafion membranes made only for limited commercial utilization in FCs. However, hydrocarbon-based membranes had a lower proton conductivity, so polymers with a high degree of sulfonation (DS) were commonly used to address this problem [48]. However, since such polymers have a high DS, they are prone to swelling, which restricts their use in PEMFC systems [49]. Based on all these experiences, researchers listed the basic requirements for an appropriate polyelectrolyte as follows: (i) high electrochemical solidity through the fuel cell's action, (ii) good water uptake, (iii) high proton conductivity with minimal resistivity losses, (iv) low penetrability to reactant substances to maximize productivity, and (v) economical and chiefly available in nature [50]. Graphene oxide (GO) and sulfonated poly(arylenethioethersulfone)-grafted graphene oxide was used to make highly sulfonated polyelectrolytes of a sulfonated poly(arylene ether sulfone) (SPAES) composite. According to Lee et al., with the presence of GO on the surface, the composite membranes containing SATS-GO showed higher proton conductivity than the pristine SPAES [17]. Another group reported improved performance with cross-linked sulfonated poly(ether ketone) (C-SPEEK) membranes for DMFC applications, using SPEEK with chloromethyl side groups (SPEEK-Cl) and poly(2,5-benzimidazole)-grafted graphene oxide (ABPBI-GO) as a cross-linker [17].

8.3.1 Electronic and Ionic Properties of Polymers

Polymers have been used as insulating materials and as a replacement for structural resources, for example wood, silica, and metal systems, since their discovery, due to their stability, negligible weight, ease of chemical modification, and reduced temperature processability. Organic doped polyacetylene was the first electrically conducting polymer documented in 1977 [51]. The p-conjugated system, which is formed by p_z-orbitals and alternating carbon–carbon bond lengths, is a distinctive electronic property of pristine (non-doped) polymer conductors. In some polymer systems such as polyaniline, nitrogen p_z-orbitals and C_6 rings form the conjugation. Several conducting polymers have a chemical structure of repeat units in their pristine form, trans- and cis-polyacetylene, polyaniline, polypyrrole, polythiophene, poly(*p*-phenylene), poly(*p*-phenylenevinylene), and poly (*p*-pyridylvinylene), as shown in Figure 8.6 [52]. Although the electronic ground states of these structures differ, the electron–phonon interaction stabilizes a two-fold degenerate insulating ground state in undoped trans-$(CH)_x$ [17]. Via the method of doping, the electrical conductivities of pristine polymers are converted from insulating to metallic, with the conductivity rising as the doping level increases. To cause an insulator-metal conversion in electronically conductive polymers, both n-type electron-donating and p-type electron-accepting dopants have been used. The doping of polymers is done by electrochemical deposition or by exposing polymers films to the vapors of the dopant vapors or solutions [53].

In a doped polymer, the electronic property is induced via dopant atoms which are positioned interstitially between chains and donate or accept a charge of the polymer backbone. Due to this mechanism, polymer and dopant form three-dimensional structures, in which the negatively or positively charged ions are introduced to the polymer chain, results in metallic conductivity behavior. The combination of phonons and electrons results in lattice distortions surrounding the doped charge in the polymer as a result of this. Charges applied to the backbone by doping the metal atoms or by photo-excitation are retained in excitons and soliton confinement states for the degenerate ground states [54]. For non-degenerate systems, the charges introduced by low doping or photo-excitation are stored as polarons or bi-polarons [55]. Since the initial report of achieving conductivity of 100 S/cm by doping with iodine or other electron donors and acceptors, incapacitated polyacetylene has been used as the prototype. However, recent developments in the dispensation of other conducting polymer structures have resulted in enhancements in their conductivity alignment in the range of 10^4–10^5 S/cm by doping, at room temperature, which behaves like a metallic conductor, as shown in Figure 8.6(b). This led to an increase in research in the chemistry of doping the semiconducting polymers to increase the high electrical conductivity and stability of the material [56].

Ionic conductivity of the polymers is the main factor that is utilized in energy storage and conversion devices. The basic requirement and determining factor for ionic membranes in fuel cells is high ionic conductivity. The type of functional groups present in a polymer, such as strong cationic and anionic groups (quaternary ammonium salts, phosphonic and sulfonic acid groups) [59] and weak groups (such as 1°, 2°, and 3° amine groups, hydroxyls, and carboxylic acids), determines the extent of ionic conductivity in ion-exchange membranes or polyelectrolytes. The porosity of polymer membranes, on the other hand, affects ionic conductivity, which would be strong in a highly porous membrane, and vice versa.

As a result, in the case of organic/inorganic composite materials, the pore structure can be fine-tuned using organic or inorganic materials and synthetic techniques [60]. These properties place polymers in an advantageous position for the design of new generation energy conversion and storage devices. The amount of water in a

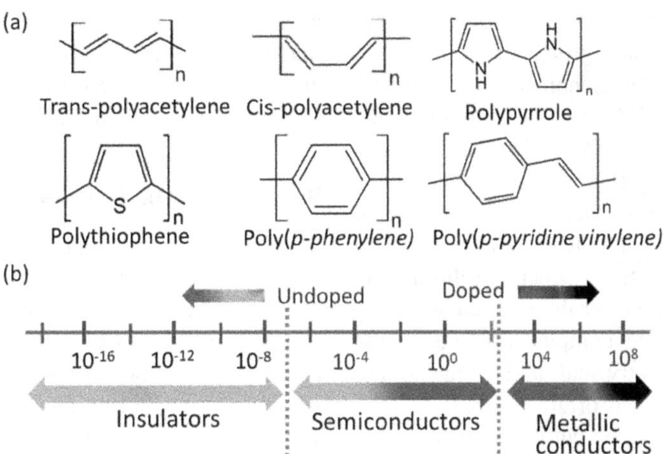

FIGURE 8.6 (a) Chemical structures of various conducting polymers [57], (b) band showing conductivity range of polymers [58].

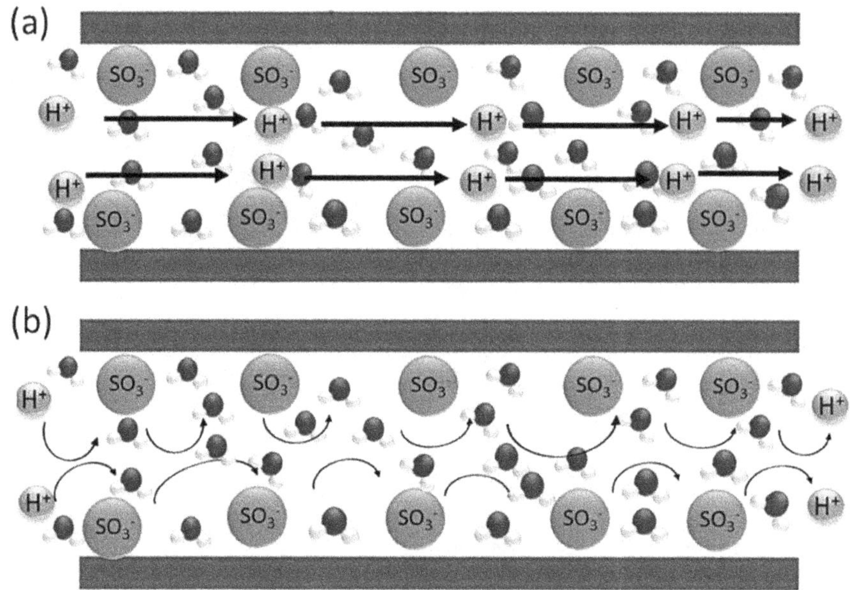

FIGURE 8.7 Ion conduction in polyelectrolyte through (a) hopping, (b) ion propagation through water channels in the fuel cell [62].

polymeric material has a big impact on ionic conductivity and channel formation. The higher the water content, the higher is the ionic conductivity and the better the propagation of protons. The lower amount of water in polymers with higher ionic conductivity is the main aspect for increasing FC stability, which will also minimize polyelectrolyte swelling and flooding. Furthermore, a polyelectrolyte should only conduct ions such as hydroxide, protons, or carbonate and should not conduct electrons [61]. In polyelectrolytes, ion conduction takes place through the hopping, and ion propagation through the hydrated water channel mechanism, as shown in Figure 8.7a and b [62].

8.3.2 Biopolymers

Monomeric units are covalently bound to form larger molecules in biopolymers. Polynucleotides, polypeptides, and polysaccharides are the three main groups of biopolymers, which are graded based on the monomers and structure formed. Polynucleotides are RNA and DNA polymers that hold 13 or more than 13 monomers of nucleotides. Proteins and poly-nucleotides, including collagen, actin, and fibrin, are polymers made up of amino acids. Polysaccharides, such as starch, cellulose, and alginate, are branched and linear polymeric carbohydrates. Other profit-making active natural polymers include bio rubbers such as lignin, suberin, isoprene (polyphenolic complex polymers), and cutin/cutan (lengthy chains of fatty acids and the melanin of complex polymers). Since their bio-origin comes from agricultural non-food crops that can be cultivated indefinitely, biopolymers are sustainable, carbon neutral, and often renewable. As a result, the use of biopolymers will result in a long-term industry [63]. In this context, cost-effective and environmentally friendly polymer electrolytes derived from renewable sources may be a viable alternative to conventional polymers for energy conversion and storage. Polysaccharides such as chitosan, chitin, agar-agar, carrageen, cellulose, and starch are examples of natural polymers considered to be the best candidates for polyelectrolyte membranes in fuel cell technology due to their abundance in the environment [64–66]. On this foundation, research on biopolymer membrane materials and systems has focused on: (i) identification of effective biopolymer materials, and (ii) increasing the processability of the membrane to be adopted as an efficient polyelectrolyte. Biopolymers such as chitosan and alginate have shown promise in fuel cells by improving proton conductivity, membrane resilience, fuel crossover, and electro-osmotic drag. Purwanto et al. reported on chitosan for direct methanol fuel cell (DMFC) applications; a montmorillonite (MMT) filler was proposed as an alternative membrane electrolyte. They discovered that as the O-MMT loadings increased, the water and methanol uptake decreased, though the ion exchange capacity (IEC) value increased [67].

8.3.3 Synthetic Polymers

Synthetic polymers can be classified into four main categories namely, thermoplastics, thermosets, elastomers, and synthetic fibers. The backbones of most synthetic polymers are made up of carbon–carbon bonds, whereas hetero chain polymers have other components added along the backbone, such as oxygen, sulfur, and nitrogen. Low-density polyethylene, high-density polyethylene, polypropylene (PP), polyvinyl chloride (PVC), polystyrene (PS), nylon, teflon, and thermoplastic polyurethane are the most widely utilized synthetic polymers [68]. Using the new Cu(I)-catalyzed

TABLE 8.2
Advantages of Polyelectrolytes over Conventional Liquid-Based Electrolytes in FCs

Properties of Polymer Electrolyte	Advantages of Polymer Electrolyte	Applications of Polymer Electrolyte
➢ Transparency ➢ Flexible and solvent-free ➢ Designed as a thin film of polymeric electrolytes ➢ Ionic conductance of polymer membrane is high ➢ Facile fabrication ➢ Wide voltage window	➢ Ensures that there are no leaks ➢ There is no internal shorting in the circuit ➢ Removes the corrosive solvents ➢ No harmful gases are produced	➢ Solid-state batteries, solar, fuel cells, and portable power sources ➢ In electric vehicles, thermoelectric generators, and thin credit cards ➢ In mobile phones, laptops, and computers

reaction "click chemistry" method for alkaline fuel cell applications, Liu et al. formed a series of cross-linked, comb-shaped, polystyrene (PS) anion exchange membranes (AEMs) with a C-16 alkyl side chain. These cross-linked AEMs showed lower water uptake which resulted in good dimensional stability.

8.4 POLYMERS AS ELECTROLYTES FOR BATTERIES, SUPERCAPACITORS, AND FUEL CELLS

A polyelectrolyte (PE) membrane is made up of salt dissolving in a high-molecular-weight polymer structure. The popularity of solid-state materials like ceramics, glass, and crystal as electrolytic media was initiated in the early 1970s; Fenton et al. were the first to implement a polyelectrolyte in 1973 [69, 70]. In recent years, PEs have shown several advantages over other liquid electrolytes in many advanced electrochemical systems, for example FCs, solar cells, supercapacitors, secondary batteries, sensors, and analog memory devices, with a large potential window as listed in Table 8.2 [71–74]. This conducting polymeric ionic phase has transfer characteristics comparable to liquid electrolytes, as well as transparency, being a solvent-free medium, of negligible weight, long-lasting stability, the capacity to form a thin film, ionic conductivity, ease of manufacture, and large operating potential windows [75]. PEs improve protection by preventing issues such as electrolyte leakage, internal shorting, corrosive solvent usage, toxic gas output, and the existence of noncombustible products on the surface of electrodes [76–78]. Polymeric electrolytes are categorized into three groups depending on their physical state and composition: (i) gel-polymer electrolytes (GPEs), (ii) solid-polymer electrolytes (SPEs) [79], and (iii) composite polymer electrolytes (CPEs) [80].

Electrochemical capacitors (ECs) can be distinguished based on their charge storage mechanism:

1. The separation of electrostatic charges on the electrode and electrolyte interface creates capacitance in electric double-layer capacitors (EDLCs) as shown in Figure 8.8(a). The electrode materials are normally made of highly porous carbon materials to improve charge storage ability.
2. Pseudo-capacitors, which depend upon fast and reversible faradaic redox reactions to store (create capacitance) the charges, as shown in Figure 8.8(b).
3. Hybrid ECs, which follow the combination of electrical double-layer (EDL) and faradaic mechanisms to store charges. The most recent advancements in this field are battery-type hybrid systems as shown in Figure 8.8(c).

The electrode, electrolyte content, cell configuration, and cell packaging are all known to influence the energy and power density of a supercapacitor. Until the advent of polyelectrolytes [81, 82], flexible devices suffered a commercial setback. Polyelectrolytes provide the much-desired intimate contact between the electrode and the electrolyte in flexible devices, lowering interfacial resistance and increasing device capacitance in the solid state [83]. Xie et al. published a novel redox-mediated gel polyelectrolyte (GPE) containing PVA, phosphoric acid, phosphomolybdic acid, and symmetric graphite carbon paper as electrodes in a solid-state carbon paper supercapacitor [17]. Similarly, a gel polyelectrolyte that is a versatile, translucent, and environmentally friendly film using carboxylated chitosan hydrochloric acid biodegradable polymer with high elasticity, has a maximum electrolyte utilization rate of 74.2 wt.% and a maximum ionic conductivity of 8.69 10^{-2} S cm^{-1} [84].

In 1970, for the first time, the use of PEs for Li-ion batteries (LIBs) with enhanced electric energy density and efficiency was observed. PEs have unusual properties in this shape, including elasticity, translucency, feather light weight, effortlessness formation, bounciness, and the capability to form good electrode/electrolyte communication. A single cell of a LIB consists of two electrodes and an electrolyte made up of salts that allow ions to migrate from the cathode to the anode and vice versa. A LIB is mainly composed of (i) a cathode, (ii) an anode, and (iii) a polyelectrolyte which is an electrolyte cum separator. A solitary battery can be composed of many electrochemical

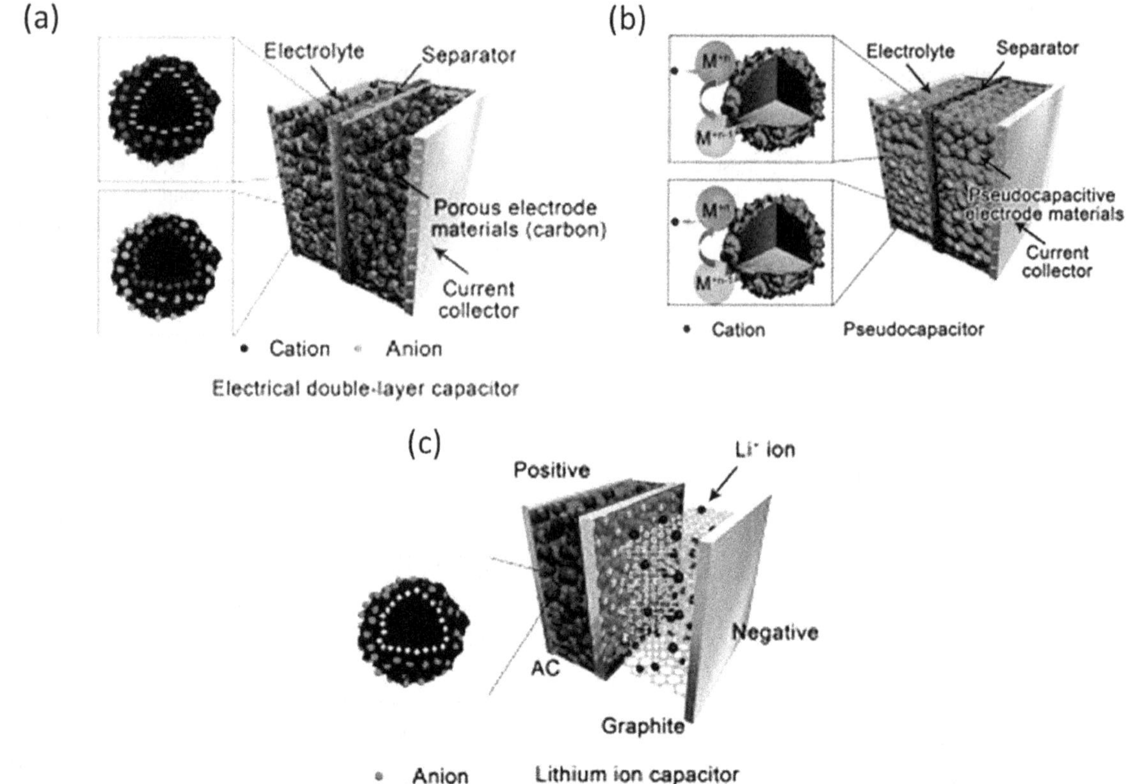

FIGURE 8.8 Figure depicting charge storage mechanism of (a) electric double-layer capacitor, (b) pseudo-capacitor, (c) lithium hybrid ion capacitor.

mono units arranged in sequences or equivalent style. The mechanism is as shown in Figure 8.9(a).

Recently Bae et al. constructed a nanostructured 3D $Li_{0.35}La_{0.55}TiO_3$ (LLTO) system derived from a hydrogel, which was used as a 3D nanofiller for an elevated performance composite polymer Li-ion electrolyte. The pre-percolating structure of the LLTO framework improved Li-ion conductivity to 8.8×10^{-5} S cm^{-1} at room temperature, as per a systematic percolation study. The material and structural architecture of this 3D nanostructured hydrogel-derived composite electrolyte represents a useful technique for fabricating high-performance composite electrolytes [85]. However, in FCs, the durability and efficiency mainly depend upon its components, as shown in Figure 8.9(b), where the polyelectrolytic membrane plays a vital role with (i) high proton conductivity and no electronic conductivity; (ii) electrochemical and chemical stability under operating conditions; (iii) adequate mechanical strength and stability; (iv) enhanced columbic performance, by use of extremely low fuel or oxygen bypass; and (v) costs of production are appropriate for the intended use [86]. Nafion, perfluorinated ionomers, partially fluorinated ionomers, non-fluorinated membranes through aromatic backbones, non-fluorinated hydrocarbons, and acid–base blend membranes are the most prominent ion exchange membranes used in FCs [86]. However, Nafion suffers from two major issues, one being high cost and the other is that methanol cross-over through these membranes in DMFCs is significantly high. To resolve these limitations, research is going on to make use of biomaterials and low-cost polymers as the composite membrane in FCs [87, 88].

8.5 OVERVIEW OF POLYMERS IN MEMBRANE-ELECTRODE ASSEMBLIES

The MEA (membrane electrode assembly) is the core of the fuel cell, so it must be designed properly, since the MEA's purpose is to effectively regulate the movement of electrons released at the negative electrode (hydrogen oxidation) to the electron-consuming reaction at the positive electrode [89, 90]. In an alkaline fuel cell, this is normally accomplished by separating the cathodic and anodic reactions using a membrane that conducts protons (H^+) and hydroxide ions. From the anode to the cathode, electrons are channeled into an external circuit. The anode and cathode reactions of polymer electrolyte fuel cells are regulated by regulating the flow and path of electrons, as shown in Equations (8.4 and 8.5).
The reaction at the anode:

$$H_2\,2H^+ + 2e^- \tag{8.4}$$

The reaction at the cathode:

$$1/2O_2 + 2H^+ + 2e^-\,H_2O \tag{8.5}$$

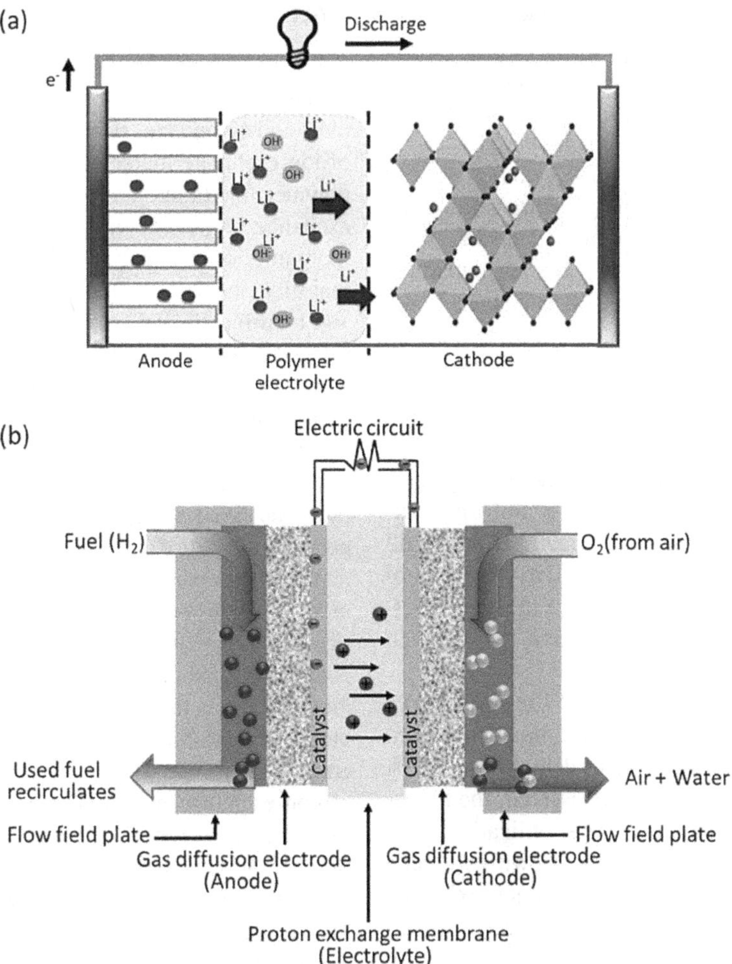

FIGURE 8.9 (a) Lithium polymer battery, (b) components of fuel cell.

To force an electron flow from the anode to the cathode, a sufficiently high electrical potential must be introduced between the two electrodes. Mitigating power losses during the electrolysis process is one of the objectives of MEA design. The properties of each portion of the MEA have been optimized by various scientists. Depending on the working and operational conditions of the FC, components with the most promising properties must always be selected as the MEA. The most commonly used polyelectrolytic membrane in the MEA is Nafion, the anode catalyst is IrO_2 [91], and carbon-supported Pt is the cathode catalyst [92]. Table 8.3 gives the list of components properties and functions in the FC.

The performance of the MEA not only depends on the individual components, but also on the entire assembly. Therefore, oxidation at the anode and reduction at the cathode of fuel and air (O_2) respectively happens at an equal rate. To achieve this, several conditions should be optimized, such as combined electrode overpotential, which must be small enough to allow for a fair rate of reaction [93]. The electron liberated from the anode due to the oxidation of fuel must be

TABLE 8.3
List of Components Involved in MEA Preparation and Their Properties in FCs

No.	Membrane Electrode Assembly Components	Properties of Components
1	Membrane	The membrane properties, such as conductance (S/cm), gas permeance, and electro-osmotic properties, are the function of temperature
2	Catalyst layer on the cathode	Catalytic behaviors of the particular material, cathode layer electrical resistance, and particle size
3	Catalyst layer on the anode	Catalytic behavior of anode, the electrical resistance of the anode diffusion layer, and particle size
4	Cathode baking layer	Mass transfer effects of products and oxidants (air)
5	Anode backing layer	Mass transfer effects of products and hydrogen fuel

available to the cathode catalyst to complete the reaction [94] and simultaneously to hydronium ions from the anodic catalyst to the cathodic catalyst across the membrane. Therefore, the MEA provides the optimal electrochemical activity and the optimal number of three-phase boundaries [95].

Factors affecting the MEA. The MEA's output is influenced by key parameters such as the glass transition temperature (G_t), pressure distribution onto and across the MEA, gas convergence at elevated pressure, gas departure, contact resistance between sub-layers, catalyst layers, ionomer material, and the sub-layers' long-lasting stability [96]. Among these factors, four important factors can be discussed:

1. Pressure distribution: During the design, engineering the assembly of a stack and a uniform, efficient pressure distribution is critical; failure to do so may result in issues of productivity in terms of gas outpouring, contact resistance of the electrodes, and cell malfunctioning/destruction of stacked elements.
2. Electrochemical active surface area: The surface area of the catalyst and its assistance are more important than the catalyst loading itself in determining MEA results. Although it is not always assured, the higher surface area results in better catalyst utilization.
3. Activation polarization loss: Also known as kinetic loss, this is caused by inefficient hydrogen oxidation at the negative (anode) and can be reduced with the usage of a highly active electro-catalyst [17].
4. Concentration polarization: At high current densities, mass transfer limitations for gaseous product diffusion and water transport between MEA sub-layer pores affect MEA efficiency. To keep the MEA dried up under moist operating status, efficient two-step transfer of electrons and protons inside the pores of the bi-layer diffusion layer and catalyst layer (CL) is needed [97].

8.6 ROLE OF POLYMERS IN FUEL CELLS

As we discussed earlier, in FCs, the polyelectrolyte (polymer membrane) allows ion transport and stops gas diffusion through it. As a result, each reaction takes place at the anode, and the cathode results in ions (hydronium ions) passing through the electrolyte. The protons can react with O_2 and electrons at the cathode after migrating across the electrolyte/polyelectrolyte as H_2 oxidizes on the anode electrode, which releases protons and electrons. The movement of charged ions across the electrolyte must be controlled by the flow of electronic charge through an external circuit to generate electricity. Hydronium ions, hydroxide ions, oxide ions, and carbonate ions are movable in electrolytes, and they are the foundation and of common interest for all forms of FCs [98].

8.6.1 Polymers as Ion Exchange Media

Ion-exchange membrane preparation techniques are closely connected to the ion exchange abilities of individual polymers and composites [99]. There are several different types of ion-exchange materials, much like resins, with distinct polymer mediums and functional units to define ion-exchange properties in the material. There is a variety of inorganic ion-exchange materials that exist [100], but the majority of them depend on zirconium phosphates, clays, and aluminosilicates. Ion-exchange membranes are generally made up of conjugated polymers and are categorized either by their application as a partitioning medium or by their content and basic structure. Ion-exchange media are categorized according to their polymer functionality: (i) cation exchange membranes, in which the membrane contains the fixed negatively charged ions but facilitates the permeability for exchange of cations; (ii) an ion-exchange membrane consisting of a static positively charged polymer that allows for selective permeability for anions; (iii) bipolar membranes, where positively and negatively charged groups are fixed together in membranes; and (iv) membranes in the atmosphere that are randomly distributed with both fixed cationic and anionic groups. The penetration of proton and hydroxide ions can be seen in the proton exchange membrane fuel cell (PEMFC) and anion exchange membrane fuel cell (AEMFC) [101]. In all cases, the membrane should be mechanically solid, with negligible contraction and expansion when switching from weak to strong ionic solutions, more stability, and stability across the pH spectrum and in the existence of an oxidation agent [102, 103]. Over time, it has become clear that hydrocarbon-based polymers, such as polyether sulfones (PESF), polyether ketones (PEK), poly(arylene ethers), polyesters, and polyimides (PI), have less chemical and oxidative stability than fluorocarbon-based polymers. Rigid sites, such as aromatic structures, are integrated directly into the polymer backbone to improve membrane stability and properties. Aromatic rings provide inflexibility, resulting in enhanced thermal and mechanical stability [66, 104].

8.6.2 Polymer Composites as Ion-Exchange Media

A variety of materials and modifying methods are available to tune the desired end properties of composite membranes. Furthermore, composite membranes can be made from a variety of sources, including precursors and particles. Nanomaterials are among the many components that are commonly used in composite fabrications, including solids, acids, silica [105], metals [106], bentonites (clays), and mixing with another polymer due to their exclusive properties such as surface activity and outstanding mechanical stability for long-lasting usage. The electronic properties of the polymers are improved by inducing effective ionic sites into them. These ionic sites are induced by different techniques such as chemical deposition, the sol–gel technique, and doping of inorganic nanomaterials into the polymer matrix [107, 108]. To enhance the effectiveness of polymer composites membranes, organic and inorganic functional groups can be used. There are different types of chemical bonds between functional groups of organic and inorganic compounds, which alter the properties of polymeric composite membranes. These types of chemical bonds are categorized into two types:

1. Van der Waals interactions and hydrogen bonding: these bonds are only limited to a weak interaction between the functional groups of organic and inorganic compounds with a polymeric chain in hybrid composite membranes.
2. Covalent/ionic interaction: In this type of interaction, functional groups, such as silanes, metal oxides, and metal alkoxides, give strong covalent/ionic bonding between the polymer chains and functional groups of chemical compounds of the composite membrane, as well as at the interface of the polymer and the functional group of inorganic compounds, which allows the covalent bonding with the polymer matrix which behaves as a compatibilizer [109, 110].

In addition to this, proton exchange membranes are classified into three categories: (i) perfluorinated composite membranes (PCMs), (ii) an acid-base complex polymer with inorganic–organic composites, and (iii) nonfluorinated inorganic–organic composites with aromatic polymers. Modified perfluorinated membranes involve hydrophilic metal oxides (hydrophilicity due to the presence of metal cations and oxygen anions), which are transition metal oxides (TiO_2, ZrO_2), SiO_2, and also hetero atoms containing polyacids (silicotungstic acid, phosphotungstic acid) and which have been studied extensively. In many cases, popular Nafion membranes with several nanomaterials of inorganic origin are generally used for enhancing the durability and productivity of the membranes. Nafion membranes with silica were produced by using different methods including grafting silica onto the polymer chain and also by using a sol-gel reaction with several silicate materials [111]. Kumari et al. reported a series of sulfonated PEEK-PEGs cross-linked and interpenetrating with SiO_2 nanocomposite ion-exchange membranes [112] with a better performance of FC applications. In another study, a new ternary nanocomposite polyelectrolyte membrane containing sulfonated PSU, MOF (NH_2MIL-53(Al)), and silica nanoparticles was investigated with significant improvement of the mechanical and thermal properties of the prepared membranes [113].

8.6.3 Polymers and Their Composites as Electrocatalysts

Carbon-supported Pt catalysts are now the most popular and practical electrocatalysts for PEM fuel cells. The corrosion of carbon support materials, on the other hand, has been reported as a major cause of catalyst failure. Other failure modes, such as Pt dissolution, sintering, and agglomeration, have also contributed to catalyst degradation, especially for platinum and (Pt)-based catalysts [29]. Multiple carbon-based and metal-based materials have been used as electrocatalysts to obtain a high-performing fuel cell catalyst. However, due to their specific optical, electrical, chemical, and mechanical properties, as well as their ease of processing and application in electrochemical energy conversion, electronically conducting polymers are acquiring more attention in the current situation [114, 115]. Polymers such as polyaniline, polypyrrole, and polythiophene have been successfully demonstrated as support for the dispersion of metallic grains [115]. These polymer backings include the possibility of large surface areas and more conductance in the wide voltage window where the polymers molecule can be oxidized, with acceptable stability of more than 72 hours in acidic electrolyte for example 1 molar hydrochloric acid. The dispersion of platinum in contact with the polymeric backings to form assisted catalysts which minimize the amount of platinum while improving the activity of electrocatalyst for oxidation of hydrogen [17]. Because of its high electrical conductance in its partly oxidized state, polyaniline (PANI) is an appealing material for use as a polymeric catalyst backing. It also has strong adhesion to certain substances, for example, CNT (carbon nanotubes) and glass fibers, and excellent chemical stability in aqueous solutions. PANI-assisted Pt alloys and platinum catalysts are an excellent strategy to decrease the cost of noble materials by increasing their performance [116]. In an attempt to prepare advanced electrocatalysts at the anode in fuel cells, new approaches using poly(2-fluoroaniline) and poly(2,3,5,6-tetrafluoroaniline) have recently centered on the alteration of PANI backings for Pt catalyst. When compared to traditional polyaniline support, the results showed that some stable fluorinated PANI increases the electrocatalytic activity of platinum against hydrogen oxidation in the fuel cell [117].

Recently Tiwari et al. synthesized SnO_2: PANI composites were used as the cathode material in an air-cathode microbial fuel cell (MFC) to improve the oxygen reduction reaction (ORR). The efficiency of SnO_2: PANI composites were tested using various SnO_2: PANI composite ratios. Similarly, polypyrrole (PPY) is used as electrocatalyst support due to its mechanical and chemical stability [118]. Not long ago, Nanostructured PPy film materials, in addition to traditional PPy films, have sparked interest because of the large surface area, high conductance, and specific charge transfer properties. Qi et al. developed blended backings for Pt catalyst consisting of PPy and PSS (polystyrene sulfonate) and analyzed as gas diffusion electrodes (GDE) with enhanced electron and hydronium ion transport, which can be utilized in manufacturing the classic PEMFC electrodes [119]. As an electrocatalyst, the Nafion membrane and polypyrrole composites were also used. Stainless steel (SS) is regarded as a worthy anode material in microbial fuel cells (MFCs) due to its low cost and high mechanical strength, according to Pu et al., who used in-situ electrochemical deposition of polypyrrole (PPy) onto stainless steel (PPy/SS) to achieve this. The corrosion resistance and power generation efficiency of the anode were significantly improved using this process benthic microbial fuel cells (BMFC) were stated by another group of researchers. By changing the surface of electrodes, according to the researchers [120]. The output of PPy induced metal oxide blended electrode materials for benthic microbial cells (BMFCs). The performances of PPy coated MnO_2, Fe_2O_3, and MnO_2-Fe_2O_3 nanocomposites were also estimated in electrochemical studies, FP(170 mW/m^2) outclassed other MFP (117.29 mW/m^2), MP (90.54 mW/m^2), and non-modified electrodes (69.19 mW/m^2) in terms of power density [121].

8.7 CHALLENGES IN DESIGNING COMPATIBLE POLYMER-BASED MEMBRANE ELECTRODE ASSEMBLIES

In contrast to commercially successful portable energy devices like lithium-ion batteries and advanced rechargeable battery concepts, direct methanol fuel cells (DMFCs) and PEMFCs with methanol solution and hydrogen gas as a fuel and as the energy transporters, respectively, provide the best scenarios for being utilized in portable systems. However, among many major challenges to obtaining a high performance in FCs, selecting an efficient, cost-effective, and stable polyelectrolytic membrane is an important step. Chlor-alkali (PFSA) copolymers, as membrane materials, also identified as Nafion membranes, have been broadly used in DMFCs and PEMFCs as the membrane for years. However, among the many drawbacks of Nafion and their composite-based membrane series is severe fuel loss, such as high methanol transport through the membrane from the negative electrode to the positive electrode, flooding, and mechanical unreliability. These flaws result in a notable decrease in fuel cell productivity due to a mixed potential at the cathode, which reduces fuel utilization efficiency and, as a result, suppresses the commercialization of FCs. These limitations still exist in many of the developed and being developed polymer-based ion-exchange membranes in FCs. This is driving extensive research toward modifying the properties of base polymers, their blends, and composites. In addition to this, research is also underway to maintain the above properties without sacrificing mechanical or chemical strength.

8.8 CONCLUSION AND FUTURE PROSPECTS

In conclusion, this chapter has provided an overview of the various aspects of the polymer-based membranes used in fuel cells. The overview of the polyelectrolytic membrane showed simple and eco-friendly sources can be utilized to accomplish high power and energy density while also overcoming the shortcomings of current synthetic polymers and modified Nafion-based membranes. For years, synthetic polymers and their composites have been highly successful as a polymer–electrolyte membrane (PEM) used in DMFCs and PEMFCs, although many studies have revealed that in DMFCs and PEMFCs methanol crossover and fuel loss is a major drawback for substantial fuel cell performance. Due to these limitations in synthetic polymer membranes, the issue of renewable resources arises. Presently, one of the queries in overcoming the conversion from synthetic and moderately performing polymers to renewable energy sources is to find a suitable stock of materials. Nevertheless, fuel cells will continue to be widely recognized as a highly efficient alternative energy generation system that works on renewable fuels. For this, improving polymer-based membrane performance and long-lasting stability are the key features for the short-term success of FCs in the world's market.

ACKNOWLEDGMENTS

S. K. Nataraj conveys many thanks to the Department of Science and Technology (Government of India) for the NANO MISSION Project (SR/NM/NT-1073/2016), DST INSPIRE (IFA12-CH-84), and DST-Technology Mission Division (DST/TMD/HFC/2K18/124G) (Government of India) for economic support, as well as the "Talent Attraction Programme Funded by the Community of Madrid Spain/Spanish (2017-T1/AMB5610)".

REFERENCES

[1] S. Li, X. Hao, A. Abudula, G. Guan, Nanostructured Co-based bifunctional electrocatalysts for energy conversion and storage: Current status and perspectives, *Journal of Materials Chemistry A* 7 (2019) 18674–18707.
[2] P. Gu, M. Zheng, Q. Zhao, X. Xiao, H. Xue, H. Pang, Rechargeable zinc–air batteries: A promising way to green energy, *Journal of Materials Chemistry A* 5 (2017) 7651–7666.
[3] A. Boudghene Stambouli, E. Traversa, Fuel cells, an alternative to standard sources of energy, *Renewable and Sustainable Energy Reviews* 6 (2002) 295–304.
[4] C. Meng, O.Z. Gall, P.P. Irazoqui, A flexible super-capacitive solid-state power supply for miniature implantable medical devices, *Biomedical Microdevices* 15 (2013) 973–983.
[5] S. Srinivasan, *Fuel cells: from fundamentals to applications*, Berlin/Heidelberg: Springer Science & Business Media, 2006.
[6] H. Xu, L. Kong, X. Wen, Fuel cell power system and high power DC–DC converter, *IEEE Transactions on Power Electronics* 19 (2004) 1250–1255.
[7] J. Walkowiak-Kulikowska, J. Wolska, H. Koroniak, Polymers application in proton exchange membranes for fuel cells (PEMFCs), *Physical Sciences Reviews* 2 (2017).
[8] A. Kirubakaran, S. Jain, R.K. Nema, A review on fuel cell technologies and power electronic interface, *Renewable and Sustainable Energy Reviews* 13 (2009) 2430–2440.
[9] B. Cook, Introduction to fuel cells and hydrogen technology, *Engineering Science & Education Journal* 11 (2002) 205–216.
[10] R.M. Ormerod, Solid oxide fuel cells, *Chemical Society Reviews* 32 (2003) 17–28.
[11] S. Mekhilef, R. Saidur, A. Safari, Comparative study of different fuel cell technologies, *Renewable and Sustainable Energy Reviews* 16 (2012) 981–989.

[12] A.B. Stambouli, Fuel cells: The expectations for an environmental-friendly and sustainable source of energy, *Renewable and Sustainable Energy Reviews* 15 (2011) 4507–4520.

[13] A. Bıyıkoğlu, RETRACTED: Review of proton exchange membrane fuel cell models, *International Journal of Hydrogen Energy* 30 (2005) 1181–1212.

[14] B.C. Ong, S.K. Kamarudin, S. Basri, Direct liquid fuel cells: A review, *International Journal of Hydrogen Energy* 42 (2017) 10142–10157.

[15] G. McLean, An assessment of alkaline fuel cell technology, *International Journal of Hydrogen Energy* 27 (2002) 507–526.

[16] S.S. Munjewar, S.B. Thombre, R.K. Mallick, Approaches to overcome the barrier issues of passive direct methanol fuel cell – Review, *Renewable and Sustainable Energy Reviews* 67 (2017) 1087–1104.

[17] J.H. Lee, Y.K. Jang, C.E. Hong, N.H. Kim, P. Li, H.K. Lee, Effect of carbon fillers on properties of polymer composite bipolar plates of fuel cells, *Journal of Power Sources 193*(2) (2009) 523–529.

[18] M. Pihlatie, *Stability of Ni-YSZ composites for solid oxide fuel cells during reduction and re-oxidation*, Otaniemi: VTT, 2010.

[19] H. Frey, T. Johann, Celebrating 100 years of "polymer science": Hermann Staudinger's 1920 manifesto, *Polymer Chemistry* 11 (2020) 8–14.

[20] H. Staudinger, Über polymerisation, *Berichte der Deutschen Chemischen Gesellschaft (A and B Series)* 53 (1920) 1073–1085.

[21] H. Shirakawa, E.J. Louis, A.G. MacDiarmid, C.K. Chiang, A.J. Heeger, Synthesis of electrically conducting organic polymers: Halogen derivatives of polyacetylene, $(CH)_x$, *Journal of the Chemical Society, Chemical Communications* (1977) 578. https://doi.org/10.1039/C39770000578

[22] H. Wu, G. Yu, L. Pan, N. Liu, M.T. McDowell, Z. Bao, Y. Cui, Stable Li-ion battery anodes by in-situ polymerization of conducting hydrogel to conformally coat silicon nanoparticles, *Nature Communications* 4 (2013) 1943.

[23] L. Hu, H. Wu, S.S. Hong, L. Cui, J.R. McDonough, S. Bohy, Y. Cui, Si nanoparticle-decorated Si nanowire networks for Li-ion battery anodes, *Chemical Communication* 47 (2011) 367–369.

[24] K.S. Weeraratne, A.A. Alzharani, H.M. El-Kaderi, Redox-active porous organic polymers as novel electrode materials for green rechargeable sodium-ion batteries, *ACS Applied Materials & Interfaces* 11 (2019) 23520–23526.

[25] X. Wu, B. Huang, R. Lv, Q. Wang, Y. Wang, Highly flexible and low capacitance loss supercapacitor electrode based on hybridizing decentralized conjugated polymer chains with MXene, *Chemical Engineering Journal* 378 (2019) 122246.

[26] D. He, Y. Rong, Z. Kou, S. Mu, T. Peng, R. Malpass-Evans, M. Carta, N.B. McKeown, F. Marken, Intrinsically microporous polymer slows down fuel cell catalyst corrosion, *Electrochemistry Communications* 59 (2015) 72–76.

[27] S. Giddey, A versatile polymer electrolyte membrane fuel cell (3 kWe) facility, *Solid State Ionics* 152–153 (2002) 363–371.

[28] S. Verhelst, T. Wallner, Hydrogen-fueled internal combustion engines, *Progress in Energy and Combustion Science* 35 (2009) 490–527.

[29] Y.-J. Wang, D.P. Wilkinson, J. Zhang, Noncarbon support materials for polymer electrolyte membrane fuel cell electrocatalysts, *Chemical Reviews* 111 (2011) 7625–7651.

[30] W. Gu, T.Y. Paul, R.N. Carter, R. Makharia, H.A. Gasteiger, Modeling of membrane-electrode-assembly degradation in proton-exchange-membrane fuel cells–local H2 starvation and start–stop induced carbon-support corrosion, *Modeling and diagnostics of polymer electrolyte fuel cells*. New York: Springer, 2009, pp. 45–87.

[31] L. Roen, C. Paik, T. Jarvi, Electrocatalytic corrosion of carbon support in PEMFC cathodes, *Electrochemical and Solid State Letters* 7 (2003) A19.

[32] R. Makharia, S. Kocha, P. Yu, M.A. Sweikart, W. Gu, F. Wagner, H.A. Gasteiger, Durable PEM fuel cell electrode materials: Requirements and benchmarking methodologies, *ECS Transactions* 1 (2006) 3.

[33] N. Mansor, T.S. Miller, I. Dedigama, A.B. Jorge, J. Jia, V. Brázdová, C. Mattevi, C. Gibbs, D. Hodgson, P.R. Shearing, C.A. Howard, F. Corà, M. Shaffer, D.J.L. Brett, P.F. McMillan, Graphitic carbon nitride as a catalyst support in fuel cells and electrolyzers, *Electrochimica Acta* 222 (2016) 44–57.

[34] Y. Nabil, S. Cavaliere, I.A. Harkness, J.D.B. Sharman, D.J. Jones, J. Rozière, Novel niobium carbide/carbon porous nanotube electrocatalyst supports for proton exchange membrane fuel cell cathodes, *Journal of Power Sources* 363 (2017) 20–26.

[35] B. Seger, A. Kongkanand, K. Vinodgopal, P.V. Kamat, Platinum dispersed on silica nanoparticle as electrocatalyst for PEM fuel cell, *Journal of Electroanalytical Chemistry* 621 (2008) 198–204.

[36] Y. Suzuki, A. Ishihara, S. Mitsushima, N. Kamiya, K.-i. Ota, Sulfated-zirconia as a support of Pt catalyst for polymer electrolyte fuel cells, *Electrochemical and Solid State Letters* 10 (2007) B105.

[37] F. Memioğlu, A. Bayrakçeken, T. Öznülüer, M. Ak, Synthesis and characterization of polypyrrole/carbon composite as a catalyst support for fuel cell applications, *International Journal of Hydrogen Energy* 37 (2012) 16673–16679.

[38] R.-C. Zhang, D. Sun, R. Zhang, W.-F. Lin, M. Macias-Montero, J. Patel, S. Askari, C. McDonald, D. Mariotti, P. Maguire, Gold nanoparticle-polymer nanocomposites synthesized by room temperature atmospheric pressure plasma and their potential for fuel cell electrocatalytic application, *Scientific Reports* 7 (2017) 1–9.

[39] A. Hermann, T. Chaudhuri, P. Spagnol, Bipolar plates for PEM fuel cells: A review, *International Journal of Hydrogen Energy* 30 (2005) 1297–1302.

[40] B.D. Cunningham, J. Huang, D.G. Baird, Review of materials and processing methods used in the production of bipolar plates for fuel cells, *International Materials Reviews* 52 (2013) 1–13.

[41] J. Huang, D.G. Baird, J.E. McGrath, Development of fuel cell bipolar plates from graphite filled wet-lay thermoplastic composite materials, *Journal of Power Sources* 150 (2005) 110–119.

[42] T.L. Smith, A.D. Santamaria, J.W. Park, K. Yamazaki, Alloy selection and die design for stamped proton exchange membrane fuel cell (PEMFC) bipolar plates, *Procedia CIRP* 14 (2014) 275–280.

[43] L. Flandin, A. Danerol, C. Bas, E. Claude, G. De-Moor, N. Alberola, Characterization of the degradation in membrane electrode assemblies through passive electrical measurements, *Journal of The Electrochemical Society* 156 (2009) B1117.

[44] A. Naji, B. Krause, P. Pötschke, A. Ameli, Hybrid conductive filler/polycarbonate composites with enhanced electrical and thermal conductivities for bipolar plate applications, *Polymer Composites* 40 (2018) 3189–3198.

[45] S. Joseph, J.C. McClure, R. Chianelli, P. Pich, P. Sebastian, Conducting polymer-coated stainless steel bipolar plates for proton exchange membrane fuel cells (PEMFC), *International Journal of Hydrogen Energy* 30 (2005) 1339–1344.

[46] K. Prater, The renaissance of the solid polymer fuel cell, *Journal of Power Sources* 29 (1990) 239–250.

[47] A. Thomas, P. Kuhn, J. Weber, M.-M. Titirici, M. Antonietti, Porous polymers: Enabling solutions for energy applications, *Macromolecular Rapid Communications* 30 (2009) 221–236.

[48] H.-S. Lee, O. Lane, J.E. McGrath, Development of multiblock copolymers with novel hydroquinone-based hydrophilic blocks for proton exchange membrane (PEM) applications, *Journal of Power Sources* 195 (2010) 1772–1778.

[49] M.-Y. Lim, K. Kim, Sulfonated poly(arylene ether sulfone) and perfluorosulfonic acid composite membranes containing perfluoropolyether grafted graphene oxide for polymer electrolyte membrane fuel cell applications, *Polymers* 10 (2018) 569.

[50] A. Alaswad, A. Palumbo, M. Dassisti, A.G. Olabi, Fuel cell technologies, applications, and state of the art. A reference guide, Reference Module in Materials Science and Materials Engineering (2016).

[51] C.K. Chiang, C.R. Fincher, Y.W. Park, A.J. Heeger, H. Shirakawa, E.J. Louis, S.C. Gau, A.G. MacDiarmid, Electrical conductivity in doped polyacetylene, *Physical Review Letters* 39 (1977) 1098–1101.

[52] E. Conwell, Transport in trans-polyacetylene, *IEEE Transactions on Electrical Insulation* EI-22 (1987) 591–627.

[53] J. Kanicki, T. Skotheim, *Handbook of conducting polymers*, New York, USA: Marcel Dekker, 1986.

[54] Y. Meng, X.J. Liu, B. Di, Z. An, Recombination of polaron and exciton in conjugated polymers, *The Journal of Chemical Physics* 131 (2009) 244502.

[55] D.K. Campbell, Solitons in Polyacetylene and Related Systems, *Dynamical Problems in Soliton Systems: Proceedings of the Seventh Kyoto Summer Institute*, Kyoto, Japan, August 27–31, 1984 30 (2013) 176.

[56] S.N. Patel, A.M. Glaudell, K.A. Peterson, E.M. Thomas, K.A. O'Hara, E. Lim, M.L. Chabinyc, Morphology controls the thermoelectric power factor of a doped semiconducting polymer, *Science Advances* 3 (2017) e1700434.

[57] S.C. Rasmussen, The path to conductive polyacetylene, *Bulletin for the History of Chemistry* 39 (2014) 64–72.

[58] T.-H. Le, Y. Kim, H. Yoon, Electrical and electrochemical properties of conducting polymers, *Polymers* 9 (2017) 150.

[59] T. Tezuka, K. Tadanaga, A. Hayashi, M. Tatsumisago, Inorganic–organic hybrid membranes with anhydrous proton conduction prepared from 3-aminopropyltriethoxysilane and sulfuric acid by the sol–gel method, *Journal of the American Chemical Society* 128 (2006) 16470–16471.

[60] E. Besson, A. Mehdi, H. Chollet, C. Réyé, R. Guilard, R.J.P. Corriu, Synthesis and cation-exchange properties of a bis-zwitterionic lamellar hybrid material, *Journal of Materials Chemistry* 18 (2008) 1193.

[61] L.A. Adams, S.D. Poynton, C. Tamain, R.C.T. Slade, J.R. Varcoe, A carbon dioxide tolerant aqueous-electrolyte-free anion-exchange membrane alkaline fuel cell, *ChemSusChem* 1 (2008) 79–81.

[62] J. Escorihuela, J. Olvera-Mancilla, L. Alexandrova, L.F. del Castillo, V. Compañ, Recent progress in the development of composite membranes based on polybenzimidazole for high temperature proton exchange membrane (PEM) fuel cell applications, *Polymers* 12 (2020) 1861.

[63] M.R. Yates, C.Y. Barlow, Life cycle assessments of biodegradable, commercial biopolymers—A critical review, *Resources, Conservation and Recycling* 78 (2013) 54–66.

[64] N.F. Ab. Rahman, L.K. Shyuan, A.B. Mohamad, A.A.H. Kadhum, Review on biopolymer membranes for fuel cell applications, *Applied Mechanics and Materials* 291–294 (2013) 614–617.

[65] S. Karthikeyan, S. Selvasekarapandian, M. Premalatha, S. Monisha, G. Boopathi, G. Aristatil, A. Arun, S. Madeswaran, Proton-conducting I-carrageenan-based biopolymer electrolyte for fuel cell application, *Ionics* 23 (2016) 2775–2780.

[66] A. Muthumeenal, S. Sundar Pethaiah, A. Nagendran, 6-Biopolymer Composites in Fuel Cells, *Biopolymer Composites in Electronics* (2017) 185–217. https://doi.org/10.1016/B978-0-12-809261-3.00006-1

[67] M. Purwanto, L. Atmaja, M.A. Mohamed, M.T. Salleh, J. Jaafar, A.F. Ismail, M. Santoso, N. Widiastuti, Biopolymer-based electrolyte membranes from chitosan incorporated with montmorillonite-crosslinked GPTMS for direct methanol fuel cells, *RSC Advances* 6 (2016) 2314–2322.

[68] K. Imato, H. Otsuka, Reorganizable and stimuli-responsive polymers based on dynamic carbon–carbon linkages in diarylbibenzofuranones, *Polymer* 137 (2018) 395–413.

[69] S. Ramesh, L.C. Wen, Investigation on the effects of addition of SiO_2 nanoparticles on ionic conductivity, FTIR, and thermal properties of nanocomposite PMMA–$LiCF_3SO_3$–SiO_2, *Ionics* 16 (2009) 255–262.

[70] D. Fenton, Complexes of alkali metal ions with poly (ethylene oxide), *Polymer* 14 (1973) 589.

[71] S. Ramesh, C.-W. Liew, E. Morris, R. Durairaj, Effect of PVC on ionic conductivity, crystallographic structural, morphological and thermal characterizations in PMMA–PVC blend-based polymer electrolytes, *Thermochimica Acta* 511 (2010) 140–146.

[72] P. Bruce, *Solid State electrochemistry*, Cambridge: *Cambridge University Press*, 1995.

[73] J.H. Kim, M.-S. Kang, Y.J. Kim, J. Won, N.-G. Park, Y.S. Kang, Dye-sensitized nanocrystalline solar cells based on composite polymer electrolytes containing fumed silica nanoparticles, *Chemical Communications* (2004) 1662.

[74] K.S. Ngai, S. Ramesh, K. Ramesh, J.C. Juan, A review of polymer electrolytes: fundamental, approaches and applications, *Ionics* 22 (2016) 1259–1279.

[75] N.N. Sa'adun, R. Subramaniam, R. Kasi, Development and characterization of poly(1-vinylpyrrolidone-co-vinyl acetate) copolymer based polymer electrolytes, *The Scientific World Journal* 2014 (2014) 1–7.

[76] M.H. Khanmirzaei, S. Ramesh, Nanocomposite polymer electrolyte based on rice starch/ionic liquid/TiO_2 nanoparticles for solar cell application, *Measurement* 58 (2014) 68–72.

[77] M. Khanmirzaei, S. Ramesh, Ionic transport and FTIR properties of lithium iodide doped biodegradable rice starch based polymer electrolytes, *International Journal of Electrochemical Science* 8 (2013) 9977–9991.

[78] K.H. Teoh, S. Ramesh, A.K. Arof, Investigation on the effect of nanosilica towards corn starch–lithium perchlorate-based polymer electrolytes, *Journal of Solid State Electrochemistry* 16 (2012) 3165–3170.

[79] A.C. Bloise, C.C. Tambelli, R.W.A. Franco, J.P. Donoso, C.J. Magon, M.F. Souza, A.V. Rosario, E.C. Pereira, Nuclear magnetic resonance study of PEO-based composite polymer electrolytes, *Electrochimica Acta* 46 (2001) 1571–1579.

[80] A. Manuel Stephan, Review on gel polymer electrolytes for lithium batteries, *European Polymer Journal* 42 (2006) 21–42.

[81] L. Dong, C. Xu, Y. Li, Z.-H. Huang, F. Kang, Q.-H. Yang, X. Zhao, Flexible electrodes and supercapacitors for wearable energy storage: a review by category, *Journal of Materials Chemistry A* 4 (2016) 4659–4685.

[82] X. Peng, L. Peng, C. Wu, Y. Xie, Two dimensional nanomaterials for flexible supercapacitors, *Chemical Society Reviews* 43 (2014) 3303.

[83] H. Wu, M. Genovese, K. Ton, K. Lian, A comparative study of activated carbons from liquid to solid polymer electrolytes for electrochemical capacitors, *Journal of The Electrochemical Society* 166 (2019) A821.

[84] H. Yang, Y. Liu, L. Kong, L. Kang, F. Ran, Biopolymer-based carboxylated chitosan hydrogel film crosslinked by HCl as gel polymer electrolyte for all-solid-sate supercapacitors, *Journal of Power Sources* 426 (2019) 47–54.

[85] J. Bae, Y. Li, J. Zhang, X. Zhou, F. Zhao, Y. Shi, J.B. Goodenough, G. Yu, A 3D nanostructured hydrogel-framework-derived high-performance composite polymer lithium-ion electrolyte, *Angewandte Chemie International Edition* 57 (2018) 2096–2100.

[86] N. Shaari, S.K. Kamarudin, Chitosan and alginate types of bio-membrane in fuel cell application: An overview, *Journal of Power Sources* 289 (2015) 71–80.

[87] L. Zhang, S.-R. Chae, Z. Hendren, J.-S. Park, M.R. Wiesner, Recent advances in proton exchange membranes for fuel cell applications, *Chemical Engineering Journal* 204–206 (2012) 87–97.

[88] P. Prapainainar, S. Maliwan, K. Sarakham, Z. Du, C. Prapainainar, S.M. Holmes, P. Kongkachuichay, Homogeneous polymer/filler composite membrane by spraying method for enhanced direct methanol fuel cell performance, *International Journal of Hydrogen Energy* 43 (2018) 14675–14690.

[89] P. Medina, M. Santarelli, Analysis of water transport in a high pressure PEM electrolyzer, *International Journal of Hydrogen Energy* 35 (2010) 5173–5186.

[90] H. Ito, T. Maeda, A. Nakano, Y. Hasegawa, N. Yokoi, C.M. Hwang, M. Ishida, A. Kato, T. Yoshida, Effect of flow regime of circulating water on a proton exchange membrane electrolyzer, *International Journal of Hydrogen Energy* 35 (2010) 9550–9560.

[91] S. Siracusano, V. Baglio, A. Stassi, R. Ornelas, V. Antonucci, A.S. Aricò, Investigation of IrO_2 electrocatalysts prepared by a sulfite-couplex route for the O_2 evolution reaction in solid polymer electrolyte water electrolyzers, *International Journal of Hydrogen Energy* 36 (2011) 7822–7831.

[92] H.-Y. Jung, S. Park, B.N. Popov, Electrochemical studies of an unsupported PtIr electrocatalyst as a bifunctional oxygen electrode in a unitized regenerative fuel cell, *Journal of Power Sources* 191 (2009) 357–361.

[93] W. Dai, H. Wang, X.-Z. Yuan, J.J. Martin, D. Yang, J. Qiao, J. Ma, A review on water balance in the membrane electrode assembly of proton exchange membrane fuel cells, *International Journal of Hydrogen Energy* 34 (2009) 9461–9478.

[94] Q. Guo, Z. Qi, Effect of freeze-thaw cycles on the properties and performance of membrane-electrode assemblies, *Journal of Power Sources* 160 (2006) 1269–1274.

[95] B. Han, J. Mo, Z. Kang, F.-Y. Zhang, Effects of membrane electrode assembly properties on two-phase transport and performance in proton exchange membrane electrolyzer cells, *Electrochimica Acta* 188 (2016) 317–326.

[96] K. Elias, H. Kurek, Principles of high performance membrane electrode assembly fabrication, *Worcester Polytechnic Institute* 188 (2014) 317–326.

[97] B. Bladergroen, H. Su, S. Pasupathi, V. Linkov, *Overview of membrane electrode assembly preparation methods for solid polymer electrolyte electrolyzer*. London: InTech, 2012.

[98] K. Haraldsson, K. Wipke, Evaluating PEM fuel cell system models, *Journal of Power Sources* 126 (2004) 88–97.

[99] E. Tooper, L. Wirth, *Ion exchange resins, Ion Exchange Technology*. Amsterdam: Elsevier, 1956, pp. 7–26.

[100] F. Helfferich, Ion-exchange kinetics.1iii. Experimental test of the theory of particle-diffusion controlled ion exchange, *The Journal of Physical Chemistry* 66 (1962) 39–44.

[101] S. Gottesfeld, D.R. Dekel, M. Page, C. Bae, Y. Yan, P. Zelenay, Y.S. Kim, Anion exchange membrane fuel cells: Current status and remaining challenges, *Journal of Power Sources* 375 (2018) 170–184.

[102] S. Heiner, *Preparation and characterization of ion exchange membranes, Membrane Science and Technology*. Amsterdam: Elsevier, 2004, pp. 89–146.

[103] K.-D. Kreuer, Proton conductivity: Materials and applications, *Chemistry of Materials* 8 (1996) 610–641.

[104] J.K. Lee, W. Li, A. Manthiram, Poly(arylene ether sulfone)s containing pendant sulfonic acid groups as membrane materials for direct methanol fuel cells, *Journal of Membrane Science* 330 (2009) 73–79.

[105] M.M. Ayad, A. Abu El-Nasr, J. Stejskal, Kinetics and isotherm studies of methylene blue adsorption onto polyaniline nanotubes base/silica composite, *Journal of Industrial and Engineering Chemistry* 18 (2012) 1964–1969.

[106] M. Khajenoori, M. Rezaei, B. Nematollahi, Preparation of noble metal nanocatalysts and their applications in catalytic partial oxidation of methane, *Journal of Industrial and Engineering Chemistry* 19 (2013) 981–986.

[107] S.M. Hosseini, S.S. Madaeni, A. Zendehnam, A.R. Moghadassi, A.R. Khodabakhshi, H. Sanaeepur, Preparation and characterization of PVC based heterogeneous ion exchange membrane coated with Ag nanoparticles by (thermal-plasma) treatment assisted surface modification, *Journal of Industrial and Engineering Chemistry* 19 (2013) 854–862.

[108] D.J. Kim, M.J. Jo, S.Y. Nam, A review of polymer–nanocomposite electrolyte membranes for fuel cell application, *Journal of Industrial and Engineering Chemistry* 21 (2015) 36–52.

[109] C.-C. Ke, X.-J. Li, S.-G. Qu, Z.-G. Shao, B.-L. Yi, Preparation and properties of Nafion/SiO_2 composite membrane derived via in situ sol-gel reaction: size controlling and size effects of SiO_2 nano-particles, *Polymers for Advanced Technologies* 23 (2012) 92–98.

[110] E. Ruiz-Hitzky, Functionalizing inorganic solids: Towards organic-inorganic nanostructured materials for intelligent and bioinspired systems, *Chemical Record* 3 (2003) 88–100.

[111] J. Joseph, C.-Y. Tseng, B.-J. Hwang, Phosphonic acid-grafted mesostructured silica/Nafion hybrid membranes for fuel cell applications, *Journal of Power Sources* 196 (2011) 7363–7371.

[112] M. Kumari, H.S. Sodaye, R.C. Bindal, Cross-linked sulfonated poly(ether ether ketone)-poly ethylene glycol/ silica organic–inorganic nanocomposite membrane for fuel cell application, *Journal of Power Sources* 398 (2018) 137–148.

[113] L. Ahmadian-Alam, H. Mahdavi, A novel polysulfone-based ternary nanocomposite membrane consisting of metal-organic framework and silica nanoparticles: As proton exchange membrane for polymer electrolyte fuel cells, *Renewable Energy* 126 (2018) 630–639.

[114] A.A. Franco, M. Guinard, B. Barthe, O. Lemaire, Impact of carbon monoxide on PEFC catalyst carbon support degradation under current-cycled operating conditions, *Electrochimica Acta* 54 (2009) 5267–5279.

[115] Rajesh, T. Ahuja, D. Kumar, Recent progress in the development of nano-structured conducting polymers/nanocomposites for sensor applications, *Sensors and Actuators B: Chemical* 136 (2009) 275–286.

[116] A. Nirmala Grace, K. Pandian, Pt, Pt–Pd and Pt–Pd/Ru nanoparticles entrapped polyaniline electrodes – A potent electrocatalyst towards the oxidation of glycerol, *Electrochemistry Communications* 8 (2006) 1340–1348.

[117] J. Niessen, U. Schröder, M. Rosenbaum, F. Scholz, Fluorinated polyanilines as superior materials for electrocatalytic anodes in bacterial fuel cells, *Electrochemistry Communications* 6 (2004) 571–575.

[118] A.F. Diaz, J.I. Castillo, J.A. Logan, W.-Y. Lee, Electrochemistry of conducting polypyrrole films, *Journal of Electroanalytical Chemistry and Interfacial Electrochemistry* 129 (1981) 115–132.

[119] Z. Qi, M. C. Lefebvre, P. G. Pickup, Electron and proton transport in gas diffusion electrodes containing electronically conductive proton-exchange polymers, *Journal of Electroanalytical Chemistry* 459 (1998) 9–14.

[120] K.-B. Pu, Q. Ma, W.-F. Cai, Q.-Y. Chen, Y.-H. Wang, F.-J. Li, Polypyrrole modified stainless steel as high performance anode of microbial fuel cell, *Biochemical Engineering Journal* 132 (2018) 255–261.

[121] O. Prakash, A. Mungray, S. Chongdar, S.K. Kailasa, A.K. Mungray, Performance of polypyrrole coated metal oxide composite electrodes for benthic microbial fuel cell (BMFC), *Journal of Environmental Chemical Engineering* 8 (2020) 102757.

9 Solid Polymer Electrolytes for Solid State Batteries

Anukul K. Thakur, Mandira Majumder, Archana Patole, and Shashikant P. Patole

Khalifa University of Science and Technology, Abu Dhabi, United Arab Emirates

CONTENTS

9.1 INTRODUCTION

The electrolyte in rechargeable batteries is a very crucial component for attaining high-performance rechargeable batteries with a long life span. An ideal electrolyte should not only exhibit suitable ionic conduction for a wide range of ambient temperatures but also should exhibit compatibility with various electrode materials and maintain a fair chemical stability. So far, concerning conductivity, optimization of commercial batteries has been attempted by implementing lithium salt-based liquid electrolytes comprising organic solvents and suitable separators. However, the safety concerns and performance decay related to liquid electrolytes, including fire, inflammable solvents, and explosions, has limited the application of lithium-ion batteries on an industrial scale [1–3]. Hence, safer electrolytes with better reliability are being searched for so as to be implemented in Li-ion batteries (LIBs). Furthermore, to reach a larger energy density and better performance of LIBs, a great demand for novel electrolytes has been generated [1].

Recently, implementing solid-state electrolytes (SSEs) has become popular to overcome the issues faced by conventional liquid electrolyte-based lithium-ion batteries. Unlike liquid electrolytes, SSEs exhibit the advantages of a broad electrochemical window, non-leakage, superior mechanical strength, and thermal stability. Attributed to these advantages, several SSEs are incorporated in lithium-ion batteries and generally branch out into polymer solid electrolytes (PSEs) and inorganic solid electrolytes (SEs). Presently, garnet oxides, superionic sodium conductors, and sulfides form the representative types of inorganic solid electrolytes delivering large room-temperature ionic conductivity almost equal to that exhibited by liquid electrolytes. However, these inorganic solid electrolytes are challenged by poor and fragile interfacial compatibility with solid electrodes [4, 5]. The incompatibility of the SE with the solid-state electrode limits their further application in commercial SSEs.

Polymers have already shown promising aspects in the energy storage field as an electrode material [6–16]. Nevertheless, PSEs are now gaining attention for their application in LIBs, attributed to their excellent mechanical properties, safety, and flexibility (Figure 9.1).

Polymers can also host ions that can traverse the free channels created by vacancies in the existing polymer host, resulting in enhanced conductivity at a temperature in which the free movement of polymer molecules is possible. The ionic conductivity of the polymers shows a dispersive liquid-like property in the solid electrolyte.

DOI: 10.1201/9781003169727-9

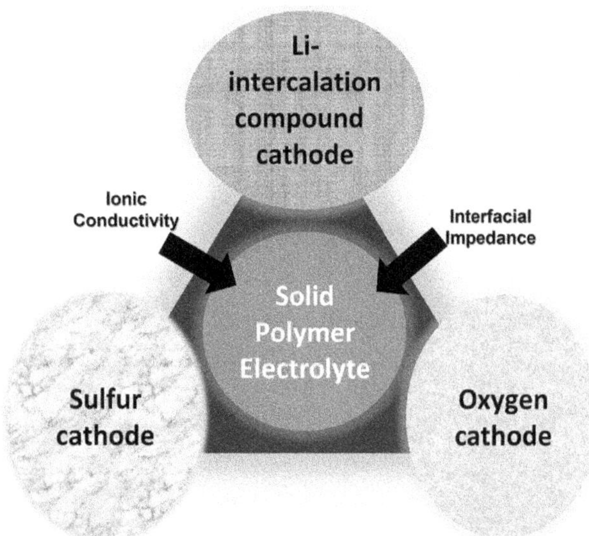

FIGURE 9.1 Chart depicting the pros and cons of various electrolytes.

Unlike their inorganic counterparts, PSEs achieved the first successful milestone of SSEs, which could be attributed to the closeness of PSEs to commercially available liquid battery separators in the manufacturing process. For the first time, electric vehicles (the Bulevar), manufactured by Bollore, were launched on the market for commercial purposes in 2011, which were powered by solid polymer lithium batteries. The timeline of the development of solid-state electrolytes based on polymers is given in Figure 9.2.

Hence, PSEs were realized as more suitable, compared to the inorganic solid electrolytes, for commercial applications in the field of soft packages, wearable devices, and power batteries. To date, several matrices of PSEs, including polyvinylidene difluoride (PVDF), polyvinylidene difluoride-hexafluoropropylene (PVDF-HFP), polyethylene oxide (PEO), polyvinyl cyanide (PAN), and polymethyl methacrylate (PMMA), have been reported [17–22].

In this chapter, we discuss in brief the current developments in the designing of PSEs, starting with a brief discussion on polymer matrices for realizing their contribution to electrolyte performance. The PSE can be classified into two: PSEs exhibiting a polymer matrix and polymer composites (PCs). In addition, detailed PSE applications in current largely attractive fields such as lithium-ion and sulfur batteries are also discussed.

9.2 POLYMER SOLID ELECTROLYTES FOR BATTERIES

The rapid development of lithium batteries, attaining all the required characteristics synchronously with single-component polymer electrolytes, is not very easy. Research on polymer frames leads to a solid base and elucidates the directions which result from the exploitation of multicomponent polymer-based electrolytes with superior performance.

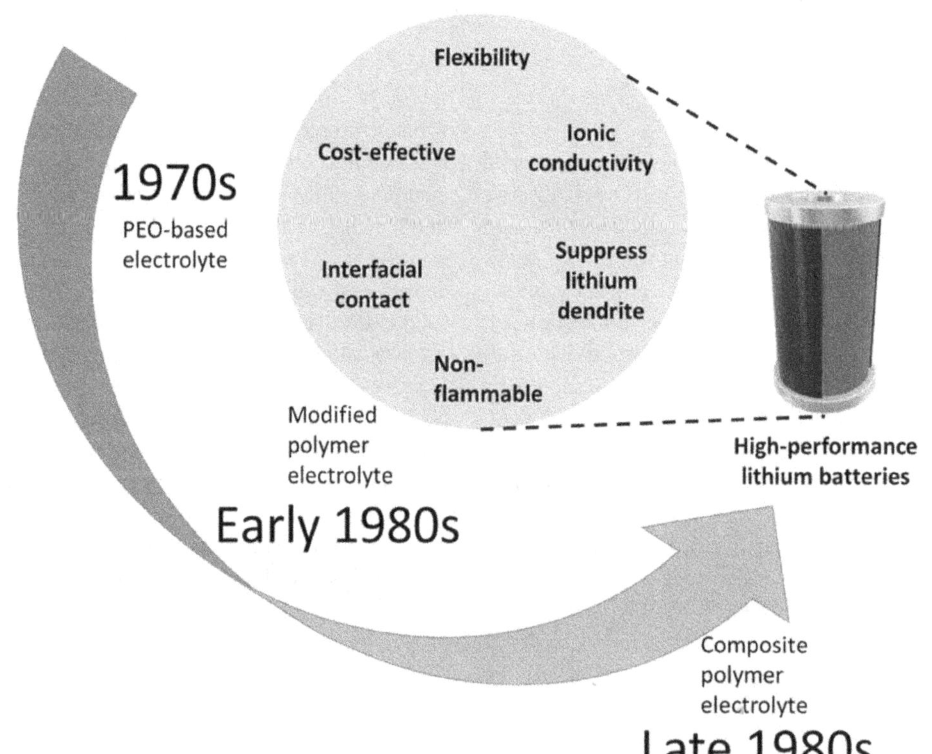

FIGURE 9.2 Timeline representing the various steps of the development of the Li-based battery implementing solid-state electrolytes and their advantages.

Figure 9.2 projects the steps of the development of solid-state electrolytes and their advantages.

9.2.1 Polyethylene Oxide (PEO)

PEO exhibiting higher molecular weight has been solely explored as a polymer composite in order to fabricate Li-polymer batteries due to its requisite mechanical properties and compatibility with the electrodes. The repeating moiety in the polymer cable permits the Li-ion complex, while adequate ion dynamics are caused by the segment's motion [4]. However, the thick packing of the chain in semi-crystalline PEO reduces the probability of effectively transferring Li-ions, leading to reduced ionic conductivity [23]. It is normally postulated that ionic transport occurs as a result of main-chain and segmental activities rapidly increasing with a decrease in the crystallinity or an increase in the temperature greater than the glass-transition temperature (T_g) [24].

Unlike conventional lithium, bis(trifluoromethanesulfonyl)imide (LiTFSI) possesses a flexible anion, CF_3-SO_2-N–SO_2-CF_3-(TFSI-), which leads to the termination of the crystallinity of PEO chains and exhibits remarkable properties like electrochemical and thermal stability. On the other hand, lithium imidodisulfuryl fluoride (LiFSI), an equivalent of LiTFSI, shows minimized viscosity associated with the melt, enhancing the Li-ion diffusion. The ionic conductivity of PEO/LiFSI has been noted to be greater than 0.001 S cm^{-1}, which is a little greater when compared to that with an LiTFSI electrolyte.

9.2.2 Polyacrylonitrile

Polyacrylonitrile (PAN), having an electron-withdrawing and polar nitrile group C≡N, displays substantial mechanics and a broad potential window [25]. Specifically, the high redox potential of PAN-based electrolytes reduces those polymers which are adaptive to the large electrical potential cathode materials, leading to a high specific energy, even though it was previously highlighted that lithium bis(trifluoromethanesulfonyl)imide could reduce the crystallinity of PAN [26]. Ionic conductivity PAN-based electrolytes does not really depend on the movement of the polymer chains. On the other hand, the existence of groups that can withdraw electrons existing on the N atom associated with the LiTFSI separates the imide anion and increases its stability, and results in the substantial delocalization of the charge corresponding to the imide anion, resulting in the dissociation of ion.

9.2.3 Polyvinylidene Difluoride

The functional group (C-F) exhibits a robust electron-withdrawing property, rendering polyvinylidene fluoride (PVDF) with a relatively large polarization and as dielectric, leading to the easy dissolution of lithium salts [27, 28]. This results in the easy transport of the ions of lithium through the PVDF/lithium salt [29]. In addition, PVDF possesses substantial electrochemical and thermal stability together with a potential of up to 5 V. PSE films comprising PVDF (thickness of 90 μm), fabricated by Nan's research group implementing different lithium salts, have been reported [30]. Amongst the three PVDF-based PSEs, the PVDF-$LiClO_4$ electrolyte exhibited the largest conductivity related to ion transport, reaching 0.0001 S cm^{-1}.

This electrolyte was regarded as a solid and exhibiting outstanding mechanical properties, unlike the usual gel electrolytes, where the dimethylformamide (DMF) molecules exist. Conversely, the PVDF-lithium bis(trifluoromethanesulfonyl)imide electrolyte exhibited high electrochemical stability, whose electrochemical window could reach a value of 4.65 V [30]. Further, PVDF-LiFSI was endowed with the least polarization, a voltage being associated with the Li stripping-plating process. Also, the PVDF-based PSE is mostly challenged by poor mechanical strength. The incorporation of HFP within the matrix of a PVDF polymer can remarkably improve the strength, but also cut down the crystallinity [29].

9.2.4 Polyacrylates

Polyacrylates (PAs) are the polymers that contain functional ester groups showing electron-donating capability, which are supposed to coordinate with cations associated with alkali salts, leading to a facile breakdown of the lithium salts [31, 32]. PAs also show more interfacial compatibilities and also project themselves as cheaper in cost [32]. These qualities project PAs as very apt as matrices for PSEs. In general, polycyanoacrylate (PCA) and polymethyl methacrylate (PMMA) are regarded as two exclusively implemented candidates. Shukla and Thakur executed infrared spectroscopy to highlight that carbon-oxygen is the only potential site in the acrylic glass framework that can form coordination bonds with the Li^+ [33]. Even though acrylic glass exhibits several advantages, as has been discussed earlier, PMMA based electrolytes project themselves as a little rigid and usually show less ionic conduction at normal temperature, which limits their commercial application. PCA has been regarded as an alternative PA-based matrix, having high binding affinity and high stability due to the existence of a functional nitrile group. Functional groups of ester and nitrile present in PCA are able to network with the ions, resulting in an increase in the dissolution of the salt and electron-donor-acceptor complex [34]. Equipped with these advantages, polymers based on PCA electrolytes exhibit a very promising attribute for Li-based batteries [34, 35].

9.3 SOLID POLYMER COMPOSITE ELECTROLYTES (SPCs)

SPCs are related to inorganic fillers seated within polymer hosts, which causes the combination of the head of both organic–inorganic materials. SPCs have garnered enhanced attention in the past few years, credited to their excellent performance in the field of lithium batteries [36]. In SPCs, the inorganic fillers are generally uniformly dispersed in a polymer framework or organized layer by layer as a separate layer, strongly attached with the film of the polymer. The increased ionic conduction, enhanced mechanical strength, together with improved structural firmness, broadens the potential window and tailors the resistance, which could be determined for PCs by means of both synthesis approaches. On the basis of the impact made to the transport of Li, inorganic packing used in PCs is categorized into active fillers and passive fillers [37, 38].

9.3.1 Inert-Polymer Filler Electrolytes

Various kinds of oxide ceramics including $ZnAl_2O_4$, [39], TiO_2, [40], SiO_2, [41] Al_2O_3, [42] ZrO_2, [43] and CeO_2, [44] are largely implemented to effect inert filler-based composite electrolytes. Weston and Steele [45] showed that the inclusion of Al_2O_3 particles in a PEO-based polymer electrolyte was capable of impactfully enhancing the strength, together with conductivity. Dissanayake et al. [46] showed the impact of the ionic conductivity of PCs on Al_2O_3 particle size and reported an increase in ionic conduction with the diminishing size of the grains. The increasing conductivity of the ions by lowering the particle size was attributed to a larger surface area as a result of the presence of minor particles acting as a surface filler which resulted in the formation of large pathways as conduction channels as a result of the developed interactions of ions and the hydroxyl group related to the shim [47].

To improve the efficiency related to the interfacial areas and the robust polymer-ceramic interactions in-between the polymers and the inert ceramics, Lin and co-workers reported the production of SiO_2 ($MUSiO_2$) particles. The synthesized particles were monodispersed through hydrolysis carried out in tetraethyl orthosilicate associated with a PEO solution [48]. This approach was able to limit the col lection of fillers, causing remarkable dispersion and large mono-dispersity, therefore improving the surface area for Lewis base interaction. Choudhury et al. [49] reported an SPCE consisting of silica grafted PEO chains, showing soft glassy rheological properties. The organization of nanoparticles varied according to the content alteration of the lithium salt in the PCs. After the lamination of the electrolyte associated with the Li-ions within the host, the PEO disappears, leading to the vanishing of pronounced aggregates in the material as examined through a scanning electron microscope (SEM) [50].

Lin and co-workers synthesized nano aluminosilicate, exhibiting a 3D tubular morphology, and implanted it as a scaffold in PEO-LiTFSI [51]. The silica surface, charged negatively related to the halloysite nanotube, resulted in splitting of the Li salt and the soaking up of the ion of Li at the exterior surface. On the other hand, anions of TFSI were attached to the internal surface holding the Al-OH groups, which were charged positively. The Lewis base interaction associated in-between the halloysite nanotube, PEO, and LiTFSI efficiently permitted the passage of the Li-ions through the 3D tubular channels, attaining a 10^{-4} S cm^{-1} value for conductivity. Also, according to the Lewis acid-base theory, Sheng et al. [52] implemented flame-retardant and cost-effective $Mg_2B_2O_5$ nanowires in a PEO-LiTFSI electrolyte to result in the improvement of the characteristics of PCs.

Various interfacial geometries of the polymer (inorganic) could be designed by altering the inorganic fillers' morphology [53, 54]. For instance, in the case of the fillers comprising the inorganic nanoparticle, though they are able to improve the conduction related to the ions, the rapid transport paths associated with them are inaccessible and small-ranged. Nanowires showing a high value of aspect ratios possess elongated transport channels. However, these are randomly dispersed and associated with the poor link between the nanowire, resulting in the restriction of the conductivity related to the ion in the electrolytes. In order to consider the recompensating effect of the interfaces, work was reported related to the investigation of a composite electrolyte created through melting followed by infiltrating a matrix of PEO-LiTFSI in the nanochannels associated with the surface-functionalized and anodized aluminum oxide (AAO). The material paved way for the advent of interfacial interaction between the inorganic fillers and polymer, causing an increase in ionic conductivity [53].

9.3.2 Active-Polymer Filler Electrolytes

Apart from cutting down the crystalline property of the polymer host by introducing a passive filler, a composite solid electrolyte as an active filler is considered to produce novel Li-ion transfer pathways, which is thought to be the important cause of conductivity augmentation [55]. For the several reported metal oxides, sulfides and ceramics have been exclusively studied [56].

9.3.2.1 Garnet-Polymer Solid Electrolytes

A typical garnet has the chemical formula of $A_3B_2(XO_4)_3$, where A, B, and X are present in the eight, six, and four coordinated cation sites, respectively [57, 58]. Different types of crystal structures are shown in Figure 9.3.

The garnet-like solid electrolyte, $Li_7La_3Zr_2O_{12}$ (LLZO), has garnered much attention from the time it was first reported [59]. Various reports are available which indicate that the highlighted material based on the polymer/LLZO composite could provide feasible mitigation for inferior electrochemical performance. ^6Li and ^7Li nuclear magnetic resonance (NMR) spectroscopy was brought in to identify an ion transport way comprising PEO-LLZO [60]. Different from the earlier case, Yang et al. [61] reported LLZO nanowires deep-seated within a PAN-$LiClO_4$ matrix, manufactured with the help of electrospinning, and proposed a unique Li-ion conduction mechanism. The designed PCs comprising LLZO nanowires exhibited augmented ionic

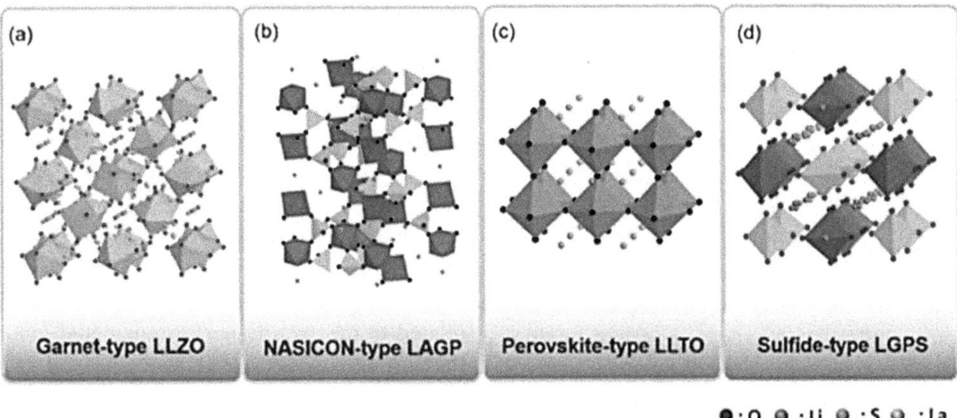

FIGURE 9.3 Crystallographic arrangement of (a) LLZO of garnet type, (b) LAGP of NASICON type, (c) LLTO of perovskite type, and (d) LGPS of thio-LISICON type. The inside sites are occupied by other elements including titanium, zirconium, germanium, and phosphorus. [58] with permission and copyright 2018 Wiley-VCH

conduction attaining a value of 0.0001 S cm^{-1} in ambient conditions. PCs associated with the nanoparticle of LLZOs, together with inert Al_2O_3 nanowires, were reported to be synthesized and studied in detail for comparison. It was observed that the ionic conductivities of the materials were less than PAN-LLZO nanowires. The authors implemented a combination of discerning ^{6}Li-isotope differentiating and NMR, which was established to indicate the conduction pathways for the Li-ions in the PEO($LiClO_4$)/LLZO, comprising 50 wt.% of the particles of LLZO, [60] to illustrate the mechanism. Recently, a solvent-free straightforward approach leading to the creation of a thin and flexible SPCE reaching a value of 80.4 wt.% was reported [62]. In the study, polytetrafluoroethylene was cast in fibers and stuck to the active $Li_{6.75}La_3Zr_{1.75}Ta_{0.25}O_{12}$ (LLZTO) in powder by milling. So interconnection in the LLZTO was easily effected with the PTFE binder to create a 3D framework. Subsequently, penetration of the LLZTO framework, which was flexible and associated with a solid electrolyte comprising succinonitrile (SN)-LiTFSI, presented a solid electrolyte exhibiting remarkably enhanced conductivity in ambient conditions.

9.3.2.2 NASICON-Polymer Electrolytes

The NASICON kind of ceramics, that is Na ionic conductors, have garnered much attention of materials scientists owing to their properties, including remarkable stability, high ionic conductivity in ambient conditions, and practical applicability [63]. As a result of the replacement of Na by Li, NASICON-type materials, it was observed that the original morphology is retained together with conversion in an inorganic filler based on Li-ions [64, 65]. In general, these materials are denoted by the chemical formula $LiM_2(PO_4)_3$, where filling of M sites is observed by Ge, Zr, or Ti. Zhai et al. [66] reported the production of SPCEs to where there was vertical alignment and connection between the NASICON-type ceramic $Li_{1+x}Al_xTi_{2-x}(PO_4)_3$ (LATP). The matrix comprising PEO was considered as the ceramic structure with high porosity to introduce flexibility in the electrolyte membrane. In the mentioned research work, the nanoparticles of LATP were first dispersed in water, followed by deposition on a substrate. The ice was sublimated after, followed by sintering of the LATP particles and hence the formation of the channels, which were vertically aligned straightaway. As a result, Li-ion passageways were created by vertical arrangement and interwoven channels. Getting advantage from the proposed ice-template approach, Liao and co-workers introduced a different NASICON like ceramic, $Li_{1.5}Al_{0.5}Ge_{1.5}(PO_4)_3$ (LAGP), effecting the creation of a vertically arranged polymer/ceramic composite electrolyte related to PEO [67]. To highlight the solid electrolyte with large, outstanding, satisfactory fracture resilience and mechanical strength simultaneously, Li et al. [68] made a composite exhibiting distinctive "brick-and-mortar" small-scale structures.

9.3.2.3 Perovskite-Polymer Electrolytes

A (La, Sr, Ca) B (Al, Ti) O_3 denotes various sites in perovskite electrolytes. Some A sites can be replaced with Li metal, leading to the formula $Li_{3x}La_{2/3-x}TiO_3$ [69]. Liu et al. [70] synthesized $Li_{0.33}La_{0.557}TiO_3$ (LLTO) nanowires using electrospinning and entrancing the nanowires (15 wt.%) into PAN associated with the enhancement corresponding to the three orders of magnitude linked to Li-ion conductivity at ambient conditions, unlike the Li salt-based polymer alone. For alleviating nanowire–nanowire cross-junctions and causing a further enhancement in the ionic conductivity of SPCE, they then synthesized an electrolyte comprising PAN showing distinctly oriented LLTO nanowires obtained through electrospinning, having interdigitally arranged Pt collectors for angle tuning [71]. For implementing the complete advantage of the rapid transmission, using the interphase at the ceramic-polymer, an $Li_{0.35}La_{0.55}TiO_3$ (LLTO-1) framework showing a 3D nanostructure with a very large area surface was fabricated by implementing a hydrogel-derived method [72].

9.3.2.4 Sulfide-Polymer Electrolytes

The replacement of oxygen with more and better polarizable sulfur can minimize the binding of Li-ions to the host framework and extend the ion transport channel. So, sulfur-based electrolytes exhibit better ion conductivity, reaching an order of magnitude of 10^{-2} S cm^{-1} in ambient conditions [73, 74]. In addition, these are easily distortable, leading to simplification of the synthesis process of the compared electrolytes based on oxides. Nevertheless, the materials based on sulfide are mostly degraded in compounds that are polar and exhibit high sensitivity to air. This makes their handling very difficult, along with exhibiting a narrow electrochemical potential window [74]. Sulfide/polymer-based composite electrolytes have shown promising aspects in eliminating these issues. A free-standing electrolyte was designed by compositing the sulfide $Li_{10}GeP_2S_{12}$ (LGPS) into a PEO matrix with the help of a simple solution-casting method [75]. Implementing a similar production method, Li et al. [76] reported the fabrication of a flexible and free-standing SPCE by introducing $Li_{10}SnP_2S_{12}$ (LSPS) into PEO, with an ionic conductivity of 10^{-4} S cm^{-1}. Attributed to their facile processability, improved ionic conductivity, and very cost-effective precursors, Li_6PS_5Cl, Li_6PS_5Br, and Li_6PS_5I have been implemented as electrolytes for solid-state batteries [77]. Li_6PS_5Cl/PEO PCs were synthesized by implementing a liquid-phase procedure [78] with 5 wt.% PEO; the PCs were able to highly limit the reactions going on at the interface and the creation of Li dendrites, leading to the exhibition of an improved ionic conductivity. Lately, a free-standing electrode exhibiting large conductance based on a composite membrane comprising $78Li_2S–22P_2S_5$ glass-ceramic sulfide was reported to have been designed using the approach of liquid-phase reactions [79].

9.4 POLYMER ELECTROLYTES FOR HOPPED-UP BATTERIES

PSEs have long been studied and implemented in various rechargeable batteries. In this section, we will discuss a few remarkable recent pieces of research that reflect the state-of-the-art of this approach. We will consider various PSEs implemented in Li-ion batteries for improved charge storage performance.

Xiaoming Sun and co-workers reported mechanical strength with good conductivity in thiol-based PSEs (Figure 9.4). The synthesis of the PSE (coded as M-PSEGDA) was carried out by cross-linking covalently poly(ethylene glycol) diacrylate (PEGDA), metal frameworks (MOFs), and pentaerythritol (3-mercaptopropionate) together with implementing multiple carbon-sulfur bonds. The PSE also exhibited a little interfacial impedance, a wide electrochemical window, and a remarkable Li^+ transference number. The symmetrical cells incorporated with thiol PSE when investigated showed huge stability in a > 54 day stability test. Moreover, a PSE with a cathode of lithium iron phosphate exhibited a capacity of more than 140 mA h g^{-1} [80].

Zhigang Xue and co-workers reported a new type of PSE comprising urea and disulfide bonds. The hydrogen bonds in-between the disulfide metathesis and urea groups render the PSE with a high degree of self-mitigating ability when there are no external stimuli present at a normal temperature, in addition to more rapid self-healing when treated at high temperatures. The entire PSE shows a large self-healing efficiency together with a trace amount of changes related to the cycling performance and ion conductivity of the lithium/$LiFePO_4$ cell as compared to the clean one (Figure 9.5) [81].

Yonghong Deng et al. synthesized a chemically in situ polymerized copolymer based PSE (PLA/PEG–PSE) showing a comb-like morphology with a large ionic conductivity comprising poly(ethylene glycol), 2-methoxyethyl methacrylate, and methyl acrylate functionalized poly(D,L-lactide) (Figure 9.6). A large ionic conductivity corresponding to a value of 0.00001 S cm^{-1} was attained at room temperature, and the highest ionic conductivity of 10^{-4} S cm^{-1} at a temperature of 60 °C was observed, related to the Li^+ transference number of 0.36 and an activation energy of 0.2 eV. The PLA/PEG-PSE showed a broad stability of 4.6 V along with an excellent lithium metal. $LiFePO_4$/PSE/Li cells integrated with a cathode and Li exhibited augmented stability with discharge capacities (149 mA h g^{-1}) and more than 82% capacity retention at a temperature of more than 50 °C. This work inspired more related work on in situ fabrication related to a vinyl-functionalized precursor that shows an ion-conducting membrane apt for application in Li-metal batteries [82].

Xiaochun Chen and co-workers reported the fabrication of a membrane by executing an imidazolium-based polymerized ionic liquid working as the scaffold for the polymer matrix. The material comprised lithium salt lithium bis(trifluoromethanesulfonyl)imide and porous fiber cloth. The ionic conduction of the electrolyte comprising 2.0 M kg^{-1} LiTFSI reached a value of 0.00001 S cm^{-1} at 30 °C and 0.0001 S cm^{-1} at 60 °C. The specific discharge capacity exhibited for $LiFePO_4$ incorporated with the produced electrolyte was ~138 mA h g^{-1}. The retained capacity for the electrode material was 90% after 250 cycles at 60 °C. LATP electrolytes were further dispersed for enhancing the conductivity in a polymer matrix resulting in an LiTFSI-LATP composite electrolyte. $LiFePO_4$ implemented with PIL-LiTFSI-LATP as a PSE exhibited remarkable rate performance and large capacity retention (~97% for 250 cycles) [83]. David Mecerreyes et al. reported the optimization of a mono-ion polymer implemented in polymer electrolytes, including ethylene carbonate, oxide, and a sulfonimide. This mono-ion copolymer was synthesized by polycondensation. Optimization of the degree of crystallinity and the conductivity of the PSE was possible by altering the monomer's stoichiometry. The optimized copolymer exhibited high ionic conductivity

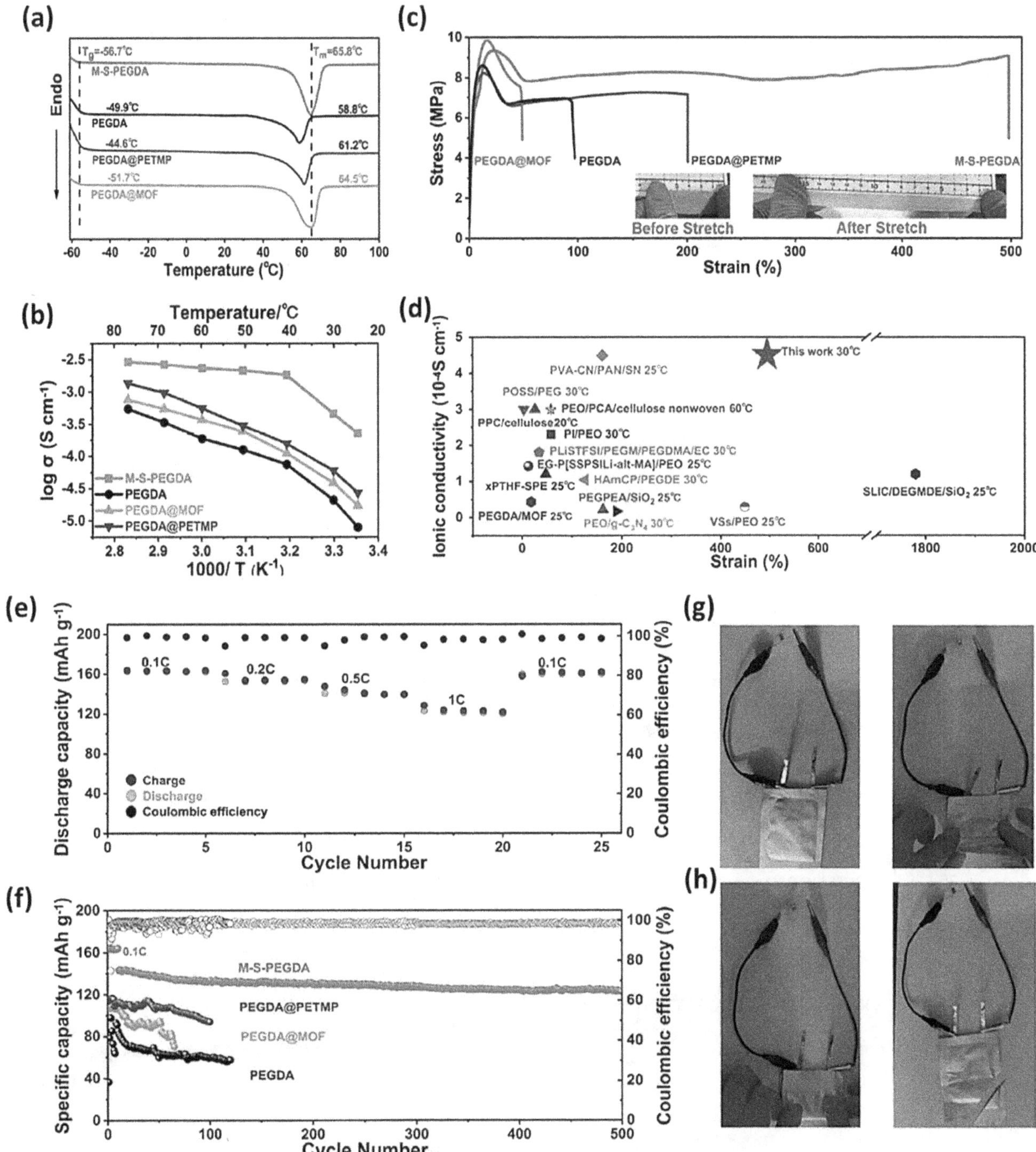

FIGURE 9.4 Various physicochemical characteristics of M-S-PEGDA. (a) DSC thermograms of the synthesized samples. (b) Stress–strain curves corresponding to the synthesized samples. (c) Drawing a comparison of M-S-PEGDA and previously reported polymer electrolytes. The real phenomenon of the stretching of M-S-PEGDA is represented in the inset figure. (d) Electrochemical performance of solid batteries based on different PSEs. (e) Capacity versus cycle number at different C-rates for Li|M-S-PEGDA|LFP full cells. (f) Cyclic stability of the various cells comprising various synthesized SPEs at 0.5 °C and at a temperature of 40 °C. (g and h) Safety and flexibility check of pouch cell comprising Li|M-S-PEGDA|LFP. [80] with permission and copyright 2020 Wiley-VCH

corresponding to a value of 10^{-4} S cm^{-1}. Interactions and mobility were studied using FTIR-ATR, NMR relaxation time measurements, a lithium pulsed field gradient, and lithium diffusion analysis. Large lithium mobility was also noted which resulted in the promotion of lithium mobility. The performance of the poly(ethylene oxide carbonate) was better than that of the analogous conventional salt in a polymer electrolyte [84].

Yi Cui and co-workers reported a PSE comprising of thick nanoporous polyimide (PI) film incorporated with

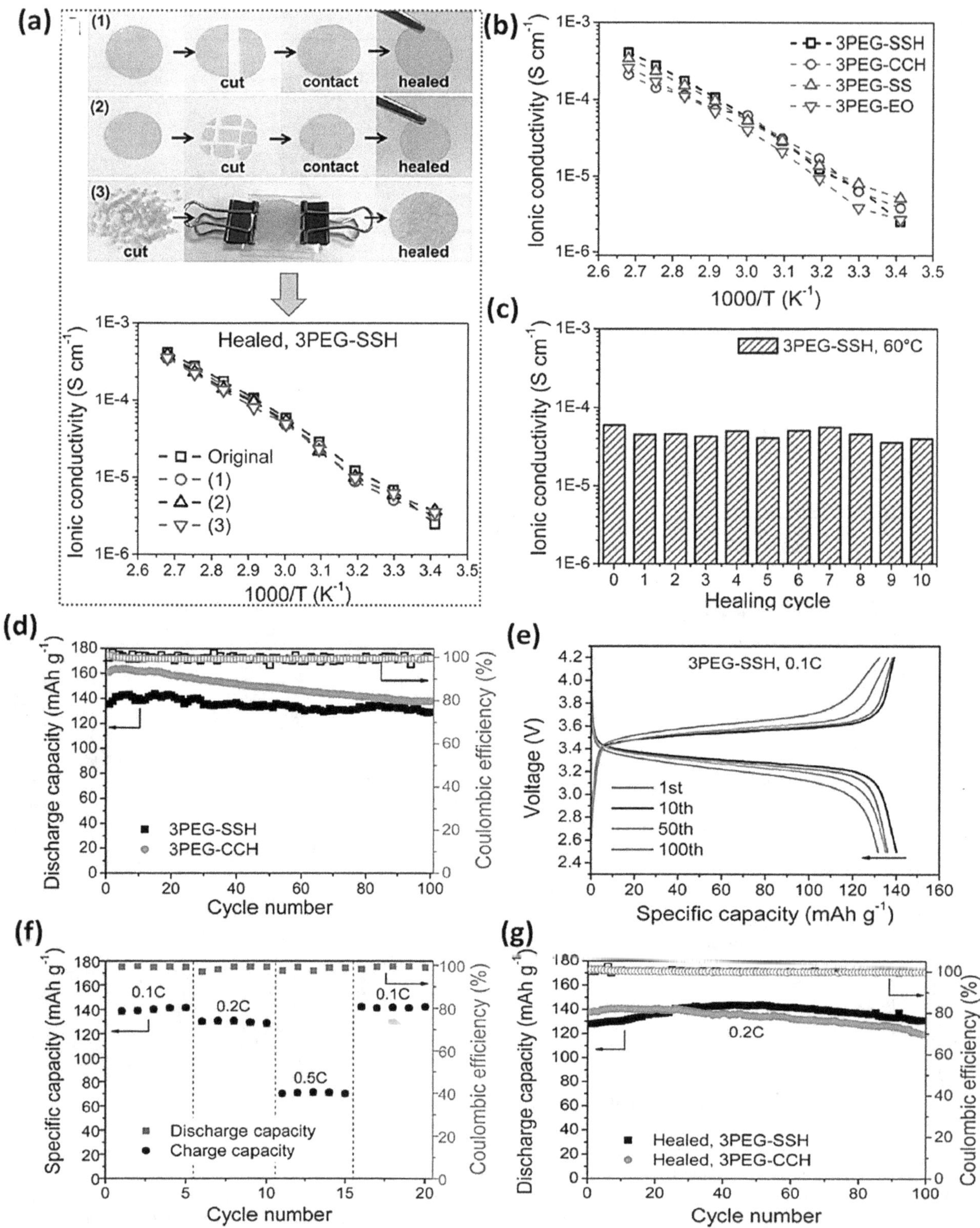

FIGURE 9.5 (a) Representation of the ionic conductivity of various solid polymer electrolytes (SPEs) cut in different forms and healed. (b) Representation of ionic conductivity, dependent on the temperature of SPEs. (c) Conductivity variation of healed 3PEG–SSH through cycles of subsequent cutting and healing. (d) Cycling stability test of cells at 0.1 C. (e) Charge–discharge profiles corresponding to 3PEG–SSH at different cycles at 0.1 C. (f) Cyclic stability of cell at various current rates. (g) Cyclic stability of cells comprising healed SPEs. [81] with permission and copyright 2020 American Chemical Society

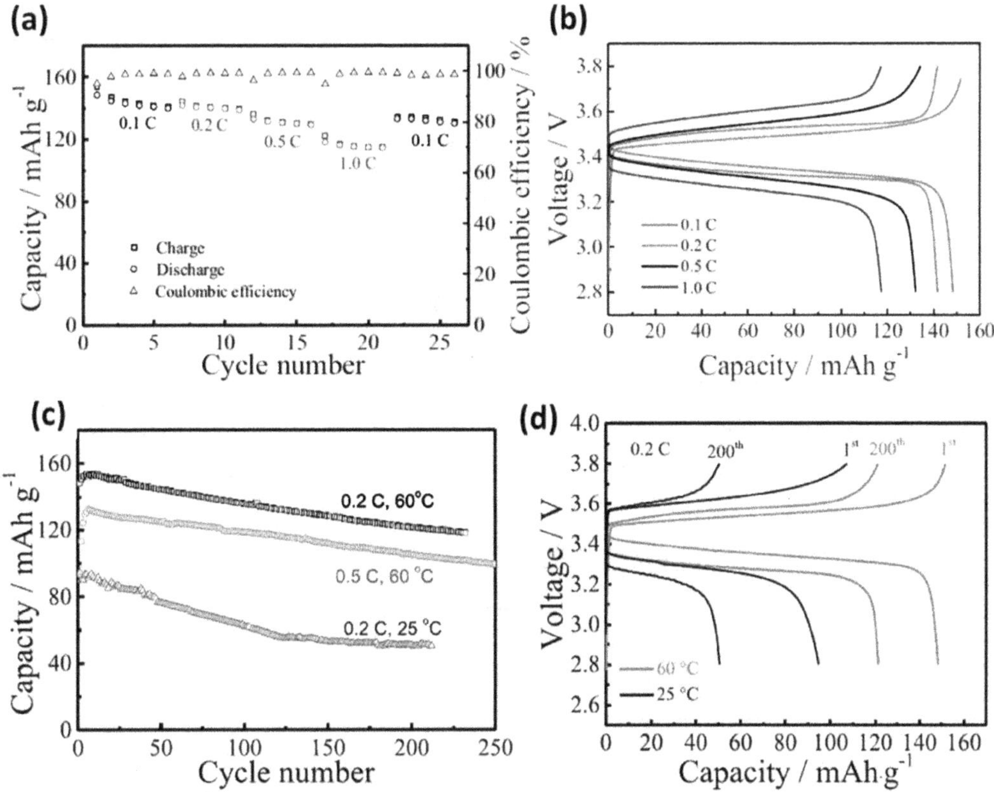

FIGURE 9.6 Representation of the rate capability corresponding to large temperatures for an LFP/SPE/Li battery. (a) Representation of the C-rate capability. (b) Representation of the charge/discharge profile at different C-rates. (c) Representation of cyclic stability (long term) of the as-prepared cells at different C-rates and temperatures. (d) The charge/discharge potentials obtained for SFP/SPE/Li cells based on PLA/PEG-SPE at two different temperatures. [82] with permission and copyright 2020 Institute of Physics

PEO/LiTFSI. The battery associated is safe to operate as PI film is non-flammable and mechanically strong, preventing fire and short-circuiting together with a long duration of cycling without any noticeable capacity change (Figure 9.7). The vertical channels enhance the ionic conductivity of the infused electrolyte. The Li-ion batteries are made up by integrating PI/PEO/LiTFSI exhibit good cycle life with 200 cycles at 0.2 C rate and endure abuse different mechanical tests [85].

Qian Zhang and co-workers reported the synthesis of PGO by incorporating different molar ratios using the monomers of oligo ethylene oxide methyl ether methacrylate (OE) and glycidyl methacrylate and its detailed investigation for application in solid Li-ion batteries. PGO-70 exhibited the highest conductivity for ions corresponding to a value of 10^{-5} S cm^{-1} containing a 50% $LiClO_4$ content. PGO (60) showed the highest value related to the transference of Li^+ (~0.5) and exhibited electrochemical stability reaching a value of 4.4 V. As a result of the hydrolysis of the epoxy groups, two hydroxyl groups in the side chains dangled as pendants. The membrane associated with ring-opening HPGO-70 exhibited a higher conductivity reaching a value of 10^{-6} S cm^{-1}, $LiClO_4$¼ 30% as compared to the corresponding PGO-70 with a conductivity of 10^{-7} S cm^{-1}, and $LiClO_4$¼ 30% at ambient temperature. The large Li^+ transport number associated with the ring-opening polymers has been reported to be 0.693 for HPGO (60). The assembled battery implementing $LiFePO_4$ as the anode and HPGO (60) acting as the PSE exhibited good cycling stability at ambient temperatures, with a capacity of 20 mA h g^{-1}. Capacity was enhanced to a value of 99 mA h g^{-1} at 70 °C during the initial cycles and further decreased rapidly to 50 mA h/g [86].

9.5 SOLID POLYMER COMPOSITE ELECTROLYTES IN RECHARGEABLE BATTERIES

The previous section highlighted work based on pristine solid polymer-based electrolytes for rechargeable batteries. However, a pristine polymer is challenged by large interfacial impedance and small ionic conductivity at ambient temperatures. These disadvantages can be mitigated by compositing the pristine polymers with other materials, resulting in SPCE. A few recently reported remarkable works on SPCE are highlighted in this section.

Liang Li and co-workers have reported the synthesis of a hybrid polymer composite consisting of chitosan drenched in liquid electrolyte. The composite exhibited a large conductivity for ions, reaching a value of 10^{-4} S cm^{-1},

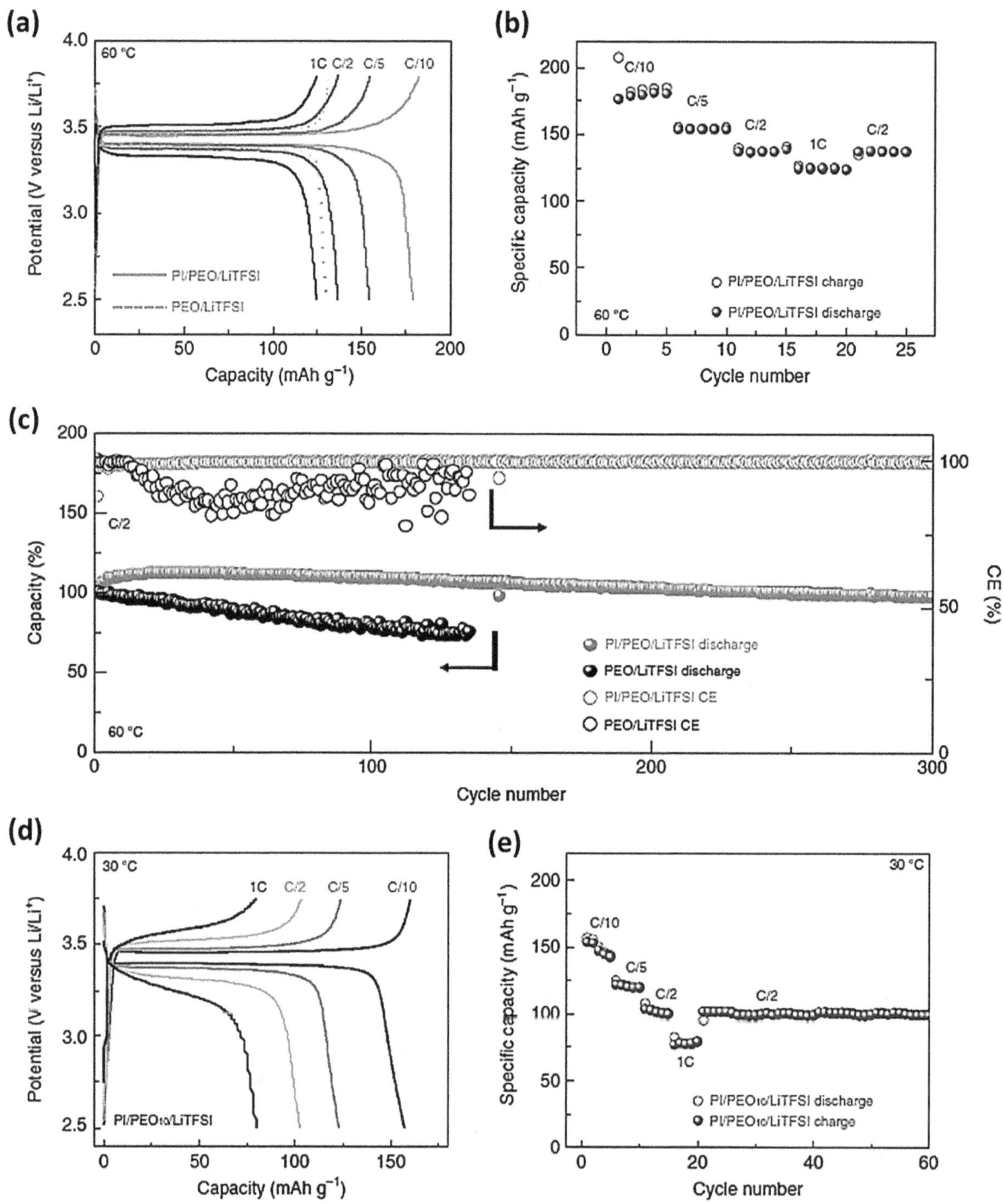

FIGURE 9.7 Electrochemical characteristics of PI/PEO/LiTFSI-implemented full cells. (a) Potential characteristics of a cell at different charging rates at 60 °C. The red dashed curve corresponds to the cell performance of Li/PEO/LiTFSI/LFP at C/10 at 60 °C. (b) Cyclic stability test for full cell at various C-rates at 60 °C. (c) Cyclic stability performance of cells at C/2 at 60 °C. (d) Potential profile and (e) cyclic stability corresponding to cell comprising Li/PI/PEO_{10}/LiTFSI/LFP at various charging rates. [85] with permission and copyright 2019 Nature

associated with little interfacial impedance and good flexibility at ambient temperatures. The $LiFePO_4$ solid-state cell fabricated with this hybrid polymer composite electrolyte maintained a large capacity, that is after 130 cycles, of 120 mA h g^{-1}. This electrolyte promises the synthesis of a novel solid-electrolyte, exhibiting large ionic conductivity and little interfacial impedance, synchronously requisite for application in rechargeable batteries [87].

Vilas G. Pol and co-workers have highlighted the fabrication of an SPCE having big room-temperature ionic conductivity corresponding to a value of 0.0001 S cm^{-1}, a broad voltage window, and significantly improved heat resistance

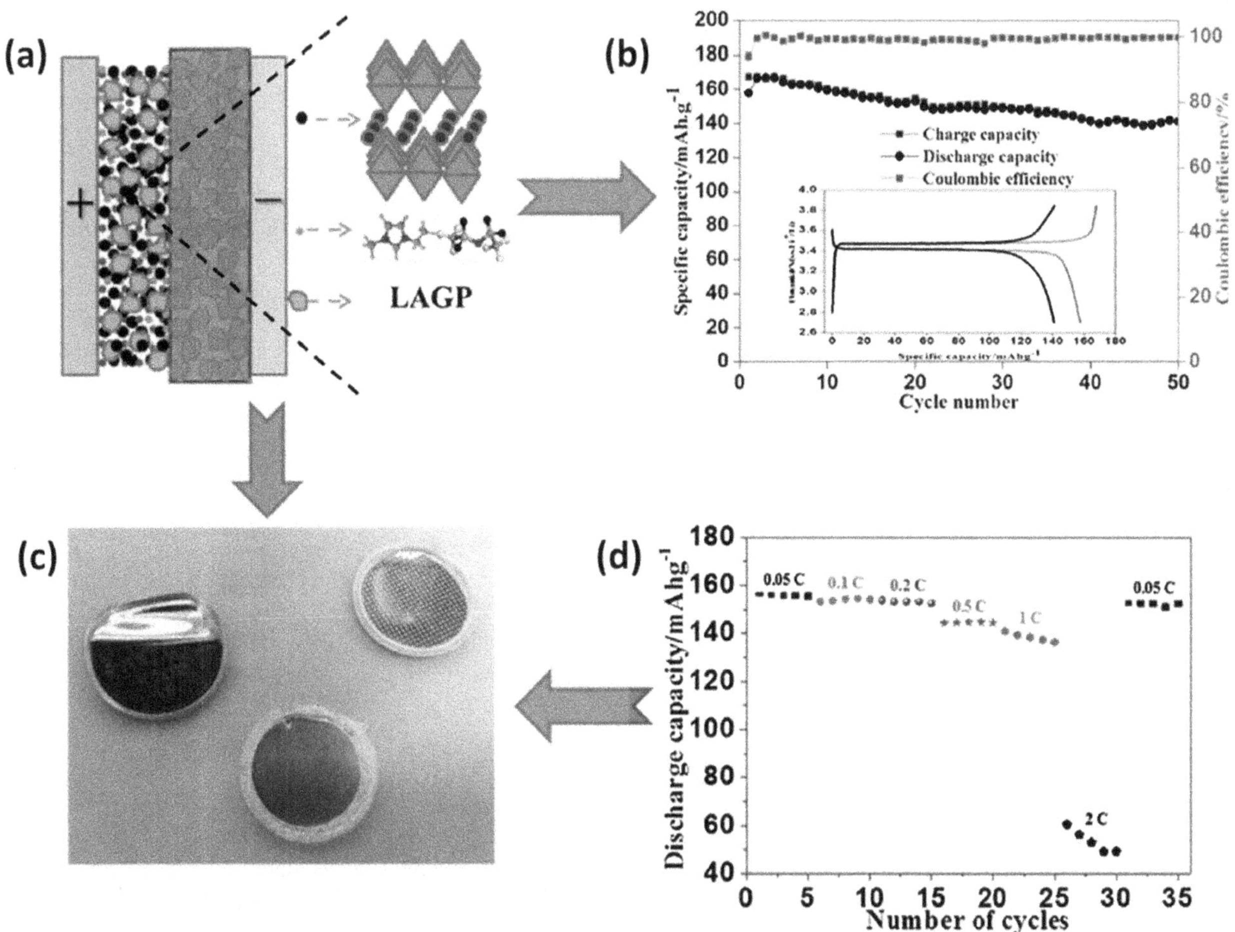

FIGURE 9.8 (a) Diagram representing $LiFePO_4$/Li-based solid-state cell. (b) Cyclic stability performance; the inset shows the charge/discharge profiles obtained for the cell. (c) Digital photographs of the button cells at the end of the cycles. (d) C-rate discharge characteristics of $LiFePO_4$/Li-based batteries. [89] with permission and copyright 2017 American Chemical Society

in a combination of a ceramic filler with PVDF and LiTFSI salt. Self-standing, mechanically robust, and scalable SPCEs distinctly connected with $LiFePO_4$ and high-voltage $iNi_{0.3}Co_{0.3}Mn_{0.3}O_2$ cathodes with Li exhibited stable cycling. More importantly, a thermal window was noticed with a little release of heat corresponding to a value of 189 J g^{-1} which is owed to the heat runaway in the cell implementing the SPCE. Unlike the former case, the electrolyte cell with liquid possessed a deteriorating thermal stability corresponding to a value of 157 °C and a discharge of 812 J g^{-1}. The increased safety of SPCEs was credited to the remarkable stability related to their thermal prospects. Importantly, this work highlights a promising SPCE along with projecting a standard for quantitative study on safety in terms of the thermal prospects of entire rechargeable batteries [88]. Qingming Guo and co-workers have reported the fabrication of SPCEs consisting of polymer P(VDF-HFP) as a scaffold in the form of a matrix and ceramic LAGP as the primary medium for ion conduction (Figure 9.8). In this work, electrochemical, chemical, and thermal stability were enhanced in the polymer significantly, suggesting that the interface in the cell was better utilized. Also, a smaller crystallinity of the polymer in the electrolyte was achieved. These property alterations indicated significant enhancement in the ionic conduction of the composite electrolyte and good compatibility with the lithium anode synchronously. The other advantages are a remarkably stable solid electrolyte and interphase, high mechanical strength resulting in suppressed growth of dendrites, and thermal safety stability. Moreover, the solid-state battery implemented with $LiFePO_4$/LPELCE/Li exhibited enhanced electrochemical properties, which shows promising potential for implementing this type of electrolyte to designs with increased safety with energy density [89].

Y. Kobayashi and co-workers reported on dense pellets with Li-lanthanum titanate, fabricated by implementing a plasma sintering method. The obtained pellets exhibited relatively higher ionic conduction, typically 0.001 S cm^{-1} at 22 °C, corresponding to an energy of activation of ~30 kJ mol^{-1}. Lithium manganese oxide was developed on the pellets by using spray casting at a temperature of 400 °C. The all-solid-state battery system, designed by implementing pellets, showed 100 cycles of charge/discharge at 60 °C [90].

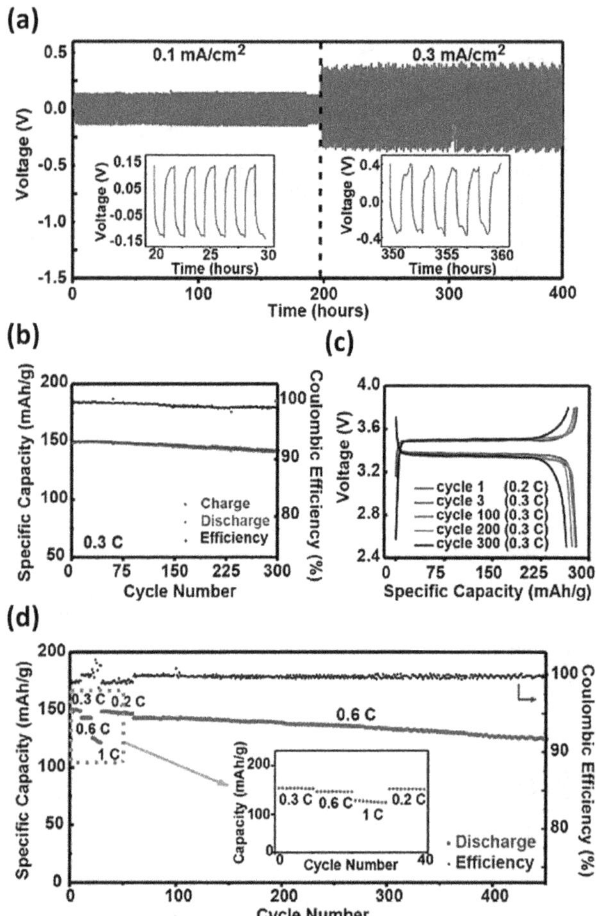

FIGURE 9.9 (a) Potential profiles and the zoomed potential traces of a cell at different densities. (b) The cycle performance of full-cell cycles. (c) Charge and discharge potentials at 0.2 C for the initial 2 cycles and for the rest of the 300 cycles at 0.3 C. (d) Li-LAGP/PEO-$LiFePO_4$ rate performance corresponding to cycles at different C-rates and a subsequent 400 cycles. [67] with permission and copyright 2019 Elsevier

Yuan Yang and co-workers reported the fabrication of ceramic nanoparticles vertically aligned and seated in a polymer, showing a good conductivity optimization of electrolytes. Solid lithium batteries were assembled by applying the ice templating method to fabricate batteries comprising a vertically aligned polymer/ceramic composite electrolyte exhibiting conductivity LAGP and PEO (Figure 9.9). Perpendicular continuous paths are made by the LAGP walls which results in rapid ion transport, while the matrix comprising of PEO imparts flexibility in the electrolyte. This solid composite exhibited a conductivity of 10^{-4} and 10^{-3} S cm^{-1} at ambient temperature and at 60 °C respectively, with a capacity retention of ~93 [67].

John B. Goodenough et al. found the Li^+ distribution-transport associated with an SPCE by investigating a NASICON–$LiZr_2(PO_4)_3$ composite linked with PEO by $^6Li \rightarrow {}^7Li$ trace-exchange nuclear magnetic resonance measurements and ^7Li relaxation time (Figure 9.10). The number of Li^+ ions associated with the environments related to the composite electrolytes is reliant on the amount of ceramic filler and Li-salt concentration. A composite electrolyte comprising of (EO)/(Li^+) with an LZP filler exhibited a high Li^+ conductivity of 10^{-4} S cm^{-1} with a little activation energy which is due to the extra Li^+ existing environment, which was mobile. Furthermore, the reaction of $LiZr_2(PO_4)_3$, lithium, and an in situ synthesized electrolyte interphase layer led to the stability of the Li/composite–electrolyte interface and tailored the resistance existing at the interphase, which resulted in solid $LiFePO_4$ [91].

Lin Zhu and co-workers reported the synthesis of $Li_{0.33}La_{0.557}TiO_3$ nanowires (LLTO-NWs) by electrospinning and calcination after exhibiting high ionic conductivity. This was integrated into LiTFSI, PPC, and PEO in order to design an SPCE. The designed SPCE contained 8 wt.% LLTO-NWs reaching a value of 10^{-5} and 10^{-4} S cm^{-1} at room temperature and 60 °C, respectively. The solid-state full cells fabricated by implementing the $LiFePO_4$/Li SPCE showed a reversible capacity of 135 mA h g^{-1} and good cycling at 0.5 C at 60 °C [92]. Furthermore, electrochemical performance exhibited by several other recent works related to solid-state electrolytes are shown in Table 9.1

9.6 CONCLUSION AND PERSPECTIVES

To mitigate the safety concerns and to cut down the cost, volume, and weight of the storage devices, the introduction of solid electrolytes is a mandate. A solid-state electrolyte is majorly challenged by ionic conductivity, which requires improvement. Various strategies have been adapted to improve the ionic conductivity of solid-state electrolytes—the various polymers which have been implemented as solid electrolytes to date. Though several advances have been achieved in designing PSEs, challenges related to many factors for application for practical purposes need to be addressed before they can be adapted for commercial lithium batteries.

In spite of the fact that cross-linked polymers have proved to be very promising candidates as PSE hosts, showing stability at high operational temperatures, large ionic conductivity, and fair mechanical strength, resulting in normal operating conditions for the associated energy storage devices, their synthesis still remains complicated. Even though the ionic conductivity for a handful of PSEs attains ~10^{-3} S cm^{-1}, this value is less than that corresponding to 0.001–0.01 S cm^{-1} exhibited by commonly used electrolytes. In addition, the potential window of this type of PSE is not broad enough for the operation of large-voltage cathodes, causing a low density of the associated batteries. Architecting room-temperature, wide potential window, polymeric frameworks with easy synthesis techniques is a mandate. Polymers have large ionic conductivity associated with their inferior mechanical properties; large-strength electro-spun and nonwoven membranes possessing porous frameworks highlights themselves as promising candidates. Secondly, the integration of inorganic fillers in polymer electrolytes

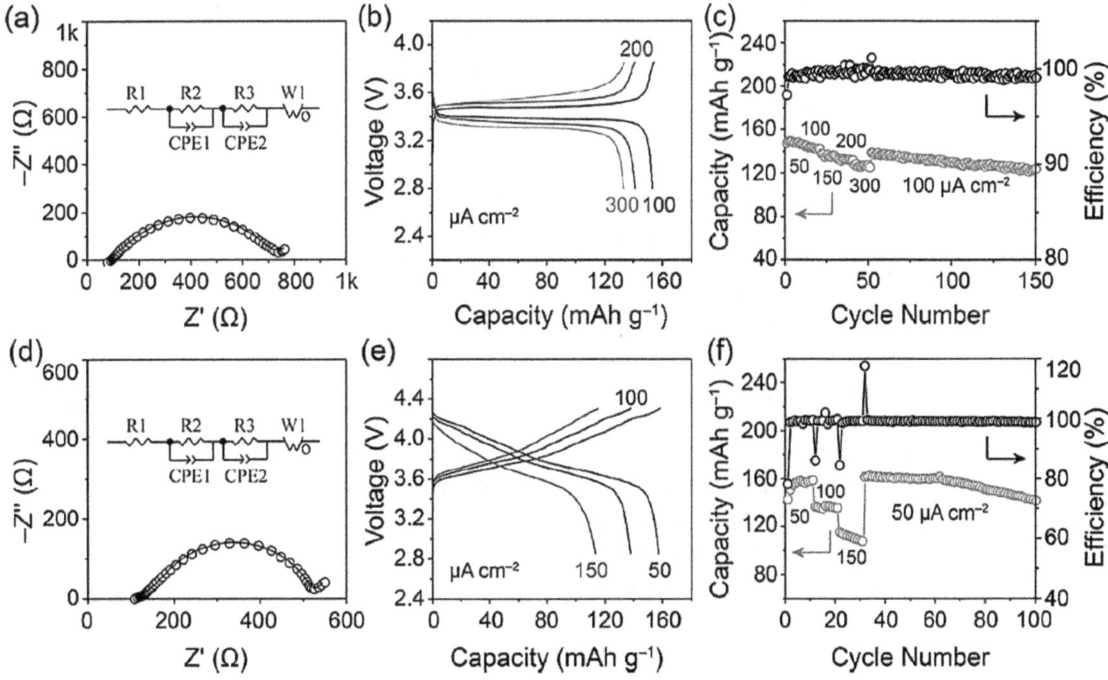

FIGURE 9.10 Electrochemical performance of Li/LiFePO$_4$ cells: (a) Nyquist plots, (b) charge–discharge potential profiles, and (c) cyclic stability and efficiency. Electrochemical performance of Li/NMC cells: (d) impedance spectroscopic analysis, (e) profiles corresponding to charge/discharge potentials, and (f) cycle performance corresponding to stability and efficiency. [91] with permission and copyright 2020 American Chemical Society

TABLE 9.1
Various PSEs and PSCEs as Electrolytes for Rechargeable Batteries

Compound	Components/Filler	Salt	Ionic Conductivity (S cm^{-1})	Reference
PEO	PEO/PEG/LiTf	LiTf	1.00×10^{-4}	[93]
PEO	PEO/PEGDME/LiTFSI	LiTFSI	3.80×10^{-4}	[94]
PEO	EC/PC/LiClO$_4$	LiClO$_4$	1.25×10^{-3}	[95]
PEO	DB/LiClO$_4$	LiClO$_4$	1.00×10^{-5}	[96]
PEO	LiTFSI/BMITFSI	LiTFSI	3.20×10^{-4}	[97]
PEO	LiTFSI/Pyr$_{24}$TFSI	Pyr$_{24}$TFSI	1.00×10^{-5}	[98]
PEO	LiPF$_6$/MMPIPF$_6$	MMPIPF$_6$	1.13×10^{-3}	[99]
PAN-LiClO$_4$-LLZO	Li$_7$La$_3$Zr$_2$O$_{12}$ nanowire	LiClO$_4$	1.31×10^{-4}	[61]
PTFE-LLZTO-SN-LiTFSI	PEO, SN, PTFE, LiTFSI, Li$_{6.75}$La$_3$Zr$_{1.75}$Ta$_{0.25}$O$_{12}$ particle	LiTFSI	1.20×10^{-4}	[62]
LAGP-SPCE	PEO, PEGDME, Li$_{1.5}$Al$_{0.5}$Ge$_{1.5}$(PO$_4$)$_3$	LiTFSI	1.25×10^{-4}	[68]
PAN-LLTO PCs	Li$_{0.33}$La$_{0.557}$TiO$_3$	LiClO$_4$	2.40×10^{-4}	[70]
PVDF-HFP	Li$_7$La$_3$Zr$_2$O$_{12}$/PVDF	PVDF-HFP	1.20×10^{-6}	[100]
Li$_{10}$SnP$_2$S$_{12}$	Li$_{10}$SnP$_2$S$_{12}$	Li$_{10}$SnP$_2$S$_{12}$	2.00×10^{-4}	[101]
PDMS/PEGDA	Li$_{1.3}$Al$_{0.3}$Ti$_{1.7}$(PO$_4$)$_3$	PDMS/PEGDA	2.40×10^{-6}	[102]

proves to be a facile approach to increase ionic conductivity and enhance the electrochemical window for PSEs. A very small contact persisting between the inorganic and organic materials results in limited well-organized interfaces causing the separation of phase and smaller Li-ion conduction pathways, which influences SPE properties positively from the fillers. The implementation of an adhesive coating in-between the inorganic–organic interface, and considerations about inorganic fillers, are favorable options to minimize the above issues. However, the complicated relationship between polymers and inorganic materials, from a case-by-case basis, shows that compositing inorganic fillers with polymers has only limited success. More research is required to elucidate the prevailing interactions in-between the host polymer and the guest materials. Furthermore, electrolytes with an inorganic polymer, layered with each component exhibiting a self-determining unceasing phase, can reduce agglomeration caused in the materials and result in perpetual Li-ion pathways, in turn effectively augmenting the overall performance of the solid-state electrolyte. In addition, they can also eliminate the contact of the cathode material with the oxidized polymer effecting high-voltage lithium batteries. Sulfide-based solid electrolytes exhibiting bulky conductivity and better mechanical strength have shown excellent performance when tested in Li-based batteries. However, these batteries are challenged by the narrow operating potential window of sulfide-based electrolytes. Further, sulfides are usually prone to attack by moisture and the indulgence of the sulfide composite electrolyte is problematic. Designing sulfide-based polymers which will be capable of tolerating large voltages at room temperature would be effective for practical application. The electrode/electrolyte interfacial properties show a critical role in determining capability and cyclic stability. To reach the adequate quality of electrolyte/electrode interfaces, the liquid electrolyte is used to wet the interface. The hot pressing approach is sometimes implemented to integrate the electrolyte-cathode framework.

It is predicted that, equipped with a newly assimilated understanding for the chemical mechanism, nano-microstructures and the operation of PSEs and PCs with additionally augmented presentations satisfying the necessities for real applications will be designed consequently to attain high-performance all-solid-state energy storage devices.

ACKNOWLEDGMENTS

AKT acknowledges SNU, Republic of Korea for post-doctoral fellowship support. SPP acknowledges Khalifa University for financial support (FSU-2018–29) and the ADEK (AARE, 2018, AARE18-136) Award for Research Excellence.

REFERENCES

[1] D. Lin, Y. Liu, Y. Cui, Reviving the lithium metal anode for high-energy batteries, *Nature Nanotechnology* 12(3) (2017) 194.

[2] A. Manthiram, J.C. Knight, S.T. Myung, S.M. Oh, Y.K. Sun, Nickel-rich and lithium-rich layered oxide cathodes: Progress and perspectives, *Advanced Energy Materials* 6(1) (2016) 1501010.

[3] A. Manthiram, B. Song, W. Li, A perspective on nickel-rich layered oxide cathodes for lithium-ion batteries, *Energy Storage Materials* 6 (2017) 125–139.

[4] L. Yue, J. Ma, J. Zhang, J. Zhao, S. Dong, Z. Liu, G. Cui, L. Chen, All solid-state polymer electrolytes for high-performance lithium ion batteries, *Energy Storage Materials* 5 (2016) 139–164.

[5] Z. Pan, L. Yao, J. Zhai, X. Yao, H. Chen, Interfacial coupling effect in organic/inorganic nanocomposites with high energy density, *Advanced Materials* 30(17) (2018) 1705662.

[6] M. Majumder, R.B. Choudhary, A.K. Thakur, I. Karbhal, Impact of rare-earth metal oxide (Eu_2O_3) on the electrochemical properties of a polypyrrole/CuO polymeric composite for supercapacitor applications, *RSC Advances* 7(32) (2017) 20037–20048.

[7] M. Majumder, R.B. Choudhary, S.P. Koiry, A.K. Thakur, U. Kumar, Gravimetric and volumetric capacitive performance of polyindole/carbon black/MoS_2 hybrid electrode material for supercapacitor applications, *Electrochimica Acta* 248 (2017) 98–111.

[8] M. Majumder, R.B. Choudhary, A.K. Thakur, U. Kumar, Augmented gravimetric and volumetric capacitive performance of rare earth metal oxide (Eu_2O_3) incorporated polypyrrole for supercapacitor applications, *Journal of Electroanalytical Chemistry* 804 (2017) 42–52.

[9] A.K. Thakur, A.B. Deshmukh, R.B. Choudhary, I. Karbhal, M. Majumder, M.V. Shelke, Facile synthesis and electrochemical evaluation of PANI/CNT/MoS_2 ternary composite as an electrode material for high performance supercapacitor, *Materials Science and Engineering: B* 223 (2017) 24–34.

[10] A.K. Thakur, R.B. Choudhary, M. Majumder, G. Gupta, In-situ integration of waste coconut shell derived activated carbon/polypyrrole/rare earth metal oxide (Eu_2O_3): A novel step towards ultrahigh volumetric capacitance, *Electrochimica Acta* 251 (2017) 532–545.

[11] A.K. Thakur, R.B. Choudhary, M. Majumder, G. Gupta, M.V. Shelke, Enhanced electrochemical performance of polypyrrole coated MoS_2 nanocomposites as electrode material for supercapacitor application, *Journal of Electroanalytical Chemistry* 782 (2016) 278–287.

[12] A.K. Thakur, R.B. Choudhary, M. Majumder, M. Majhi, Fairly improved pseudocapacitance of PTP/PANI/TiO_2 nanohybrid composite electrode material for supercapacitor applications, *Ionics* 24(1) (2018) 257–268.

[13] M. Majumder, R.B. Choudhary, A.K. Thakur, C.S. Rout, G. Gupta, Rare earth metal oxide (RE_2O_3; RE= Nd, Gd, and Yb) incorporated polyindole composites: Gravimetric and volumetric capacitive performance for supercapacitor applications, *New Journal of Chemistry* 42(7) (2018) 5295–5308.

[14] M. Majumder, R.B. Choudhary, A.K. Thakur, Hemispherical nitrogen-doped carbon spheres integrated with polyindole as high performance electrode material for supercapacitor applications, *Carbon* 142 (2019) 650–661.

[15] R.B. Choudhary, M. Majumder, A.K. Thakur, Two-dimensional exfoliated MoS2 flakes integrated with polyindole for supercapacitor application, *Chemistry Select* 4(23) (2019) 6906–6912.

[16] M. Majumder, A.K. Thakur, M. Bhushan, D. Mohapatra, Polyaniline integration and interrogation on carbon nano-onions empowered supercapacitors, *Electrochimica Acta* 370 (2021) 137659.

[17] J.R. Harding, C.V. Amanchukwu, P.T. Hammond, Y. Shao-Horn, Instability of poly(ethylene oxide) upon oxidation in lithium–air batteries, *The Journal of Physical Chemistry C* 119(13) (2015) 6947–6955.

[18] J.-H. Park, J.-H. Cho, W. Park, D. Ryoo, S.-J. Yoon, J.H. Kim, Y.U. Jeong, S.-Y. Lee, Close-packed SiO_2/poly(methyl methacrylate) binary nanoparticles-coated polyethylene separators for lithium-ion batteries, *Journal of Power Sources* 195(24) (2010) 8306–8310.

[19] X. Zhang, T. Liu, S. Zhang, X. Huang, B. Xu, Y. Lin, B. Xu, L. Li, C.-W. Nan, Y. Shen, Synergistic Coupling between $Li_{6.75}La_3Zr_{1.75}Ta_{0.25}O_{12}$ and Poly(vinylidene fluoride) induces

high ionic conductivity, mechanical strength, and thermal stability of solid composite electrolytes, *Journal of the American Chemical Society* 139(39) (2017) 13779–13785.

[20] S. Slane, M. Salomon, Composite gel electrolyte for rechargeable lithium batteries, *Journal of Power Sources* 55(1) (1995) 7–10.

[21] T. Liu, Z. Chang, Y. Yin, K. Chen, Y. Zhang, X. Zhang, The PVDF-HFP gel polymer electrolyte for Li-O_2 battery, *Solid State Ionics* 318 (2018) 88–94.

[22] T.-C. Wen, W.-C. Chen, Gelled composite electrolyte comprising thermoplastic polyurethane and poly(ethylene oxide) for lithium batteries, *Journal of Power Sources* 92(1–2) (2001) 139–148.

[23] R. Tan, R. Gao, Y. Zhao, M. Zhang, J. Xu, J. Yang, F. Pan, Novel organic-inorganic hybrid electrolyte to enable LiFePO(4) quasi-solid-state Li-ion batteries performed highly around room temperature, *ACS Appl Mater Interfaces* 8(45) (2016) 31273–31280.

[24] J. Lopez, D.G. Mackanic, Y. Cui, Z. Bao, Designing polymers for advanced battery chemistries, *Nature Reviews Materials* 4(5) (2019) 312–330.

[25] P. Hu, J. Chai, Y. Duan, Z. Liu, G. Cui, L. Chen, Progress in nitrile-based polymer electrolytes for high performance lithium batteries, *Journal of Materials Chemistry A* 4(26) (2016) 10070–10083.

[26] J. Li, X. Huang, L. Chen, X-ray diffraction and vibrational spectroscopic studies on PAN-LiTFSI polymer electrolytes, *Journal of The Electrochemical Society* 147(7) (2000) 2653.

[27] H. Wang, Z. Zeng, P. Xu, L. Li, G. Zeng, R. Xiao, Z. Tang, D. Huang, L. Tang, C. Lai, Recent progress in covalent organic framework thin films: Fabrications, applications and perspectives, *Chemical Society Reviews* 48(2) (2019) 488–516.

[28] J. Zhang, J. Yang, T. Dong, M. Zhang, J. Chai, S. Dong, T. Wu, X. Zhou, G. Cui, Aliphatic polycarbonate-based solid-state polymer electrolytes for advanced lithium batteries: Advances and perspective, *Small* 14(36) (2018) 1800821.

[29] Y. Zhu, S. Xiao, Y. Shi, Y. Yang, Y. Hou, Y. Wu, A composite gel polymer electrolyte with high performance based on Poly(Vinylidene Fluoride) and polyborate for lithium ion batteries, *Advanced Energy Materials* 4(1) (2014) 1300647.

[30] X. Zhang, S. Wang, C. Xue, C. Xin, Y. Lin, Y. Shen, L. Li, C.-W. Nan, Self-suppression of lithium dendrite in all-solid-state lithium metal batteries with Poly(vinylidene difluoride)-based solid electrolytes, *Advanced Materials* 31(11) (2019) 1806082.

[31] M.S. Su'ait, A. Ahmad, H. Hamzah, M.Y.A. Rahman, Effect of lithium salt concentrations on blended 49% poly(methyl methacrylate) grafted natural rubber and poly(methyl methacrylate) based solid polymer electrolyte, *Electrochimica Acta* 57 (2011) 123–131.

[32] H. Zhang, J. Zhang, J. Ma, G. Xu, T. Dong, G. Cui, Polymer electrolytes for high energy density ternary cathode material-based lithium batteries, *Electrochemical Energy Reviews* 2(1) (2019) 128–148.

[33] N. Shukla, A.K. Thakur, Role of salt concentration on conductivity optimization and structural phase separation in a solid polymer electrolyte based on PMMA-LiClO 4, Ionics 15(3) (2009) 357–367.

[34] Y. Cui, J. Chai, H. Du, Y. Duan, G. Xie, Z. Liu, G. Cui, Facile and reliable in situ polymerization of poly(ethyl cyanoacrylate)-based polymer electrolytes toward flexible lithium batteries, *ACS Applied Materials & Interfaces* 9(10) (2017) 8737–8741.

[35] J. Chai, J. Zhang, P. Hu, J. Ma, H. Du, L. Yue, J. Zhao, H. Wen, Z. Liu, G. Cui, A high-voltage poly(methylethyl α-cyanoacrylate) composite polymer electrolyte for 5 V lithium batteries, *Journal of Materials Chemistry A* 4(14) (2016) 5191–5197.

[36] Y. Liu, B. Xu, W. Zhang, L. Li, Y. Lin, C. Nan, Composition modulation and structure design of inorganic-in-polymer composite solid electrolytes for advanced lithium batteries, *Small* 16(15) (2020) 1902813.

[37] J. Wan, J. Xie, D.G. Mackanic, W. Burke, Z. Bao, Y. Cui, Status, promises, and challenges of nanocomposite solid-state electrolytes for safe and high performance lithium batteries, *Materials Today Nano* 4 (2018) 1–16.

[38] W. Wang, E. Yi, A.J. Fici, R.M. Laine, J. Kieffer, Lithium ion conducting poly(ethylene oxide)-based solid electrolytes containing active or passive ceramic nanoparticles, *The Journal of Physical Chemistry C* 121(5) (2017) 2563–2573.

[39] L. Wang, W. Yang, J. Wang, D.G. Evans, New nanocomposite polymer electrolyte comprising nanosized $ZnAl_2O_4$ with a mesopore network and PEO-$LiClO_4$, *Solid State Ionics* 180(4–5) (2009) 392–397.

[40] S.H. Chung, Y. Wang, L. Persi, F. Croce, S.G. Greenbaum, B. Scrosati, E. Plichta, Enhancement of ion transport in polymer electrolytes by addition of nanoscale inorganic oxides, *Journal of Power Sources* 97 (2001) 644–648.

[41] C.-W. Nan, L. Fan, Y. Lin, Q. Cai, Enhanced ionic conductivity of polymer electrolytes containing nanocomposite SiO_2 particles, *Physical Review Letters* 91(26) (2003) 266104.

[42] S.N. Banitaba, D. Semnani, E. Heydari-Soureshjani, B. Rezaei, A.A. Ensafi, Electrospun polyethylene oxide-based membranes incorporated with silicon dioxide, aluminum oxide and clay nanoparticles as flexible solvent-free electrolytes for lithium-ion batteries, *JOM* 71(12) (2019) 4537–4546.

[43] G. Derrien, J. Hassoun, S. Sacchetti, S. Panero, Nanocomposite PEO-based polymer electrolyte using a highly porous, super acid zirconia filler, *Solid State Ionics* 180(23–25) (2009) 1267–1271.

[44] F. Zhou, X. Zhao, H. Xu, C. Yuan, CeO_2 spherical crystallites: Synthesis, formation mechanism, size control, and electrochemical property study, *The Journal of Physical Chemistry C* 111(4) (2007) 1651–1657.

[45] J.E. Weston, B.C.H. Steele, Effects of inert fillers on the mechanical and electrochemical properties of lithium salt-poly(ethylene oxide) polymer electrolytes, *Solid State Ionics* 7(1) (1982) 75–79.

[46] M.A.K.L. Dissanayake, P.A.R.D. Jayathilaka, R.S.P. Bokalawala, I. Albinsson, B.E. Mellander, Effect of concentration and grain size of alumina filler on the ionic conductivity enhancement of the $(PEO)_9LiCF_3SO_3$:Al_2O_3 composite polymer electrolyte, *Journal of Power Sources* 119–121 (2003) 409–414.

[47] W. Tang, S. Tang, X. Guan, X. Zhang, Q. Xiang, J. Luo, High-performance solid polymer electrolytes filled with vertically aligned 2D materials, *Advanced Functional Materials* 29(16) (2019) 1900648.

[48] D. Lin, W. Liu, Y. Liu, H.R. Lee, P.-C. Hsu, K. Liu, Y. Cui, High ionic conductivity of composite solid polymer electrolyte via in situ synthesis of monodispersed SiO_2 nanospheres in poly(ethylene oxide), *Nano Letters* 16(1) (2016) 459–465.

[49] S. Choudhury, S. Stalin, Y. Deng, L.A. Archer, Soft colloidal glasses as solid-state electrolytes, *Chemistry of Materials* 30(17) (2018) 5996–6004.

[50] D. Lin, P.Y. Yuen, Y. Liu, W. Liu, N. Liu, R.H. Dauskardt, Y. Cui, A silica-aerogel-reinforced composite polymer electrolyte with high ionic conductivity and high modulus, *Advanced Materials* 30(32) (2018) 1802661.

[51] Y. Lin, X. Wang, J. Liu, J.D. Miller, Natural halloysite nanoclay electrolyte for advanced all-solid-state lithium-sulfur batteries, *Nano Energy* 31 (2017) 478–485.

[52] O. Sheng, C. Jin, J. Luo, H. Yuan, H. Huang, Y. Gan, J. Zhang, Y. Xia, C. Liang, W. Zhang, $Mg_2B_2O_5$ nanowire enabled multifunctional solid-state electrolytes with high ionic conductivity, excellent mechanical properties, and flame-retardant performance, *Nano Letters* 18(5) (2018) 3104–3112.

[53] X. Zhang, J. Xie, F. Shi, D. Lin, Y. Liu, W. Liu, A. Pei, Y. Gong, H. Wang, K. Liu, Vertically aligned and continuous nanoscale ceramic–polymer interfaces in composite solid polymer electrolytes for enhanced ionic conductivity, *Nano Letters* 18(6) (2018) 3829–3838.

[54] Q. Zhang, S. Sun, W. Liu, P. Leng, X. Lv, Y. Wang, H. Chen, S. Ye, S. Zhuang, L. Wang, Integrating TADF luminogens with AIE characteristics using a novel acridine–carbazole hybrid as donor for high-performance and low efficiency roll-off OLEDs, *Journal of Materials Chemistry C* 7(31) (2019) 9487–9495.

[55] M. Dirican, C. Yan, P. Zhu, X. Zhang, Composite solid electrolytes for all-solid-state lithium batteries, *Materials Science and Engineering: R: Reports* 136 (2019) 27–46.

[56] Y. Liu, P. He, H. Zhou, Rechargeable solid-state Li–Air and Li–S batteries: Materials, construction, and challenges, *Advanced Energy Materials* 8(4) (2018) 1701602.

[57] M.P. O'Callaghan, A.S. Powell, J.J. Titman, G.Z. Chen, E.J. Cussen, Switching on fast lithium ion conductivity in garnets: The structure and transport properties of $Li_{3+x}Nd_3Te_{2-x}Sb_xO_{12}$, *Chemistry of Materials* 20(6) (2008) 2360–2369.

[58] G. Xi, M. Xiao, S. Wang, D. Han, Y. Li, Y. Meng, Polymer-based solid electrolytes: material selection, *Design, and Application, Advanced Functional Materials* 31 (2020) 2007598.

[59] R. Murugan, V. Thangadurai, W. Weppner, Fast lithium ion conduction in garnet-type $Li_7La_3Zr_2O_{12}$, *Angewandte Chemie International Edition* 46(41) (2007) 7778–7781.

[60] A. Sakuda, A. Hayashi, M. Tatsumisago, Recent progress on interface formation in all-solid-state batteries, *Current Opinion in Electrochemistry* 6(1) (2017) 108–114.

[61] T. Yang, J. Zheng, Q. Cheng, Y.-Y. Hu, C.K. Chan, Composite polymer electrolytes with $Li_7La_3Zr_2O_{12}$ garnet-type nanowires as ceramic fillers: Mechanism of conductivity enhancement and role of doping and morphology, *ACS Applied Materials & Interfaces* 9(26) (2017) 21773–21780.

[62] T. Jiang, P. He, G. Wang, Y. Shen, C.-W. Nan, L.-Z. Fan, Solvent-free synthesis of thin, flexible, nonflammable garnet-based composite solid electrolyte for all-solid-state lithium batteries, *Advanced Energy Materials* 10(12) (2020) 1903376.

[63] J.B. Goodenough, H.-P. Hong, J.A. Kafalas, Fast Na^+-ion transport in skeleton structures, *Materials Research Bulletin* 11(2) (1976) 203–220.

[64] V. Epp, Q. Ma, E.-M. Hammer, F. Tietz, M. Wilkening, Very fast bulk Li ion diffusivity in crystalline $Li_{1.5}Al_{0.5}Ti_{1.5}(PO_4)_3$ as seen using NMR relaxometry, *Physical Chemistry Chemical Physics* 17(48) (2015) 32115–32121.

[65] H. Aono, E. Sugimoto, Y. Sadaoka, N. Imanaka, G.Y. Adachi, Ionic conductivity of solid electrolytes based on lithium titanium phosphate, *Journal of the Electrochemical Society* 137(4) (1990) 1023.

[66] H. Zhai, P. Xu, M. Ning, Q. Cheng, J. Mandal, Y. Yang, A flexible solid composite electrolyte with vertically aligned and connected ion-conducting nanoparticles for lithium batteries, *Nano Letters* 17(5) (2017) 3182–3187.

[67] X. Wang, H. Zhai, B. Qie, Q. Cheng, A. Li, J. Borovilas, B. Xu, C. Shi, T. Jin, X. Liao, Rechargeable solid-state lithium metal batteries with vertically aligned ceramic nanoparticle/polymer composite electrolyte, *Nano Energy* 60 (2019) 205–212.

[68] A. Li, X. Liao, H. Zhang, L. Shi, P. Wang, Q. Cheng, J. Borovilas, Z. Li, W. Huang, Z. Fu, M. Dontigny, K. Zaghib, K. Myers, X. Chuan, X. Chen, Y. Yang, Nacre-inspired composite electrolytes for load-bearing solid-state lithium-metal batteries, *Advanced Materials* 32(2) (2020) 1905517.

[69] S. Stramare, V. Thangadurai, W. Weppner, Lithium lanthanum titanates: A review, *Chemistry of Materials* 15(21) (2003) 3974–3990.

[70] W. Liu, N. Liu, J. Sun, P.-C. Hsu, Y. Li, H.-W. Lee, Y. Cui, Ionic conductivity enhancement of polymer electrolytes with ceramic nanowire fillers, *Nano Letters* 15(4) (2015) 2740–2745.

[71] W. Liu, S.W. Lee, D. Lin, F. Shi, S. Wang, A.D. Sendek, Y. Cui, Enhancing ionic conductivity in composite polymer electrolytes with well-aligned ceramic nanowires, *Nature Energy* 2(5) (2017) 1–7.

[72] J. Bae, Y. Li, J. Zhang, X. Zhou, F. Zhao, Y. Shi, J.B. Goodenough, G. Yu, A 3D nanostructured hydrogel-framework-derived high-performance composite polymer lithium-ion electrolyte, *Angew Chem International Edition in English* 57(8) (2018) 2096–2100.

[73] N. Kamaya, K. Homma, Y. Yamakawa, M. Hirayama, R. Kanno, M. Yonemura, T. Kamiyama, Y. Kato, S. Hama, K. Kawamoto, A lithium superionic conductor, *Nature Materials* 10(9) (2011) 682–686.

[74] Y.S. Jung, D.Y. Oh, Y.J. Nam, K.H. Park, Issues and challenges for bulk-type all-solid-state rechargeable lithium batteries using sulfide solid electrolytes, *Israel Journal of Chemistry* 55(5) (2015) 472–485.

[75] Y. Zhao, C. Wu, G. Peng, X. Chen, X. Yao, Y. Bai, F. Wu, S. Chen, X. Xu, A new solid polymer electrolyte incorporating $Li_{10}GeP_2S_{12}$ into a polyethylene oxide matrix for all-solid-state lithium batteries, *Journal of Power Sources* 301 (2016) 47–53.

[76] X. Li, D. Wang, H. Wang, H. Yan, Z. Gong, Y. Yang, Poly(ethylene oxide)–$Li_{10}SnP_2S_{12}$ composite polymer electrolyte enables high-performance all-solid-state lithium sulfur battery, *ACS Applied Materials & Interfaces* 11(25) (2019) 22745–22753.

[77] S. Yubuchi, S. Teragawa, K. Aso, K. Tadanaga, A. Hayashi, M. Tatsumisago, Preparation of high lithium-ion conducting Li_6PS_5Cl solid electrolyte from ethanol solution for all-solid-state lithium batteries, *Journal of Power Sources* 293 (2015) 941–945.

[78] J. Zhang, C. Zheng, J. Lou, Y. Xia, C. Liang, H. Huang, Y. Gan, X. Tao, W. Zhang, Poly(ethylene oxide) reinforced Li_6PS_5Cl composite solid electrolyte for all-solid-state lithium battery: Enhanced electrochemical performance, mechanical property and interfacial stability, *Journal of Power Sources* 412 (2019) 78–85.

[79] Y. Zhou, S. Wu, Y. Ma, H. Zhang, X. Zeng, F. Wu, F. Liu, J.E. Ryu, Z. Guo, Recent advances in organic/composite phase change materials for energy storage, *ES Energy & Environment* 9 (2020) 28–40.

[80] H. Wang, Q. Wang, X. Cao, Y. He, K. Wu, J. Yang, H. Zhou, W. Liu, X. Sun, Thiol-branched solid polymer electrolyte featuring high strength, toughness, and lithium ionic conductivity for lithium-metal batteries, *Advanced Materials* 32(37) (2020) 2001259.

[81] Y.H. Jo, S. Li, C. Zuo, Y. Zhang, H. Gan, S. Li, L. Yu, D. He, X. Xie, Z. Xue, Self-healing solid polymer electrolyte facilitated by a dynamic cross-linked polymer matrix for lithium-ion batteries, *Macromolecules* 53(3) (2020) 1024–1032.

[82] M. Zaheer, H. Xu, B. Wang, L. Li, Y. Deng, An in situ polymerized comb-like PLA/PEG-based solid polymer electrolyte for lithium metal batteries, *Journal of The Electrochemical Society* 167(7) (2019) 070504.

[83] F. Ma, Z. Zhang, W. Yan, X. Ma, D. Sun, Y. Jin, X. Chen, K. He, Solid polymer electrolyte based on polymerized ionic liquid for high performance all-solid-state lithium-ion batteries, *ACS Sustainable Chemistry & Engineering* 7(5) (2019) 4675–4683.

[84] L. Meabe, N. Goujon, C. Li, M. Armand, M. Forsyth, D. Mecerreyes, Single-ion conducting poly(Ethylene Oxide Carbonate) as solid polymer electrolyte for lithium batteries, *Batteries & Supercaps* 3(1) (2020) 68–75.

[85] J. Wan, J. Xie, X. Kong, Z. Liu, K. Liu, F. Shi, A. Pei, H. Chen, W. Chen, J. Chen, X. Zhang, L. Zong, J. Wang, L.-Q. Chen, J. Qin, Y. Cui, Ultrathin, flexible, solid polymer composite electrolyte enabled with aligned nanoporous host for lithium batteries, *Nature Nanotechnology* 14(7) (2019) 705–711.

[86] W. Yao, Q. Zhang, F. Qi, J. Zhang, K. Liu, J. Li, W. Chen, Y. Du, Y. Jin, Y. Liang, N. Liu, Epoxy containing solid polymer electrolyte for lithium ion battery, *Electrochimica Acta* 318 (2019) 302–313.

[87] S. Ai, T. Wang, T. Li, Y. Wan, X. Xu, H. Lu, T. Qu, S. Luo, J. Jiang, X. Yu, D. Zhou, L. Li, A Chitosan/Poly(ethylene oxide)-based hybrid polymer composite electrolyte suitable for solid-state lithium metal batteries, *Chemistry Select* 5(10) (2020) 2878–2885.

[88] S. Zhang, Z. Li, Y. Guo, L. Cai, P. Manikandan, K. Zhao, Y. Li, V.G. Pol, Room-temperature, high-voltage solid-state lithium battery with composite solid polymer electrolyte with in-situ thermal safety study, *Chemical Engineering Journal* 400 (2020) 125996.

[89] Q. Guo, Y. Han, H. Wang, S. Xiong, Y. Li, S. Liu, K. Xie, New class of LAGP-based solid polymer composite electrolyte for efficient and safe solid-state lithium batteries, *ACS Applied Materials & Interfaces* 9(48) (2017) 41837–41844.

[90] Y. Kobayashi, H. Miyashiro, T. Takeuchi, H. Shigemura, N. Balakrishnan, M. Tabuchi, H. Kageyama, T. Iwahori, All-solid-state lithium secondary battery with ceramic/polymer composite electrolyte, *Solid State Ionics* 152–153 (2002) 137–142.

[91] N. Wu, P.-H. Chien, Y. Li, A. Dolocan, H. Xu, B. Xu, N.S. Grundish, H. Jin, Y.-Y. Hu, J.B. Goodenough, Fast Li+ conduction mechanism and interfacial chemistry of a NASICON/polymer composite electrolyte, *Journal of the American Chemical Society* 142(5) (2020) 2497–2505.

[92] L. Zhu, P. Zhu, S. Yao, X. Shen, F. Tu, High-performance solid PEO/PPC/LLTO-nanowires polymer composite electrolyte for solid-state lithium battery, *International Journal of Energy Research* 43(9) (2019) 4854–4866.

[93] L. Niedzicki, M. Kasprzyk, K. Kuziak, G.Z. Zukowska, M. Armand, M. Bukowska, M. Marcinek, P. Szczeciński, W. Wieczorek, Modern generation of polymer electrolytes based on lithium conductive imidazole salts, *Journal of Power Sources* 192(2) (2009) 612–617.

[94] H. Wang, D. Im, D.J. Lee, M. Matsui, Y. Takeda, O. Yamamoto, N. Imanishi, A composite polymer electrolyte protect layer between lithium and water stable ceramics for aqueous lithium-air batteries, *Journal of the Electrochemical Society* 160(4) (2013) A728–A733.

[95] L. Fan, Z. Dang, C.-W. Nan, M. Li, Thermal, electrical and mechanical properties of plasticized polymer electrolytes based on PEO/P(VDF-HFP) blends, *Electrochimica Acta* 48(2) (2002) 205–209.

[96] M.S. Michael, M.M.E. Jacob, S.R.S. Prabaharan, S. Radhakrishna, Enhanced lithium ion transport in PEO-based solid polymer electrolytes employing a novel class of plasticizers, *Solid State Ionics* 98(3) (1997) 167–174.

[97] J.-W. Choi, G. Cheruvally, Y.-H. Kim, J.-K. Kim, J. Manuel, P. Raghavan, J.-H. Ahn, K.-W. Kim, H.-J. Ahn, D.S. Choi, C.E. Song, Poly(ethylene oxide)-based polymer electrolyte incorporating room-temperature ionic liquid for lithium batteries, *Solid State Ionics* 178(19) (2007) 1235–1241.

[98] K. Vignarooban, M.A.K.L. Dissanayake, I. Albinsson, B.E. Mellander, Effect of TiO_2 nano-filler and EC plasticizer on electrical and thermal properties of poly(ethylene oxide) (PEO) based solid polymer electrolytes, *Solid State Ionics* 266 (2014) 25–28.

[99] T. Sutto, Hydrophobic and hydrophilic interactions of ionic liquids and polymers in solid polymer gel electrolytes, *Journal of the Electrochemical Society* 154 (2007) P101–P107.

[100] W. Zhang, J. Nie, F. Li, Z.L. Wang, C. Sun, A durable and safe solid-state lithium battery with a hybrid electrolyte membrane, *Nano Energy* 45 (2018) 413–419.

[101] J. Ju, Y. Wang, B. Chen, J. Ma, S. Dong, J. Chai, H. Qu, L. Cui, X. Wu, G. Cui, Integrated interface strategy toward room temperature solid-state lithium batteries, *ACS Applied Materials & Interfaces* 10(16) (2018) 13588–13597.

[102] X. Liu, S. Peng, S. Gao, Y. Cao, Q. You, L. Zhou, Y. Jin, Z. Liu, J. Liu, Electric-field-directed parallel alignment architecting 3D lithium-ion pathways within solid composite electrolyte, *ACS Applied Materials & Interfaces* 10(18) (2018) 15691–15696.

10 Polymer Batteries

Shishir Kumar Singh
Department of Physics, I.Sc., Banaras Hindu University, Varanasi, India

Dimple Dutta
Chemistry Division, Bhabha Atomic Research Centre, Mumbai, India

Rajendra Kumar Singh
Department of Physics, I.Sc., Banaras Hindu University, Varanasi, India

CONTENTS

10.1 INTRODUCTION

In modern times, the consumption of fossil fuel-based energy has increased rapidly due to the improvement in living standards as well as the increase in population. Therefore, to overcome the energy crisis, efforts have focused on the development of clean energy technologies, which include suitable energy storage devices for an efficient energy supply. Nowadays, fossil fuel-based vehicles are being replaced by hybrid-electric vehicles (HEVs). In order to achieve this, rechargeable batteries, one of the efficient energy storage devices, are considered to be the next generation main power sources for transportation and portable electronic devices [1,2]. Currently, conventional battery technologies such as nickel-metal hydride (Ni-MH), nickel-cadmium (Ni-Cd), and valve-regulated lead-acid (VR-LA) batteries are being replaced by rechargeable lithium-ion batteries (R-LIBs), because they have a high gravimetric/volumetric energy storage capability with a high-rate capability. But R-LIBs suffer from safety issues such as thermal rupture due to the use of carbonate-based ionic solvents. Nowadays, many researchees have focused on non-volatility and liquid free electrolytes like polymer electrolytes, because these overcomes problems associated with commercial organic liquid electrolytes. Therefore, rechargeable polymer batteries are emerging as one of the most promising candidates to provide long cycling stability as well as removing safety concerns. Figure 10.1 shows that the widely used rechargeable batteries with high revenue contributions are rechargeable lithium-ion batteries (~37%), starting, lighting, and ignition (SLI) batteries (~20%), alkaline batteries (~15%), and lead-acid batteries (~8%) [3].

10.1.1 Rechargeable Batteries

Batteries are basically a combination of electrochemical cells connected in series, parallel, or both. These batteries convert electrochemical energy into electrical energy through redox reaction, and vice versa, and are used as an electric power

DOI: 10.1201/9781003169727-10

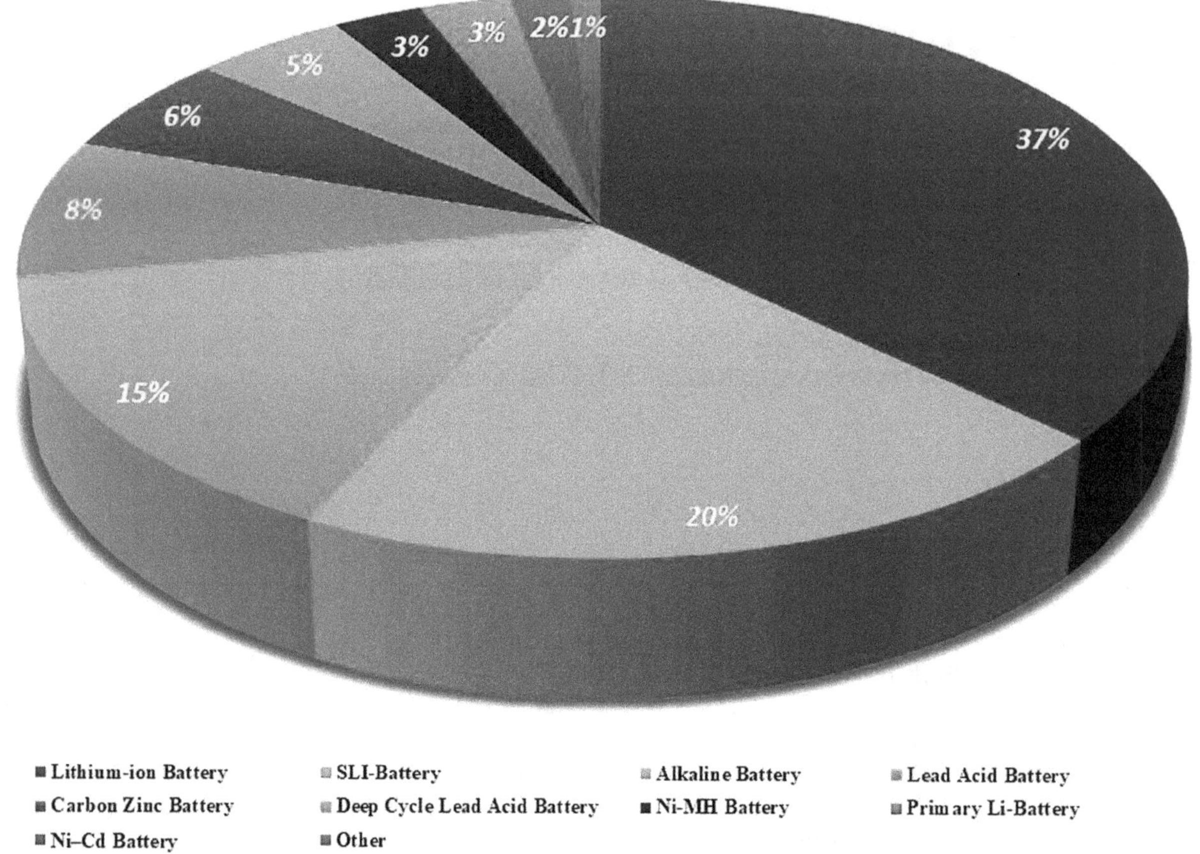

FIGURE 10.1 Different battery chemistries used in revenue generation.

source. Batteries are mainly divided into two groups: primary (non-rechargeable) and secondary (rechargeable) batteries. In primary batteries (i.e., alkaline batteries), the chemical reactions aren't reversible so when the chemicals are used up the cell stops producing electricity. So, a primary battery is used once and discarded. In contrast to this, a secondary battery (like lead-acid, Ni-Cd, lithium, sodium, and lithium sulphur) can be charged and discharged several times as the chemistry is reversible [4]. Rechargeable batteries consist of an anode (negative electrode), electrolyte, and cathode (positive electrode). During charging, the cathode supplies electrons through the external circuit by oxidation reactions, and the anode gains electrons by reduction reactions. In discharging, the anode supplies electrons through the external load by oxidation reactions, and the cathode gains electrons through reduction reactions. During the charging–discharging process, the electrolyte is used as an electronic separator and also provides the ionic medium for the transportation of ions between the electrode Figure 10.2 illustrates the energy density of various rechargeable batteries.

10.1.1.1 Lead-Acid Battery

In 1859 Gaston Planté invented the lead-acid battery. In this battery, lead (Pb), used as an anode, and lead dioxide (PbO_2), used as a cathode, are dipped into an aqueous solution of sulfuric acid (H_2SO_4) which is used as an electrolyte. It has a very low self-discharge rate and is very inexpensive and simple to manufacture. Due to these reasons, they are preferred for applications in light-weight electric vehicles, home inverters, uninterrupted power supply (UPS), and so on. But they suffer from some limitations such as slow charging, low specific energy (35 W h kg^{-1}), low volumetric–gravimetric energy density, toxicity, and a limited cycle life [3,5]. The redox reaction of a lead-acid battery is:

$$Pb_{(s)} + PbO_{2(s)} + 2H_2SO_{4(aq)} \rightleftarrows 2PbSO_{4(s)} + 2H_2O_{(l)} \tag{10.1}$$

10.1.1.2 Nickel-Cadmium (Ni-Cd) Battery

This battery was invented by Waldemar Jungner in 1899. With few technological improvements, Ni-Cd became the most preferable battery in the 1980s due to its ultra-fast charging with low temperature performance. In an Ni-Cd battery, there is NiOOH (cathode), Cd (anode), and a KOH aqueous solution which is used as an electrolyte. The Ni-Cd battery shows better specific energy (50 W h kg^{-1}) as compared to the lead-acid battery. But they have some drawbacks, such as a lower specific energy density, self-discharging, a memory effect, and high toxicity due to cadmium (Cd). Most importantly an Ni-Cd battery can't be disposed of in landfills [3,6]. The following reaction occurs in an Ni-Cd battery:

$$2NiOOH + Cd + 2H_2O \rightleftarrows Ni(OH)_2 + Cd(OH)_2 \tag{10.2}$$

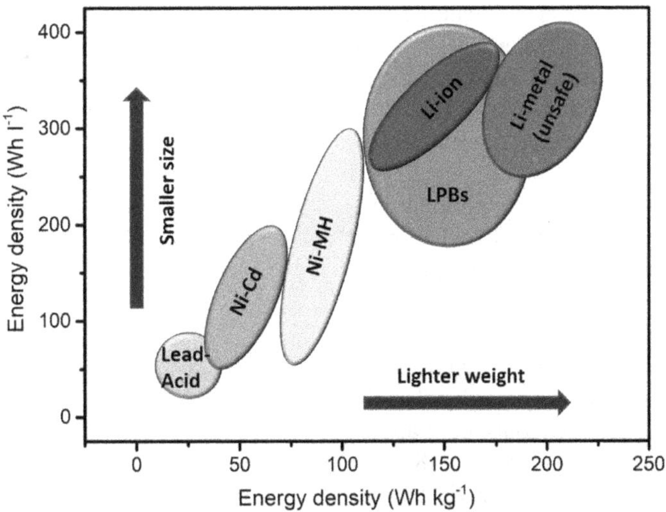

FIGURE 10.2 Ragone plot of various rechargeable batteries.

10.1.1.3 Nickel-Metal Hydride Battery (Ni-MH Battery)

The Ni-MH battery was available from the 1980s through leading battery manufacturers. It achieved a higher specific capacity than an Ni-Cd battery (~40%) but with less memory effect, and so is suitable under a wide temperature range with very low toxicity. It fulfills the requirement for cheap and low-end applications. But it has some limitations, such as a high intolerance to overcharging, self-discharging, a limited-service life, requires a complex charging algorithm, and low coulombic efficiency (~65%) [3,6]. The redox reaction which occurs in an Ni-MH battery is:

$$2NiOOH + H_2 \rightleftarrows 2Ni(OH)_2 \quad (10.3)$$

10.1.1.4 Rechargeable Lithium Batteries (R-LBs)

The technology of R-LBs was developed by J. Goodenough, S. Whittingham, R. Yazami, and A. Yoshino and then launched by Sony in 1991. Nowadays, the market of lithium rechargeable batteries has become more dominant in specific applications in electronic devices and electric vehicles, because they have achieved a high energy/power density compared to all other rechargeable batteries with high cell voltage (~3.6 V). A lithium battery has mainly three components in which a cathode is used as a positive electrode ($LiFePO_4$, $LiCoO_2$, $LiMn_2O_4$, $LiMnPO_4$, etc.), an anode is used as a negative electrode (Li-metal, graphite, hard carbon, $Li_4Ti_5O_{12}$, etc.), and a lithium ion-conducting-based electrolyte placed between both electrodes (cathode and anode) provides ion conduction as well as the electronic insulating medium [4]. During the discharge process, Li^+-ions flow from anode to cathode via the electrolyte, while electrons flow through the external load. However, during the charging process, Li^+-ions flow from the cathode to the anode via the electrolyte, while electrons move through the external circuit. The self-discharging of lithium batteries is less than 2% per month in contrast to 20% for Ni-Cd batteries and 30% for Ni-MH batteries. However, it needs a battery management system to protect it from abuse, and also some improvement is required such as safety with cyclic stability [7]. On the basis of lithium chemistry, rechargeable lithium batteries are broadly of two types:

1. *Rechargeable lithium-ion battery*: In this type of battery, the anode used is a lithium intercalated material (graphite, hard carbon, $Li_4Ti_5O_{12}$, etc.). This type of anode material improved cyclability with safety but it has some limitations such as low cell potential and suffers from a high charge transfer rate [8]. The configuration of the rechargeable lithium-ion battery is: LiC_6/Electrolyte/Cathode
2. *Rechargeable lithium-metal battery (R-LMB)*: In the R-LMB, the anode used is of lithium metal, because the theoretical capacity of lithium metal (~3860 mA h g^{-1}) is extremely high compared to lithium intercalated materials (LiC_6 ~ 372 mA h g^{-1}, $Li_4Ti_5O_{12}$ ~ 175 mA h g^{-1}), and it shows the lowest electrochemical potential (~−3.04 V w.r.t. a hydrogen electrode) in comparison to other elements. Because of these beneficial properties, it has some unique advantages such as a high operating voltage as well as a high specific energy/power density [8]. The configuration of the rechargeable lithium metal battery is:Li(metal)/Electrolyte/Cathode

10.2 BATTERY COMPONENTS AND PARAMETERS

➢ *Cathode (positive electrode)*: The cathode gains electrons through the external circuit during discharging and it produces electricity with electrochemical reactions. The cathode is a good oxidizing agent and provides stability and working potential when it comes in contact with the electrolyte. The material used in a positive electrode is called the active material. The specific capacity of the battery is dependent on active material. Some active materials are $LiFePO_4$, $LiCoO_2$, $LiMn_2O_4$, $LiMnPO_4$, LNMC, and LNCA [8].

- *Anode (negative electrode)*: The anode releases electrons during discharging through the external load and produces electricity. Such anodes are more preferred which have interesting properties such as good conductivity, low cost, high coulombic output (A h g^{-1}), high stability, and ease of fabrication. The material used in the negative electrode is called the anode material. In cell assembling, the ion storage capacity of the anode material should always be higher than the cathode material. Some active materials are graphite, hard carbon, silicon, $Li_4Ti_5O_{12}$, and Li-metal [8].
- *Electrolyte*: This is an ion-conducting material and has negligible electronic conduction that provides a suitable medium to transfer ions inside the cell (from positive to negative electrode, and vice versa) during the electrochemical reaction and that also protects against short-circuiting. Typically, it is liquid in nature, a solution of salts, alkalis, or acids to impart ionic conductivity. It should have some beneficial properties such as it should be unreactive with the active material, and its physical and chemical properties should not change with temperature and voltage variation. Apart from these, the electrolyte also should have high Li^+-ion conductivity with a wide electrochemical stability window. Nowadays, many rechargeable batteries are fabricated using solid electrolytes, especially polymer-based electrolytes, which are ionic conductors.
- *Separator*: This is just like a physical barrier between the electrodes to prevent short circuiting and must be chemically inert as well as permeable to the ions. Polymer-based solid electrolytes avoid the use of additional separators.
- *Specific capacity*: The capacity of the battery depends on the total amount of electric charge stored in the electrode materials. It is expressed in ampere hours (A h). The specific capacity of the battery is dependent on the total amount of charge per unit weight of cathode material contained in the electrode (mA h g^{-1}). The specific theoretical capacity (C_{specific}) of the cathode material is:

$$C_{\text{specific}} = \frac{x \times F}{nM} \tag{10.4}$$

Here, x is the total number of electrons released in a redox reaction, F is the Faraday constant (96,485 C mol^{-1}), n is the number atoms of the active material, and M is the molecular weight of the cathode material. As an example, the theoretical capacity of $LiCoO_2$ cathode material calculated by using equation (10.4) can be given as:

$$LiCoO_2 \rightleftarrows Li^+ + e^- + CoO_2 \tag{10.5}$$

Here, the molecular weight of the $LiCoO_2$ cathode is 97.9 g mol^{-1}, the number of total electrons transferred is $x = 1$ from the $LiCoO_2$ molecule, and the number of atoms of active material is $n = 1$ which is 1 in the redox reaction. So, the C_{specific} of $LiCoO_2$ is 274 mA h g^{-1}.

- *Specific energy density*: The energy density of a battery is the energy stored by the electrode material per unit weight or per unit volume. It can also be defined in the form of capacity and voltage as:

$$\text{Energy density}\left(\text{W h kg}^{-1}\right) = \text{Capacity} \times \text{Cell voltage}.$$

The high specific energy density of a battery means that it is able to store a lot of energy in a small mass of active material.

- *Power density*: The power density of the battery is the power stored by the battery per unit weight of active material. It can be denoted in terms of voltage, current, and energy density:

$$\text{Power density} = \text{Energy density/Time} = \text{Current} \times \text{Voltage}$$

The high-power density of the battery indicates that the system can release large amounts of energy. It can be expressed as watts per kilogram (W kg^{-1}).

- *Current-rate (C-rate)*: The charge rate (or discharge rate) of the battery is often expressed as the C-rate. A 1C rate means the battery requires a fixed current to be fully charged or discharged in 1 hour. Similarly, a C/10 rate shows that a battery can be charged or discharged at a particular current in 10 hours. The value of the required current for charging as 1C (one hour or $h = 1$) is calculated by the following equation:

$$1C = \text{Theoretical capacity of active material}\left(\frac{\text{mA h}}{\text{g}}\right) \times \text{weight of active material per unit area of electrode}\left(\text{g}\right) \tag{10.6}$$

- *Charging*: In the charging process of a battery, the current flows in the reverse direction and the oxidation reaction occurs on the cathode and the reduction reaction occurs on the anode.

  The reaction occurring during the charging process is:

$$\text{On cathode: } LiFePO_4 \rightarrow Li_{1-x}FePO_4 + xLi^+ + xe^- \tag{10.7}$$

$$\text{On anode: } Li^+ + e^- \rightarrow Li \tag{10.8}$$

- *Discharging*: During the discharging process, the battery is connected with the load, during which the flow of electrons occurs from the anode to the cathode, and ions move inside the battery through the electrolyte and the reduction reaction occurs on the cathode.

The reaction occurring during the discharging process is:

$$\text{On cathode: } Li_{1-x}FePO_4 + xLi^+ + xe^- \rightarrow LiFePO_4 \quad (10.9)$$

$$\text{On anode: } Li \rightarrow Li^+ + e^- \quad (10.10)$$

- *Cyclability*: The cycle life or cyclability of a battery is a very important parameter because high cyclability indicates the durability of the battery which is important for electronic devices and electric vehicles. The cyclability of the battery is related to the number of charge/discharge cycles with almost constant specific capacity. The cyclability of a battery depends on many important factors such as the chemistry of the battery, operating temperature/voltage, compatibility of the electrode with the electrolyte, and the depth of discharge.
- *Coulombic efficiency*: The coulombic efficiency is indicative of the reversibility of the electrochemical reactions on the electrodes. It gives, after a full cycle, the ratio of the total amount of charge released from the battery during discharge and the total amount of charge received into the battery during charge. The total amount of charge released during charging a battery (Q_{charge}) is larger than the charge gained during discharge ($Q_{discharge}$). The coulombic efficiency (η) is calculated by:

$$\eta = \frac{Q_{discharge}}{Q_{charge}} \times 100\% \quad (10.11)$$

A 100% coulombic efficiency indicates that the total number of deintercalated lithium-ions from the cathode is equal to the intercalated lithium-ions entering the cathode. Generally, the coulombic efficiency of a battery is less than 100% because initial cycles lose some of the charge due to unwanted irreversible side reactions [9].

- *Memory effect*: This is an effect observed in rechargeable batteries which forces them to store less charge. If a battery is repeatedly charged after its partial discharge then the battery cannot achieve its highest capacity. This is known as the memory effect.

10.3 ELECTROLYTES FOR RECHARGEABLE LITHIUM BATTERIES

The electrochemical performance of a lithium battery mainly depends on the ion-conducting electrolyte because it provides the medium for Li^+-ion conduction between both electrodes during the redox reaction. For efficient lithium batteries, electrolytes should have high ionic conductivity, negligible electronic conductivity, a wide range of working potential, high thermal/chemical stability, non-volatility, non-flammability, be environmentally friendly, and have wettability. According to the physical state, the electrolytes are divided into two main categories.

10.3.1 Liquid Electrolytes

The commercial liquid electrolytes are composed of salt $LiPF_6$ (lithium hexa-fluorophosphate) dissolved in carbonate-based organic solvents (some commercial electrolytes are: 1 M $LiPF_6$ in EC: DEC, 1 M $LiPF_6$ in EC: DMC, and 1 M $LiPF_6$ in EC: DEC: DMC) [10]. So, they exhibit high ionic conductivity, low viscosity, and also provide better contact electrodes with low interface resistance. Nowadays, the safety of batteries is one of the key concerns due to the use of commercial liquid electrolytes, because the carbonate-based ones have some drawbacks, such as high volatility/flammability, leakage problems, electrode corrosion, and CO_2 gas evolution when the battery is working at high potential or after increasing the battery temperature. However, salt $LiPF_6$ has many concerns, such as being easily decomposed (Reaction 10.12) and being highly reactive when water is present in the electrolyte (Reaction 10.13):

$$LiPF_6 \rightarrow LiF + PF_5 \quad (10.12)$$

$$LiPF_6 + H_2O \rightarrow LiF + POF_3 + 2HF \quad (10.13)$$

In Reactions (10.12) and (10.13), some highly reactive and harmful products are formed such as HF, POF_3, and PF_5. So, electrode degradation is due to the presence of HF gas which has an effect on the cycling stability of the battery when electrolyte degradation occurs due to the presence of another compound—POF_3 (Reaction 10.14)—and when CO_2 gas is released, which is the main reason for battery explosion:

$$CH_3CH_4O_3 + POF_3 \rightarrow CH_2FCH_2OPF_2O + CO_2(\text{gas}) \quad (10.14)$$

Furthermore, lithium deposition at the interface of the lithium metal anode and the electrolyte during the charging–discharging process leads to the growth of spiky electron-conducting dendrite crystals from the anode to the cathode. After long cycling, these electron-conducting dendrites penetrate the separator and come to the contact of the cathode which causes electrical short-circuit [11] (see Figure 10.3).

10.3.2 Solid Electrolytes

Solid electrolytes (SEs) conduct ions at room temperature, and their applications are of fundamental and practical interest. In device applications, superionic conductors (SICs) or

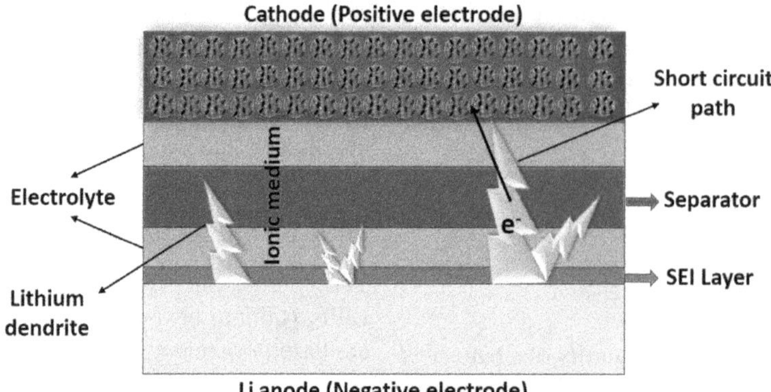

FIGURE 10.3 Schematic diagram of the growth of lithium dendrites in rechargeable lithium batteries.

solid electrolytes (SEs) are a good replacement for conventional organic electrolytes because SEs overcome problems such as leakage/portability issues, electrode corrosion during charging–discharging, and gas evolution during heating associated with liquid electrolytes [12,13]. Generally, solid electrolytes are classified into two main groups.

10.3.2.1 Inorganic Solid Electrolytes

Conventional ionic solids such as alkali and silver halides (AgI, AgBr, AgF, etc.) have very low ionic conductivity (~10^{-10} to 10^{-12} S cm^{-1}), because of which they are not suitable for solid-state devices. Therefore, inorganic solid electrolytes are synthesized by mixing a poor ion conductor with insulating material. Inorganic solid electrolytes are designed on certain structural criteria:

1. The size of the ions should be suitable for conduction;
2. Inorganic solid electrolytes should be single-ion conductors (the cationic transference number should be unity);
3. Negligible or very low electronic conductivity is required;
4. No concentration gradient inside the cell during operation.

Due to the occurrence of these properties, inorganic solid electrolytes are used in portable devices like batteries, sensors, supercapacitors, and fuel cells [14,15]. On the basis of physical and chemical properties, inorganic solid electrolytes are mainly classified in two groups.

10.3.2.1.1 Crystalline or Polycrystalline Solid Electrolytes

Ion transport behavior in crystalline solid electrolytes depends on their crystalline structure, which is related to the ion hopping mechanism in the crystal structures. Generally, crystalline materials possess two sub-lattices and a rigid matrix enclosing a sub-lattice, which provides an easy path for ion conduction within the crystalline framework [16,17]. This crystalline framework is divided into two types:

1. *Soft framework crystals*: The soft framework materials are characterized by:

   a. Ionic bonding;
   b. Polarizability of heavy ions (e.g., Ag^+, Cu^+);
   c. A sharp phase transition;
   d. Low Debye temperature.

   Some examples of this class of materials are AgI, CuI, $RbAg_4I_5$, and Ag_2HgI_2 [16].
2. *Hard framework crystals*: The hard framework solids are characterized by:

   a. Covalent bonding with low polarizability of mobile ions;
   b. High Debye temperature;
   c. Generally, less order–disorder transitions.

   Some examples of this class of materials are NASICONS and Na-β-alumina [18].

10.3.2.1.2 Glassy Solid Electrolytes

The glassy solid electrolytes are synthesized by quenching a molten mixture of a glass former, metal oxide glass modifier with ionic dopant salts. Glassy solid electrolytes have some unique advantages with respect to crystalline/polycrystalline solid electrolytes such as the absence of grain boundary conduction and a wide range of compositions with ease of preparation into different shapes and sizes. In general, glassy solid electrolytes can be expressed as MX-M_2O-A_xO_y. Here, MX-dopant ionic salts (AgI, NaI, CuI, PbI_2, CdI_2, etc.), M_2O as a glass modifier (Ag_2O, Cu_2O, etc.), and A_xO_y as a glass former (B_2O_3, MoO_3, P_2O_5, SiO_2, As_2O_5, etc.) [19,20]. The ionic conductivity of some glassy solid electrolytes are shown in Table 10.1.

10.3.2.2 Polymer Electrolytes (PEs)

Polymer electrolytes are considered as the most probable next generation electrolytes for enhancing the electrochemical performance of batteries. They are classified in the following categories as shown in Figure 10.4.

TABLE 10.1
The Ionic-Conductivity of Crystalline/Polycrystalline and Glassy Solid Electrolytes

Inorganic Solid Electrolytes	Ionic Conductivity (σ) S cm^{-1}	Conducting Ions	References
Crystalline/Polycrystalline Solid Electrolytes			
MAg_4I_5 (M = Rb, K, NH_4)	~0.20 (22 °C)	Ag^+	[21]
AgI-Ag_3PO_4	~6.0 × 10^{-2} (RT)	Ag^+	[22]
$Li_{0.34}La_{0.51}TiO_{2.94}$	~1.0 × 10^{-3} (27 °C)	Li^+	[23]
$Li_{3.25}$Ge0.25$P_{0.7}5S_4$	~2.0 × 10^{-3} (25 °C)	Li^+	[24]
Glassy Solid Electrolytes			
75Ag-$_{27}Ag_2SeO_4$	~2.2 × 10^{-2} (25°C)	Ag^+	[25]
$75Li_2S$-$20P_2S_5$-$5P_2S_3$	~6.4 × 10^{-4} (25°C)	Li^+	[26]
CuI-Cu_2O-P_2O_5	~1.0 × 10^{-3} (25°C)	Cu^+	[27]
$60Na_2S$-SiO_2-B_2O_3	~2.0 × 10^{-3} (27°C)	Na^+	[28]

10.3.2.2.1 Solid Polymer Electrolytes (SPEs)

In SPEs, alkali metal salts are incorporated into high molecular weight polymers like polyvinylidene fluoride (PVdF), polyethylene oxide (PEO), polyvinyl acetate (PVA), and polymethymethacrylate (PMMA) (see Figure 10.5). The ion conducting property of SPEs by complexing alkali metal salts with polar polymer PEO was reported the first time by Wright et. al (1973). However, Armand and his co-workers in 1979 for the first time recognized the potential applications of these materials. After that, not only polymer PEO but different polymer host matrices such as PVdF, PMMA, and PVdF-HFP were used for the preparation of SPEs. Nowadays, a huge number of ionic salts used in PEs consisting of monovalent, divalent, and transition metal ions (e.g., Li^+, Na^+, Ag^+, H^+, K^+, Mg^{2+}) have been reported [29,30].

A few important criteria of the formation of SPEs can be expressed as: [31]

1. The polymer should have a large number of polar groups (e.g., hydroxyl (–OH) group, carboxyl (–C=O–OH) group, and carbonyl (>C=O) group) and a flexible chain.
2. The polymer should have a low T_g (glass transition temperature).
3. The polymer should have low cohesive energy to facilitate the higher dissociation of salts.

Currently, SPEs have become fast-growing electrolytes in current research interest for application in electrochemical portable devices due to their having some important properties such as a wide electrochemical stability window, high mechanical/thermal stability, and a thin film-forming ability. But some limitations are associated with SPEs such as:

- Most SPEs have very low ionic conductivity (~0.001 to 0.01 mS cm^{-1} at RT);
- A high interface resistance;
- A low lithium-ion conductivity.

SPEs have better mechanical stability and safety for device application but they possess a low ionic-conductivity value at room temperature. The ionic conductivity of some SPEs are given in Table 10.2. Therefore, in order to enhance ionic conductivity, different approaches are adopted:

1. *Co-polymerization*: In the co-polymerization method, another polymer is linked with the original host matrix and provides an additional polymer chain, which reduces the glass transition temperature (T_g) as well as the percentage of crystallinity of the host polymer matrix and it also enhances ionic conductivity [39]. Liang et al. [40] reported that the high ionic conducting comb-like copolymer Poly(AN-co-PEGMEM) was prepared by the co-polymerization of acrylonitrile and poly(ethylene glycolmethyl methacrylate).

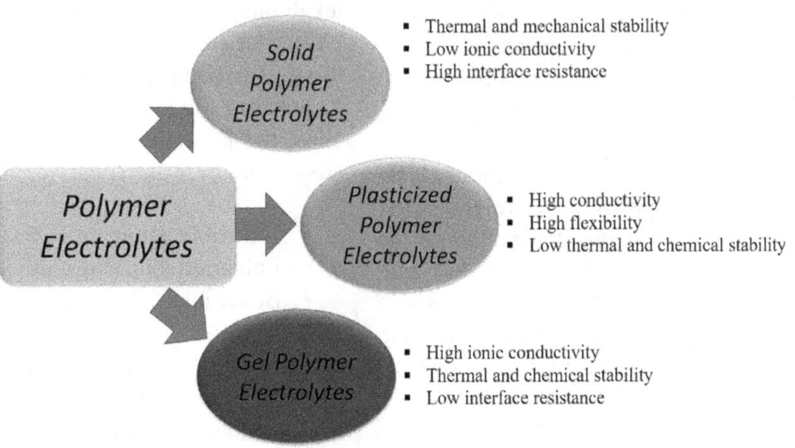

FIGURE 10.4 Classification of the polymer electrolytes for polymer batteries.

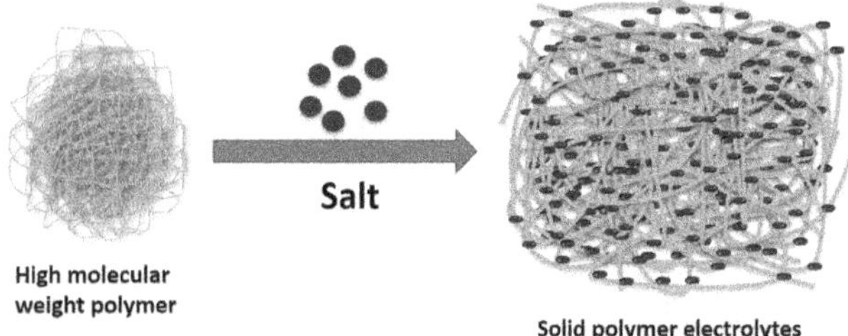

FIGURE 10.5 Schematic diagram of solid polymer electrolytes.

TABLE 10.2
The Ionic Conductivity of SPEs

SPEs	Ionic Conductivity (σ) S cm^{-1}	Temperature (°C)	References
PEO + $LiN(CF_3SO_3N)_2$	~1.0×10^{-4}	RT	[32]
PEO + $LiAsF_6$	~1.4×10^{-4}	25	[33]
PVA–PVdF–$LiClO_4$	~3.0×10^{-5}	25	[34]
PVA–LiBOB,	~1.0×10^{-4}	25	[35]
PEO + $Mg(ClO_4)_2$	~1.3×10^{-7}	27	[36]
PEG + $LiClO_4$	~2.0×10^{-7}	20	[37]
PEO + $NaPF_6$	~5.7×10^{-6}	30	[38]

2. *Blending of polymer*: Polymer blending processes are effective ways to reduce the crystalline phase of polymers which enhance the ionic conductivity results in the system. The blended polymers are the physical mixtures of two or more host polymers which are not interlinked by a covalent bond in which the T_g of the host polymer is low, such as poly methymethacrylate (PMMA), polyvinyl chloride (PVC), and polyvinylpyrrolidone (PVP) [41]. When two or more polymers are completely miscible at a segmental level, they form a single homogeneous phase, that is, a single melting point, a single crystallinity point, and a single glass transition temperature. Some blend polymer electrolytes are shown in Table 10.3.
3. *Composite polymer electrolytes (CPEs)*: This is another effective way to enhance the ionic-conductivity of polymer electrolytes. The CPEs manually consist of three components: polymer host matrix, dopant ionic salt, and an inorganic filler (like SiO_2, ZrO_2, $LiAlO_2$, Al_2O_3, TiO_2, fumed silica, and $BaTiO_3$). These nano-fillers are doped in the salt-polymer system to obtain the nano-composite solid polymer electrolytes. After uniform dispersion of the inorganic nanofiller in the salt-polymer system, increased amorphicity with better mechanical and electrochemical properties of the PEs can be achieved [45]. Some CPEs with ionic conductivity are shown in Table 10.4.

10.3.2.2.2 Plasticized Polymer Electrolytes

Recently, much research work has focussed on enhancing the ionic conductivity of PEs at ambient temperatures. Apart from the advantage of the polymer host system and dopant alkali metal salts there is a need to change the structure of the host polymer matrix which would improve the ionic conductivity of the PE system. Many PEs have high glass transition temperatures and also have low amorphicity which provide low ionic conduction at RT. One of the effective approaches to increase the total ionic-conductivity of PEs is the use of organic plasticizers in the PE system. The plasticized PEs used as a hybrid electrolyte have combined beneficial natures such as providing a liquid-like nature as well as safety and a flexible shape. These hybrid electrolytes consist of a host polymer matrix, an inorganic alkali metal salt providing conducting ions, and an organic plasticizer (e.g., ethylene carbonate, propylene carbonate, di-ethyl carbonate, and di-methyl carbonate) which is used to enhance the amorphous nature of the system [50,51]. Now, the plasticized polymer electrolytes exhibit several advantages over the SPEs:

- A flexible nature as well as being mechanical stable;
- Enhanced ionic conductivity;
- Better interface contact with electrodes.

As these organic plasticizers enhance the ionic conductivity of PEs they suffer from less thermal/chemical stability, have a reduced environmentally hazardous nature, and can't operate at high temperatures because of highly volatility/flammability and toxicity [52]. Currently, many researchers

TABLE 10.3
The Ionic-Conductivity of Blend PEs

Blend Polymer Electrolytes	Ionic Conductivity (σ) S cm^{-1}	Temperature (°C)	References
PEMA-PVC-PC-$LiClO_4$	~1.9×10^{-4}	30	[42]
PAN-PMMA-$LiClO_4$	~5.6×10^{-4}	30	[43]
PVA–PVdF–$LiClO_4$	~3.0×10^{-5}	25	[33]
PEO–PMMA–$LiCF_3SO_3$	~1.01×10^{-5}	30	[44]

TABLE 10.4
The Ionic-Conductivity of Composite PEs

Composite Polymer Electrolytes	Ionic Conductivity (σ) S cm^{-1}	Temperature (°C)	References
PEO–$LiBF_4$–TiO_2,	~1.0×10^{-5}	25	[46]
PEO–NH_4SCN-Mg–Zn ferrite	~8.0×10^{-6}	RT	[47]
PEO–$LiClO_4$–Al_2O_3	~1.0×10^{-5}	30	[45]
PVC–PEMA–PC–$LiClO_4$-TiO_2	~7.2×10^{-3}	28	[48]
PVA:NaI + SiO_2	~3.8×10^{-3}	RT	[49]

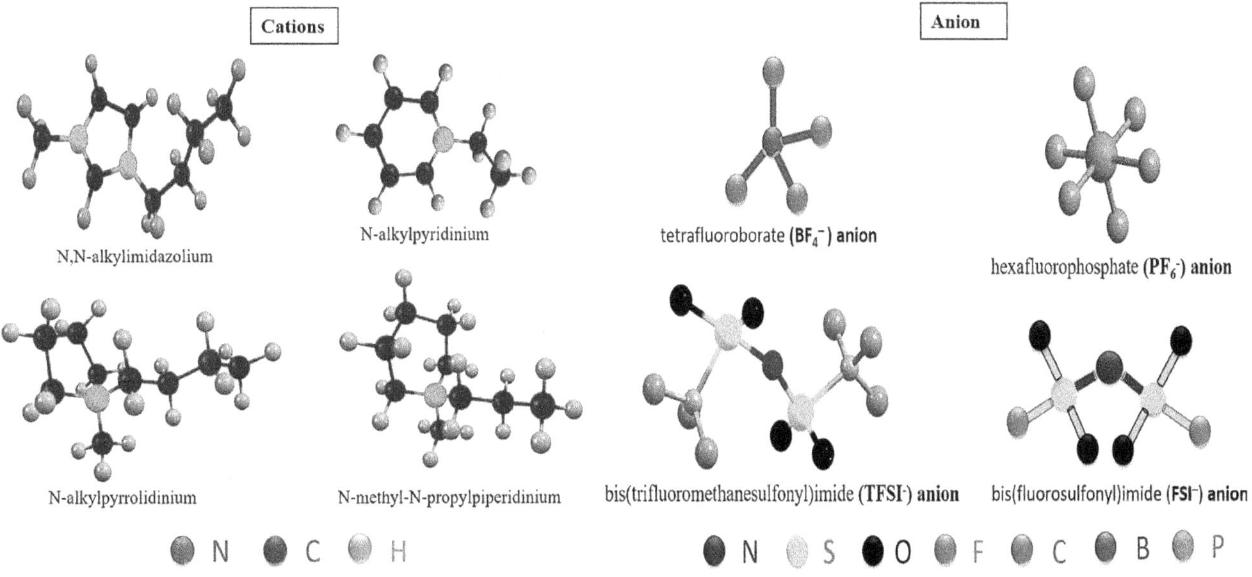

FIGURE 10.6 Some common cations/anions of room-temperature ionic liquids.

are discovering an alternative approach for achieving thermal/chemical stable polymer electrolytes with high ionic-conductivity. In this approach, room-temperature ionic liquids (R-ILs) are used in the place of low molecular weight organic plasticizers. R-ILs are molten salts generally, which are composed of asymmetric self-dissociated bulky organic cations and organic/inorganic anions. R-ILs have very low ionic bonding between asymmetric bulky cations and anions with respect to other ionic salts (e.g., NaI, NaCl, KCl); because of having low lattice energy they easily dissociate into cations and anions [53]. The most common anions/cations of ionic liquids are shown in Figure 10.6. These cations and anions can be tailored in millions of possible ways to make suitable R-ILs for many applications. Due to this unique nature, they are also called "designer solvents."

So, due to some of their important properties, R-ILs are considered as an ideal ionic solvent:

- ✓ High ionic conductivity due to mobility of free mobile anions/cations (~0.1–0.01 S cm^{-1});
- ✓ Wide electrochemical stability window (EWS) (~4.0–6.0 V);
- ✓ Low melting point ($T_m < 100$ °C);
- ✓ High thermal stability (~400–500 °C);
- ✓ Low glass transition temperature (T_g < room temperature);
- ✓ Wide liquid range;
- ✓ Negligible vapor pressure;
- ✓ Easily recyclable;
- ✓ High flame resistance;
- ✓ More ennvironmentally friendly than other organic solvents due to high decomposition temperature.

Therefore, because of the high ionic conductivity as well as the plasticizing nature of R-ILs, they are an attractive ionic solvent for the preparation of novel ionic liquid incorporated polymer electrolytes/ionic liquid-based gel polymer electrolytes [54,55].

10.3.2.2.3 Ionic Liquid-Based Gel Polymer Electrolytes (IL-GPEs)

Conventional polymer electrolytes are obtained by immobilizing an organic liquid electrolyte (salt solution in carbonate based organic solvents like DEC, EC, PC, and DMC) in some neutral host polymer matrix. But these polymer electrolytes suffer from problems such as gas evolution (like CO_2, CO, and HF) when the electrolyte is decomposed, due to its volatile nature and its easily being vaporized when the device gets heated. Also, the organic liquid electrolytes degrade and form stable inorganic precipitations (such as LiF, Li_2O, and Li_2CO_3) on the anode surface when the device is operated at a high working voltage or high temperature [56]. To overcome the above problems, a R-IL is incorporated into a host polymer matrix with lithium salts. R-ILs act as a plasticizer due to which the amorphicity and ionic conductivity of PEs (see Figure 10.7) gets increased. Because of the presence of large and asymmetric cations/anions of IL in the polymer-salt system, the segmental motion of the polymer chain increases, which enhances ion diffusion in the electrolyte system. This obtained electrolyte is known as an "IL-GPE." Some IL-GPEs with ionic conductivity are shown in Table 10.5.

Nowadays, IL-GPEs are one of the most attractive electrolytes for rechargeable lithium polymer battery (LPBs) applications. IL-GPEs are better than conventional carbonate-based electrolytes, brittle crystalline/polycrystalline solid electrolytes, and other types of solid electrolytes because they exhibit free-standing/flexible, good thermal/chemical stability, high ionic-conductivity with wide ESW, and better mechanical stability. Some important

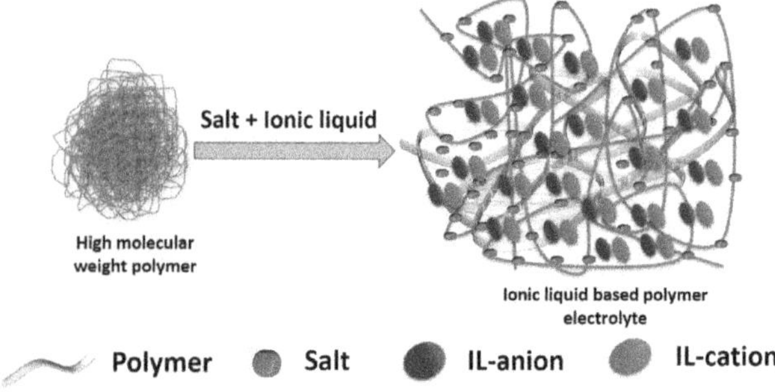

FIGURE 10.7 Schematic diagram of IL-GPE.

TABLE 10.5
Some IL-PEs and Their Ionic Conductivity

IL-PEs	Ionic Conductivity (σ) S cm^{-1}	Temperature (°C)	References
PEO + LiTFSI + PyR_{13}-TFSI	~1.0×10^{-4}	30	[57]
PEO + LiTFSI + PyR_{14}TFSI	~3.4×10^{-4}	RT	[58]
PEO + LiTFSI + BMPyTFSI	~6.9×10^{-4}	40	[59]
PEO + $LiClO_4$ + $BMIMPF_6$	~8.8×10^{-6}	30	[60]
PVdF-HFP + LiTFSI + 50 wt.% PYR_{14}TFSI	~4.0×10^{-4}	RT	[61]
PVdF-HFP + LiTFSI + xEMIMTFSI (x = 2)	~4.27×10^{-3}	30	[62]
PVdF-HFP + LiTFSI + 70 wt.% PYR_{13}FSI	~3.9×10^{-3}	30	[9]

characteristics of IL-GPEs are shown in Table 10.6. Furthermore, some features of LPBs with IL-GPEs still need to be improved for enhancing electrochemical performance:

- Interface resistance between electrode/electrolyte needs to be low;
- Cathode (active) materials deliver maximum capacity (nearly the theoretical value);
- High cyclability and coulombic efficiency (above 95%) should be achieved.

So, safety concerns related to LPBs can be addressed by IL-based GPEs but interface resistance of the cell must be minimized [63]. This can be done by some improvements:

- The ionic and lithium-ion conductivity of IL containing PEs should be increased;
- Cell impedance should be minimized;
- A stable and thin solid electrolyte interface (SEI) layer should form in-between the interface of the electrolyte and anode.

TABLE 10.6
Properties of IL-GPEs for Application in LPBs

Characteristics	Effect on Battery Performance
Ionic conductivity ~10^{-3}–10^{-4} S cm^{-1} at RT	Increased current and energy density
Wide electrochemical stability window (ESW) up to 4.5 V	Battery performs at higher voltage
Abundant nontoxic constituent	Reducing manufacturing cost and safety issues
Mechanical and thermal stability	High temperature operation
Chemically unreactive with electrode materials	Long-term cycling stability

10.4 ELECTROCHEMICAL CHARACTERIZATIONS OF IL-GPES FOR LPBS

Lithium-ion conductivity, the growth of lithium-dendrite, ESW, and galvanostatic charge–discharge measurement are some important measurements of IL-GPEs for efficient LPBs.

10.4.1 Ionic and Lithium-Ion Conductivity of IL-GPEs

In rechargeable batteries, IL-GPEs are used as an electrotonically insulating separator as well as providing an ion-conducting path between cathode and anode or vice versa. The total ionic-conductivity of the PEs depends on some important factors like the fixability of polymer chains (or the amorphicity of polymer system). When the flexibility of the polymer chain is increased, the mobility of free ions in the polymer electrolyte is enhanced due to the large intake of ILs in the host polymer matrix. The effect of IL concentration in a polymer-salt system is reported by Singh et al. [9] and Srivastava et al. [64] in Figure 10.8a and b, respectively. Singh et al. [9] showed that the total ionic-conductivity of a PVdF-HFP + 20 wt.% LiTFSI system is ~5.70×10^{-6} S cm^{-1} at room temperature but after IL doping in a polymer salt system, due to the increased free mobile ions and also the increased amorphicity of the solid polymer electrolyte, the total ionic conductivity obtained was ~3.9×10^{-3} S cm^{-1} for 70 wt.% IL containing PE. On the other hand, Srivastava et al. [64] synthesized an IL incorporated blend polymer-salt system [PVdF-HFP + PMMA] + 20 wt.% LiTFSI + X wt.% EMIMTFSI; for $0 \leq x \leq 70$ (BGPEs). They also found that the ionic-conductivity of BGPEs increased with the IL concentration and that the activation energy value decreased with IL concentration (Figure 10.8c and d). This happened due to the presence of the higher loading of IL which increased the mobility of charge carriers as well as reducing the crystallinity of the PEs. M. Safa et al. [65] reported the ionic and thermal properties of an IL-salt doped polymeric ionic liquid electrolyte. The 80 wt.% IL-salt containing a polymeric ionic liquid electrolyte [{PDAD-MATFSI + (LiTFSI + EMIMTFSI);

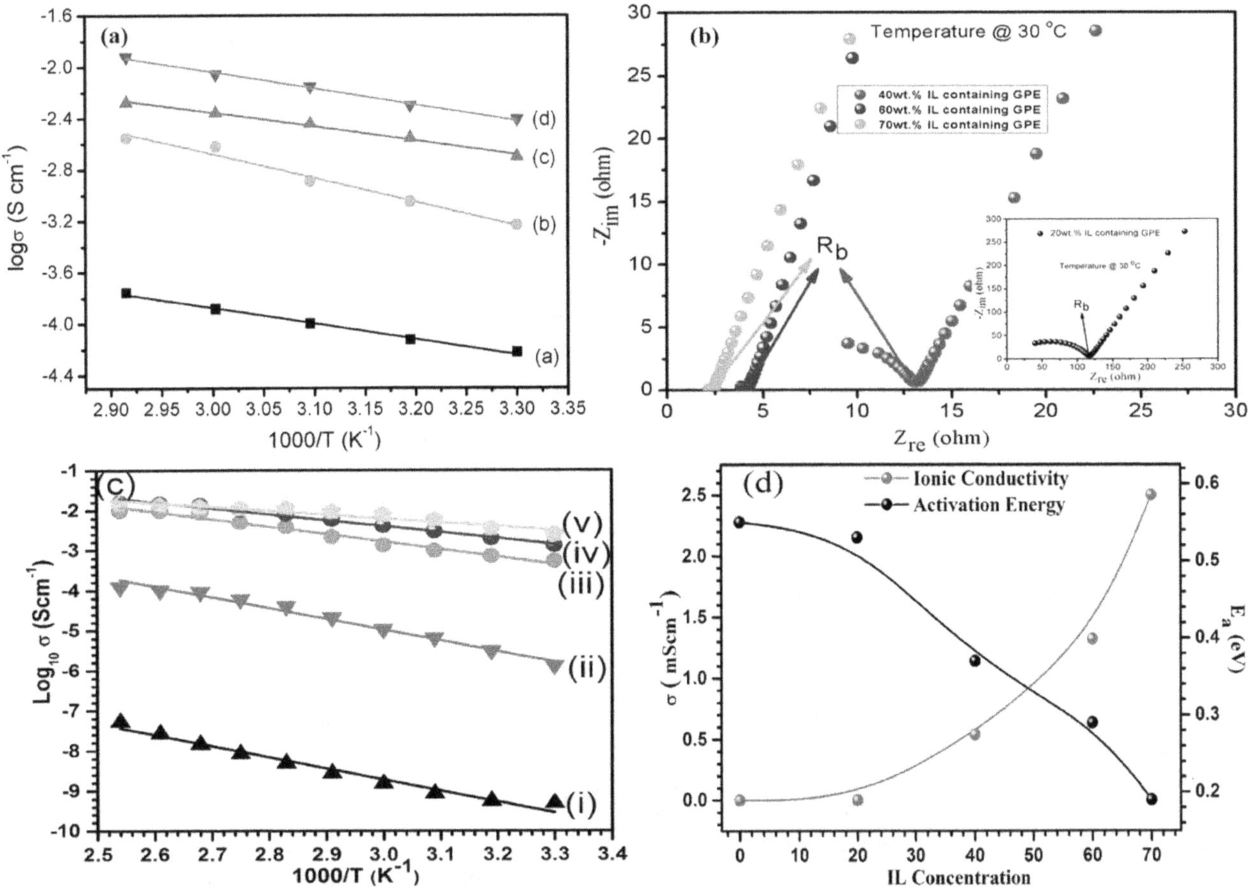

FIGURE 10.8 (a and b) Conductivity and impedance spectra of IL-GPEs at different concentrations of $PYR_{13}FSI$ ionic liquid (a = 20, b = 40, c = 60, and d = 70 wt.%) [9] (c and d) Conductivity and activation energy of the BGPEs at different weight percentages of EMIMTFSI ionic liquid (I = 0, II = 20, III = 40, IV = 60, and V = 70 wt.%) [64].

20:80}] showed the highest ionic-conductivity (~3.3 × 10^{-3} S cm^{-1} at 25 °C). The polymeric IL and IL pair contained a high loading IL-salt solution that allowed the enchaining of the ionic conductivity.

Furthermore, in an IL-GPE system, many types of ions such as cations/anions of ionic liquid and salt are present. For efficient LPB applications, our main focus is to determine the lithium transference number which shows the mobility of lithium-ions in the electrolyte system. The lithium transference number of IL-GPEs is calculated by using a combined AC/DC technique, in which a constant DC voltage is applied across the symmetric cell (Li/IL-GPE/Li) and the resulting current is recorded with respect to time. Finally, the cell resistances before/after polarization are measured by AC impedance spectroscopy (see Figure 10.9a). The lithium transference number (t_{Li^+}) is measured by the Bruce–Vincent formula: [66]

$$t_{Li^+} = \frac{I_S\left(\Delta V - I_0 R_0^{sei+ct}\right)}{I_0\left(\Delta V - I_S R_S^{sei+ct}\right)} \tag{10.15}$$

In this equation, I_S is the steady-state current and I_0 is the initial current; ΔV is the constant DC voltage; R_0^{sei+ct} and R_0^{sei+ct} respectively are the before/after polarization electrolyte resistance.

The lithium-ion conductivity is one of the most important parameters of polymer electrolytes and is dependent on the diffusion of the Li^+-ion (lithium transference number) and total ionic conductivity in an IL-based PE system. The lithium-ion conductivity (σ_{Li^+}) of IL-based PEs is measured by: [61]

$$\sigma_{Li^+} = t_{Li^+} \times \sigma_T \tag{10.16}$$

where σ_{Li^+} is the lithium-ion conductivity, and t_{Li^+} and σ_T are the lithium transference number and total ionic conductivity.

The DC polarization curves and AC impedance along with the circuit are shown in Figure 10.9a. The circuit shows that the first intercept at high frequency of the impedance curve indicates bulk resistance (R_b); the asymmetric semicircle shows the passive film resistance, that is, the combined combination of the solid electrolyte interface resistance (R_{sei}) and the charge transfer resistance (R_{ct}) with a constant phase element (CPE) [66]. The CPE has both components resistive and capacitive due to which it is more effective than a capacitor component. Li et al. [61]

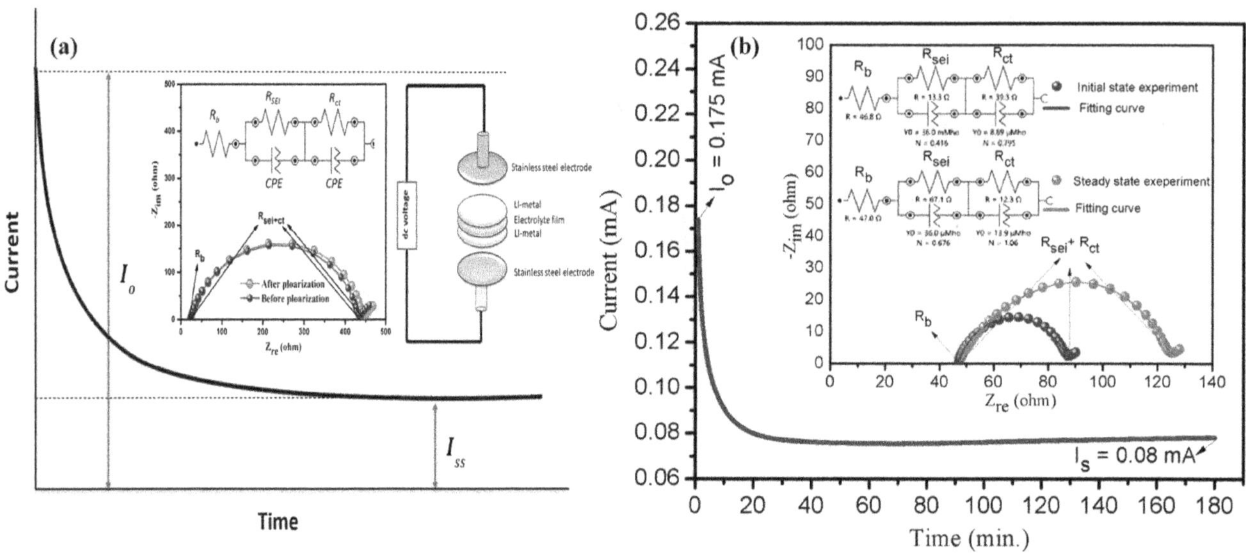

FIGURE 10.9 (a) Schematic diagram of combined AC/DC technique (b) DC polarization curve of symmetrical cell Li/70 wt.% IL containing GPE/Li [9].

investigated a PVdF-HFP + LiTFSI + *X* wt.% PYR_{14}TFSI polymer electrolyte. The t_{Li^+} of polymer electrolytes is increased by increasing the concentration of IL up to 33.3 wt.% of PYR_{14}TFSI, because at high IL concentration the ionic cluster is formed around the lithium-ion. Due to which PYR_{14}^+ cations are absorbed by this ionic cluster and form a large ionic atmosphere, thereby diffusion of lithium-ion is decreased in the electrolyte. So, the maximum t_{Li^+} was found to be 0.8 for 33.3 wt.% PYR_{14}TFSI containing PE, which means 80% of ionic conductivity is related to Li^+-ions. The Li^+-ion conductivity is ~2.2×10^{-5} for a polymer electrolyte without IL. The maximum lithium ion-conductivity ~2.8×10^{-4} was achieved when 33.3 wt.% PYR_{14}TFSI was incorporated in a PVdF-HFP-LiTFSI system. In another study, Singh et al. [9] showed that t_{Li^+} and the σ_{Li^+} of a polymer-salt system (PVdF-HFP + 20 wt.% LiTFSI) increased with a PYR_{13}FSI IL concentration. The t_{Li^+} and σ_{Li^+} of 70 wt.% IL containing GPE was reported at 0.43 (which means 43% of conductivity is related to Li^+-ions) and ~1.6×10^{-3} S cm^{-1} at 30 °C (see Figure 10.9b). Furthermore, the t_{Li^+} and σ_{Li^+} of IL-GPEs are shown in Table 10.7.

10.4.2 Solid Electrolyte Interface (SEI)

The SEI is a protective passivation layer that is formed between the anode and electrolyte interface which is related to a thin inorganic layer and a thick porous organic layer. It is mainly dependent on the composition of the electrolyte and anode material used. This SEI layer is formed in a few initial charge–discharge cycles when the electrolyte directly comes in contact with the lithium metal anode. The SEI layer consists of a combination of two distinct insoluble layers and shows an electronic insulator nature though it has an ionic conducting nature:

- A thin dense inorganic layer near the anode, which is related to inorganic species (e.g., LiF, Li_2O, Li_2CO_3);

TABLE 10.7
Ionic-Conductivity, Transference Number, Li^+-ion Conductivity, and Electrochemical Stability Window of Some IL-GPEs

IL-GPEs	Ionic Conductivity (σ) S cm^{-1} (RT)	Lithium Transference Number (t_{Li^+})	ESW (V)	Lithium-Ion Conductivity ((σ_{Li+}))	Reference
PVdF-HFP + LiTFSI + 33.3 wt.% PYR_{14}TFSI	~3.5×10^{-4}	0.80	~4.8	~2.8×10^{-4}	[61]
PEO + LiFSI + 7.5 wt.% EMIMFSI	~2.8×10^{-4}	0.28	~3.8	~8.0×10^{-5}	[67]
PVdF-HFP + LiTFSI + 70 wt.% PYR_{13}FSI	~3.9×10^{-3}	0.43	~4.3	~1.6×10^{-3}	[9]
{PDAD-MATFSI + (LiTFSI + EMIMTFSI); 20:80}	~3.3×10^{-3}	0.41	~5.0	—	[65]
PEO + LiTFSI + 30 wt.% BMPImTFSI	~2.5×10^{-5}	0.41	~5.2	~1.0×10^{-5}	[68]
PVdF-HFP + LiTFSI + 40 wt.% EMIMFSI	~3.8×10^{-4}	0.39	~4.7	~1.5×10^{-4}	[69]
[P(VdF-HFP) + PMMA] + 70 wt.% EMIMTFSI + 20 wt.% LiTFSI	~2.5×10^{-3}	0.43	~4.3	~1.1×10^{-3}	[64]

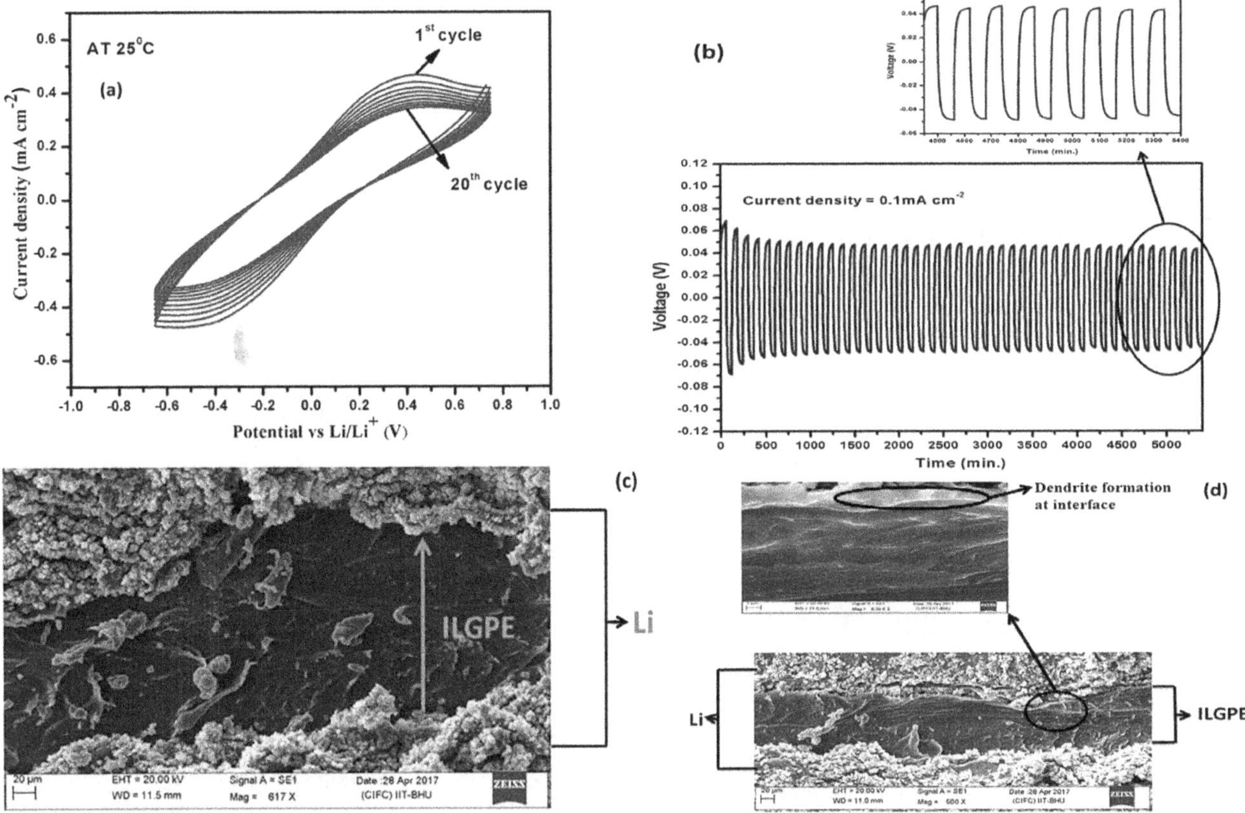

FIGURE 10.10 (a and b) CVs and voltage vs. time profile of Li/40 wt.% IL containing GPE/Li. Cross-section images of symmetrical cell (c) and fresh cell (d) after cycling [69].

- A thick porous organic layer near the electrolyte, which is related to organic species (e.g., polyolephines, $EMIM^+$, PYR_{13}^+).

So, the uniform and porous formation of the SEI layer is better for battery performance because the thin SEI layer allows the transfer of Li^+-ions, avoids the growth of Li electro-deposition, and prevents any further electrolyte reduction [61]. The thickness of the SEI layer is a critical factor for long term battery performance, because the thicker SEI layer increases the internal resistance of the battery, due to which less transportation of Li^+-ions occurs at the interface. The thickness of a suitable SEI layer has been reported to be around 10 to 100 nm [70].

However, Li^+-ion batteries suffer from low coulombic efficiency, due to the growth of uncontroled dead-lithium depositions on the surface of the Li-metal anode because lithium is electrochemically unstable with organic-based liquid electrolytes. Therefore, during charging–discharging, continuously growing Li dendrite breaks the SEI layer, and the surface area of the Li anode is increased due to which side reactions are increased between the Li anode and electrolyte [61]. These are one of the key concerns for battery failure (see Figure 10.3). However, optimized IL-based PEs can be used to prevent the growth of Li dendrite as well as provide stability to the SEI layer, thereby the electrolyte does not directly come in contact with the Li metal and thus can avoid further side reactions occurring between the Li-metal anode and electrolyte. So, the compatibility of electrolytes with Li-metal and the mechanical stability of the SEI layer is evaluated by using Li plating-stripping. In this process, the growth of dead lithium on the lithium surface is known by applying a constant areal current density (e.g., 10^{-1} mA cm^{-2}) across the Li/IL-GPE/Li symmetric cell and a corresponding voltage is recorded against time.

$$\text{In stripping: } Li_s \rightarrow Li^+ + e^{-1} \tag{10.17}$$

$$\text{In plating: } Li^+ + e^{-1} \rightarrow Li_s \tag{10.18}$$

Singh et al. [69] discussed the CVs and the lithium plating–stripping cycling profile of a symmetric cell (Li/40 wt.% IL containing GPE/Li) which is shown in Figure 10.10. From this figure, it can be clearly seen that a stable plating-stripping voltage and constant coulombic efficiency of lithium plating-stripping is observed after complete cycling, which indicates the reversibility of the redox reaction and a stable and uniform SEI layer on the lithium anode and electrolyte which are shown in SEM images.

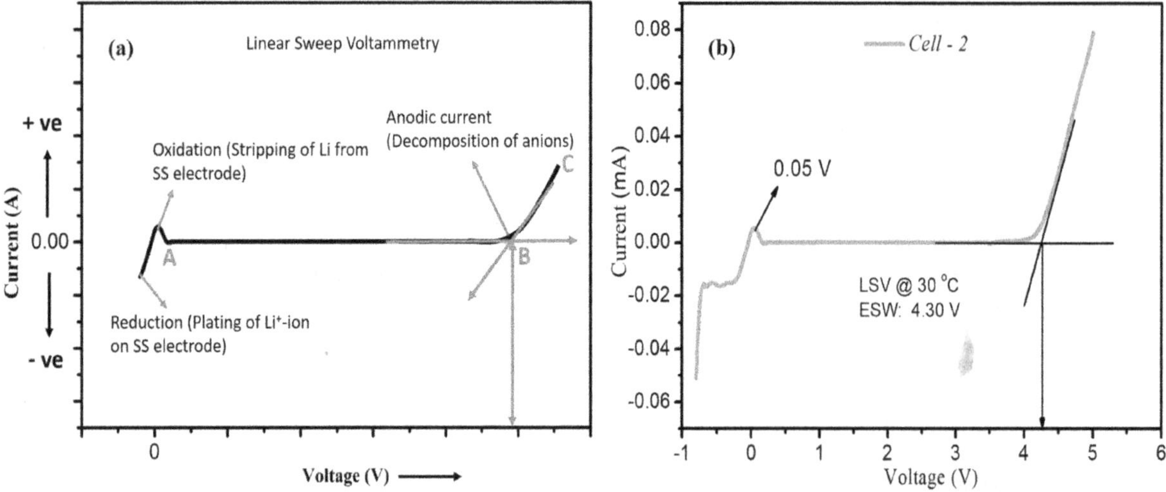

FIGURE 10.11 (a) Schematic diagram of LSV curve (b) LSV curve of cell Li/70 wt.% IL containing GPE/SS at 0.1 mV s^{-1} [9].

10.4.3 Electrochemical Stability Window (ESW)

The ESW of electrolytes is the most important study used in high voltage battery applications. The ESW is the electrochemically unreactive potential range, which means in this region the electrolyte is neither reduced nor oxidized. Before the charging–discharging operation of the cell, the ESW of the optimized composition of IL-GPE is checked because the battery operates within this voltage region. The linear sweep voltammetry (LSV) technique is used for the measurement of the ESW of electrolytes. In this technique, the sample is placed between the Li electrode (reference/counter-electrode) and stainless steel (SS; working electrode), and a constant scan rate is applied across the electrode [61].

The linear sweep voltammetry curve indicates the electrolyte decomposition at the interface of the inert SS electrode (Figure 10.11). In this curve, the negative current shows the reduction of Li^+-ions at the stainless steel electrode, which is called a plating process, and a small positive current indicates the oxidation of lithium from the stainless-steel electrode, and this process is called stripping. But after a certain voltage is started, the oxidation of anions (such as $TFSI^-$ and FSI^-) present in the electrolyte occurs, due to which the anodic current increases rapidly (point B to C). However, this anodic current is not reversible, which means that degradation of the electrolyte started above this voltage. When the anodic/cathodic current is not present in-between the AB region, this indicates that it is not involved in any type of electrochemical reaction in that region. Therefore, the ESW of an electrolyte should be high for high voltage Li-battery applications. Singh et al. [9] developed IL-GPE, (PVdF-HFP + X wt.% $PYR_{13}FSI$) + 20 wt.% LiTFSI for an Li-polymer battery and an ESW of 70 wt.% $PYR_{13}FSI$ incorporated GPE is found to be ~4.3 V vs. Li/Li^+ (Figure 10.11b). In another study, Srivastava et al. [64] developed a blend polymer system using P(VdF-HFP)/PMMA polymer (3:1 ratio), 20 wt.% LiTFSI salt and EMIMTFSI ionic liquid. A 70 wt.% IL containing BGPE showed electrochemical stability up to ~4.3 V. Also, Gupta et al. [68] used the PEO based polymer system with pyridinium based IL and prepared GPE, PEO + 20% LiTFSI + 30% IL which was stable up to ~5.2 V. Therefore, the electrochemical performance of rechargeable batteries (such as the charge–discharge) is recorded within this voltage region and, finally, it was concluded that the electrochemical stability of IL based GPE is suitable for high voltage battery applications. The ESW of some IL-GPEs are also shown in Table 10.7.

10.4.4 Charge–Discharge Performance of Lithium Batteries Using IL-GPEs

The charging–discharging process indicates the de-intercalation–intercalation of Li^+-ion from/in a positive electrode (cathode) through a redox reaction (see Figure 10.12). Mainly, three types of charging–discharging processes are used for LPBs:

1. The charging–discharging of a cell with constant voltage (CV) where the current varies with respect to time with fixed upper and lower cut-off currents.
2. The charging–discharging of a cell with constant current (CC) where the voltage varies with respect to time. The upper and lower cut-off voltages of the charging and discharging curve remain within an electrochemical stability window.
3. A hybrid process (CC–CV) of the charging–discharging of a cell, in which initially the cell charges/discharges with a constant current within the upper/lower voltage limits and the cell charges/discharges with a constant voltage up to the upper/lower limits of the current.

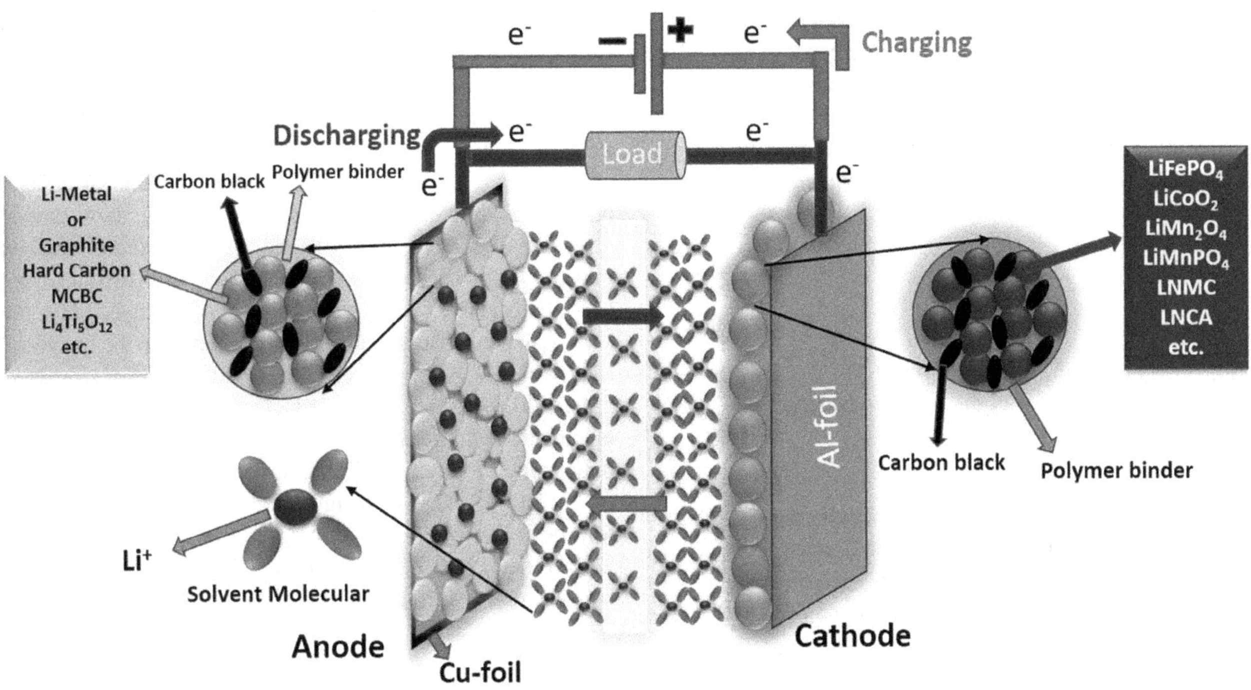

FIGURE 10.12 Schematic diagram of rechargeable lithium battery.

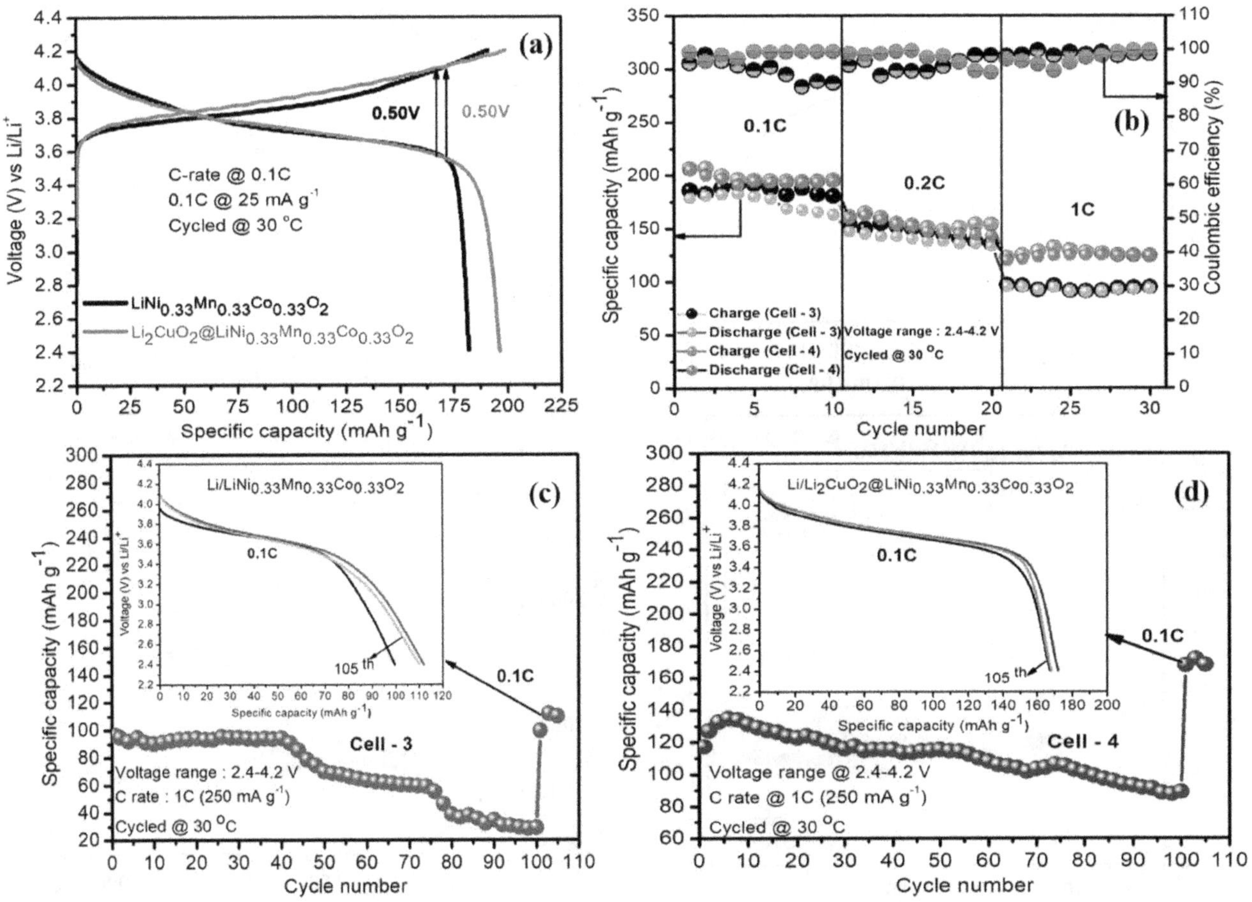

FIGURE 10.13 (a–d) Electrochemical performance of 70 wt.% pyrrolidinium-FSI based gel polymer electrolytes with $LiNi_{0.33}Mn_{0.33}Co_{0.33}O_2$ and $Li_2CuO_2@LiNi_{0.33}Mn_{0.33}Co_{0.33}O_2$ cathodes, respectively [9].

The constant current method is preferred for the charging–discharging of the battery. The specific capacity (*Q*) of the battery is calculated by the multiplication of current (*i*) and the total time (*t*) taken during the charging/discharging of the battery to the upper/lower voltage. The specific capacity of the battery is obtained by the following equations:

$$\text{Capacity}\left(Q\right) = i\left(\text{A}\right) \times t\left(\text{h}\right) \text{A h} \tag{10.19}$$

$$\text{Specific capacity} = \frac{Q\left(\text{A h}\right)}{\text{Weight of active material}\left(\text{g}\right)}\left(\frac{\text{A h}}{\text{g}}\right) \tag{10.20}$$

In the CC method, the charging/discharging current is related to the current-rate (C-rate). The C-rate is defined as a current when the battery is fully charged/discharged in a fixed time with a constant current. For example, 0.1C, 0.2C, 0.5C, and 1C are the current rates at which the battery will fully charge–discharge in 10, 5, 2, and 1 hours respectively.

Recently, IL-GPEs have been frequently used in rechargeable Li-polymer batteries, because they have some advantages such as higher energy/power densities, and safer and long cyclic-lives. Because IL-GPEs have several beneficial advantages, like high lithium-ion conductivity, they provide a stable SEI passive layer between the anode and electrolyte interface, enhanced cycling stability, and are thermally and chemically stable. The electrochemical performance of some Li-polymer batteries has already been studied in the literature. Singh et al. [9] prepared the (PVdF-HFP + 70 wt.% $PYR_{13}FSI$) + 20 wt.% LiTFSI electrolyte and reported its electrochemical performance in a rechargeable Li-polymer battery in (Li/70 wt.% $PYR_{13}FSI$ incorporated GPE/LNMC or Li_2CuO_2@LNMC) configurations (see Figure 10.13a and b). They obtained a maximum specific discharge capacity of Li/70 wt.% $PYR_{13}FSI$ incorporated GPE/Li_2CuO_2@LNMC is ~196 mA h g^{-1} at 0.1C, whereas Li/70 wt.% $PYR_{13}FSI$

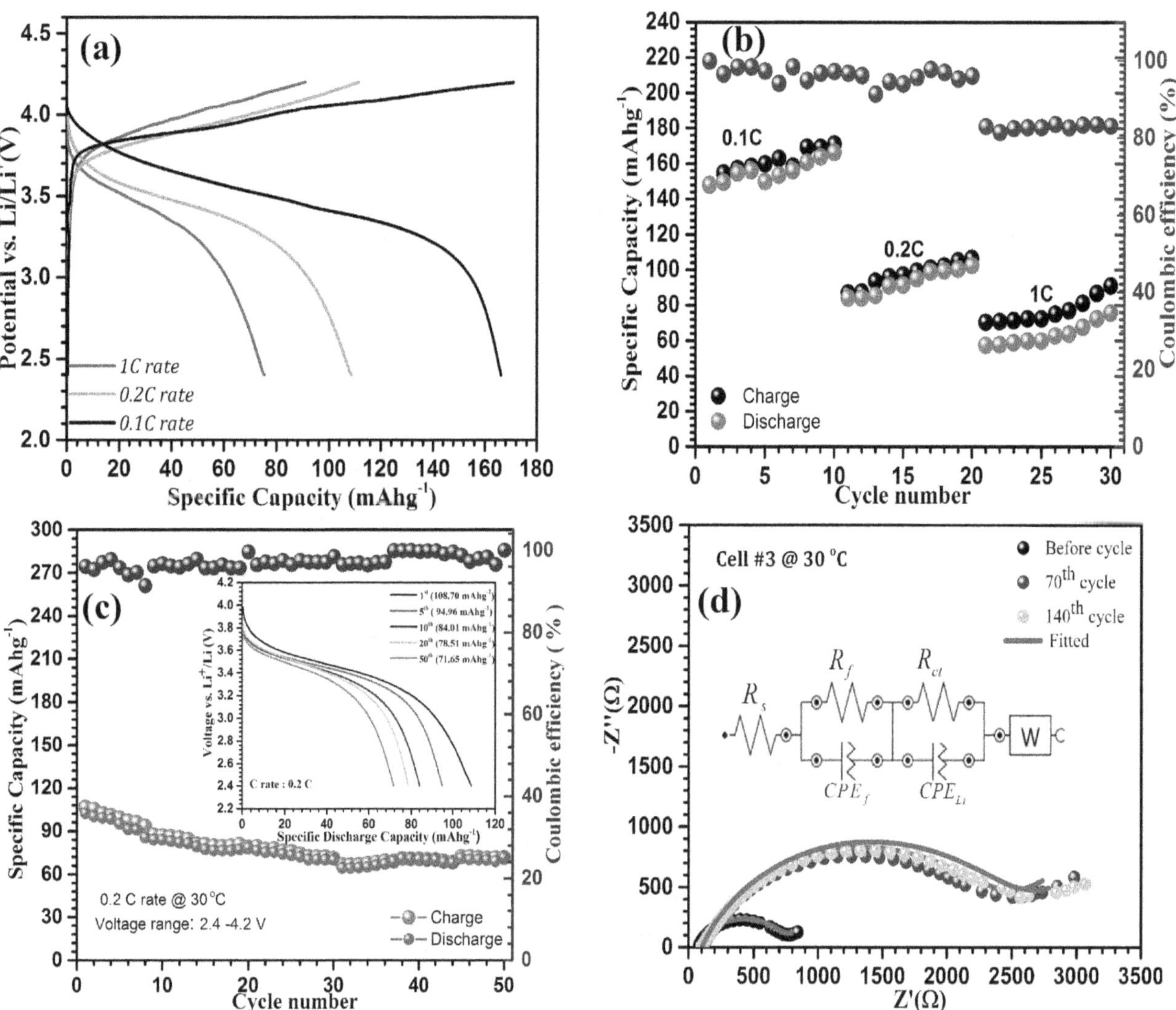

FIGURE 10.14 (a) Charge–discharge curves at different C-rates (b) rate performance (c) cycling stability (at 0.2C) (d) interface resistance after cycling of lithium cell (Li/70 wt.% EMIMTFSI containing BGPE/$Li_{1.2}Ni_{0.6}Mn_{0.1}Co_{0.1}O_2$) [64].

TABLE 10.8
The Specific Discharge Capacity of Li-Polymer Batteries Using IL-GPEs

IL-GPEs	Lithium Battery	Temp. (°C)	C-Rate	Specific Discharge Capacity (mA h g^{-1})	Reference
PEO_{20} + LiTFSI + Pyr_{13}TFSI	Li/$LiFePO_4$	RT	0.1C	~115	[71]
PEO_{20} + LiTFSI + $Pip_{1,1}O_1$TFSI	Li/$LiFePO_4$	RT	0.05C	~120	[72]
$PEO_{20}LiTFSI_2$ [Pyr_{14}TFSI]$_4$	Li/$LiFePO_4$	40	0.1C	~140	[73]
PEO_{20} + LiTFSI + [Pyr_{13}TFSI]	Li/$LiFePO_4$	40	0.05C	~125	[74]
{PDAD-MATFSI + (LiTFSI + EMIMTFSI); 20:80}	Li/$LiFePO_4$	RT	0.1C	~169.3	[65]
PVdF-HFP + LiTFSI + 33.3 wt.% PYR_{14}TFSI	Li/$LiFePO_4$	RT	0.1C	~122.6	[61]
PVdF-HFP + LiTFSI + 70 wt.% PYR_{13}FSI	Li/Li_2CuO_2@LNMC	RT	0.1C	~196	[9]
PVdF-HFP + LiTFSI + 40 wt.% EMIMFSI	Li/$LiFePO_4$	50	0.1C	~160.3	[69]
[P(VdF-HFP) + PMMA] + 70 wt.% EMIMTFSI + 20 wt.% LiTFSI	Li/$Li_{1.2}Ni_{0.6}Mn_{0.1}Co_{0.1}O_2$	30	0.1C	~166	[64]

TABLE 10.9
Challenges and Solutions to Developing Safe and Efficient IL-GPEs for LPB

Challenges	Solutions
Low gravimetric and volumetric capacity in rechargeable Li^+-ion batteries (R-LIBs).	Li-metal is used as an anode due to its high theoretical capacity (~3860 mA h g^{-1}).
R-LIBs face safety problems by using carbonate-based liquid electrolytes, because liquid electrolytes are highly reactive with lithium metal and suffer from leakage problems.	To shortcut the problem solid electrolytes can be used in place of liquid electrolytes, because solid electrolytes provide thermodynamic stability with lithium metal.
High interface resistance is formed between the electrode/electrolyte, also solid electrolytes cannot provide flexibility in the battery.	The solid polymer electrolytes provide better electrode/electrolyte contact at the interface as well as flexibility in the battery.
The solid polymer electrolytes provide mechanical/thermal stability but show low ionic/σ_{Li^+} (~10^{-7}–10^{-4} S cm^{-1} at RT).	The ionic-conductivity is improved using low molecular weight organic plasticizers in the polymer electrolyte.
Plasticized PEs suffer some drawbacks such as poor thermal/chemical stability, are hazardous for the environment, and can't operate at high temperatures, due to the flammable/toxic nature of organic plasticizers.	To overcome these problems, room temperature ionic liquids are used in place of organic plasticizers and are incorporated in the polymer-salt system. The IL-GPEs show good thermal/electrochemical stability with high ionic conductivity, and the battery can operate at high temperatures due to the non-flammable nature of R-ILs. Polymer batteries minimize safety concerns, and also provide long cycle stability.

incorporated GPE/LNMC has a discharge capacity ~182 mA h g^{-1}. After cycling, the specific discharge capacity of Li/Li_2CuO_2@LNMC is found to be ~90 mA h g^{-1} at 1C (~69% of the highest discharge capacity), while an Li/LNMC cell delivered ~29 mA h g^{-1} specific discharge capacity at 1C, that is ~30% capacity retention of the maximum discharge capacity (Figure 10.13c and d). An Li_2CuO_2 modified LNMC cathode provided long cyclic stability at a high current rate when compared to a pristine LNMC cathode because the Li_2CuO_2 coating layer minimized unwanted chemical reactions occurring between the electrolyte–cathode interface; it also enhanced the diffusion of lithium-ions. In another study, Safa et al. [65] reported that the specific discharge capacity of Li/$LiFePO_4$ cells with an IL-PE {PDAD-MATFSI + (LiTFSI + EMIMTFSI); 20:80} electrolyte was ~169.3 (99.9% of the theoretical capacity of $LiFePO_4$; ~170 mA h g^{-1}) at 0.1C; ~166.3 (97.8%) at 0.2C; ~153.8 (90.5%) at 0.5C; ~126.8 (74.6%) at 1C; and ~71.4 mA h g^{-1} (42.0%) at 2C, respectively. Also, Srivastava et al. [64] developed 70 wt.% EMIMTFSI containing BGPE and used an Li-rich NMC ($Li_{1.2}Ni_{0.6}Mn_{0.1}Co_{0.1}O_2$) cathode and obtained a maximum discharge capacity of ~166 mA h g^{-1} at 0.1C. The cell Li/70 wt.% EMIMTFSI containing BGPE/$Li_{1.2}Ni_{0.6}Mn_{0.1}Co_{0.1}O_2$) retained 67% of maximum discharge capacity up to the 50th cycle and maintained coulombic efficiency (~97%) up to the 140th cycle (Figure 10.14a–c). They also investigated the interface resistance of the cell before and after cycling and observed the increase in the impedance of the cell from 641 to 2380 Ω after 140 cycles (Figure 10.14d). Furthermore, other reports on the specific discharge capacity of rechargeable Li-polymer batteries are shown in Table 10.8. Finally, it can be concluded that the IL-GPEs are suitable for high performance battery applications due to not only high ionic-conductivity, flexibility, and thermal/chemical stability, but also due to the fact that these electrolytes provide a long cycle life, and are safer to battery applications.

10.5 SUMMARY

For high gravimetric and volumetric capacity in rechargeable Li-polymer batteries, Li-metal is the best anode material, which is the reason for their high specific capacity compared to other anodic elements present in the periodic table. Furthermore, in a rechargeable Li-polymer battery, electrolytes play an important role in enhancing the electrochemical performance of the battery. Also, the operating voltage of rechargeable Li-polymer batteries is related to the nature of the electrode–electrolyte and its chemical potential. Typically, carbonate-based liquid electrolytes have less thermal/chemical stability than PEs when the device is working at high voltage and temperatures. But PEs possess limitations of low ionic-conductivity as compared to carbonate-based liquid electrolytes. Therefore, many approaches to enhance the ionic conductivity of PEs have been made, such as the blending of PEs, composite polymer electrolytes, and the use of organic plasticizers in PEs. On the other hand, organic plasticizers are thermally unstable as well as highly reactive with a lithium metal anode. Therefore, ILs are used as an alternative plasticizer due to having beneficial properties, which overcome problems associated with organic liquid electrolytes. These IL-GPEs provide high ionic conductivity, a flexible/freestanding thin film, high thermal stability, a wide operating voltage, and safety, and thereby are suitable for enhancing the durability of LPBs (Table 10.9).

ACKNOWLEDGMENTS

R. K. Singh gratefully acknowledges the financial support from BRNS-DAE Mumbai, DST, and SERB New Delhi, India.

REFERENCES

1. Wilberforce, T., Z. El Hassan, F. N. Khatib, A. Makky, A. Baroutaji, J. G. Carton and A. G. Olabi. 2017. Developments of electric cars and fuel cell hydrogen electric cars. *Int. J. Hydrog. Energy*. 42: 25695–25734.
2. He, H., X. Zhang, R. Xiong, Y. Xu and H. Guo. 2012. Online model-based estimation of state-of-charge and open-circuit voltage of lithium-ion batteries in electric vehicles. *Energy* 39: 310–318.
3. Buchmann, I. 2017. A Handbook on Rechargeable Batteries for Non-Engineers. *Battery University BU-103: Global Battery Markets*. ISBN 978-0968211847
4. Tarascon, J. M. and M. Armand. 2001. Issues and challenges facing rechargeable lithium batteries. *Nature* 414: 359–367.
5. Salameh, Z. M., M. A. Casacca and W. A. Lynch. 1992. A mathematical model for lead-acid batteries. *IEEE T Energy Conver*. 7(1): 93–98.
6. Linden, D. 1984. *Handbook of Batteries and Fuel Cells*, McGraw-Hill Book Co, New York. 1075.
7. Abada, S., G. Marlair, A. Lecocq, M. Petit, V. Sauvant-Moynot and F. Huet. 2016. Safety focused modeling of lithium-ion batteries: A review. *J. Power Sources*. 306: 178–192.
8. Crompton, T. R. 2000. *Battery Reference Book*, 'Third Edition. Newnes, Oxford.
9. Singh, S. K., D. Dutta and R. K. Singh. 2020. Enhanced structural and cycling stability of Li_2CuO_2-coated $LiNi_{0.33}Mn_{0.33}Co_{0.33}O_2$ cathode with flexible ionic liquid-based gel polymer electrolyte for lithium polymer batteries. *Electrochim. Acta*. 343: 136122.
10. Zinigrad, E., L. Larush-Asraf, J. S. Gnanaraj, M. Sprecher and D. Aurbach. 2005. On the thermal stability of $LiPF_6$. *Thermochim. Acta*. 438: 184–191.
11. Wang, A., S. Kadam, H. Li, S. Shi and Y. Qi. 2018. Review on modeling of the anode solid electrolyte interphase (SEI) for lithium-ion batteries. *NpjComput. Mater*. 4(1): 1–26.
12. Kim, J. K., L. Niedzicki, J. Scheers, C. R. Shin, D. H. Lim, W. Wieczorek, P. Johansson, J. H. Ahn, A. Matic and P. Jacobsson. 2013. Characterization of N-butyl-N-methyl-pyrrolidinium bis(trifluoromethanesulfonyl)imide-based polymer electrolytes for high safety lithium batteries. *J. Power Sources*. 224: 93–98.
13. Hagenmuller, P. and W. V. Gool. 1978. *Solid Electrolytes*, Academic Press, New York.
14. Yamane, H., S. Kikkawa and M. Koizumi. 1987. Preparation of lithium silicon nitrides and their lithium ion conductivity. *Solid State Ionics*. 25: 183–191.
15. Thangadurai, V., S. Narayanan and D. Pinzaru. 2014. Garnet-type solid-state fast Li ion conductors for Li batteries: Critical review. *Chem. Soc. Rev*. 43: 4714–4727.
16. Chandra, S. 1981. *Superionic Solids: Principles and Applications*, North Holland, Amsterdam.
17. Takahashi, T. ed., 1989. *High conductivity solid ionic conductors: recent trends and applications*. World Scientific, Singapore.
18. Yao, Y. F. Y. and J. T. Kummer. 1967. Ion exchange properties of and rates of ionic diffusion in beat-alumina. *J. Inorg. Nucl. Chem*. 29: 2453–2466.
19. Geller, S., W. L. Roth and G. D. Mahan. 1976. *Superionic Conductors*, Springer-Verlag US, Boston. 171–182.
20. Varsamis, C. P. E., A. Vegiri and E. I. Kamitsos. 2002. Molecular dynamics investigation of lithium borate glasses: Local structure and ion dynamics. *Phys. Rev. B*. 65: 104203–104214.
21. Owens, B. B. and G. R. Argue. 1967. High-conductivity solid electrolytes-MAg_4I_5. *Science*. 157: 308–310.
22. Machida, N., S. Nishida, T. Shigematsu, H. Sakai, M. Tatsumisago and T. Minami. 2000. Mechano-chemical synthesis of a silver ion conductor in the system AgI-Ag_3PO_4. *Solid State Ionics*. 136: 381–386.
23. Stramare, S., V. Thangadurai and W. Weppner. 2003. Lithium lanthanum titanates: A review. *Chem. Mater*. 15: 3974–3990.
24. Murayama, M., R. Kanno, M. Irie, S. Ito, T. Hata, N. Sonoyama and Y. Kawamoto. 2002. Synthesis of new lithium ionic conductor thio-LISICON - Lithium silicon sulfides system. *J. Solid State Chem*. 168: 140–148.
25. Scrosati, B., Magistris, A., Mari, C.M. and Mariotto, G, 2012. *Fast Ion Transport in Solids North Holland*, Springer Science & Business Media, Amsterdam. 405.
26. Minami, K., A. Hayashi, S. Ujiie and M. Tatsumisago. 2009. Structure and properties of $Li_2S–P_2S_5–P_2S_3$ glass and glass–ceramic electrolytes. *J. Power Sources*. 189: 651–654.
27. Liu, C. and C. A. Angell. 1984. Fast Cu^+ ion conducting phosphate iodide-glasses. *Solid State Ionics*. 13: 105–109.
28. Otto, K. 1966. Electrical conductivity of SiO_2-B_2O_3 glasses containing Lithium or Sodium. *Phys. Chem. Glasses*. 7: 29–37.

29. Fenton, D. E., J. M. Parker and P. V. Wright. 1973. Complexes of alkali metal ions with poly(ethylene oxide). *Polymer*. 14: 589.
30. Armand, M. B., J. M. Chabagno, M. J. Duclot, P. Vashistha, J. N. Mundy and G. K. Shenopy. 1979. *Fast Ion Transport in Solids*. Elsevier North Holland, New York.
31. Ratner, M. A., J. R. Maccallum and C. A. Vincent. 1987. *Polymer Electrolytes Review*-1, Elsevier Applied Science. London.
32. Magistris, A., P. Mustarelli, E. Quartarone and C. Tomasi. 2000. Transport and thermal properties of $(PEO)_n$-$LiPF_6$ electrolytes for super-ambient applications. *Solid State Ionics*. 136–137: 1241–1247.
33. Reddy, C. V. S., G. P. Wu, C. X. Zhao, W. Jin, Q. Y. Zhu, W. Chen and S. Mho. 2007. Mesoporous silica (MCM-41) effect on (PEO + $LiAsF_6$) solid polymer electrolyte. *Curr. Appl. Phys*. 7: 655–661.
34. Rajendran, S., M. Sivakumar, R. Subadevi and M. Nirmala. 2004. Characterization of PVA–PVdF based solid polymer blend electrolytes. *Physica B: Condens. Matter*. 348: 73–78.
35. Noor, S. A. M., P. M. Bayley, M. Forsyth and D. R. Macfarlane. 2013. Ionogels based on ionic liquids as potential highly conductive solid state electrolytes. *Electrochim. Acta*. 91: 219–226.
36. Dissanayake, M. A. K. L., L. R. A. K. Bandara, L. H. Karaliyadda, P. A. R. D. Jayathilaka and R. S. P. Bokalawala. 2006. Thermal and electrical properties of solid polymer electrolyte PEO_9 $Mg(clo_4)_2$ incorporating nano-porous Al_2O_3 filler. *Solid State Ionics*. 177: 343–346.
37. Wieczorek, W., P. Lipka, G. Zukowska and H. Wycislik. 1998. Ionic interactions in polymeric electrolytes based on low molecular weight poly(ethylene glycols). *J. Phys. Chem. B*. 102: 6968–6974.
38. Hashmi, S. A. and S. Chandra. 1995. Experimental investigations on a sodium-ion-conducting polymer electrolyte based on poly(ethylene oxide) complexed with $NaPF_6$. *Mater. Sci. Engg. B*. 34: 18–26.
39. Gray, F. M. 1997. *Polymer Electrolytes*, Royal Society of Chemistry. Cambridge, Cambridge, UK
40. Liang, Y. H., C. C. Wang and C. Y. Chen. 2008. Comb-like copolymer-based gel polymer electrolytes for lithium ion conductors. *J. Power Sources*. 176: 340–346.
41. Utracki, L. A. (Ed.). 2002. *Polymer Blend Handbook*, vol. I, Kluwer Academic Publishers, Dordrecht, The Netherlands.
42. Rajendran, S. and P. Sivakumar, 2008. An investigation of PVdF/PVC-based blend electrolytes with EC/PC as plasticizers in lithium battery applications. *J. Phys. Condens. Matter*. 403: 509–516.
43. Flora, X. H., M. Ulaganathan and S. Rajendran. 2012. Influence of Lithium Salt Concentration on PAN-PMMA Blend Polymer Electrolytes. *Int. J. Electrochem. Sci*. 7: 7451–7462.
44. Sengwa, R. J., P. Dhatarwal and S. Choudhary. 2015. Effects of plasticizer and nanofiller on the dielectric dispersion and relaxation behaviour of polymer blend based solid polymer electrolytes. *Curr. Appl. Phys*. 15: 135–143.
45. Croce, F., G. B. Appetecchi, L. Persi and B. Scrosati. 1998. Nanocomposite polymer electrolytes for lithium batteries. *Nature*. 394: 456–458.
46. Kumar, B. and L. G. Scanlon. 1994. Polymer-ceramic composite electrolytes. *J. Power Sources*. 52: 261–268.
47. Kumar, Y., S. A. Hashmi and G. P. Pandey. 2011a. Ionic liquid mediated magnesium ion conduction in poly(ethylene oxide) based polymer electrolyte. *Electrochim. Acta*. 56: 3864–3873.
48. Pradeepa, P., S. Edwinraj and M. R. Prabhu. 2015. Effects of ceramic filler in poly(vinyl chloride)/poly(ethyl methacrylate) based polymer blend electrolytes. *Chinese Chemical Letters*. 26: 1191–1196.
49. Kulshrestha, N., B. Chatterjee and P. N. Gupta. 2014. Structural, thermal, electrical, and dielectric properties of synthesized nanocomposite solid polymer electrolytes. *High Perform. Polym*. 26: 677–688.
50. Appetecchi, G. B., F. Croce and B. Scrosati. 1995. Kinetics and stability of the lithium electrode in poly(methylmethacrylate)-based gel electrolytes. *Electrochim. Acta*. 40: 991–997.
51. Qian, X., N. Gu, Z. Cheng, X. Yang, E. Wang and S. Dong. 2002. Plasticizer effect on the ionic conductivity of PEO-based polymer electrolyte. *Mater. Chem. Phys*. 74: 98–103.
52. Kalhoff, J., G. G. Eshetu, D. Bresser and S. Passerini. 2015. Safer electrolytes for lithium-ion batteries: State of the art and perspectives. *ChemSusChem*. 8(13): 2154–2175.
53. MacFarlane, D. R., M. Forsyth, P. C. Howlett, M. Kar, S. Passerini, J. M. Pringle, H. Ohno, M. Watanabe, F. Yan, W. Zheng and S. Zhang. 2016. Ionic liquids and their solid-state analogues as materials for energy generation and storage. *Nat. Rev. Mater*. 1(2): 15005.
54. Lahiri, A. and F. Endres. 2017. Electrodeposition of nanostructured materials from aqueous, organic and ionic liquid electrolytes for Li-ion and Na-ion batteries: A comparative Review. *J. Electrochemical Society*. 164: D597–D612.
55. Macfarlane, D. R., N. Tachikawa, M. Forsyth, J. M. Pringle, P. C. Howlett, G. D. Elliott, J. H. Davis, M. Watanabe, P. Simon and C. A. Angell. 2014. Energy applications of ionic liquids To cite this version. *Energy Environ. Sci*. 7: 232–250.
56. Chawla, N., N. Bharti and S. Singh. 2019. Recent advances in non-flammable electrolytes for safer lithium-ion batteries. *Batteries*. 5: 1–26.
57. Shin, J.-H., W. A. Henderson and S. Passerini. 2003. Ionic liquids to the rescue overcoming the ionic conductivity limitations of polymer electrolytes. *Electrochem. Commun*. 5(12): 1016–1020.
58. Wetjen, M., G.-T. Kim, M. Joost, G. B. Appetecchi, M. Winter and S. Passerini. 2014. Thermal and electrochemical properties of PEO-LiTFSI-Pyr14TFSI-based composite cathodes, incorporating 4 V-class cathode active materials. *J. Power Sources*. 246, 846–857.
59. Cheng, H., C. Zhu, B. Huang, M. Lu and Y. Yang. 2007. Synthesis and electrochemical characterization of PEO-based polymer electrolytes with room temperature ionic liquids. *Electrochim. Acta*. 52(19): 5789–5794.
60. Chaurasia, S. K., R. K. Singh and S. Chandra. 2011a. Dielectric relaxation and conductivity studies on (PEO:$LiClO_4$) polymer electrolyte with added ionic liquid [BMIM][PF_6]: Evidence of ion-ion interaction. *J Polym. Sci. Pol. Phys*. 49(4): 291–300.
61. Li, L., J. Wang, P. Yang, S. Guo, H. Wang, X. Yang, X. Ma, S. Yang and B. Wu. 2013. Preparation and characterization of gel polymer electrolytes containing N-butyl-N-methylpyrrolidinium bis (trifluoromethanesulfonyl) imide ionic liquid for lithium ion batteries. *Electrochim. Acta*. 88: 147–156.
62. Yang, P., W. Cui, L. Li, L. Liu and M. An. 2012. Characterization and properties of ternary P (VdF-HFP)-LiTFSI-EMITFSI ionic liquid polymer electrolytes. *Solid State Sci*. 14(5): 598–606.

63. Li, Q. and H. Ardebili. 2016. Flexible thin-film battery based on solid-like ionic liquid-polymer electrolyte. *J. Power Sources*. 303: 17–21.
64. Srivastava, N., S. K. Singh, H. Gupta, D. Meghnani, R. Mishra, R. K. Tiwari, A. Patel, A. Tiwari and R. K. Singh. 2020. Electrochemical performance of Li-rich NMC cathode material using ionic liquid based blend polymer electrolyte for rechargeable Li-ion batteries. *J. Alloys Compd.* 843: 155615.
65. Safa, M., A. Chamaani, N. Chawla and B. El-Zahab. 2016. Polymeric ionic liquid gel electrolyte for room temperature lithium battery applications. *Electrochim. Acta.* 213: 587–593.
66. Evans, J., C. A. Vincent and P. G. Bruce. 1987. Electrochemical measurement of transference numbers in polymer electrolytes. *Polymer*. *28*: 2324–2328.
67. Balo, L., H. Gupta, S. K. Singh, V. K. Singh, S. Kataria and R. K. Singh. 2018. Performance of EMIMFSI ionic liquid based gel polymer electrolyte in rechargeable lithium metal batteries. *J Ind Eng Chem*. 65: 137–145.
68. Gupta, H., L. Balo, V. K. Singh, S. K. Singh, A. K. Tripathi, Y. L. Verma and R. K. Singh. 2017. Effect of temperature on electrochemical performance of ionic liquid based polymer electrolyte with Li/$LiFePO_4$ electrodes. *Solid State Ionics*. 309: 192–199.
69. Singh, S. K., L. Balo, H. Gupta, V. K. Singh, A. K. Tripathi, Y. L. Verma and R. K. Singh. 2018. Improved electrochemical performance of EMIMFSI ionic liquid based gel polymer electrolyte with temperature for rechargeable lithium battery. *Energy*. 150: 890–900.
70. Osada, I., H. de Vries, B. Scrosati and S. Passerini. 2016a. Ionic-liquid-based polymer electrolytes for battery applications. *Angew*. 55(2): 500–513.
71. An, Y., X. Cheng, P. Zuo, L. Liao and G. Yin. 2011. The effects of functional ionic liquid on properties of solid polymer electrolyte. *Mater. Chem. Phys*. 128(1-2): 250–255.
72. Philippe, B., R. Dedryvere, M. Gorgoi, H. Rensmo, D. Gonbeau and K. Edstrom. 2013. Improved performances of nanosilicon electrodes using the salt LiFSI: a photoelectron spectroscopy study. *J. Am. Chem. Soc*. 135(26): 9829–9842.
73. Appetecchi, G. B., G. T. Kim, M. Montanino, F. Alessandrini and S. Passerini, 2011. Room temperature lithium polymer batteries based on ionic liquids. *J. Power Sources*. 196(16): 6703–6709.
74. Qian, J., W. A. Henderson, W. Xu, P. Bhattacharya, M. Engelhard, O. Borodin and J.-G. Zhang. 2015. High rate and stable cycling of lithium metal anode. *Nat. Commun*. 6: 6362.

11 Polymer Semiconductors

Moises Bustamante-Torres
Yachay Tech University, Urcuqui City, Ecuador
National Autonomous University of Mexico, Mexico City, Mexico

Jocelyne Estrella-Nuñez
Yachay Tech University, Urcuqui City, Ecuador

Odalys Torres and Sofía Abad-Sojos
ELTE Eötvös Loránd University, Budapest, Hungary

Bryan Chiguano-Tapia
Yachay Tech University, Urcuqui City, Ecuador

Emilio Bucio
National Autonomous University of Mexico, Mexico City, Mexico

CONTENTS

DOI: 10.1201/9781003169727-11

11.1 INTRODUCTION

Semiconductor polymers are the most promising functional polymers (Schenning and Meijer 2001), with a huge range of potential applications. Polymeric semiconductors promise a new paradigm in electronic device fabrication. The ability to utilize roll-to-roll processing or other existing printing technologies to produce large-area, flexible, and stretchable devices has the potential to reduce manufacturing costs drastically (McBride et al. 2018).

During the past few years, polymer semiconductors have had wide demand in many fields such as medicine, the military, and chiefly electronics. They present the attractive characteristics of low production costs, high extinction coefficients, wide absorption bands, and high-quality flexibility and stability films in different environments. In 1987, the first electronic device based on semiconductors was the thin-film diode (Heeger 2010).

Semiconductor polymers have been widely studied since that time. There are several ways to synthesize semiconductor polymers. This depends on the goals: to increase the power conversion efficiency (PCE), the hole and electron mobilities, the cost of synthesis, and so on. Among the most studied building block units can be found diketopyrrolopyrroles (DPPs), isoindigo (IID), imides, and benzodithiophene (BDT). All these units have been used for the development of organic field-effect transistors (OFETs), organic thin-film transistors (OTFTs), organic photovoltaics (OPVs), amongst others.

11.2 SEMICONDUCTING POLYMERS

The attractiveness of semiconducting polymers lies in the possibility of processing them from solution, applying conventional printing techniques for a controlled film formation (Hoppe and Sariciftci 2006). One can find materials showing very different electrical properties among polymers: insulators, dielectrics, semiconductors, photoconductors, metal-like conductors, and even superconductors (Glowacki et al. 2012). Polyaniline was the first polymer used as a photothermal agent; meanwhile, in terms of its optical properties, polydiacetylene has the same optical absorption property as melanin (Deng, Liu and Li 2019). Figure 11.1 illustrates some molecular structures of semiconducting polymers.

11.3 SYNTHESIS OF SEMICONDUCTORS POLYMERS

During the past few years, polymer semiconductors have presented attractive characteristics. Improved film-forming properties can enhance the properties and applications of polymers. Organic semiconductor materials based on heterocyclic monomers like aniline, pyrrole, and thiophene have been widely used in technological applications in organic light-emitting diodes (OLEDs), organic field-effect transistors (OFETs), organic thin-film transistors (OTFTs), organic photovoltaics (OPVs), rechargeable batteries, and so on (Bouabdallah et al. 2019; Sun, Guo and Facchetti 2020; Qu, Qi and Huang 2021).

11.3.1 Building Block Selection

Polymer semiconductors as essential components for OTFTs and OPVs have been widely studied since 1983 and 1993, respectively (Ebisawa, Kurokawa and Nara 1983; Sariciftci et al. 1993). Under an electrical bias, for a polymer semiconductor, the charge carriers depend on the highest occupied molecular orbital (HOMO) and the lowest unoccupied molecular orbital (LUMO) levels. The nature of conjugated polymers allows their being divided into n-type polymers or p-type polymers, conforming to the electron-donor or electron-acceptor role in the conjugated structure. An n-type semiconductor with low LUMO levels favors the transport of negative electrons. Nevertheless, a p-type semiconductor with significant HOMO levels transports

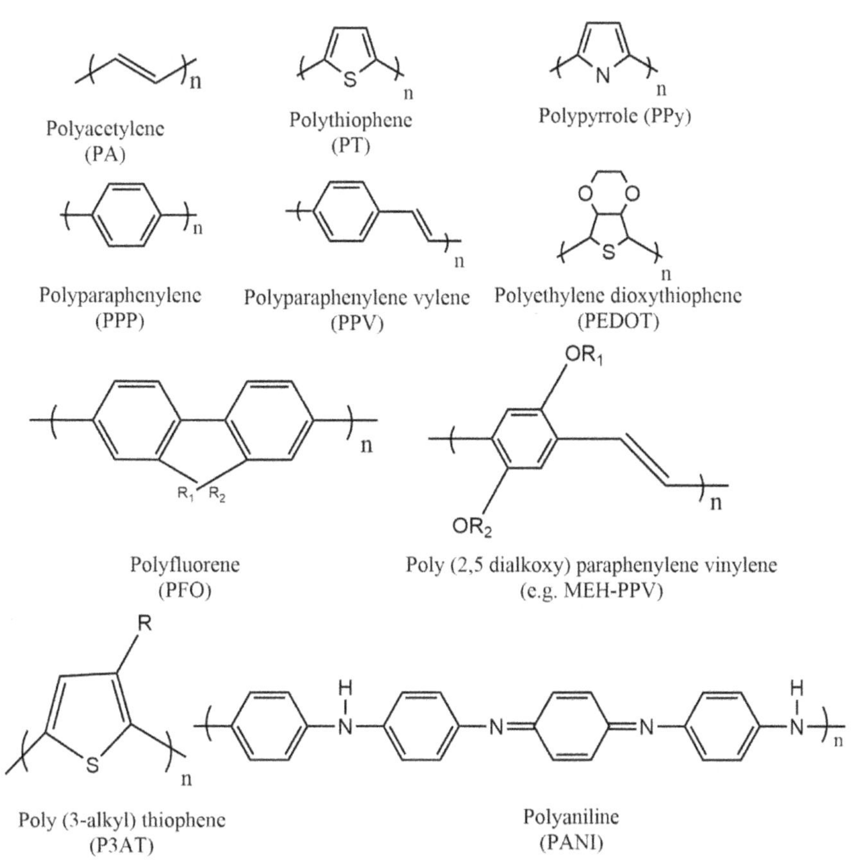

FIGURE 11.1 Molecular structures of some important semiconducting polymers.

positive electron holes. The polymer's π-conjugated building blocks govern the HOMO and LUMO levels attributed to polymer semiconductors. Energy levels would decrease when there is an electron deficiency or electron-accepting moiety; it increases when electron-rich or electron-giving moiety increases. Be that as it may, regardless of whether a moiety is an electron donor or acceptor depends on its adjoining moiety (Yuen et al. 2011; He, Hong and Li 2014).

11.3.1.1 Acceptor Building Blocks

A good choice of repetitive units can be obtained as electron-deficient or electron-rich conjugated polymers, allowing the balance of holes and electrons. Furthermore, the absorption and emission wavelengths can be controlled by regulating the bandgap. Nowadays, polymers with an alternating organization of donor-acceptor (D-A) moieties have become of increased interest (Müllen and Pisula 2015). D-A polymers have an interspersed arrangement of electron-deficient and electron-rich moieties. Employing the hybridization of the molecular orbitals of D-A polymers, small bandgaps can be obtained. This delivers a useful mechanism to enhance the power conversion efficiency (PCE) of OPVs (Chen and Cao 2009; Cheng, Yang and Hsu 2009; Boudreault, Najari and Leclerc 2011). Until now, diketopyrrolopyrroles (DPPs), isoindigo (IID), imides, and so on have been used as acceptor molecules in conjugated polymers. Suitable donors paired with different acceptor energy can make polymers with the appropriate behavior. DPPs have a planar conjugated bicyclic lactam unit, which induces the HOMO and LUMO level because it has a more significant electron-withdrawing capability. Figure 11.2 displays a typical route for DPP synthesis. To improve the reaction rate of DPP polymers, it is recommended that the succinate concentration and temperature are enhanced.

Furthermore, the addition of a succinic ester at a high temperature is necessary. The DPP body has distinct contiguous aromatic structures because of the different aromatic nitrile materials used for the synthesis. In this way, the energy levels, the steric hindrance, and the molecule's planarity between the DPP and neighboring aromatic units can be adjusted (Shi et al. 2017).

DPP derivative acceptors and their use in the fabrication of OTFTs and OPVs has been extensively researched (Y. Li et al. 2013). During recent years, DPP-based polymers have shown increasing mobility values from ~0.1 to 12 $cm^2\ V^{-1}\ s^{-1}$ (I. Kang et al. 2013). Figure 11.3 shows some of the diketopyrrolo[3,4-c]pyrrole derivatives that have been synthesized to make OTFTs. The D-A polymer A_1 (Lu et al. 2013) used in OTFTs showed a hole mobility of 0.034 $cm^2\ V^{-1}\ s^{-1}$, and for OPVs, it reached a 5.1% PCE. By introducing a para quinoid ring into A_1 an A_2 polymer was obtained. Researchers report several A_2-based polymers that increase the electron mobilities by varying the building blocks units. The polymer PA_21 (Cui, Yuen and Wudl 2011)

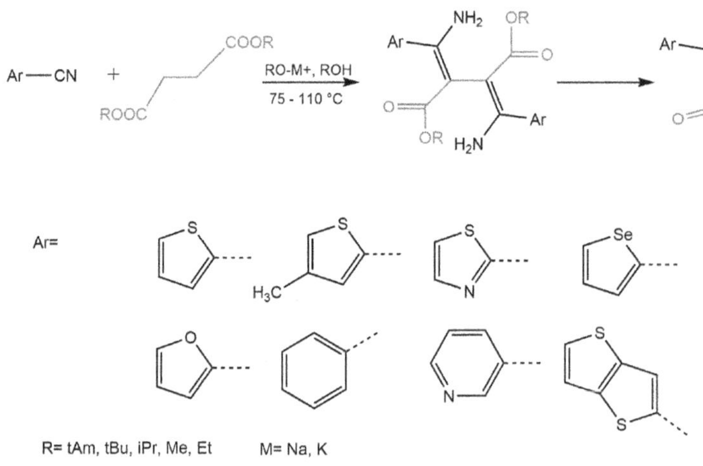

FIGURE 11.2 General way for symmetric DPP synthesis.

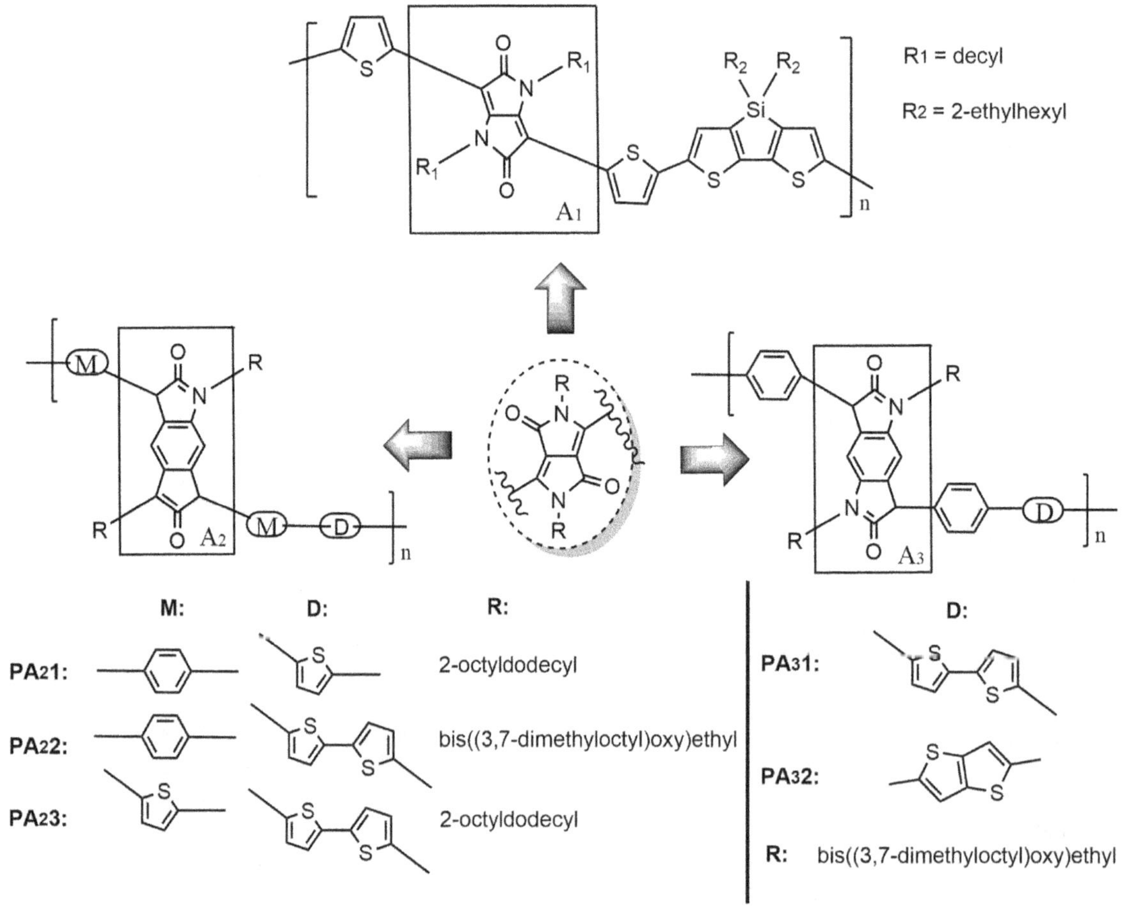

FIGURE 11.3 Polymers synthesized from DPP-derived building blocks.

reported for OTFT applications showed hole and electron mobilities of the order of 10^{-3} cm^2 V^{-1} s^{-1} and ambipolar charge transport behavior. From an acetal substituent, the addition of bithiophene to the A_2 polymer obtained a D-A polymer (PA_22) with higher solubility (Hong et al. 2012). PA_22 showed a hole mobility of 0.03 cm^2 V^{-1} s^{-1} and unipolar p-type behavior. PA_21 and PA_22 polymers are bounded by two sterically hindered benzene rings that produce a torsion in the backbone of the polymer, which causes a reduction in the effective π-conjugation length. Two thiophene units were utilized to synthesize the PA_23 (Rumer et al. 2013) polymer to get over this problem. This new polymer reached ambipolar charge transport behavior with a large hole and electron mobilities of 0.2 and 0.1 cm^2 V^{-1} s^{-1}, respectively. The electron-deficient pyrazine units were used to synthesize the PA_31 and PA_32 (Hong et al. 2013)

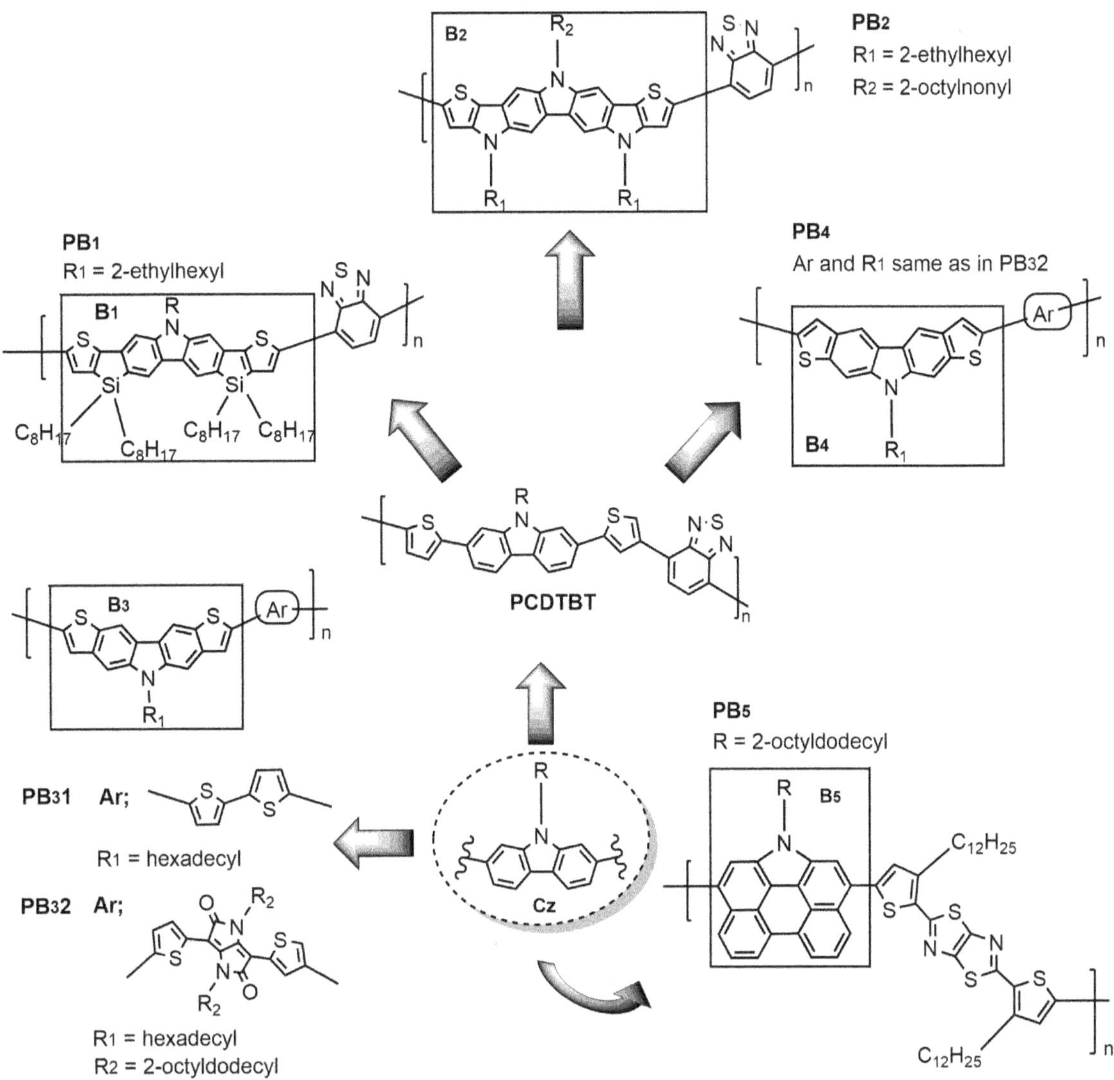

FIGURE 11.4 Polymer synthesized from Cz-derived building blocks.

polymers from an A_1 building block. They were used to develop OTFTs exhibiting ambipolar behavior, and good hole and electron mobilities (He, Hong and Li 2014).

11.3.1.2 Donor Building Blocks

Figure 11.4 shows the 2,7-carbazole (Cz) derivatives. The PCDTBT (Chu et al. 2011) is a well-known Cz-polymer for developing OPVs that reach a PCE of over 7%. The capabilities of PCDTBT building blocks were studied by bridging silicon and nitrogen to carbazole-thiophene structures; two polymers (PB_1 and PB_2) were obtained (J.-S. Wu et al. 2012). This proves that the PB_1-polymer shows a higher PCE of 5.2% than the PB_2-polymer, which presented a 2.6% PCE. A novel p-type B_3 building block with an enlarged aromatic system was created from thiophene rings in the carbazole (Y. Chen et al. 2012). Polymer $PB_3$1 presented 0.39 cm^2 V^{-1} s^{-1} hole mobility in OTFTs. Moreover, the $PB_3$2 polymer showed excellent mobility at 1.36 cm^2 V^{-1} s^{-1} (Deng et al. 2012). However, this polymer's observed high mobility was due to the insertion of a DPP unit, which is known as a good n-type building block (Y. Li et al. 2013). An isomer of B_3-building blocks was studied (B_4). However, the obtained PB_4-polymer presented lower mobility (below 10^{-3} cm^2 V^{-1} s^{-1}). This lower mobility could be due to the less effective conjugation across the B_4-building-block unit. A B_5-building-block unit is a structure of carbazole combined with a biphenyl. The obtained PB_5-polymer showed a mild appearance in the hole mobility of 0.13 cm^2 V^{-1} s^{-1} for OTFTs and a PCE of 3.2% for OPVs (He, Hong and Li 2014).

11.3.2 Backbone Halogenation

Currently, solution-processed bulk heterojunction (BHJ) solar cells have shown a notable acceleration in efficiency, achieving PCE amounts above 10%. This development is thanks to the enhancement of device architectures and the optimization of semiconducting polymers. As an effect, several polymers' optoelectronic properties were improved, allowing them to reach high efficiencies. The synthesis of D-A polymers by a convenient choice of electron-acceptor or electron-donor moieties allows adjusting the frontier molecular orbital (FMO) energy levels. Furthermore, the nature and location of the solubilizing alkyl side-chains, the backbone planarity, and the molecular weight influence the polymer optoelectronic properties (Leclerc et al. 2016).

11.3.2.1 Fluorination

Moieties such as halogens, cyano groups, and diimide are the usual electron-deficient groups utilized for acceptor semiconductors in OFETs. The unique properties of fluorine atoms call researchers' attention to making a fluorine substitution on conjugated polymers (Shi et al. 2017). The development of more efficient semiconducting polymers has recently involved the fluorination of the conjugated polymer backbone by using fluorine atoms. Such as the case of the PffBT4T-2OD (Liu et al. 2014) polymer, which holds a record PCE for single-junction solar cells. The PDTP-DFTBT (You et al. 2013) copolymer is one more compelling backbone fluorination case, which has a 1.38 eV band and reaches a PCE of at least 8% in single-junction devices. Another use for fluorination is the development of highly performing molecular semiconductors. The *p*-DTS(FBTTH$_2$)$_2$ (Van Der Poll et al. 2012) derivative is currently one of the best-fluorinated molecules for BHJ devices. Figure 11.5 shows the above-mentioned fluorinated polymers.

11.3.2.2 Synthesis of Fluorinated Conjugated Polymers

In contrast with the non-fluorinated counter-derivatives, the synthesis of many fluorinated conjugated polymers needs, at any rate, one extra chemical step, taking into account that fluorine atoms have to be introduced into the building-block unit before the functionalization step (Son et al. 2011). Figure 11.6 shows the fluorination building-block synthesis of two well-known units in the conjugated polymers area. It can be appreciated that the synthesis requires a minimum of three and one additional steps for the fluorinated derivatives of the thiophene electron-donating group (Sakamoto, Komatsu and Suzuki 2010; Fei et al. 2015) and the 2,1,3-benzothiadiazole electron-accepting group (Zhang et al. 2012; Chen et al. 2014), respectively.

Interesting enhancements have reached these fluorinated building-block units and their commercial availability; they are more expensive than their non-fluorinated analogs due to the specific synthesis pathway. Another chemical problem occurs with the regioregularity when it is used on asymmetric mono-fluorinated compounds. Traditional polycondensation methods do not provide regioselectivity control when using asymmetric monomers. The monomer direction's lack of management causes a random regioselectivity, which provokes a lower structural assembly in solid-state and inferior charge transport properties. To solve random regioselectivity a different stepwise chemical approach was proposed which involves synthesizing symmetric monomers from asymmetrical fragments. This strategy enables polymer synthesis utilizing mono-fluorine aryl groups with well-controlled regioregularity and well-defined monomer alteration (Leclerc et al. 2016).

11.3.2.3 Chlorination

Just as with fluorination, chlorination joins to the backbone of the polymer semiconductor, a chlorine atom. Because a

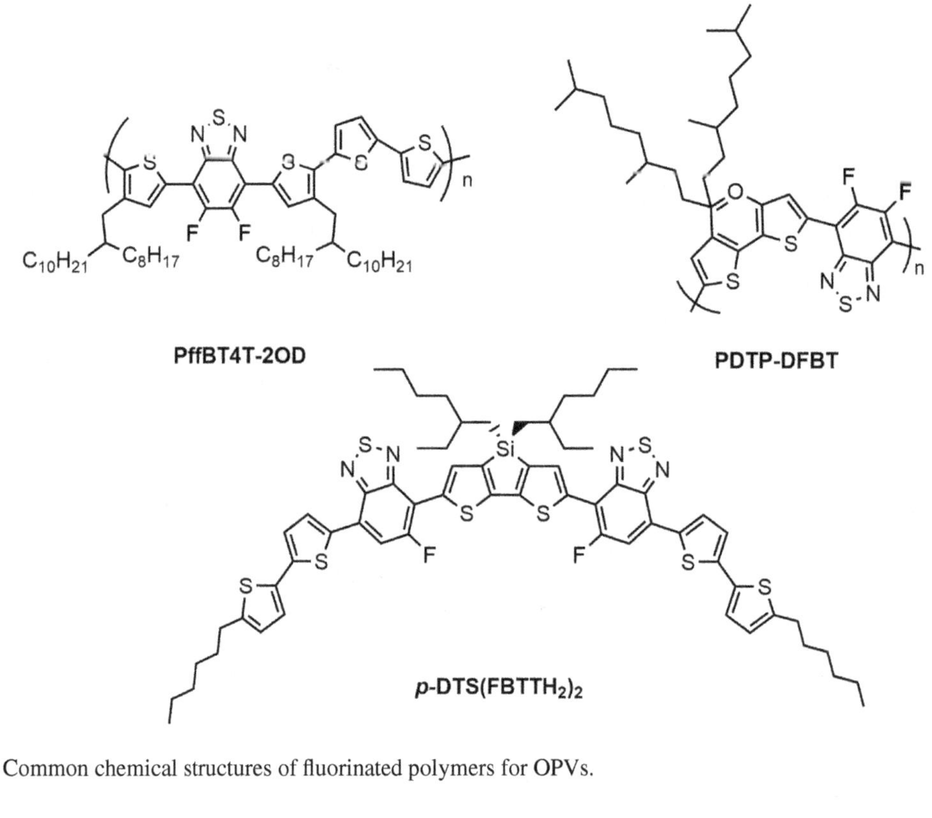

PffBT4T-2OD

PDTP-DFBT

p-DTS(FBTTH$_2$)$_2$

FIGURE 11.5 Common chemical structures of fluorinated polymers for OPVs.

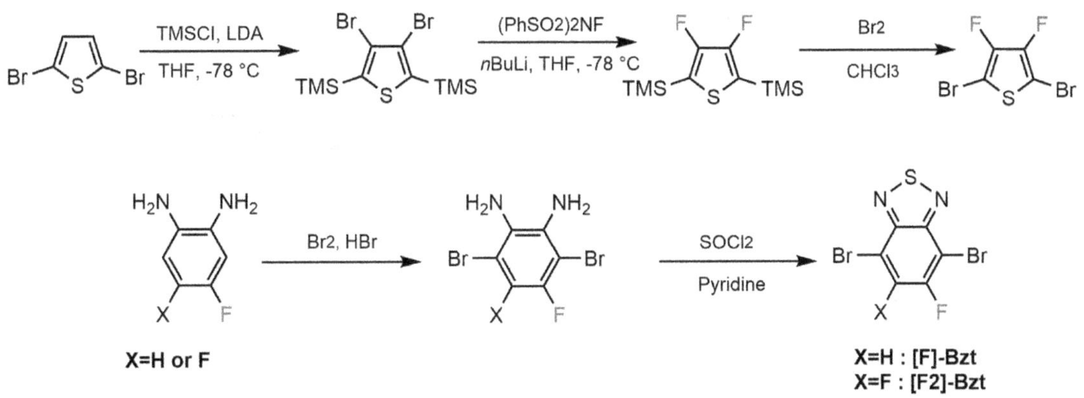

FIGURE 11.6 Synthesis of fluorinated units.

chlorine atom is more significant than a fluorine atom, it may bring steric hindrance effects (Shi et al. 2017). Experiments that compare both the fluorination and chlorination effects in IDD-based polymers have demonstrated that polymers that hold a fluorinated IID core showed a more significant OFET efficiency but a lower OPV yield than polymers that have a chlorinated core (Zheng et al. 2015).

A new polymer donor PBDB-T-SF presented a 13.1% PCE when mixed with the IT-4F acceptor unit (Zhao et al. 2017). Recently, impressive photovoltaic performances have been evaluated as elucidating the effect of chlorination on the thienyl side-chains of benzodithiophene (BDT). Figure 11.7 shows the PBDB-T-2Cl chlorinated polymer synthesized by substituting the fluorinated 2F-BDT in PBDB-T-SF with the chlorinated 2Cl-BDT. The changes in the properties of the polymer provoked by the addition of chlorine atoms were studied. This change was used as a strategy to get an easy and cost-effective synthesis of the 2Cl-BDT polymer. Figure 11.7 illustrates that fluorination carries out four steps while chlorination only carries out one, which makes the procedure cheaper. Density functional theory (DFT) studies of PBDB-T-2Cl exhibited that chorine atom substitution slightly reduced the HOMO energy levels; on the side-chains there was not a significant influence on the conformation of the conjugated backbone. Chlorination has been facilitated to improve the molecular weight, π–π interactions, and the formation of effective pathways for the charge transport in the polymers, which in turn produce higher charge carrier mobility and photovoltaic specifications in the OPVs (S. Zhang et al. 2018; Kini, Jeon and Moon 2020).

11.3.3 Side-Chain Engineering

π-conjugated polymers have poor solubility due to their rigid backbones. Side-chains such as alkyl chains, oligo (ethylene glycol) chains, siloxane-terminated side chains, carbosilane chains, and fluoroalkyl chains are introduced into the backbone of the polymer to solve the problem of solubility. The side-chains conduct different molecular packing distances, crystallinity, and interchain interactions, which finally results in the performance of the other OFETs. In side-chain engineering, the polymer properties are affected by the form, length, density, distribution, and branching position of the side-chains (Shi et al. 2017). One of the best ways to increase the conjugated polymers' solubility is by adding alkyl chains to the backbone. The research into different linear alkyl side-chains on OFETs shows that the more ordered side-chains improve the length; however, the charge transporting reduces as the alkyl chain length increases (Park et al. 2016).

On the other hand, branched side-chains can afford better solubility for polymers than linear side-chains. Despite this, they have traditionally been seen to interfere with charge transport in polymer films because they are insulating and bulky enough to hinder effective intermolecular interactions. It has recently been demonstrated that well-adjusted, bulky, long-branched side-chains can increase polymers' charge-carrier mobilities (Lei, Dou and Pei 2012; Zhang et al. 2013). As was mentioned, DPP-based D-A polymers can afford higher a charge-carrier despite their relatively low crystalline nature (Bürgi et al. 2008; J. Li et al. 2012). Experiments with DPP-based polymers showed that more branched side-chains presented considerably more mobility than polymers with a linear chain due to the enhanced solubility. Figure 11.8 shows the chemical structures of these compounds. The high self-organization properties during the solution process enabled the obtaining of high-mobility polymeric semiconductors suitable for OFETs. This was possible thanks to the improved intermolecular interactions with extraordinarily short π–π stacking distances with the polymers' good solubility (I. Kang et al. 2013).

In addition to alkyl side-chains, siloxane-terminated, carbosilane, and fluoroalkyl side-chains in conjugated polymers can also be found. Figure 11.9a displays the synthetic route of the brominated IID monomer with siloxane-terminated side-chains. Siloxane-terminated side-chains are more bendable due to the longer length and angle of the Si-O and Si-O-Si bonds, respectively (Mei et al. 2011). Consequently, an IID-based polymer with siloxane-terminated side-chains obtains better solubility, a reduced π-stacking distance, and a consistent crystalline length

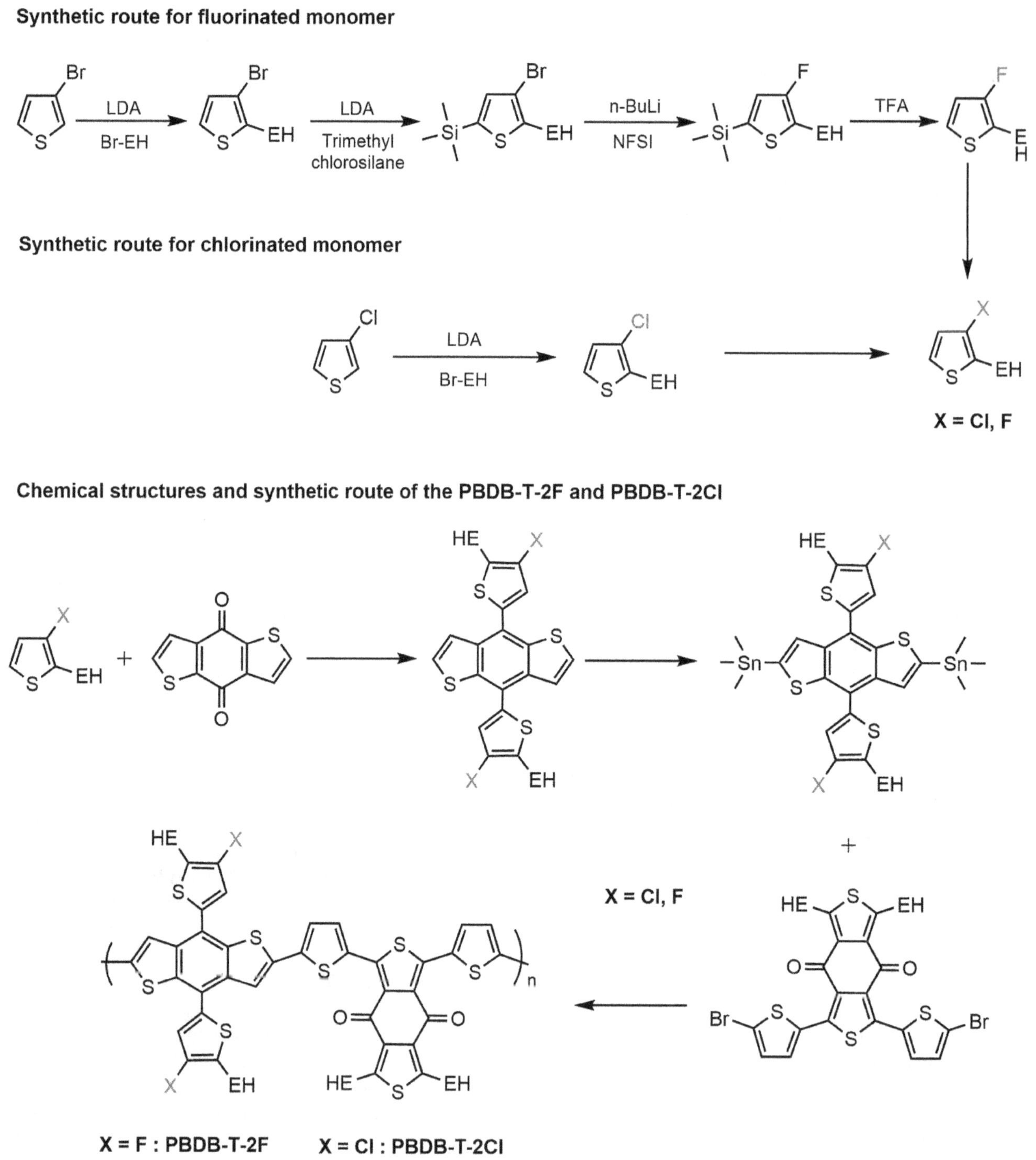

FIGURE 11.7 Synthetic routes for fluorinated and chlorinated monomers, and the synthetic routes for PDBT-T-2F and PDBT-T-2Cl.

which provides higher charge mobility than with the branched alkyl side-chains (Shi et al. 2017). In comparison, carbosilane chains also have a silicon element in their structure. Introducing carbosilane side-chains (Figure 11.9b) into IID-based polymers promises the acquisition of high mobility, stretchability, and mechanical stability. These novel carbosilane side-chains could become high-performance semiconductors (Wu et al. 2016).

Last but not least important, side-chain engineering with fluoroalkyl chains was developed. As was mentioned above, fluorine atoms are adjacent to the conjugated backbone and directly adjust the conformation of the backbone and energy levels. Fluoroalkyl chains have several properties which include rigidity, thermal stability, hydrophobicity, chemical and oxidative resistance, and self-assembly ability (Shi et al. 2017). This kind of side-chain has been studied, especially for acceptor semiconductors. Research with fluoroalkyl chains has produced very good results by improving the hole mobility, the microstructure of the polymer, and also they induce the strong self-organization of the backbone.

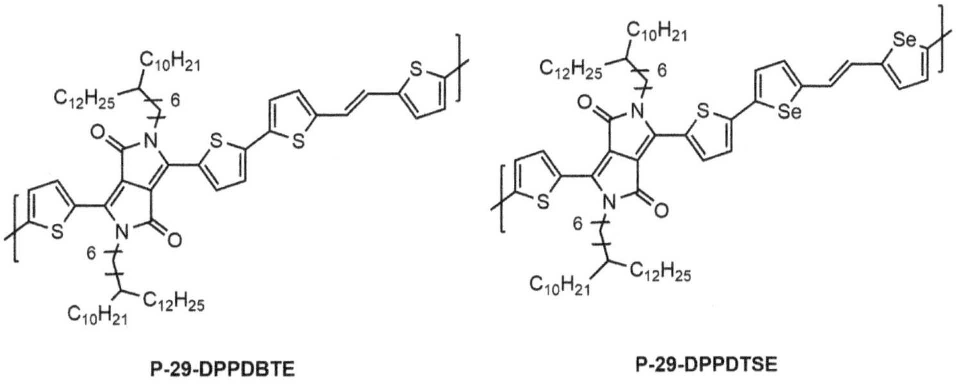

P-29-DPPDBTE

P-29-DPPDTSE

FIGURE 11.8 Chemical structures of P-29-DPPDTSE and P-29-DPPDBTE semiconductor polymers.

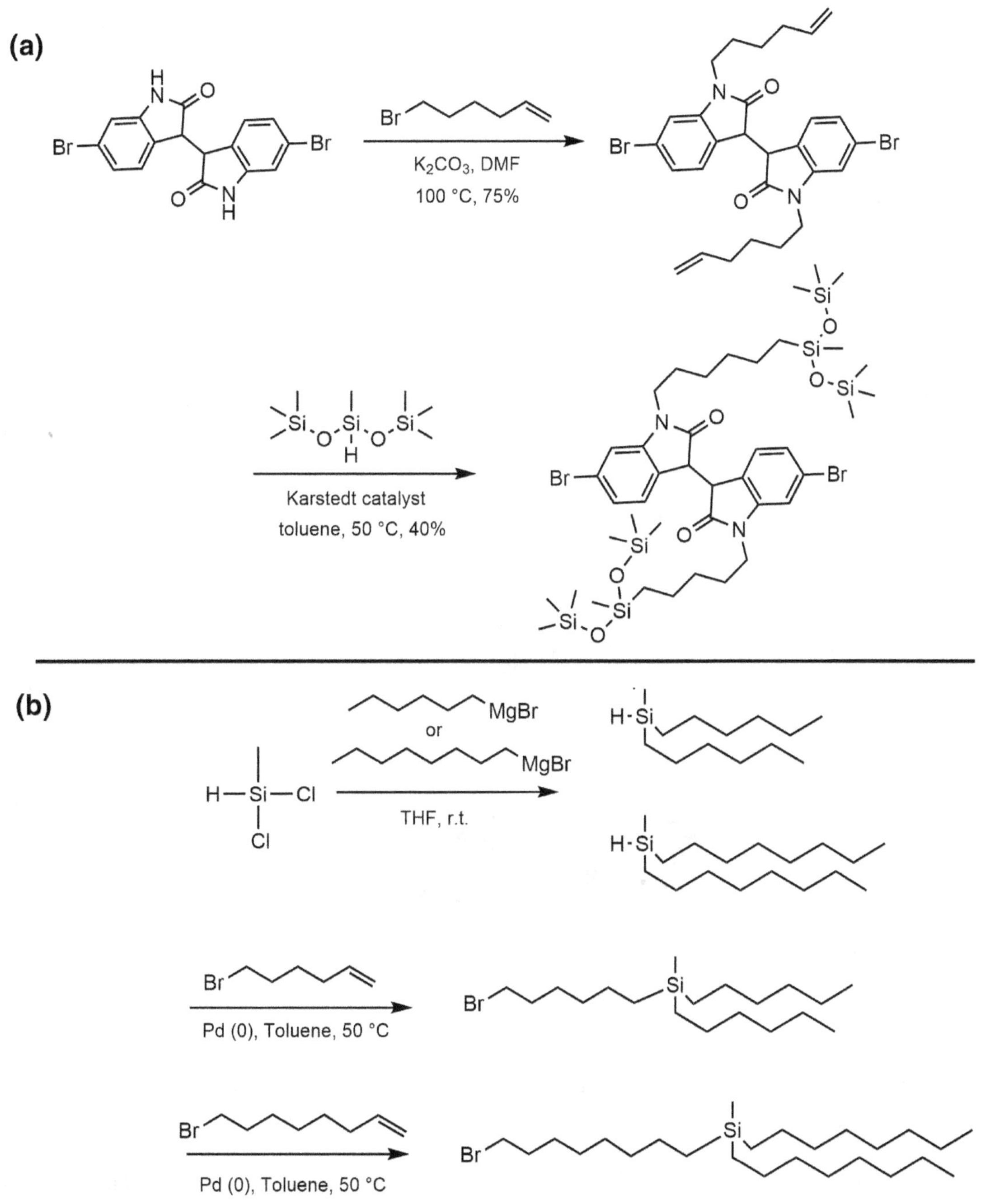

(a)

K$_2$CO$_3$, DMF
100 °C, 75%

Karstedt catalyst
toluene, 50 °C, 40%

(b)

THF, r.t.

Pd (0), Toluene, 50 °C

Pd (0), Toluene, 50 °C

FIGURE 11.9 (a) Synthetic route of brominated IID monomer with siloxane-terminated side chains. (b) Synthesis of carbosilane chains.

Fluoroalkyl side-chains for OPVs need to be further investigated (Homyak et al. 2015; Kang et al. 2016).

11.3.4 Random Copolymerization

Random copolymers are made up of two or more different monomers with an utterly random repeat unit sequence. A polymer that is obtained this way is also known as a statistical copolymer. Currently, the usual way to make a polymerization process of conjugated polymers is by Stille polycondensation (Carsten et al. 2011), Suzuki polycondensation (Sakamoto et al. 2009), or direct arylation polymerization (Bura, Blaskovits and Leclerc 2016). A general scheme of the polymerization processes is shown in Figure 11.10.

Previously, researchers focused their efforts on studying regular conjugated polymers to release their high charge transporting properties. The HOMO and LUMO levels can often be controlled for regular conjugated polymers by choosing different pairs of charge donor and acceptor monomers. Problems can appear due to the complexity and difficulty of designing monomers with different structures. Hence, random copolymerization has been used as an easy way to synthesize polymer semiconductors to regulate HOMO and LUMO energy levels, crystallinity, solubility, molecular packing, functionality, among other properties (Shi et al. 2017).

HD-PPTV and HD-PPPV (Figure 11.11) are regular polymers that showed ambipolar behavior and unipolar p-type behavior, respectively (Wu, Kim and Jenekhe 2011). On the other hand, a random copolymer PPTPV, as shown in Figure 11.11, was endowed with ambipolar behavior between HD-PPTV and HD-PPPV when it was composed of 50% of each regular polymer. Other characterization studies of the random copolymer showed photophysical property and crystallinity behavior between regular polymers (Kim et al. 2014).

Other random copolymerization experiments demonstrated an enhancement in crystallization temperatures, the polymer's solubility, mechanical features through the formation of intra- and intermolecular hydrogen bonding, and modification of the film structure. This allowed obtaining polymers that exhibit stretchable and healable properties (Park et al. 2016). Researchers have also used random copolymerization to verify a polymer's hydrogen bonds (Yao et al. 2015).

11.4 PROPERTIES

11.4.1 Electronic Properties

Most polymers are good electrical insulators; however, they can have the mobile charge carriers redistributed according to the applied electric field's intensity and period. In some cases, the direct current conductivity applied under high electric fields may have resulted from impurities (Glowacki et al. 2012). Charge carriers are called electrons or holes. The nature and role of charge carriers will depend on the direct relationship between the structure, morphology, and transport of the studied system (Jaiswal and Menon 2006). The continuous development of organic semiconducting polymers plays a vital role in the manufacture of various electronic devices such as OTFTS, OPVs,

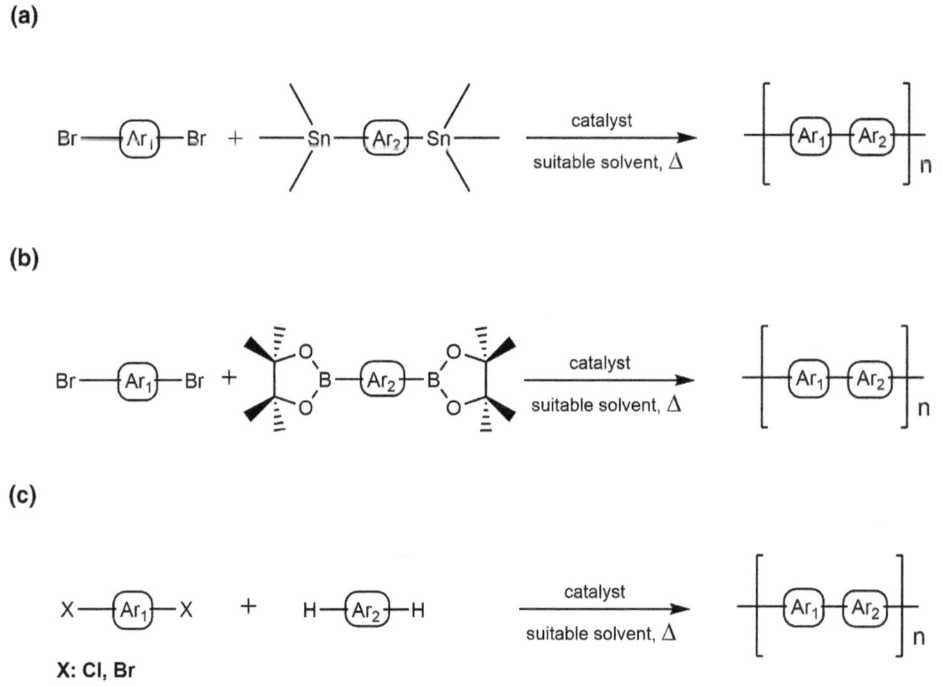

(a)

Br–Ar₁–Br + –Sn–Ar₂–Sn– —catalyst, suitable solvent, Δ→ [–Ar₁–Ar₂–]n

(b)

Br–Ar₁–Br + (pinacol)B–Ar₂–B(pinacol) —catalyst, suitable solvent, Δ→ [–Ar₁–Ar₂–]n

(c)

X–Ar₁–X + H–Ar₂–H —catalyst, suitable solvent, Δ→ [–Ar₁–Ar₂–]n

X: Cl, Br

FIGURE 11.10 Polymerization conjugated polymer methodologies: (a) Stille-coupling reaction; (b) Suzuki–Miyaura coupling reaction; and (c) direct-arylation reaction.

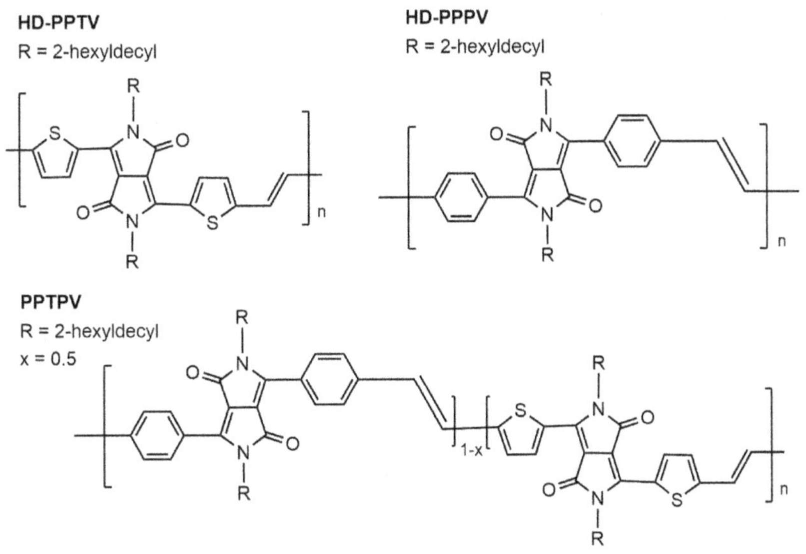

FIGURE 11.11 Chemical structure of regular copolymer HD-PPTV and HD-PPPV and random copolymer PPTPV.

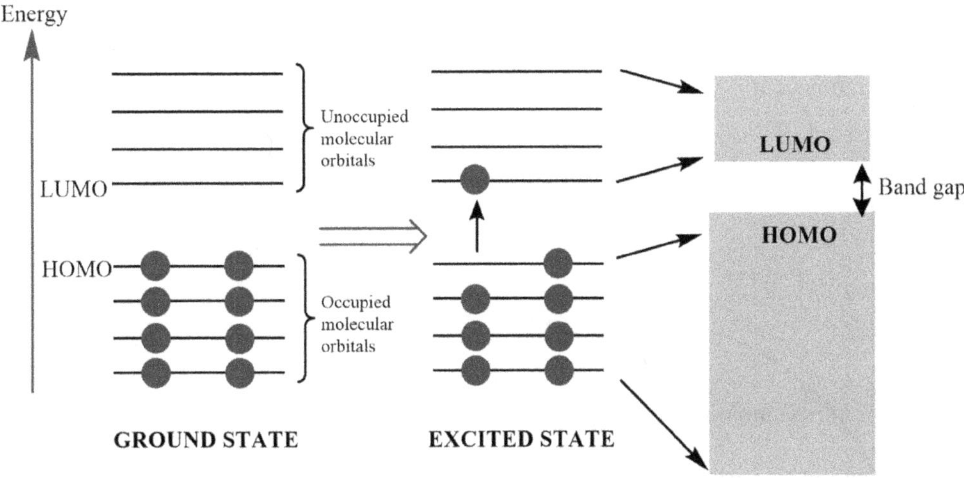

FIGURE 11.12 Representation of the LUMO and HOMO energy levels.

and organic light-emitting diodes (OLEDs) (Sirringhaus 2014; Paterson et al. 2018).

Conjugated polymers are structured by a series of simple and double bonds. Electrons in the bound state are not available for electrical conduction. However, π electrons' delocalization in the carbon chain gives them their semiconducting transport nature (Jaiswal and Menon 2006). In semiconductor polymers, the electronic properties are associated with the π-conjugation in the polymer chains, resulting in rigid and semicrystalline materials (Wang G et al. 2018a).

As is shown in Figure 11.12, the difference between the LUMO and HOMO energy levels is associated with the optical bandgap. LUMO and HOMO correspond to the lowest and highest occupied molecular orbital, respectively. The HOMO level can be raised by increasing the donor strength while the LUMO level remains similar (Hendriks et al. 2014). The bandgap in these quickly modifies the changes in the chemical structure.

The polymer backbone formed by the π-conjugated structural units of the electron donor and acceptor will enhance the semiconducting polymer's performance (Mikie and Osaka 2019). The dipole–dipole interchain interactions reduce the π–π stacking distances, facilitating charge carrier mobility (Tsao et al. 2011). Alternatively, it can occur through the mixing of molecular orbitals (HOMO and LUMO). Deep HOMO levels are related to chemical stability in air and high voltages, and in the case of deep LUMO levels, it is related to stable electron transport in the air (Liu et al. 2016).

Electronic applications require high current densities, which imply a high concentration of charge carriers; however, polymers are materials with low charge mobility (less than 10^{-2}–10^{-1}cm^2 V^{-1} s^{-1}) that can be improved by heavy doping or high levels of the charge injection (Arkhipov et al. 2003). Charge carriers can be produced by electrodes, adjacent organic materials, or generated through photoexcitation (Zhao and Zhan 2011). In previous studies, the polymer synthesized by Sista et al. (2010) had greater electron delocalization and a lower band gap. The extended electron delocalization was obtained through the substitution of phenylenthynul groups on the benzene ring (Sista et al. 2010). A similar example was developed by Hundt et al. (2009) who worked on polymers with a fused benzodithiphene core and phenylethynyl substituents, which produce a lower bandgap and a fluorescent response (Hundt et al. 2009).

11.4.2 Charge Carrier Mobility

Charge carrier mobility (μ) can be explained as the speed at which carriers pass through the material in a given direction under an applied field (Wadsworth et al. 2019). The mobility of a material has its units of velocity per unit field and is dependent on the electric field and the temperature (Zhan et al. 2010). This dependence is based on different models, such as carrier hopping, trapping/detrapping, and tunneling. These characteristics are determinant in the system's transport regimen (dispersive/nondispersive, polaronic, and correlative) (Jaiswal and Menon 2006). The first semiconducting polymers produced were stable under environmental conditions, but they lack structural order, which provokes the inhibition of their carrier mobilities. The following studies were focused on the microstructure and its relation with their electrical properties (Holliday, Donaghey and McCulloch 2013). Figure 11.13 illustrates the progress of the principal semiconducting polymers and their improvement in mobility. Recent advances obtain a charge carrier mobility beyond those of amorphous silicon, even with values exceeding 10 $cm^2\ V^{-1}\ s^{-1}$ (Sirringhaus 2014).

Solvation and the use of flexible branched linkers have a critical impact on polymer charge transport. The influence of the insertion of flexible linker nonconjugated branches in conjugated polymers has been studied; one study case is represented in Figure 11.14. It is reported that alkyl branching agents can enhance polymer semiconductors' solvent interaction while having the most negligible impact on charge carrier mobility (Wang et al. 2018b). It has also been studied how crystals with high ordering and long-range interconnectivity exhibit a high carrier charge mobility (McBride et al. 2018). One of the most studied polymers is DPP, which has a hole transporting predominant mechanism.

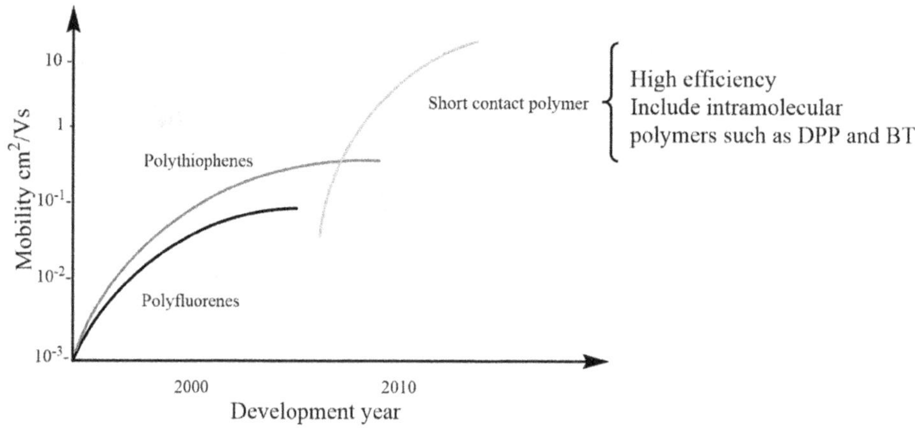

FIGURE 11.13 Progress in charge carrier mobility based on semiconducting polymer types.

FIGURE 11.14 Schematic structure of the polymer and its respective flexible linkers, which were analyzed in previous studies.

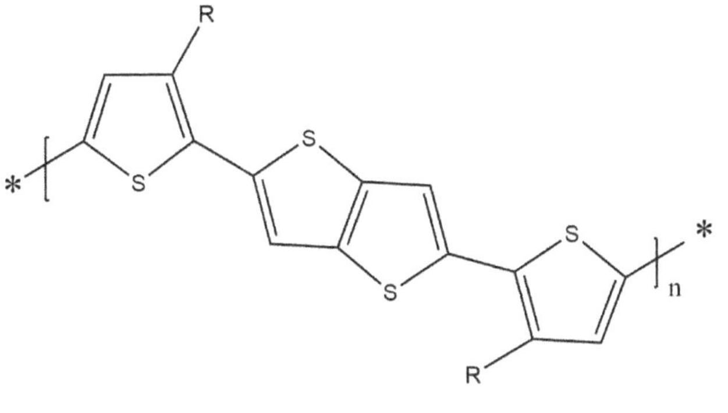

FIGURE 11.15 Chemical structure of poly(2,5-bis(3-alkulthiophen-2-yl) thieno[3,2-b]thiophene).

Previous work reported on the thieno[3,2-b]thiophene semiconducting polymer in the liquid crystal phase. The extended planar π electron system allows a close intermolecular π–π distance through which the carrier mobility can be promoted (McCulloch et al. 2006). Figure 11.15 shows that the structure of poly(2,5-bis(3-chain thiophen-2-yl) thieno[3,2-b]thiophene) with a low-energy backbone conformation promotes highly ordered crystal domain formation.

Electronic devices need to improve semiconductors' stretchability because existing ones usually sacrifice their charge transport mobility to achieve stretchability. Conjugated polymers developed as softer semiconductors are characterized by their high charge carrier mobility, like poly-Si (Nielsen, Turbiez, and McCulloch 2012), but their stretchability remains poor. The nanoconfinement of polymers plays an essential role in improving stretchability, where the modulus of the conjugated polymer is reduced to avoid early cracks. The results related to semiconductor films are encouraging, which were stretched to 100% strain without affecting mobility (Xu et al. 2017).

11.4.2.1 Intrinsic Charge Trapping

Electron traps are located in the bandgap with a Gaussian distribution, and they are associated with intrinsic defects like kinks in the polymer backbone, environmental conditions, and impurities (Kaake, Barbara and Zhu 2010; Nicolai et al. 2012). Charge trapping is a phenomenon that has been a challenge in the performance of various devices. When the electron current is considerably smaller than the hole current, it could be explained by trap-limited conduction. An additional charge trapping agent can be added by introducing metal nanoparticles or an extra buffer layer, thereby enhancing the charge trapping ability, retention force, and durability (Murari et al. 2016). Figure 11.16 shows the structure of naphthalene diimide-alt-biselenophene (PNDIBS) and its phenyl end-capped derivative (ePNDIBS) studied by Murari et al. (2016) by their charge trapping capability. The first n-type OFET memory transistor based on a fullerene semiconductor and an electron-trapping polymer was reported. The method used consists of charge trapping and de-trapping, and modulating the transistor channel's conductivity (Dao, Matsushima and Murata 2012).

There are semiconducting bipolar polymers, which have the intrinsic characteristics of transporting electrons and holes, thereby providing the charge-trapping layer with opportunities to trap both of them (Zhou et al. 2013).

11.4.2.2 Light Polymers

Conjugated polymers are related to the electroluminescence phenomenon, which can be explained by the light emission that occurs when current flows through electron-hole recombination (De Boer and Facchetti 2008). The development of polymer light-emitting diodes (PLEDs) is essential for different applications. In these materials, the luminescence produced is given by holes and electrons injected from the device contacts. The luminescence obtained presents a characteristic color related to the energy difference between the excited and ground states (De Boer and Facchetti 2008).

11.4.3 Charge Carrier Transport

Charge transport is explained as continuous electron-transfer reactions between neutral and charged molecules or polymeric repeating units (Facchetti 2007).

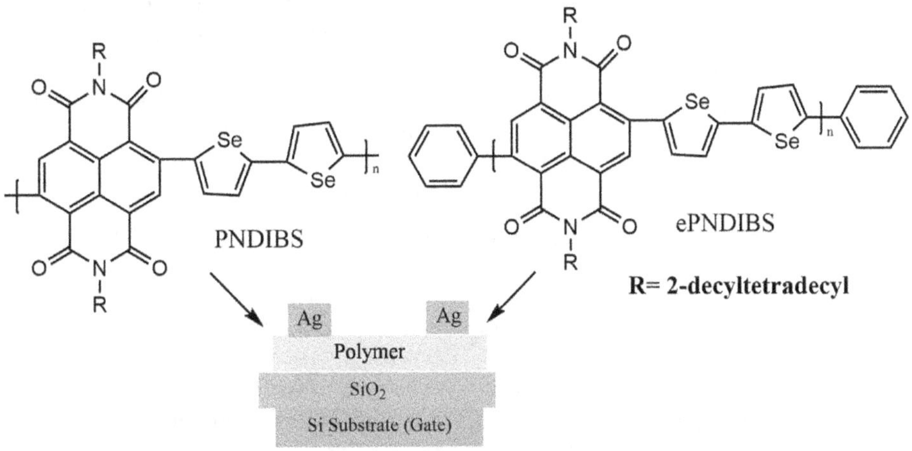

FIGURE 11.16 Molecular structures of PNIBS and ePNIBS where their intrinsic charge trapping was studied.

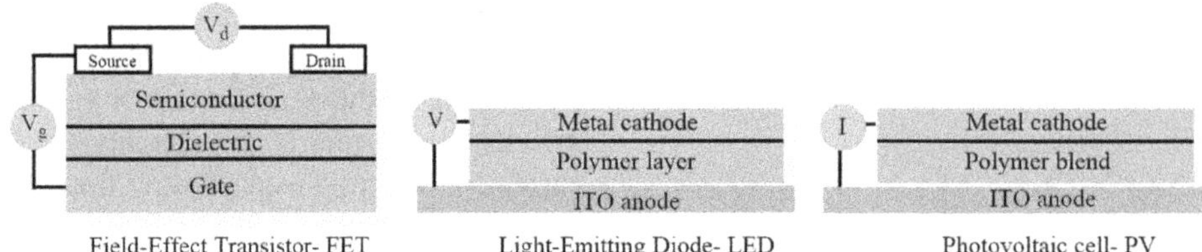

FIGURE 11.17 Schematic diagram of various semiconductor-based devices.

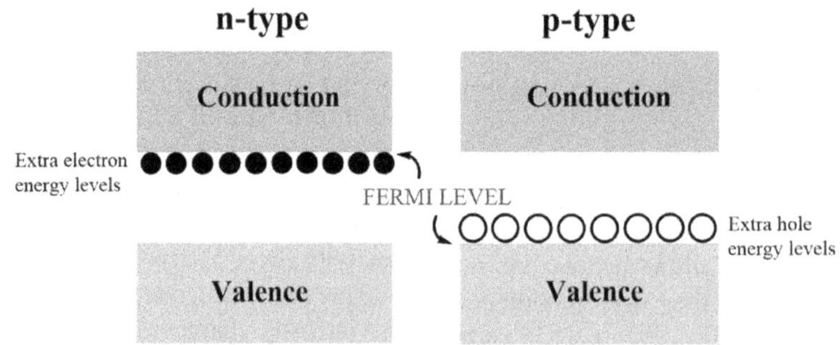

FIGURE 11.18 Schematic diagram of n- and p-type polymeric materials.

Holliday, Donaghey, and McCulloch (2013) reported two methods to improve charge transport. The first mode is to use bridging units to reduce conformational high-energy backbone disorder, and the second mode is to induce coplanarity by using extended aromatic structures (Holliday, Donaghey and McCulloch 2013). Both present effective results.

The electron transporting semiconducting polymers are commonly applied in developing organic p-n junctions, PVs, FETs (Dimitrakopoulos and Malenfant 2002), LEDs, and complementary logic circuits. According to the previous report, it was synthesized semiconducting polymers used in LED applications which are composed of naphthalene diimide (NDI) and per diimide (PDI) acceptor parts (Yan et al. 2009; Zhan et al. 2010). Kim et al. proved that FETs are highly resistant under tensile stress, and that the change of its carrier mobility is negligible (Kim et al. 2017). Some structures of device applications are schematized in Figure 11.17.

11.4.3.1 Single-Layer (SL) Transportation

Single-layer transport occurs when the photosensitive layer collects holes and electrons, and there is a low-sheet resistance path for photo-generated current (Kolesov et al. 2016). The single layer is composed of the desired semiconducting polymer. This may result in a simplified architecture, which can achieve long-life high power efficiency (Kotadiya, Blom and Wetzelaer 2019).

The charge carriers are classified as hole (HT) or electron (ET) transport. Figure 11.18 illustrates how n- and p-type polymeric materials work. Some of the semiconducting polymers can be doped with n or p-type dopants which have an influence on their electrical conductivity and at the same time in their linear optical properties, which are associated with the creation of a charge transfer band when a charge is added of taken from the polymer backbone (Kajzar and Sujan 2016).

11.4.3.2 n-Type (Electron-Transporting)

In order to obtain an acceptable performance, electron transport materials must have the following characteristics: high electron affinity (greater than 3 eV, but not significantly more than 5 eV), proper intermolecular electron orbital overlap (suitable for high mobility) as shown in Figure 11.18, as well as high air stability (Zhan et al. 2010). The other fundamental aspects are:

- Suitable HOMO/LUMO levels which minimize the barrier for electron injection and hole blocking;
- High electron mobility to promote the charge recombination zone and to improve the exciton generation rate;
- An easy process to be used in different techniques like spin coating or printing;
- High T_g to generate an amorphous phase as well as to avoid light scattering.

11.4.3.3 p-Type (Hole-Transporting)

The study of hole transport is essential in many applications like nanoelectronics or photovoltaics (Bescond 2015). Hole transportation velocity is related to hole mobility (Zhao and Zhan 2011). Some p-type polymers present high hole mobility, higher than 1 $cm^2 V^{-1} s^{-1}$, so similar to amorphous silicon. Conjugated polymers are organic p-type semiconductors with a direct bandgap (1.5–3 eV). Usually, luminescent polymers are p-type materials that accept and transport holes due to their relatively high HOMO level (Zhao and Zhan 2011). Hole mobility can be enhanced by solution-processable semiconductors based on the addition of the molecular acid $B(C_6F_5)_3$. This compound's action mole is based on its dual role as a p dopant and a microstructure modifier (Panidi et al. 2017). Organic semiconductors with an extremely high optical absorption coefficient in the order of 107 cm help thin organic solar cells (Skompska 2010). Moreover, there are ambipolar materials that present analogous hole and electron mobility (Zhan et al. 2010).

11.4.4 Intra- and Interchain Charge Transport

Different routes are needed to approach the intra- and interchain charge transport of polymeric chains. To determine the charge transport systems with or without crystalline phases, the study of the different contributions of electronic processes at various length scales is needed (Noriega et al. 2013). Polymeric chains are bound through weak interactions that are characterized by different degrees of freedom. The presence of aggregates in the polymeric chain can generate energetic disorder, which induces electronic localization, limiting the charge transport in conjugated high mobility (Delongchamp and Kline 2012).

These kinds of interactions are helpful in the modeling of new materials where we can obtain predictive behaviors. In hole mobility, some parameters are commonly used in transport modeling (Gali et al. 2017), like polymer chain length, polymeric internal space, conformational freedom, and chain orientation based on the applied field.

Previous studies use quantum mechanical techniques to determine the charge mobility of self-assembled P3HT molecules. A hopping transport model and the parameters of charge mobility, torsional angle, and intermolecular distance were taken into account. A higher rate of charge transfer through intra-chain than along π–π interchain was obtained (Lan, Yang and Yang 2009).

11.4.5 Optical Properties

Hybrid polymer-semiconductor materials are an alternative for photovoltaic cells due to the advantage of having two components that are enabled to absorb visible light (Skompska 2010).

Some materials can absorb photons with energy equal to or greater than the bandgap energy. They form bound electron-hole pairs, which diffuse in the material and dissociate into individual charges. These pairs will diffuse in the internal electric field generated by the entire device.

Polymer emeraldine polyaniline (PANI) is an economical material, and its optical gap of ~1.8 eV allows the capture of a large part of the visible solar spectrum. Its optical properties are applied for the discoloration of synthetic solutions by adsorption and photocatalysis (Belabed et al. 2018).

11.4.6 Mechanical Properties of Organic Semiconductors

Conjugated polymers have mechanical characteristics like mechanically flexible, stretchability, and solution-processability, making it feasible to apply them in electronic

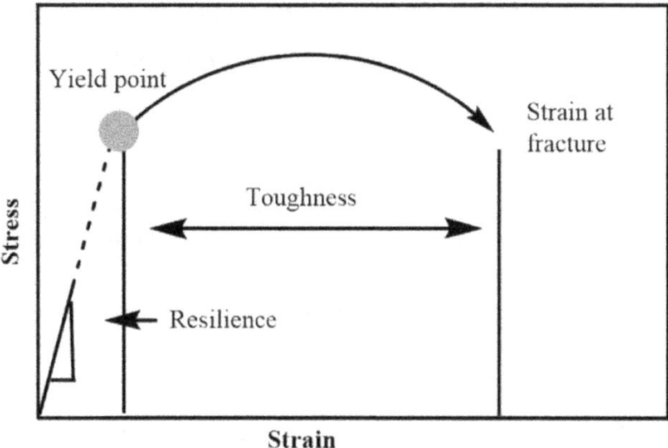

FIGURE 11.19 Hypothetical stress–strain curve.

products (Lei et al. 2017). In the case of polymer films it is necessary to measure mechanical parameters, such as the elastic modulus (Young's modulus) which plays an essential role in the material's durability (Root et al. 2017). Some critical mechanical properties can be deduced by stress–strain curves like the one represented in Figure 11.19.

There are numerous techniques to determine mechanical properties. Two prominent ones have been used previously to characterize semiconductors for application in soft electronics. Most of the studies developed are based on the improvement of stretchable semiconductor properties without significant electrical performance influence. Some techniques used consist of side-chain engineering (Oh et al. 2016), the incorporation of dynamic bonding, nanoconfinement, and soft cross-linkers which were used to reduce the tensile modulus to improve fracture strain (Wang et al. 2016). The film was characterized over water (FOW method) and an elastomer (FOE method) (Rodriquez et al. 2017). Mung et al. report the practical tuning of mechanical properties by incorporating non-conjugated spacers, which improves stretchability (Mun et al. 2018). Other work reports on the use of polymer acceptors (PPAs), which produce tie molecules and polymer entanglements, dissipating substantial mechanical strain energy with sizeable plastic deformation (Choi et al. 2019).

11.4.7 Physical Properties

Usually, air instability is related to the sensitivity of free radical anions from H_2O or O_2 in the air, which can suppress electron transport processes. To improve environmental stability, the mobile electrons in polymers should be kinetically or thermodynamically resistant to trapping, and the LUMO level should be below 4 eV (Zhao and Zhan 2011). It is reported that the ionization potential has a relation to oxidative stability. The use of electron-rich polythiophenes could be related to easy oxidation in the air due to their high HOMO energies, which can be improved by using phenyl rings in the place of thiophene (Facchetti 2007). The use of heteroatoms in conjugated polymer backbones has been reported which improves air stability.

11.5 POLYMER SEMICONDUCTOR CHARACTERIZATION TECHNIQUES

11.5.1 Physicochemical Characterization Techniques

Polymer semiconductors could be characterized by physicochemical techniques to define various physical and chemical properties such as size, shape, composition, stability, purity, and other essential features. Therefore, microscopy, spectroscopy, and X-ray-based characterization techniques are applied to study a range of materials. Here is a summary of the most common physicochemical techniques used for polymer semiconductor characterization.

11.5.1.1 Microscopy Based Characterization Techniques

Scanning electron microscopy (SEM) is used to obtain information about morphology (surface), chemical composition, and polymer structure. SEM works at the nanometer or micrometer scale with 1–2 million times magnification. In this technique, a high-energy electron beam emits primary electrons onto the surface of the sample material. SEM provokes the emission of secondary electrons from each spot of the material. A detector identifies the secondary electrons producing an electronic signal. The amplification of this signal allows the generation of 2D or 3D high-contrast images. Field emission scanning electron microscopy (FESEM) uses a field emission gun and in-lens detectors, usually comparable with SEM. These features contribute to obtaining clear and less electrostatically distorted images with high magnification. FESEM allows the observation of structures on the polymer material's surface with a resolution around 0.5 nm. This technique avoids damage to sensitive samples (Cik, Foo and Nor 2015; Inkson 2016; Semnani 2017; Mayeen et al. 2018).

Transmission electron microscopy (TEM) uses the transmitted electrons to create a 2D image offering valuable information about the polymer material's internal microstructure with a 50 million times magnification. TEM allows the evaluation of nanostructures such as thin films, particles, and fibers. This microscopic technique uses electromagnetic lenses placed before and after the material so as to characterize it. This technique possesses greater control of the electron beam compared with SEM (Williams and Carter 2009). SEM and TEM images are widely used to characterize polymer semiconductors (Chiu et al. 2003; Deschler et al. 2015; Giusto et al. 2020).

Atomic force microscopy (AFM) provides essential information about the physical properties of the sample. AFM measures size, morphology, surface area, volume, and other physical properties such as intermolecular forces, electric forces, conductivity forces, magnetic forces, resistance, and surface potential. AFM allows identifying materials with particle sizes from 1 to 8 μm. High-resolution 3D nanoscale images can be generated showing the topology of the material. This technique allows the generation of images from polymers and other biological samples. AFM is cheaper, easy to operate, and requires less space than SEM and TEM (Ray 2013).

Near-field scanning optical microscopy (NSOM) is a method that can exceed the far-field resolution limit. In NSOM, laser light passes across an optical fiber and generates a light-field emission. This light emission is highly confined in a tip aperture located very close to the material under investigation. In the NSOM technique, the resolution does not work in the function of the diffraction limit; instead, it works in the aperture size, which is around 10–100 nm in diameter. NSOM produces high-resolution topographic, optical, transmission, and fluorescence images. The resolution of the images generated is approximately 50 nm and 20 times better than conventional light microscopes. To enhance the resolution and contrast, the aperture should be as close as possible to the polymer semiconductor's surface (Barbara, Adams and O'Connor 1999; Frisbie 2003; Lin et al. 2014).

11.5.1.2 Spectroscopy Based Characterization Techniques

Raman scattering (RS) is a method that uses light scattering to establish the chemical and morphological structure of polymeric materials. RS measures the inelastic scattering photons from the incident light in a non-invasive way without sample preparation. RS has a submicron spatial resolution, confocal light collection, and excellent depth resolution (Popović et al. 2010).

Mass spectrometry (MS) is a widely used analytical technique that measures the mass-to-charge ratios of atoms and molecules. MS requires an external device to introduce the sample that could be gas chromatography, liquid chromatography, or capillary electrophoresis. MS produces ions in the gas phase then, and a mass analyzer separates the ions according to their mass-to-charge ratios. Finally, the detector counts the ions (De Hoffmann 2005). MS provides high precision, accuracy, selectivity, and sensibility of detection to determine molecular weight (Gmoshinskii et al. 2013).

Ultraviolet-visible spectroscopy (UV-Vis) is an analytic technique to characterize the size, concentration, functional groups, aggregation state, and conjugation of materials. UV-Vis light is emitted to a sample and the transmittance of the light is measured, obtaining an absorbance spectrum at different wavelengths. Additionally, this technique confirms the identity of a compound based on the absorption profiles (Sapsford et al. 2011; Venkatachalam 2016).

Dynamic light scattering (DLS) is a technique that can detect size, shape, structure, aggregation state, and conformation of particles at a submicron to 1 nm scale. In this technique, non-invasive visible laser light is emitted to suspended particles or macromolecules in solution. The scattering light obtained from the incident light is analyzed to detect the size of the dissolved particles. The scattered laser light is recorded and quantified by a proton-counting detector (Falke and Betzel 2019).

Electron energy loss spectroscopy (EELS) uses a beam to emit fast electrons that interact with the sample material. Some of these electrons can undergo inelastic interactions with the sample. EELS allows the understanding of the chemical elements, stoichiometry, energy levels, electronic structure, and dielectric constants, among other characteristics (Hirai 2003; Verbeeck et al. 2005).

11.5.1.3 X-ray Based Characterization Techniques

The use of various microscopy, spectroscopy, and X-ray based characterization techniques permits the obtaining of a wide variety of information from surfaces, interfaces, thin films, and multilayers (Stoev and Sakurai 1999). There are some X-ray based techniques for characterizing materials such as X-ray diffraction (XRD), X-ray fluorescence (XRF), X-ray photoelectron spectroscopy (XPS), and grazing incidence X-ray diffraction (GIXRD). An ample number of X-ray based characterization techniques have been used for the physicochemical characterization of polymer semiconductors (Pron et al. 2010; Zhang et al. 2014; Wang, Gasperini and Bao 2018**c**).

XRD characterizes the crystalline planes of thin-film materials on an atomic scale following Bragg's diffraction law. This technique allows for the determining of crystalline phases, orientation, structural properties measuring the thickness of thin-film layers, and the atomic arrangement of small crystal or grain regions (Caminade, Laurent and Majoral 2005; Sapsford et al. 2011).

XRF determines the concentrations and the type of elements present in a material. XRF is a non-destructive method that uses a primary X-ray emission to excite the sample and generate a secondary fluorescent X-ray emission. This secondary emission of the sample is measured to characterize its chemical structure. Additionally, the fluorescent X-ray emission is a unique characteristic (fingerprint) of each element present in the sample (Feng, Zhang, and Yu 2020).

XPS based on the photoelectric effect is used to analyze the chemical characteristics of the surface materials due to a short range of exciting photoelectrons from the sample (Tiede et al. 2008). GIXRD analyzes the phases or crystalline fractions in thin organic layers, films, and organic field transistors (Zhang et al. 2014).

11.5.2 Electrical and Optical Polymer Semiconductor Characterization Techniques

The electrical and optical methods determine the optoelectrical features of materials. These properties, such as bipolar transistor gain, switching properties (powering devices), and its applicability in solar cells, is essential to understand the material's suitability. New characterization techniques have been reported for polymer semiconductors.

11.5.2.1 Recombination Lifetime Characterization

This is an optical analysis based on the induced absorption of a free carrier. This technique consists mainly of using a laser beam for energizing electron-hole pairs. The primary purpose is to estimate the low defect densities of semiconductors (De Laurentis and Irace 2014).

11.5.2.2 Deep Level Transient Spectroscopy (DLTS)

This is a thermal transient scanning technique that operates on a megahertz frequency. DLTS outperforms previous techniques due to its improved sensitivity and observable range of trap depths. It studies the Schottky barrier, which consists of a boundary for electrons formed by potential energy at a metal-semiconducting material junction or the polymer's deep center, and its state changes. The spectroscopic nature of this test makes it suitable for the deep-level characterization of the polymer. Also, it gives energy values, capture rates, and concentration of majority and minority carrier traps. Finally, together with a thermal transmission process, we can obtain a complete spectroscopic analysis of the semiconductor's bandgap. It is mainly used for characterization of photovoltaic materials (Khan and Masafumi 2015).

11.5.2.3 Fourier Transform Infrared Spectroscopy

This analysis technique recognizes chemical bonds from any molecule through infrared absorption spectroscopy. The testing results in a distinctive profile of a sample, like a fingerprint. The results are applied in the screening and scanning of other samples. These results serve to identify functional groups and help characterize the formation of covalent bonds (Yakuphanoglu et al. 2006).

11.5.2.4 Ellipsometric Spectroelectrochemistry

This characterization technique is an *in situ* UV-Vis-NIR transmission spectroscopy-ellipsometry technique. This technique measures the dielectric tensor of the polymer through a cuvette. The cuvette is a highly controlled electrochemical environment for elliptical measurements. Later, a series of oxidation steps are carried out along with an increase in the electrochemical potential. A precise measure of the resonant energies and orientation is obtained depending on the oscillatory force of the exciton-polaron interchain resonances (Cobet et al. 2018).

Moreover, for the analysis of the electrical properties of organic transistors, it is widely used in semiconductor parameter analyzers such as HP 4155B and HP 4192A, analyzers used for F8T2 and pBTTT (Noh et al. 2007), and the Keithley 4200SCS semiconductor parameter analyzer used for water-stable PTFT (Roberts et al. 2008).

11.6 DEVICES BASED ON ORGANIC POLYMER SEMICONDUCTORS

11.6.1 Hybrid Organic–Inorganic Materials

The basis of hybrid nanomaterials is a combination of organic and inorganic components. Among the inorganic components, we can find oxides, sulfides, salts, and nonmetallic elements. Among the organics, we can find organic pollutants, polymers, biomolecules, substances of pharmaceutical origin, among others. Its molecular combination gives rise to new polymeric materials with high performance and functionality. The appearance of hybrid material models began with cubic siloxane, the smallest particle obtained from silica. These copolymers were prepared through a hydroxylation reaction with a propargyl group known as hybrid macromolecules (Chujo 1996).

The process of building hybrid materials is developed at different levels:

1. The construction of the molecules from the union of covalent bonds or a π complexation.
2. It is assembled on a nanoscopic scale where intermolecular interactions (hydrogen bonds, electrostatic and dispersion interactions) play the most crucial role.
3. Micro-structuring of the hybrid material through cooperative interactions of the different modules.

The process itself and the infinite varieties of possible interactions can give rise to lots of materials with particular properties for application (Table 11.1) (Ananikov 2019). The application of hybrid materials can explore high potential areas such as enzyme stabilization discovery and incorporation in nano, micro-sized hybrid materials in nanoscale machines and devices (Hwang and Gu 2012; Cobo et al. 2015).

The technologies previously known as ""soft chemistry" nowadays have aroused intense interest in academia and industry. These techniques originated in solid-state chemistry areas such as organic, organometallic, polymer chemistry, and supramolecular chemistry, and are now used for polymerization reactions to generate hybrid materials from polymer molecules and nanoparticles (Nicole, Rozes and Sanchez 2010). Today the research focuses on combining the purest obtained materials to produce functional nanocomposites.

About its structure, the inorganic part of a hybrid material can easily vary from monometallic species to inorganic nano-sized particles and their extended phases (Gomez-Romero 2001; Rao, Cheetham and Thirumurugan 2008; Schubert 2011). Organic compounds allow the direct incorporation of metals into the polymeric structure to give rise to chains and polymer frameworks coordinated with metals. This metallic coordination ensures the dynamic selectivity of reactants within complex mixtures due to the flexibility of metal-ligand coordinate bonds that are formed (Koppens et al. 2014; Gobbi, Orgiu and Samorì 2018).

These new hybrid organic–inorganic materials give rise to nano-responsive molecules and nanoparticles that can be used in innovative materials that respond to stimulation, nanoscale biological machines and devices, catalyst enhancement, drug delivery systems, photovoltaics, energy research, conversion, light generation, and biocompatible materials (Table 11.2) which make them one of the most versatile synthesized tools of the century that essentially tries to mimic natural structures and fulfilling highly complex tasks with high efficiency (Prosa et al. 2020)

11.6.2 Polymeric Field-Effect Thin-Film Transistors (PTFTs)

PTFTs are field-effect devices with organic or polymeric thin-film conductors used as a channel material for their activation. An insulator is necessary to avoid the effect of the environmental conditions on the polymer semiconductor's chemical structure. Field-effect mobility will be affected by the material used as an insulator (Parashkov et al. 2004). A PTFT is made up of three electrodes (source, drain, and gate) arranged in an organic semiconductor polymer layer and a dielectric gate layer, as shown in Figure 11.20. For its operation, an electric field passes through the source-drain electrodes. The transistor is activated when the voltage (Vo) is applied to the gate electrode, inducing current (Ic) flow from the source electrode to the drain. When the voltage Vo = 0, then the transistor is off. The current flow Ic is modulated by the voltage. The characterization of the transistor is given in current–voltage (output–transfer) plots. These plots help show essential parameters such as the mobility (μ) of the field-effect transistor, current on/off ratio, threshold voltage, and subthreshold swing (Facchetti 2011).

TABLE 11.1
Structural Types of Hybrid Materials and Their Applications

Structural Types	Applications	Reference
Nano, micro-sized hybrid materials	– Nanoscale machine devices – Biopolymers and biomacromolecules – Responsive bioconjugate materials – Crossbreed responsible polymers	Hwang and Gu (2012), Cobo et al. (2015)
Organic–inorganic hybrid materials between nanomaterials and nanocomposites	– Solar cells – Photoactive devices – Photocatalysis – High-performance electrochemical capacitors – Gas sensing and capture	Ananikov (2019)

TABLE 11.2
Hybrid Materials: Composition, Function, and Fields of Interest

Material	Function	Fields	Reference
Improved PdHx materials	Absorption of hydrogen by palladium	Catalysis and Electrochemistry	Ananikov (2019)
Highly ordered PdIn intermetallic nanostructures	Catalytic properties in diphenylacetylene Hydrogenation	Organic chemistry, General and inorganic chemistry	Mashkovsky et al. (2018)
Aluminum oxide (inorganic) and polymethyl methacrylate (organic)	High yield strength and fracture toughness	Material sciences and Electrochemistry	Munch et al. (2008)
Metallic nanoparticles, "nano salt" structures	Transient simulation of nucleation, high polarization degree, reversibly switching between two solid states	Functional and polymer materials, nanomaterials	Yang et al. (2019)
Catalytic cocktails (monometallic species, metal clusters, and metal nanoparticles)/organic–inorganic nanoscale hybrid systems	Higher activity, better selectivity, and improved stability	Organic chemistry, and catalysis	Eremin and Ananikov (2017)
Nanostructures and carbon materials, and heteroatom-doped derivatives	Suitable properties for energy storage, electrocatalysis, optoelectronic devices, and photoactive layers	Advanced materials and polymer advanced technologies	Bounioux, Katz and Yerushalmi-Rozen (2012), Koppens et al. (2014), Gobbi, Orgiu and Samorì (2018)
Carbon materials with metal-containing particles under microwave irradiation	High-performance catalytic systems	Organic chemistry, metallurgy, material sciences, and nanoparticles	Pentsak et al. (2018)
Cellulose nanocrystals grafted with titanium dioxide for long term release	Drug delivery nanocomposite, compensation of triclosan, antibacterial activity facing *E. coli* and *S. aureus*	Biotechnology and medicine	Ananikov (2019)
Polyphenols from tea extracts for the synthesis of gold nanoparticles (AuNPs)	Double function reducing and capping agents	Structural chemistry	Alegria et al. (2018)
Polyethyleneimine attached to iron oxide nanoparticles for on-site reduction of an Au inorganic salt	Supramagnetic properties, a plasmonic response in dark field microscope, dual image probe, cytotoxicity reduction	Organic chemistry, composite nanomaterials, dual imaging	Yoon et al. (2018)
Nanocrystalline semiconductor matrix	Photosensitivity shifted to low energies (longer wavelengths)	Nanomaterials and electrochemistry	Rumyantseva et al. (2018)
Supramolecular organogels (N-benzylbispidinols in aromatic solvents)	Poly/multifunctional chelating agent, possible use in luminescence gels and metal ion recognition gels	Medicine, organic chemistry, physics, petrochemistry, and geology	Medved'ko et al. (2019)

The type of polymer chain and the substitution positions of the conjugated polymers are crucial for the molecular packaging and the morphology of the transistor. These properties will be reflected in the device's efficiency and performance (Lei, Dou and Pei 2012). The active polymeric thin layer's thickness is also essential and produces changes in the voltage threshold (Vt). There is high mobility in the transistor when the thickness of the layer is not enormous. The electrons' carrier mobility will decrease with a thick layer (Reséndiz et al. 2010). There is a strong correlation between the mobility of the field effect, the solvent used as a dielectric, and the insulator's dielectric constant. These characteristics can be tested through X-ray diffractometry and electron spectroscopy (Parashkov et al. 2004). One of the critical parameters based on an outstanding performance of a PTFT is the nature of its semiconductor to conduct holes, electrons, or both under various gate bias conditions (Facchetti 2011).

11.6.2.1 Classification of the Transistors Based on Semiconductor Polymers

11.6.2.1.1 p-Channel Polymer Transistors: Ability to Conduct Holes

Most of the polymers that have worked optimally for PTFTs are p-channeled. To improve hole-transporting polymers, a balance is made among the HOMO energy levels, which should be between −5 and −5.5 eV in normal conditions. As a result, the E_{HOMO} presents an inverse behavior to the ionization potential. This condition gives rise to easy air-induced oxidation, and the acceptor sites quickly begin to complicate the environmental stability of the PTFT. Then, the current flow is excellent, and high mobility values can be obtained. The most common problem for these transistors is high voltage thresholds. The addition of substituents improves their solubility and stability (Facchetti 2011). The positioning of the alkyl groups as part of the side chain of

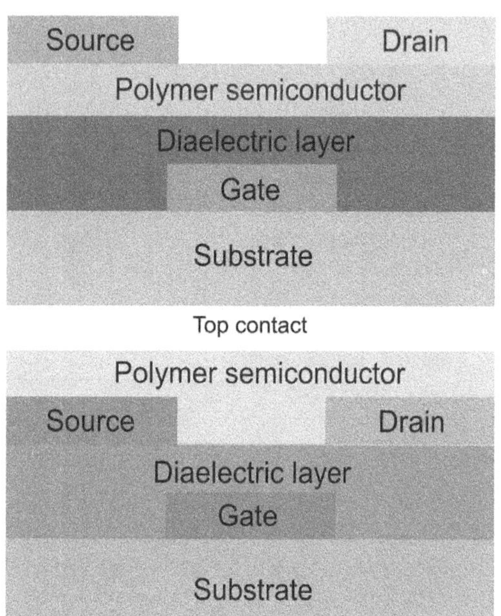

FIGURE 11.20 Top and bottom contact structure of polymer thin film transistor.

the polymer allows the transistors to have greater mobility (Lei, Dou and Pei 2012).

It is not easy to maintain regularity in films under air conditions due to the increase in the carriers' density. Therefore, new polymers are designed based on three structural approaches to improve their self-assemblage and their ability to form thin films:

- The presence of long chains of alkyl sidechains for a better processibility in solution;
- Molecular self-assembly facilitated and induced thanks to its structural stability;
- Achieve a delicate balance of transistor functionality and oxidative doping stability by controlling the conjugation extension. (Facchetti 2011)

The most promising semiconductor candidates are poly triarylamines, poly indolocarbazoles, and thiophene-based polymers.

11.6.2.1.2 n-Channel Polymer Transistors: Ability to Conduct Electrons

The behavior of these polymers corresponds to the energy levels of the lowest molecular orbital (LUMO). The LUMO orbital must be low enough for the injection or generation, extraction, and stable charge transport of the electron. The main limiting factors in the construction of devices with n-channel polymers are a steric hindrance and the lack of solubilizing power in some electron-withdrawing groups (EWGs) that are strong, such as cyano (CN) groups or fluorocarbons. These types of polymers are functionalized using strong heteroarenes (Sun, Guo and Facchetti 2020).

11.6.2.1.2.1 Imide-Functionalized n-Type Polymers

Functionalization with imide groups improves the electron-withdrawing and attachment capacity of many polymer chains, especially in n-type polymers (Sun et al. 2018). The great success of rylene diimide as a semiconductor increased interest in n-channel polymers. Naphthalene diimide (NDI) has a deep-positioned LUMO level. These NDIs could be used in a wide range of optoelectronic devices, so their optimal molecular weight, the number of alkyl groups they should have, film processing, and device engineering have improved.

The crystallinity of a compound reflects better mobility, as in the case of semi-fluoroalkyl with a highly crystalline backbone and side-chain structure. This feature allows them to self-organize efficiently in a planar backbone structure. The high mobility and affinity of NDI polymers and their crystalline structure also exhibit powerful aggregation, enhanced crystal structure, as well as a changed bend morphology. These types of PTFTs are excellent for photovoltaic devices (Sun, Guo and Facchetti 2020).

11.6.2.1.2.2 Amide-Functionalized n-Type Polymers

The compounds diketopyrrolopyrrole (DPP), isoindigo (IID), and naphthalenebisamide (NBA) usually present as ambipolar. Electron-withdrawing groups (EGWs) can be introduced to balance them, or they can be embedded to have electron-deficient functionalities and to build a ladder backbone of type A-A. Some compounds have been synthesized with a rigid double-bond-locked confirmation to eliminate rotation along the conjugated backbone. These polymers' limited mobility and solubility require more research work so as to be used as stable n-type polymeric thermoelectrical materials in the future (Sun, Guo and Facchetti 2020).

11.6.2.1.2.3 B–N Embedded Polymers

B is for boron (partially negatively charged) and N is for nitrogen (partially positively charged). In this B–N coordination, a pair of electrons are donated from nitrogen to the boron p orbital, allowing the transport of electrons. They have good light absorption and tunable LUMO levels. However, LUMO levels are not low enough for the efficient injection of electrons and/or stable transport. They are amorphous and have a poor structure, making it difficult for electrons to pass. They are also very unstable and difficult to purify due to their easy bind with the Lewis bases. Their functionalization with alkyl amides and phenyl rings can help improve their properties and applicability for PTFTs (Sun, Guo and Facchetti 2020).

11.6.2.1.2.4 Cyano-Functionalized Polymers

The π-conjugated systems are suitable transistors due to their long uniform areas with isotropic charge transport. They have easy control of their rheological properties in solutions and a very low variation in device-device performance. These characteristics are essential for the printing process with which most transistors are formed. Baeg, Caironi, and Noh (2013) establish that for its proper application, the polymer must present two characteristics. Firstly, it has a π-conjugated backbone formed or connected to unsaturated units that results in extended π-orbitals along the polymer chain. The second one concerns the functionalization of the polymer core with solubilizing substituents such as the electron-withdrawing CN groups.

The polydispersity index and molecular weight are also crucial for adequate solubility, rheological properties, morphology, and thin-film formation. They are limited by their sizable steric hindrance and their lack of solubility. Fused-ring electron acceptors are upright for CN-functionalization. They usually have a very long side-chain. This characteristic improves their stability but much remains to be done for a correct optimization of structure for industrial purposes (Baeg, Caironi and Noh 2013; Sun, Guo and Facchetti 2020).

11.6.2.2 Ambipolar Polymeric Semiconductors

Most organic semiconductors are either n-channel or p-channel. However, these kinds of polymers are limited by their exclusive characteristics. That is why researchers are increasingly interested in ambipolar compounds. They have a lower bandgap (<2 eV) compared to classical unipolar semiconductors. These semiconductors may be the origin of the understanding of traps in hydroxyls, silanol, and carboxyl groups, which make up the surface of many dielectrics. They would also originate the production of complementary-like circuits, reducing the complexity of the manufacturing of transistors. Transistors could combine the switching and light functions in a single bifunctional device. They could be used for light sensing (LS-OTFTs) or light emission (Facchetti 2011). Semiconductors are just one of the critical materials for the development of PTFTs. New solutions, printable compounds, dielectric gates, and protective encapsulants must also be studied for an optimal and effective transistor design.

11.6.3 Current Techniques of Transistor Fabrication

11.6.3.1 Inkjet Printing

The vast majority of PTFTs were manufactured using the inkjet printing technique. This technique prints each part of the transistor in an aqueous dispersion conductive polymer. For the microelectronic industry, inkjet printing technology benefits from creating polymers in solution on a large scale (Kawase et al. 2003). Also, it has experimented with improving the resolution, overlap capacitances, and limitations in the film's thickness. These enhanced conditions would improve the speed and operating voltages of PTFTs for small-scale production. Using a well-structured self-aligned gate is imperative to enhance the switching speed in our printed transistors. The minimum parameters to produce PFTFs on a small scale are established. The information on downscaling of the process helps develop ambient intelligent devices giving PTFTs a higher range of industrial applications (Noh et al. 2007).

11.6.3.2 Push Coating

Polymeric conductors have the advantage of good processibility in solution. That is the origin of their cheap manufacture and electronic flexibility under normal conditions. On the other hand, their affinity with the substrate is not good; liquid handling on very hydrophobic surfaces is complicated. To improve this characteristic, the push coating system compresses one microliter of the polymer in solution with a viscoelastic stamp on the substrate's surface. This structure causes the retention of the solvent in the stamp and minimizes its deformation. Besides, it helps the stamp peeling process even after a long time. This new technique is low cost, uses a small amount of polymer in solution, is preserved for a long time, and improves field-effect mobility (Ikawa et al. 2012).

11.6.3.3 Improvements in PTFT Structure

11.6.3.3.1 Low Voltage PTFTs on Plastic (Ion-Gel Gate)

An ion-gel-composed dielectric gate can be manufactured to work in combination with a semiconductor polymer. The ion-gel polarization response is more excellent than that reported in solid-state polymer electrolytes. This ion-gel gate can work at high frequencies, which is advantageous when building organic electronics (Lee et al. 2007). The ion-gel dielectrics have a large capacitance, a high capacity to be printed, and an ultimate frequency to generate a response. Its high polarizability allows the synthesis of more simplified transistors since the gate electrode's alignment can be more flexible. Optimizing both the material and the gate should improve the performance of future transistors (Cho et al. 2008).

11.6.3.3.2 Self-Encapsulation

This technique is based on the phase separation of the polymeric mixtures by managing the energy control of the substrate's surface. It is possible to generate flat perpendicular layers on the surface of the film. This process helps the polymer to be encapsulated with insulating material altogether within the transistor. The process does not compromise the equipment's operation, and the exposure of the semiconductor to environmental conditions is minimal because they are vertically segregated. Furthermore, the transistor's higher stability can be obtained at room temperature (Arias, Endicott and Street 2006).

11.6.3.3.3 Nucleic Agents

The solidification process of the polymers, or even when in a solution, can be regulated through additives, including nucleating agents. There are commercially available nucleating agents that, without altering the semiconductors, assist in the melt, solution, or solid-state polymer process. Heterogeneous nucleation makes possible the production of thin-film transistors with uniform electrical characteristics on a large scale. Their electronic and optoelectronic performance must be controlled through the microstructural components of the active film layer to be applied in photovoltaic devices. Nucleated agents can help form a huge variety of materials based on nucleating agent concentration (Treat et al. 2013).

With the materials available nowadays, it is possible to consider industrial production. Many have good stability, long lifetimes, are cheap to produce, structurally stable, and some can even be used in a complementary way due to their ambipolar nature for more complete devices. The study of the other components that supplement the semiconductor polymer is now indispensable for taking the next step towards producing better optoelectronic devices.

11.6.4 Sensors

Research around new organic semiconductors has opened the opportunity to use them in 0/1/2-dimensional technology. Electronic devices that fold easily and inexpensively can be obtained to produce clean and eco-friendly technologies, including sensors. The high conductivity of charge carriers gives rise to technologies such as nano/sub-microstructures that are good to be used in sensors that can be applied in electronics, optoelectronics, and photonics (Neupane et al. 2019). The emerging field of functional organic compounds such as carbon materials makes carbon nanotubes and other structures an excellent choice for sensors (Baughman, Zakhidov and De Heer 2002). Much more investigation must be done to establish suitable quality sensors to be produced at the industrial level. Besides, we have some technologies already established around them.

11.6.4.1 Chemical Sensors

OTFTs are an excellent option for chemical and biological sensors. They are cheap, highly compatible, flexible, large-area substrates, effortless to process, and very tunable. These sensors have been proven effective in aqueous media and aqueous solutions with low trinitrobenzene, cysteine, methyl phosphonic acid, and glucose concentrations. The drain current is sensitive to certain compounds; sensitivity to alcohols as analytes has been tested to parts per billion. The dielectric interface of the sensor can be affected by small ion molecules that diffuse into the semiconductor. Therefore, this sensor can be used in aqueous sensing (Roberts et al. 2008). Non-metallic nanotubes can be used for transport and thermal power for sensor applications. Differences in temperature between them cause existing voltages between joints. Substances with high sensitivity are affected by a large amount of charge that is injected. However, very little material is required, and we get tiny-sized sensing nanotube elements (Baughman, Zakhidov and De Heer 2002).

11.6.4.2 Metal-Organic Frameworks as Chemical Sensors

Metal-organic frameworks (MOFs) are crystalline structures within materials with high porosity, diversity, tunability, and structural diversity. MOF "nodes" are composed of metallic cations or cation clusters. They are connected multitopically to the "linkers" that are the ions of the molecules. These can be used as chemical sensors due to suitable properties and high reusability. The porosity of the material allows a higher concentration of the analyte in the structure than on the outside. Stronger binding to the analyte also reflects lower detection limits. However, the specificity with which it binds to analytes is still being studied. The most intuitive process currently used is size exclusion (molecular sieving). Depending on the MOF's porosity, the smaller particles have a greater possibility of being absorbed while large ones are excluded. Therefore, the viable analyte for a specific MOF is determined by its topological properties (nodes and linker structure, appendices, orientation, interpenetration, and interweaving of the framework) to improve its quick response, as summarized in Table 11.3. MOFs are very suitable for forming a copy of themselves, and so this catenating property can be avoided, and other

TABLE 11.3
Requirements for a Quick Response of a Chemical MOF Sensor

Important Aspect for Chemical Sensor Designing (Sources of Selectivity)	How to Improve This Requirement
Catenation	Sterically hindered strut design Templating MOF assembly under conditions of high dilution Bulky linker appendages added during MOF assembly
Pore dimension	Remotion of the non-structural ligands or replacement of node-coordinated solvent molecules using smaller or larger ligands
Chemically specific interactions of the adsorbate tendency to absorbate MOF corner	Post-synthetic modifications: alteration or addition of functional groups, pore modifying, and distribution at coordinatively unsaturated metal sites at MOFs
Reversive bonding reactions in open metal MOF sites	Selectively sensing structures and functionalities previously investigated
Sorption kinetics (physisorption) and thermodynamics	Increasing MOF aperture sites

characteristics of the material structure must be considered for the correct functioning of the sensor (Kreno et al. 2011).

11.6.4.3 Gas Sensors

One of the most critical environmental problems is gaseous pollutants such as nitrogen oxide, sulfur, and many other toxic gases that represent a threat to human life. Measuring the concentration of these pollutants is imperative for human health, and, consequently, gas sensors provide us with an essential alternative for this problem (Adhikari and Majumdar 2004). Phthalocyanine sensors exhibit resistivity to different types of gases. Many conductive polymers can be synthesized from phthalocyanine precursors (Pc). For example, PbPc and ZnPc, as doped polymeric conductors, have been proved to have greater sensitivity than those assembled with transition metals for the sensing of NO_2, ammonia, and H_2S. Besides, they possess high sensitivity to highly flammable gases like H_2, CO, and CH_4 at room temperature. However, for gases such as CL_2, F_2, and BF_3 which have low conductance and a prolonged recovery time, high temperatures (170 °C) are required for optimal performance. More research is needed to establish materials that work optimally for gas sensors (Miasik, Hooper and Tofield 1986).

New sensors based on high-performance conjugated polymers are currently under development, which involves improving the compound's stability until these sensors can manufacture smart microelectronic devices. For this, structural tailoring is studied in detail. A very recent example is DPP-based polymer semiconductors. A modification of the polymer backbone was performed to improve its sensitivity to gases and their electronic stability in the air (Mukhopadhyaya et al. 2020). Today's gas sensors are susceptible to the presence of fuels and other toxic gases in the air, but not to all of them. The need for sensors with a more excellent detection range is essential, and as a result, a new material with significant sensitivity to low concentrations at room temperature is required.

11.6.5 Organic Photovoltaics

Recently, polymer-based photovoltaics are under study due to their beneficial and exciting characteristics. These low-cost photovoltaic devices are produced through the chemical manipulation of organic polymers and polymer-based materials. The materials used to produce organic solar cells contain a delocalized pi-electron system that can absorb sunlight photons, create photogenerated charge carriers, and transport them. To introduce extrinsic charge carriers into organic materials, it is necessary to have a doping step. Organic donor-acceptor bilayer devices work similarly to classical p-n junctions. This means that the electron transfer occurs through the junction from the positive charge donor layer (p-type) to the negative charge acceptor layer (n-type) (Wöhrle and Meissner 1991; Lane et al. 2000; Clarke and Durrant 2010).

Organic thin films through vacuum evaporation and solution processing techniques are pre-prepared to produce organic solar cells. Donor and acceptor components are dissolved in a standard or mixture solvent, creating blends. These blends could be deposited using printing or coating techniques such as inkjet printing or spin-coating. Some soluble monomers need a post-deposition polymerization to achieve a final insoluble thin-film semiconducting form (Trivedi and Nalwa 1997). Thin films are deposition by vacuum evaporation and reduce contaminants like water and oxygen (Hoppe et al. 2004). Donor-acceptor networks are created or induced through molecular doping or evaporation techniques. Organic solar cells are located between two contact layers (electrodes) (Hiramoto, Fujiwara and Yokoyama 1991; Geens et al. 2002; Günes, Neugebauer and Sariciftci 2007).

Organic solar cells are capable of performing the conversion of light into electricity in the following order: the incident solar light in the form of a photon is absorbed by the solar cell. The photon provokes the formation of an excited state, bonding the electron with a hole pair, forming an exciton. The produced exciton diffuses to a region of the solar cell to be later dissociated. In the dissociation step, the charge separation occurs. Finally, this charge is transported through the organic semiconductor to the electrodes (Nunzi 2002).

Organic photovoltaic devices constitute a sustainable, less expensive, lightweight, flexible, and environmentally friendly alternative to inorganic semiconductors. The main concerns about polymer-based photovoltaics are the low power conversion efficiency and short lifetime. The power conversion efficiency of a polymeric photovoltaic is above 8%, meanwhile in the case of inorganic photovoltaics it is 25% (Bounioux, Katz and Yerushalmi-Rozen 2012).

11.6.5.1 Carbon Nanotubes for Organic Photovoltaics

The exploration of new materials integrates organic semiconductors with inorganic nanoparticles such as carbon nanotubes. This interaction generates a hybrid material that enhances the functionality of semiconductors. Carbon nanotubes contain desirable characteristics to be used in a semiconductor, such as natural abundance, structural and chemical stability, and high and controllable aspect ratios. The dimensions and diameters of carbon nanotubes can be controlled and defined during synthesis and purification processes. The work function of carbon nanotubes is determined through the energy appropriate to take one electron from the Fermi to the vacuum level outside the surface. This is a crucial characteristic to consider during semiconductor manufacturing (Arranz-Andrés and Blau 2008).

The electron mobility within a semiconducting carbon nanotube can exceed 10^5 cm^2 V^{-1} s^{-1} at room temperature; this characteristic makes carbon nanotubes of potential use in electronic devices (Arnold et al. 2009). A single cylinder of graphene forms single-wall carbon nanotubes (SWCNTs) with a diameter around 0.4–3 nm with a work function of approximately 4.8–4.9 eV. Meanwhile, multi-wall carbon nanotubes (MWCNTs) are formed by several SWCNTs of different radii, arranged with a typical diameter between 1.4 and 100 nm. MWCNTs have a work function of around 4.3–4.4 eV (Ago et al. 1999; Bounioux, Katz and Yerushalmi-Rozen 2012).

Since the discovery of photoinduced charge transfer between organic conjugated polymers and carbon nanotubes, significant investigation is underway to combine these materials into an efficient photovoltaic organic cell that surpasses inorganic cell features (Kymakis, Alexandrou, and Amaratunga 2003; Hatton et al. 2007). To obtain an efficient charge separation between a conjugated polymer (donor material) and carbon nanotube (acceptor material) it is necessary to consider the LUMO and HOMO. The LUMO of the donor should be located above the LUMO acceptor; simultaneously, the HOMO level of the acceptor should be located below the HOMO of the donor (Ago et al. 1999; Arranz-Andrés and Blau 2008).

Ferguson et al. (2010) assert that the excited state of poly(3-hexylthiophene) (P3HT) is quenched in the presence of SWCNTs. To reach this conclusion, they performed two spectroscopy techniques: steady-state photoluminescence and transient microwave conductivity. Investigations observed long-lived charge carriers generated in P3HT and short-lived carriers confined in carbon nanotubes. The results suggest that P3HT:SWCNT composites can be optimized for their possible use as an organic photovoltaic active layer with good photoconversion efficiency.

11.7 CONCLUSION

Semiconductors are given by π-bonds, which are derived from bond and anti-bond energy levels. The principal characteristics of semiconductor polymers are based on their electronic properties which depend on charge transport. Two types of materials are defined based on charge transport. First, n-type materials are related to charge mobility through electrons, and second p-type semiconductors are associated with hole transporting. Most of the properties are correlated to polymer structure, morphology, transport, and other weak interactions related to inter- and intra-chain interactions. These polymers are significantly present during the manufacturing of different electronic devices, requiring some mechanical characteristics, such as good flexibility, stretchability, and processability. Nowadays, several methods of synthesizing polymers can be found. However, the best method depends on the goal to be achieved. Building block selection allows the focusing of the synthesis of the semiconductor polymer by selecting the p-type or n-type unit. The backbone halogenation enhances the efficiency of n-type-based semiconducting polymers. Better properties can be obtained, such as the solubility of polymers used in side-chain engineering. Moreover, a good way to control the HOMO and LUMO energetic levels is by working with random copolymerization. In addition, this allows us to increase or improve other polymer properties such as solubility and crystallinity. Currently, a variety of characterization methods are employed to understand the characteristics of the surface, interface, thin film, and multilayers of polymers that are compatible with organic semiconductors. Semiconducting polymer materials can provide a variety of applications in several devices. Some of these applications are transistors, sensors, solar cells, and conductive polymers, such as hybrid materials and carbon nanotubes. Polymers used as semiconductors are environmentally friendly alternatives to inorganic semiconductors.

REFERENCES

Adhikari, Basudam and Majumdar, Sarmishtha. 2004. "Polymers in Sensor Applications". *Progress in Polymer Science* 29 (7): 699–766. doi:10.1016/j.progpolymsci.2004.03.002.

Ago, Hiroki et al. 1999. "Work Functions and Surface Functional Groups of Multiwall Carbon Nanotubes". *The Journal of Physical Chemistry B* 103 (38): 8116–8121. doi:10.1021/jp991659y.

Alegria, Elisabete et al. 2018. "Effect of Phenolic Compounds on the Synthesis of Gold Nanoparticles and its Catalytic Activity in the Reduction of Nitro Compounds". *Nanomaterials* 8 (5): 6–10. MDPI AG. doi:10.3390/nano8050320.

Ananikov, Valentine P. 2019. "Organic–Inorganic Hybrid Nanomaterials". *Nanomaterials* 9 (9): 1197. MDPI AG. doi:10.3390/nano9091197.

Arias, Ana Claudia, Endicott, Fred and Street, Robert A. 2006. "Surface-Induced Self-Encapsulation of Polymer Thin-Film Transistors". *Advanced Materials* 18 (21): 2900–2904. doi:10.1002/adma.200600623.

Arkhipov, Vladimir et al. 2003. "Charge carrier mobility in doped semiconducting polymers". *Applied Physics Letters* 82 (19): 3245–3247. doi:10.1063/1.1572965.

Arnold, Michael S. et al. 2009. "Broad Spectral Response Using Carbon Nanotube/Organic Semiconductor/C60 Photodetectors". *Nano Letters* 9 (9): 3354–3358. doi:10.1021/nl901637u.

Arranz-Andrés, Javier and Blau, Werner J. 2008. "Enhanced Device Performance Using Different Carbon Nanotube Types in Polymer Photovoltaic Devices". *Carbon* 46 (15): 2067–2075. Elsevier BV. doi:10.1016/j.carbon.2008.08.027.

Baeg, Kang-Jun, Caironi, Mario and Noh, Yong-Young. 2013. "Toward Printed Integrated Circuits Based on Unipolar or Ambipolar Polymer Semiconductors". *Advanced Materials* 25 (31): 4210–4244. doi:10.1002/adma.201205361.

Barbara, Paul F., Adams, D. M. and O'Connor, Donald B. 1999. "Characterization of Organic Thin Film Materials with Near-Field Scanning Optical Microscopy (NSOM)". *Annual Review of Materials Science* 29 (1): 433–469. doi:10.1146/annurev.matsci.29.1.433.

Baughman, Ray H., Zakhidov, Anvar A. and De Heer, Walt A. 2002. "Carbon Nanotubes--the Route Toward Applications". *Science* 297 (5582): 787–792. doi:10.1126/science.1060928.

Belabed, Chemseddin et al. 2018. "Optical and Dielectric Properties for the Determination of Gap States of the Polymer Semiconductor: Application to Photodegradation of Organic Pollutants". *Optik* 160: 218–226. doi:10.1016/j.ijleo.2018.01.109.

Bescond, Mark. 2015. "Quantum Transport in Semiconductor Nanowires". *Semiconductor Nanowires* 173–202. doi:10.1016/b978-1-78242-253-2.00006-2.

Bouabdallah, Daho et al. 2019. "Synthesis and Characterization of Semiconductor Polymer Doped with $FeCl_3$ and I2". *Semiconductors* 53 (12): 1656–1664. doi:10.1134/s1063782619160073.

Boudreault, Pierre-Luc T., Najari, Ahmed and Leclerc, Mario. 2011. "Processable Low-Bandgap Polymers for Photovoltaic Applications†". *Chemistry of Materials* 23 (3): 456–469. doi:10.1021/cm1021855.

Bounioux, Céline, Katz, Eugene A. and Yerushalmi-Rozen, Rachel. 2012. "Conjugated Polymers - Carbon Nanotubes-Based Functional Materials for Organic Photovoltaics: A Critical Review". *Polymers for Advanced Technologies* 23 (8): 1129–1140. doi:10.1002/pat.3054.

Bura, Thomas, Blaskovits, J. Terence and Leclerc, Mario. 2016. "Direct (Hetero)arylation Polymerization: Trends and Perspectives". *Journal of the American Chemical Society* 138 (32): 10056–10071. doi:10.1021/jacs.6b06237.

Bürgi, Lukas et al. 2008. "High-Mobility Ambipolar Near-Infrared Light-Emitting Polymer Field-Effect Transistors". *Advanced Materials* 20 (11): 2217–2224. doi:10.1002/adma.200702775.

Caminade, Anne-Marie, Laurent, Régis and Majoral, Jean-Pierre. 2005. "Characterization of Dendrimers". *Advanced Drug Delivery Reviews* 57 (15): 2130–2146. doi:10.1016/j.addr.2005.09.011.

Carsten, Bridget et al. 2011. "Stille Polycondensation for Synthesis of Functional Materials". *Chemical Reviews* 111 (3): 1493–1528. doi:10.1021/cr100320w.

Chen, Junwu and Cao, Yong. 2009. "Development of Novel Conjugated Donor Polymers for High-Efficiency Bulk-Heterojunction Photovoltaic Devices". *Accounts of Chemical Research* 42 (11): 1709–1718.. doi:10.1021/ar900061z.

Chen, Yagang et al. 2012. "Novel Conjugated Polymers Based on Dithieno[3,2-b:6,7-b]carbazole for Solution Processed Thin-Film Transistors". *Macromolecular Rapid Communications* 33 (20): 1759–1764. doi:10.1002/marc.201200330.

Chen, Zhenhui et al. 2014. "Low Band-Gap Conjugated Polymers with Strong Interchain Aggregation and Very High Hole Mobility Towards Highly Efficient Thick-Film Polymer Solar Cells". *Advanced Materials* 26 (16): 2586–2591. doi:10.1002/adma.201305092.

Cheng, Yen-Ju, Yang, Sheng-Hsiung and Hsu, Chain-Shu. 2009. "Synthesis of Conjugated Polymers for Organic Solar Cell Applications". *Chemical Reviews* 109 (11): 5868–5923. doi:10.1021/cr900182s.

Chiu, Jiann-Jong. et al. 2003. "Organic Semiconductor Nanowires for Field Emission". *Advanced Materials* 15 (16): 1361–1364. doi:10.1002/adma.200304918.

Cho, Jeong Ho et al. 2008. "Printable Ion-Gel Gate Dielectrics for Low-Voltage Polymer Thin-Film Transistors on Plastic". *Nature Materials* 7 (11): 900–906. doi:10.1038/nmat2291.

Choi, Joonhyeong et al. 2019. "Influence of Acceptor Type and Polymer Molecular Weight on the Mechanical Properties of Polymer Solar Cells". *Chemistry of Materials* 31 (21): 9057–9069. doi:10.1021/acs.chemmater.9b03333.

Chu, Ta-Ya et al. 2011. "Morphology Control in Polycarbazole Based Bulk Heterojunction Solar Cells and Its Impact on Device Performance". *Applied Physics Letters* 98 (25): 253301. doi:10.1063/1.3601474.

Chujo, Yoshiki. 1996. "Organic-Inorgaic Hybrid Materials." *Current Opinion in Solid State and Materials Science* 1 (6): 806–811. doi:10.1016/S1359-0286(96)80105-7.

Cik, Rohaida Che, Foo, Choo Thye and Nor, Azillah Fatimah. (2015). Field Emission Scanning Electron Microscope (FESEM) Facility in BTI. NTC 2015: Nuclear Technical Convention 2015, Malaysia

Clarke, Tracey M. and Durrant, James R. 2010. "Charge Photogeneration in Organic Solar Cells". *Chemical Reviews* 110 (11): 6736–6767. doi:10.1021/cr900271s.

Cobet, Christoph et al. 2018. "Ellipsometric Spectro-electrochemistry: An in Situ Insight in the Doping of Conjugated Polymers". *The Journal of Physical Chemistry C* 122 (42): 24309–24320. doi:10.1021/acs.jpcc.8b08602.

Cobo, Isidro et al. 2015. "Smart Hybrid Materials by Conjugation of Responsive Polymers to Biomacromolecules". *Nature Materials* 14 (2): 143–159. doi:10.1038/nmat4106.

Cui, Weibin, Yuen, Jonathan and Wudl, Fred. 2011. "Benzodipyrrolidones and their Polymers". *Macromolecules* 44 (20): 7869–7873. doi:10.1021/ma2017293.

Dao, Toan Thanh, Matsushima, Toshinori and Murata, Hideyuki. 2012. "Organic Nonvolatile Memory Transistors Based on Fullerene and an Electron-Trapping Polymer". *Organic Electronics* 13 (11): 2709–2715. doi:10.1016/j.orgel.2012.07.041.

De Boer, Bert and Facchetti, Antonio. 2008. "Semiconducting Polymeric Materials". *Polymer Reviews* 48 (3): 423–431. doi:10.1080/15583720802231718.

De Hoffmann, Edmond. 2005. "Mass Spectrometry". *Kirk-Othmer Encyclopedia of Chemical Technology*. doi:10.1002/0471238961.1301191913151518.a01.pub2.

De Laurentis, Martina and Irace, Andrea. 2014. "Optical Measurement Techniques of Recombination Lifetime Based on the Free Carriers Absorption Effect". *Journal of Solid State Physics* 2014: 1–19. doi:10.1155/2014/291469.

DeLongchamp, Dean M. and Kline, R. Joseph. 2012. "Characterization of Order and Orientation in Semiconducting Polymers". *Organic Electronics II*: 27–66. doi:10.1002/9783527640218.ch2.

Deng, Yunfeng et al. 2012. "Donor–Acceptor Conjugated Polymers with Dithienocarbazoles as Donor Units: Effect of Structure on Semiconducting Properties". *Macromolecules* 45 (21): 8621–8627. doi:10.1021/ma301864f.

Deng, Lin et al. 2019. "Hybrid nanocomposites for imaging-guided synergistic theranostics". *Nanomaterials For Drug Delivery And Therapy*, 117–147. doi:10.1016/b978-0-12-816505-8.00017-5

Deschler, Felix et al. 2015. "Imaging of Morphological Changes and Phase Segregation in Doped Polymeric Semiconductors". *Synthetic Metals* 199: 381–387. doi:10.1016/j.synthmet.2014.11.037.

Dimitrakopoulos, Christos and Malenfant, Patrick. 2002. "Organic Thin Film Transistors for Large Area Electronics". *Advanced Materials* 14 (2): 99–117. doi:10.1002/1521-4095(20020116)14:2<99::aid-adma99>3.0.co;2-9.

Ebisawa, Fumihiro, Kurokawa, Takeshi and Nara, Shigetoshi. 1983. "Electrical Properties of Polyacetylene/Polysiloxane Interface". *Journal of Applied Physics* 54 (6): 3255–3259. doi:10.1063/1.332488.

Eremin, Dmitry B. and Ananikov, Valentine P. 2017. "Understanding Active Species in Catalytic Transformations: From Molecular Catalysis to Nanoparticles, Leaching, "Cocktails" of Catalysts and Dynamic Systems". *Coordination Chemistry Reviews* 346: 2–19. doi:10.1016/j.ccr.2016.12.021.

Facchetti, Antonio. 2007. "Semiconductors for Organic Transistors". *Materials Today* 10 (3): 28–37. doi:10.1016/s1369-7021(07)70017-2.

Facchetti, Antonio. 2011. "π-Conjugated Polymers for Organic Electronics and Photovoltaic Cell Applications†". *Chemistry of Materials* 23 (3): 733–758. doi:10.1021/cm102419z.

Falke, Sven and Betzel, Christian. 2019. "Dynamic Light Scattering (DLS)". *Radiation in Bioanalysis*: 173–193. doi:10.1007/978-3-030-28247-9_6.

Fei, Zhuping et al. 2015. "Influence of Backbone Fluorination in Regioregular Poly(3-alkyl-4-fluoro)thiophenes". *Journal of the American Chemical Society* 137 (21): 6866–6879. doi:10.1021/jacs.5b02785.

Feng, Xin, Zhang, Huihua and Yu, Peiqiang. 2020. "X-ray Fluorescence Application in Food, Feed, and Agricultural Science: A Critical Review". *Critical Reviews in Food Science and Nutrition*: 1–11. doi:10.1080/10408398.2020.1776677.

Ferguson, Andrew J. et al. 2010. "Photoinduced Energy and Charge Transfer in P3HT:SWNT Composites". *The Journal of Physical Chemistry Letters* 1 (15): 2406–2411. doi:10.1021/jz100768f.

Frisbie, C. Daniel. 2003. "Scanning Probe Microscopy". *Encyclopedia of Physical Science and Technology*: 469–484. doi:10.1016/b0-12-227410-5/00675-x.

Gali, Sai Manoj et al. 2017. "Energetic Fluctuations in Amorphous Semiconducting Polymers: Impact on Charge-Carrier Mobility". *The Journal of Chemical Physics* 147 (13): 134904. doi:10.1063/1.4996969.

Geens, W et al. 2002. "Organic Co-Evaporated Films of a PPV-Pentamer and C60: Model Systems for Donor/Acceptor Polymer Blends". *Thin Solid Films* 403–404: 438–443. doi:10.1016/s0040-6090(01)01585-1.

Giusto, Paolo et al. 2020. "Shine Bright Like a Diamond: New Light on an Old Polymeric Semiconductor". *Advanced Materials* 32 (10): 1908140. doi:10.1002/adma.201908140.

Glowacki, Ireneusz et al. 2012. "Conductivity Measurements". *Polymer Science: A Comprehensive Reference*: 847–877. doi:10.1016/b978-0-444-53349-4.00058-3.

Gmoshinskii, Ivan V. et al. 2013. "Nanomaterials and Nanotechnologies: Methods of Analysis and Control". *Russian Chemical Reviews* 82 (1): 48–76. doi:10.1070/rc2013v082n01abeh004329.

Gobbi, Marco, Orgiu, Emanuele and Samorì, Paolo. 2018. "When 2D Materials Meet Molecules: Opportunities and Challenges of Hybrid Organic/Inorganic van der Waals Heterostructures". *Advanced Materials* 30 (18): 1706103. doi:10.1002/adma.201706103.

Gómez-Romero, Pedro. 2001. "Hybrid Organic-Inorganic Materials—In Search of Synergic Activity". *Advanced Materials* 13 (3): 163–174. doi:10.1002/1521-4095(200102)13:3<163::aid-adma163>3.0.co;2-u.

Grozema, Ferdinand C. and Siebbeles, Laurens D. A. 2011. "Charge Mobilities in Conjugated Polymers Measured by Pulse Radiolysis Time-Resolved Microwave Conductivity: From Single Chains to Solids". *The Journal of Physical Chemistry Letters* 2 (23): 2951–2958. doi:10.1021/jz201229a.

Günes, Serap, Neugebauer, Helmut and Sariciftci, Niyazi Serdar. 2007. "Conjugated Polymer-Based Organic Solar Cells". *Chemical Reviews* 107 (4): 1324–1338. doi:10.1021/cr050149z.

Hatton, Ross A. et al. 2007. "A Multi-Wall Carbon Nanotube–Molecular Semiconductor Composite for Bi-Layer Organic Solar Cells". *Physica E: Low-dimensional Systems and Nanostructures* 37 (1–2): 124–127. doi:10.1016/j.physe.2006.07.001.

He, Yinghui, Hong, Wei and Li, Yuning. 2014. "New Building Blocks for π-Conjugated Polymer Semiconductors for Organic Thin Film Transistors and Photovoltaics". *Journal of Material Chemistry*. 2 (41): 8651–8661. doi:10.1039/c4tc01201a.

Heeger, Alan J. 2010. "Semiconducting Polymers: The Third Generation". *Chemical Society Reviews* 39 (7): 2354. doi:10.1039/b914956m.

Hendriks, Koen H. et al. 2014. "Small-Bandgap Semiconducting Polymers with High Near-Infrared Photoresponse". *Journal of the American Chemical Society* 136 (34): 12130–12136. doi:10.1021/ja506265h.

Hirai, Hisako. 2003. "Electron Energy-Loss Spectroscopy and Its Applications to Characterization of Carbon Materials". *Carbon Alloys*: 239–256. doi:10.1016/b978-008044163-4/50015-2.

Hiramoto, Masahiro, Fujiwara, Hiroshi and Yokoyama, Masaaki. 1991. "Three-Layered Organic Solar Cell with a Photoactive Interlayer of Codeposited Pigments". *Applied Physics Letters* 58 (10): 1062–1064. doi:10.1063/1.104423.

Holliday, Sarah, Donaghey, Jenny E. and McCulloch, Iain. 2013. "Advances in Charge Carrier Mobilities of Semiconducting Polymers Used in Organic Transistors". *Chemistry of Materials* 26 (1): 647–663. doi:10.1021/cm402421p.

Homyak, Patrick et al. 2015. "Effect of Pendant Functionality in Thieno[3,4-b]thiophene-alt-benzodithiophene Polymers for OPVs". *Chemistry of Materials* 27 (2): 443–449. doi:10.1021/cm503334h.

Hong, Wei et al. 2012. "Synthesis and Thin-Film Transistor Performance of Benzodipyrrolinone and Bithiophene Donor-Acceptor Copolymers". *Journal of Materials Chemistry* 22 (41): 22282. doi:10.1039/c2jm34867e.

Hong, Wei et al. 2013. "Dipyrrolo[2,3-b:2′,3′-e]pyrazine-2,6(1H,5H)-dione Based Conjugated Polymers for Ambipolar Organic Thin-Film Transistors". *Chemical Communication* 49 (5): 484–486. doi:10.1039/c2cc37266e.

Hoppe, Harald et al. 2004. "Modeling of Optical Absorption in Conjugated Polymer/Fullerene Bulk-Heterojunction Plastic Solar Cells". *Thin Solid Films* 451–452: 589–592. doi:10.1016/j.tsf.2003.11.173.

Hoppe, Harald and Sariciftci, Serdar. 2006. "Nanostructure and Nanomorphology Engineering in Polymer Solar Cells". *Nanostructured Materials for Solar Energy Conversion*: 277–318. doi:10.1016/b978-044452844-5/50011-1.

Hundt, Nadia et al. 2009. "Polymers Containing Rigid Benzodithiophene Repeating Unit with Extended Electron Delocalization". *Organic Letters* 11 (19): 4422–4425. doi:10.1021/ol901786z.

Hwang, Ee Taek and Gu, Man Bock. 2012. "Enzyme Stabilization by Nano/Microsized Hybrid Materials". *Engineering in Life Sciences* 13 (1): 49–61. doi:10.1002/elsc.201100225.

Ikawa, Mitsuhiro et al. 2012. "Simple Push Coating of Polymer Thin-Film Transistors". *Nature Communications* 3 (1): 1–8. doi:10.1038/ncomms2190.

Inkson, Beverley J. 2016. "Scanning Electron Microscopy (SEM) and Transmission Electron Microscopy (TEM) for Materials Characterization". *Materials Characterization Using Nondestructive Evaluation (NDE) Methods*: 17–43. doi:10.1016/b978-0-08-100040-3.00002-x.

Jaiswal, Manu and Menon, Reghu. 2006. "Polymer Electronic Materials: A Review of Charge Transport". *Polymer International* 55 (12): 1371–1384. doi:10.1002/pi.2111.

Kaake, Loren, Barbara, Paul and Zhu, Xiaoyang. 2010. "Intrinsic Charge Trapping in Organic and Polymeric Semiconductors: A Physical Chemistry Perspective". *The Journal of Physical Chemistry Letters* 1 (3): 628–635. doi:10.1021/jz9002857.

Kajzar, Francois and Sujan, Ghosh Kumer. 2016. "Organic Conductors and Semiconductors, Optical Properties of". *Reference Module in Materials Science and Materials Engineering*. doi:10.1016/b978-0-12-803581-8.02420-6.

Kang, Boseok et al. 2016. "Side-Chain-Induced Rigid Backbone Organization of Polymer Semiconductors through Semifluoroalkyl Side Chains". *Journal of the American Chemical Society* 138 (11): 3679–3686. doi:10.1021/jacs.5b10445.

Kang, Il et al. 2013. "Record High Hole Mobility in Polymer Semiconductors via Side-Chain Engineering". *Journal of the American Chemical Society* 135 (40): 14896–14899. doi:10.1021/ja405112s.

Kawase, Takeo et al. 2003. "Inkjet printing of polymer thin film transistors". *Thin Solid Films* 438–439: 279–287. doi:10.1016/s0040-6090(03)00801-0

Khan, Aurangzeb and Masafumi, Yamaguchi. 2015. "Deep Level Transient Spectroscopy: A Powerful Experimental Technique for Understanding the Physics and Engineering of Photo-Carrier Generation, Escape, Loss and Collection Processes in Photovoltaic Materials". *Solar Cells-New Approaches and Reviews*. doi:10.5772/59419.

Kim, Ki-Hyun et al. 2014. "Determining Optimal Crystallinity of Diketopyrrolopyrrole-Based Terpolymers for Highly Efficient Polymer Solar Cells and Transistors". *Chemistry of Materials* 26 (24): 6963–6970. doi:10.1021/cm502991d.

Kim, Min Je et al. 2017. "Structure–Property Relationships of Semiconducting Polymers for Flexible and Durable Polymer Field-Effect Transistors". *ACS Applied Materials & Interfaces* 9 (46): 40503–40515. doi:10.1021/acsami.7b12435.

Kini, Gururaj P., Jeon, Sung Jae and Moon, Doo Kyung. 2020. "Design Principles and Synergistic Effects of Chlorination on a Conjugated Backbone for Efficient Organic Photovoltaics: A Critical Review". *Advanced Materials* 32 (11): 1906175. doi:10.1002/adma.201906175.

Kolesov, Vladimir A. et al. 2016. "Solution-Based Electrical Doping of Semiconducting Polymer Films Over a Limited Depth". *Nature Materials* 16 (4): 474–480. doi:10.1038/nmat4818.

Koppens, F. H. L. et al. 2014. "Photodetectors Based on Graphene, Other Two-Dimensional Materials and Hybrid Systems". *Nature Nanotechnology* 9 (10): 780–793. doi:10.1038/nnano.2014.215.

Kotadiya, Naresh B., Blom, Paul W. M. and Wetzelaer, Gert-Jan A. H. 2019. "Efficient and Stable Single-Layer Organic Light-Emitting Diodes Based on Thermally Activated Delayed Fluorescence". *Nature Photonics* 13 (11): 765–769. doi:10.1038/s41566-019-0488-1.

Kreno, Lauren E. et al. 2011. "Metal–Organic Framework Materials as Chemical Sensors". *Chemical Reviews* 112 (2): 1105–1125. doi:10.1021/cr200324t.

Kymakis, E., Alexandrou, I. and Amaratunga, G. A. J. 2003. "High Open-Circuit Voltage Photovoltaic Devices from Carbon-Nanotube-Polymer Composites". *Journal of Applied Physics* 93 (3): 1764–1768. doi:10.1063/1.1535231.

Lan, Yi-Kang, Yang, Cheng Han and Yang, Hsiao-Ching. 2009. "Theoretical Investigations of Electronic Structure and Charge Transport Properties in Polythiophene-Based Organic Field-Effect Transistors". *Polymer International* 59 (1): 16–21. doi:10.1002/pi.2683.

Lane, P et al. 2000. "Electroabsorption Studies of Phthalocyanine/Perylene Solar Cells". *Solar Energy Materials and Solar Cells* 63 (1): 3–13. doi:10.1016/s0927-0248(00)00013-1.

Leclerc, Nicolas et al. 2016. "Impact of Backbone Fluorination on π-Conjugated Polymers in Organic Photovoltaic Devices: A Review". *Polymers* 8 (1): 11. doi:10.3390/polym8010011.

Lee, Jiyoul et al. 2007. "Ion Gel Gated Polymer Thin-Film Transistors". *Journal of the American Chemical Society* 129 (15): 4532–4533. doi:10.1021/ja070875e.

Lei, Ting et al. 2017. "Biocompatible and Totally Disintegrable Semiconducting Polymer for Ultrathin and Ultralightweight Transient Electronics". *Proceedings of the National Academy of Sciences* 114 (20): 5107–5112. doi:10.1073/pnas.1701478114.

Lei, Ting, Dou, Jin-Hu and Pei, Jian. 2012. "Influence of Alkyl Chain Branching Positions on the Hole Mobilities of Polymer Thin-Film Transistors". *Advanced Materials* 24 (48): 6457–6461. doi:10.1002/adma.201202689.

Li, Jun et al. 2012. "A Stable Solution-Processed Polymer Semiconductor with Record High-Mobility for Printed Transistors". *Scientific Reports* 2 (1). doi:10.1038/srep00754.

Li, Yuning et al. 2013. "High Mobility Diketopyrrolopyrrole (DPP)-Based Organic Semiconductor Materials for Organic Thin Film Transistors and Photovoltaics". *Energy & Environmental Science* 6 (6): 1684. doi:10.1039/c3ee00015j.

Lin, Ping-Chang et al. 2014. "Techniques for Physicochemical Characterization of Nanomaterials". *Biotechnology Advances* 32 (4): 711–726. doi:10.1016/j.biotechadv.2013.11.006.

Liu, Chang et al. 2016. "Low Bandgap Semiconducting Polymers for Polymeric Photovoltaics". *Chemical Society Reviews* 45 (17): 4825–4846. doi:10.1039/c5cs00650c.

Liu, Yuhang et al. 2014. "Aggregation and Morphology Control Enables Multiple Cases of High-Efficiency Polymer Solar Cells". *Nature Communications* 5 (1): 1–8. doi:10.1038/ncomms6293.

Lu, Shaofeng et al. 2013. "3,6-Dithiophen-2-yl-diketopyrrolo[3,2-b]pyrrole (isoDPPT) as an Acceptor Building Block for Organic Opto-Electronics". *Macromolecules* 46 (10): 3895–3906. doi:10.1021/ma400568b.

Mashkovsky, Igor et al. 2018. "Highly-Ordered PdIn Intermetallic Nanostructures Obtained from Heterobimetallic Acetate Complex: Formation and Catalytic Properties in Diphenylacetylene Hydrogenation". *Nanomaterials* 8 (10): 4–8. doi:10.3390/nano8100769.

Mayeen, Anshida et al. 2018. "Morphological Characterization of Nanomaterials". *Characterization of Nanomaterials*: 335–364. doi:10.1016/b978-0-08-101973-3.00012-2.

McBride, Michael et al. 2018. "Process-Structure-Property Relationships for Design of Polymer Organic Electronics Manufacturing". *13th International Symposium on Process Systems Engineering (PSE 2018)*: 2467–2472. doi:10.1016/b9780444-64241-7.50406-7.

McCulloch, Iain et al. 2006. "Liquid-Crystalline Semiconducting Polymers with high Charge-Carrier Mobility". *Nature Materials* 5 (4): 328–333. doi:10.1038/nmat1612.

Medved'ko, Alexey et al. 2019. "Supramolecular Organogels Based on N-Benzyl, N′-Acylbispidinols". *Nanomaterials* 9 (1): 1–17. doi:10.3390/nano9010089.

Mei, Jianguo et al. 2011. "Siloxane-Terminated Solubilizing Side Chains: Bringing Conjugated Polymer Backbones Closer and Boosting Hole Mobilities in Thin-Film Transistors". *Journal of the American Chemical Society* 133 (50): 20130–20133. doi:10.1021/ja209328m.

Miasik, Jan J., Hooper, Alan and Tofield, Bruce C. 1986. "Conducting Polymer Gas Sensors". *Journal of the Chemical Society, Faraday Transactions 1: Physical Chemistry in Condensed Phases* 82 (4): 1117–1126. doi:10.1039/f19868201117

Mikie, Tsubasa and Osaka, Itaru. 2019. "Ester-Functionalized Naphthobispyrazine as an Acceptor Building Unit for Semiconducting Polymers: Synthesis, Properties, and Photovoltaic Performance". *Macromolecules* 52 (10): 3909–3917. doi:10.1021/acs.macromol.9b00521

Mukhopadhyaya, Tushita et al. 2020. "Design and Synthesis of Air-Stable p-Channel-Conjugated Polymers for High Signal-to-Drift Nitrogen Dioxide and Ammonia Sensing". *ACS Applied Materials & Interfaces* 12 (19): 21974–21984. doi:10.1021/acsami.0c04810.

Müllen, Klaus and Pisula, Wojciech. 2015. "Donor–Acceptor polymers". *Journal Of The American Chemical Society* 137(30), 9503–9505. doi:10.1021/jacs.5b07015

Mun, Jaewan et al. 2018. "Effect of Nonconjugated Spacers on Mechanical Properties of Semiconducting Polymers for Stretchable Transistors". *Advanced Functional Materials* 28 (43): 1804222. doi:10.1002/adfm.201804222.

Munch, Etienne et al. 2008. "Tough, Bio-Inspired Hybrid Materials". *Science* 322 (5907): 1516–1520. doi:10.1126/science.1164865.

Murari, Nishit M. et al. 2016. "Organic Nonvolatile Memory Devices Utilizing Intrinsic Charge-Trapping Phenomena in An n-Type Polymer Semiconductor". *Organic Electronics* 31: 104–110. doi:10.1016/j.orgel.2016.01.015.

Neupane, Guru Prakash et al. 2019. "2D Organic Semiconductors, the Future of Green Nanotechnology". *Nano Materials Science* 1 (4): 246–259. doi:10.1016/j.nanoms.2019.10.002.

Nicolai, Herman Theunis et al. 2012. "Unification of Trap-Limited Electron Transport in Semiconducting Polymers". *Nature Materials* 11 (10): 882–887. doi:10.1038/nmat3384.

Nicole, Lionel, Rozes, Laurence and Sanchez, Clément. 2010. "Integrative Approaches to Hybrid Multifunctional Materials: From Multidisciplinary Research to Applied Technologies". *Advanced Materials* 22 (29): 3208–3214. doi:10.1002/adma.201000231.

Nielsen, Christian B., Turbiez, Mathieu and McCulloch, Iain. 2012. "Recent Advances in the Development of Semiconducting DPP-Containing Polymers for Transistor Applications". *Advanced Materials* 25 (13): 1859–1880. doi:10.1002/adma.201201795.

Noh, Yong-Young et al. 2007. "Downscaling of Self-Aligned, All-Printed Polymer Thin-Film Transistors". *Nature Nanotechnology* 2 (12): 784–789. doi:10.1038/nnano.2007.365.

Noriega, Rodrigo et al. 2013. "A General Relationship Between Disorder, Aggregation and Charge Transport in Conjugated Polymers". *Nature Materials* 12 (11): 1038–1044. doi:10.1038/nmat3722.

Nunzi, Jean-Michel. 2002. "Organic Photovoltaic Materials and Devices". *Comptes Rendus Physique* 3 (4): 523–542. doi:10.1016/s1631-0705(02)01335-x.

Oh, Jin Young et al. 2016. "Intrinsically Stretchable and Healable Semiconducting Polymer for Organic Transistors". *Nature* 539 (7629): 411–415. doi:10.1038/nature20102.

Panidi, Julianna et al. 2017. "Remarkable Enhancement of the Hole Mobility in Several Organic Small-Molecules, Polymers, and Small-Molecule: Polymer Blend Transistors by Simple Admixing of the Lewis Acid p-Dopant B(C6F5)3". *Advanced Science* 5 (1): 1700290. doi:10.1002/advs.201700290.

Parashkov, Radoslav et al. 2004. "All-Organic Thin-Film Transistors Made of poly(3-butylthiophene) Semiconducting And Various Polymeric Insulating Layers". *Journal of Applied Physics* 95 (3): 1594–1596. doi:10.1063/1.1636524.

Park, Won-Tae et al. 2016. "Effect of Donor Molecular Structure and Gate Dielectric on Charge-Transporting Characteristics for Isoindigo-Based Donor-Acceptor Conjugated Polymers". *Advanced Functional Materials* 26 (26): 4695–4703. doi:10.1002/adfm.201504908.

Paterson, Alexandra F. et al. 2018. "Recent Progress in High-Mobility Organic Transistors: A Reality Check". *Advanced Materials* 30 (36): 1801079. doi:10.1002/adma.201801079.

Pentsak, Evgeniy et al. 2018. "Systematic Study of the Behavior of Different Metal and Metal-Containing Particles under the Microwave Irradiation and Transformation of Nanoscale and Microscale Morphology". *Nanomaterials* 9 (1): 19. doi:10.3390/nano9010019.

Popović, Zoran V. et al. 2010. "Raman Scattering on Nanomaterials and Nanostructures". *Annalen der Physik* 523 (1–2): 62–74. doi:10.1002/andp.201000094.

Pron, Adam et al. 2010. "Electroactive Materials for Organic Electronics: Preparation Strategies, Structural Aspects and Characterization Techniques". *Chemical Society Reviews* 39 (7): 2577. doi:10.1039/b907999h.

Prosa, Mario et al. 2020. "Nanostructured Organic/Hybrid Materials and Components in Miniaturized Optical and Chemical Sensors". *Nanomaterials* 10 (3): 480. doi:10.3390/nano10030480.

Qu, Dunshuai, Qi, Ting and Huang, Hui. 2021. "Acceptor–Acceptor-Type Conjugated Polymer Semiconductors". *Journal of Energy Chemistry* 59: 364–387. doi:10.1016/j.jechem.2020.11.019.

Rao, C N R, Cheetham, A K and Thirumurugan, A. 2008. "Hybrid Inorganic–Organic Materials: A New Family in Condensed Matter Physics". *Journal of Physics: Condensed Matter* 20 (8): 083202. doi:10.1088/0953-8984/20/8/083202.

Ray, Suprakas S. 2013. "Structure and Morphology Characterization Techniques". *Clay-Containing Polymer Nanocomposites*: 39–66. doi:10.1016/b978-0-444-59437-2.00003-x.

Reséndiz, Luis Martín et al. 2010. "Effect of Active Layer Thickness on the Electrical Characteristics of Polymer Thin Film Transistors". *Organic Electronics* 11 (12): 1920–1927. doi:10.1016/j.orgel.2010.09.002.

Roberts, Mark E. et al. 2008. "Water-Stable Organic Transistors and Their Application in Chemical and Biological Sensors". *Proceedings of the National Academy of Sciences* 105 (34): 12134–12139. doi:10.1073/pnas.0802105105.

Rodriquez, Daniel et al. 2017. "Comparison of Methods for Determining the Mechanical Properties of Semiconducting Polymer Films for Stretchable Electronics". *ACS Applied Materials & Interfaces* 9 (10): 8855–8862. doi:10.1021/acsami.6b16115.

Root, Samuel E. et al. 2017. "Mechanical Properties of Organic Semiconductors for Stretchable, Highly Flexible, and Mechanically Robust Electronics". *Chemical Reviews* 117 (9): 6467–6499. doi:10.1021/acs.chemrev.7b00003.

Rumer, Joseph W. et al. 2013. "BPTs: Thiophene-Flanked Benzodipyrrolidone Conjugated Polymers for Ambipolar Organic Transistors". *Chemical Communications* 49 (40): 4465–4467. doi:10.1039/c3cc40811f.

Rumyantseva, Marina et al. 2018. "Photosensitive Organic-Inorganic Hybrid Materials for Room Temperature Gas Sensor Applications". *Nanomaterials* 8 (9): 1–16. doi:10.3390/nano8090671.

Sakamoto, Junji et al. 2009. "Suzuki Polycondensation: Polyarylenes à la Carte". *Macromolecular Rapid Communications* 30 (9–10): 653–687. doi:10.1002/marc.200900063.

Sakamoto, Youichi, Komatsu, Shingo and Suzuki, Toshiyasu. 2010. "ChemInform Abstract: Tetradecafluorosexithiophene: The First Perfluorinated Oligothiophene.". *ChemInform* 32 (37): no–no. doi:10.1002/chin.200137121.

Sapsford, Kim E. et al. 2011. "Analyzing Nanomaterial Bioconjugates: A Review of Current and Emerging Purification and Characterization Techniques". *Analytical Chemistry* 83 (12): 4453–4488. doi:10.1021/ac200853a.

Sariciftci, Niyazi Serdar et al. 1993. "Semiconducting Polymers (as donors) and Buckminsterfullerene (as acceptor): Photo-induced Electron Transfer and Heterojunction Devices". *Synthetic Metals* 59 (3): 333–352. doi:10.1016/0379-6779(93)91166-y.

Schenning, Albert and Meijer, Bert. 2001. "Functional Conjugated Polymers, Molecular Design of: Architecture". *Encyclopedia of Materials: Science and Technology*: 3400–3407. doi:10.1016/b0-08-043152-6/00607-0.

Schubert, Ulrich. 2011. "Cluster-Based Inorganic–Organic Hybrid Materials". *Chemical Society Reviews* 40 (2): 575–582. doi:10.1039/c0cs00009d.

Semnani, Dariush. 2017. "Geometrical Characterization of Electrospun Nanofibers". *Electrospun Nanofibers*: 151–180. doi:10.1016/b978-0-08-100907-9.00007-6.

Shi, Longxian et al. 2017. "Design and Effective Synthesis Methods for High-Performance Polymer Semiconductors in Organic Field-Effect Transistors". *Materials Chemistry Frontiers* 1 (12): 2423–2456. doi:10.1039/c7qm00169j.

Sirringhaus, Henning. 2014. "25th Anniversary Article: Organic Field-Effect Transistors: The Path Beyond Amorphous Silicon". *Advanced Materials* 26 (9): 1319–1335. doi:10.1002/adma.201304346.

Sista, Prakash et al. 2010. "Synthesis and Electronic Properties of Semiconducting Polymers Containing Benzodithiophene with Alkyl Phenylethynyl Substituents". *Macromolecules* 43 (19): 8063–8070. doi:10.1021/ma101709h.

Skompska, Magdalena. 2010. "Hybrid Conjugated Polymer/Semiconductor Photovoltaic Cells". *Synthetic Metals* 160 (1–2): 1–15. doi:10.1016/j.synthmet.2009.10.031.

Son, Hae Jung et al. 2011. "Synthesis of Fluorinated Polythienothiophene-co-benzodithiophenes and Effect of Fluorination on the Photovoltaic Properties". *Journal of the American Chemical Society* 133 (6): 1885–1894. doi:10.1021/ja108601g.

Stoev, Krassimir N. and Sakurai, Kenji. 1999. "Review on Grazing Incidence X-ray Spectrometry and Reflectometry". *Spectrochimica Acta Part B: Atomic Spectroscopy* 54 (1): 41–82. doi:10.1016/s0584-8547(98)00160-8.

Sun, Huiliang et al. 2018. "Imide-Functionalized Polymer Semiconductors". *Chemistry – A European Journal* 25 (1): 87–105. doi:10.1002/chem.201803605.

Sun, Huiliang, Guo, Xugang and Facchetti, Antonio. 2020. "High-Performance n-Type Polymer Semiconductors: Applications, Recent Development, and Challenges". *Chem* 6 (6): 1310–1326. doi:10.1016/j.chempr.2020.05.012.

Tiede, Karen et al. 2008. "Detection and Characterization of Engineered Nanoparticles in Food and the Environment". *Food Additives & Contaminants: Part A* 25 (7): 795–821. doi:10.1080/02652030802007553.

Treat, Neil D. et al. 2013. "Microstructure Formation in Molecular and Polymer Semiconductors Assisted by Nucleation Agents". *Nature Materials* 12 (7): 628–633. doi:10.1038/nmat3655.

Trivedi, D. C. and Nalwa, H. S.. 1997. "Handbook of Organic Conductive Molecules and Polymers." *Wiley, New York* 2 (1997): 505.

Tsao, Hoi Nok et al. 2011. "Ultrahigh Mobility in Polymer Field-Effect Transistors by Design". *Journal of the American Chemical Society* 133 (8): 2605–2612. doi:10.1021/ja108861q.

Van Der Poll, Thomas S. et al. 2012. "Non-Basic High-Performance Molecules for Solution-Processed Organic Solar Cells". *Advanced Materials* 24 (27): 3646–3649. doi:10.1002/adma.201201127.

Venkatachalam, Sridevi. 2016. "Ultraviolet and Visible Spectroscopy Studies of Nanofillers and Their Polymer Nanocomposites". *Spectroscopy of Polymer Nanocomposites*: 130–157. doi:10.1016/b978-0-323-40183-8.00006-9.

Verbeeck, Johann. et al. 2005. "Electron Energy Loss Spectrometry". *Encyclopedia of Analytical Science*: 324–331. doi:10.1016/b0-12-369397-7/00605-1.

Wadsworth, Andrew et al. 2019. "Modification of Indacenodithiophene-Based Polymers and Its Impact on Charge Carrier Mobility in Organic Thin-Film Transistors". *Journal of the American Chemical Society* 142 (2): 652–664. doi:10.1021/jacs.9b09374.

Wang, Feifei et al. 2018a. "Incorporation of Heteroatoms in Conjugated Polymers Backbone toward Air-Stable, High-Performance n-Channel Unencapsulated Polymer Transistors". *Chemistry of Materials* 30 (15): 5451–5459. doi:10.1021/acs.chemmater.8b02359.

Wang, Ging-Ji Nathan et al. 2016. "Inducing Elasticity through Oligo-Siloxane Crosslinks for Intrinsically Stretchable Semiconducting Polymers". *Advanced Functional Materials* 26 (40): 7254–7262. doi:10.1002/adfm.201602603.

Wang, Ging-Ji Nathan et al. 2018b. "Nonhalogenated Solvent Processable and Printable High-Performance Polymer Semiconductor Enabled by Isomeric Nonconjugated Flexible Linkers". *Macromolecules* 51 (13): 4976–4985. doi:10.1021/acs.macromol.8b00971.

Wang, Ging-Ji Nathan, Gasperini, Andrea and Bao, Zhenan. 2018c. "Stretchable Polymer Semiconductors for Plastic Electronics". *Advanced Electronic Materials* 4 (2): 1700429. doi:10.1002/aelm.201700429.

Williams, David B. and Carter, C. Barry. 2009. "Scattering and Diffraction". *Transmission Electron Microscopy*: 23–38. doi:10.1007/978-0-387-76501-3_2.

Wöhrle, Dieter and Meissner, Dieter. 1991. "Organic Solar Cells". *Advanced Materials* 3 (3): 129–138. doi:10.1002/adma.19910030303.

Wu, Hung-Chin et al. 2016. "Isoindigo-Based Semiconducting Polymers Using Carbosilane Side Chains for High Performance Stretchable Field-Effect Transistors". *Macromolecules* 49 (22): 8540–8548. doi:10.1021/acs.macromol.6b02145.

Wu, Jhong-Sian et al. 2012. "Dithienocarbazole-Based Ladder-Type Heptacyclic Arenes with Silicon, Carbon, and Nitrogen Bridges: Synthesis, Molecular Properties, Field-Effect Transistors, and Photovoltaic Applications". *Advanced Functional Materials* 22 (8): 1711–1722. doi:10.1002/adfm.201102906.

Wu, Pei-Tzu, Kim, Felix Sunjoo and Jenekhe, Samson A. 2011. "New Poly(arylene vinylene)s Based on Diketopyrrolopyrrole for Ambipolar Transistors". *Chemistry of Materials* 23 (20): 4618–4624. doi:10.1021/cm202247a.

Xu, Jie et al. 2017. "Highly Stretchable Polymer Semiconductor Films Through the Nanoconfinement Effect". *Science* 355 (6320): 59–64. doi:10.1126/science.aah4496.

Yakuphanoglu, Fahrettin et al. 2006. "Electrical and Optical Properties of an Organic Semiconductor Based on Polyaniline Prepared by Emulsion Polymerization and Fabrication of Ag/Polyaniline/n-Si Schottky Diode". *The Journal of Physical Chemistry B* 110 (34): 16908–16913. doi:10.1021/jp060445v.

Yan, Bing. 2012. "Recent Progress in Photofunctional Lanthanide Hybrid Materials". *RSC Advances* 2 (25): 874–882. doi:10.1039/c2ra20976d.

Yan, He et al. 2009. "A High-Mobility Electron-Transporting Polymer for Printed Transistors". *Nature* 457 (7230): 679–686. doi:10.1038/nature07727.

Yang, Fut et al. 2019. "A Hybrid Material that Reversibly Switches Between Two Stable Solid States". *Nature Materials* 18 (8): 874–882. doi:10.1038/s41563-019-0434-0.

Yao, Jingjing et al. 2015. "Significant Improvement of Semiconducting Performance of the Diketopyrrolopyrrole–Quaterthiophene Conjugated Polymer through Side-Chain Engineering via Hydrogen-Bonding". *Journal of the American Chemical Society* 138 (1): 173–185. doi:10.1021/jacs.5b09737.

Yoon, Gyu et al. 2018. "Synthesis of Iron Oxide/Gold Composite Nanoparticles Using Polyethyleneimine as a Polymeric Active Stabilizer for Development of a Dual Imaging Probe". *Nanomaterials* 8 (5): 300. doi:10.3390/nano8050300.

You, Jingbi et al. 2013. "A Polymer Tandem Solar Cell with 10.6% Power Conversion Efficiency". *Nature Communications* 4 (1). doi:10.1038/ncomms2411.

Yuen, Jonathan D. et al. 2011. "High Performance Weak Donor–Acceptor Polymers in Thin Film Transistors: Effect of the Acceptor on Electronic Properties, Ambipolar Conductivity, Mobility, and Thermal Stability". *Journal of the American Chemical Society* 133 (51): 20799–20807. doi:10.1021/ja205566w.

Zhan, Xiaowei et al. 2010. "Rylene and Related Diimides for Organic Electronics". *Advanced Materials* 23 (2): 268–284. doi:10.1002/adma.201001402.

Zhang, Fengjiao et al. 2013. "Critical Role of Alkyl Chain Branching of Organic Semiconductors in Enabling Solution-Processed N-Channel Organic Thin-Film Transistors with Mobility of up to 3.50 cm 2 V–1 s–1". *Journal of the American Chemical Society* 135 (6): 2338–2349. doi:10.1021/ja311469y.

Zhang, Lei et al. 2014. "Oligothiophene Semiconductors: Synthesis, Characterization, and Applications for Organic Devices". *ACS Applied Materials & Interfaces* 6 (8): 5327–5343. doi:10.1021/am4060468.

Zhang, Shaoqing et al. 2018. "Over 14% Efficiency in Polymer Solar Cells Enabled by a Chlorinated Polymer Donor". *Advanced Materials* 30 (20): 1–7. doi:10.1002/adma.201800868.

Zhang, Yong et al. 2012. "Significant Improved Performance of Photovoltaic Cells Made from a Partially Fluorinated Cyclopentadithiophene/Benzothiadiazole Conjugated Polymer". *Macromolecules* 45 (13): 5427–5435. doi:10.1021/ma3009178.

Zhao, Wenchao et al. 2017. "Molecular Optimization Enables over 13% Efficiency in Organic Solar Cells". *Journal of the American Chemical Society* 139 (21): 7148–7151. doi:10.1021/jacs.7b02677.

Zhao, Xingang and Zhan, Xiaowei. 2011. "Electron Transporting Semiconducting Polymers in Organic Electronics". *Chemical Society Reviews* 40 (7): 3728. doi:10.1039/c0cs00194e.

Zheng, Yu-Qing et al. 2015. "Effect of Halogenation in Isoindigo-Based Polymers on the Phase Separation and Molecular Orientation of Bulk Heterojunction Solar Cells". *Macromolecules* 48 (16): 5570–5577. doi:10.1021/acs.macromol.5b01074.

Zhou, Ye et al. 2013. "Nonvolatile Multilevel Data Storage Memory Device from Controlled Ambipolar Charge Trapping Mechanism". *Scientific Reports* 3 (1). doi:10.1038/srep02319.

12 Polymer Organic Photovoltaics

Sahidul Islam
Trivenidevi Bhalotia College, Raniganj, West Bengal, India
The University of Burdwan, Golapbag, West Bengal, India

Ujjwal Mandal
The University of Burdwan, Golapbag, West Bengal, India

CONTENTS

12.1 INTRODUCTION

Solar energy is one of the most naturally abundant sources of renewable energy. Photosynthesis is the most efficient natural process for converting sunlight into the form of chemical energy (Scholes et al. 2011). In human society electrical energy is one of the easiest transmittable forms. Therefore, effort for the conversion of sunlight into electrical energy is very important. A great development in the conversion efficiency of electromagnetic radiation into electrical energy has been achieved over the last 50 years (Ginley et al. 2008). Organic photovoltaics (OPVs) is a promising field for producing low cost and clean energy, and harvesting photons from sunlight (Su et al. 2012). In OPVs a thin layer (100 nm) of organic photoactive material is sandwiched between two electrodes. The absorption of light in an organic layer generates excitons (a tightly bound electron-hole pair). The dissociation of excitons into electrons and holes (charge carriers) by choosing an appropriate material and collecting charges in electrodes in suitably designed device architecture is required for the generation of electricity (Cusumano et al. 2020). An easy fabrication process of organic semiconductors involves the vacuum sublimation method, printing, and spray-coatings from solutions or by using vapors of the appropriate organic materials (Brabec 2004). Thin films of organic semiconductors absorb light very efficiently (absorption coefficient ~10^5 cm^{-1}). Using a double pass and reflecting metal electrode, a 100 nm thick organic layer can absorb 90% of the incident electromagnetic radiation. Therefore, a small amount of organic material is required for making a photovoltaic device; its architecture must be thin, light weight, easily portable, and easy to install (Dou et al. 2013). The extensive research on OPVs has been done for over two decades and the power conversion efficiency (PCE) achieved till now is 13% (Hedley et al. 2016). The highest occupied molecular orbital (HOMO) of active organic material contains two electrons of opposite spin. Upon irradiation by a photon, one of the electrons goes to the lowest unoccupied molecular orbital (LUMO), conserving the spin multiplicity, thus a singlet exciton is generated (Chen et al. 2017). The binding energy in a typical singlet exciton was found to be between 0.1 and 0.5 eV (Xie and Wu 2020). To split the exciton into electron and hole, a donor–acceptor heterojunction is necessary where the HOMO and LUMO of the acceptors are at lower energy compared to that of the donors (Facchetti 2013). In this situation, in the excited state, the electron transfer from donor to acceptor and hole, and from acceptor to donor, has become energetically feasible. Experimentally the electron transfer

DOI: 10.1201/9781003169727-12

rate (K_{ET}) was found to depend exponentially on the distance (R_{DA}) between electron donor and acceptor:

$$K = K_0 e^{-\beta R_{DA}}$$

Here K_0 represents the electron transfer rate when the donor and acceptor are at close proximity; the β parameter is the attenuation factor; and β depends on the electronic properties and orientations of donor and acceptor and the bridge between them (Albinsson and Martensson 2008). The excitons are transported to the heterojunction by means of Foster resonance energy transfer and diffusion (Feron et al. 2012). Two types of heterojunctions are mainly used in OPVs. In the first kind, the bilayer heterojunction (Lee et al. 2015) is formed when donor and acceptor materials are deposited one on another; and in the second kind a bulk heterojunction is formed when donor and acceptor materials are mixed (Scharber and Sariciftci 2013). The harvesting of generated excitons is limited within the combination of diffusion length and FRET distance form the heterojunction in the bilayer entity. Both the FRET distance and exciton diffusion distance is less than 20 nm (Mikhnenko et al. 2015). Here "light harvesting" means the generation of excitons upon irradiation by sunlight. The exciton generation is largely favored by the roughness of the interface between donor and acceptor materials (Cnops et al. 2014). When polycrystalline films are subjected to vacuum sublimation or donor and acceptor layers are allowed to deposit one by one, roughness in the active layer occurs (Ayzner et al. 2009). With the increase of exciton diffusion, PCE increases. For planar heterojunction, the light harvesting can be increased by using multilayer architecture where the FRET between two adjacent layers of donor and acceptor materials is efficient (Ayzner et al. 2009). Alternatively, light harvesting can also be increased by increasing the interface area of the heterojunctions. The bulk heterojunction nanostructure is formed by using donor and acceptor blend. Blends can be deposited from a solution of the same or different solvents for donor and acceptor or by vacuum evaporation of donor and acceptor materials (Wang et al. 2017). The diffused excitons, after reaching the bulk heterojunction, splits, generating electrons and holes. The nascent electrons and holes must overcome the coulombic force of attraction to become separate charge carrying species. The transportation of holes and electrons in the acceptor and donor phase establish clearly the charge transport pathway in the bulk heterojunction (Casalegno et al. 2017). It has been found that the blend morphology should be appropriate to dissociating the excitons into carrier species (Alam et al. 2013). For the transport of carriers to the respective electrodes, the donor and acceptor must form a continuous and uninterrupted pathway. Despite the continuity in the pathway, the carriers must go through the labyrinth pathway due to the morphological complexity. In the transport pathway, electrons and holes can encounter each other, resulting in the formation of geminate electron hole pairs which may be converted into excitons or again separate into charge carriers. The morphology in the heterojunctions largely affects the non-geminate electron–hole pairing rate (Lakhwani et al. 2014). The progress in PCE depends on the selection of the efficiently light absorbing donor and acceptor organic materials, the charge transport, the active layer morphology, the suitability of the transporting layer, and device architecture (Guerrero and Garcia-Belmonte 2016).

12.2 MATERIALS FOR ORGANIC PHOTOVOLTAICS

The electrical properties of organic materials are different as compared to inorganic materials.

The electrical conduction of organic materials involves the migration of π electrons in conjugated systems. There are several advantages, such as ease of chemical tailoring and modulation of the optical band gap, in organic materials. They are low cost, soft, flexible, and light weight. Hence OPV devices can be inexpensive. Organic polymers having a delocalizable π-electron cloud are used in the construction of OPV devices. Some examples of such types of polymers are *p*-phenylenevinylens (PPVs), polyacetylene (PA), polythiophens (PTs), C_{60}, C_{61}, 3-4-ethylenedioxythiophene (PEDOT), 6-6-phenyl-C_{61}-butyric acid methyl ester (PCBM), poly 3-hexyl thiophene (P3HT), and poly styrene-sulfonate (PSS).

12.3 PROCESSING OF OPV CELLS

For photovoltaic applications, semiconducting organic materials can be easily fabricated on less-expensive substances such as plastic, glass, and metal foils. The molecular nature of organic materials permits a solution-based fabrication process. In many cases the purification of organic materials is done by a sublimation process under multiple rounds of a ground zero gradient before the fabrication of OPVs. The research community uses mainly two film deposition techniques: (i) vacuum thermal evaporation (Kovacik et al. 2010) and (ii) spin coating (Wantana et al. 2017). There are many other techniques for small molecules and polymers. In order to grow the thin films using small organic molecules, vacuum thermal evaporation (VTE) is appropriate. Polymers are not suitable as they decompose below the temperature of their evaporation. The major advantages of the VTE process are the ability to produce high-quality film and incorporate inorganic and organic dielectrics in the same system. The film thickness can be finely controlled and multilayer structures can be formed (Eslamian 2016). For the purpose of doping, VTE is preferably exploited, although in OPV doping organic light emitting diodes (OLEDs) depend extensively on it.

The organic polymers cannot be processed with a VTE technique. The general techniques of the processing of organic materials are involved with methods that are solution based such as spin-coating.

In the spin-coating method, the solution of organic materials is spread over the rotating substrate by centrifugal force and after evaporation of the solvent, there is formation of a thin film having a thickness of 20–300 nm. Using a large variety of solutions, thin films having well defined thickness and uniformity can be achieved. Beside these two main methods of device fabrication, there are some other popular laboratory scale and industrial scale production techniques. One such important technique is inkjet printing. This is a solution-based technique exploited to develop very small (e.g., 5 μm) non-lithographic patterning and high throughput fabrications.

In this method a very reduced amount of solution is wasted. However, the major challenge in this technique is the ink formulation and nozzle alignment which is essential for small patterns (de Gans et al. 2004). Another solution-based technique suitable for a large deposition surface and throughput process is spray deposition. With the help of an atomizer or pumping the liquid through an ultrasonic nozzle, spray is allowed to form. With the inert gas flow, the spray is brought to the surface of the substrate and allowed to deposit. The main disadvantage in the spray deposition technique is the difficulty in forming good quality films with uniform thickness (Vak et al. 2007). Another solution processing method which is comparable with the VTE technique is organic vapor phase deposition (OVPD). The organic material is evaporated and the vapor is carried by an inert gas flow which is then directed to the surface of the substrate for deposition. The deposition rate of the material depends on the flow rate of the inert gas, the temperature, and the pressure, therefore in the OVPD technique controlling of the film thickness is difficult (Shtein et al. 2003). Another relevant technique is organic vapor jet printing (OVJP). This technique involves the spraying of the vapor of organic material through a printing nozzle, then carrying it, by the flow of inert gas, towards the substrate; printing occurs without shadow masking (Sun et al. 2005).

12.4 THE BASIC OPERATIONAL PROCESS OF OPV

In a simple OPV there are four basic steps in the operational pathway: (i) absorption of photons (generating excitons);(ii) diffusion of excitons; (iii) charge transfer (dissociation of excitons); and (iv) collection of charges (Blom et al. 2007). The inorganic photovoltaic materials upon absorption of photons directly produce carriers; contrarily the organic materials produce excitons which are diffused to heterojunctions and not influenced by an externally applied electric field (Lungenschmied et al. 2007). In an effective OPV all the generated excitons migrate to the heterojunctions and dissociate there to produce carriers. The diffusion time of the excitons must be shorter than the life time. The exciton life time (τ_e) is the sum of the reciprocal of all the decay rates. The efficiency of an OPV directly depends on the migration of excitons to the heterojunction within time τ_e. The distance (L_e) crossed by an exciton while diffusing to the heterojunction is given by $L_e = \sqrt{D_e \tau_e}$, where D_e represents the diffusion coefficient of the exciton (Forrest 2015). Charge transfer between donor and acceptor material occurs over a very short distance. The dissociation of excitons at the donor–acceptor heterojunction to form the carriers sets up the bottleneck step. The carriers must overcome the coulombic force of attraction, hence recombination occurs. The difference in the electron affinities between donor and acceptor material is the main driving force for the dissociation of excitons, and the process becomes thermodynamically and kinetically favorable. In the donor–acceptor interface excitons decay to form a radical cation and a radical anion:

$$D + A + h\gamma \rightarrow D* + A\left(\text{or } D + A*\right) \rightarrow D^{\bullet+}A^{\bullet-})$$

Here D and A stand for donor and acceptor respectively.

After exciton dissociation, the generated electron and hole remain in close proximity. The large chemical potential gradient is the driving force for the migration of the charges towards the electrodes (Zhang et al. 2019). The charge is collected by placing a transparent conductive oxide (TCO) (e.g., indium tin oxide (ITO)) and metal electrodes on two different sides in the OPV architecture. The PCE of the OPV depends on the overall external quantum efficiency (η_{EQE}) which in turn is associated with the efficiency of each of the steps (Myers and Xue 2012):

$$\eta_{EQE} = \eta_A \eta_{ED} \eta_{CT} \eta_{CC}$$

Here, η_A, η_{ED}, η_{CT}, η_{CC} is the efficiency of the photon absorption (generation of excitons), the diffusion of excitons, the charge transfer (dissociation of excitons), and the collection of charge respectively. The internal quantum efficiency η_{IQE} can be defined as:

$$\eta_{IQE} = \eta_{ED} \eta_{CT} \eta_{CC}$$

Thus, $\eta_{EQE} = \eta_A \eta_{IQE}$.

12.5 CURRENT DENSITY (*J*)–VOLTAGE (*V*) CHARACTERISTICS FOR OPVS

The performance of OPVs can be understood by studying the current density–voltage characteristics. Three depicted curves in the current density–voltage curves are (i) the dark current density; (ii) the generated photocurrent; and (iii) the total current, that is the sum of the dark current and the generated photo current under the condition of photo illumination. Separate study of the photo current is not necessary as the dark current is always present. To separately study the photo current, sophisticated techniques are required (Myers et al. 2009). The current density at zero applied bias is called the short circuit current density (J_{sc}) and the

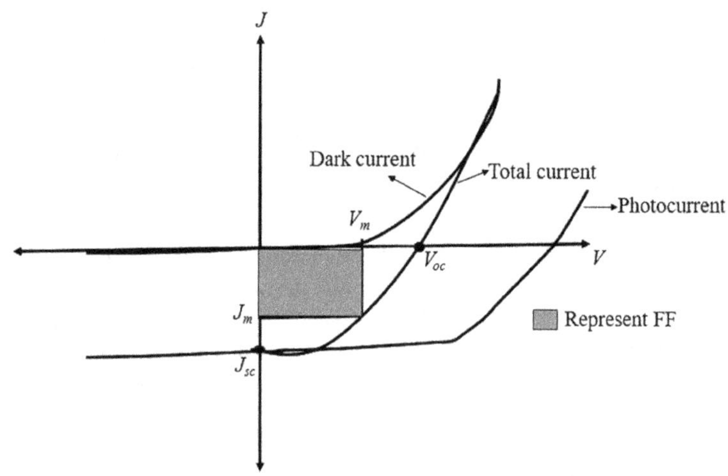

FIGURE 12.1 Current density–voltage characteristics.

J_m is the maximum current density and V_m is the maximum voltage at the power point maximum (P_m) in the fourth quadrant.

$P_m = |J_m V_m|$, i.e., the enclosed area in the rectangle in Figure 12.1 which represents the obtained maximum power.

The fill factor (FF) is defined as $FF = \frac{J_m V_m}{J_{sc} V_{sc}}$, i.e., the ratio of maximum power obtained to the ideally obtainable maximum power under illuminated conditions.

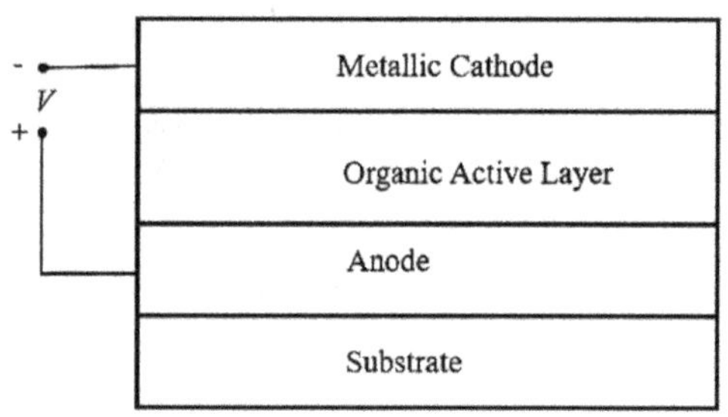

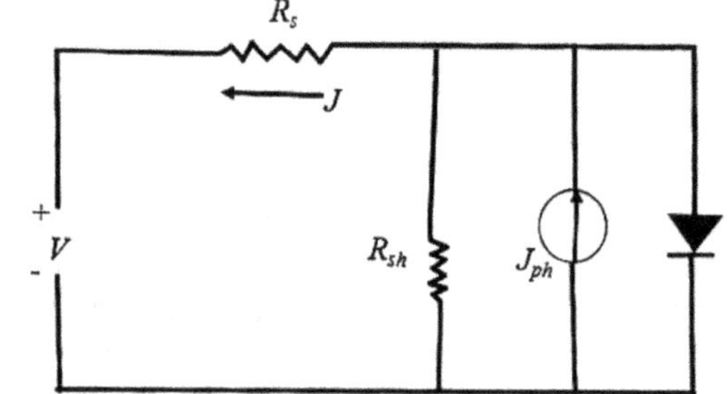

FIGURE 12.2 Circuit representation of OPVs.

voltage at which the total current is zero is called the open circuit voltage (V_{oc}). These two points are shown in the current density voltage curves (Figure 12.1).

The ideal value of the fill factor is unity. In reality FF ≤ 0.7 for modern and optimized OPV devices (Green et al. 2011).

The total power conversion efficiency can be defined as:

$$\eta_p = \frac{J_{sc} V_{sc} FF}{P_0}$$

Here P_0 represents the intensity of optical power.

Conventionally, the external quantum efficiency (EQE) can be defined, at a certain wavelength, as the ratio of the number of electrons generated to the number of photons absorbed in the process:

$$\eta_{EQE} = \frac{J_{sc} / q}{P_0 \lambda / hc}$$

Here q is the electronic charge.

Back calculation of J_{sc} can be done by calculating the EQE using the reference 1 sun AM1.5G illumination and the equation:

$$J_{sc} = \int_{\lambda_1}^{\lambda_2} \frac{q\lambda}{hc} EQE(\lambda) S(\lambda) d\lambda$$

Here, the power intensity of reference AM1.5G is $S(\lambda)$. The difference between the experimentally obtained value and the obtained value from integration must be very much less for the correct calibration of the EQE and systems for J–V measurements. With the help of an equivalent circuit, a photovoltaic device can be represented (Figure 12.2). A diode is placed parallel to the current source which represents photocurrent generation. There are two kinds of resistors in the circuit: a series resistor R_s, which represents the intrinsic resistant of the active organic layer and shunt resistor and R_{sh}, a parallel resistant placed to represent current leakage. An expression of current voltage relation can be developed from this circuit which is known as a Shockley equation (Stanojević et al. 2020):

$$J = J_s \left\{ \exp\left[\frac{q(V - JR_s)}{nKT} \right] - 1 \right\} + \frac{q(V - JR_s)}{R_{sh}} + J_{ph}$$

Here, J_s stands for the diode's reverse bias saturation current density, J_{ph} for the generated photocurrent density, n for the ideality factor, and the rest of the terms have the usual meanings.

It is clear from the Shockley equation that an increase of J is favored by a decrease of R_s and an increase of R_{sh}. The decrease of FF and J_{sc} favor an increase of R_s. FF and V_{oc} decrease with a decrease of R_{sh}. The generated photocurrent in an ideal OPV (where $R_s = 0$ and $R_{sh} = \infty$) linearly increases with the increase of the intensity of the incident photons. When the total current $J = 0$, then

$$V_{OC} = \frac{nKT}{q} \ln\left[\left| \frac{J_{ph}}{J_s} \right| + 1 \right]$$

Here V_{oc} is called the open circuit voltage. We can see that V_{oc} logarithmically depends on J_{ph} and so also on P_0 (Zheng and Xue 2010). Assuming no current loss, η_p should logarithmically increase with the incident power as V_{oc} increases, and the value of J_{sc}/P_0 can be fixed to constant.

Since the series and shunt resistant in an actual OPV is finite, the values of J_{sc}, FF, and V_{oc} deviate from their respective ideal values. An increase of the intensity of illumination, increases the concentration of free carriers, so the bimolecular recombination rate also increases. Therefore, with an increase of the illumination intensity, the J_{sc}/P_0 ratio is decreased:

$$R_{BM} = \gamma (np - n_i p_i) \propto np$$

$$\gamma = \frac{q}{\varepsilon \varepsilon_0} (\mu_n + \mu_p)$$

The highest value of η_p for an OPV device is achieved at the condition of illuminating power intensity when the increase in V_{oc} exceeds the decrease in J_{sc}/P_0 attributed to bimolecular recombination (Myers and Xue 2012).

12.6 SMALL MOLECULE-BASED OPVs

OPV devices made earlier were mainly based on organic small molecules. The major limitation in their operation was less efficiency in the bottleneck step, that is exciton dissociation into charge carriers in the donor–acceptor heterojunction. To get rid of this problem, several architectural changes were made, one of them being the mixing of donor and acceptor materials. The photocurrent was found to increase as a greater number of excitons reached the donor–acceptor interface for dissociation, though that caused another limitation in charge transport properties, hence the photocurrent increase becomes limited (Uchida et al. 2004). Another modification is the use of a planar mixed heterojunction, in which a layer consists of mixed donor and acceptor materials is sandwiched between two layers where one layer acts as a neat donor and the other one as a neat acceptor (Xue et al. 2005). This design was mainly aimed at improvement in the efficiency of exciton dissociation as well as charge transport. In the earlier modification a significant improvement in photocurrent J_{sc} was found but the fill factor decreased, largely due to the poor morphology of the mixed layer (Rostalski and Meissner 2000). While preparing the mixed materials, poor morphology becomes very common, though it can be overcome if the phase segregation in the co-evaporated material is controlled by allowing it to deposit on hot substrates or by thermal annealing (Zheng et al. 2009). The phase segregation causes the easy transport of free carriers to their collection electrodes, hence η_{CC} increases. Thus, the planar mixed-architecture become an efficient choice for small molecule based OPV devices. The incorporation of an electron transporting thin layer between the cathode and acceptor is another modification in small organic molecule-based devices. The addition of a thin layer prevents the quenching of excitons. A typical transporting layer is made of bathocuproine (BCP). A BCP layer in small molecule OPVs is commonly known as an exciton blocker (Patil et al. 2018). The major contribution in the enhancement in efficiency of OPVs comes from the increase of absorption of photons, the open circuit voltage, and the length of the diffusion of excitons. One example of the advancement is the changing of 3,4,9,10-perylene tetracarboxylic-bis-benzimidazole (PTCBI) (exciton diffusion length ~3 nm) into C_{60} (exciton diffusion length ~40 nm) (Peumans and Forrest 2001). The absorption coefficient of C_{60} in the region of short wavelengths is largely enhanced in combination with phthalocyanines compared to PTCBI. The use of C_{70} further increases the efficiency, as it strongly absorbs in the blue–green region of electromagnetic spectra. Hence the current use of C_{60} and C_{70} as standard acceptors in small organic molecule-based devices is rational (Pfuetzner et al. 2009).

12.7 POLYMER-BASED OPVs

In the primitive stage of development, organic polymers such as polythiophenes (Glenis et al. 1984), polyacetylenes (Kanicki and Fedorko 1984), and polyvinylenes (Marks et al. 1994), having conjugated π-electron clouds, were used to make single layer Schottky diodes. The performance of these devices was very low. Considerable performance was observed for the first time when a bilayer structure made of organic polymers like MEH-PPV was used as a donor and C_{60} as a standard acceptor material (Sariciftci et al. 1993). Later a polymer trivially known as PC61BM, a derivative of C_{60}, and in some more efficient devices polymers like PC71BM, a derivative of C_{70}, was used in OPVs (Hummelen et al. 1995). The main focus then shifted to the choice of appropriate donor materials. Polymers like

MEH-PPV (Yu et al. 1995) and MDMO-PPV (Wienk et al. 2003) were then used in devices as donor materials, resulting in a considerable enhancement in performance. Despite having a high quantum efficiency, these earlier devices lost attention as it was found that with the increase of illumination intensity, strong bimolecular recombination of the carriers occurred and the film morphology was poor. A study on the donor polymer poly(3-hexylthiophene) (P3HT) has been conducted, widely replacing the PPVs or MDMO-PPV because of the higher absorption coefficient in a wide range of electromagnetic spectra (Ma et al. 2007). The extensive morphological study of P3HT: PCBM film increased efficiency by up to 5%. After the study of donor–acceptor composition optimization and a morphological study of heterojunction, the PTB7 polymer was used along with PC71BM and the efficiency was found to increase by up to 7.4% (Liang et al. 2010). The highest efficiency of the commercially developed polymer based OPVs by Konarka has been reported as 8.3% (Jannat et al. 2013).

12.8 HYBRID ORGANIC–INORGANIC PHOTOVOLTAIC DEVICES

Recently emphasis has been given to inorganic–organic hybrid material-based photovoltaic fabrication, aiming at an ease in processibility and commercial feasibility. In hybrid OPVs, inorganic material-based nanocrystals are used along with organic materials where the favorable properties of inorganic materials, including environmental compatibility, rapid transport of charge, and easy band gap tuning as a function of size, are incorporated in the device. It was found from research study that exciton dissociation is favored at the organic/inorganic nanocrystal interface. The light absorption can be modulated by changing the physical parameters of nanocrystals, such as shape and size. An earlier developed P3HT/CdSe nanorod active layer hybrid cell had the efficiency of 1.7% (Huynh et al. 2002). When a nanorod is used instead of a nanosphere, the interparticle barrier is decreased and the transport of the charge becomes feasible. A similar effect can also be observed by using a nano tetrapod instead of a nanosphere (Sun and Greenham. 2006). For hybrid cell fabrication, mostly cadmium salt nanocrystals are used; other salts used for efficient device fabrication are silicon (Liu et al. 2011) and ZnO (Beek et al. 2006). An advantage of using Pb based nano material is the low band gap, hence the efficient absorption of light in the near infrared region. Until now the highest efficiency of a hybrid solar cell (3.8%) was found when a Pb based nano material, PbS, was used along with a low-band gap polymer PDTPBT (Seo et al. 2011).

12.9 TANDEM ORGANIC PHOTOVOLTAIC DEVICES

In most of organic systems, there is a fundamental tradeoff between η_A and η_{IQE}. One of the methods to get rid of this is the stacking of multiple devices with high η_{IQE} in series; the resulting architecture is called a tandem cell (Meng et al. 2018). The highest efficiency of a tandem cell was found for inorganic photovoltaics using different materials in the constituent sub-cells, thus covering a very wide range of solar spectra. This concept can be extended to organic solar cells. Even without increasing the absorption of photons, a tandem architecture can increase device efficiency by increasing the output voltage. In tandem cells, the lowest current among the sub-cells is taken, and the voltage simply follows additivity. In a tandem cell of organic material-based sub-cells the most important focus needs to be given to the carrier recombination zone (CRZ), where two sub-cells are connected. The recombination of electrons and holes from the two closely placed sub-cells occurs in the CRZ. The first CRZ layer in a tandem cell was developed using a 15 nm thickness gold layer. The electrical conductivity of the gold layer is very high. Because of the semi-transparency of the gold layer, the incident photons are largely blocked (Hiramoto et al. 1990). A better CRZ layer was made by Yakimov and Forrest by depositing silver nanoparticles on a layer of high conducting small molecules in a vacuum. This CRZ was used in CuPc/PTCBI (η_P = ~2.5%) and CuPc/C_{60} based tandem devices (η_P = ~5.7%). It is difficult to fabricate tandem cells where polymers are used in sub-cells (Yakimov and Forrest 2002). While fabricating the architecture of tandem cells, protection of the front cell is very important while preparing the back cell. Hence the front cell has to be insoluble in the solvent used in preparing the back cell, or it needs to be protected during fabrication (Janssen et al. 2007). In a tandem cell, the front cell can be the polymer cell and the nanoparticle coated small molecule-based cell can be the back cell, or by engineering the CRZ in such a way that the front cell becomes isolated using nanoparticles or solvent resisting organic layers (Kim et al. 2007). Major difficulties in fabricating tandem cells arise because of the requirement of choosing appropriate materials and the complex architecture.

12.10 EFFECTS OF TEMPERATURE ON OPV CELLS

In OPV, with the increase of temperature, the mobility of the charge carriers increases. In organic materials, phonons considerably assist the delocalization of charge carrying species. The probability of finding a sufficiently energized phonon is less at low temperature. Therefore, the photocurrent at low temperature is reduced. With the increase of temperature, the short circuit current density (J_{sc}) increases, then proceeds towards saturation, and then decreases while the open circuit voltage (V_{oc}) linearly decreases. The fill factor (FF) which depends on the short circuit current and open circuit voltage follows their variations. Thus, with increase of temperature, the PCE of OPV first increases then tends to decrease, passing through the maximum value (Belhocine et al. 2012).

12.11 FUNDAMENTAL LIMITATIONS OF OPVs

The main limitations of OPVs is much reduced PCE and a poor life time. Degradation of organic materials used in OPVs is faster compared to inorganic materials such as cadmium telluride or silicon. Two types of degradation mainly occur in OPVs: (i) extrinsic degradations and (ii) intrinsic degradations. It is found that aerial oxygen and moisture or water cause extrinsic degradation; photo-induced burn-in is the reason for intrinsic degradation. To protect the OPV from oxygen and water damage, the device needs to be well encapsulated by barrier materials having very little permeability to oxygen and water. To improve stability in operating temperatures, materials of glass at temperatures above 100 °C need to be used. The exact mechanism of photo-induced burning is still not well established. However, from several studies it has been found that the use of densely packed and well-ordered film of pure materials reduces photo-induced burning (Gupta et al. 2013).

12.12 FUTURE DEVELOPMENT REGARDING THE PCE ENHANCEMENT OF OPVs

The improvement in the performance of OPVs can be done by improving the absorption of photons. One possibility is increasing the thickness of the organic layer and shifting the absorption spectra towards longer wavelengths. The increase in layer thickness will increase the absorption of photons but this will affect the lifetime and transport of the carriers. With the increase of thickness of the active layer, the transit time of the carriers increases and becomes larger than the lifetime, resulting in a larger probability of charge recombination, hence the increase of layer thickness is trivial to the increase of the device performance (Ou et al. 2016). Without considerable loss of current the thickness of the layer of polymer such as P3HT can be increased from 100 up to 500 nm. The absorption of electromagnetic radiation in OPVs occurs in the UV region, up to ~650 nm, depending on the choice of organic materials. The external quantum efficiency can even be 70%, meaning that a very large number of adsorbed photons contribute to exciton generation. The shifting of the absorption wavelength towards the infrared region increases photon intensity, hence using a polymer of low band gap will increase the PCE. Other aspects of PCE improvement are the development of new active materials and device architecture optimizations (Ulum et al. 2019).

12.13 STATUS OF THE OPV INDUSTRY

The criteria for the commercial entry of OPVs are device performance, lifetime, processing method, environmental impact, and production cost. Fabrication for laboratory scale production, for the purpose of research, is done by coating the organic materials on a small glass surface using the spin coating technique, and in most of the cases toxic solvents are used. For mass scale production of OPVs, it is necessary to have a roll-to-roll fabrication technique, a green solvent, and flexible substrates with large surface areas. Many academic researchers are trying to innovate the processing of OPVs, and several scientific leaders are assessing the commercial aspects of OPV based technologies. The first company commercially producing OPV devices is Konarka Technologies. Many other companies like Solarmer energy, Plextronics, and Polyera have started commercial-scale polymer-based device production, and the Global Photonic Energy Corporation, Mitsubishi Chemical, and Heliatek have manufactured small molecule-based devices. OPV device manufacturing companies maintain good relationships and terms with academia and encourage research on high performing organic materials in laboratories. Until now the full fledged commercialization of OPV based devices has been quite challenging.

12.14 CONCLUSIONS

The field of organic semiconductors has grown tremendously in the last two decades and a considerable endeavor by researchers from various academic disciplines, such as chemistry, material science, physics, and electrical engineering, has been observed. Mainly the research has concerned the enhancement in power conversion efficiency, durability, and stability of devices.

Research on progress mainly includes: (i) the complete understanding of the conversion of solar energy to electricity, that is conversion of photons to electrons; (ii) the synthesis of suitable organics; (iii) the optimization of composition morphology; and (iv) innovation in device architecture. To achieve a high PCE, it is required to have a substrate with a large surface area and an appropriate technique for the roll-to-roll printing of organic materials from solutions. Collaboration between researchers and engineers is required to solve the emerging problems in processes for increasing device life time and architectural stability. The commitment of research groups is mainly centralized in synthesizing new organic materials having good absorption coefficients in a wide range of electromagnetic spectra, being stable to photo-induced burn-in and photobleaching, and having good solubility. The role of the engineers is on innovating a new feasible fabrication process, device architecture optimization, designing various electrodes, and so on. Thus, by the various combined efforts, one day the full-fledged commercialization of OPV devices may become a reality.

ACKNOWLEDGMENTS

This work was financially supported by UGC-BSR research start-up grant (Ref. No. F.30-383/2017(BSR)) dated 15 December 2017 and DST-SERB (File no EEQ/2018/000964). SI thanks the Department of Chemistry, Trivenidevi Balotia College, for being supportive. The authors thank the Department of Chemistry, The University of Burdwan.

REFERENCES

Alam, M.A., Ray, B., Khan, M.R., and Dongaonkar, S. 2013. The essence and efficiency limits of bulk-heterostructure organic solar cells: A polymer-to-panel perspective. *J. Mater. Res.* 28(4): 541–557.

Albinsson, B., and Martensson, J. 2008. Long-range electron and excitation energy transfer in donor-bridge-acceptor systems. *J. Photochem. Photobiol. C* 9(3): 138–155.

Ayzner, A.L., Tassone, C.J., Tolbert, S.H., and Schwartz, B.J. 2009. Reappraising the need for bulk heterojunctions in polymer–fullerene photovoltaics: The role of carrier transport in all solution-processed P3HT/PCBM bilayer solar cells. *J. Phys. Chem.* C 113(46): 20050–20060.

Beek, W.J.E., Wienk, M.M., and Janssen, R.A.J. 2006. Hybrid solar cells from regioregular polythiophene and zno nanoparticles. *Adv. Funct. Mater.* 16: 1112–1116.

Belhocine, N.F., Djediga, H., Boughias, O., and Said, B.M. 2012. Effect of temperature on the organic solar cells parameters. *J. Energ. Power Eng.* 6: 921–924.

Blom, P.W.M., Mihailetchi, V.D., Koster, L.J.A., and Markov, D.E. 2007. Device physics of polymer: Fullerene bulk heterojunction solar cells. *Adv. Mater.* 9: 1551–1566.

Brabec, C.J. 2004. Organic photovoltaics: Technology and market. *Sol. Energy Mater. Sol. Cells* 83(2–3): 273–292.

Casalegno, M., Pastore, R., Idé, J., Po, R., and Raos, G. 2017. Origin of charge separation at organic photovoltaic heterojunctions: A mesoscale quantum mechanical view. *J. Phys. Chem. C* 121(31): 16693–16701.

Chen, R., Tang, Y., Wan, Y., Chen, T., Zheng, C., Qi, Y., and Huang, W. 2017. Promoting singlet/triplet exciton transformation in organic optoelectronic molecules: Role of excited state transition configuration. *Scientific Rep.*, 7(1): 1–11.

Cnops, K., Rand, B.P., Cheyns, D., Verreet, B., Empl, M.A., and Heremans, P. 2014. 8.4% efficient fullerene-free organic solar cells exploiting long-range exciton energy transfer. *Nat. Commun.* 5: 3406.

Cusumano, P., Arnone, C., Giambra, M.A., and Parisi, A. 2020. Donor/acceptor heterojunction organic solar cells. *Electronics* 9(1): 1–8.

de Gans, B.J.; Duineveld, P.C., and Schubert, U.S. 2004. Inkjet printing of polymers: State of the art and future developments. *Adv. Mater.* 16: 203–213.

Dou, L., You, J., Hong, Z., Xu, Z., Li, G., Street, R.A., and Yang, Y. 2013. 25th anniversary article: A decade of organic/polymeric photovoltaic research. *Adv. Mater.* 25(46): 6642–6671.

Eslamian, M. 2016. Inorganic and organic solution-processed thin film devices. *Nanomicro Lett.* 9(1): 1–23.

Facchetti, A. 2013. Polymer donor–polymer acceptor (all-polymer) solar cells. *Mat. Today* 16(4): 123–132.

Feron, K., Belcher, W., Fell, C., and Dastoor, P. 2012. Organic solar cells: understanding the role of förster resonance energy transfer. *Int. J. Molecular Sci.* 13(12): 17019–17047.

Forrest, S.R. 2015. Excitons and the lifetime of organic semiconductor devices. *Pfhilos. Trans. R. Soc. A* 373(2044): 20140320–20140320.

Ginley, D., Green, M.A., and Collins, R. 2008. Solar energy conversion toward 1Terawatt. *MRS Bull.* 33(4): 355–364.

Glenis, S., Horowitz, G., Tourillon, G., and Garnier, F. 1984. Electrochemically grown polythiophene and poly(3-methylthiophene) organic photovoltaic cells. *Thin Solid Films* 111: 93–103.

Green, M.A. Emery, K., Hishikawa, Y., and Warta, W. 2011. Solar cell efficiency tables (version 37). *Prog. Photovolt: Res. Appl.* 19: 84–92.

Guerrero, A., and Garcia-Belmonte, G. 2016. Recent advances to understand morphology stability of organic photovoltaics. *Nano-Micro Lett.*, 9(1): 1–16.

Gupta, S.K., Dharmalingam, K., Pali, L.S., Rastogi, S., Singh, A., and Garg, A. 2013. Degradation of organic photovoltaic devices: A review. *Nanomater. Energy* 2(1): 42–58.

Hedley, G.J., Ruseckas, A., and Samuel, I.D.W. 2016. Light harvesting for organic photovoltaics. *Chem. Rev.* 117(2): 796–837.

Hiramoto, M., Suezaki, M., and Yokoyama, M. 1990. Effect of thin gold interstitial-layer on the photovoltaic properties of tandem organic solar cell. *Chem. Lett.* 19: 327–330.

Hummelen, J.C. Knight, B.W., LePeq, F., Wudl, F., Yao, J., and Wilkins, C.L. 1995. Preparation and characterization of fulleroid and methanofullerene derivatives. *J. Org. Chem.* 60: 532–538.

Huynh, W.U., Dittmer, J.J., and Alivisatos, A.P. 2002. Hybrid nanorod-polymer solar cells. *Science* 295: 2425–2427.

Jannat, A., Rahman, M.F., and Khan, M.S.H. 2013. A review study of organic photovoltaic cell. *Int. J. Sci. Eng. Res.* 4: 1–6.

Janssen, A.G.F., Riedl, T., Hamwi, S., Johannes, H.H., and Kowalsky, W. 2007. Highly efficient organic tandem solar cells using an improved connecting architecture. *Appl. Phys. Lett.* 91: 073519–073523.

Kanicki, J., and Fedorko, P. 1984. Electrical and photovoltaic properties of trans-polyacetylene. *J. Phys. D: Appl. Phys.* 17: 805–817.

Kim, J.Y., Lee, K., Coates, N.E., Moses, D., Nguyen, T.Q., Dante, M., and Heeger, A.J. 2007. Efficient tandem polymer solar cells fabricated by all-solution processing. *Science* 317: 222–225.

Kovacik, P., Sforazzini, G., Cook, A.G., Willis, S.M., Grant, P.S., Assender, H.E., and Watt, A.A. R. 2010. Vacuum-deposited planar heterojunction polymer solar cells. *ACS Appl. Mater. Interfaces* 3(1): 11–15.

Lakhwani, G., Rao, A., and Friend, R.H. 2014. Bimolecular recombination in organic photovoltaics. *Annu. Rev. Phys. Chem.* 65(1): 557–581.

Lee, J., Jung, Y.K., Lee, D.Y., Jang, J.W., Cho, S., Son, S., and Park, S.H. 2015. Enhanced efficiency of bilayer polymer solar cells by the solvent treatment method. *Synthetic Metals* 199: 408–412.

Liang, Y., Xu, Z., Xia, J., Tsai, S., Wu, Y., Li, G., Ray, C., and Yu, L. 2010. For the bright future: Bulk heterojunction polymer solar cells with power conversion efficiency of 7.4%. *Adv. Mater.* 22: E135–E138.

Liu, C.Y., Holman, Z.C., and Kortshagen, U.R. 2011. Hybrid solar cells from P3HT and silicon nanocrystals. *Nano Lett.* 9: 449–452.

Lungenschmied, C., Dennler, G., Neugebauer, H., Sariciftci, S.N., Glatthaar, M., Meyer, T., and Meyer, A. 2007. Flexible, long-lived, large-area, organic solar cells. *Sol. Energy Mater. Sol. Cells* 91: 379–384.

Ma, W., Gopinathan, A., and Heeger, A.J. 2007. Nanostructure of the interpenetrating networks in poly(3- hexylthiophene)/fullerene bulk heterojunction materials: Implications for charge transport. *Adv. Mater.* 19: 3656–3659.

Marks, R.N., Halls, J.J.M., Bradley, D.D.C., Friend, R.H., and Holmes, A.B. 1994. The photovoltaic response in poly(p-phenylene vinylene) thin-film devices. *J. Phys.: Condens Matter.* 6: 1379–1394.

Meng, L., Zhang, Y., Wan, X., Li, C., Zhang, X., Wang, Y., and Chen, Y. 2018. Organic and solution-processed tandem solar cells with 17.3% efficiency. *Science*, 361: 1094–1098

Mikhnenko, O.V., Blom, P.W.M., and Nguyen, T.Q. 2015. Exciton diffusion in organic semiconductors. *Energ. Environ. Sci.* 8(7): 1867–1888.

Myers, J.D., and Xue, J. 2012. Organic semiconductors and their applications in photovoltaic devices. *Polym. Rev.* 52(1): 1–37.

Myers, J.D., Tseng, T.K., and Xue, J. 2009. Photocarrier behavior in organic heterojunction photovoltaic cells. *Org. Elec.* 10: 1182–1186.

Ou, Q.D., Li, Y.Q., and Tang, J.X. 2016. Light manipulation in organic photovoltaics. *Adv. Sci.* 3(7): 1600123–1600148.

Patil, B.R., Ahmadpour, M., Sherafatipour, G., Qamar, T., Fernández, A.F., Zojer, K., and Madsen, M. 2018. Area dependent behavior of bathocuproine (BCP) as cathode interfacial layers in organic photovoltaic cells. *Sci. Rep.* 8(1): 1–9.

Peumans, P., and Forrest, S.R. 2001. Very-high-efficiency double-heterostructure copper phthalocyanine/C [sub 60] photovoltaic cells. *Appl. Phys. Lett.* 79: 126–128.

Pfuetzner, S., Meiss, J., Petrich, A., Riede, M., and Leo, K. 2009. Improved bulk heterojunction organic solar cells employing C [sub 70] fullerenes. *Appl. Phys. Lett.* 94: 223307–223320.

Rostalski, J., and Meissner, D. 2000. Monochromatic versus solar efficiencies of organic solar cells. *Sol. Energy Mater. Sol. Cells.* 61: 87–95.

Sariciftci, N.S. Braun, D., Zhang, C., Srdanov, V.I., Heeger, A.J., Stucky, G., and Wudl, F. 1993. Semiconducting polymer-buckminsterfullerene heterojunctions: Diodes, photodiodes, and photovoltaic cells. *Appl. Phys. Lett.* 62: 585–587.

Scharber, M.C., and Sariciftci, N.S. 2013. Efficiency of bulk-heterojunction organic solar cells. *Prog. Polymer Sci.* 38(12): 1929–1940.

Scholes, G.D., Fleming, G.R., Olaya-Castro, A., and van Grondelle, R. 2011. Lessons from nature about solar light harvesting. *Nat. Chem.* 3(10): 763–774.

Seo, J., Cho, M.J., Lee, D., Cartwright, A.N., and Prasad, P.N. 2011. Efficient heterojunction photovoltaic cell utilizing nanocomposites of lead sulfide nanocrystals and a low-band gap polymer. *Adv. Mater.* 23: 3984–3988.

Shtein, M., Peumans, P., Benziger, J.B., and Forrest, S.R. 2003. Micropatterning of small molecular weight organic semiconductor thin films using organic vapor phase deposition. *J. Appl. Phys.* 93: 4005–4016.

Stanojević, M., Gojanović, J., Matavulj, P., and Živanović, S. 2020. Organic solar cell physics analyzed by Shockley diode equation. *Opt. Quant. Electr.* 52(7). 3–10.

Su, Y.W., Lan, S.C., and Wei, K.H. 2012. Organic photovoltaics. *Mat. Today* 15(12): 554–562.

Sun, B., and Greenham, N.C. 2006. Improved efficiency of photovoltaics based on CdSe nanorods and poly(3-hexylthiophene) nanofibers. *Phys. Chem. Chem. Phys.* 8: 3557–3560.

Sun, Y., Shtein, M., Forrest, S.R. 2005. Direct patterning of organic light-emitting devices by organicvapor jet printing. *Appl. Phys. Lett.* 86: 113504–113508.

Uchida, S. Xue, J., Rand, B.P., and Forrest, S.R. 2004. Organic small molecule solar cells with a homogeneously mixed copper phthalocyanine: C [sub 60] active layer. *Appl. Phys. Lett.* 84: 4218–4220.

Ulum, M.S., Sesa, E., Kasman, W., and Belcher, W. 2019. The effect of active layer thickness on P3HT: PCBM nanoparticulate organic photovoltaic device performance. *J. Phys.: Conf. Ser.* 1242: 012025.

Vak, D., Kim, S.S., Jo, J., Oh, S.H., Na, S.I., Kim, J., and Kim, D.Y. 2007. Fabrication of organic bulk heterojunction solar cells by a spray deposition method for low-cost power generation. *Appl. Phys. Lett.* 91: 081–102.

Wang, Z., Zhou, Y., Miyadera, T., Chikamatsu, M., and Yoshida, Y. 2017. Constructing nanostructured donor/acceptor bulk heterojunctions via interfacial templates for efficient organic photovoltaics. *ACS Appl. Mat. Interfaces* 9(50): 43893–43901.

Wantana, K., Aniwat, P., Bunlue, S., Alongkot, T., Anusit, K., and Pisist, K. 2017. Study of thin film coating technique parameters for low cost organic solar cells fabrication. *Mat. Today: Proc.* 4(5): 6626–6632.

Wienk, M.M. Kroon, J.M. Verhees, W.J.H., Knol, J., Hummelen, J.C., van Hal, P.A., and Janssen, R.A.J. 2003. Efficient methano[70]fullerene/MDMO-PPV bulk heterojunction photovoltaic cells. *Angew. Chem. Int. Ed.* 42: 3371–3375.

Xie, Y., and Wu, H. 2020. Balancing charge generation and voltage loss toward efficient norfullerene organic solar cells. *Mat. Today Adv.* 5: 1–8.

Xue, J., Rand, B.P., Uchida, S., and Forrest, S.R. 2005. A hybrid planar-mixed molecular heterojunction photovoltaic cell. *Adv. Mater.* 17: 66–71.

Yakimov, A., and Forrest, S.R. 2002. High photovoltage multiple-heterojunction organic solar cells incorporating interfacial metallic nanoclusters. *Appl. Phys. Lett.* 80: 1667–1669.

Yu, G., Gao, J., Hummelen, J.C., Wudl, F., and Heeger, A.J. 1995. Polymer photovoltaic cells: Enhanced efficiencies via a network of internal donor–acceptor heterojunctions. *Science* 270: 1789–1791.

Zhang, Z., Fang, W.H., Long, R., and Prezhdo, O.V. 2019. Exciton dissociation and suppressed charge recombination at 2D perovskite edges: Key roles of unsaturated halide bonds and thermal disorder. *J. Am. Chem. Soc.* 141: 15557–15566.

Zheng, Y., and Xue, J. 2010. Organic photovoltaic cells based on molecular donor–acceptor heterojunctions. *Polym. Rev.* 50: 420–453.

Zheng, Y., Pregler, S.K. Myers, J.D. Ouyang, J. Sinnott, S.B., and Xue, J. 2009. Computational and experimental studies of phase separation in pentacene [sub 60] mixtures. *J. Vac. Sci. Technol. B.* 27: 169–179.

13 Polymers and Their Composites for Wearable Electronics

Svetlana Jovanović and Dragana Jovanović
Vinča Institute of Nuclear Sciences, University of Belgrade, National Institute of the Republic of Serbia, Belgrade, Serbia

CONTENTS

13.1 WEARABLE ELECTRONICS: DEFINITION AND DRIVING FORCES

In the age of fast technological development, a new field of the electronic industry—wearable electronics (WEs)—finds its way to meet modern humans' needs. By implementing both electronic and computing elements into unique devices, these new products are becoming a part of everyday life and a fast-growing market (Ghahremani Honarvar and Latifi 2017).

These products are connecting textiles/clothing and accessories such as watches, glasses, fitness bands, or earrings with sensors that can detect and transfer different signals for the human body to read as information. Due to fast development in the last decade, WEs are becoming an important segment of medicine, academic life, as well as wide consumer markets.
These devices can measure:

- Heart rate, pulse, blood pressure, blood oxygen saturation (like the oximeter), and electrophysiological signals such as the electrocardiograph, electromyogram, and electroencephalogram;
- Body temperature, physical motions, skin and breath moisture;
- Biochemical parameters such as blood sugar and electrolytes.

(Farandos et al. 2015, Kanoun et al. 2015, Harito et al. 2020, Wang et al. 2020)

Depending on the way they can be worn, there are:

- Accessories such as smartwatches, smart glasses, smart earrings, and headsets;
- Electronics impregnated in clothing;
- Electronic devices implanted in the user's body, or tattooed on the surface of the skin, or subdermally placed in the deeper tissue.

What are WEs actually? The simplest definition would be: devices attached to our clothes, fashion accessories, or other means with sensors. These sensors track motion; brain, heart, or muscle activity; skin conductivity; hydration; or other parameters.

These devices can function as sensors for luminescence, energy harvesting, and thermo-electricity generation (Choi et al. 2019, Harito et al. 2020, Huang et al. 2017, Zhou et al. 2016). Among different applications, the one in healthcare seems to be the most important and groundbreaking, due to the assumption that WEs could lead to transformation.

DOI: 10.1201/9781003169727-13

Thus, the necessity of flexible sensing electronics is increasing every day. All components or wearable devices must be flexible and stretchable, mechanically robust, lightweight, as well as chemically stable, inert and, biocompatible (Fan et al. 2019, Qu et al. 2016). These characteristics are key factors in producing safe and conformable wearable devices. To achieve these demands, flexible and stretchable materials are used. Some of the most important properties of materials for WEs are:

- Softness and stretchability;
- High electrical conductivity;
- Good thermoelectric properties.

The main classes of materials used in the production of these devices are: conductive polymers, carbon materials, piezoelectric elastomer polymers, chitin and chitosan, and rare earth nanoparticles as well as composites of these materials. All these materials will be analyzed in this chapter.

13.2 CONDUCTIVE POLYMERS AND THEIR COMPOSITES IN WES

Different conductive polymers have been investigated for application in WEs. Due to biocompatibility, the electrical properties that can be tuned by doping and de-doping, and the sensitivity of ionic molecules (Harito et al. 2020), these polymers have been largely investigated as a different component of WE. Conductive polymers have been investigated as:

- Thermoelectric generators (Fan and Ouyang 2019);
- Sensors for health monitoring applications (del Agua et al. 2018);
- Connectors (Guo et al. 2016);
- Actuators (Zhou et al. 2016).

The drawbacks of conductive polymers for WE application are poor mechanical properties, such as low stretchability and flexibility, hard processability, and brittleness.

To improve the mechanical behavior of conductive polymer poly(3,4-ethylenedioxythiophene) (PEDOT), the adding of an elastomer to create a more elastic polymer was performed (Kim et al. 2020). Kim et al. produced a poly(3,4-ethylenedioxythiophene)-polyurethane-ionic liquid composite. The material showed an astonishing conductivity at above 140 S cm^{-1}, the capability to stretch over 600% of its starting length, and an elasticity of <7 MPa. This material was prepared for application in thermoelectric generators which convert body heat into electricity. The development of this kind of electricity generator is very attractive for WEs and would allow the assembling of self-powered devices. The even higher electrical conductivity of PEDOT was achieved at above 1000 S cm^{-1}, with thermopower between 10 and 50 $\mu V\ K^{-1}$ and thermal conductivity below 1 W m^{-1} K^{-1} (Fan and Ouyang 2019, Yano et al. 2019). These properties have been explained by the weak vibration of the lattice.

Due to good thermoelectric properties, PEDOT: polystyrene sulfonate (PEDOT:PSS) was most often investigated between conductive polymers (Fan et al. 2019). This copolymer was deposited on cotton (Zhang and Cui 2017), silk (Ryan et al. 2017), polyethylene terephthalate (PET) fabric (Guo et al. 2016), and cellulose/polyester cloth (Manjakkal et al. 2020). Although PEDOT:PSS shows advantages due to its conductivity and thermoelectric properties, its application in WEs is limited due to unfavorable mechanical properties such as hardness and brittleness of thick polymer films.

Apart from conducting polymers, a very large number of other polymer-based materials were developed including ionic gels, stimulus-response polymers, liquid crystalline polymers, and piezoelectric materials (Harito et al. 2020, Wang et al. 2018).

To improve mechanical properties such as elasticity and flexibility as well as electrical properties, the composites with carbon-based nanomaterials are investigated.

13.3 COMPOSITES OF CARBON-BASED NANOMATERIALS AND POLYMERS FOR WEARABLE ELECTRONICS

In the following part of the chapter carbon-based nanomaterials—graphene, graphene oxide (GO), carbon nanotubes (CNTs), graphene nanoribbons (GNRs), and graphene quantum dots (GQDs)—are discussed. Their structure, properties, as well as composites with polymers are presented to shed light on their significance in WEs.

Carbon-based nanomaterials have superior electrical conductivity (10^4–10^5 S m^{-1}) (Stankovich et al. 2006), flexibility, stability upon stress, stretchability (Lee, Kim, and Ahn 2015), and biocompatibility (Bhattacharya et al. 2016), and they can be chemically functionalized (Bardhan 2017), and thus have been widely exploited in sensing and energy wearable devices.

13.3.1 Graphene

Graphene was discovered in 2004 and ever since has attracted tremendous scientific attention (Novoselov et al. 2004). It is composed of sp^2 hybridized C atoms organized in a honeycomb-like crystalline lattice. This lattice is presented in Figure 13.1. Due to the fact that three electrons build covenant bonds with three neighboring C atoms, one electron per C atom is free and placed in a non-hybridized p-orbital. These p-orbitals overlap and create a unique cloud where electrons can freely move. This delocalization of electrons is the reason for the unique and extraordinary graphene properties, such as electrical (10^8 S m^{-1}) and heat conductivity (5000 W m^{-1} K^{-1}), very low resistivity (~1.0 $\mu\Omega$ cm), as well as the unique and ultra-high electron mobility of 15,000–200,000 $cm^2\ V^{-1}\ s^{-1}$ due to the ballistic transport

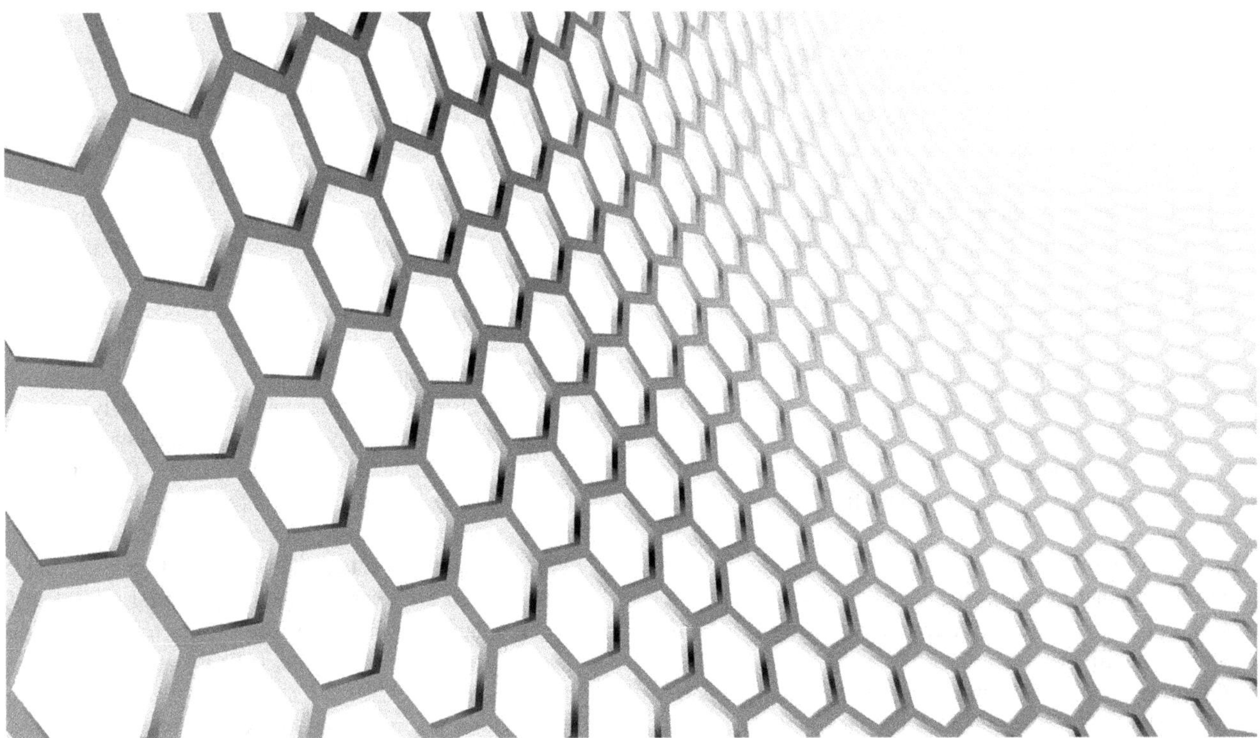

FIGURE 13.1 Hexagonal graphene sheet structure. Adapted from www.freeimages.com with permission.

of carriers, ambipolar properties, and conductivity (Basu, Basu, and Bhattacharyya 2010, Bolotin et al. 2008, Marinho et al. 2012). Graphene is a semiconductor with zero bandgap. With only one atom thickness, the highest possible specific surface area between carbon nano/materials has been measured (2600 m^2 g^{-1}), while the theoretical capacitance was calculated to be 550 F g^{-1} (Xia et al. 2009). Due to its excellent mechanical properties such as Young's modulus in the range 0.5–1 TPa, huge tensile strength at around 130 GPa, a spring constant between 1 and 5 N m^{-1}, a breaking strength of 42 N m^{-1}, and huge stretching ability (Akinwande et al. 2017, Geim and Novoselov 2007, Lee et al. 2008), graphene is an attractive and appropriate material for the production of components for wearable/stretchable electronic devices. In the visible part of the spectrum, the transparency of the single graphene layer is around 97% (Yen et al. 2014), and it is electrically sensitive to biological molecules. All these properties make graphene a good candidate for WEs.

Graphene can be produced in the form of a layer, using scotch tape exfoliation (Novoselov et al. 2004, Novoselov et al. 2005), chemical vapor deposition (CVD) (Ani et al. 2018, Lin et al. 2018, Mattevi, Kim, and Chhowalla 2011), rapid thermal annealing (Prekodravac et al. 2015, Prekodravac et al. 2016, Prekodravac et al. 2017), epitaxial growth (Kumar et al. 2016), and arc discharge (Subrahmanyam et al. 2009). Other methods lead to obtaining graphene in large amounts and a support-free state: Hummer's oxidation of graphite (Hummers and Offeman 1958, Jovanovic et al. 2020), electrochemical exfoliation (Marković et al. 2017), microwave-assisted exfoliation (Wang et al. 2014), and others.

Due to its electrical conductivity, elasticity, thermal conductivity, light weight, and electron mobility, different composites based on graphene and polymers have been produced.

One of the possible applications of a graphene–polymer nanocomposite is the production of smart textiles (Tehrani et al. 2019). One more exceptional graphene property is that it can be easily incorporated into natural or synthetic textiles (Molina 2016). By combing the polymers and graphene, mechanical properties such as strength, flexibility, and toughness can be improved. Graphene can be incorporated in textile fibers *during* synthesis or by *post-synthetic* impregnation. Approaches to achieving this are based on the deposition of graphene stabilized with a polymer onto a textile, such as:

- Polyvinylpyrrolidone (PVP) to a polymer solution (polyvinyl alcohol, PVA) or a melted phase. (Das et al. 2013)
- Graphene oxide and reduced GO (rGO) were mixed with polycaprolactone (PCL). (Ramazani and Karimi 2015)
- A conductive polymer, polyaniline/polyethylene oxide (PANI/PEO), was functionalized with graphene; electrical conductivity was enhanced from 2.4×10^{-3} to 9.9×10^{-1} S cm^{-1} by adding graphene. (Moayeri and Ajji 2015)
- A conductive polymer polypyrrole (PPy) and a graphene fiber were produced, starting from monomer pyrrole and GO, where the chemical reduction was conducted after producing fibers of PPy/GO. (Ding et al. 2014)

- PEDOT with graphene core–shell fibers showed a higher capacitance and a larger current density than pure graphene flakes. (Meng et al. 2017)
- A graphene and PEDOT:PSS composite (Qu et al. 2016) was produced in the form of a hollow fiber, with highly improved mechanical and electronic properties, appropriate for the production of flexible powering textiles for portable and wearable electronic devices.
- A composite based on poly(vinylidene-difluoride) (PVDF) and graphene functionalized with PSS was produced by mixing polymer monomers and functionalized graphene (*in situ*) (Sankar et al. 2020). The electrical conductivity of the PVDF/PSS-HE was 0.3 S cm^{-1}, which was two orders of magnitude higher when compared to the graphene-free sensor.

After the composites were produced, electrospinning, melt spinning, or other methods were used to obtained fibers for textile production (Chen, Ma, et al. 2016, Ding et al. 2014, Nilsson et al. 2013).

Graphene-polymer composites can be deposited onto a textile surface. The advantages of such an approach are the avoiding of a problem with polymer synthesis and a simple, economically favorable route that can be easily implemented at a large scale. Such examples are:

- Cotton fibers were impregnated with the dispersion of thermoplastic polyurethane and graphene, creating electrically conductive materials for WEs (Cataldi et al. 2017). Both conductive (~10 Ω sq^{-1}) and flexible textile were produced with outstanding resilience against weight-pressed severe folding as well as laundry cycles, solar irradiation, and humidity.
- GO was deposited on different fibers: nylon, cotton, and polyester, and later converted into rGO (Yun et al. 2013). The functionalization was achieved with bovine serum albumin (BSA) via electrostatic self-assembly. Amphiphilic protein albumin was adhesive and caused an increase in the GO sheet adsorption onto a polymer. To convert GO into rGO, a chemical reduction at reduced temperature was applied. Electrical conductivity above 1000 S m^{-1} was maintained after a large number of repetitive bending cycles, temperatures changing from low to high, and washing.
- The dispersion of GO in dimethylformamide (DMF) was coated onto nonwoven polypropylene (PP), then reduced into rGO with hydroiodic acid (HI). This produced a highly conductive textile with 35.6 S m^{-1} at 5.2% of graphene (Pan et al. 2017). Stability against washing conditions was tested and the material was stable.
- A flexible strain sensor was produced from cotton fabric where GO was deposited by vacuum filtration (Ren et al. 2017). To convert GO into rGO, a hot press at 180 °C for 60 min was used, leading to an increase of the C/O ratio from 1.77 to 3.72. Sheet resistance was 0.9 kΩ sq^{-1}. A strain sensor activity was kept even after 400 deformation cycles.
- Due to numerous defects in the structure of rGO, the other strategy is to transfer graphene produced in chemical vapor deposition (CVD) on polymers such as polypropylene, nylon, PP, and poly(lactic acid) (Neves et al. 2017, Torres Alonso et al. 2018). Although the process of graphene transfer can cause defect formation, the sheet resistance values were 600 Ω sq^{-1} which proved that graphene preserved its electrical conductivity.
- Post-synthetic deposition of rGO onto a polymer surface PEDOT:PSS was achieved using a layer by layer (LbL) technique (Shathi et al. 2020). This material was used to produce a sports bra. The electrical conductivity of the graphene-coated textile was 400 ± 5.0 kΩ, a sheet resistance of 50 kΩ, and an impedance of 40–150 Hz over an electric potential of 1.0 mV. This material offered the opportunity to record high-quality electrocardiograms (ECGs) and a pulse rate response.
- Graphene was produced by a CVD method and transferred onto a PE support and then to cotton fabric using an adhesive layer and laminating graphene from a PE sheet (Ergoktas et al. 2020). The produced WEs were investigated as adaptive optical textiles.
- Chitosan, polyethylene glycol (PEG), and glycerol were used as binding and dispersive agents or as matrices for producing stable graphene dispersion (Pereira et al. 2020). The flax fabric was then impregnated with graphene/polymer by the dip-pad-dry method. Electrical conductivity measured on the impregnated fabric was around 0.04 S m^{-1}. The gauge factor (GF) was 1.89 when 0.5% graphene nanoplatelets (GNPs) were used, which is explained by piezoresistive behavior.
- Graphene was deposited on fabric using spray coating techniques (Sadanandan et al. 2020, Samanta and Bordes 2020, Wang et al. 2016). GO water dispersion at a concentration between 0.05 and 0.5 weight% was sprayed onto PET fabric (Samanta and Bordes 2020) and dried at room temperature for 24 h followed by annealing at 200 °C for 2 h, and reduction with NaBH4. Electrical conductivity was between 0.3 and 2.4 mS m^{-1} and the maximum stress was 80 MPa, indicating very good mechanical and electrical properties.

WE devices for monitoring heart rate, arterial oxygen saturation (SpO_2), and respiratory rate were prepared using CVD-produced graphene (Polat et al. 2019). This sensor operated with ambient light which reduced power consumption.

Thus, the combination of graphene with polymer fibers is a promising approach for enhancing both the electrical and mechanical properties of polymers.

13.3.2 Carbon Nanotubes

Carbon nanotubes were discovered in 1991 by Iijima (1991). They are cylindrical structures that are made of sp^2 hybridized C atoms. The most simple definition is that they are graphene rolled up into a cylinder. Depending on the number of graphene sheets, they can be single (SWCNTs) or multi-wall carbon nanotube (MWCNTs). The diameter of SWCNTs is usually from 0.8 to 3.3 nm (Chen et al. 2015, Iijima and Ichihashi 1993), while the length is up to 14 (Sugime et al. 2021) and 55 cm (Zhang et al. 2013). In Figure 13.2, the structure of SWCNTs (a and b) as well as MWCNTs (c and d) are presented.

Considering they are made from only C atoms, they are highly hydrophobic and tend to form van der Waals interactions between each other. These interactions are very weak, between 0.5 and 1 kcal/mol (Moondra et al. 2018), but in the case of CNTs they are formed along the very large free surface, with the specific surface area of an individual tube calculated in the range of 900–1000 $m^2\ g^{-1}$ (Peigney et al. 2001). Thus, nanotubes are highly entangled in very stable structures, named bundles. Bundles are stable, not only due to their van der Waals interactions, but also due to the high flexibility and elasticity of the tubes. When bundles are formed, the surface area is reduced to 50 $m^2\ g^{-1}$ or lower (Peigney et al. 2001). They are the main obstacle in the application of CNTs in different fields; they block the formation of uniform CNT film, the transport of nanotubes through the cell membrane, and cause other practical issues.

To overcome this problem, the detached nanotubes were obtained through two main different approaches:

1. The covalent functionalization of CNT with different functional groups. (Georgakilas et al. 2002, Kleut et al. 2012, Tagmatarchis and Prato 2004, Vazquez and Prato 2010)
2. Non-covalent functionalization. (Jovanovic et al. 2009, Markovic et al. 2009)

Also, the combination of both covalent and non-covalent functionalization (Jovanović et al. 2009) is one of the approaches for unbundling CNTs, where mild structural modification improved the non-covalent functionalization of CNTs by using a lower amount of stabilizing polymer.

CNTs show exquisite mechanical and other properties: a strength between 11 and 63 GPa, a superior Young's modulus in the range 270–950 GPa (Yu et al. 2000), an extremely high thermal conductivity of 1000 $W\ m^{-1}\ K^{-1}$ at room temperature (Kim et al. 2001), a lightweight and low density of 1.3 to 1.4 $g\ cm^{-3}$ (Collins and Avouris 2000), a high aspect or length-to-diameter ratio of 132,000,000:1, a specific surface area around $10^3\ m^2$ (Zaman et al. 2015), and high conductivity of 10^6 to 10^7 S m^{-1} (Wang and Weng 2018). Due to these properties, CNTs are investigated solely or in polymer composites for WEs (Di et al. 2016, Lima et al. 2018, Olejník et al. 2020). In this chapter, only the composites of CNT with polymers will be discussed.

One of the first e-textiles based on CNTs was produced by Shim et al. (2008). They coated a cotton yarn with SWCNTs, MWCNTs, polyelectrolytes, Nafion-ethanol, and PSS in water. Cotton was dipped into the solution and dried. The obtained CNT-cotton yarn showed electrical resistivity of 15 Ω/cm. The authors coated the cotton yarn with both nanotubes and anti-albumin, which resulted in the creating of a biosensor for albumin detection.

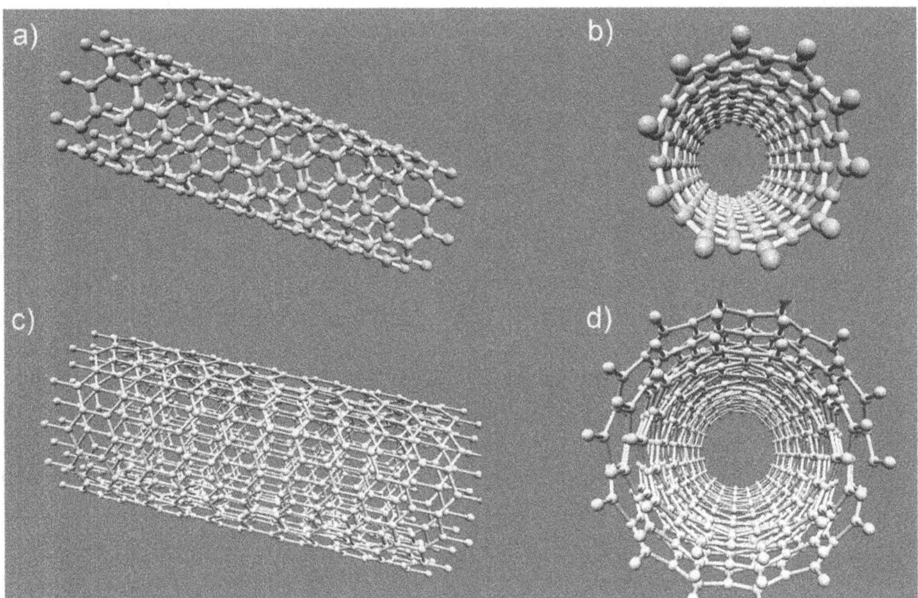

FIGURE 13.2 The structure of SWCNTs (a and b) and MWCNTs (c and d). Software Nanotube Modeler © JCrystalSoft, 2005–2018 was used for drawing.

An example of WEs based on a CNT/polymer composite is polypyrrole on CNT/cotton fiber (Lima et al. 2018). A CNT/polymer composite was prepared in several steps:

1. Cotton fibers were washed with surfactant and alcohol.
2. MWCNTs were oxidized with strong acids to incorporate carboxy and hydroxyl functional groups in an NT structure which made them more dispersible in polar organic solvents.
3. MWCNTs were sonicated (100 mg) with sodium dodecylbenzene sulphonate (SDBS, 100 mg) in 100 mL water and sonicated with cotton fiber, followed by drying.

In the next phase polypyrrole was polymerized on cotton/CNTs in two approaches:

4. Chemical polymerization by adding pyrrole, $FeCl_3$, and HCl, monomers were polymerized on cotton/CNT fibers.
5. Interfacial polymerization, where pyrrole was dispersed in a hexane solution, while $FeCl_3$ and camphor sulfonic acid was in a water solution, and cotton/CNT fiber was immersed in the water phase.

The produced material showed both good bactericidal, electrochemical, and heating properties. The electrical conductivity of 10.44 S cm^{-1}, the capacitance of 30 F g^{-1}, and a 100% bacterial colony reduction tested on *Staphylococcus aureus* showed that the produced material can be used in smart electronic devices for heat and microbial treatment on various body parts and for the storage of energy (Lima et al. 2018).

High electronic conductivity (57 S cm^{-1}) was also reported for a composite prepared of SWCNTs with a fluorinated copolymer and an ionic liquid (Sekitani et al. 2008). The produced elastic conductor showed a remarkable elongation of 134%. The mechanical properties of the polymer were also improved by adding modified MWCNTs during the process of polymer synthesis (Velasco-Santos et al. 2003). The storage modulus was increased by 1135%, while the glass transition temperature was 40 °C higher compared to the polymer without nanotubes.

A composite of CNTs and a polydimethylsiloxane (PDMS) elastomer was developed for application in WEs (Kim et al. 2018). This composite was produced by the sonication of CNTs in isopropyl alcohol, followed by the addition of methyl group-terminated PDMS (MEP), the base of the Sylgard 184 silicone elastomer kit (PDMS-A, viscosity of 3500 cSt) was added and sonicated, and then a curing agent. At the end the IPA was evaporated by heating. In this way, a highly homogeneous dispersion of the composite based on CNTs and PDMS was produced. The proposed method is a five-step approach, that takes 6 to 8 hours while the price of the final product is only \$1–5 per gram of the composite. The sheet resistance of the composite was below 20 Ω sq^{-1}, the tensile stress was 3.65 MPa, the flexibility more than 90°, and the elasticity above 45% yield strain. The material was used for creating sensors for monitoring brain, heart, and muscle signals via EEG, ECG, and EMG, respectively.

Strain sensors were produced based on CNTs and a polymer and printed on the different supports (Kanoun et al. 2015). These sensors were able to measure the strain locally and internally, depending on their geometry.

A composite based on MWCNTs and poly(octadecyl acrylate) was produced using a reversible addition-fragmentation chain-transfer (RAFT) reaction (Wang et al. 2020). The composite showed a switch-like electronic response with a positive temperature coefficient of resistance values of 7496.53% ± 3950.58% K^{-1} at 42.0 °C. This composite showed a large potential for wearable temperature sensors and other thermometric applications.

Simultaneous chemical reduction of both GO and acid oxidized MWCNTs was conducted with hydrazine and the produced material was cast onto the thermoplastic spandex-based membranes (Kong et al. 2018). The produced material showed super-stretchability (387% of elongation at the break) and electrical conductivity of 49.5 S cm^{-1}. The strain sensitivity of the material was dependent on the hybrids' content and showed great promise for stretchable energy storage/conversion devices in WEs.

The stretchable CNT/polymer sensor was produced to overcome the problems with the nonlinear response to strain variation (Fu et al. 2019). A serpentine-shaped MWCNT/PMDS nanocomposite was encapsulated in a PMDS matrix to achieve a uniform response. The elasticity of the polymer was improved by adding MWCNTs from 2 to 5 MPa of Young's modulus of elasticity, as well as density (0.97 to 1.13 g cm^{-3}), while the resistance was 607 kΩ. The sensor showed better response linearity.

To construct a fully stretchable and comfortable device for wearing on the skin or wardrobe, components such as connectors must be stretchable too. This problem was solved by Dou et al. (2020). The phases of the synthetic approach are presented in Figure 13.3. They used PDMS as a matrix, and both liquid metals and CNTs for creating a conductive network. The produced composite showed stretchability of 40% tensile strain, stabile electrical conductivity (10 S m^{-1}), and a change of resistance in deformation ($\Delta R/R$) of 16%.

A composite based on a binary polymer polyvinylidene fluoride/poly(3,4-ethylenedioxythiophene)–poly(styrene sulfonate) (PVDF/PEDOT:PSS) and SWCNTs was used for manufacturing a wearable sensor (Aziz and Chang 2018). The sensors with high linearity and a stable response in the range 25–100 °C and under different bending angles were produced. This material showed sensitivity to bending movement of 90 kΩ per degree angle and a stabile response from 0° to 120°.

CNTs were also used for producing flexible, stretchable, and wearable power systems which are highly significant for WEs. An example of a flexible and stretchable power

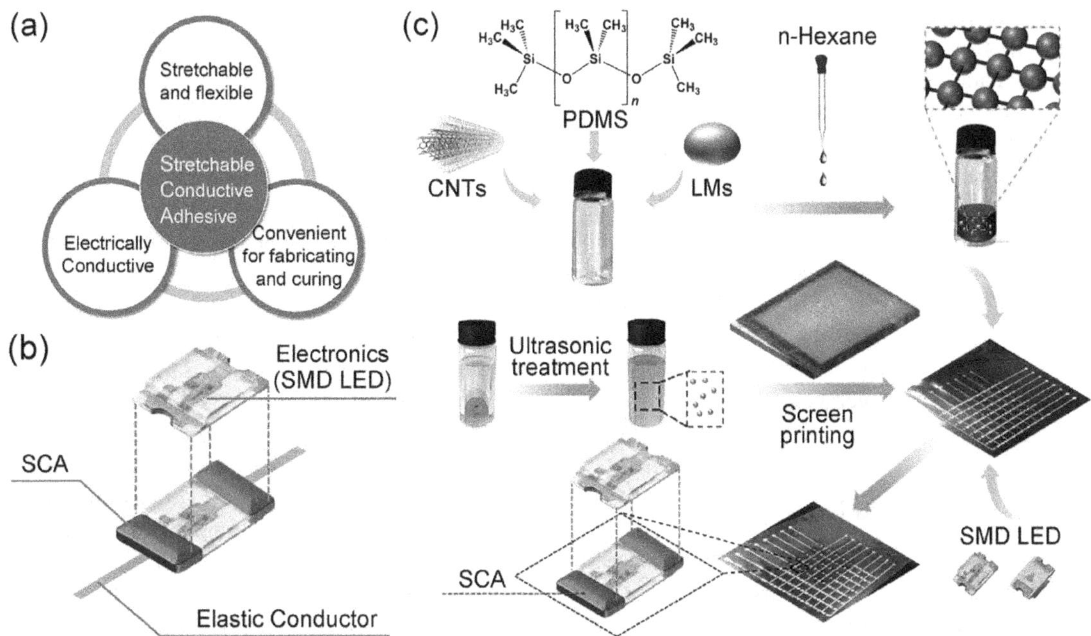

FIGURE 13.3 Stages of synthesis stretchable WE device. Dou et al. (2020) with permission from Elsevier.

supply is supercapacitors based on CNTs (Yang et al. 2013). The elastic fiber was coated with electrolyte and wrapped with CNTs. This supercapacitor showed a power density of 421 W kg^{-1} at a 1 A g^{-1} charge–discharge current density and a mass-energy density of 0.363 W h kg^{-1}.

Composites based on CNTs and polymers were largely investigated for WEs, as sensors and cathodes, in energy harvesting and storage. However, the fiber-shape of CNTs is a reason for possible health issues such as damage of DNA molecules, inflammatory skin reactions, as well as damages of the skin or organs in the form of interstitial fibrosis. Owing to the potential toxicity of CNTs, GNRs are now in focus for application in WEs.

13.3.3 Graphene Nanoribbons

Graphene nanoribbons are strips of graphene with a width below 50 nm (Celis et al. 2016, Bang et al. 2018). They show semiconducting properties due to quantum confinement and edge effects (Bang et al. 2018). The size of the energy bandgap is inversely proportional to the GNR width (Han et al. 2007). GNRs can be produced by top-down methods, such as longitudinal unzipping of nanotubes (Jovanović et al. 2014, Kosynkin et al. 2009), gamma rays cutting out graphene sheets (Marković et al. 2016, Tošić et al. 2012), and plasma etching on CNTs (Jiao et al. 2009); and bottom-up methods, such as chemical vapor deposition (Chen, Zhang, et al. 2016), epitaxy (Miettinen, Nevius, and Conrad 2019), and chemical synthetic approaches (Miettinen, Nevius, and Conrad 2019).

One approach to produce e-textiles using GNRs is to deposit a water dispersion of nanoribbons stabilized with SDS onto cotton fabric (Gan et al. 2015). The fabric was dipped inside GNR/SDS dispersion and dried. This step was done several times. Good adhesion of GNR onto cotton fabric was achieved with the assistance of SDS. The cotton fabric's OH groups and COOH or OH groups of GNR were formed with strong hydrogen bonds, while SDS increased the affinity of GNR to the fabric due to enhancing the dispersibility of the ribbons. GNR improved the thermal stability of the cotton fabric, increased tensile stress from 7.6 to 12.1 MPa, Young's modulus from 13.2 to 21.7, while tensile strain lowered from 105.7 to 56.3%. Electrical resistivity was 80 Ω for coated cotton, while the relative resistivity of the material was not changed largely upon bending (Gan et al. 2015).

Nanofibers based on GNRs and polyacrylonitrile (PAN) (Matsumoto et al. 2013) were produced. Both the mechanical and electrical properties of the material were improved. The conductivity was 165 S cm^{-1} while both tensile strain and elasticity were improved significantly (Matsumoto et al. 2013).

Similarly, graphene oxide nanoribbon (GONR)/PAN composites were used to produce films by electrospinning (Matsumoto et al. 2015). In the next step, twisting and carbonization improved the mechanical and electrical properties of the fibers. It was detected that GNRs were highly oriented in the fibers, and at a lower weight fraction the nanoribbons enhanced the mechanical characteristics of the nanofiber yarns. The authors have suggested the application of this material for flexible and wearable electronic devices and implantable medical devices.

GNRs were deposited onto Kevlar fibers using polyurethane as an interlayer (Xiang et al. 2012) and conductive fiber was produced. GNR/Kevlar fiber showed a conductivity of 20 S cm^{-1}. The authors suggested the application of this material as conductive wires in WEs and battery-heated armor due to its electrical and mechanical properties.

A new form of GNR, obtained from penta graphene, can change properties from ferromagnetic to antiferromagnetic if both electric field and bending strain are applied (He, Wang, and Zhang 2017). So-called penta GNRs showed a high potential for application in computational WEs and others.

Other research showed that fibers can be produced from vertically oriented GNTs with a high length specific capacitance of 3.2 mF cm^{-1} and a voltammetric capacitance of 234.8 F cm^{-1} at 2 mV s^{-1} (Sheng et al. 2017). This material was used as the negative electrode in an all-solid-state supercapacitor. The device showed an excellent volumetric capacitance of 12.8 F cm^{-3} and 5.7 mW h cm^{-3}, very good cycling stability, as well as tolerance to mechanical deformation. Thus it can be considered for application in WEs.

13.3.4 Graphene Quantum Dots

Graphene quantum dots are very small, circular pieces of graphene sheet functionalized with a large number of polar functional groups such as carboxyl, hydroxyl, epoxy, and ethoxy (Jovanović 2019, Li et al. 2019a). With a diameter below 100 nm, a height between 0.5 nm and a few nm, and a bandgap around 2 eV, they are semiconductive materials with strong and stable photoluminescence (Jovanovic et al. 2015, Jovanović et al. 2020, Ye et al. 2015). Due to a large number of oxygen-containing functional groups, GQDs are dispersible in water and biocompatible (Henna and Pramod 2020, Jovanović et al. 2020).

Nanofibers based on polyvinylidene fluoride (PVDF) and GQDs were produced by the electrospinning method (Choi et al. 2019). This material was investigated as a triboelectric nanogenerator (TENG) and as a wearable mechanical energy harvesting system. Nanofibers were produced by electrospinning, as schematically presented in Figure 13.4.

The material was prepared by dissolving 15 wt.% of PVDF powder in *N,N*-dimethylformamide at room temperatures for 24 h. GQDs doped with N atoms were produced from GO/polyethyleneimine composite powders, heated for 1 h. at 300 °C followed by filtration. The obtained dots were less than 5 nm and showed an extraordinary photoluminescence efficiency of 90%. To produce the PVDF/GQD composite NF, ethanol dispersion of GQDs was added to a polymer solution in different concentrations of 2.5 to 10 vol%. The applied voltage was 19 kV and electrospun material was collected onto Al sheets, while the TENG device was produced by depositing on Al thin film. The material showed strong photoluminescence centered at 453 nm (2.75 eV) due to the following electron transitions: $C\pi^* \rightarrow C\pi$, $N\pi^* \rightarrow C\pi$, and $O\pi^* \rightarrow C\pi$. It was observed that the maximum power output of the TENG device increased from 35 to 97 μW when the added amount of GQDs was higher. This effect was assigned to the β-phase formation and detrimental effect. With a higher GQD content, the β-phase of the polymer was larger. Negatively charged, conductive GQDs were embedded in the dielectric polymer which trapped the charge and stored it. Charges can be transferred from Al to polymer, but in the presence of GQDs electrical charges were generated between dots and polymers. The authors demonstrated the TENG device efficiency by connecting it with commercial light-emitting diode (LED) bulbs and proved that the device was working.

One more flexible power source based on GQDs was produced by electrodeposition (Li et al. 2019b). The ternary complex was obtained by growing graphene hydrogel (GH) on carbon fibers (CFs) and then nitrogen-doped GQDs were electrodeposited (Li et al. 2019b). In this a flexible asymmetric supercapacitor was assembled. The potential window was 2 V, and the energy density was 3.6 mW h cm^{-3} at

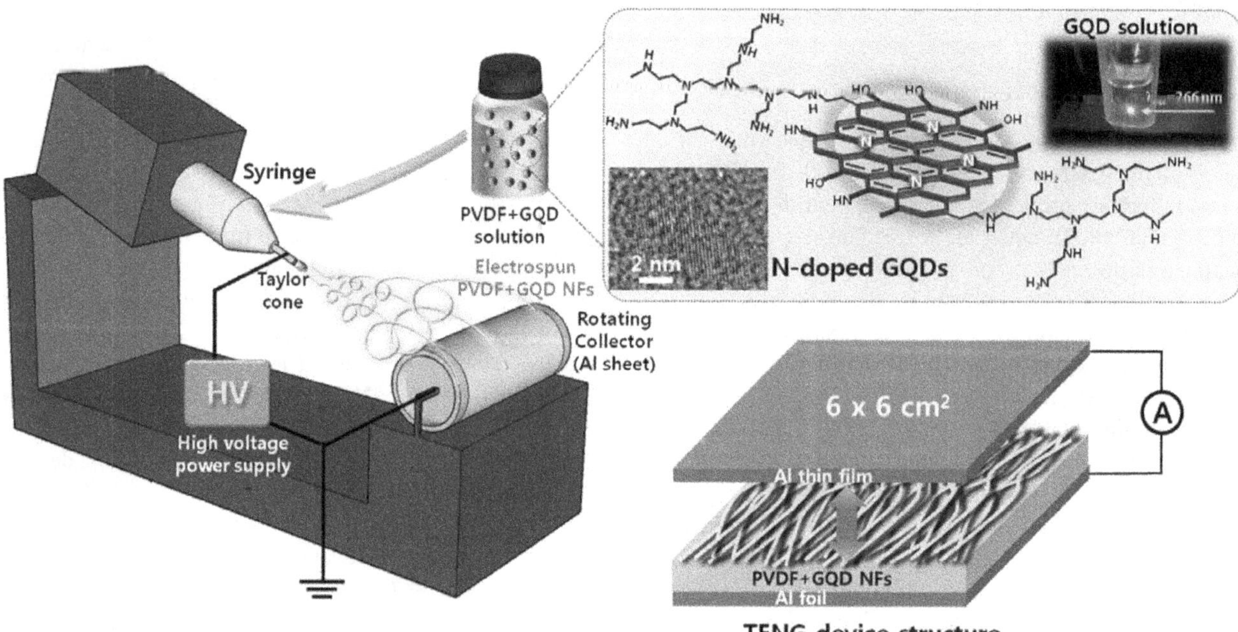

FIGURE 13.4 Production of PVDF nanofiber doped with GQDs. Choi et al. (2019) with permission from Elsevier.

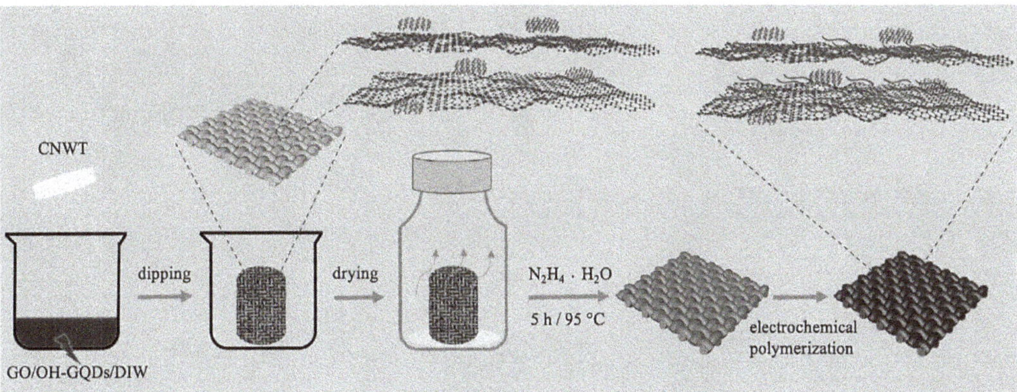

FIGURE 13.5 Deposition of GQDs/GO on textile. Shao et al. (2020) with permission from Elsevier.

35.6 mW cm^{-3}. It was predicted that the specific volumetric and gravimetric capacitance might achieve 6.6 F cm^{-3} and 36.9 F g^{-1} at a value of current density of 35.6 mA cm^{-3}. Another composite based on PET, graphene and GQDs, was assembled in a micro-flexible supercapacitor, with high transparency and a storage capacity of 9.09 µF cm^{-2} (Lee et al. 2016). One more GQD based flexible energy storage device was produced using N doped GQDs (Li et al. 2017). The specific capacitance was 461 mF cm^{-2}. One more composite based on the natural polymer lignin sulfate and GQDs was used for electrode production, showing an excellent specific capacitance of 451.7 F g^{-1} as well as good mechanical properties (Xu et al. 2020).

A GQD/PVA composite showed significant potential for application as non-volatile memory cells for flexible electronics due to bipolar non-volatile switchings (Ivanov et al. 2019). The authors prepared a composite of partially fluorinated graphene with GQDs on PVA (named PFG/PVA) using 2D printing. Resistive switchings, unipolar threshold switchings, and bipolar non-volatile ones were detected. The advantages of composites with graphene and GQDs are chemical stability, transparency, and thin-film production at room temperature on flexible and solid substrates using 2D printing. When the thickness of the PVA layer was reduced, and PFG increased, the unipolar threshold was switched to bipolar. This behavior was maintained for 1 year in the temperature range 80 to 300 K. The resistive switchings were explained by current path formation where fluorinated graphene was placed on top of PVA. These current paths included both traps in polymer and GQDs.

Figure 13.5 presents GQDs/GO composited onto textiles. Here GQD functionalized with OH groups were deposited together with GO onto a compressed non-woven towel (CNWT) using the "dyeing and drying" technique (Shao et al. 2020). In this way, 0.09 mg cm^{-2} of GQDs was deposited onto CNWT.

The presence of GQDs improves the electrochemical properties of the composite. Additionally, PANI nanoparticles were electrochemically deposited on the composite surface which led to an areal capacitance of 195 mF cm^{-1} at 0.1 mA cm^{-2} and capacity retention of 96.5% after 6000 cycles at a current density of 5 mA cm^{-2}. This material can be used as an electrode in e-textiles.

A promising self-powered electricity generator and sensor was produced using GQDs (Huang et al. 2017). This nanogenerator was triggered with moisture and delivered a voltage up to 0.27 V at 70% humidity variation and a power density of 1.86 mW cm^{-2}. The device was fabricated by depositing GQD on an Au interdigital electrode onto the PET substrate. Thanks to economically favorable synthesis and the harvesting of ambient energy, the produced power system is considered an ecologically friendly approach to supplying power.

A composite based on GQDs, PANI, and PET showed the ability to detect ammonia but at the same time exhibit flexibility, low cost, and is wearable (Gavgani et al. 2016). The conductivity of PANI was 32.8 S cm^{-1} at a 10 nA applied current while PANI/GQDs showed increased conductivity of 95.8 S cm^{-1}. It was suggested that the mechanism for ammonia detection was based on an acid-based doping/de-doping process, the carrier's mobility, and the swelling process.

A stretchable photodetector based on GQDs and wrinkled poly(dimethylsiloxane) showed high sensitivity and stretchability (Chiang et al. 2016). Graphene layers were placed onto the rippled, wavy polymer, which enhanced the formation of charge carriers in GQDs as a result of photon reflection between the ripples, as presented in Figure 13.6.

13.4 CHITIN AND ITS DERIVATIVES FOR WES

The natural polymers chitin and chitosan were investigated for WEs (Pottathara et al. 2020). Biocompatibility and flexibility are the main advantages of these polymers.

A composite based on chitin and silk fibroin was produced for application in contact-lens-type ocular sensors, a film type wireless heater, and a transparent display (Hong et al. 2018). The authors assembled a wireless, ocular glucose sensor (Farandos et al. 2015), a wireless heater for a WE (Jang et al. 2017) with chitin and silver nanofibers, and a composite thin film for a toughened glass (Choi et al. 2017).

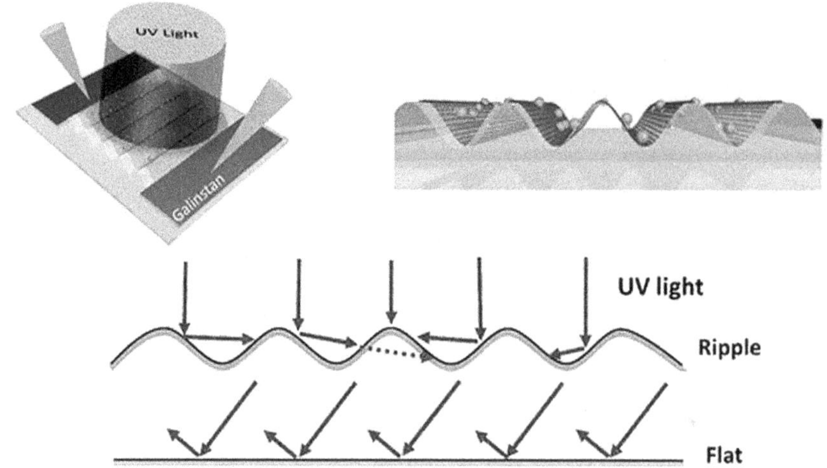

FIGURE 13.6 The mechanism of an action stretchable GQD photodetector. Chiang et al. (2016) reprinted with permission from ACS Applied Materials & Interfaces, copyright 2021 American Chemical Society.

Negatively charged chitin nanofibers, with a diameter of 5 nm and a length of 1200 nm, were produced from crab shells (You et al. 2018). These fibers were used for the stabilization of reduced GO flakes in a water solution and composite production. The composite had a modulus of elasticity of 10.6 GPa and an impressive electric conductivity of 163.1 S m^{-1}, an areal capacity of 78.6 mF cm^{-2} at a scan rate of 5 mV s^{-1}, and excellent cycling stability (over 84%) after 2500 cycles at a scan rate of 100 mV s^{-1} (You et al. 2018). Due to these properties, the composite is considered a promising candidate for assembling wearable circuits and biodegradable electronics.

A filament produced from chitin, cellulose nanofibers, and CNTs showed an electrical conductivity of 2056 S m^{-1} (Zhang et al. 2020). This filament showed the following mechanical properties: a tensile strain of 125 ± 13 MPa, a strain at break of 10.5 ± 3%, and a Young's modulus of 6.3 ± 2 GPa. The authors suggested the application of this material in WEs due to their electrical and mechanical characteristics.

A composite based on chitin, PANI, and ZnO nanoparticles was used for producing a thin film with a solution casting method (Andre et al. 2020). The composite showed the ability to sense ethanol in a range of 20–100 ppm, while the limit of detection was 127 ppm. The material showed high tensile strength (>6 MPa) but low elongation at break (<3%), an electrical resistance of 10^6 Ω, and high hydrophilicity with a contact angle of 25°.

Considering their biocompatibility and degradability, it is expected that composites based on these natural polymers will be largely investigated for WEs.

13.5 PIEZOELECTRIC ELASTOMERS AND THEIR COMPOSITES FOR WEs

Piezoelectric elastomer composites are investigated mostly as potential nanogenerators that convert vibrational or mechanical energy directly into electricity (Hu et al. 2019, Mokhtari et al. 2020). These composites are simple and compact, which makes them ideal for application in WEs.

Poly(vinylidene fluoride) (PVDF) is a piezoelectric polymer, suitable for WE energy application thanks to its high flexibility, sensitivity, ductility, low price, and biocompatibility (Guo et al. 2020). To create PVDF fibers, the most often used technique is electrospinning (Forouharshad et al. 2019). The electroactivity of a PVDF fiber is highly dependent on the fiber thickness due to changes in the ratio between the α- and β-phase (Forouharshad et al. 2019). The percentage of the β-phase can vary from 45 to 90% (Guo et al. 2020). By adding additives such as inorganic salt (like LiCl), the formation of the β-phase can be improved and lead to an enhancement in voltage output of 5 V at 100 Hz (Forouharshad et al. 2019). Using this approach, efficient energy harvesting textiles can be produced.

Moghadam et al. prepared a membrane of PVDF fibers and zirconium-based metal-organic frameworks (MOFs) for pulse monitoring in the radial artery, as shown in Figure 13.7 (Moghadam, Hasanzadeh, and Simchi 2020). They succeeded in increasing the piezoelectric coefficient by 3.4-fold, while flexibility was preserved. The improvement in piezoelectric properties was assigned to higher crystallinity and the larger β-phase content (75%). The voltage output of the composite was 568 ± 76 mV.

A numerously modified PVDF was prepared to improve mechanical and electrical properties, and these polymers and composites were investigated for application in WEs as energy harvesting components, actuators, and physical, chemical, and bio-sensors (Guo et al. 2020). The piezoelectric coefficient of a PVDF is still less than for inorganic piezoelectric materials. Thus, in the future scientists will continue to optimize synthetic procedures towards more efficient material with better skin affinity and biocompatibility, improved vertical piezoelectric coefficients, and other properties (Guo et al. 2020).

Apart from PVDF and its composites, polyolefins such as polyethylene (PE) or polypropylene (PP) are a significant

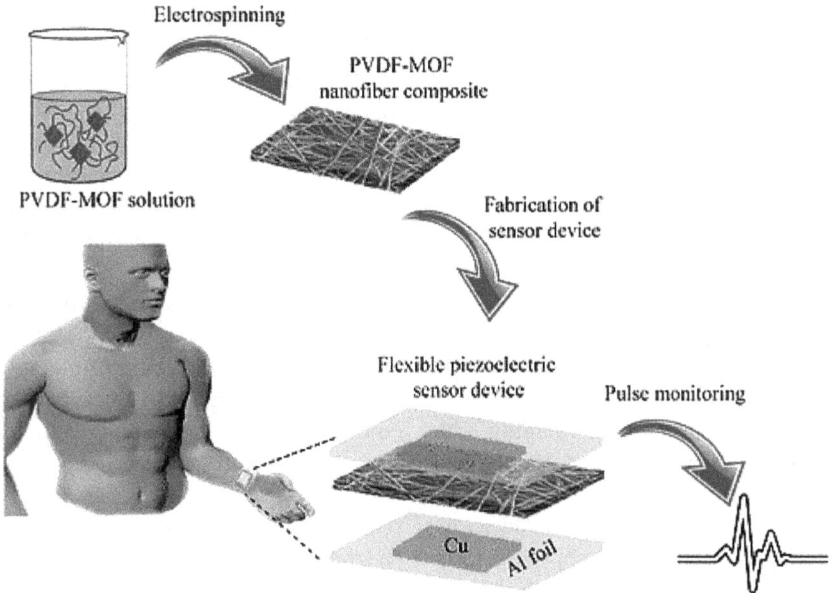

FIGURE 13.7 Schematic presentation of synthesis and application of PVDF-MOF composite. Moghadam, Hasanzadeh, and Simchi (2020) reprinted with permission from ACS Applied Nano Materials, copyright American Chemical Society.

group of polymers with piezoelectric properties, as well as polydimethylsiloxane (PDMS), poly(*p*-xylylene), and so on (Kaczmarek et al. 2019). Biopolymers with piezoelectric properties are collagen, chitin, α-keratin, and wood cellulose (Kaczmarek et al. 2019).

Piezoelectric polymers, their derivates, and composites show a high potential for application in modern medicine and microelectronics. Their advantages over classical piezoelectric ceramics and inorganic material are biocompatibility, flexibility, and low cost. Thus, further investigation of these materials will lead to producing more efficient electricity generators, sensors, and actuators.

13.6 RARE-EARTH-BASED COMPOSITES

Rare-earth-based materials/composites are crucial to modern electronics and electrical engineering. So far, tens of thousands of rare-earth-based compounds have been synthesized and characterized with properties that are sufficient for technological applications (Gavrilović et al. 2016, Jovanović 2020, Labrador-Páez et al. 2017, del Rosal et al. 2016). Depending on the application, the synthesized materials should have precisely defined functional characteristics, such as brightness, resolution, distribution spectral energies, and emission lifetimes. These properties are a consequence of certain structural and morphological characteristics of compounds, among which the most important are uniform particle size distribution, spherical morphology, high purity and homogeneous distribution activator ions, and the absence of agglomerates (Gavrilović et al. 2014, Hirano and Ohmori 2015, Yang, Yao, and Lin 2004).

Activator (rare-earth) ions are formed when electrons are removed from 6s orbitals and left as optically active 4f orbitals within filled 5s and 5p shells. The lowest bond energy in an atom (ion) has 4f electrons and represents valence electrons, responsible for the luminescent (optical) properties of these elements. The f-orbitals (inner orbitals) are the key to both chemical and spectroscopic properties of rare-earth ions. In general, luminescent ion spectra consist of groups or several sharp lines, which all correspond to one characteristic transition. Using the rare-earth ions to represent a key point in the technological development of optical devices and electronics, due to narrow lines in the spectrum of the f–f transition, the luminescence can be reduced to the visible part of the spectrum, which results in high efficiency and light intensity (Shionoya, Yen, and Yamamoto 2007). Rare-earth ions have a large number of excited energy levels that allow them to be absorbed and emit photons in the range from ultraviolet to infrared. Each ion gives a different color to the visible part of the spectrum which is a consequence of their characteristic energy transitions between their excited and basic electronic state. Activator ions are surrounded by ions of the matrix (polymer) and form luminescent centers which take place in excitation and emission processes.

Based on an enormous effort by scientists, everyday life cannot be imagined without devices based on luminescence. Due to their superior physicochemical properties, rare-earth-based composites, as a kind of luminescent material, are suitable candidates for a wide variety of applications in everyday life and wearable electronics: the next-generation computer chips, electroluminescent lamps, television cathode ray tubes (CRTs), ductile and flat-panel displays with high-definition of resolution 0 (FPD), plasma displays, electrochromic display devices, screens with liquid crystals (LCDs), electroluminescent screens with thin-film (TFEL), field emission screens (FEDs), batteries with high energy density, fluorescent and compact fluorescent lamps, diagnostic devices used in medicine (X-ray topography and positron emission tomography), solid-state lasers and laser diodes, magnets and sensors with high power and high sensitivity, safety markers, sensing thermometers, automobiles with high fuel efficiency, aerospace components, longer-lasting satellites, infrared detectors with quantum counters, and so on (Feldmann et al. 2003, Rapaport et al. 2006). Thus, applications can be classified as: light sources (fluorescent lamps, LED lasers), displays (surface emission displays, cathode ray tubes, plasma displays), and detector systems (tomography screens, scintillators). Also, recently, the possibility of high penetration near-infrared fluorescence imaging with rare-earth-based material as a new generation of versatile, biomedical nanomaterials was reported (Naczynski et al. 2013, del Rosal et al. 2016). The most important factors which should be taken into account for applying rare-earth-based materials/composites are the efficiency and lifetime of their luminescence, emission color, quenching effects, physical and chemical stability, reproducibility, availability of materials, environmental impact, and price (Yen 1998).

In recent years, there has been a great interest in the development of new synthetic composites with a unique combination of properties. Rare-earth ions in combination with different kinds of ligands/polymers present a new kind of material for practical use and potential applications. The formation of a complex between rare-earth ions and organic ligands allows larger absorption coefficients which could be used in optoelectronic applications such as amplifiers, fibers, and waveguides (Sun et al. 2005).

Most of the polymer matrices in previous reports used polymers such as PVA, poly(methyl methacrylate), PS, polyacrylic acid, and their copolymers. PVA is a widely used water-soluble polymer with good mechanical properties, remarkable physical and chemical stability, good natural degradation, and biocompatibility. Luminescent complexes/composites can be prepared by the incorporation of rare-earth elements into the polymer matrix. During the sensitization, the luminescence of the rare-earth ions in the complexes can be additionally enhanced by using synergistic agents/ligands, which provide a protective layer around the complex, which reduces the probability of energy transfer from the complex to the solvent. The luminescent efficiency is dependent on their structure, kind of ligand, the synergy between organic ligand and rare-earth ions, a combination of more earth ions, their molar ratio and concentration, as well as the electronic configuration of used rare-earth ions (Wei et al. 2020). There are some examples of PVA and rare composites and complexes composed of bidentate Schiff-base-ligand-functionalized polysulfone and Eu(III) ions (Ali 2019, Frontera et al. 2010, Gao, Zhang, and Don 2015, Hamdalla, Hanafy, and Bekheet 2015).

In the last two decades, much attention has been paid to the metal-organic frameworks (MOFs) identified as a class of porous polymeric materials consisting of metal ions/rare-earth ions linked together by organic bridging ligands. MOFs containing trivalent rare-earth ions have a lot of advantages, such as increased brightness and emission quantum yield, enhanced absorption cross-sections, broadened excitation ranges, allow excitation by cheap and safe light-emitting diodes (LEDs), as well as good photostability under UV and blue irradiation (Rocha et al. 2011). There are some examples of nanoparticles, near-infrared (NIR) and up-conversion emitting MOFs, as well as dual magnetic-photoluminescent systems with trivalent rare-earth ions: $[Gd(bdc)1.5(H_2O)_2]$ nanorods were prepared in the cationic CTAB/isooctane/1-hexanol/water microemulsion system, $Ln_2(bdc)_3(H_2O)_4$ (Ln^{3+} = Eu^{3+}, Tb^{3+}) in size of 4–5 nm prepared using an organic polymer, $[(Er_xYb_{1-x})_2(pvdc)_3(H_2O)_2]$ (x = 0.32, 0.58, 0.70 and 0.81), $[Nd_2(H_2O)_4(pdca)(Hbdc)_2]$, and $[Dy(hpd)_3Mn_{1.5}(H_2O)_6]$ (Kerbellec et al. 2008, Rieter et al. 2006, White et al. 2009, Yang et al. 2006, Zhao et al. 2006).

13.7 CONCLUSION

This chapter has shown that polymers and their composites have a large potential application in WEs. The most promising are materials that combine polymers and carbon-based nanomaterials due to their potential to improve the electrical, mechanical, and chemical properties of polymers. But special attention should be paid to the cytotoxicity of these composites, due to the ability of carbon-based nanomaterials to produce toxic species upon visible irradiation. Considering that WEs are aimed to be worn when exposed to sunlight, the analysis of biocompatibility in these exact conditions is still missing.

ACKNOWLEDGMENTS

This work was financially supported by the Ministry of Education, Science and Technological Development of the Republic of Serbia (Grant No. 451-03-68/2022-14/200017).

REFERENCES

Akinwande, Deji, Christopher J. Brennan, J. Scott Bunch, Philip Egberts, Jonathan R. Felts, Huajian Gao, Rui Huang, Joon-Seok Kim, Teng Li, Yao Li, Kenneth M. Liechti, Nanshu Lu, Harold S. Park, Evan J. Reed, Peng Wang, Boris I. Yakobson, Teng Zhang, Yong-Wei Zhang, Yao Zhou, and Yong Zhu. 2017. A review on mechanics and mechanical properties of 2D materials—Graphene and beyond. *Extreme Mechanics Letters* 13:42–77.

Ali, F. M. 2019. Synthesis and characterization of a novel erbium doped poly(vinyl alcohol) films for multifunctional optical materials. *Journal of Inorganic and Organometallic Polymers and Materials* 30:2418–2429.

Andre, Rafaela S., Danilo M. dos Santos, Luiza A. Mercante, Murilo H. M. Facure, Sergio P. Campana-Filho, Luiz H. C. Mattoso, and Daniel S. Correa. 2020. Nanochitin-based composite films as a disposable ethanol sensor. *Journal of Environmental Chemical Engineering* 8 (5):104163.

Ani, M. H., M. A. Kamarudin, A. H. Ramlan, E. Ismail, M. S. Sirat, M. A. Mohamed, and M. A. Azam. 2018. A critical review on the contributions of chemical and physical factors toward the nucleation and growth of large-area graphene. *Journal of Materials Science* 53 (10):7095–7111.

Aziz, Shahid, and Seung-Hwan Chang. 2018. Smart-fabric sensor composed of single-walled carbon nanotubes containing binary polymer composites for health monitoring. *Composites Science and Technology* 163:1–9.

Bang, Kyuhyun, Sang-Soo Chee, Kangmi Kim, Myungwoo Son, Hanbyeol Jang, Byoung Hun Lee, Kwang Hyeon Baik, Jae-Min Myoung, and Moon-Ho Ham. 2018. Effect of ribbon width on electrical transport properties of graphene nanoribbons. *Nano Convergence* 5 (1):7.

Bardhan, Neelkanth M. 2017. 30 years of advances in functionalization of carbon nanomaterials for biomedical applications: A practical review. *Journal of Materials Research* 32 (1):107–127.

Basu, Joydeep, Jayanta Basu, and Tarun Bhattacharyya. 2010. The evolution of graphene-based electronic devices. *International Journal of Smart and Nano Materials* 1:201–223.

Bhattacharya, Kunal, Sourav P. Mukherjee, Audrey Gallud, Seth C. Burkert, Silvia Bistarelli, Stefano Bellucci, Massimo Bottini, Alexander Star, and Bengt Fadeel. 2016. Biological interactions of carbon-based nanomaterials: From coronation to degradation. *Nanomedicine: Nanotechnology, Biology and Medicine* 12 (2):333–351.

Bolotin, Kirill, K. J. Sikes, M. Zhifang Jiang, Klima, G. Fudenberg, James Hone, Phaly Kim, and H. L. Stormer. 2008. Ultrahigh electron mobility in suspended graphene. *Solid State Communications* 146:351–355.

Cataldi, Pietro, Luca Ceseracciu, Athanassia Athanassiou, and Ilker S. Bayer. 2017. Healable cotton–graphene nanocomposite conductor for wearable electronics. *ACS Applied Materials & Interfaces* 9 (16):13825–13830.

Celis, Arlensiú, Maya Narayanan Nair, A. Taleb-Ibrahimi, E. Conrad, Claire Berger, W. Heer, and A. Tejeda. 2016. Graphene nanoribbons: Fabrication, properties and devices. *Journal of Physics D: Applied Physics* 49:143001.

Chen, Junjun, Xiangju Xu, Lijie Zhang, and Shaoming Huang. 2015. Controlling the diameter of single-walled carbon nanotubes by improving the dispersion of the uniform catalyst nanoparticles on substrate. *Nano-Micro Letters* 7 (4):353–359.

Chen, Shaohua, Wujun Ma, Hengxue Xiang, Yanhua Cheng, Shengyuan Yang, Wei Weng, and Meifang Zhu. 2016. Conductive, tough, hydrophilic poly(vinyl alcohol)/graphene hybrid fibers for wearable supercapacitors. *Journal of Power Sources* 319:271–280.

Chen, Zongping, Wen Zhang, Carlos-Andres Palma, Alberto Lodi Rizzini, Bilu Liu, Ahmad Abbas, Nils Richter, Leonardo Martini, Xiao-Ye Wang, Nicola Cavani, Hao Lu, Neeraj Mishra, Camilla Coletti, Reinhard Berger, Florian Klappenberger, Mathias Kläui, Andrea Candini, Marco Affronte, Chongwu Zhou, Valentina De Renzi, Umberto del Pennino, Johannes V. Barth, Hans Joachim Räder, Akimitsu Narita, Xinliang Feng, and Klaus Müllen. 2016. Synthesis of graphene nanoribbons by ambient-pressure chemical vapor deposition and device integration. *Journal of the American Chemical Society* 138 (47):15488–15496.

Chiang, Chia-Wei, Golam Haider, Wei-Chun Tan, Yi-Rou Liou, Ying-Chih Lai, Rini Ravindranath, Huan-Tsung Chang, and Yang-Fang Chen. 2016. Highly stretchable and sensitive photodetectors based on hybrid graphene and graphene quantum dots. *ACS Applied Materials & Interfaces* 8 (1):466–471.

Choi, Geon-Ju, Seong-Ho Baek, Sang-Seok Lee, Firoz Khan, Jae Hyun Kim, and Il-Kyu Park. 2019. Performance enhancement of triboelectric nanogenerators based on polyvinylidene fluoride/graphene quantum dot composite nanofibers. *Journal of Alloys and Compounds* 797:945–951.

Choi, Gwang-Mun, Jungho Jin, Dahye Shin, Yun Hyeok Kim, Ji-Hoon Ko, Hyeon-Gyun Im, Junho Jang, Dongchan Jang, and Byeong-Soo Bae. 2017. Flexible hard coating: Glass-like wear resistant, yet plastic-like compliant, transparent protective coating for foldable displays. *Advanced Materials* 29 (19):1700205.

Collins, P. G., and P. Avouris. 2000. Nanotubes for electronics. *Scientific American* 283 (6):62–69.

Das, Sriya, Ahmed S. Wajid, Sanjoy K. Bhattacharia, Michael D. Wilting, Iris V. Rivero, and Micah J. Green. 2013. Electrospinning of polymer nanofibers loaded with non-covalently functionalized graphene. *Journal of Applied Polymer Science* 128 (6):4040–4046.

del Agua, Isabel, Daniele Mantione, Usein Ismailov, Ana Sanchez-Sanchez, Nora Aramburu, George G. Malliaras, David Mecerreyes, and Esma Ismailova. 2018. DVS-Crosslinked PEDOT:PSS free-standing and textile electrodes toward wearable health monitoring. *Advanced Materials Technologies* 3 (10):1700322.

del Rosal, Blanca, Alberto Pérez-Delgado, Elisa Carrasco, Dragana J. Jovanović, Miroslav D. Dramićanin, Goran Dražić, Ángeles Juarranz de la Fuente, Francisco Sanz-Rodriguez, and Daniel Jaque. 2016. Neodymium-based stoichiometric ultrasmall nanoparticles for multifunctional deep-tissue photothermal therapy. *Advanced Optical Materials* 4 (5):782–789.

Di, Jiangtao, Xiaohua Zhang, Zhenzhong Yong, Yong Zhang, Da Li, Ru Li, and Qingwen Li. 2016. Carbon-nanotube fibers for wearable devices and smart textiles. *Advanced Materials* 28: 10529–10538.

Ding, Xiaoteng, Yang Zhao, Chuangang Hu, Yue Hu, Zelin Dong, Nan Chen, Zhipan Zhang, and Liangti Qu. 2014. Spinning fabrication of graphene/polypyrrole composite fibers for all-solid-state, flexible fibriform supercapacitors. *Journal of Materials Chemistry A* 2 (31):12355–12360.

Dou, Jiabin, Lixue Tang, Lei Mou, Rufan Zhang, and Xingyu Jiang. 2020. Stretchable conductive adhesives for connection of electronics in wearable devices based on metal-polymer conductors and carbon nanotubes. *Composites Science and Technology* 197:108237.

Ergoktas, M. Said, Gokhan Bakan, Pietro Steiner, Cian Bartlam, Yury Malevich, Elif Ozden-Yenigun, Guanliang He, Nazmul Karim, Pietro Cataldi, Mark A. Bissett, Ian A. Kinloch, Kostya S. Novoselov, and Coskun Kocabas. 2020. Graphene-enabled adaptive infrared textiles. *Nano Letters* 20 (7):5346–5352.

Fan, Xi, Wanyi Nie, Hsinhan Tsai, Naixiang Wang, Huihui Huang, Yajun Cheng, Rongjiang Wen, Liujia Ma, Feng Yan, and Yonggao Xia. 2019. PEDOT:PSS for flexible and stretchable electronics: Modifications, strategies, and applications. *Advanced Science* 6 (19):1900813.

Fan, Zeng, and Jianyong Ouyang. 2019. Thermoelectric properties of PEDOT: PSS. *Advanced Electronic Materials* 5 (11):1800769.

Farandos, Nicholas M., Ali K. Yetisen, Michael J. Monteiro, Christopher R. Lowe, and Seok Hyun Yun. 2015. Smart lenses: Contact lens sensors in ocular diagnostics *Advance Healthcare Material* 4 (6):785–785.

Feldmann, Claus, Thomas Jüstel, Cees R Ronda, and Peter Schmidt. 2003. Inorganic luminescent materials: 100 years of research and application. *Journal of Advanced Functional Materials* 13 (7):511–516.

Forouharshad, Mahdi, Simon G. King, Wesley Buxton, Philip Kunovski, and Vlad Stolojan. 2019. Textile-compatible, electroactive polyvinylidene fluoride electrospun mats for energy harvesting. *Journal of Macromolecular Chemistry* 220 (24):1900364.

Frontera, Patrizia, Concetta Busacca, Vincenza Modafferi, Pierluigi Antonucci, and Massimiliano Lo Faro. 2010. Preparation of PVA/Sm (NO3) 3-Sm_2O_3 composites nanofibers by electrospinning technique. *Advances in Science and Technology* 71:22–27.

Fu, Xiang, Ahmed M. Al-Jumaily, Maximiano Ramos, Ata Meshkinzar, and Xiyong Huang. 2019. Stretchable and sensitive sensor based on carbon nanotubes/polymer composite with serpentine shapes via molding technique. *Journal of Biomaterials Science, Polymer Edition* 30 (13):1227–1241.

Gan, Lu, Songmin Shang, Chun Wah Marcus Yuen, and Shouxiang Jiang. 2015. Graphene nanoribbon coated flexible and conductive cotton fabric. *Composites Science and Technology* 117:208–214.

Gao, Baojiao, Dandan Zhang, and Tintin Don. 2015. Preparation and photoluminescence properties of polymer–rare-earth complexes composed of bidentate Schiff-base-ligand-functionalized polysulfone and Eu (III) Ion. *Journal of Physical Chemistry C* 119 (29):16403–16413.

Gavgani, Jaber Nasrollah, Amirhossein Hasani, Mohammad Nouri, Mojtaba Mahyari, and Alireza Salehi. 2016. Highly sensitive and flexible ammonia sensor based on S and N co-doped graphene quantum dots/polyaniline hybrid at room temperature. *Sensors and Actuators B: Chemical* 229:239–248.

Gavrilović, T. V., D. J. Jovanović, V. Lojpur, and M. D. Dramićanin. 2014. Multifunctional Eu3+- and Er3+/Yb3+-doped GdVO4 nanoparticles synthesized by reverse micelle method. *Scientific Reports* 4:4209.

Gavrilović, Tamara V., Dragana J. Jovanović, Krisjanis Smits, and Miroslav D. Dramićanin. 2016. Multicolor upconversion luminescence of GdVO4:Ln^{3+}/Yb^{3+} (Ln^{3+} = Ho^{3+}, Er^{3+}, Tm^{3+}, Ho^{3+}/Er^{3+}/Tm^{3+}) nanorods. *Dyes and Pigments* 126:1–7.

Geim, A. K., and K. S. Novoselov. 2007. The rise of graphene. *Nature Materials* 6 (3):183–191

Georgakilas, V., K. Kordatos, M. Prato, D. M. Guldi, M. Holzinger, and A. Hirsch. 2002. Organic functionalization of carbon nanotubes. *Journal of the American Chemical Society* 124 (5):760–761.

Ghahremani Honarvar, Mozhdeh, and Masoud Latifi. 2017. Overview of wearable electronics and smart textiles. *The Journal of The Textile Institute* 108 (4):631–652.

Guo, S., X. Duan, M. Xie, K. C. Aw, and Q. Xue. 2020. Composites, fabrication and application of polyvinylidene fluoride for flexible electromechanical devices: A review. *Micromachines (Basel)* 11 (12):1076.

Guo, Yang, Michael T. Otley, Mengfang Li, Xiaozheng Zhang, Sneh K. Sinha, Gregory M. Treich, and Gregory A. Sotzing. 2016. PEDOT:PSS "Wires" Printed on Textile for Wearable Electronics. *ACS Applied Materials & Interfaces* 8 (40):26998–27005.

Hamdalla, Taymour A., Taha A. Hanafy, and Ashraf E. Bekheet. 2015. Influence of erbium ions on the optical and structural properties of polyvinyl alcohol. *Journal of Specroscopy* 2015:204867.

Han, Melinda Y., Barbaros Özyilmaz, Yuanbo Zhang, and Philip Kim. 2007. Energy band-gap engineering of graphene nanoribbons. *Physical Review Letters* 98 (20):206805.

Harito, Christian, Listya Utari, Budi Riza Putra, Brian Yuliarto, Setyo Purwanto, Syed Z. J. Zaidi, Dmitry V. Bavykin, Frank Marken, and Frank C. Walsh. 2020. Review—The development of wearable polymer-based sensors: Perspectives. *Journal of the Electrochemical Society* 167 (3):037566.

He, C., X. F. Wang, and W. X. Zhang. 2017. Coupling effects of the electric field and bending on the electronic and magnetic properties of penta-graphene nanoribbons. *Physical Chemistry Chemical Physics* 19 (28):18426–18433.

Henna, T. K., and K. Pramod. 2020. "Chapter 16 - Biocompatibility of graphene quantum dots and related materials." In *Handbook of Biomaterials Biocompatibility*, edited by Masoud Mozafari, 353–367. Woodhead Publishing.

Hirano, Masanori, and Toshiaki Ohmori. 2015. Direct formation and luminescence of nanocrystals in the system Eu2Sn2O7–Gd2Sn2O7 complete solid solutions. *Journal of the American Ceramic Society* 98 (12):3726–3732.

Hong, Moo-Seok, Gwang-Mun Choi, Joohee Kim, Jiuk Jang, Byeongwook Choi, Joong-Kwon Kim, Seunghwan Jeong, Seongmin Leem, Hee-Young Kwon, Hyun-Bin Hwang, Hyeon-Gyun Im, Jang-Ung Park, Byeong-Soo Bae, and Jungho Jin. 2018. Biomimetic chitin–silk hybrids: An optically transparent structural platform for wearable devices and advanced electronics. *Advanced Functional Materials* 28 (24):1705480.

Hu, Sanming, Zhijun Shi, Weiwei Zhao, Li Wang, and Guang Yang. 2019. Multifunctional piezoelectric elastomer composites for smart biomedical or wearable electronics. *Composites Part B: Engineering* 160:595–604.

Huang, Yaxin, Huhu Cheng, Gaoquan Shi, and Liangti Qu. 2017. Highly efficient moisture-triggered nanogenerator based on graphene quantum dots. *ACS Applied Materials & Interfaces* 9 (44):38170–38175.

Hummers, William S., and Richard E. Offeman. 1958. Preparation of graphitic oxide. *Journal of the American Chemical Society* 80 (6):1339–1339.

Iijima, Sumio. 1991. Helical microtubules of graphitic carbon. *Nature* 354 (6348):56–58.

Iijima, Sumio, and Toshinari Ichihashi. 1993. Single-shell carbon nanotubes of 1-nm diameter. *Nature* 363 (6430):603–605.

Ivanov, A. I., N. A. Nebogatikova, I. A. Kotin, S. A. Smagulova, and I. V. Antonova. 2019. Resistive switching effects in fluorinated graphene films with graphene quantum dots enhanced by polyvinyl alcohol. *Nanotechnology* 30 (25):255701.

Jang, Jiuk, Byung Gwan Hyun, Sangyoon Ji, Eunjin Cho, Byeong Wan An, Woon Hyung Cheong, and Jang-Ung Park. 2017. Rapid production of large-area, transparent and stretchable electrodes using metal nanofibers as wirelessly operated wearable heaters. *NPG Asia Materials* 9 (9):e432–e432.

Jiao, Liying, Li Zhang, Xinran Wang, Georgi Diankov, and Hongjie Dai. 2009. Narrow graphene nanoribbons from carbon nanotubes. *Nature* 458 (7240):877–880.

Jovanović, D.J. 2020. "Chapter 6. Syntheses, crystal structures, photoluminescence properties and applications of Ln2+/Ln3+ -doped vanadate based phosphors." In *Spectroscopy of Lanthanide Doped Oxide Materials*, edited by Vijay Pawade Sanjay Dhoble, Hendrik Swart, Vibha Chopra, Woodhead Publishing, Elsevier Ltd.

Jovanović, S. 2019. "Graphene Quantum Dots—A New Member of the Graphene Family: Structure, Properties, and Biomedical Applications." In *Handbook of Graphene Set, I-VIII*, edited by Alexander N. Chaika Edvige Celasco, Tobias Stauber, Mei Zhang, Cengiz Ozkan, Cengiz Ozkan, Umit Ozkan, Barbara Palys, Sulaiman Wadi Harun, 267–299, Wiley

Jovanović, S., T. Da Ross, A. Ostric, D. Tošić, J. Prekodravac, Z. Marković, and B. Todorović Marković. 2014. Raman spectroscopy of graphene nanoribbons synthesized by longitudinal unzipping of multiwall carbon nanotubes. *Physica Scripta* T162:014023.

Jovanovic, S., Z. Markovic, D. Kleut, N. Romcevic, M. M. Cincovic, M. Dramicanin, and B. T. Markovic. 2009. Functionalization of single wall carbon nanotubes by hydroxyethyl cellulose. *Acta Chimica Slovenica* 56 (4):892–899.

Jovanović, S. P., Z. M. Marković, D. N. Kleut, N. Z. Romčević, V. S. Trajković, M. D. Dramićanin, and B. M. Todorović Marković. 2009. A novel method for the functionalization of γ-irradiated single wall carbon nanotubes with DNA. *Nanotechnology* 20 (44):445602.

Jovanovic, S. P., Z. Syrgiannis, Z. M. Markovic, A. Bonasera, D. P. Kepic, M. D. Budimir, D. D. Milivojevic, V. D. Spasojevic, M. D. Dramicanin, V. B. Pavlovic, and B. M. Todorovic Markovic. 2015. Modification of structural and luminescence properties of graphene quantum dots by gamma irradiation and their application in a photodynamic therapy. *ACS Applied Materials & Interfaces* 7 (46):25865–25874.

Jovanovic, Svetlana, Olaf C. Haenssler, Milica Budimir, Duška Kleut, Jovana Prekodravac, and Biljana Todorovic Markovic. 2020. Reduction of graphene oxide and graphene quantum dots using nascent hydrogen: The investigation of morphological and structural changes. *Journal of Resolution and Discavery* 5 (1):1.

Jovanović, Svetlana P., Zois Syrgiannis, Milica D. Budimir, Dusan D. Milivojević, Dragana J. Jovanovic, Vladimir B. Pavlović, Jelena M. Papan, Malte Bartenwerfer, Marija M. Mojsin, Milena J. Stevanović, and Biljana M. Todorović Marković. 2020. Graphene quantum dots as singlet oxygen producer or radical quencher: The matter of functionalization with urea/thiourea. *Materials Science and Engineering: C* 109:110539.

Kaczmarek, H., B. Królikowski, E. Klimiec, M. Chylińska, and D. Bajer. 2019. Advances in the study of piezoelectric polymers. *Russian Chemical Reviews* 88 (7):749–774.

Kanoun, O., C. Müller, A. Benchirouf, A. Sanli, C. Gerlach, and A. Bouhamed. 2015. "Carbon Nanotube Polymer Composites for High Performance Strain Sensors." *1st Workshop on Nanotechnology in Instrumentation and Measurement (NANOFIM)*, University of Salento, Lecce, Italy, 24–25 July 2015.

Kerbellec, Nicolas, Laure Catala, Carole Daiguebonne, Alexandre Gloter, Odile Stephan, Jean-Claude Bünzli, Olivier Guillou, and Talal Mallah. 2008. Luminescent coordination nanoparticles. *New Journal of Chemistry* 32 (4):584–587.

Kim, Jeong Hun, Ji-Young Hwang, Ha Ryeon Hwang, Han Seop Kim, Joong Hoon Lee, Jae-Won Seo, Ueon Sang Shin, and Sang-Hoon Lee. 2018. Simple and cost-effective method of highly conductive and elastic carbon nanotube/polydimethylsiloxane composite for wearable electronics. *Scientific Reports* 8 (1):1375.

Kim, Nara, Samuel Lienemann, Ioannis Petsagkourakis, Desalegn Alemu Mengistie, Seyoung Kee, Thomas Ederth, Viktor Gueskine, Philippe Leclère, Roberto Lazzaroni, Xavier Crispin, and Klas Tybrandt. 2020. Elastic conducting polymer composites in thermoelectric modules. *Nature Communications* 11 (1):1424.

Kim, P., L. Shi, A. Majumdar, and P. L. McEuen. 2001. Thermal transport measurements of individual multiwalled nanotubes. *Phys Rev Lett* 87 (21):215502e.

Kleut, D., S. Jovanović, Z. Marković, D. Kepić, D. Tošić, N. Romčević, M. Marinović-Cincović, M. Dramićanin, I. Holclajtner-Antunović, V. Pavlović, G. Dražić, M. Milosavljević, and B. Todorović Marković. 2012. Comparison of structural properties of pristine and gamma irradiated single-wall carbon nanotubes: Effects of medium and irradiation dose. *Materials Characterization* 72:37–45.

Kong, Qingning, Zhonglin Luo, Yanbin Wang, and Biaobing Wang. 2018. Fabrication of super-stretchable and electrical conductive membrane of spandex/multi-wall carbon nanotube/reduced graphene oxide composite. *Journal of Polymer Research* 25 (11):231.

Kosynkin, Dmitry V., Amanda L. Higginbotham, Alexander Sinitskii, Jay R. Lomeda, Ayrat Dimiev, B. Katherine Price, and James M. Tour. 2009. Longitudinal unzipping of carbon nanotubes to form graphene nanoribbons. *Nature* 458 (7240):872–876.

Kumar, B., M. Baraket, M. Paillet, J. R. Huntzinger, A. Tiberj, A. G. M. Jansen, L. Vila, M. Cubuku, C. Vergnaud, M. Jamet, G. Lapertot, D. Rouchon, A. A. Zahab, J. L. Sauvajol, L. Dubois, F. Lefloch, and F. Duclairoir. 2016. Growth protocols and characterization of epitaxial graphene on SiC elaborated in a graphite enclosure. *Physica E: Low-dimensional Systems and Nanostructures* 75:7–14.

Labrador-Páez, L., D. J. Jovanović, M. I. Marqués, K. Smits, S. D. Dolić, F. Jaque, H. E. Stanley, M. D. Dramićanin, J. García-Solé, P. Haro-González, and D. Jaque. 2017. Unveiling molecular changes in water by small luminescent nanoparticles. *Small* 13 (30):28605131.

Lee, Changgu, Xiaoding Wei, Jeffrey W. Kysar, and James Hone. 2008. Measurement of the elastic properties and intrinsic strength of monolayer graphene. *Science* 321 (5887):385–388.

Lee, Keunsik, Hanleem Lee, Yonghun Shin, Yeoheung Yoon, Doyoung Kim, and Hyoyoung Lee. 2016. Highly transparent and flexible supercapacitors using graphene-graphene quantum dots chelate. *Nano Energy* 26:746–754.

Lee, Seung-Mo, Jae-Hyun Kim, and Jong-Hyun Ahn. 2015. Graphene as a flexible electronic material: Mechanical limitations by defect formation and efforts to overcome. *Materials Today* 18 (6):336–344.

Li, Meixiu, J. Tao Chen, Justin Gooding, and Jingquan Liu. 2019a. Review of carbon and graphene quantum dots for sensing. *ACS Sensors* 4 (7):1732–1748.

Li, Zhen, Yanfeng Li, Liang Wang, Ling Cao, Xiang Liu, Zhiwen Chen, Dengyu Pan, and Minghong Wu. 2017. Assembling nitrogen and oxygen co-doped graphene quantum dots onto hierarchical carbon networks for all-solid-state flexible supercapacitors. *Electrochimica Acta* 235:561–569.

Li, Zhen, Junjie Wei, Jing Ren, Xiaomin Wu, Liang Wang, Dengyu Pan, and Minghong Wu. 2019b. Hierarchical construction of high-performance all-carbon flexible fiber supercapacitors with graphene hydrogel and nitrogen-doped graphene quantum dots. *Carbon* 154:410–419.

Lima, Ravi M. A. P., Jose Jarib Alcaraz-Espinoza, Fernando A. G. da Silva, and Helinando P. de Oliveira. 2018. Multifunctional wearable electronic textiles using cotton fibers with polypyrrole and carbon nanotubes. *ACS Applied Materials & Interfaces* 10 (16):13783–13795.

Lin, Li, Bing Deng, Jingyu Sun, Hailin Peng, and Zhongfan Liu. 2018. Bridging the gap between reality and ideal in chemical vapor deposition growth of graphene. *Chemical Reviews* 118 (18):9281–9343.

Manjakkal, Libu, Abhilash Pullanchiyodan, Nivasan Yogeswaran, Ensieh S. Hosseini, and Ravinder Dahiya. 2020. Nanofibers-based piezoelectric energy harvester for self-powered wearable technologies. *Polymers* 32 (24):1907254.

Marinho, Bernardo, Marcos Ghislandi, Evgeniy Tkalya, and Cor Koning. 2012. Electrical conductivity of compacts of graphene, multi-wall carbon nanotubes, carbon black, and graphite powder. *Powder Technology - Powder Technol* 221:351–358.

Markovic, Z., S. Jovanovic, D. Kleut, N. Romcevic, V. Jokanovic, V. Trajkovic, and B. Todorovic-Markovic. 2009. Comparative study on modification of single wall carbon nanotubes by sodium dodecylbenzene sulfonate and melamine sulfonate superplasticiser. *Applied Surface Science* 255 (12):6359–6366.

Marković, Z., S. Jovanović, M. Milosavljević, I. Holclajtner-Antunović, and B. Todorovic-Marković. 2016. "Graphene nanoribbons synthesis by gamma irradiation of graphene and unzipping of multiwall carbon nanotubes." In *Graphene Science Handbook: Fabrication Methods*, edited by Guoxin Zhang and Xiaoming Sun. Boca Raton, FL: CRC Press, 361–374.

Marković, Zoran M., Danka M. Matijašević, Vladimir B. Pavlović, Svetlana P. Jovanović, Ivanka D. Holclajtner-Antunović, Zdenko Špitalský, Matej Mičušik, Miroslav D. Dramićanin, Dušan D. Milivojević, Miomir P. Nikšić, and Biljana M. Todorović Marković. 2017. Antibacterial potential of electrochemically exfoliated graphene sheets. *Journal of Colloid and Interface Science* 500:30–43.

Matsumoto, Hidetoshi, S. Imaizumi, S. Masuda, Y. Konosu, Minoru Ashizawa, and Akihiko Tanioka. 2015. Graphene nanoribbon as promising filler of composite fibers and textiles. *Chemical Fibers International* 65:237–239.

Matsumoto, Hidetoshi, Shinji Imaizumi, Yuichi Konosu, Minoru Ashizawa, Mie Minagawa, Akihiko Tanioka, Wei Lu, and James M. Tour. 2013. Electrospun composite nanofiber yarns containing oriented graphene nanoribbons. *ACS Applied Materials & Interfaces* 5 (13):6225–6231.

Mattevi, Cecilia, Hokwon Kim, and Manish Chhowalla. 2011. A review of chemical vapour deposition of graphene on copper. *Journal of Materials Chemistry* 21 (10):3324–3334.

Meng, Yuning, Lin Jin, Bin Cai, and Zhenling Wang. 2017. Facile fabrication of flexible core–shell graphene/conducting polymer microfibers for fibriform supercapacitors. *RSC Advances* 7 (61):38187–38192.

Miettinen, A., M. S. Nevius, and E. H. Conrad. 2019. The growth and structure of epitaxial graphene nanoribbons, 1–29. In *Graphene Nanoribbons*, edited by Luis Brey, Pierre Seneor, and Antonio Tejeda. Bristol: IOP Publishing.

Moayeri, Ali, and Abdellah Ajji. 2015. Fabrication of polyaniline/poly(ethylene oxide)/non-covalently functionalized graphene nanofibers via electrospinning. *Synthetic Metals* 200:7–15.

Moghadam, Bentolhoda Hadavi, Mahdi Hasanzadeh, and Abdolreza Simchi. 2020. Self-powered wearable piezoelectric sensors based on polymer nanofiber–metal–organic framework nanoparticle composites for arterial pulse monitoring. *ACS Applied Nano Materials* 3 (9):8742–8752.

Mokhtari, Fatemeh, Mahnaz Shamshirsaz, Masoud Latifi, and Javad Foroughi. 2020. Nanofibers-based piezoelectric energy harvester for self-powered wearable technologies. *Polymers* 12 (11):2697.

Molina, J. 2016. Graphene-based fabrics and their applications: A review. *RSC Advances* 6 (72):68261–68291.

Moondra, Shruti, Rahul Maheshwari, Neha Taneja, Muktika Tekade, and Rakesh K. Tekadle. 2018. "Chapter 6 - Bulk Level Properties and its Role in Formulation Development and Processing." In *Dosage Form Design Parameters*, edited by Rakesh K. Tekade, 221–256. Academic Press.

Naczynski, D. J., M. C. Tan, M. Zevon, B. Wall, J. Kohl, A. Kulesa, S. Chen, C. M. Roth, R. E. Riman, and P. V. Moghe. 2013. Rare-earth-doped biological composites as in vivo shortwave infrared reporters. *Nature Communications* 4 (1):2199.

Neves, Ana I. S., Daniela P. Rodrigues, Adolfo De Sanctis, Elias Torres Alonso, Maria S. Pereira, Vitor S. Amaral, Luis V. Melo, Saverio Russo, Isabel de Schrijver, Helena Alves, and Monica F. Craciun. 2017. Towards conductive textiles: Coating polymeric fibres with graphene. *Scientific Reports* 7 (1):4250.

Nilsson, Erik, Henrik Oxfall, Wojciech Wandelt, Rodney Rychwalski, and Bengt Hagström. 2013. Melt spinning of conductive textile fibers with hybridized graphite nanoplatelets and carbon black filler. *Journal of Applied Polymer Science* 130 (4):2579–2587.

Novoselov, K. S., A. K. Geim, S. V. Morozov, D. Jiang, Y. Zhang, S. V. Dubonos, I. V. Grigorieva, and A. A. Firsov. 2004. Electric field effect in atomically thin carbon films. *Science* 306 (5696):666.

Novoselov, K. S., D. Jiang, F. Schedin, T. J. Booth, V. V. Khotkevich, S. V. Morozov, and A. K. Geim. 2005. Two-dimensional atomic crystals. *Proceedings of the National Academy of Sciences of the United States of America* 102 (30):10451.

Olejník, Robert, Stanislav Goňa, Petr Slobodian, Jiří Matyáš, Robert Moučka, and Romana Daňová. 2020. Polyurethane-carbon nanotubes composite dual band antenna for wearable applications. *Polymers* 12 (11):2759.

Pan, Qin, Eunkyoung Shim, Behnam Pourdeyhimi, and Wei Gao. 2017. Highly conductive polypropylene–graphene nonwoven composite via interface engineering. *Langmuir* 33 (30):7452–7458.

Peigney, Alain, Christophe Laurent, Emmanuel Flahaut, Revathi Bacsa, and A. Rousset. 2001. Specific surface area of carbon nanotubes and bundles of carbon nanotubes. *Carbon* 39:507–514.

Pereira, Pedro, Diana P. Ferreira, Joana C. Araújo, Armando Ferreira, and Raul Fangueiro. 2020. The potential of graphene nanoplatelets in the development of smart and multifunctional ecocomposites. *Polymers* 12 (10). Accessed 2020/09//. doi:10.3390/polym12102189.

Polat, Emre O., Gabriel Mercier, Ivan Nikitskiy, Eric Puma, Teresa Galan, Shuchi Gupta, Marc Montagut, Juan José Piqueras, Maryse Bouwens, Turgut Durduran, Gerasimos Konstantatos, Stijn Goossens, and Frank Koppens. 2019. Flexible graphene photodetectors for wearable fitness monitoring. *Science Advances* 5 (9):7846.

Pottathara, Yasir Beeran, Hanuma Reddy Tiyyagura, Zakiah Ahmad, and Sabu Thomas. 2020. "Chapter 3 - Chitin and chitosan composites for wearable electronics and energy storage devices." In *Handbook of Chitin and Chitosan*, edited by Sreerag Gopi, Sabu Thomas and Anitha Pius, 71–88. Elsevier.

Prekodravac, J., Z. Marković, S. Jovanović, M. Budimir, D. Peruško, I. Holclajtner-Antunović, V. Pavlović, Z. Syrgiannis, A. Bonasera, and B. Todorović-Marković. 2015. The effect of annealing temperature and time on synthesis of graphene thin films by rapid thermal annealing. *Synthetic Metals* 209:461–467.

Prekodravac, J., Z. Marković, S. Jovanović, I. Holclajtner-Antunović, V. Pavlović, and B. Todorović-Marković. 2016. Raman spectroscopy study of graphene thin films synthesized from solid precursor. *Optical and Quantum Electronics* 48 (2):1–6.

Prekodravac, J. R., Z. M. Marković, S. P. Jovanović, I. D. Holclajtner-Antunović, D. P. Kepić, M. D. Budimir, and B. M. Todorović-Marković. 2017. Graphene quantum dots and fullerenol as new carbon sources for single–layer and bi–layer graphene synthesis by rapid thermal annealing method. *Materials Research Bulletin* 88:114–120.

Qu, Guoxing, Jianli Cheng, Xiaodong Li, Demao Yuan, Peining Chen, Xuli Chen, Bin Wang, and Huisheng Peng. 2016. A fiber supercapacitor with high energy density based on hollow graphene/conducting polymer fiber electrode. *Advanced Materials* 28 (19):3646–3652.

Ramazani, Soghra, and Mohammad Karimi. 2015. Aligned poly(ε-caprolactone)/graphene oxide and reduced graphene oxide nanocomposite nanofibers: Morphological, mechanical and structural properties. *Materials Science and Engineering: C* 56:325–334.

Rapaport, Alexandra, Janet Milliez, Michael Bass, Arlete Cassanho, and Hans Jenssen. 2006. Review of the properties of up-conversion phosphors for new emissive displays. *Journal of Display Technology* 2 (1):68.

Ren, Jiesheng, Chaoxia Wang, Xuan Zhang, Tian Carey, Kunlin Chen, Yunjie Yin, and Felice Torrisi. 2017. Environmentally-friendly conductive cotton fabric as flexible strain sensor based on hot press reduced graphene oxide. *Carbon* 111:622–630.

Rieter, William J, Kathryn ML Taylor, Hongyu An, Weili Lin, and Wenbin Lin. 2006. Nanoscale metal– organic frameworks as potential multimodal contrast enhancing agents. *Journal of the American Chemical Society* 128 (28):9024–9025.

Rocha, Joao, Luís D Carlos, Filipe A Almeida Paz, and Duarte Ananias. 2011. Luminescent multifunctional lanthanides-based metal–organic frameworks. *Journal of Chemical Society Reviews* 40 (2):926–940.

Ryan, Jason D., Desalegn Alemu Mengistie, Roger Gabrielsson, Anja Lund, and Christian Müller. 2017. Machine-washable PEDOT:PSS dyed silk yarns for electronic textiles. *ACS Applied Materials & Interfaces* 9 (10):9045–9050.

Sadanandan, Kavya, Agnes Bacon, Dong-Wook Shin, Saad Alkhalifa, S. Russo, Monica Craciun, and Ana Neves. 2020. Graphene coated fabrics by ultrasonic spray coating for wearable electronics and smart textiles. *Journal of Physics: Materials* 4:014004.

Samanta, Archana, and Romain Bordes. 2020. Conductive textiles prepared by spray coating of water-based graphene dispersions. *RSC Advances* 10 (4):2396–2403.

Sankar, Vetrivel, Ashwin Nambi, Vivek Nagendra Bhat, Debadatta Sethy, Krishnan Balasubramaniam, Sumitesh Das, Mriganshu Guha, and Ramaprabhu Sundara. 2020. Waterproof flexible polymer-functionalized graphene-based piezoresistive strain sensor for structural health monitoring and wearable devices. *ACS Omega* 5 (22):12682–12691.

Sekitani, T., Y. Noguchi, K. Hata, T. Fukushima, T. Aida, and T. Someya. 2008. A rubberlike stretchable active matrix using elastic conductors. *Science* 321 (5895):1468–1472.

Shao, Feng, Nantao Hu, Yanjie Su, Hong Li, Bin Li, Cheng Zou, Gang Li, Zhi Yang, and Yafei Zhang. 2020. PANI/Graphene quantum dots/graphene co-coated compressed non-woven towel for wearable energy storage. *Synthetic Metals* 270:116571.

Shathi, Mahmuda Akter, Minzhi Chen, Nazakat Ali Khoso, Md Taslimur Rahman, and Bidhan Bhattacharjee. 2020. Graphene coated textile based highly flexible and washable sports bra for human health monitoring. *Materials & Design* 193:108792.

Sheng, Lizhi, Tong Wei, Yuan Liang, Lili Jiang, Liangti Qu, and Zhuangjun Fan. 2017. Vertically oriented graphene nanoribbon fibers for high-volumetric energy density all-solid-state asymmetric supercapacitors. *Small* 13 (22):1700371.

Shim, Bong Sup, Wei Chen, Chris Doty, Chuanlai Xu, and Nicholas A. Kotov. 2008. Smart electronic yarns and wearable fabrics for human biomonitoring made by carbon nanotube coating with polyelectrolytes. *Nano Letters* 8 (12):4151–4157.

Shionoya, Shigeo, William M Yen, and Hajime Yamamoto. 2007. *Phosphor Handbook*: CRC Press.

Stankovich, Sasha, Dmitriy A. Dikin, Geoffrey H. B. Dommett, Kevin M. Kohlhaas, Eric J. Zimney, Eric A. Stach, Richard D. Piner, SonBinh T. Nguyen, and Rodney S. Ruoff. 2006. Graphene-based composite materials. *Nature* 442 (7100):282–286.

Subrahmanyam, K. S., L. S. Panchakarla, A. Govindaraj, and C. N. R. Rao. 2009. Simple method of preparing graphene flakes by an arc-discharge method. *The Journal of Physical Chemistry C* 113 (11):4257–4259.

Sugime, Hisashi, Toshihiro Sato, Rei Nakagawa, Tatsuhiro Hayashi, Yoku Inoue, and Suguru Noda. 2021. Ultra-long carbon nanotube forest via in situ supplements of iron and aluminum vapor sources. *Carbon* 172:772–780.

Sun, Li-Ning, Hong-Jie Zhang, Qing-Guo Meng, Feng-Yi Liu, Lian-She Fu, Chun-Yun Peng, Jiang-Bo Yu, Guo-Li Zheng, and Shu-Bin Wang. 2005. Near-infrared luminescent hybrid materials doped with lanthanide (Ln) complexes (Ln= Nd, Yb) and their possible laser application. *The Journal of Physical Chemistry B* 109 (13):6174–6182.

Tagmatarchis, Nikos, and Maurizio Prato. 2004. Functionalization of carbon nanotubes via 1,3-dipolar cycloadditions. *Journal of Materials Chemistry* 14 (4):437–439.

Tehrani, Farshad, Mara Beltrán-Gastélum, Karan Sheth, Aleksandar Karajic, Lu Yin, Rajan Kumar, Fernando Soto, Jayoung Kim, Joshua Wang, Shemaiah Barton, Michelle Mueller, and Joseph Wang. 2019. Laser-induced graphene composites for printed, stretchable, and wearable electronics. *Advence Materials Technologies* 4 (8):1900162.

Torres Alonso, Elias, Daniela P. Rodrigues, Mukond Khetani, Dong-Wook Shin, Adolfo De Sanctis, Hugo Joulie, Isabel de Schrijver, Anna Baldycheva, Helena Alves, Ana I. S. Neves, Saverio Russo, and Monica F. Craciun. 2018. Graphene electronic fibres with touch-sensing and light-emitting functionalities for smart textiles. *NPJ Flexible Electronics* 2 (1):25.

Tošić, D., Z. Marković, M. Dramićanin, I. Holclajtner Antunović, S. Jovanović, M. Milosavljević, J. Pantić, and B. Todorović Marković. 2012. Gamma ray assisted fabrication of fluorescent oligographene nanoribbons. *Materials Research Bulletin* 47 (8):1996–2000.

Vazquez, E., and M. Prato. 2010. Functionalization of carbon nanotubes for applications in materials science and nanomedicine. *Pure and Applied Chemistry* 82 (4):853–861.

Velasco-Santos, Carlos, Ana L. Martínez-Hernández, Frank T. Fisher, Rodney Ruoff, and Victor M. Castaño. 2003. Improvement of thermal and mechanical properties of carbon nanotube composites through chemical functionalization. *Chemistry of Materials* 15 (23):4470–4475.

Wang, Alexander J., Surendra Maharjan, Kang-Shyang Liao, Brian P. McElhenny, Kourtney D. Wright, Eoghan P. Dillon, Ram Neupane, Zhuan Zhu, Shuo Chen, Andrew R. Barron, Oomman K. Varghese, Jiming Bao, and Seamus A. Curran. 2020. Poly(octadecyl acrylate)-grafted multiwalled carbon nanotube composites for wearable temperature sensors. *ACS Applied Nano Materials* 3 (3):2288–2301.

Wang, Bin, Jinzhang Liu, Yi Zhao, Yan Li, Wei Xian, Mojtaba Amjadipour, Jennifer MacLeod, and Nunzio Motta. 2016. Role of graphene oxide liquid crystals in hydrothermal reduction and supercapacitor performance. *ACS Applied Materials & Interfaces* 8 (34):22316–22323.

Wang, Haifei, Ziya Wang, Jian Yang, Chen Xu, Qi Zhang, and Zhengchun Peng. 2018. Ionic gels and their applications in stretchable electronics. *Macromolecular Rapid Communications* 39:1800246.

Wang, Xiaobo, Heqing Tang, Shuangshuang Huang, and Lihua Zhu. 2014. Fast and facile microwave-assisted synthesis of graphene oxide nanosheets. *RSC Advances* 4 (104):60102–60105.

Wang, Yang, and George J. Weng. 2018. "Electrical conductivity of carbon nanotube- and graphene-based nanocomposites." In *Micromechanics and Nanomechanics of Composite Solids*, edited by Shaker A. Meguid and George J. Weng, 123–156. Cham: Springer International Publishing.

Wei, Yifan, Zhengquan Fu, Hao Zhao, Ruiqi Liang, Chengyu Wang, Di Wang, and Jian Li. 2020. Preparation of PVA fluorescent gel and luminescence of europium sensitized by Terbium (III). *Journal of Polymers* 12 (4):893.

White, Kiley A., Demetra A. Chengelis, Kristy A. Gogick, Jack Stehman, Nathaniel L. Rosi, and Stéphane Petoud. 2009. Near-infrared luminescent lanthanide MOF barcodes. *Journal of the American Chemical Society* 131 (50):18069–18071.

Xia, Jilin, Fang Chen, Jinghong Li, and Nongjian Tao. 2009. Measurement of the quantum capacitance of graphene. *Nature Nanotechnology* 4 (8):505–509.

Xiang, Changsheng, Wei Lu, Yu Zhu, Zhengzong Sun, Zheng Yan, Chi-Chau Hwang, and James M. Tour. 2012. Carbon nanotube and graphene nanoribbon-coated conductive kevlar fibers. *ACS Applied Materials & Interfaces* 4 (1):131–136.

Xu, Lanshu, Chen Cheng, Chunli Yao, and Xiaojuan Jin. 2020. Flexible supercapacitor electrode based on lignosulfonate-derived graphene quantum dots/graphene hydrogel. *Organic Electronics* 78:105407.

Yang, Jin, Qi Yue, Guo-Dong Li, Jun-Jun Cao, Guang-Hua Li, and Jie-Sheng Chen. 2006. Structures, photoluminescence, up-conversion, and magnetism of 2D and 3D rare-earth coordination polymers with multicarboxylate linkages. *Inorganic Chemistry* 45 (7):2857 2865.

Yang, Ping, Guang-Qing Yao, and Jian-Hua Lin. 2004. Energy transfer and photoluminescence of BaMgAl10O17 co-doped with Eu2+ and Mn2+. *Optical Materials* 26 (3):327–331.

Yang, Zhibin, Jue Deng, Xuli Chen, Jing Ren, and Huisheng Peng. 2013. A highly stretchable, fiber-shaped supercapacitor. *Angewandte Chemie International Edition* 52 (50):13453–13457.

Yano, Hirokazu, Kazuki Kudo, Kazumasa Marumo, and Hidenori Okuzaki. 2019. Fully soluble self-doped poly(3,4-ethylenedioxythiophene) with an electrical conductivity greater than 1000 S cm^{-1}. *Science Advances* 5 (4):9492.

Ye, Ruquan, Zhiwei Peng, Andrew Metzger, Jian Lin, Jason Mann, Kewei Huang, Changsheng Xiang, Xiujun Fan, Errol Samuel, Lawrence Alemany, Angel Martí, and James Tour. 2015. Bandgap engineering of coal-derived graphene quantum dots. *ACS Applied Materials & Interfaces* 7: 7041–7048.

Yen, Wen-Chun, Yu-Ze Chen, Chao-Hui Yeh, Jr-Hau He, Po-Wen Chiu, and Yu-Lun Chueh. 2014. Direct growth of self-crystallized graphene and graphite nanoballs with Ni vapor-assisted growth: From controllable growth to material characterization. *Scientific Reports* 4 (1):4739.

Yen, William M. 1998. *Phosphor Handbook* CRC Press.

You, J., L. Zhu, Z. Wang, L. Zong, M. Li, X. Wu, and C. Li. 2018. Liquid exfoliated chitin nanofibrils for re-dispersibility and hybridization of two-dimensional nanomaterials. *Chemical Engineering Journal* 344:498–505.

Yu, Min-Feng, Oleg Lourie, Mark J. Dyer, Katerina Moloni, Thomas F. Kelly, and Rodney S. Ruoff. 2000. Strength and breaking mechanism of multiwalled carbon nanotubes under tensile load. *Science* 287 (5453):637.

Yun, Yong Ju, Won Hong, Wan-Joong Kim, Yongseok Jun, and Byung Hoon Kim. 2013. A novel method for applying reduced graphene oxide directly to electronic textiles from yarns to fabrics. *Advanced Materials (Deerfield Beach, Fla.)* 25.

Zaman, Asaduz, Taslim Ur Rashid, Mubarak A. Khan, and Mohammed Mizanur Rahman. 2015. Preparation and characterization of multiwall carbon nanotube (MWCNT) reinforced chitosan nanocomposites: Effect of gamma radiation. *BioNanoScience* 5 (1):31–38.

Zhang, Kaitao, Lukas Ketterle, Topias Järvinen, Shu Hong, and Henrikki Liimatainen. 2020. Conductive hybrid filaments of carbon nanotubes, chitin nanocrystals and cellulose nanofibers formed by interfacial nanoparticle complexation. *Materials & Design* 191:108594.

Zhang, Rufan, Yingying Zhang, Qiang Zhang, Huanhuan Xie, Weizhong Qian, and Fei Wei. 2013. Growth of half-meter long carbon nanotubes based on Schulz–Flory distribution. *ACS Nano* 7 (7):6156–6161.

Zhang, Yalei, and Yue Cui. 2017. Cotton-based wearable PEDOT:PSS electronic sensor for detecting acetone vapor. *Flexible and Printed Electronics* 2 (4):042001.

Zhao, Bin, Hong-Ling Gao, Xiao-Yan Chen, Peng Cheng, Wei Shi, Dai-Zheng Liao, Shi-Ping Yan, and Zong-Hui Jiang. 2006. A promising MgII-ion-selective luminescent probe: Structures and properties of Dy–Mn polymers with high symmetry. *Chemistry A European Journal* 12 (1):149–158.

Zhou, Jian, Matthieu Mulle, Yaobin Zhang, Xuezhu Xu, Er Qiang Li, Fei Han, Sigurdur T. Thoroddsen, and Gilles Lubineau. 2016. High-ampacity conductive polymer microfibers as fast response wearable heaters and electromechanical actuators. *Journal of Materials Chemistry C* 4 (6):1238–1249.

14 Polymer-Based Organic Electronics

Sonali Verma, Bhavya Padha, Prerna, and Sandeep Arya
University of Jammu, Jammu and Kashmir, India

CONTENTS

14.1 INTRODUCTION

Nowadays, our daily life depends highly on different electronic devices. Much of these are based on silicon (Si), a technology which permits a quick switching process required to execute information processing. On the other hand, organic electronics, as the name suggests, make use of organic materials like polymers. Here, we will discuss organic electronics in the field of science and technology where polymers demonstrate unique electronic properties [1–3]. In this field, the most popular organic devices are optical actuators, such as light-emitting diodes, organic solar cells, and organic sensors. There is a large variety of materials appropriate for this field but organic electronics comprising polymers is an opportunity to prepare stretchable systems [4]. Organic electronics highly contributes to developing a smart society owing to various advantages such as device flexibility, cost effectiveness, and simple manufacturing by printing methods. Rapid progress has been reported worldwide on different organic materials for integration in devices like organic transistors, photovoltaics, and sensors [5–10].

For developing novel functional materials for commercial utilization, they are first designed and then integrated with each other, such when an organic sensor comprises a sensor, a data acquisition system, or a power supply along with a data communication function. Many studies have been done on different organic electronics but much less work on its electrical connectivity for flexible systems. Although flexible and stretchable electrical components have been designed, fully flexible devices remain unfeasible until the electric connectivity among the components attains equal flexibility. Thus, it is essential to study bonding techniques for flexible organic electronics by making use of organic soft materials like conductive polymers. The development in organic electronics depends highly on the structural and fundamental development of polymers. In 1976, Heeger, MacDiarmid, and Shirakawa reported for the first time conducting polymers [11].

The solution or melt processability of polymers facilitates various cost effective and consumer-friendly materials that have changed our day-to-day life. Further, polymers with optoelectronic properties are considered as high class materials because of their multistep synthesis as well as comparatively high cost. If properly soluble, they can be spun into fiber as well as printed into thin film, and then used in various applications where an optoelectronic property is needed. Among various polymers, the class of conjugate polymers plays an important role in organic electronics. In this chapter, we will discuss the integration of conjugated polymers in organic electronics. The chapter concentrates on different conjugated polymers and their performances in organic devices.

DOI: 10.1201/9781003169727-14

14.2 HISTORY OF CONJUGATED POLYMERS

Although the concept of conjugated polymers was introduced many years ago, their expansion and utilization as a particular research field has only taken place recently. The chronological progress of conjugated polymers has been extensively reported in the literature [12–16]. Conjugated polymers came into existence in 1834, but until the 1970s very limited study was done on them. The work of Labes et al. on inorganic polymers served as a motivation to the significant achievements made by MacDiarmid, Heeger, and Shirakawa [17]. Their observations resulted in an increased popularity of the first generation of conjugated polymers (Figure 14.1a) [18] that are relatively modest in comparison to the attractive designs of recent conjugated polymers (PM6 in Figure 14.1b). Polymers like polyaniline (PANI), polypyrrole (PPy), and poly(paraphenylenevinylene) (PPV) were considered to be high-tech polymers during the early stages of this period; however, the information garnered from investigating these simple structures is not sufficient. On the other hand, these simple polymers can be easily synthesized by using commercially available precursor materials, like aniline for PANI or thiophene for polythiophene [19]. However, this structural simplicity results in fewer strategies for synthesizing monomers and polymers as the development of organic electronics depends highly on advancements in synthesis techniques. During the 1980s, various significant results were achieved from the combined work of Heeger and Wudl [20,21]. Revolutionary studies, like the spectroscopic investigation of poly (alkylthiophenes) (P3AT) as well as the preparation of the first narrow bandgap polymer (Figure 14.1b), were the result of collaborative work [22–25]. This association results in various significant discoveries that encouraged the researchers of different fields of science to explore this underdeveloped field of conjugate polymers. From 1970 to 1990, "conjugated polymer" was just a theoretical term as, at that time, these materials had not yet been practically utilized. Then, the general technique for characterizing conjugated polymers was based on determining the conductivity in the doped state. Heeger et al. reported conductivity values for

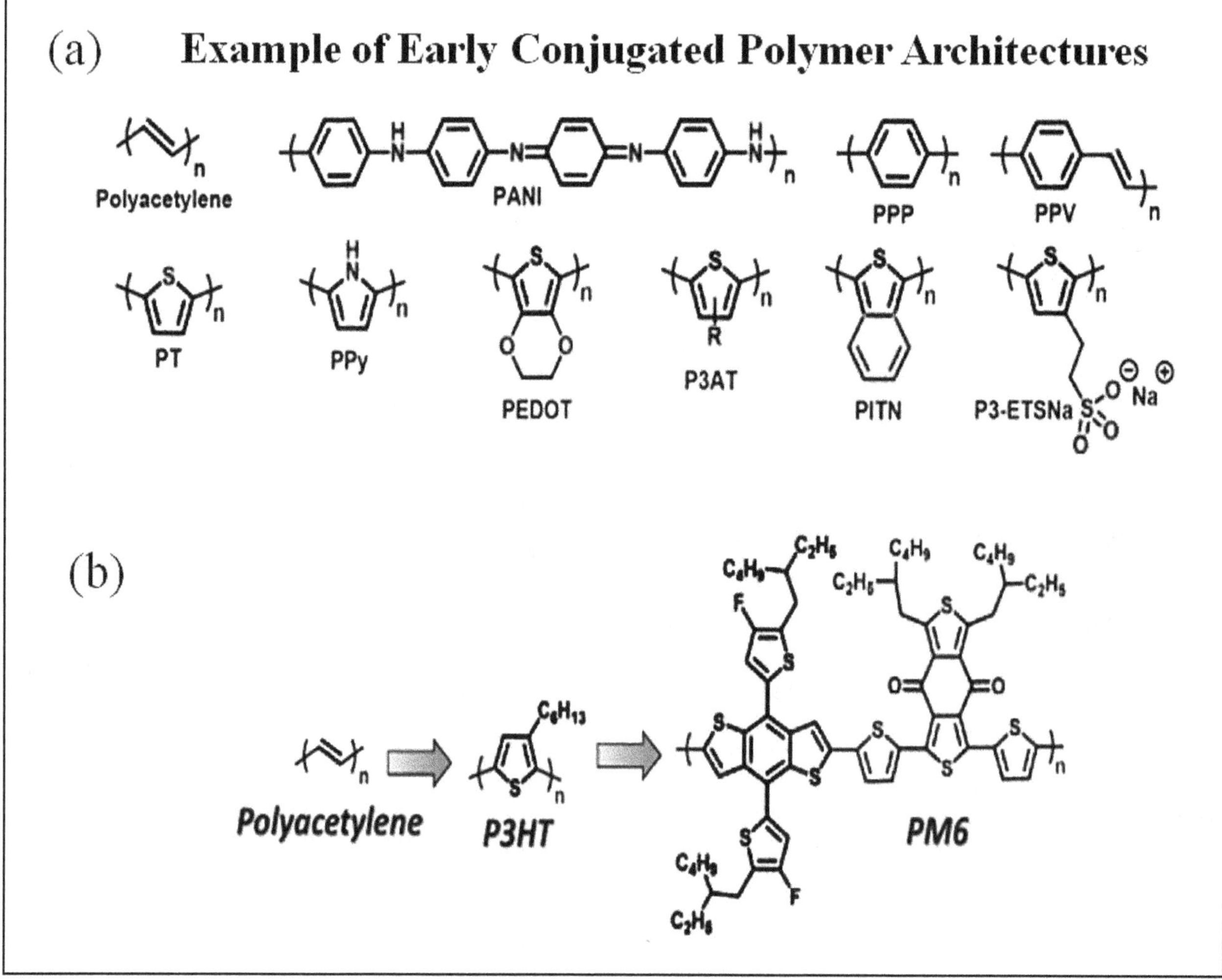

FIGURE 14.1 (a) Some early conjugated polymers; (b) from simple repeat units of polyacetylene to the synthetic undertaking of synthesizing regioregular P3HT, and to the morphological complexity of 2D-polymers like PM6. [18] with permission andcopyright Elsevier.

polythiophene, polyacetylene, and PITN as 14, 220, and 0.4 S/cm, respectively [23,24,26].

In 1989, the material which gained much attention as well as finding new applications from its description in a patent by Bayer A.G. was poly (3,4-ethylenedioxythiophene) (PEDOT) [27,28] which was synthesized via chemical oxidation with anhydrous iron (III) chloride and which resulted in a conductivity of 2.3 S/cm. However, from 1990 to 2000, the integration of conjugated polymers into organic devices became very popular. In particular, during the 1990s, the implementation of conjugated polymers in various organic devices like organic photovoltaics (OPVs), organic field effect transistors (OFETs), and organic light emitting diodes (OLEDs) or in organic lasers gainedmuch interest for researchers due to the conjugated polymers providing low cost alternatives in synthesis along with superior mechanical properties in comparison to their inorganic counterparts [29,30]. During this time, modification in polymer synthesis techniques resulted in the formation of new structures as well as their consequent utilization in organic electronic open doors for the next 10–15 years (from 2000 to the present).

14.3 CONJUGATED POLYMERS

Progress in this field increased tremendously following knowledge that their electric conductivity is enhanced significantly via electrochemical doping [31]. Alan Heeger, Alan MacDiarmid, and Hideki Shirakawa won the Nobel Prize in Chemistry (2000) for their important discovery in the field of conjugated polymers. After that, various research teams devoted much attention to such polymers in order to take advantage of their exceptional optical as well as semiconductor characteristics that open doors in the fields of plastic electronics and photonics [32]. Conjugated polymers as semiconductors have gained extensive academic as well as industrial interest for optoelectronic application. Specifically, their utilization in polymer light emitting diodes (PLEDs), polymer solar cells (PSCs), and organic field effect transistors (OFETs) provides many opportunities for resolving various energy related problems [33]. In contrast to inorganic materials, conjugated polymers offer various merits. More significantly, soluble polymer semiconductors are easily processed as well as printed, overcoming the conventional photolithography required for patterning and boosted large-scale processing of printed electronics [34].

Conjugated polymers have many applications in electronics owing to their semiconductor nature, and electrochemical as well as optical properties. They are appropriate substitutes for inorganic counterparts due to various advantages such as: being economical; solution processable, hence permitting web processing; as well as being able to be used in devices with high flexibility [35]. Moreover, it is possible to utilize conjugated polymers in applications like OPVs, OFETs, as well as OLEDs [36–39] by extensively optimizing their physical (solubility and crystallinity) as well as electronic (charge transport and light absorption) properties as a result of synthetic tunability. Furthermore, due to their synthetic modifications, they also show application in electrochromic devices, organic lasers, as well as in chemical sensors [40–43]. Initially, knowledge about conjugated polymers revolved only around their fundamental properties, like conductivity and electrochemistry, and it took many years to understand how a small change in the physical properties influences the performance according to a particular application. Thus, the structural alteration as well as the progress of conjugated polymers was a slow but continuous process over decades [44]. The development of conjugated polymers has brought improvements in synthetic techniques, permitting the synthesis of novel designs. The structural alteration of these polymers was highly application driven and explains how polymer structure influences device performance. The significant modifications in the structure of conjugated polymers arethe advancement of polymerization parameters to synthesize polymers with good morphological reliability, like regioregular P3HT, and modification in electronic properties and the integration of side chains for enhancing solubility. Moreover, the dimensionality of conjugated polymers highly influences its optoelectronic properties. The different dimensionalities of conjugated polymers and their performances in organic electronics are discussed below in detail.

14.4 ONE-DIMENSIONAL (1D) CONJUGATED POLYMERS

The majority of these exhibit 1D-conjugated backbones. Initially focus was mainly on polyphenylenevinylenes (PPVs) and poly-paraphenylenes (PPPs), but after some time, different polyphenylene derivatives like ladder-type poly-para-phenylenes (LPPP) as well as shorter bridged stepladder polymers gained in interest. Moreover, polyfluorenes (PFs) and polycarbazoles (PCz) have been extensively investigated for application in organic devices such as PLEDs and PSCs. Conjugated polymers comprising thiophene building blocks show applications in OFETs. All such building blocks that served as donors are integrated with the appropriate acceptors for the development of donor-acceptor (D-A) copolymers that result in fast advancements in charge transportation as well as in photovoltaic polymer materials.

14.4.1 Conjugated Polyphenylenes

Forthe last 20 years, conjugated polyphenylenes have been a highly investigated category of polymers showing application in organic electronics [45]. Poly-para-phenylene (PPP1), as shown in Scheme 14.1, and its derivatives were introduced in the 1960s and show various significant applications [46,47]. Unfortunately, poly-para-phenylene without any substitution is insoluble as well as inflexible which results in its limited applications. The solubility of PPP can

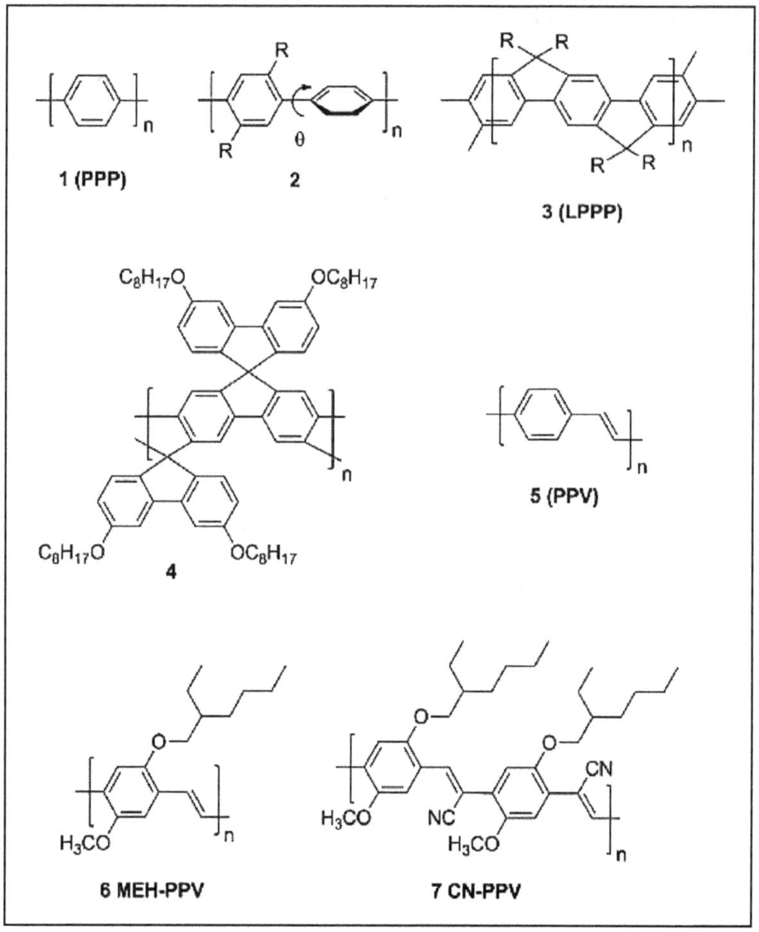

SCHEME 14.1 Linear and ladder-type polyphenylenes and polyphenylenevinylenes. [46] with permission andc Elsevier.

be enhanced via incorporating specific dopants into the main chain that also results in increased torsion among the adjoining repeating units and hence disturbs the conjugation [48,49]. Thus, for enhancing the conjugation, various techniques have been adopted that include the utilization of phenylenvinylenes in the polymer main chain. These techniques proved to be very effective in restricting rotation among the repeating units and hence enhance the solubility as well as the stability, and also modify the electrochemical characteristics. The optoelectronic properties of these conjugated polyphenylenes have been discussed in detail. Moreover, in the past decade, both PFs and PCz demonstrate great potential for organic electronic applications, so their discussion also becomes important.

14.4.1.1 Linear and Ladder-Type Polyphenylenes

PPP1 is thermally stable as well as exhibiting a wide band gap (3.5 eV) and hence serves as an active material in blue PLEDs [50] though it is not solution processable. However, the side chains incorporated into the phenylenes (2) enhance the solubility, while the emission shifts to the UV region. Thus for obtaining visible color emission, the adjoining phenylene units are planarized by means of a methine bridging. Regarding this, fully ladderized poly-para-phenylenes (3) (LPPPs) have been introduced [51,52]. Also, such a fully planar arrangement of the polymer backbone strongly enhances the conjugation that was revealed via the bathochromic shift in optical absorption. The photoluminescence spectra of 3 (LPPP) shows a red-shift in comparison to the non-bridged PPP, though both the photoluminescence as well as electroluminescence shifted to 600 nm [53] as a result of excimer formation and ketone defects [54]. Further, for overcoming the emission in the yellow–green region, stepladder copolymers with dialkylated phenylenes introduced into the backbone of LPPPs were designed that suppress excimer production as well as showing blue emission in the photoluminescence and electroluminescence spectrum due to deformed configuration [55]. Another derivative of LPPP (4) comprising spiro-fluorene was also designed in which no low energy emission was detected [56]. Another linear conjugated polymer obtained through phenylenes is polyphenylenevinylene (PPV) that is an extremely popular semiconductor polymer exhibiting wide application in PLEDs as well as in PSCs. Initially in 1990, the non-substituted PPV 5 was synthesized via solution-processing and served as an electroluminescent active material in OLEDs showing green–yellow emission [57].

Later on, the focus shifted towards enhancing the solubility of PPV via incorporating side chains into the backbone [58] that resulted not only in solution-processable

PPV based polymers but that also tunes the color emission. One of the highly renowned PPVs with high solubility is poly[2-methoxy-5- (2 -ethylhexyloxy)-1,4-phenylene vinylene] (6) that was also the first conjugated polymer used in the fabrication of a flexible "plastic" orange LED with an external quantum efficiency (EQE) of 1% [59]. Usually, the derivatives of PPV serve as the donor units. Via incorporating electron-withdrawing cyano groups, excellent internal efficiencies (up to 4%) were attained with polymer 7 as the emitting layer [60].

14.4.1.2 Stepladder Polyphenylenes with Bridging Atoms

For preparing large-area LEDs, a polymer having an emission in the blue region is required. Linear polyphenylenes, apart from PPP, LPPP, or PPV, do not fulfill the above criteria. Thus, more interest developed in designing stepladder polymers, termed as "partially ladderized PPPs," in which the neighboring phenylene units were planarized via methine groups (Scheme 14.2) or bridge-head atoms like silicon and phosphorous. Moreover, the carbon bridged stepladder polyphenylenes, owing to their comparatively facile synthesis, have fewer defects as well as superior photophysical properties when compared with the completely ladderized polymers that had become the main center of focus. Additionally, the absorption as well as the fluorescence characteristics of the stepladder polyphenylenes is enhanced via altering the size of the bridging unit. In particular, the highest emission peak of polymer 8 (Scheme 14.2) is around 420 nm, whereas polymers 9, 10, and 11 show a red shift from 430 to about 440–445 nm (Figure 14.2a) [61]. Now before moving ahead, it is important to first discuss PFs, one of the most popular categories of this field.

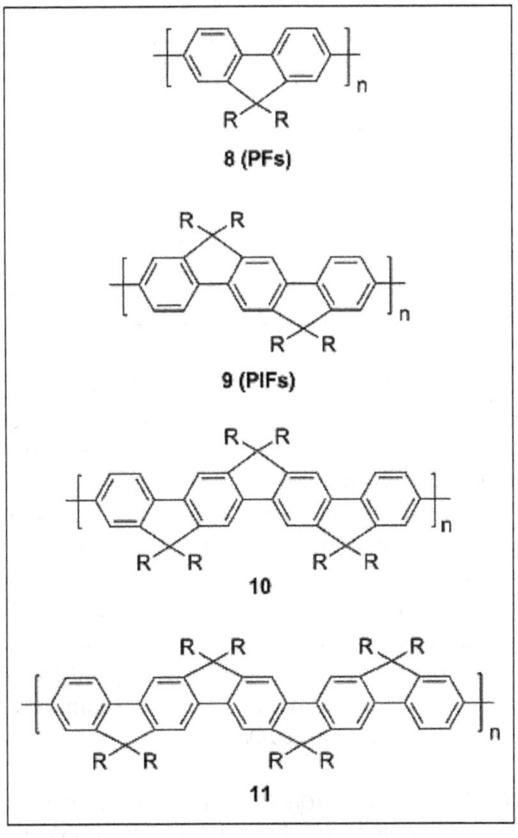

SCHEME 14.2 Stepladder polyphenylenes having methine groups. [46] with permission andcopyright Elsevier.

14.4.1.2.1 Polyfluorenes (PFs)

PFs are considered as glitter in this field owing to their attractive application in different organic optoelectronics like LEDs, FETs, OPVs, biosensors, and lasers [62]. PFs, including coplanar fluorene as a repeating unit, reveal that every adjoining phenylene pair in a PPP is flattened through methylene groups [63]. Moreover, the facile functionalization of fluorene at positions 2 and 7 allows the simple formation of polymerizable monomers, leading to PF homo- and copolymers that fulfill the necessary conditions for application in organic electronics. During the last decade, various fluorene-based polymers were extensively explored.

14.4.1.2.1.1 PFs: Polymers for Blue PLEDs These are very promising for blue PLEDs owing to their structure tunability, leading to easy functionalization as well as excellent properties, like high chemical and thermal stability, along with a good fluorescence quantum yield [64]. On the other hand, soluble PFs with an alkyl substitution at position 9 show some limitations, such as electroluminescence efficiency and insensitive emission to human eyes. Figure 14.2b shows the relationship between eye sensitivity and color wavelength and color luminosity [46]. The human eye is highly sensitive to yellow–green emissions; however, a decrease in sensitivity is observed on moving toward the red and blue regions. Particularly in the blue region, the human eye shows sensitivity in the range of 440–450 nm [65–67] though PFs usually show emission at 420 nm. Consequently, for good visibility, PFs must emit in longer wavelengths. Furthermore, for ensuring good color purity, Commission Internationale de l'Éclairage (CIE) coordinates should only be in the range of 450–485 nm. One of the solutions to overcome this problem is to design new polyphenylenes with a chemical structure in-between PFs and LPPPs, such as ladderized polytetraphenylene (10) and polypentaphenylene (11). The emission of PFs can also be tuned to the pure blue region via introducing a relatively low energy material that traps the excitons and which turns out to be an emitter. Such a system is known as a "host-guest"/"host-dopant," in which the PF acts as a host and guest (dopant) and is incorporated as a low energy unit.

14.4.1.2.1.2 PFs: Hosts for Red, Green, Blue, as well as for White PLEDs PFs serve as outstanding hosts for the realization of high performance red-green-blue (RGB) and white PLEDs via chemical doping or physical blending due to their outstanding photoluminescence efficiency, wide band gap, as well as superior charge carrier transportation characteristics. Here, the focus is on a host/dopant system (chemical variations) in which a rapid Förster

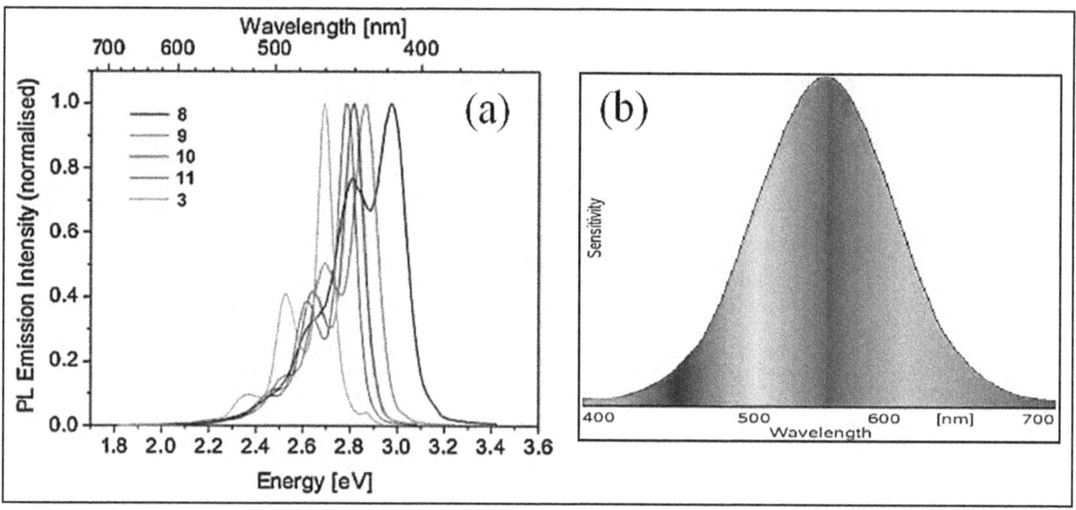

FIGURE 14.2 (a) Fluorescence of stepladder polyphenylenes 8, 9, 10, 11, and LPPP 3. (b) Light sensitivity curve of human eyes with respect to colors. [46] with permission andcopyright Elsevier.

TABLE 14.1
Performance of Some Different Color Based PF Polymers

Color	V_{onset} (V)	B_{max}(cd/m²)	LE_{max}(cd/A)	PE_{max} (lm/W)	EQE (%)	CIE1931 (x, y)	Reference
Red							
	~3.5	~10,000	8.3	7.5	—	0.63, 0.35	[78]
	10.6	~850	1.45	—	2.54	0.66, 0.34	[74]
Green							
	4.8	21,595	7.43	2.96	—	0.26, 0.58	[72]
	12	—	0.9	—	0.6	0.36, 0.56	[69]
Blue							
	4.4	4080	0.63	0.19	1.21	0.19, 0.14	[75]
	6.0	1600	0.19	0.06	0.20	0.17, 0.12	[76]
	3.3	6539	3.43	—	2.42	0.15, 0.16	[68]
White							
	3.5	18,480	12.8	8.5	5.4	0.31, 0.36	[77]
	3.1	6770	2.96	2.33	1.37	0.34, 0.35	[78]
	5.8	12,040	18.0	8.38	6.36	0.33, 0.35	[79]

energy transfer occurs that results in a highly efficient emission [68,69]. Simultaneously, the color stability is enhanced by the trapped excitons produced on the PF segment. The performance of some different colors based on PF polymers is illustrated in Table 14.1. The copolymer comprising benzothiadiazole (BTZ) in the PF backbone shows application in OLEDs for green emission. Such a polymer based highly stable device exhibits a low value of turn-on voltage and luminous efficiency (LE) of 10.5 cd/A at 6600 cd/m² at 4.85V [70]. Further, if the contents of the BTZ material are tuned, a peak efficiency as high as 10.9 cd/A at 22,800 cd/cm² is achieved [71]. Moreover, by introducing a naphthalimide derivative, a highest LE of 7.45 cd/A at (0.26, 0.58) is attained [72]. Red light-emitting polymers are designed via introducing small-molecule red emitters in PFs. The attachment of two alkyl substituents at thiophene units results in an enhanced fluorescence quantum yield, though the electroluminescence efficiency and CIE are the same as those of the parent polymer [80]. In addition, benzoselenadiazole and naphthoselenadiazole have also been reported as narrow band gap units for achieving red emissions [81,82]. In particular, Jian Yang et al. [83] reported a polymer showing an emission at 657 nm and an EQE of 3.1%. The dyes of perylene attached with a PF backbone in the form of end-capping groups, as pendant side groups or as co monomers in the backbone, also help in achieving red emissions. From such polymers, the highest electroluminescence efficiency obtained is 1.6 cd/A [69].

14.4.1.2.1.3 PFs: Electron Rich Material for PSCs A PF has been reported as an able candidate for polymer solar cells (PSCs). By the copolymerization of fluorene units using aromatic groups, the absorptions of the obtained

TABLE 14.2
Performance of Some PF-Based PSCs

Short-Circuit Current Density (mA/cm²)	Open-Circuit Voltage (V)	Fill Factor	Power Conversion Efficiency (%)	Reference
2.0	0.78	0.50	0.78	[84]
4.66	1.04	0.46	2.2	[84]
~6	~1	0.63	3.7	[86]
5.35	0.52	0.50	1.38	[87]
8.88	0.59	0.42	2.2	[88]

polymers show a red shift. In such copolymers, the fluorene groups enhance the solubility by incorporating solubilizing groups at position 9 without disturbing the delocalization of the electrons in the entire copolymer main chain. Usually, they serve as donors that are blended with fullerene derivatives like [6, 6]-phenyl-C_{61}-butyric acid methyl ester (PCBM) for fabricating bulk-heterojunction solar cells. Table 14.2 demonstrates the performance of different PF-based PSCs.

14.4.1.2.2 C-Bridged Stepladder Polyphenylenes

In addition to polyfluorenesand polyindenofluorenes, ladderizedpolytetra- and polypentaphenylenes are the well-known C-bridged polyphenylenes. Initially, their homopolymers with alkyl substitution were designed on the basis of bathochromic shift in emissions of the PF to the blue region due to the fact that their chemical structure lies in the space between the PF and LPPP. Consequently, various ladderized multi-phenylene monomer derivatives were synthesized in order to study them. In this regard, copolymers obtained from indenofluorene (IF) became very popular among researchers.

14.4.1.2.3 PCz and Heteroatom Bridged Stepladder Polyphenylenes

The chemical as well as the electronic properties of a polymer can also be modified via incorporating N atoms in the place of methylene groups into polyphenylene. The most simple N-bridged polyphenylenes is PCz which is among the highly explored organic semiconductors exhibiting excellent photoconductive features along with the potential of forming charge transfer complexes [89]. Some different stepladder polyphenylenes with incorporated N-bridges [90,91] were also synthesized for applications in PSCs and PLEDs.

In addition, some other heteroatoms such as Si, Ge, S, and P were also used for replacing the methine bridges. Dibenzosilole (or silafluorene) having an Si atom at the bridging position is comparable to fluorene. Si exhibits analogous electronic properties to that of C but, in Si, no oxidation takes place for forming a ketone. As a result, it is expected that dibenzosilole-based homopolymers will not show an extra band corresponding to the green region as in the case of PF. In order to investigate this, poly(2,7-dibenzosilole) having alkyl substitutions has been investigated; the electroluminescence emission analogous to PF was observed with no degradation even at 200°C while under the same conditions; its PF analog results in a strong green emission [92]. Further, for preparing a blue light-emitting copolymer, the introduction of dibenzosilole into the PF backbone is very effective [93]. The introduction of 3,6-dibenzosilole into polydioctylfluorene not only restrains the formation of supplementary emission bands but also improves the color purity as well as the LE. The device shows an LE of 2.02 cd/A and an EQE of 3.34% with CIE coordinates of (0.16, 0.07) [93].

14.4.2 Polycyclic Aromatic Hydrocarbon (PAH)-Based Conjugated Polymers

PAHs comprise fused aromatic rings without any substituent. PAHs, owing to their outstanding electronic and optoelectronic properties along with the potential for organic devices attract researchers all around the world [94]. Figure 14.3 illustrates some common PAHs [95]. Among them, the simple PAHs, for example phenanthrene and pyrene, were used for synthesizing conjugated polymers. Some additional PAHs having methylene bridges were also synthesized for constructing new semiconductor polymers. The optical and electronic characteristics of such PAH-based polymers were properly scrutinized and it was observed that some also showed potential in PSCs and PLEDs. Further, for synthesizing poly(2,7) or poly(3,6)-phenathrylene polymers, the polymerization of phenanthrene at different positions is very effective. Moreover, the attachment of alkyl or aryl groups at positions 9 and 10 of phenanthrene enhances its solubility as well as processability. First, with the attachment of alkyl, strong aggregation was revealed via large bathochromical shifting in fluorescence; such aggregation would then be overcome by substituting an aryl group. The aryl substituted polymers show deep blue electroluminescence as well as high color stability in PLEDs [96].

Additionally, triphenylene, a small discotic PAH, served as a monomer for synthesizing homo-and co-polymers for PLEDs [97,98]. As a result of the twisted phenyl rings around a triphenylene core, π-π stacking does not occur in the polymer, leading to an analogous photoluminescence spectrum in both the film and solution. Similarly, another PAH named pyrene, owing to its excellent properties, also attracts researchers of various scientific fields, making this PAH very popular in fundamental as well as in photochemical research [99]. For synthesizing pyrene-based polymers, pyrene can be linked into the main chains via three ways: (i) 2, 7-linked, (ii) 1, 6-linked, and (iii) 1, 3-linked (Figure 14.4).

Further, for electrophilic substitution, positions 1, 3, 6, and 8 are highly reactive, whereas the remaining two positions are not directly accessible. Considering this fact, polypyrenylene, with substitution at positions 2 and 7, having four aryl

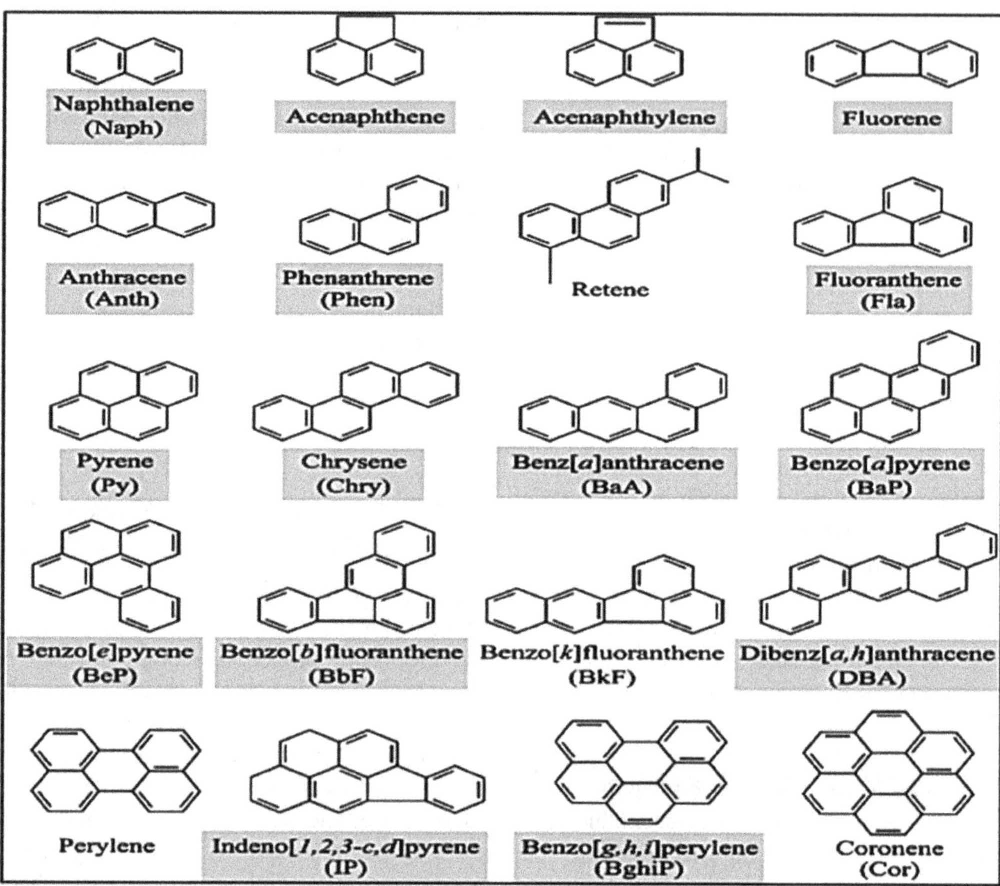

FIGURE 14.3 Examples of some polycyclic aromatic hydrocarbons (PAHs) [95].Creative Commons Attribution (CC BY) license.

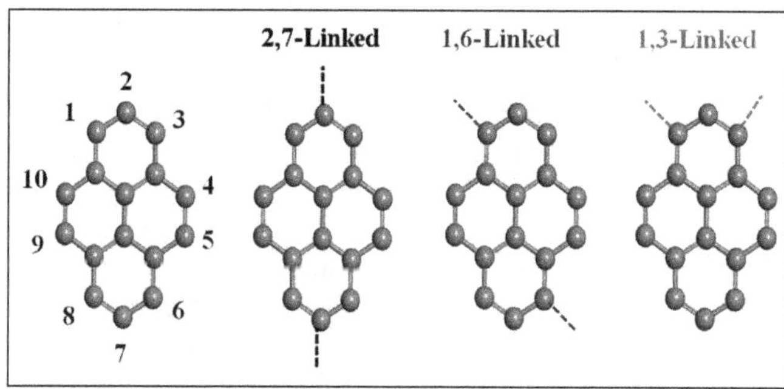

FIGURE 14.4 The numbered positions of pyrene and different options to extend pyrene into polymers.

substituents, was prepared via incorporating dopants before annulation of pyrene. Such a polymer shows some interesting blue fluorescence phenomena [100]. In the case of 2,7-linked poly(tetraalkoxypyrenylene), the maximum fluorescence in both solutions as well as in films must be nearly analogous, such that the alkoxy substituent hinders aggregation, inducing a red shift in fluorescence spectra. However, in LEDs, a prominent red shift comes into view.

14.4.3 Thiophene-Containing Conjugated Polymers

Thiophene-based conjugated polymers have gained ample research interest as an active material in the field of organic electronics. As a result of various studies on thiophene over a long time, it is now possible to tune its electronic properties via different attractive alterations in its structure. Polymers comprising thiophene moieties exhibit fascinating optical, electronic, and redox properties along with a unique self-assembling property. Furthermore, the excellent polarizability of S-atoms in thiophenes highly contributes to stabilize the conjugated chains as well as to enhance charge carrier transportation, essential for electro-optic applications [101]. Thiophene as well as its benzo- and thieno-fused derivatives, owing to its high electron-donating nature, is very efficient in constructing donor-accepter polymers for utilization in organic devices. Additionally, via copolymerization or homopolymerization with donor materials, the obtained polymers show tunable properties due to the extended π-conjugation system. Interestingly, the thiophene-based polymers show hole transport properties and are used as p-type semiconductors for OFETs and OPVs. Firstly, we discuss polythiophene derivatives taking into consideration their importance, followed by a discussion on thiophene-based polymers comprising different fused-thiophene groups.

14.4.4 Polythiophenes and Their Derivatives

Initially, in the 1980s, study was mainly based upon the unsubstituted polythiophenes for OFET devices, demonstrating mobilities of around 10^{-5} cm^2/Vs [102]. For improving the solubility as well as the film-forming ability, there is a need to incorporate various side chains into the polythiophene backbone that encourages studies on polythiophene derivatives. One such well-known polymer material showing application in OFETs and OPVs is poly(3-hexylthiophene) (P3HT). The regioregularity of this polymer greatly contributes to the molecular configuration as well as device performance [103]. The regioregular P3HT demonstrates mobilities in the range of 0.05–0.2 cm^2/Vs via an edge-on orientation [104]. Further, via different optimization techniques, it is possible to enhance the performance of P3HT in OPVs and PCEs by 6.5% which is recorded by making use of proper acceptors [105]. Additionally, by synthesizing novel polymers with elongated but few chains, the stability of polythiophene-based FETs can be enhanced. A high hole mobility of ~0.14 cm^2/Vs was achieved by solution-processed regioregular polythiophene with the partial replacement of thiophene rings via alkyl substitution [106]. On the other hand, a selenophene-containing polymer shows a relatively low mobility of 0.02 cm^2/Vs [107]. Initially, the selenophene-based polymers were introduced for organic photovoltaic applications as an alternative to polythiophenes, but it was observed that such polymers generally demonstrate ambipolar charge transportation [108]. The introduction of big aromatic units like naphthalene enhances the torsion of polythiophene backbones that results in improved stability; however, it sacrifices the carrier mobilities [109]. Today, the 3,4-substituted polyalkylthiophenes have gained great interest. Further, in comparison to regioregular monosubstituted poly(3-alkylthiophenes), the polymer 146a-c [46] shows analogous transistor mobilities of 0.17 cm^2/Vs along with improved environmental stability. The power conversion efficiencies of such polymers reached 4.2% in polymer:fullerene bulk heterojunction solar cells [110]. The advanced studies on these polymers revealed that the extent of the main chain twisting increases the ionization potential as well as the open-circuit voltage (Voc) while maintaining the short-circuit current (Isc). Such molecular design offers a facile approach for tuning the twisting degree in the main chain for enhancing performance [111].

14.4.5 Thienoacene-Containing Conjugated Polymers

Thieno[3,2-b]thiophene incorporated into polythiophene main chains results in a low highest occupied molecular orbital (HOMO) level with excellent stability because of the relatively higher resonance stabilization energy in comparison to the single thiophene ring [112]. In comparison to P3HT, a lower HOMO level of about 0.3 eV is shown by thieno[3,2- b]thiophene-based polythiophenes. The transistors formed using such polymers show charge mobilities of around 0.2–0.7 cm^2/Vs; on the other hand one of the multi-thienoacene-based polymers offers greater mobilities of about 1.0 cm^2/V [112–114]. Via varying the dip-coating speed, it is possible to have full control of the monolayer as well as of the resultant microstructure of multi-thienoacene-based polymers [46] formed on the surface. It was also observed that the polymer chains are uniaxially oriented at a low dip-coating speed that yields definite structural anisotropy along with good carrier mobility of 1.3 cm^2/Vs in the direction of alignment [115]. Further, for developing p-type semiconductors, some other thienoacenes with extended conjugations were also introduced into the polythiophene backbone. In particular, trithienoacene polymers having alkyl substitution at the thiophene rings demonstrate high hole mobility of 0.3 cm^2/Vs [116] which is significantly higher in comparison to some other reported polymers (1.7×10^{-3} cm^2/Vs [117] and 0.05–0.06 cm^2/Vs [118]). Jun Li et al. [118] via blending the synthesized polymer with PC71BM and investigating it in a PSC device obtained an excellent PCE of 3.2%. Among different tetrathienoacene polymers, the polymer with bithiophene exhibits a superior mobility of 0.33 cm^2/Vs [119] while, on the other hand, for pentathienoacene-based polymers, mobilities of about 2.3×10^{-3} cm^2/Vs were achieved [117]. Moreover, the variation in field-effect behavior of such polymers is due to the C2 symmetry of various thienoacenes [120].

14.4.6 Naphthodithiophene-Containing Conjugated Polymers

One of the recently introduced building blocks for offering a highly stiff planar main chain when incorporated into a polythiophene polymer is naphthodithiophene (NDT). When compared with benzothienobenzothiophene (BTBT), the structural arrangement of NDT is highly effective in avoiding twisting among the adjoining thiophene rings, and hence preserves π-stacking as well as maintaining carrier mobility. However, BTBT served as small-molecule semiconductors in OFETs exhibits a highly twisted backbone with no OFET response. In particular, the [1,2-b:5,6-b'] NDT copolymer shows good mobilities of greater than 0.5 cm^2/Vs due to strong π-stacking [121]. Moon Chan Hwang et al. [122] reported a disk shaped NDT isomer that was copolymerized using thiophene and bithiophene for obtaining p-type material. Further, as reviewed by Xin Guo et al.

[46], in comparison to polymer 164a a comparatively much more ordered intermolecular structural arrangement was obtained for polymer 164b. FET results show mobilities of 0.01 and 0.076 cm^2/Vs for polymers 164a and 164b with current on/off ratios of 105 and 106, respectively.

14.4.7 Donor-Acceptor (D-A) Polymers

D-A conjugated polymers consist of electron donor and acceptor units. The electron push/pull ability and molecular arrangement in donor and acceptor units as well as their interactions highly influence the optoelectronic characteristics of such polymers. Thus, it is imperative to carefully choose the donor and acceptor groups before developing such polymers. The D-A polymers with a low bandgap gained much attention as active materials for OPVs, OLEDs, as well as for electrochromic devices [123–126]. Typical D-A polymers are composed of electron-deficient N-containing aromatic heterocycles, for example 2,1,3-benzothiadiazole, whereas electron-rich derivatives are usually thiophene, fluorine, and so on [127–129]. Thus donors are electron rich units usually comprising thiophene, and acceptors are electron deficient units mostly comprising N-heterocycles, as shown in Figure 14.5 [130]. The incorporation of donor and acceptor units in one polymer system effectively contributes in modifying the characteristics for utilization in organic devices. Considering OFETs, the intra- and intermolecular interactions among donors and acceptors in the polymer results in the strong π-stacking of polymer chains favoring charge carrier transportation. Moreover, in such donor-acceptor systems, it is possible to tune the HOMO and LUMO energy levels and hence band gaps by selecting appropriate electron-rich and electron-deficient units due to the fact that HOMO is usually a donor unit and LUMO an acceptor unit. Such evidence is very helpful in controlling the Voc and Jsc values for obtaining high power conversion efficiencies in PSC devices.

Further, the vinylene-linked D-A conjugated polymers (VDAs) having a vinylene linkage among the donor and acceptor units are low-bandgap polymers. These vinylene linkages effectively planarize the backbone via eradicating the torsion interactions among donor and acceptor rings, consequently decreasing the bandgap as a result of the extended conjugation length. Additionally, these vinylene groups partially increase the solubility of the polymer by offering rotational flexibility. Jianguo Mei et al. [131] report a simple method for synthesizing VDA polymers with no imperfections. The synthesized polymer was scrutinized as a donor unit in photovoltaic devices after blending with PCBM as an acceptor unit. PCEs of ~0.2–0.3% were achieved with an I_{sc} of 1.6 $mA/cm^2$and a V_{oc} of 0.61 V.

14.5 TWO-DIMENSIONAL (2D) CONJUGATED POLYMERS

In addition to a 1D system, higher dimensional polymer systems are very popular among researchers globally. Although the term "two-dimensional (2D) polymers" has been extensively employed there is still no proper definition for such a category of polymer. Various techniques have been adopted for synthesizing 2D polymers and some of them are covalent chemistry synthesis, hydrogen-bonded networks, and a supramolecular self-assembly that results in a layered structure. Among these techniques, limited focus

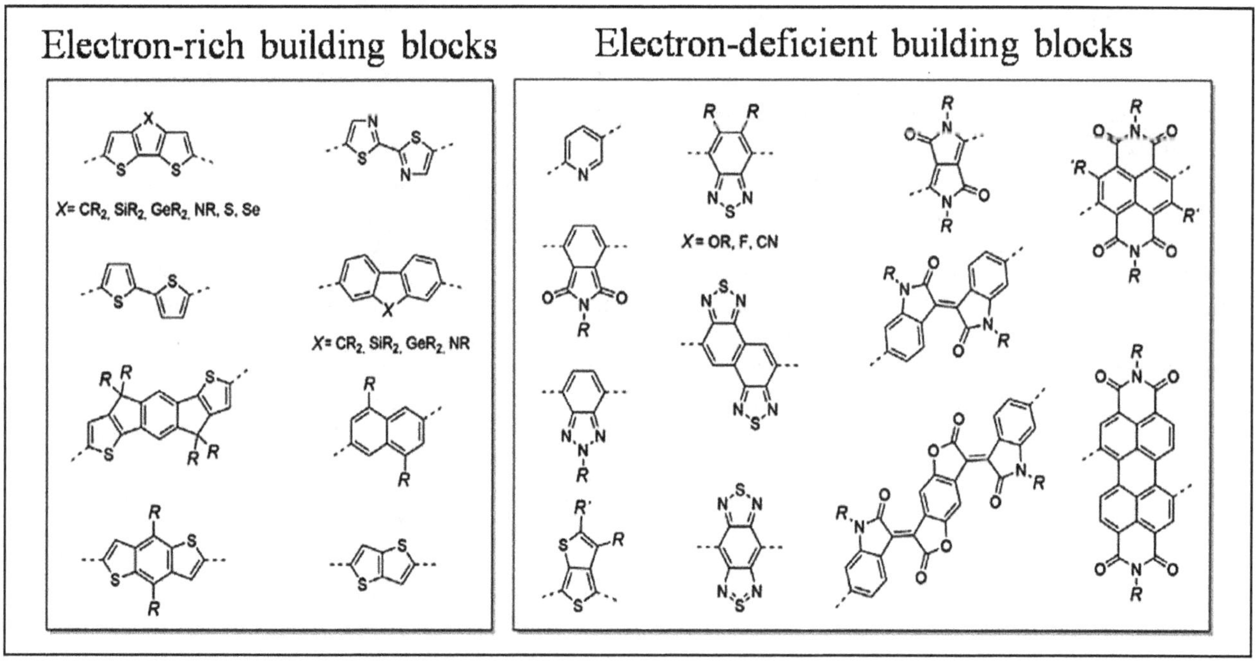

FIGURE 14.5 Left: electron-rich building blocks, right: electron-deficient building blocks. Reproduced from [130] with permission. Copyright 2021, Elsevier.

has been devoted to covalent chemistry. Various superstructures are formed in a solid state as a result of increased dimensionalities in π-conjugated systems, leading to multidirectional charge transportation. Graphene, a single and extremely thin layer of graphite that is one-atom thick, is one of the highly preferred 2D polymers. Its amazing thermal, electronic, as well as mechanical properties make graphene a highly potential candidate for multiple applications.

14.5.1 Conjugated Macrocycles

In the past two decades, macrocycles with rigid and fully π-conjugated backbones,, because of their useful features and ability to serve as units for discotic liquid crystals, guest-host complexes, as well as for 3D nanomaterials, have become very popular. Furthermore, as discrete molecular entities, macrocycles exhibit some exceptional properties that are highly desirable in organic electronics [132]. These conjugated macrocycles can be synthesized by using various building blocks such as thiophene, acetylene, benzene, and pyridine [132]. For a better understanding of these conjugated macrocycles, some of them with the interesting self-assembling property and potential in organic electronics are discussed below.

By using triphenylene, triangular-shaped macrocycles can be obtained via the covalent linking of three units of triphenylene at positions 7 and 10. These macrocycles, possessing promising emissive properties, were then investigated as active materials in OLEDs, demonstrating high device stability. Luminance maxima of 200 cd/m^2 at 6.5V, aV_{oc} of 3.9V, as well as a narrow deep blue emission was achieved via a solution-processable OLED comprising TPBI as an electron-transportation layer and macrocycles as an active material. These results show that such triphenylene-based macrocycles exhibit great potential for application in OLEDs [133]. Similarly, for OFET application, various other macrocyclic oligothiophenes as well as their π-extended derivatives have been synthesized. For example, a neutral meso-substituted tetrathia annulene aromatic macrocycle was synthesized and a good hole mobility of 0.63 cm^2/Vs was observed because of its highly crystalline nature [134]. Similarly, a benzothiophene-cornered rectangular thiopheneethynylene macrocycle possessing a diameter of up to 2 nm was also reported as a potential material for OFET application. A film obtained via spin-coating demonstrates a maximum hole mobility of 7.3×10^{-3} cm^2/Vs [135].

14.5.2 Two-Dimensional (2D) D-A Polymers

Further, for designing organic materials for application in organic devices, the idea of 2D D-A materials was introduced with more delocalization of electrons perpendicular to the polymer backbone. In the literature, two categories of 2D D-A polymers are mentioned. The first is the main-chain donor and side-chain acceptor (mD-sA) polymer in which a conjugated backbone that served as a donor unit is attached to a conjugated pendent chain. In the second category, the electron-rich and electron-deficient units are in the backbone, analogous to 1D D-A polymers; however, the donor/acceptor unit consists of aromatic substitutions that remain in conjugation with the main chain, resulting in a 2D arrangement. On comparing 1D mD–mA with that of 2D mD–sA, the latter exhibits various attractive properties, like high hole transportation, isotropic charge transportation, and internal charge transfer. In 2009, the first two mD-sA materials were introduced, having a fluorene-triarylamine copolymer main chain joined with diethylthiobarbituric acid or styrylthiophene π-bridged malononitrile as an acceptor unit that results in a JSC of 9.62 mA/cm^2, a Voc of 0.99 V, an FF of 0.50, along with a PCE of ~4.7% [136].

An additional 2D D–A polymer design was obtained by making use of benzo[1,2-b:4,5-b']dithiophene (BDT) as a donor. Such a design maintains the donor/acceptor moieties in the main chain; however, this results in a 2D arrangement via joining the conjugated side chains with the backbone. Moreover, such aromatic substitutions can efficiently enhance charge transportation. Additionally, these substituted aromatic side chains have sufficient donor units for lowering the HOMO levels and enhancing the value of Voc. Furthermore, D-A polymers consisting of tris(thienylenevinylene) substitutions have been investigated; a remarkable enhancement in PCEs was observed when compared with the polymers without conjugated side chains. For OPV devices, an I_{sc} of 22.6 mA/cm^2 has so far been obtained. Such polymers exhibit great potential for the development of novel 2D D-A materials [137].

14.6 FUTURE SCOPE AND CONCLUSIONS

This chapter has revolved around various categories of conjugated polymers classified on the basis of their molecular structures and their application in organic electronics. Without getting into details of the synthesis techniques, the performance of 1D and 2D conjugated polymers in organic electronics have been discussed. PFs and PCz belonging to the family of conjugated polyphenylenes are considered as the most important polymers. PFs gained great attention as potential semiconductors for application in optoelectronics, particularly as luminescence materials for RGB as well as for white electroluminescence. On the other hand, carbazole-based polymers not only serve as luminescence and host materials, they also exhibit potential application in PSCs. On the same note, the stepladder conjugated polymers consisting of C- or heteroatom bridging also exhibit the potential for utilization in electro-optical applications, though these polymers, after a first analysis, did not become the highly demanded polymers for future applications. For application in PLEDs, PAH-based polymers have also been scrutinized as semiconductors. Another material that plays a very crucial role in organic electronics is thiophene containing polymers that have gained much attention of researchers working in this field. Here, the thiophene- and fused thiophene-based polymers which usually serve as p-type semiconductors have also been discussed for OFET applications. Further, π-conjugated and rigid macrocycle possessing self-assembled supramolecular arrangements were also discussed on the basis of their potential to be used in OLEDs and OFETs. In the 2D polymer category, graphene is one of the potential candidates for various organic electronic devices. In addition, the D-A approach proves to be very promising for synthesizing highly efficient conjugated polymers that show great potential in OFET and PSC applications. In such an approach, the donor units are the electron-rich building blocks that are combined with appropriate acceptor units for obtaining high performance D-A polymers. Here, various donor as well as acceptor units have been listed and the performances of some D-A polymers with different dimensionalities were also discussed. It was also observed that the maximum values of FET mobilities and the PCEs of PSCs were achieved using D-A copolymers. It is also not wrong to say that these polymers take the foremost position in the race of conjugated polymers for application in organo-electronic applications. In addition, the research on designing and synthesizing new donor and acceptor units still continues in academic as well as industrial laboratories. Moreover, the focus is also on the proper choice of the donor and acceptor units as well as their appropriate grouping in polymers. As a result, the enhancement of molecular weights as well as the precise refinement of the polymers will be the main concern for further enhancing the performances of polymers in the future.

REFERENCES

[1] Xu, Y., Liu, C., Khim, D. and Noh, Y.Y., 2015. Development of high-performance printed organic field-effect transistors and integrated circuits. *Physical Chemistry Chemical Physics*, 17(40), pp. 26553–26574.

[2] Teixeira da Rocha, C., Haase, K., Zheng, Y., Löffler, M., Hambsch, M. and Mannsfeld, S.C., 2018. Solution coating of small molecule/polymer blends enabling ultralow voltage and high-mobility organic transistors. *Advanced Electronic Materials*, 4(8), p. 1800141.

[3] Allard, S., Forster, M., Souharce, B., Thiem, H. and Scherf, U., 2008. Organic semiconductors for solution-processable field-effect transistors (OFETs). *Angewandte Chemie International Edition*, 47(22), pp. 4070–4098.

[4] Rogers, J.A., Someya, T. and Huang, Y., 2010. Materials and mechanics for stretchable electronics. *Science*, 327(5973), pp. 1603–1607.

[5] Cheng, P., Li, G., Zhan, X. and Yang, Y., 2018. Next-generation organic photovoltaics based on non-fullerene acceptors. *Nature Photonics*, 12(3), pp. 131–142.

[6] Assadi, M.K., Bakhoda, S., Saidur, R. and Hanaei, H., 2018. Recent progress in perovskite solar cells. *Renewable and Sustainable Energy Reviews*, 81, pp. 2812–2822.

[7] Kweon, O.Y., Lee, M.Y., Park, T., Jang, H., Jeong, A., Um, M.K. and Oh, J.H., 2019. Highly flexible chemical sensors based on polymer nanofiber field-effect transistors. *Journal of Materials Chemistry C*, 7(6), pp. 1525–1531.

[8] Li, H., Shi, W., Song, J., Jang, H.J., Dailey, J., Yu, J. and Katz, H.E., 2018. Chemical and biomolecule sensing with organic field-effect transistors. *Chemical Reviews*, 119(1), pp. 3–35.

[9] Menšík, M., Toman, P., Bielecka, U., Bartkowiak, W., Pfleger, J. and Paruzel, B., 2018. On the methodology of the determination of charge concentration dependent mobility from organic field-effect transistor characteristics. *Physical Chemistry Chemical Physics*, 20(4), pp. 2308–2319.

[10] Haase, K., Teixeira da Rocha, C., Hauenstein, C., Zheng, Y., Hambsch, M. and Mannsfeld, S.C., 2018. High-mobility, solution-processed organic field-effect transistors from C8-BTBT: Polystyrene blends. *Advanced Electronic Materials*, 4(8), p. 1800076.

[11] Shirakawa, H., Louis, E.J., MacDiarmid, A.G., Chiang, C.K. and Heeger, A.J., 1977. Synthesis of electrically conducting organic polymers: Halogen derivatives of polyacetylene, (CH) x. *Journal of the Chemical Society, Chemical Communications*, (16), pp. 578–580.

[12] Rasmussen, S.C., 2011. Electrically conducting plastics: Revising the history of conjugated organic polymers. In *100+ Years of Plastics: Leo Baekeland and Beyond* (pp. 147–163). American Chemical Society.

[13] Heeger, A.J., 2010. Semiconducting polymers: The third generation. *Chemical Society Reviews*, 39(7), pp. 2354–2371.

[14] Swager, T.M., 2017. 50th anniversary perspective: Conducting/semiconducting conjugated polymers: A personal perspective on the past and the future. *Macromolecules*, 50(13), pp. 4867–4886.

[15] Rasmussen, S.C., 2018. Early history of conjugated polymers: From their origins to the handbook of conducting polymers. In *Handbook of Conducting Polymers*, 4th ed., Reynolds, J.R., Skotheim, T., Thompson, B., Eds. Marcel Dekker.

[16] Rasmussen, S.C., 2020. Conjugated and conducting organic polymers: The first 150 years. *ChemPlusChem*, 85(7), pp. 1412–1429.

[17] Labes, M.M., Love, P. and Nichols, L.F., 1979. Polysulfur nitride-a metallic, superconducting polymer. *Chemical Reviews*, 79(1), pp. 1–15.

[18] Pankow, R.M. and Thompson, B.C., 2020. The development of conjugated polymers as the cornerstone of organic electronics. *Polymer*, 207, p. 122874.

[19] Selvan, T., Spatz, J.P., Klok, H.A. and Möller, M., 1998. Gold–polypyrrole core–shell particles in diblock copolymer micelles. *Advanced Materials*, 10(2), pp. 132–134.

[20] Bendikov, M., Martin, N., Perepichka, D.F. and Prato, M., 2011. Fred Wudl: Discovering new science through making new molecules. *Journal of Materials Chemistry*, 21(5), pp. 1292–1294.

[21] Zhang, Q., Perepichka, D.F. and Bao, Z., 2018. Fred Wudl's fifty-year contribution to organic semiconductors. *Journal of Materials Chemistry C*, 6(14), pp. 3483–3484.

[22] Hotta, S., Rughooputh, S.D.D.V., Heeger, A.J. and Wudl, F., 1987. Spectroscopic studies of soluble poly (3-alkylthienylenes). *Macromolecules*, 20(1), pp. 212–215.

[23] Kobayashi, M., Chen, J., Chung, T.C., Moraes, F., Heeger, A.J. and Wudl, F., 1984. Synthesis and properties of chemically coupled poly (thiophene). *Synthetic Metals*, 9(1), pp. 77–86.

[24] Chen, S.A. and Lee, C.C., National Science Council, 1996. Method for preparing processable polyisothianaphthene. U.S. Patent 5, 510, 457.

[25] Patil, A.O., Ikenoue, Y., Wudl, F. and Heeger, A.J., 1987. Water soluble conducting polymers. *Journal of the American Chemical Society*, 109(6), pp. 1858–1859.

[26] Chiang, C.K., Fincher Jr, C.R., Park, Y.W., Heeger, A.J., Shirakawa, H., Louis, E.J., Gau, S.C. and MacDiarmid, A.G., 1977. Electrical conductivity in doped polyacetylene. *Physical Review Letters*, 39(17), p. 1098.

[27] Jonas, F., Heywang, G. and Schmidtberg, W., 1989. Preparation of (alkylenedioxy) thiophene polymers for use as antistatic agents. *DE*, 3813589, p.A1.

[28] Elschner, A., Kirchmeyer, S., Lovenich, W., Merker, U. and Reuter, K., 2010. *PEDOT: Principles and Applications of An Intrinsically Conductive Polymer*. CRC Press.

[29] Thompson, B.C. and Fréchet, J.M., 2008. Polymer–fullerene composite solar cells. *Angewandte Chemie International Edition*, 47(1), pp. 58–77.

[30] Otieno, F., Airo, M., Ranganathan, K. and Wamwangi, D., 2016. Annealed silver-islands for enhanced optical absorption in organic solar cell. *Thin Solid Films*, 598, pp. 177–183.

[31] Nalwa, H. S., Ed, 1997. *Handbook of Organic Conductive Molecules and Polymers*. John Wiley & Sons: Chichester, Vols. 1–4.

[32] Farchioni, R., Grosso, G. and Vignolo, P., 2001. Recursive algorithms for polymeric chains. In *Organic Electronic Materials* (pp. 89–125). Springer, Berlin, Heidelberg.

[33] Bian, L., Zhu, E., Tang, J., Tang, W. and Zhang, F., 2012. Recent progress in the design of narrow bandgap conjugated polymers for high-efficiency organic solar cells. *Progress in Polymer Science*, 37(9), pp. 1292–1331.

[34] Kola, S., Sinha, J. and Katz, H.E., 2012. Organic transistors in the new decade: Toward n-channel, printed, and stabilized devices. *Journal of Polymer Science Part B: Polymer Physics*, 50(15), pp. 1090–1120.

[35] Søndergaard, R.R., Hösel, M. and Krebs, F.C., 2013. Roll-to-Roll fabrication of large area functional organic materials. *Journal of Polymer Science Part B: Polymer Physics*, 51(1), pp. 16–34.

[36] Luo, G., Ren, X., Zhang, S., Wu, H., Choy, W.C., He, Z. and Cao, Y., 2016. Recent advances in organic photovoltaics: Device structure and optical engineering optimization on the nanoscale. *Small*, 12(12), pp. 1547–1571.

[37] Günes, S., Neugebauer, H. and Sariciftci, N.S., 2007. Conjugated polymer-based organic solar cells. *Chemical Reviews*, 107(4), pp. 1324–1338.

[38] Ostroverkhova, O., 2016. Organic optoelectronic materials: Mechanisms and applications. *Chemical Reviews*, 116(22), pp. 13279–13412.

[39] Mei, J., Diao, Y., Appleton, A.L., Fang, L. and Bao, Z., 2013. Integrated materials design of organic semiconductors for field-effect transistors. *Journal of the American Chemical Society*, 135(18), pp. 6724–6746.

[40] Beaujuge, P.M. and Reynolds, J.R., 2010. Color control in π-conjugated organic polymers for use in electrochromic devices. *Chemical Reviews*, 110(1), pp. 268–320.

[41] McQuade, D.T., Pullen, A.E. and Swager, T.M., 2000. Conjugated polymer-based chemical sensors. *Chemical Reviews*, 100(7), pp. 2537–2574.

[42] Lin, P. and Yan, F., 2012. Organic thin-film transistors for chemical and biological sensing. *Advanced Materials*, 24(1), pp. 34–51.

[43] McGehee, M.D. and Heeger, A.J., 2000. Semiconducting (conjugated) polymers as materials for solid-state lasers. *Advanced Materials*, 12(22), pp. 1655–1668.

[44] B. Schmatz, R.M. Pankow, B.C. Thompson, J.R. Reynolds, 2019. Perspective on the advancements in conjugated polymer synthesis, design, and functionality over the past ten years. In *Conjugated Polymers: Perspective, Theory, and New Mater* (pp. 107–148). CRC Press.

[45] Grimsdale, A.C. and Müllen, K., 2007. Oligomers and polymers based on bridged phenylenes as electronic materials. *Macromolecular Rapid Communications*, 28(17), pp. 1676–1702.

[46] Guo, X., Baumgarten, M. and Müllen, K., 2013. Designing π-conjugated polymers for organic electronics. *Progress in Polymer Science*, 38(12), pp. 1832–1908.

[47] Akiyama, M., Iwakura, Y., Shiraishi, S. and Imai, Y., 1966. Poly-p-phenylene-1, 2, 4-oxadiazole from p-cyanobenzonitrile oxide by a solid-state cycloaddition polymerization. *Journal of Polymer Science Part B: Polymer Letters*, 4(5), pp. 305–308.

[48] Vahlenkamp, T. and Wegner, G., 1994. Poly (2, 5-dialkoxy-p-phenylene) s—synthesis and properties. *Macromolecular Chemistry and Physics*, 195(6), pp. 1933–1952.

[49] Yang, Y., Pei, Q. and Heeger, A.J., 1996. Efficient blue polymer light-emitting diodes from a series of soluble poly (paraphenylene) s. *Journal of Applied Physics*, 79(2), pp. 934–939.

[50] Grem, G., Leditzky, G., Ullrich, B. and Leising, G., 1992. Realization of a blue-light-emitting device using poly (p-phenylene). *Advanced Materials*, 4(1), pp. 36–37.

[51] Scherf, U. and Müllen, K., 1991. Polyarylenes and poly (arylenevinylenes), 7. A soluble ladder polymer via bridging of functionalized poly (p-phenylene)-precursors. Die Makromolekulare Chemie, Rapid *Communications*, 12(8), pp. 489–497.

[52] Scherf, U. and Muellen, K., 1992. Poly (arylenes) and poly (arylenevinylenes). 11. A modified two-step route to soluble phenylene-type ladder polymers. *Macromolecules*, 25(13), pp. 3546–3548.

[53] Grem, G., Paar, C., Stampfl, J., Leising, G., Huber, J. and Scherf, U., 1995. Soluble segmented stepladder poly (p-phenylenes) for blue-light-emitting diodes. *Chemistry of Materials*, 7(1), pp. 2–4.

[54] Romaner, L., Heimel, G., Wiesenhofer, H., Scandiucci de Freitas, P., Scherf, U., Brédas, J.L., Zojer, E. and List, E.J., 2004. Ketonic defects in ladder-type poly (p-phenylene) s. *Chemistry of Materials*, 16(23), pp. 4667–4674.

[55] Grüner, J., Hamer, P.J., Friend, R.H., Huber, H.J., Scherf, U. and Holmes, A.B., 1994. A high efficiency blue-light-emitting diode based on novel ladder poly (p-phenylene) s. *Advanced Materials*, 6(10), pp. 748–752.

[56] Wu, Y., Zhang, J., Fei, Z. and Bo, Z., 2008. Spiro-bridged ladder-type poly (p-phenylene) s: Towards structurally perfect light-emitting materials. *Journal of the American Chemical Society*, 130(23), pp. 7192–7193.

[57] Burroughes, J.H., Bradley, D.D., Brown, A.R., Marks, R.N., Mackay, K., Friend, R.H., Burns, P.L. and Holmes, A.B., 1990. Light-emitting diodes based on conjugated polymers. *Nature*, 347(6293), pp. 539–541.

[58] Gilch, H.G. and Wheelwright, W.L., 1966. Polymerization of α-halogenated p-xylenes with base. Journal of Polymer Science Part A–1: *Polymer Chemistry*, 4(6), pp. 1337–1349.

[59] Gustafsson, G., Cao, Y., Treacy, G.M., Klavetter, F., Colaneri, N. and Heeger, A.J., 1992. Flexible light-emitting diodes made from soluble conducting polymers. *Nature*, 357(6378), pp. 477–479.

[60] Greenham, N.C., Moratti, S.C., Bradley, D.D.C., Friend, R.H. and Holmes, A.B., 1993. Efficient light-emitting diodes based on polymers with high electron affinities. *Nature*, 365(6447), pp. 628–630.

[61] Laquai, F., Mishra, A.K., Ribas, M.R., Petrozza, A., Jacob, J., Akcelrud, L., Muellen, K., Friend, R.H. and Wegner, G., 2007. Photophysical Properties of a Series of Poly (ladder-type phenylene) s. *Advanced Functional Materials*, 17(16), pp. 3231–3240.

[62] Xie, L.H., Yin, C.R., Lai, W.Y., Fan, Q.L. and Huang, W., 2012. Polyfluorene-based semiconductors combined with various periodic table elements for organic electronics. *Progress in Polymer Science*, 37(9), pp. 1192–1264.

[63] Li, C., Liu, M., Pschirer, N.G., Baumgarten, M. and Mullen, K., 2010. Polyphenylene-based materials for organic photovoltaics. *Chemical Reviews*, 110(11), pp. 6817–6855.

[64] Neher, D., 2001. Polyfluorene homopolymers: Conjugated liquid-crystalline polymers for bright blue emission and polarized electroluminescence. *Macromolecular Rapid Communications*, 22(17), pp. 1365–1385.

[65] Hashim, A.A., Majid, M.A. and Mustafa, B.A., 2013, December. Legibility of web page on full high definition display. In *2013 International Conference on Advanced Computer Science Applications and Technologies* (pp. 521–524). IEEE.

[66] Setayesh, S., Marsitzky, D. and Müllen, K., 2000. Bridging the Gap between Polyfluorene and Ladder-Poly-p-phenylene: Synthesis and Characterization of Poly-2, 8-indenofluorene. *Macromolecules*, 33(6), pp. 2016–2020.

[67] Jacob, J., Sax, S., Piok, T., List, E.J., Grimsdale, A.C. and Müllen, K., 2004. Ladder-type pentaphenylenes and their polymers: Efficient blue-light emitters and electron-accepting materials via a common intermediate. *Journal of the American Chemical Society*, 126(22), pp. 6987–6995.

[68] Guo, X., Cheng, Y., Xie, Z., Geng, Y., Wang, L., Jing, X. and Wang, F., 2009. Fluorene-based copolymers containing dinaphtho-s-indacene as new building blocks for high-efficiency and color-stable blue LEDs. *Macromolecular Rapid Communications*, 30(9–10), pp. 816–825.

[69] Ego, C., Marsitzky, D., Becker, S., Zhang, J., Grimsdale, A.C., Müllen, K., MacKenzie, J.D., Silva, C. and Friend, R.H., 2003. Attaching perylene dyes to polyfluorene: Three simple, efficient methods for facile color tuning of light-emitting polymers. *Journal of the American Chemical Society*, 125(2), pp. 437–443.

[70] Guo, X., Qin, C., Cheng, Y., Xie, Z., Geng, Y., Jing, X., Wang, F. and Wang, L., 2009. White electroluminescence from a phosphonate-functionalized single-polymer system with electron-trapping effect. *Advanced Materials*, 21(36), pp. 3682–3688.

[71] Hou, Q., Zhou, Q., Zhang, Y., Yang, W., Yang, R. and Cao, Y., 2004. Synthesis and electroluminescent properties of high-efficiency saturated red emitter based on copolymers from fluorene and 4, 7-di (4-hexylthien-2-yl)-2, 1, 3-benzothiadiazole. *Macromolecules*, 37(17), pp. 6299–6305.

[72] Liu, J., Tu, G., Zhou, Q., Cheng, Y., Geng, Y., Wang, L., Ma, D., Jing, X. and Wang, F., 2006. Highly efficient green light emitting polyfluorene incorporated with 4-diphenylamino-1, 8-naphthalimide as green dopant. *Journal of Materials Chemistry*, 16(15), pp. 1431–1438.

[73] Shu, C.F., Dodda, R., Wu, F.I., Liu, M.S. and Jen, A.K.Y., 2003. Highly efficient blue-light-emitting diodes from polyfluorene containing bipolar pendant groups. *Macromolecules*, 36(18), pp. 6698–6703.

[74] Vak, D., Chun, C., Lee, C.L., Kim, J.J. and Kim, D.Y., 2004. A novel spiro-functionalized polyfluorene derivative with solubilizing side chains. *Journal of Materials Chemistry*, 14(8), pp. 1342–1346.

[75] Liu, J., Shao, S.Y., Chen, L., Xie, Z.Y., Cheng, Y.X., Geng, Y.H., Wang, L.X., Jing, X.B. and Wang, F.S., 2007. White electroluminescence from a single polymer system: Improved performance by means of enhanced efficiency and red-shifted luminescence of the blue-light-emitting species. *Advanced Materials*, 19(14), pp. 1859–1863.

[76] Chen, L., Li, P., Cheng, Y., Xie, Z., Wang, L., Jing, X. and Wang, F., 2011. White electroluminescence from star-like single polymer systems: 2, 1, 3-Benzothiadiazole derivatives dopant as orange cores and polyfluorene host as six blue arms. *Advanced Materials*, 23(26), pp. 2986–2990.

[77] Wu, W., Inbasekaran, M., Hudack, M., Welsh, D., Yu, W., Cheng, Y., Wang, C., Kram, S., Tacey, M., Bernius, M. and Fletcher, R., 2004. Recent development of polyfluorene-based RGB materials for light emitting diodes. *Microelectronics Journal*, 35(4), pp. 343–348.

[78] Herguth, P., Jiang, X., Liu, M.S. and Jen, A.K.Y., 2002. Highly efficient fluorene-and benzothiadiazole-based conjugated copolymers for polymer light-emitting diodes. *Macromolecules*, 35(16), pp. 6094–6100.

[79] Hou, Q., Xu, Y., Yang, W., Yuan, M., Peng, J. and Cao, Y., 2002. Novel red-emitting fluorene-based copolymers. *Journal of Materials Chemistry*, 12(10), pp. 2887–2892.

[80] Yang, R., Tian, R., Yan, J., Zhang, Y., Yang, J., Hou, Q., Yang, W., Zhang, C. and Cao, Y., 2005. Deep-red electroluminescent polymers: Synthesis and characterization of new low-band-gap conjugated copolymers for light-emitting diodes and photovoltaic devices. *Macromolecules*, 38(2), pp. 244–253.

[81] Yang, R., Tian, R., Hou, Q., Yang, W. and Cao, Y., 2003. Synthesis and optical and electroluminescent properties of novel conjugated copolymers derived from fluorene and benzoselenadiazole. *Macromolecules*, 36(20), pp. 7453–7460.

[82] Yang, J., Jiang, C., Zhang, Y., Yang, R., Yang, W., Hou, Q. and Cao, Y., 2004. High-efficiency saturated red emitting polymers derived from fluorene and naphthoselenadiazole. *Macromolecules*, 37(4), pp. 1211–1218.

[83] Huo, L., Hou, J., Chen, H.Y., Zhang, S., Jiang, Y., Chen, T.L. and Yang, Y., 2009. Bandgap and molecular level control of the low-bandgap polymers based on 3, 6-dithiophen-2-yl-2, 5-dihydropyrrolo [3, 4-c] pyrrole-1, 4-dione toward highly efficient polymer solar cells. *Macromolecules*, 42(17), pp. 6564–6571.

[84] Svensson, M., Zhang, F., Veenstra, S.C., Verhees, W.J., Hummelen, J.C., Kroon, J.M., Inganäs, O. and Andersson, M.R., 2003. High-performance polymer solar cells of an alternating polyfluorene copolymer and a fullerene derivative. *Advanced Materials*, 15(12), pp. 988–991.

[85] Gadisa, A., Mammo, W., Andersson, L.M., Admassie, S., Zhang, F., Andersson, M.R. and Inganäs, O., 2007. A new donor–acceptor–donor polyfluorene copolymer with balanced electron and hole mobility. *Advanced Functional Materials*, 17(18), pp. 3836–3842.

[86] Becerril, H.A., Miyaki, N., Tang, M.L., Mondal, R., Sun, Y.S., Mayer, A.C., Parmer, J.E., McGehee, M.D. and Bao, Z., 2009. Transistor and solar cell performance of donor–acceptor low bandgap copolymers bearing an acenaphtho [1, 2-b] thieno [3, 4-e] pyrazine (ACTP) motif. *Journal of Materials Chemistry*, 19(5), pp. 591593.

[87] Zhang, F., Mammo, W., Andersson, L.M., Admassie, S., Andersson, M.R. and Inganäs, O., 2006. Low-bandgap alternating fluorene copolymer/methanofullerene heterojunctions in efficient near-infrared polymer solar cells. *Advanced Materials*, 18(16), pp. 2169–2173.

[88] Morin, J.F., Leclerc, M., Ades, D. and Siove, A., 2005. Polycarbazoles: 25 years of progress. *Macromolecular Rapid Communications*, 26(10), pp. 761–778.

[89] Li, Y., Wu, Y. and Ong, B.S., 2006. Polyindolo [3, 2-b] carbazoles: A new class of p-channel semiconductor polymers for organic thin-film transistors. *Macromolecules*, 39(19), pp. 6521–6527.

[90] Mishra, A.K., Graf, M., Grasse, F., Jacob, J., List, E.J. and Müllen, K., 2006. Blue-emitting carbon-and nitrogen-bridged poly (ladder-type tetraphenylene) s. *Chemistry of Materials*, 18(12), pp. 2879–2885.

[91] Chan, K.L., McKiernan, M.J., Towns, C.R. and Holmes, A.B., 2005. Poly (2, 7-dibenzosilole): A blue light emitting polymer. *Journal of the American Chemical Society*, 127(21), pp. 7662–7663.

[92] Wang, E., Li, C., Mo, Y., Zhang, Y., Ma, G., Shi, W., Peng, J., Yang, W. and Cao, Y., 2006. Poly (3, 6-silafluorene-co-2, 7-fluorene)-based high-efficiency and color-pure blue light-emitting polymers with extremely narrow band-width and high spectral stability. *Journal of Materials Chemistry*, 16(42), pp. 4133–4140.

[93] Feng, X., Pisula, W. and Müllen, K., 2009. Large polycyclic aromatic hydrocarbons: Synthesis and discotic organization. *Pure & Applied Chemistry*, 81(12).

[94] Felemban, S., Vazquez, P. and Moore, E., 2019. Future trends for in situ monitoring of polycyclic aromatic hydrocarbons in water sources: The role of immunosensing techniques. *Biosensors*, 9(4), p. 142.

[95] Balakrishnan, K., Datar, A., Zhang, W., Yang, X., Naddo, T., Huang, J., Zuo, J., Yen, M., Moore, J.S. and Zang, L., 2006. Nanofibril self-assembly of an arylene ethynylene macrocycle. *Journal of the American Chemical Society*, 128(20), pp. 6576–6577.

[96] Saleh, M., Baumgarten, M., Mavrinskiy, A., Schäfer, T. and Müllen, K., 2010. Triphenylene-based polymers for blue polymeric light emitting diodes. *Macromolecules*, 43(1), pp. 137–143.

[97] Saleh, M., Park, Y.S., Baumgarten, M., Kim, J.J. and Müllen, K., 2009. Conjugated triphenylene polymers for blue OLED devices. *Macromolecular Rapid Communications*, 30(14), pp. 1279–1283.

[98] Figueira-Duarte, T.M. and Mullen, K., 2011. Pyrene-based materials for organic electronics. *Chemical Reviews*, 111(11), pp. 7260–7314.

[99] Kawano, S.I., Yang, C., Ribas, M., Baluschev, S., Baumgarten, M. and Müllen, K., 2008. Blue-emitting poly (2, 7-pyrenylene) s: Synthesis and optical properties. *Macromolecules*, 41(21), pp. 7933–7937.

[100] Mishra, A., Ma, C.Q. and Bauerle, P., 2009. Functional oligothiophenes: Molecular design for multidimensional nanoarchitectures and their applications. *Chemical Reviews*, 109(3), pp. 1141–1276.

[101] Tsumura, A., Koezuka, H. and Ando, T.J.A.P.L., 1986. Macromolecular electronic device: Field-effect transistor with a polythiophene thin film. *Applied Physics Letters*, 49(18), pp. 1210–1212.

[102] Sirringhaus, H., Brown, P.J., Friend, R.H., Nielsen, M.M., Bechgaard, K., Langeveld-Voss, B.M.W., Spiering, A.J.H., Janssen, R.A., Meijer, E.W., Herwig, P. and De Leeuw, D.M., 1999. Two-dimensional charge transport in self-organized, high-mobility conjugated polymers. *Nature*, 401(6754), pp. 685–688.

[103] Chang, J.F., Sun, B., Breiby, D.W., Nielsen, M.M., Sölling, T.I., Giles, M., McCulloch, I. and Sirringhaus, H., 2004. Enhanced mobility of poly (3-hexylthiophene) transistors by spin-coating from high-boiling-point solvents. *Chemistry of Materials*, 16(23), pp. 4772–4776.

[104] Zhao, G., He, Y. and Li, Y., 2010.6.5% Efficiency of polymer solar cells based on poly (3-hexylthiophene) and indene-C60 bisadduct by device optimization. *Advanced Materials*, 22(39), pp. 4355–4358.

[105] Ong, B.S., Wu, Y., Liu, P. and Gardner, S., 2004. High-performance semiconducting polythiophenes for organic thin-film transistors. *Journal of the American Chemical Society*, 126(11), pp. 3378–3379.

[106] Kong, H., Chung, D.S., Kang, I.N., Park, J.H., Park, M.J., Jung, I.H., Park, C.E. and Shim, H.K., 2009. New selenophene-based semiconducting copolymers for high performance organic thin-film transistors. *Journal of Materials Chemistry*, 19(21), pp. 3490–3499.

[107] Chen, Z., Lemke, H., Albert-Seifried, S., Caironi, M., Nielsen, M.M., Heeney, M., Zhang, W., McCulloch, I. and Sirringhaus, H., 2010. High mobility ambipolar charge transport in polyselenophene conjugated polymers. *Advanced Materials*, 22(21), pp. 2371–2375.

[108] Chung, D.S., Park, J.W., Kim, S.O., Heo, K., Park, C.E., Ree, M., Kim, Y.H. and Kwon, S.K., 2009. Alternating copolymers containing bithiophene and dialkoxynaphthalene for the applications to field effect transistor and photovoltaic cell: performance and stability. *Chemistry of Materials*, 21(22), pp. 5499–5507.

[109] Ko, S., Verploegen, E., Hong, S., Mondal, R., Hoke, E.T., Toney, M.F., McGehee, M.D. and Bao, Z., 2011.3, 4-Disubstituted polyalkylthiophenes for high-performance thin-film transistors and photovoltaics. *Journal of the American Chemical Society*, 133(42), pp. 16722–16725.

[110] Osaka, I., Akita, M., Koganezawa, T. and Takimiya, K., 2012. Quinacridone-based semiconducting polymers: Implication of electronic structure and orientational order for charge transport property. *Chemistry of Materials*, 24(6), pp. 1235–1243.

[111] McCulloch, I., Heeney, M., Bailey, C., Genevicius, K., MacDonald, I., Shkunov, M., Sparrowe, D., Tierney, S., Wagner, R., Zhang, W. and Chabinyc, M.L., 2006. Liquid-crystalline semiconducting polymers with high charge-carrier mobility. *Nature Materials*, 5(4), pp. 328–333.

[112] Chabinyc, M.L., Toney, M.F., Kline, R.J., McCulloch, I. and Heeney, M., 2007. X-ray scattering study of thin films of poly (2, 5-bis (3-alkylthiophen-2-yl) thieno [3, 2-b] thiophene). *Journal of the American Chemical Society*, 129(11), pp. 3226–3237.

[113] Li, Y., Wu, Y., Liu, P., Birau, M., Pan, H. and Ong, B.S., 2006. Poly (2, 5-bis (2-thienyl)-3, 6-dialkylthieno [3, 2-b] thiophene) s—High-Mobility Semiconductors for Thin-Film Transistors. *Advanced Materials*, 18(22), pp. 3029–3032.

[114] Wang, S., Kiersnowski, A., Pisula, W. and Müllen, K., 2012. Microstructure evolution and device performance in solution-processed polymeric field-effect transistors: The key role of the first monolayer. *Journal of the American Chemical Society*, 134(9), pp. 4015–4018.

[115] Li, J., Qin, F., Li, C.M., Bao, Q., Chan-Park, M.B., Zhang, W., Qin, J. and Ong, B.S., 2008. High-performance thin-film transistors from solution-processed dithienothiophene polymer semiconductor nanoparticles. *Chemistry of Materials*, 20(6), pp. 2057–2059.

[116] He, M., Li, J., Sorensen, M.L., Zhang, F., Hancock, R.R., Fong, H.H., Pozdin, V.A., Smilgies, D.M. and Malliaras, G.G., 2009. Alkylsubstituted thienothiophene semiconducting materials: structure− property relationships. *Journal of the American Chemical Society*, 131(33), pp. 11930–11938.

[117] Li, J., Tan, H.S., Chen, Z.K., Goh, W.P., Wong, H.K., Ong, K.H., Liu, W., Li, C.M. and Ong, B.S., 2011. Dialkyl-substituted dithienothiophene copolymers as polymer semiconductors for thin-film transistors and bulk heterojunction solar cells. *Macromolecules*, 44(4), pp. 690–693.

[118] Fong, H.H., Pozdin, V.A., Amassian, A., Malliaras, G.G., Smilgies, D.M., He, M., Gasper, S., Zhang, F. and Sorensen, M., 2008. Tetrathienoacene copolymers as high mobility, soluble organic semiconductors. *Journal of the American Chemical Society*, 130(40), pp. 13202–13203.

[119] He, M., Li, J., Tandia, A., Sorensen, M., Zhang, F., Fong, H.H., Pozdin, V.A., Smilgies, D.M. and Malliaras, G.G., 2010. Importance of C 2 symmetry for the device performance of a newly synthesized family of fused-ring thiophenes. *Chemistry of Materials*, 22(9), pp. 2770–2779.

[120] Osaka, I., Abe, T., Shinamura, S., Miyazaki, E. and Takimiya, K., 2010. High-mobility semiconducting naphthodithiophene copolymers. *Journal of the American Chemical Society*, 132(14), pp. 5000–5001.

[121] Hwang, M.C., Jang, J.W., An, T.K., Park, C.E., Kim, Y.H. and Kwon, S.K., 2012. Synthesis and characterization of new thermally stable poly (naphthodithiophene) derivatives and applications for high-performance organic thin film transistors. *Macromolecules*, 45(11), pp. 4520–4528.

[122] Bundgaard, E. and Krebs, F.C., 2007. Low band gap polymers for organic photovoltaics. *Solar Energy Materials and Solar Cells*, 91(11), pp. 954–985.

[123] Kulkarni, A.P., Zhu, Y. and Jenekhe, S.A., 2008. Photodegradation of emissive conjugated copolymers and oligomers containing thienopyrazine. *Macromolecules*, 41(2), pp. 339–345.

[124] Gunbas, G.E., Durmus, A. and Toppare, L., 2008. Could green be greener? novel donor–acceptor-type electrochromic polymers: towards excellent neutral green materials with exceptional transmissive oxidized states for completion of RGB color space. *Advanced Materials*, 20(4), pp. 691–695.

[125] Beaujuge, P.M., Ellinger, S. and Reynolds, J.R., 2008. Spray processable green to highly transmissive electrochromics via chemically polymerizable donor–acceptor heterocyclic pentamers. *Advanced Materials*, 20(14), pp. 2772–2776.

[126] Kitamura, C., Tanaka, S. and Yamashita, Y., 1996. Design of narrow-bandgap polymers: Syntheses and properties of monomers and polymers containing aromatic-donor and o-quinoid-acceptor units. *Chemistry of Materials*, 8(2), pp. 570–578.

[127] Van Mullekom, H.A.M., Vekemans, J.A.J.M. and Meijer, E.W., 1998. Band-gap engineering of donor–acceptor-substituted π-conjugated polymers. *Chemistry–A European Journal*, 4(7), pp. 1235–1243.

[128] Zhu, Y., Champion, R.D. and Jenekhe, S.A., 2006. Conjugated donor– acceptor copolymer semiconductors with large intramolecular charge transfer: Synthesis, optical properties, electrochemistry, and field effect carrier mobility of thienopyrazine-based copolymers. *Macromolecules*, 39(25), pp. 8712–8719.

[129] Sommer, M., 2021. Development of conjugated polymers for organic flexible electronics. In *Organic Flexible Electronics* (pp. 27–70). Woodhead Publishing.

[130] Mei, J., Heston, N.C., Vasilyeva, S.V. and Reynolds, J.R., 2009. A facile approach to defect-free vinylene-linked benzothiadiazole-thiophene low-bandgap conjugated polymers for organic electronics. *Macromolecules*, 42(5), pp. 1482–1487.

[131] Iyoda, M., Yamakawa, J. and Rahman, M.J., 2011. Conjugated macrocycles: Concepts and applications. *Angewandte Chemie International Edition*, 50(45), pp. 10522–10553.

[132] Schwab, M.G., Qin, T., Pisula, W., Mavrinskiy, A., Feng, X., Baumgarten, M., Kim, H., Laquai, F., Schuh, S., Trattnig, R. and W.E.J. List, 2011. Molecular triangles: Synthesis, self-assembly, and blue emission of Cyclo-7, 10-tris-triphenylenyl macrocycles. *Chemistry–An Asian Journal*, 6(11), pp. 3001–3010.

[133] Singh, K., Sharma, A., Zhang, J., Xu, W. and Zhu, D., 2011. New sulfur bridged neutral annulenes: Structure, physical properties and applications in organic field-effect transistors. *Chemical Communications*, 47(3), pp. 905–907.

[134] Ie, Y., Hirose, T. and Aso, Y., 2009. Synthesis, properties, and FET performance of rectangular oligothiophene. *Journal of Materials Chemistry*, 19(43), pp. 8169–8175.

[135] Huang, F., Chen, K.S., Yip, H.L., Hau, S.K., Acton, O., Zhang, Y., Luo, J. and Jen, A.K.Y., 2009. Development of new conjugated polymers with donor– π-bridge– acceptor side chains for high performance solar cells. *Journal of the American Chemical Society*, 131(39), pp. 13886–13887.

[136] Zhou, E., Cong, J., Hashimoto, K. and Tajima, K., 2012. Introduction of a conjugated side chain as an effective approach to improving donor–acceptor photovoltaic polymers. *Energy & Environmental Science*, 5(12), pp. 9756–9759.

[137] Ma, Z., Ding, J., Zhang, B., Mei, C., Cheng, Y., Xie, Z., Wang, L., Jing, X. and Wang, F., 2010. Red-emitting efficiency. *Advanced Functional Materials*, 20(1), pp. 138–146.

15 Polymers and Their Composites for Thermoelectric Applications

Yu Guo and Yanfei Xu
University of Massachusetts Amherst, Amherst, United States

CONTENTS

15.1 INTRODUCTION: BACKGROUND AND MOTIVATION

Directly recovering wasted heat and converting it into electrical power by thermoelectric devices offer new potential for providing cleaner forms of energy [1, 2]. Heat can be converted into electricity by thermoelectric generators [3, 4]. These generators are generally constructed using numerous thermocouples. These thermocouples are connected electrically in series but thermally in parallel [5–7]. A thermocouple is made of two different materials with temperature-dependent properties; thus, an electrical voltage will be generated when there is a temperature gradient across the thermocouple [8, 9]. The dimensionless figure of merit (*ZT*) of materials plays a key role in determining thermoelectric conversion efficiency. The dimensionless figure of merit (*ZT*) of materials is defined as $ZT = (S^2\sigma/\kappa)T$, where S is the Seebeck coefficient (or thermopower), σ is the electrical conductivity, κ is the thermal conductivity, and T is the absolute temperature [5, 6, 10, 11]. A thermoelectric device must have an average *ZT* higher than 1 in the application's temperature range to be competitive [11].

In the early 2000s, engineering both phonon transport and electron transport in inorganic materials led to a remarkable breakthrough in the dimensionless figure of the merit of thermoelectric materials, which could further lead to high power generation efficiency [11–13]. So far, alloys of bismuth, antimony, and telluride have shown a high *ZT* of ~1 [10, 11, 13], which are promising materials for future thermoelectric applications. However, such alloys are rare and are manufactured using expensive processes, limiting the thermoelectric applications. Materials with low cost and easy processing properties are desired for thermoelectric applications.

Polymers used as organic thermoelectric materials have drawn lots of researchers' attention, because of their unparalleled and desired properties including light weight, low cost, easy processing, and mechanical flexibility [14]. However, compared with bismuth, antimony, and telluride alloys, organic materials generally have low *ZT* values [5, 14–19], which were in a range between 10^{-6} and 10^{-3} in the 1990s [18, 20]. In 2010, enlightening research on decoupling the Seebeck coefficient, thermal conductivity, and electrical conductivity opened up promising fields in organic thermoelectrics [21, 22]; a *ZT* of ~0.02 was achieved at room temperature [22]. In 2015, a *ZT* of ~0.5 was reported [17, 23]. Despite the relatively low *ZT* of organic thermoelectrics compared to that of inorganic thermoelectrics [24], the unique advantages of high scalability and low cost enabled polymers to be competitive candidate materials in thermoelectric applications.

In this chapter, we will discuss the current progress in developing polymers and polymer-based composites for thermoelectric generators. We will review how critical factors determine the performance of thermoelectrics. We will then discuss design strategies for improving the performance of thermoelectrics by decoupling electrical conductivity, the Seebeck coefficient, and thermal conductivity. Finally, to achieve better thermoelectric properties in polymer-based materials, we will address perspectives and unanswered questions.

DOI: 10.1201/9781003169727-15

15.2 FUNDAMENTALS OF THERMOELECTRICS AND KEY PARAMETERS

15.2.1 Conversion Efficiency (η) of Thermoelectric Devices

Thermoelectric device efficiency is directly related to the *ZT* of thermoelectric materials and the Carnot efficiency. The power generation efficiency of a thermoelectric device is expressed by Equation 15.1 [25]:

$$\eta = \frac{T_h - T_c}{T_h} \frac{\sqrt{1+Z\bar{T}} - 1}{\sqrt{1+Z\bar{T}} + T_c / T_h} \tag{15.1}$$

For refrigeration, the coefficient of performance is calculated by Equation 15.2 [3]:

$$\text{COP} = \frac{T_c}{T_h - T_c} \frac{\sqrt{1+Z\bar{T}} - T_h / T_c}{\sqrt{1+Z\bar{T}} + 1} \tag{15.2}$$

where T_h and T_c are the hot-side and cold-side temperature of the thermoelectric materials, respectively, and $\bar{T}$ is the average temperature of T_h and T_c. Thus, to achieve high efficiency of a thermoelectric device, it is important to use thermoelectric materials with a high *ZT* value.

From Equations (15.1) and (15.2), the efficiency of thermoelectric devices can be reduced by a low *ZT* value. Currently, the commercial thermoelectric devices based on bismuth, telluride, and related alloys show a *ZT* of ~1 and have been used for low-power cooling [26]. Materials with a high *ZT* are required for high-power cooling applications [10, 27].

15.2.2 The Dimensionless Figure of Merit (*ZT*) of Thermoelectric Materials

Equation (15.3) shows how to calculate the *ZT* value in thermoelectric materials:

$$ZT = \frac{S^2\sigma}{\kappa} T \tag{15.3}$$

where the Seebeck coefficient (or thermopower) is expressed by *S*, and the electrical conductivity and thermal conductivity are expressed as σ and κ. κ_l and κ_e are the lattice thermal conductivity and electronic thermal conductivity. The absolute temperature is *T*. To achieve a high *ZT* value in polymer-based materials, a high Seebeck coefficient, high electrical conductivity, and low thermal conductivity are needed [28].

15.2.3 Seebeck Coefficient (*S*) of Thermoelectric Materials

The Seebeck effect is caused by an asymmetrical distribution of the electron transport under a temperature gradient [14] that leads to an electromotive force. As shown in

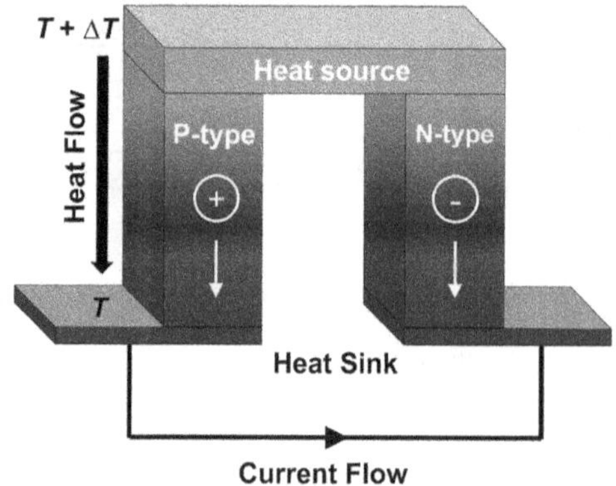

FIGURE 15.1 Schematic diagram of the Seebeck effect for power generation.

Figure 15.1, it is the fundamental effect for a thermoelectric generator.

The electronic density of states (DOS) near the Fermi level (E_F) influence the *S* [18, 35, 36]. Based on the Mott relationship, *S* for degenerate semiconductors is calculated by:

$$S = \frac{\pi^2 k_B^2 T}{3q} \left(\frac{\mathrm{d} \ln \sigma(E)}{\mathrm{d}E} \right) \Bigg|_{E=E_F} \tag{15.4}$$

where k_B is the Boltzmann constant, *q* is the elementary charge, and $\sigma(E)$ is the electrical conductivity contributed by mobile charge carriers with energy *E*. The *S* for nondegenerate semiconductors is given by:

$$S = \frac{k_B}{q} \left[\frac{(E - E_F)}{k_B T} + A \right] \tag{15.5}$$

where *A* is the heat of transport constant. The doping level of semiconductors plays a key role in controlling the position of E_F and determining the magnitude of thermoelectric power generation. Thus, from Equation (15.3), maximizing electrical conductivity is crucial for achieving a high *ZT*—in the case of κ_e contributed by electron transport remaining negligible while keeping the asymmetry of DOS above and below E_F [14, 15, 18].

15.2.4 Electrical Conductivity (σ) of Thermoelectric Materials

Electrical conductivity, σ, in materials can be calculated by Equation (15.6). Elementary charge is expressed by *q*; free carrier concentration is expressed by *n*; charge carrier mobility is expressed by μ. Typically, electrical conductivity for organic materials is ~10^{-6} S cm^{-1} [18].

$$\sigma = nq\mu \tag{15.6}$$

15.2.5 Thermal Conductivity (κ) of Thermoelectric Materials

Thermal conductivity (κ) plays a key role in thermoelectric materials' *ZT*. The thermal conductivity is estimated by:

$$\kappa = \kappa_e + \kappa_l \tag{15.7}$$

where κ_e is the electronic thermal conductivity and κ_l is the lattice thermal conductivity. Based on the Wiedemann–Franz law, the κ_e of a material is proportional to its electrical conductivity, which is estimated by [37]:

$$\kappa_e = L\sigma T \tag{15.8}$$

where *L* is the Lorentz number. Conducting polymers do not completely follow the Wiedemann–Franz law, because of the charge lattice coupling [19]. For conducting polymers, given the typical electrical conductivity of 10^{-6} S cm^{-1}, the electronic thermal conductivity κ_e can be estimated to be ~10^{-10} W m^{-1} K^{-1}, which is negligible compared to κ_l. Thus, the thermal conductivity in polymers is probably independent of the doping level [18, 38].

15.3 FUNDAMENTAL STUDIES ON IMPROVING THERMOELECTRIC PERFORMANCE

To achieve better thermoelectric performance, fundamental studies have been performed, especially for inorganic materials. Nevertheless, the fundamentals of heat and charge transport in polymer-based thermoelectric materials are not well understood [17]. The effects of various parameters such as carrier concentration, electronic thermal conductivity, and lattice thermal conductivity on the *ZT* of nanostructured thermoelectric materials have been addressed [25]. Mahan et al. reported the desired electronic structure for high-performance thermoelectric materials and explained the relationships between the *ZT* value and transport distribution [39]. Mahan et al. also suggested that a higher carrier mobility along the applied electric field and narrower energy carrier distributions could lead to a higher *ZT* of thermoelectric materials.

15.4 DEVELOPMENT OF POLYMER-BASED THERMOELECTRIC MATERIALS

Polymers have unique properties and advantages, such as their low thermal conductivity, possibilities for achieving high electrical conductivity, and possibilities of balancing an incompatible Seebeck coefficient and electrical conductivity [15]. These unique properties offer new possibilities for developing polymer-based thermoelectric generators with higher thermoelectric efficiency [15].

According to Equation (15.3), a high Seebeck coefficient, high electrical conductivity, and low thermal conductivity in polymers are required to maximize *ZT* [25]. In pristine semiconducting polymers, the low electrical conductivity ranging from 10^{-3} to 10^{-8} S cm^{-1} leads to a low power factor (less than 1 μW m^{-1} K^{-2}). Figure 15.2a shows the chemical structures of several semiconducting polymers for thermoelectric materials that include polyacetylene (PA) [40–44], polypyrrole (PPy) [45–47], polythiophene (PT) [48, 49], poly(3hexylthiophene) (P3HT) [50, 51], poly(3,4-ethylenedioxythiophene) (PEDOT) [52, 53], and polyaniline (PANI) [54]. Doping can improve electrical conductivity and the power factor of polymers [17, 55]. Semiconducting polymers can be oxidized (p-doped) [56] or be reduced (n-doped) [14, 16, 17, 57]. For example, a low power factor in pristine polyacetylene is 0.17 μW m^{-1} K^{-2} [44]. After doping by iodine, the measured power factor value in the doped polyacetylene was 400 μW m^{-1} K^{-2} [44]. However, thermoelectric applications using polyacetylene are limited, because it is unstable in air [17]. Figure 15.2 shows the chemical structures of selected semiconducting polymers (Figure 15.2a) and dopants (Figure 15.2b) for thermoelectric devices. Common dopants include poly(styrene sulfonate) (PSS), tosylate (Tos), 2,3,5,6-tetrafluoro-7,7,8,8-tetracyanoquinodimethane (F_4TCNQ), and nitrosonium tetrafluoroborate ($NOBF_4$).

To develop stable doped polymers for thermoelectric devices with high power generation efficiency, dopants with different molecular structures have been developed [18, 58]. Efficient charge transfer relies on the favorable alignment of the frontier electronic levels of the dopant in relation to those of semiconducting polymers [5, 17]. Due to the need for the chemical stability of doped polymers, more p-type dopants are known than n-type dopants [5]. However, new types of air-stable n-type dopants have been developed [5, 16, 59–61]. In addition to the conventional doping method by chemical redox reactions, electrochemical doping is an alternative approach for optimizing the thermoelectric performance of polymers. Electrochemical doping methods enable the fine tuning of the doping levels [62]. Park et al. optimized the doping level of PEDOT electrochemically and achieved a high power factor of 1270 μW m^{-1} K^{-2} in the doped samples [63].

Dopant concentration influences electrical transport in polymers [5]. Electrical conductivity in doped semiconducting polymers nonlinearly increases, when there is an increasing concentration of dopants [64, 65]. Temperature-dependent electrical conduction in doped polymers is based on temperature assisted processes, according to the Arrhenius equation. Electrical conductivity is directly proportional to $\exp(-E_A/k_BT)$. The activation energy is E_A [66, 67]. Temperature-dependent electrical conduction can also be observed in polycrystalline materials, due to barriers at the boundaries of domains. The electrical conductivity is estimated by Equation 15.9 [67–72]:

$$\sigma = \sigma_0 \exp\left[-\left(\frac{T_0}{T}\right)^{\gamma}\right] \tag{15.9}$$

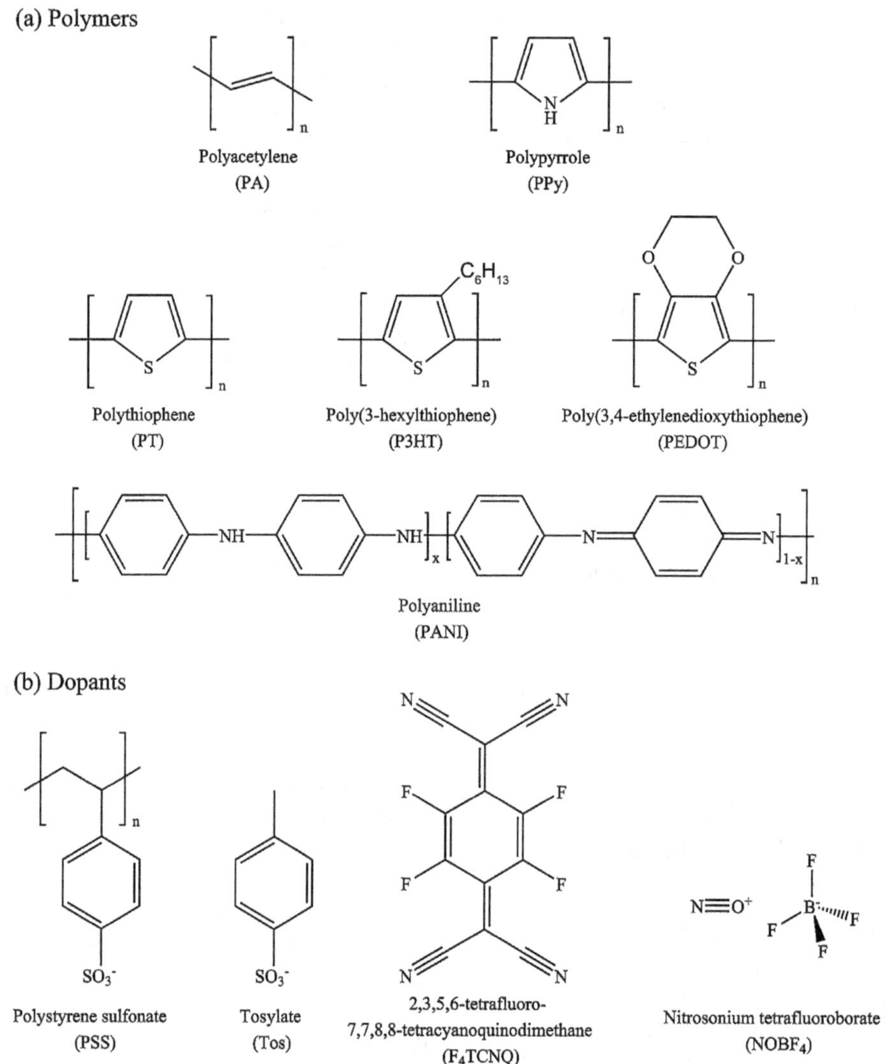

FIGURE 15.2 (a) Chemical structures of selected semiconducting polymers. (b) Chemical structures for common dopants.

where γ is a constant related to the dimensionality of the system and T_0 is the characteristic temperature. Low electrical conductivity can also be measured, which is due to the phase-separated domains in nonuniform polymers [73].

Dopant size also influences electrical transport in polymers. A bulky dopant molecule can lead to distortions in polymer chain packing, which hinders efficient electrical transport in polymers [66, 74–77]. For example, compared with Na^+ as a counterion to the sulfonic acid, K^+ leads to low electrical conductivity with a fivefold decrease in polymers [78]. Understanding the relationships between electrical transport property and nano- and microstructure is important for developing next-generation thermoelectric materials [5].

Compared with inorganic materials, the Seebeck coefficient and charge transport in polymers are different, due to the complex dependence on carrier concentration and microstructures [5]. The Seebeck coefficient is calculated by the Equation 15.10 [79]:

$$S = -\frac{k_B}{q}\int \frac{E - E_F}{k_B T}\frac{\sigma(E)}{\sigma}\,dE \tag{15.10}$$

The Seebeck coefficient can take on different temperature-dependent forms, which are related to the transport mechanisms in polymers. For lightly doped polymers (in a homogeneous system), the Seebeck coefficient is generally not temperature-dependent and is consistent with hopping behavior [80]. In heavily doped polymers, the Seebeck coefficient increases with temperature [20, 81–83]. In a homogeneous polymer system, transport models have usually assumed that carriers are distributed homogeneously and have the same charge transport mechanism [56, 84]. Heterogeneous treatment effects on electrical conductivity in a heterogeneous system have been investigated in view of mixed transport and are suggested by the electrical conductivity and Seebeck coefficient values [85]. The total electrical resistance ($\rho = \sigma^{-1}$) in the system is the sum of the contributions of the ordered regions (σ_m) and disordered regions (σ_d) [5, 86]:

$$\sigma^{-1} = f_m\sigma_m^{-1} + f_d\sigma_d^{-1} \quad (15.11)$$

where f is a general geometrical factor, and the subscripts m and d represent ordered and disordered (amorphous) domains, respectively. As there are different thermal and electrical transport mechanisms in heterogeneous regions, the electrical conductivity and Seebeck coefficient are weighted differently in the ordered (crystalline) and disordered (amorphous) phases [5, 20].

The mobility is low in polymers and there is also energetic disorder. Thus, it is challenging to directly measure carrier concentration. However, carrier concentration in well-ordered polymer semiconductors could be estimated by Hall coefficient measurements [72]. Because of the difficulty of direct measurement of carrier concentration and the lack of a robust model, it is relatively simple to analyze experimental measurements of the Seebeck coefficient, which is a function of electrical conductivity and can be analyzed by Equation (15.10). An empirical relationship between the electrical conductivity and the Seebeck coefficient emerges (Figure 15.3), where $S \propto \sigma^{-1/4}$ (Figure 15.3a) and power factor $S^2\sigma \propto \sigma^{1/2}$ (Figure 15.3b) [5, 49, 85]. The origin of such an empirical relationship is not fully understood, which calls for further investigation.

In addition to the doping of semiconducting polymers, the relationships between thermoelectric properties and polymer structures have been investigated to improve thermoelectric performance [17, 89–91]. Monomer structures can influence the charge transport and thermoelectric properties of polymers [14]. Strong π–π stacking and strong intermolecular coupling are responsible for the efficient charge transfer and high electrical conductivity [92]. Common electrically conductive polymers have low thermal conductivity of the order 0.1 W m^{-1} K^{-1}, which indicates further Seebeck coefficient enhancements could be achieved by increasing doped polymer mobility [93, 94].

Side chains influence the thermoelectric properties of polymers [78]. Shorter side chains lead to more ordered morphologies, which may enhance carrier mobility [78]. The side chain structures could also influence polymer processability, electrical conductivity, and thermal stability [95]. P3HT with aliphatic side chains is commonly processed using chlorinated solvents. However, commonly used dopants such as F_4TCNQ have limited solubility in chlorinated solvents, which leads to a suboptimal nanostructure and relatively low room-temperature electrical conductivity (<10 S cm^{-1}) in P3HT based composites. In comparison, a thiophene, p($g_4$2T-T), with polar tetraethylene glycol side chains could be dissolved in more polar solvents such as dimethylformamide, which offers better solubility for polar molecular dopants such as F_4TCNQ. As a result, F4TCNQ doped p($g_4$2T-T) achieved an electrical conductivity up to 100 S cm^{-1} and excellent thermal

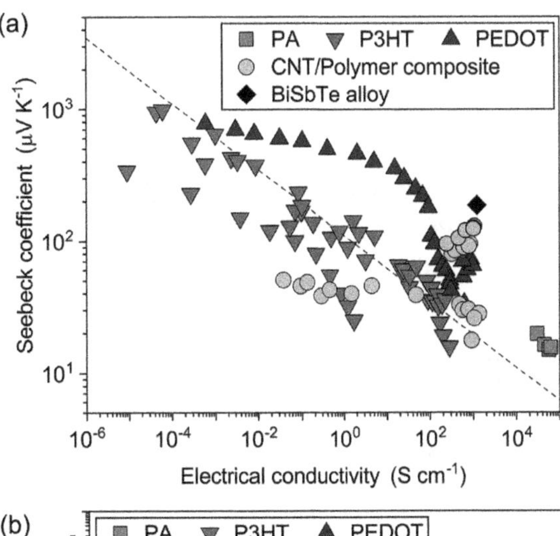

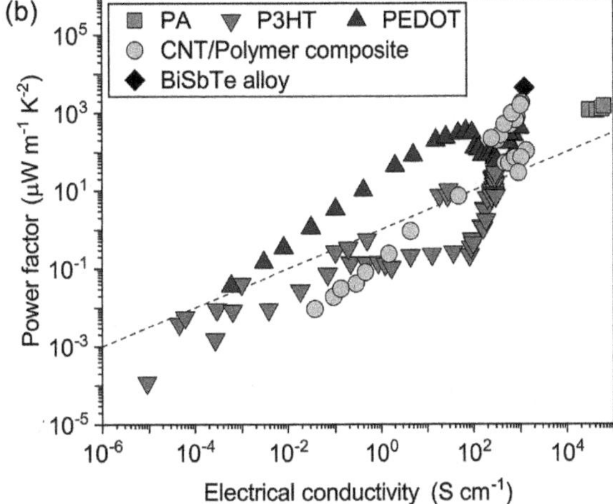

FIGURE 15.3 (a) Seebeck coefficient (S) and (b) power factor ($S^2\sigma$) as functions of electrical conductivity (σ) for selected polymer-based thermoelectrics. The plots include the thermoelectric properties of polyacetylene (PA) [83], poly(3,4-ethylenedioxythiophene) (PEDOT) [53, 55], poly(3hexylthiophene) (P3HT) [20, 49, 65], and carbon nanotube (CNT)/polymer composites [21, 87, 88]. For reference, the Seebeck coefficient and power factor for an inorganic bismuth, antimony, and telluride (BiSbTe) alloy are also included [11]. The empirical relationships $S \propto \sigma^{-1/4}$ (a) and $S^2\sigma \propto \sigma^{1/2}$ (b) are indicated by the dashed lines.

stability up to 150 °C [95]. Great effort has been made in investigating the thermoelectric property–polymer structure relationships and developing polymers for thermoelectric generators. However, the theory for a better design of polymers with high thermoelectric performance remains lacking [17].

Polymer morphology is critical to optimizing thermoelectric properties [5]. Polymer fibers show a high thermal conductivity of ~30 W m^{-1} K^{-1} along the backbone direction, but a low thermal conductivity of ~0.1 W m^{-1} K^{-1} in other directions, demonstrating the complexity of anisotropy in thermal conductivity [96]. The tuning of the spatial distribution of the electronic density of states is another possible way to increase the Seebeck coefficient [5]. Field-effect-modulated Seebeck coefficients larger than 200 μV K^{-1} at room temperature have been reported for nearly disorder-free polymers [97]. The density of states can also be tailored in disordered polymers, which enable the simultaneous enhancement in both the Seebeck coefficient and electrical conductivity [98].

15.5 DEVELOPMENT OF POLYMER-BASED COMPOSITES FOR THERMOELECTRIC MATERIALS

Polymer-based composites have demonstrated high thermoelectric properties, thanks to unique effects from the combined advantages of both polymers and fillers [99–103]. Polymers have low thermal conductivity, light weight, easy processability, mechanical flexibility, and low cost. Nanofillers can enhance electrical transport in nanocomposites and optimize the balance between the heat and charge transports. Adding highly electrically conductive nanofillers into polymers can enhance electrical conductivity without reducing the Seebeck coefficient, which leads to high power factors [22, 62, 87, 88, 100, 101]. Moreover, the interfaces between fillers and polymers create numerous boundaries, which hinder phonon transport and result in low thermal conductivities in polymer-based composites [9, 104]. Additionally, nanostructures could further improve the thermoelectric properties due to the change in the density of states and the quantum confinement effects [105].

Nanostructured fillers have been developed, which were added into polymers and which improved the thermoelectric performance of polymer-based composites [17, 99–103, 106]. Fillers can be either inorganic or organic [14, 107]. We mainly discuss nanostructured carbon fillers including carbon nanotubes (CNTs) and graphene, which can increase the electrical conductivity of polymer-based composites due to their high electrical conductivity [108–113]. Table 15.1 shows the thermoelectric properties of selected polymer-based composites. Yu et al. reported polymer-based composites consisting of 30 wt.% PEDOT:PSS, 60 wt.% carbon nanotubes, and 10 wt.% polyvinyl acetate [87]. The Seebeck coefficient of such composites is ~40 μV K^{-1} [87]. Composites composed of semiconducting polymers and carbon nanofillers have also been developed [87, 102, 114, 115]. For example, Yao et al. [115] prepared SWCNT/PANI nanocomposites through the *in situ* polymerization of aniline using SWCNTs as templates. It was observed that the PANI grew along the SWCNTs, forming ordered chain structures.

Post-treatments have been also investigated to enhance the electrical conductivity of polymer-based composites [99, 103]. The high power factor of ~151 μW m^{-1} K^{-2} has been observed [116]. Chemical treatments on thermoelectric polymer composites could contribute to electrical conductivity enhancements without improving the Seebeck coefficient. Notably, formic acid treated polymer-based composites show a Seebeck coefficient lower than that of the untreated ones [117]. Because of heat conduction and joule heating, the power conversion efficiency can be limited by parasitic heat losses in thermoelectric devices [5].

TABLE 15.1
Thermoelectric Properties of Selected Carbon-Based Filler/Polymer Composites at Room Temperature

Polymers	Fillers	Loading (wt.%)	S (μV K^{-1})	σ (S cm^{-1})	PF (μW m^{-1} K^{-2})	κ (W m^{-1} K^{-1})	ZT	Reference
P3HT	SWCNTs	60	31.1	2760	267	—	—	[118]
PEDOT:PSS	SWCNTs	6.7	59	≈ 1330	464	—	—	[117]
PEDOT:PSS	SWCNTs	13	≈ 18.3	570.4	19	—	—	[119]
PEDOT:PSS	SWCNTs	35	25	400	25	0.4	0.02	[22]
PEDOT:PSS	DWCNTs	20	43.7	780	151	—	—	[116]
PEDOT:PSS	Graphene	3	26.8	637	45.7	—	—	[120]
PEDOT:PSS	Graphene	3	22.4	1067	53.3	0.3	0.05	[121]
PEDOT:PSS	Graphene	—	28.1	1096	57.9	—	—	[122]
PEDOT:PSS	Graphene	16	31.8	50.8	5.2	—	—	[123]
PANI	SWCNTs	41.4	40	125	20	1.5	0.004	[115]
PANI	SWCNTs	64	≈ 47	769	≈ 170	0.43	0.12	[124]
PANI	SWCNTs	65	38.9	1440	217	0.44	—	[125]
PANI	DWCNTs	30	61	610	220	0.7[a]	≈ 0.1	[100]
PANI	MWCNTs	0.93	79.8	14.1	9	0.27	0.01	[126]
PANI	MWCNTs	84.2	28.6	61.47	5	≈ 0.5	—	[127]
PANI	Graphene	48	26	814	55	—	—	[128]
PPy	SWCNTs	37.5	22.2	399	19.7	—	—	[129]
PPy	Graphene	66.7	26.9	41.6	3.01	—	—	[130]

Note: poly(3hexylthiophene) (P3HT), poly(3,4-ethylenedioxythiophene):poly(styrene sulfonate) (PEDOT:PSS), polyaniline (PANI), polypyrrole (PPy), single-walled carbon nanotubes (SWCNTs), double-walled carbon nanotubes (DWCNTs), multi-walled carbon nanotubes (MWCNTs)
a Calculated thermal conductivity.

15.6 SUMMARY AND OUTLOOK

Over recent decades, studies on the thermoelectric properties of polymer-based materials have led to breakthroughs in both the fundamental understanding of thermoelectric mechanisms and device demonstrations. However, there is still much room for material design and thermoelectric property enhancement in polymers and polymer-based composites. Because of the complicated structures in polymer-based materials, there are still challenges in gaining a deep understanding of thermoelectric behavior in polymers and translating this behavior into high-performance devices. Continued efforts are needed to explore the relationships between polymer structures and thermoelectric properties.

REFERENCES

1. Bell, L.E., Cooling, heating, generating power, and recovering waste heat with thermoelectric systems. *Science*, 2008. **321**(5895): pp. 1457–1461.
2. Goldsmid, H.J., *Thermoelectric Refrigeration*. Springer, 1964.
3. Mao, J., G. Chen, and Z. Ren, Thermoelectric cooling materials. *Nature Materials*, 2020. 20: pp. 454–461
4. Tritt, T.M., Thermoelectric phenomena, materials, and applications. *Annual Review of Materials Research*, 2011. **41**(1): pp. 433–448.
5. Russ, B., et al., Organic thermoelectric materials for energy harvesting and temperature control. *Nature Reviews Materials*, 2016. **1**(10): p. 16050.
6. Crispin, X., Thermoelectrics: Carbon nanotubes get high. *Nature Energy*, 2016. **1**(4): p. 16037.
7. Chen, H., et al., Thermal conductivity of polymer-based composites: Fundamentals and applications. *Progress in Polymer Science*, 2016. **59**: pp. 41–85.
8. Tritt, T.M., *Semiconductors and Semimetals, Recent Trends in Thermoelectric Materials Research: Part One to Three*. Academic, San Diego, CA, 2001. 69–71.
9. Chen, Z., X. Zhang, and Y. Pei, Manipulation of phonon transport in thermoelectrics. *Advanced Materials*, 2018. **30**(17): p. 1705617.
10. Rowe, D.M., *CRC Handbook of Thermoelectrics*. CRC Press, Boca Raton.
11. Poudel, B., et al., High-thermoelectric performance of nanostructured bismuth antimony telluride bulk alloys. *Science*, 2008. **320**(5876): pp. 634–638.
12. Cahill, D.G., et al., Nanoscale thermal transport. II. 2003–2012. *Applied Physics Reviews*, 2014. **1**(1): p. 011305.
13. Venkatasubramanian, R., et al., Thin-film thermoelectric devices with high room-temperature figures of merit. *Nature*, 2001. **413**(6856): pp. 597–602.
14. Shi, X.-L., J. Zou, and Z.-G. Chen, Advanced thermoelectric design: From materials and structures to devices. *Chemical Reviews*, 2020. **120**(15): pp. 7399–7515.
15. Goel, M. and M. Thelakkat, Polymer thermoelectrics: Opportunities and challenges. *Macromolecules*, 2020. **53**(10): pp. 3632–3642.
16. Wang, Y., et al., Flexible thermoelectric materials and generators: Challenges and innovations. *Advanced Materials*, 2019. **31**(29): p. 1807916.
17. Wang, H. and C. Yu, Organic thermoelectrics: Materials preparation, performance optimization, and device integration. *Joule*, 2019. **3**(1): pp. 53–80.
18. Chen, Y., Y. Zhao, and Z. Liang, Solution processed organic thermoelectrics: Towards flexible thermoelectric modules. *Energy & Environmental Science*, 2015. **8**(2): pp. 401–422.
19. Zhang, Q., et al., Organic thermoelectric materials: Emerging green energy materials converting heat to electricity directly and efficiently. *Advanced Materials*, 2014. **26**(40): pp. 6829–6851.
20. Zhang, Q., et al., What to expect from conducting polymers on the playground of thermoelectricity: Lessons learned from four high-mobility polymeric semiconductors. *Macromolecules*, 2014. **47**(2): pp. 609–615.
21. Yu, C., et al., Thermoelectric behavior of segregated-network polymer nanocomposites. *Nano Letters*, 2008. **8**(12): pp. 4428–4432.
22. Kim, D., et al., Improved thermoelectric behavior of nanotube-filled polymer composites with poly(3,4-ethylenedioxythiophene) poly(styrenesulfonate). *ACS Nano*, 2010. **4**(1): pp. 513–523.
23. Wang, H., et al., Thermally driven large n-type voltage responses from hybrids of carbon nanotubes and poly(3,4-ethylenedioxythiophene) with tetrakis(dimethylamino)ethylene. *Advanced Materials*, 2015. **27**(43): pp. 6855–6861.
24. Zhao, L.-D., et al., Ultralow thermal conductivity and high thermoelectric figure of merit in SnSe crystals. *Nature*, 2014. **508**(7496): pp. 373–377.
25. Snyder, G.J. and E.S. Toberer, Complex thermoelectric materials. *Nature Materials*, 2008. **7**(2): pp. 105–114.
26. Lan, Y., et al., Enhancement of thermoelectric figure-of-merit by a bulk nanostructuring approach. *Advanced Functional Materials*, 2010. **20**(3): pp. 357–376.
27. Tritt, E.T.M., *Recent Trends in Thermoelectric Materials Research: Part Three*. Academic Press, San Diego, CA. 2001.
28. Zhang, X. and L.-D. Zhao, Thermoelectric materials: Energy conversion between heat and electricity. *Journal of Materiomics*, 2015. **1**(2): pp. 92–105.
29. Heremans, J.P., et al., Enhancement of thermoelectric efficiency in PbTe by distortion of the electronic density of states. *Science*, 2008. **321**(5888): p. 554.
30. Pei, Y., et al., Convergence of electronic bands for high performance bulk thermoelectrics. *Nature*, 2011. **473**(7345): pp. 66–69.
31. Liu, X., et al., Low electron scattering potentials in high performance Mg2Si0.45Sn0.55 based thermoelectric solid solutions with band convergence. *Advanced Energy Materials*, 2013. **3**(9): pp. 1238–1244.
32. Harman, T.C., et al., Quantum dot superlattice thermoelectric materials and devices. *Science*, 2002. **297**(5590): p. 2229.
33. Hong, M., et al., n-type $Bi_2Te_{3-x}Se_x$ nanoplates with enhanced thermoelectric efficiency driven by wide-frequency phonon scatterings and synergistic carrier scatterings. *ACS Nano*, 2016. **10**(4): pp. 4719–4727.
34. Yang, L., et al., High-performance thermoelectric Cu_2Se nanoplates through nanostructure engineering. *Nano Energy*, 2015. **16**: pp. 367–374.
35. Rowe, D.M., *Thermoelectrics Handbook: Macro to Nano*. 2005: CRC Press, Boca Raton.
36. MacDonald, D.K.C., *Thermoelectricity: An Introduction to the Principles*. 2006: Dover Publications, Mineola.
37. Chen, G., *Nanoscale Energy Transport and Conversion: A Parallel Treatment of Electrons, Molecules, Phonons, and Photons*. 2005: Oxford University Press, Oxford.
38. Gao, X., et al., Theoretical studies on the thermopower of semiconductors and low-band-gap crystalline polymers. *Physical Review B*, 2005. **72**(12): p. 125202.

39. Mahan, G.D. and J.O. Sofo, The best thermoelectric. *Proceedings of the National Academy of Sciences*, 1996. **93**(15): p. 7436.
40. Moses, D. and A. Denenstein, Experimental determination of the thermal conductivity of a conducting polymer: Pure and heavily doped polyacetylene. *Physical Review B*, 1984. **30**(4): pp. 2090–2097.
41. Park, Y.W., et al., Metallic nature of heavily doped polyacetylene derivatives: Thermopower. *Physical Review B*, 1984. **30**(10): pp. 5847–5851.
42. Park, Y.W., et al., Conductivity and thermoelectric power of the newly processed polyacetylene. *Synthetic Metals*, 1989. **28**(3): pp. D27–D34.
43. Zuzok, R., et al., Thermoelectric power and conductivity of iodine-doped "new" polyacetylene. *The Journal of Chemical Physics*, 1991. **95**(2): pp. 1270–1275.
44. Park, Y.W., Structure and morphology: Relation to thermopower properties of conductive polymers. *Synthetic Metals*, 1991. **45**(2): pp. 173–182.
45. Maddison, D.S., R.B. Roberts, and J. Unsworth, Thermoelectric power of polypyrrole. *Synthetic Metals*, 1989. **33**(3): pp. 281–287.
46. Maddison, D.S., J. Unsworth, and R.B. Roberts, Electrical conductivity and thermoelectric power of polypyrrole with different doping levels. *Synthetic Metals*, 1988. **26**(1): pp. 99–108.
47. Kemp, N.T., et al., Thermoelectric power and conductivity of different types of polypyrrole. *Journal of Polymer Science Part B: Polymer Physics*, 1999. **37**(9): pp. 953–960.
48. Kaloni, T.P., et al., Polythiophene: From fundamental perspectives to applications. *Chemistry of Materials*, 2017. **29**(24): pp. 10248–10283.
49. Glaudell, A.M., et al., Impact of the doping method on conductivity and thermopower in semiconducting polythiophenes. *Advanced Energy Materials*, 2015. **5**(4): pp. 1401072.
50. Hynynen, J., et al., Enhanced electrical conductivity of molecularly p-doped poly(3-hexylthiophene) through understanding the correlation with solid-state order. *Macromolecules*, 2017. **50**(20): pp. 8140–8148.
51. Hynynen, J., D. Kiefer, and C. Müller, Influence of crystallinity on the thermoelectric power factor of P3HT vapour-doped with F4TCNQ. *RSC Advances*, 2018. **8**(3): pp. 1593–1599.
52. Groenendaal, L., et al., Poly(3,4-ethylenedioxythiophene) and its derivatives: Past, present, and future. *Advanced Materials*, 2000. **12**(7): pp. 481–494.
53. Bubnova, O., et al., Optimization of the thermoelectric figure of merit in the conducting polymer poly(3,4-ethylenedioxythiophene). *Nature Materials*, 2011. **10**(6): pp. 429–433.
54. Wang, H., et al., Facile charge carrier adjustment for improving thermopower of doped polyaniline. *Polymer*, 2013. **54**(3): pp. 1136–1140.
55. Kim, G.H., et al., Engineered doping of organic semiconductors for enhanced thermoelectric efficiency. *Nature Materials*, 2013. **12**(8): pp. 719–723.
56. Heeger, A.J., Semiconducting and metallic polymers: The fourth generation of polymeric materials. *The Journal of Physical Chemistry B*, 2001. **105**(36): pp. 8475–8491.
57. Lu, Y., J.-Y. Wang, and J. Pei, Strategies to enhance the conductivity of n-type polymer thermoelectric materials. *Chemistry of Materials*, 2019. **31**(17): pp. 6412–6423.
58. Dubey, N. and M. Leclerc, Conducting polymers: Efficient thermoelectric materials. *Journal of Polymer Science Part B: Polymer Physics*, 2011. **49**(7): pp. 467–475.
59. Guo, S., et al., n-doping of organic electronic materials using air-stable organometallics. *Advanced Materials*, 2012. **24**(5): pp. 699–703.
60. Qi, Y., et al., Solution doping of organic semiconductors using air-stable n-dopants. *Applied Physics Letters*, 2012. **100**(8): p. 083305.
61. Du, Y., et al., Flexible thermoelectric materials and devices. *Applied Materials Today*, 2018. **12**: pp. 366–388.
62. Petsagkourakis, I., et al., Thermoelectric materials and applications for energy harvesting power generation. *Science and Technology of Advanced Materials*, 2018. **19**(1): pp. 836–862.
63. Park, T., et al., Flexible PEDOT electrodes with large thermoelectric power factors to generate electricity by the touch of fingertips. *Energy & Environmental Science*, 2013. **6**(3): pp. 788–792.
64. Pingel, P. and D. Neher, Comprehensive picture of p-type doping of P3HT with the molecular acceptor F4TCNQ. *Physical Review B*, 2013. **87**(11): p. 115209.
65. Xuan, Y., et al., Thermoelectric properties of conducting polymers: The case of poly(3-hexylthiophene). *Physical Review B*, 2010. **82**(11): p. 115454.
66. Walzer, K., et al., Highly efficient organic devices based on electrically doped transport layers. *Chemical Reviews*, 2007. **107**(4): pp. 1233–1271.
67. Wang, Z.H., et al., Three dimensionality of "metallic" states in conducting polymers: Polyaniline. *Physical Review Letters*, 1991. **66**(13): pp. 1745–1748.
68. Heeger, A.J., et al., Solitons in conducting polymers. *Reviews of Modern Physics*, 1988. **60**(3): pp. 781–850.
69. Yoon, C.O., et al., Hopping transport in doped conducting polymers in the insulating regime near the metal-insulator boundary: Polypyrrole, polyaniline and polyalkylthiophenes. *Synthetic Metals*, 1995. **75**(3): pp. 229–239.
70. Paloheimo, J., et al., Conductivity, thermoelectric power and field-effect mobility in self-assembled films of polyanilines and oligoanilines. *Synthetic Metals*, 1995. **68**(3): pp. 249–257.
71. van de Ruit, K., et al., Quasi-one dimensional in-plane conductivity in filamentary films of PEDOT:PSS. *Advanced Functional Materials*, 2013. **23**(46): pp. 5778–5786.
72. Wang, S., et al., Hopping transport and the Hall effect near the insulator–metal transition in electrochemically gated poly(3-hexylthiophene) transistors. *Nature Communications*, 2012. **3**(1): p. 1210.
73. Epstein, A.J., et al., Inhomogeneous disorder and the modified Drude metallic state of conducting polymers. *Synthetic Metals*, 1994. **65**(2): pp. 149–157.
74. Winokur, M., et al., X-ray scattering from sodium-doped polyacetylene: Incommensurate-commensurate and order-disorder transformations. *Physical Review Letters*, 1987. **58**(22): pp. 2329–2332.
75. Winokur, M.J., et al., Structural evolution in iodine-doped poly(3-alkylthiophenes). *Macromolecules*, 1991. **24**(13): pp. 3812–3815.
76. Duong, D.T., et al., The chemical and structural origin of efficient p-type doping in P3HT. *Organic Electronics*, 2013. **14**(5): pp. 1330–1336.
77. Cochran, J.E., et al., Molecular interactions and ordering in electrically doped polymers: blends of PBTTT and F4TCNQ. *Macromolecules*, 2014. **47**(19): pp. 6836–6846.
78. Mai, C.-K., et al., Side-chain effects on the conductivity, morphology, and thermoelectric properties of self-doped narrow-band-gap conjugated polyelectrolytes. *Journal of the American Chemical Society*, 2014. **136**(39): pp. 13478–13481.

79. Fritzsche, H., A general expression for the thermoelectric power. *Solid State Communications*, 1971. **9**(21): pp. 1813–1815.
80. Park, Y., et al., Semiconductor-metal transition in doped $(CH)_x$: Thermoelectric power. *Solid State Communications*, 1979. **29**: pp. 747–751.
81. Reghu, M., et al., Counterion-induced processibility of polyaniline: Transport at the metal-insulator boundary. *Physical Review B*, 1993. **47**(4): pp. 1758–1764.
82. Yoon, C.O., et al., Transport near the metal-insulator transition: Polypyrrole doped with PF6. *Physical Review B*, 1994. **49**(16): pp. 10851–10863.
83. Nogami, Y., et al., On the metallic states in highly conducting iodine-doped polyacetylene. *Solid State Communications*, 1990. **76**(5): pp. 583–586.
84. Sirringhaus, H., 25th anniversary article: Organic field-effect transistors: The path beyond amorphous silicon. *Advanced Materials*, 2014. **26**(9): pp. 1319–1335.
85. Kaiser, A.B., Thermoelectric power and conductivity of heterogeneous conducting polymers. *Physical Review B*, 1989. **40**(5): pp. 2806–2813.
86. Kaiser, A.B., Electronic transport properties of conducting polymers and carbon nanotubes. *Reports on Progress in Physics*, 2000. **64**(1): pp. 1–49.
87. Yu, C., et al., Light-weight flexible carbon nanotube based organic composites with large thermoelectric power factors. *ACS Nano*, 2011. **5**(10): pp. 7885–7892.
88. Cho, C., et al., Completely organic multilayer thin film with thermoelectric power factor rivaling inorganic tellurides. *Advanced Materials*, 2015. **27**(19): pp. 2996–3001.
89. Yue, R., et al., Electrochemistry, morphology, thermoelectric and thermal degradation behaviors of free-standing copolymer films made from 1,12-bis(carbazolyl)dodecane and 3,4-ethylenedioxythiophene. *Polymer Journal*, 2011. **43**(6): pp. 531–539.
90. Yue, R., et al., Facile electrosynthesis and thermoelectric performance of electroactive free-standing polythieno[3,2-b]thiophene films. *Journal of Solid State Electrochemistry*, 2011. **15**(3): pp. 539–548.
91. Chen, S., et al., Systematic study on chemical oxidative and solid-state polymerization of poly(3,4-ethylenedithiathiophene). *Journal of Polymer Science Part A: Polymer Chemistry*, 2012. **50**(10): pp. 1967–1978.
92. Li, H., et al., Modification of the poly(bisdodecylquaterthiophene) structure for high and predominantly nonionic conductivity with matched dopants. *Journal of the American Chemical Society*, 2017. **139**(32): pp. 11149–11157.
93. Aïch, R.B., et al., Electrical and thermoelectric properties of poly(2,7-carbazole) derivatives. *Chemistry of Materials*, 2009. **21**(4): pp. 751–757.
94. Wakim, S., et al., Charge transport, photovoltaic, and thermoelectric properties of poly(2,7-carbazole) and poly(indolo[3,2-b]carbazole) derivatives. *Polymer Reviews*, 2008. **48**(3): pp. 432–462.
95. Kroon, R., et al., Polar side chains enhance processability, electrical conductivity, and thermal stability of a molecularly p-doped polythiophene. *Advanced Materials*, 2017. **29**(24): p. 1700930.
96. Wang, X., et al., Thermal conductivity of high-modulus polymer fibers. *Macromolecules*, 2013. **46**(12): pp. 4937–4943.
97. Venkateshvaran, D., et al., Approaching disorder-free transport in high-mobility conjugated polymers. *Nature*, 2014. **515**(7527): pp. 384–388.
98. Sun, J., et al., Simultaneous increase in Seebeck coefficient and conductivity in a doped poly(alkylthiophene) blend with defined density of states. *Macromolecules*, 2010. **43**(6): pp. 2897–2903.
99. McGrail, B.T., A. Sehirlioglu, and E. Pentzer, Polymer composites for thermoelectric applications. *Angewandte Chemie International Edition*, 2015. **54**(6): pp. 1710–1723.
100. Wang, H., et al., Simultaneously improving electrical conductivity and thermopower of polyaniline composites by utilizing carbon nanotubes as high mobility conduits. *ACS Applied Materials & Interfaces*, 2015. **7**(18): pp. 9589–9597.
101. Wang, H., S.-I. Yi, and C. Yu, Engineering electrical transport at the interface of conjugated carbon structures to improve thermoelectric properties of their composites. *Polymer*, 2016. **97**: pp. 487–495.
102. Wang, L., et al., Polymer composites-based thermoelectric materials and devices. *Composites Part B: Engineering*, 2017. **122**: pp. 145–155.
103. Yao, H., et al., Recent development of thermoelectric polymers and composites. *Macromolecular Rapid Communications*, 2018. **39**(6): p. 1700727.
104. Vineis, C.J., et al., Nanostructured thermoelectrics: Big efficiency gains from small features. *Advanced Materials*, 2010. **22**(36): pp. 3970–3980.
105. Dresselhaus, M.S., et al., New directions for low-dimensional thermoelectric materials. *Advanced Materials*, 2007. **19**(8): pp. 1043–1053.
106. Chen, R., et al., Thermoelectrics of nanowires. *Chemical Reviews*, 2019. **119**(15): pp. 9260–9302.
107. Zhang, Y. and G.D. Stucky, Heterostructured approaches to efficient thermoelectric materials. *Chemistry of Materials*, 2014. **26**(1): pp. 837–848.
108. Hewitt, C.A. and D.L. Carroll, Carbon nanotube-based polymer composite thermoelectric generators, in *Polymer Composites for Energy Harvesting, Conversion, and Storage*. 2014, American Chemical Society, Washington, DC. pp. 191–211.
109. Zong, P.-A., et al., Graphene-based thermoelectrics. *ACS Applied Energy Materials*, 2020. **3**(3): pp. 2224–2239.
110. Blackburn, J.L., et al., Carbon-nanotube-based thermoelectric materials and devices. *Advanced Materials*, 2018. **30**(11): p. 1704386.
111. Pei, Y., H. Wang, and G.J. Snyder, Band engineering of thermoelectric materials. *Advanced Materials*, 2012. **24**(46): pp. 6125–6135.
112. Jiang, Q., et al., Recent advances, design guidelines, and prospects of flexible organic/inorganic thermoelectric composites. *Materials Advances*, 2020. **1**(5): p. 1038–1054.
113. Kaiser, A.B. and V. Skákalová, Electronic conduction in polymers, carbon nanotubes and graphene. *Chemical Society Reviews*, 2011. **40**(7): pp. 3786–3801.
114. Zhang, K., Y. Zhang, and S. Wang, Enhancing thermoelectric properties of organic composites through hierarchical nanostructures. *Scientific Reports*, 2013. **3**(1): p. 3448.
115. Yao, Q., et al., Enhanced thermoelectric performance of single-walled carbon nanotubes/polyaniline hybrid nanocomposites. *ACS Nano*, 2010. **4**(4): pp. 2445–2451.
116. Lee, W., et al., Improving the thermoelectric power factor of CNT/PEDOT: PSS nanocomposite films by ethylene glycol treatment. *RSC Advances*, 2016. **6**(58): pp. 53339–53344.
117. Hsu, J.-H., et al., Origin of unusual thermoelectric transport behaviors in carbon nanotube filled polymer composites after solvent/acid treatments. *Organic Electronics*, 2017. **45**: pp. 182–189.
118. Hong, C.T., et al., Effective doping by spin-coating and enhanced thermoelectric power factors in SWCNT/P3HT hybrid films. *Journal of Materials Chemistry A*, 2015. **3**(23): pp. 12314–12319.

119. Hu, X., G. Chen, and X. Wang, An unusual coral-like morphology for composites of poly(3,4-ethylenedioxythiophene)/carbon nanotube and the enhanced thermoelectric performance. *Composites Science and Technology*, 2017. **144**: pp. 43–50.
120. Yoo, D., J. Kim, and J.H. Kim, Direct synthesis of highly conductive poly(3,4-ethylenedioxythiophene):poly(4-styrenesulfonate) (PEDOT:PSS)/graphene composites and their applications in energy harvesting systems. *Nano Research*, 2014. **7**(5): pp. 717–730.
121. Xiong, J., et al., Liquid exfoliated graphene as dopant for improving the thermoelectric power factor of conductive PEDOT: PSS nanofilm with hydrazine treatment. *ACS Applied Materials & Interfaces*, 2015. **7**(27): pp. 14917–14925.
122. Park, C., et al., Large-scalable RTCVD graphene/PEDOT: PSS hybrid conductive film for application in transparent and flexible thermoelectric nanogenerators. *RSC Advances*, 2017. **7**(41): pp. 25237–25243.
123. Xu, K., G. Chen, and D. Qiu, Convenient construction of poly(3,4-ethylenedioxythiophene)–graphene pie-like structure with enhanced thermoelectric performance. *Journal of Materials Chemistry A*, 2013. **1**(40): pp. 12395–12399.
124. Yao, Q., et al., Abnormally enhanced thermoelectric transport properties of SWNT/PANI hybrid films by the strengthened PANI molecular ordering. *Energy & Environmental Science*, 2014. **7**(11): pp. 3801–3807.
125. Wang, L., et al., Engineered molecular chain ordering in single-walled carbon nanotubes/polyaniline composite films for high-performance organic thermoelectric materials. *Chemistry – An Asian Journal*, 2016. **11**(12): pp. 1804–1810.
126. Zhang, K., et al., Thermoelectric properties of porous multi-walled carbon nanotube/polyaniline core/shell nanocomposites. *Nanotechnology*, 2012. **23**(38): p. 385701.
127. Meng, C., C. Liu, and S. Fan, A promising approach to enhanced thermoelectric properties using carbon nanotube networks. *Advanced Materials*, 2010. **22**(4): pp. 535–539.
128. Wang, L., et al., PANI/graphene nanocomposite films with high thermoelectric properties by enhanced molecular ordering. *Journal of Materials Chemistry A*, 2015. **3**(13): pp. 7086–7092.
129. Liang, L., et al., Large-area, stretchable, super flexible and mechanically stable thermoelectric films of polymer/carbon nanotube composites. *Journal of Materials Chemistry C*, 2016. **4**(3): pp. 526–532.
130. Han, S., et al., Morphology and thermoelectric properties of graphene nanosheets enwrapped with polypyrrole. *RSC Advances*, 2014. **4**(55): pp. 29281–29285.

16 Polymeric Materials for Hydrogen Storage

Seyyedeh Fatemeh Hosseini, Shiva Mohajer, and Mir Saeed Seyed Dorraji
University of Zanjan, Zanjan, Iran

CONTENTS

16.1 INTRODUCTION

There are many worries about the adverse effects of rising levels of CO_2 and other greenhouse gases on the environment and climate. It is noteworthy that increasing energy demand is the main reason for increasing CO_2 emissions in the atmosphere. In recent decades, heat and electricity generated by burning carbon-based fuels have led to an increase in the dependence of today's world, especially developing countries, on fossil fuels. As a result, with the rapid decline of non-renewable fossil fuel reserves along with possible global warming, seeking to meet clean and renewable energy sources is imperative (McKeown et al., 2007; Schlapbach and Züttel, 2011). For this purpose, hydrogen (H_2) gas has been proposed as an attractive option to reduce dependence on petroleum products, due to its being pollution-free, of large abundance, and of high calorimetric value. Moreover, H_2 has a high gravimetric energy density (142 kJ g^{-1}) which is about three times greater than that of petrol (47 k Jg^{-1}).

However, H_2 is an abundant element in nature; it is mainly found in combination with oxygen as water or organic compounds. There are many ways to produce H_2, for instance, steam reforming of fossil fuel, electrochemical cycling, water electrolysis, direct splitting, and the fermentation of water photocatalytically or thermochemically. It is important that the use of renewable energy sources (e.g., solar and wind) in H_2 production leads to a significant reduction in CO_2 emissions. At the same time, the use of fossil fuels as non-renewable energy sources will increase the need for carbon capture and sequestration to reduce CO_2 emissions (Züttel, 2003).

Generally, in its production cycle, H_2 must first be produced and then stored before consumption. However, the low volumetric energy density of H_2 in ambient conditions has made it unusable for various actual applications. Due to this, the lack of a reliable H_2 storage system remains a major technical barrier to industrialization, widespread use, and the uptake of molecular H_2, especially in transportation

DOI: 10.1201/9781003169727-16

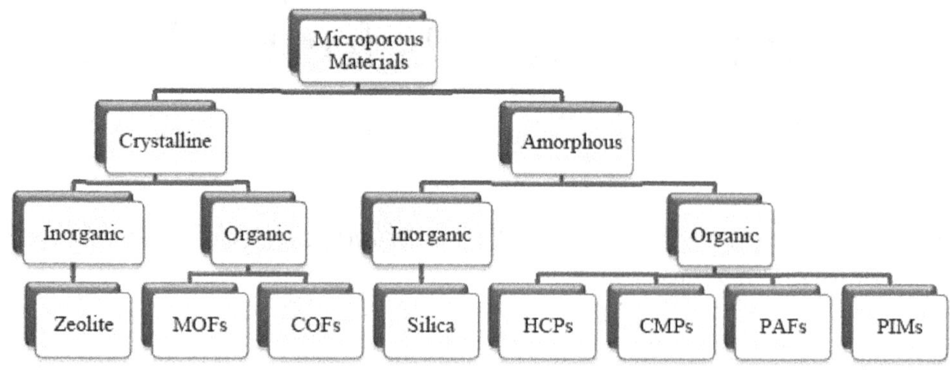

FIGURE 16.1 Cataloguing of microporous materials in terms of their construction and components.

Source: Ramimoghadam et al. (2016) with permission.

Notes: COFs: covalent organic frameworks, HCPs: hyper-cross-linked polymers, MOFs: metal organic frameworks, PAFs: porous aromatic frameworks, CMPs: conjugated microporous polymers.

systems. In an H_2 storage system, benchmarks such as reversible uptake and the release rate of H_2, safety, cost, volume, and the weight of the system should be considered (Jena, 2011; Principi et al., 2009). H_2 can be stored by three possible methods: gas compression (high-pressure gas cylinders), liquefaction (liquid H_2 in cryogenic tanks at 20 K), and storage in solid-state materials (Iulianelli and Basile, 2014; Jenkins, 2013). In particular, the reversible storage of H_2 in ambient conditions using lightweight, safe, and inexpensive solids can potentially provide the highest H_2 density along with the highest efficiency in practical H_2 storage. This method can be divided into physisorption, the physical adsorption of H_2 on the surface of porous materials by van der Waals forces, and chemisorption, the chemical bonding of H_2 with adsorbents. However, chemisorption allows the adsorption of large amounts of H_2 gas; it is not reversible and requires a high temperature to desorb H_2 gas. Meanwhile, physisorption benefits complete reversibility, fast adsorption/desorption cycles, high stability, and high energy efficiency. As a result, H_2 uptake through physisorption has become an active research field. Of course, it is worth noting that the weak interaction and relatively low adsorption enthalpy (4–7 kJ mol^{-1}) of physisorption increase the need for cryogenic temperatures (about 77 K) and high pressures to appropriate storage capacity.

On the other hand, it is well known that H_2 uptake is controlled by the structural properties of the adsorbent material. In other words, pores larger than the kinetic diameter of H_2 (0.289 nm) lead to unemployed void volume at the center of the pores and thus reduce hydrogen uptake. According to the goals set by the U.S. Department of Energy (DOE), it wasecessary to develop materials that could store 6.5 wt.% and 60 g/L of H_2 by 2050. Moreover, increasing the surface area of the adsorbent and the enthalpy of adsorption leads to an increase in H_2 adsorption capacity, and also reduces the need for cooling in the physisorption-based H_2 storage systems (Cheng et al., 2008; David, 2005; Ramimoghadam et al., 2016; Tedds et al., 2011; Züttel, 2003). Consequently, microporous materials (pores smaller than 2 nm) with low densities and high surface areas are ideal for H_2 storage applications. For this, Figure 16.1 illustrates the classification of microporous materials in terms of their construction and components. While inorganic microporous materials are more industrially important, organic ones have control over surface performance. Among these materials, highly porous polymeric materials have been found to be promising adsorbents for H_2 storage, which also helps in achieving the DOE goals. The H_2 uptake in polymeric materials occurs through a physisorption mechanism and can potentially increase system efficiency because of their structural flexibility, thermal and chemical stabilities, chemical homogeneity, and lightweight. This chapter seeks to introduce and identify the polymeric materials used in H_2 storage systems with particular emphasis on their mechanism and synthesis (Ramimoghadam et al., 2016; Tedds et al., 2011).

16.2 HYDROGEN STORAGE MEASUREMENT

Testing material in a prototype storage unit to determine how well it acts as an H_2 store is probably the best way, but it is not practical for finding new materials. Therefore, to ascertain the H_2 sorption properties of materials that have the potential to store it, laboratory gas sorption measurement methods are commonly utilized. In this type of measurement, knowing a material's engineering properties is essential for the plan of the storage unit and also for all stages of the process, from material discovery to the development of the storage unit. These measurement methods can be generally divided into three categories: temperature-programmed techniques, volumetric, and gravimetric (Keller and Staudt, 2005; Rouquerol et al., 2013).

The amount of H_2 adsorbed is usually determined by gravimetric or manometric methods. In the manometric

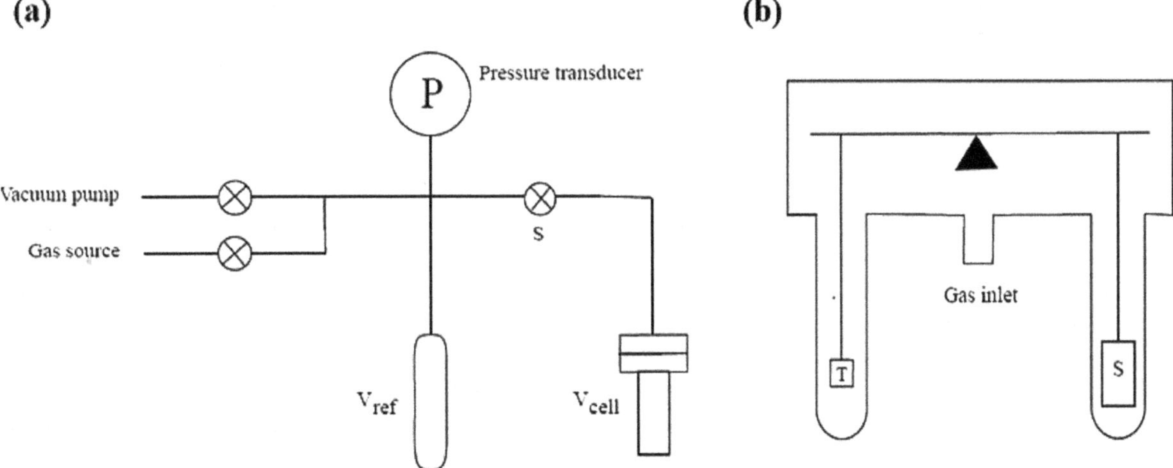

FIGURE 16.2 Schematic diagram of (a) simplified manometric device and (b) gravimetric device using a symmetric microbalance with both the S (sample) and the T (tare weight) suspended in the gas. (Webb and Gray, 2014a, with permission; Webb and Gray, 2014b, with permission.)

method to calculate the amount of adsorption, step by step, a certain amount of H_2 is prepared and inserted into the sample cell, by means of an equation of state for H_2 and a measurement of pressure, temperature, and volume (see Figure 16.2a). The amount of H_2 in the gas phase is computed at each point and any lost H_2 is ascribed to the adsorbed phase. Nevertheless, to ascertain the number of moles in the gas phase, the amount of available volume needs to be recognized. The available volume is the volume difference between the vacant sample cell and the sample cell filled with the sample and the adsorbent, which is called the "dead volume" or "void." Measuring the combined sample and the adsorbent volume is naturally difficult, and sample swelling can complicate it.

In contrast, in the gravimetric technique, the amount of H_2 adsorbed is determined directly and based on the weight measured by a micro-balance (see Figure 16.2b). In this method, there is no need to calculate the amount of introduced gas; also the error that existed in the manometric technique is avoided. However, the micro-balance reading must be corrected in a way similar to the manometric technique, which is required to know the volume of the sample and adsorbent (Broom et al., 2016).

Both of these methods have advantages and disadvantages (Broom, 2011), but to calculate the capacity or absolute adsorption, they both use the determination of the sample volume and the adsorbed phase. In this case, the problem of knowing the adsorbent volume as a function of uptake, temperature, and pressure for materials amenable to helium pycnometry is reduced. In manometric calculations, if the adsorbent volume (V_{ad}) is ignored, it is assumed that it is zero and that both techniques have used only the sample volume to estimate the adsorbed amount. Lastly, the amount of H_2 adsorption is determined by the quantity of gas that will occupy the adsorbent volume. The determined value is later recognized as excess capacity or adsorption.

16.3 POLYMER-BASED HYDROGEN STORAGE SYSTEMS

16.3.1 Organic Polymers

Polymers of intrinsic microporosity (PIMs), covalent organic frameworks (COFs), and hyper-cross-linked polymers (HCPs) are the three key categories of microporous organic polymers that have lately appeared as potential H_2 stores (Makowski et al., 2009). COFs are crystalline organic analogs of metal organic frameworks. HCPs and PIMs are non-crystalline materials that have an amorphous structure. These two are more similar to activated carbon than crystalline materials such as metal organic frameworks, COFs, and zeolites.

Polymers of intrinsic microporosity are rigid macromolecules with fused-ring components. These materials create a microporous network and have high BET (Brunauer–Emmett–Teller) surface areas (500–1100 $m^2\ g^{-1}$) (Budd et al., 2007). Among macromolecules in polymeric materials, HCPs have a high BET surface area, a high degree of microporosity, and a highly rigid network, due to having a high density of covalent chemical bonds and crosslinks (Wood et al., 2007). On the other hand, covalent organic frameworks are crystalline materials composed of organic monomers and light elements such as B, H, O, C, and Si that have strong covalent bonds (Yu et al., 2017). Due to their high surface area and low density, these materials have attracted a lot of attention for use in gas storage (Ding and Wang, 2013; Xiang and Cao, 2013).

H_2 storage capacities for trip-PIM (a triptycene-based polymer) have been reported at 1.0 MPa and 77 K, up to 2.7 wt.% (Budd et al., 2007). Presented capacities for HPCs at 77 K and 1.5 MPa, for example for a polymer based on [4, 40-bis (chloromethyl)-1, 10-biphenyl] (BCMBP), is higher, reaching 3.68 wt.% (Wood et al., 2007). Even though the uptake of these materials is not excellent, because they are made of light elements, they are noteworthy as a possible option for storage, and the development and preparation of new materials may result in a polymer with a high storage capacity.

16.3.1.1 Polymers of Intrinsic Microporosity (PIM)

PIMs are non-crystalline organic microporous polymers based on phthalocyanine, which was first reported in the 1990s (McKeown, 2012). They include the features of a microporous material that has interconnected holes with a diameter of less than 2 nm (Mason et al., 2014). These glassy polymers are made of fused ring structures that prevent them from rotating freely, resulting in a very fixed and rigid shape. This shape and rigidity result in the creation of a continuous network of interconnected intermolecular voids during preparation, which eventually leads to intrinsic microporosity. Since other polymers tend to increase the interaction between the components and decrease the unoccupied space, PIMs are distinguished by their intrinsic microporosity. In addition, those with fused cyclic structures have a high internal molecular free volume. Most importantly, when the polymer structure fails to help the orientation of co-planar moleculars, the available free volume for guest molecules is created (Long and Swager, 2001). In other words, these materials produce solids with a lot of free volumes that are interconnected and provide up to 2000 m^2 g^{-1} of available internal surface area.

PIMs combine the benefits of microporous materials and polymers. They have the same high surface area of microporous materials and the synthetic diversity of organic polymers (McKeown et al., 2007). Like other microporous organic materials, PIMs are made of lightweight elements and have properties such as relatively high thermal stability and high chemical stability. In addition, these materials exhibit several remarkable properties over other microporous organic polymers, containing solution processability, which is not present in porous aromatic frameworks, conjugated microporous polymers, HCPs, and, unlike boroxine-based covalent organic frameworks subjected to air, they maintain their microporosity in wet chemical conditions (Xu and Tan, 2014). One of the other properties of these polymers is their chemical homogeneity, which is a distinct advantage over carbon based materials (McKeown et al., 2006).

The use of PIMs in a wide range of applications such as heterogeneous catalysis, membrane separations, H_2 storage, and the adsorption of organic compounds is promising. Research has shown that the H_2 solubility coefficients for PIM-1 and PIM-7 are more significant than other polymers reported so far (Budd et al., 2007).

16.3.1.2 Synthesis

Since polymer chains are usually aromatic in the structure of PIMs and must be perfectly rigid to prevent freedom of rotation (Wu et al., 2012), the reaction used to synthesize these materials must meet these requirements to lead the intrinsic microporosity. As shown in Table 16.1, three methods used to synthesize these materials include dibenzodioxane reaction, imidization, and amidization. The most common route between these three methods is dibenzodioxane reaction (Van Mullekom et al., 2001).

In the dibenzodioxane method, the polymer is synthesized by mixing an aryl halide and a catechol. Equal amounts of both monomers are combined in a dimethylformamide solution together with a certain amount of potassium carbonate at 50–60 °C for 24–72 h (Budd, Ghanem, et al., 2004b). This reaction is sometimes referred to as the "low temperature method." Similarly, Song et al. (2008) developed a method called the "high-temperature method," in which a combination of two components was quickly stirred in dimethylacetamide and toluene at 155 °C for 8 minutes. This method has a much shorter reaction time, but control of the low temperature method is easier.

In this process, each couple of halide and its neighbor hydroxyl can be regarded as a functional group. The functionality is described by the number of the functional groups so that if the functionality is 2, the PIM is soluble, and if it is more than 2, the PIM is non-soluble (Ramimoghadam et al., 2016).

The microporosity produced in network PIMs has a strong network of covalent bonds that is also usual in other microporous materials. Although there is no network structure in soluble PIMs, these materials have microporosity due to the presence of free volumes that are partially connected.

Imidization is the second method used to produce these materials, which is almost traditional, and its products are commonly referred to as PIM-PIs. This method is based on the reaction between two monomers that have anhydride and diamine functional groups. The reason for the high use of polyimides in commercial applications, especially for electrical devices, is due to properties such as high thermal stability and a low dielectric constant (Carter et al., 2000). However, due to their significant intrinsic microporosity, their applications have expanded to gas separation membranes and energy storage. Microporosity can be provided in polyimides by using a methyl group that owns an aromatic diamine next to the amine functional group with which it is required to form a bond of C-N, which can rotate (McKeown, 2012). In addition, the solubility of linear aromatic polyimides is lower than that of the contorted aromatic polyimides synthesized by using a spiro-bifluorone moiety (Weber et al., 2007).

Amidization is the third way used to produce PIMs with the common name PIM-PAs. In this method, a condensation reaction occurs between two aromatic substances with the functional groups carboxyl and diamine. Due to polyamides having properties similar to polyimides, such as physical and mechanical ones, these materials can also be used significantly in commercial applications. The methods of preparation, surface areas, and the reaction that takes place to produce PIMs are given in Table 16.1. The preparation of the polyimides, concentrating on their intrinsic microporosity and the spiro-center polyamides, was reported by Weber et al. (2007). In this work, the

TABLE 16.1
Some Synthetic Techniques, Surface areas of BET, and Reaction Diagrams to Synthesize Polymers of Intrinsic Microporosity

Synthesis Method	Surface Area (m^2 g^{-1})	Monomer 1 + Monomer 2→ Polymer
Dibenzodioxane reaction	120–1760	Catechol + Aryl halide → PIM
Imidization reaction (PIM-PIs)	551–1407	Anhydride monomer + Diamine monomer → PIM- PIs
Amidization reaction (PIM-PAs)	50–156	Carboxylic monomer + Diamine monomer → PIM- PAs

Source: Ramimoghadam et al., 2016, with permission.

processing technique was particularly effective on the microporosity of these two materials. Polymers with a high surface area were produced using low polarity solvents.

16.3.1.3 Characterization

As mentioned above, such properties as low intrinsic density (Cote et al., 2005), thermal stability, synthetic reproducibility, chemical stability, and chemical homogeneity are among the attractive features of PIMs. McKeown et al. (2006) studied the possibility of tailoring the micropore structure of PIMs using selection of monomer precursors. To investigate this possibility, cyclotricatechylene (CTC) was combined with a network PIM through the benzodioxane-forming reaction between tetrafluoroterephthalonitrile and CTC. The CTC-network-PIM structure (Figure 16.3) was compared with two materials, a HATN-network-PIM and PIM-1 (Budd, Elabas, et al., 2004a; Budd et al., 2003). In this work, the structure of synthesized polymers was confirmed by elemental analysis, FT-IR analysis, ^{1}H NMR analysis, and ^{13}C NMR analysis. Gel permeation chromatography determined that each sample hadmolecular weight M_w of 230,000 g mol^{-1}. Nitrogen adsorption at 77 K was used to determine the BET surface area and for each sample the results were in the region of 800 m^2 g^{-1} (Figure 16.4a). Moreover, analyses of data obtained from nitrogen adsorption at low pressure using the Horváth–Kawazoe method (Horváth and Kawazoe, 1983) showed a pore size distribution in the 0.6–0.7 nm range for each sample (Figure 16.4b).

In this work, gravimetric and volumetric analyses were used to obtain sorption isotherms for each sample. The isotherms show that at relatively low pressures, the three PIMs each adsorb considerable amounts of H_2 (1.4–1.7 wt.%); in fact, the maximum amount of adsorption occurs below 1 bar.

At 1 bar and 77 K, the H_2 uptake for a CTC network is more than for an HATN-network; for an HATN-network it is more than a PIM-1; and interestingly the opposite is true for nitrogen adsorption (Figure 16.4a). Moreover, according to the measurements, the maximum number of H_2 molecules adsorbed per fused ring of the polymeric repeat unit is almost 0.5 H_2.

Polak-Kraśna et al. (2017) studied the detailed characterization of PIM-1 by considering its gas adsorption, mechanical, and chemical properties. In this work, at first, the polymer was cast into films which were then analyzed to determine the amount of H_2 adsorption at −196 °C and high pressures up to 17 MPa, showing the uptake behavior and maximum excess adsorbed capacity of the material.

The measured Young's modulus of the polymer film was 1.26 GPa and the tensile strength was 31 MPa, while the average storage modulus was greater than 960 MPa. The film was found to be thermally stable down to −150 °C. These findings show that PIM-1 has enough elasticity to endure the

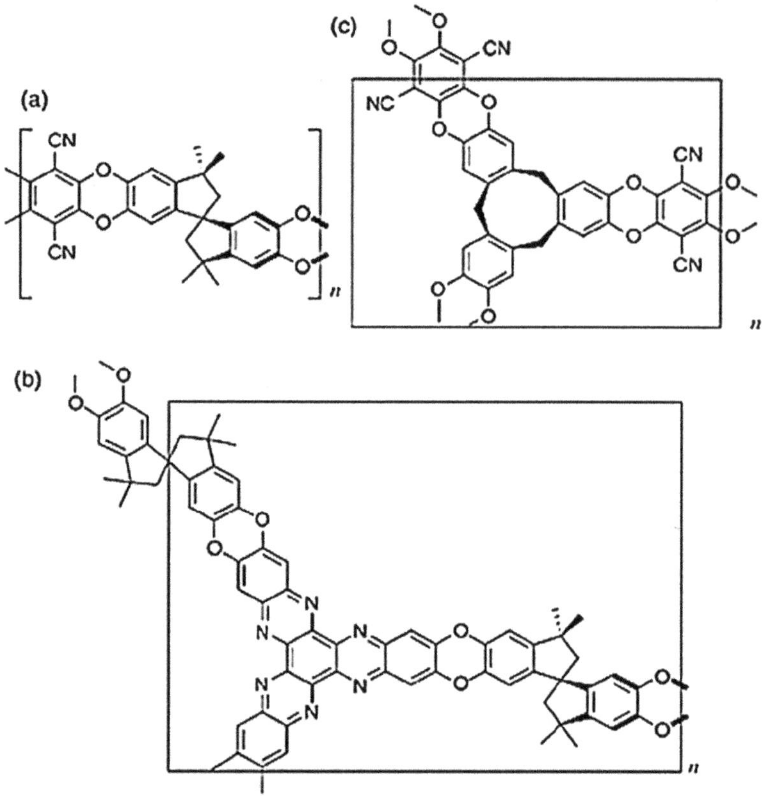

FIGURE 16.3 Structure of (a) PIM-1, (b) HATN-network-PIM, and (c) CTC-network-PIM. (McKeown et al., 2006, with permission.)

elastic deformations happening in high-pressure H_2 storage tanks. Moreover, it has the appropriate thermal stability to be used in the temperature range essential for gas storage uses.

16.3.1.4 Hydrogen Uptake and BET Surface Area

According to studies by Ramimoghadam et al. (2016), Figure 16.5 shows the H_2 uptake vs. the surface area of the BET of non-network PIMs, network PIMs, and HCPs. As can be seen, there is a linear relationship between H_2 uptake and the BET surface area of PIMs. The non-network PIMs have a lower H_2 uptake and surface areas than network PIMs. The line slope of all these polymers shows a correlation between H_2 uptake and the surface area of 0.5 wt.% per 500 m^2 g^{-1} at 77 K and 1 bar, which matches favorably the Chahine Rule (Bénard and Chahine, 2001) for carbon-based composites (1 wt.% per 500 m^2 g^{-1} at 77 K and 30–50 bar) according to the pressure difference of H_2.

For comparison, the results obtained by Hirscher (2011) about the connection between the BET (specific surface area) and the H_2 uptake of some MOFs are shown in Figure 16.6. The relationship between these two features for metal organic frameworks has been employed to induce the importance of the BET surface area (Panella et al., 2006). There is a good correlation between data obtained from the BET surface area and crystallography in MOFs (Walton and Snurr, 2007), but not for carbon materials (Kaneko and Ishii, 1992). According to what can be seen in Figure 16.5, the correctness of the BET method for PMIs has been questioned (Thomas, 2009).

Even though the BET method is widely applied to measure the surface area of numerous materials, in the case of microporous materials, this method should be used with caution. Factors such as the difficulty in distinguishing the mono-layer, multi-layer, and so on, limit the use of this method in the field of microporous materials (Thommes et al., 2015). The assumptions of an energetically homogeneous surface and no primary micropores being filled are generally invalid for microporous composites (Thomas, 2009). Because of pore filling rather than multilayer adsorption happening, nitrogen adsorption methods are more effective than BET methods for determining the pore capacity of microporous materials (McKeown et al., 2007). In the case of PIMs, H_2 simply accesses pores, which are not accessible to nitrogen molecules.

The relationship between H_2 uptake and micropore volume of some PIMs is shown in Figure 16.7 (Ramimoghadam et al., 2016) and it can be seen that the H_2 uptake capacity of these materials is also linearly related to the pore volume. In the case of carbon materials, similar results have been developed in which the volume of micropores is calculated using different theories (Gogotsi et al., 2005; Texier-Mandoki et al., 2004).

16.3.2 Nanoporous Organic Polymers

Nanoporous organic polymers include the lighter-weight elements of the periodic table, such as C, B, O, and N, have very low densities, and show certain possible benefits over

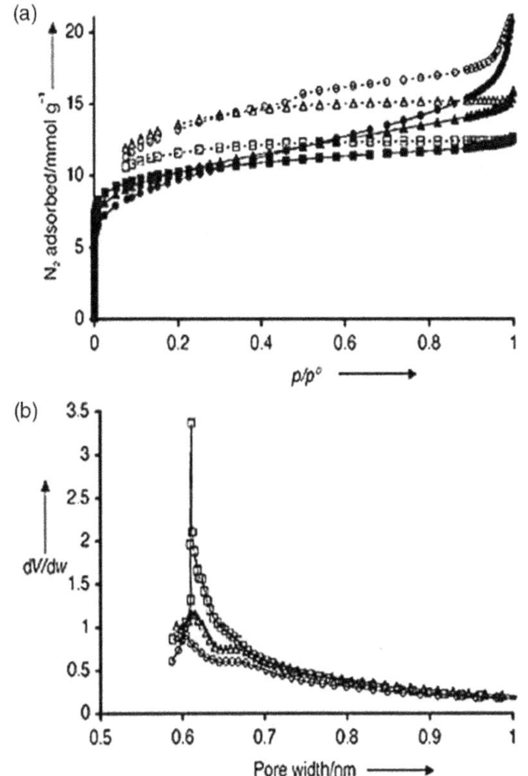

FIGURE 16.4 (a) Nitrogen adsorption and desorption isotherms at 77 K (filled symbols and open symbols, respectively); (b) apparent micropore size distributions (dV/dw, the unit is mL g^{-1} nm^{-1}) calculated by the Horváth–Kawazoe method (carbon slit-pore model) for a CTC-network-PIM (□), PIM-1 (**O**), and an HATN-network-PIM (Δ). (McKeown et al., 2006, with permission.)

inorganic-based porous materials such as widely available preparation techniques (Yu et al., 2017).

Nanoporous polymers are divided into three wide classes: cross-linked polymers which have surface areas of up to about 1000 m^2 g^{-1} and which have been applied broadly as separation media, hyper-cross-linked polymers with achievable surface areas of more than 2000 m^2 g^{-1}, and polymers with intrinsic microporosity and with surface areas up to 1000 m^2 g^{-1} (Germain et al., 2008).

These materials have attracted a lot of attention in various fields, including electrodes, chromatography, and H_2 storage (Ghanem et al., 2007; Lee et al., 2006). Recently, nanoporous polymers have shown potential as adsorbents for H_2 storage. In nanoporous materials, the capacity and temperature at which H_2 is adsorbed and desorbed are determined by the pore volume, surface area, and enthalpy of adsorption, respectively.

16.3.2.1 Synthesis

Hyper-cross-linking is one of the methods used to synthesize nanoporous polymers. This method consists of the dissolving/swelling of a non-cross-linked or slightly cross-linked polymer, then "immobilizing" the pores in their solvated state via a secondary cross-linking procedure, resulting in pores that last even after the solvent has been removed.

Figure 16.8 shows the approach most commonly applied for the synthesis of hyper-cross-linked polymers.

Typically, these polymers are synthesized using a method reported by Ahn et al. (2006). In this method, two precursors are used to prepare the desired polymer. Through

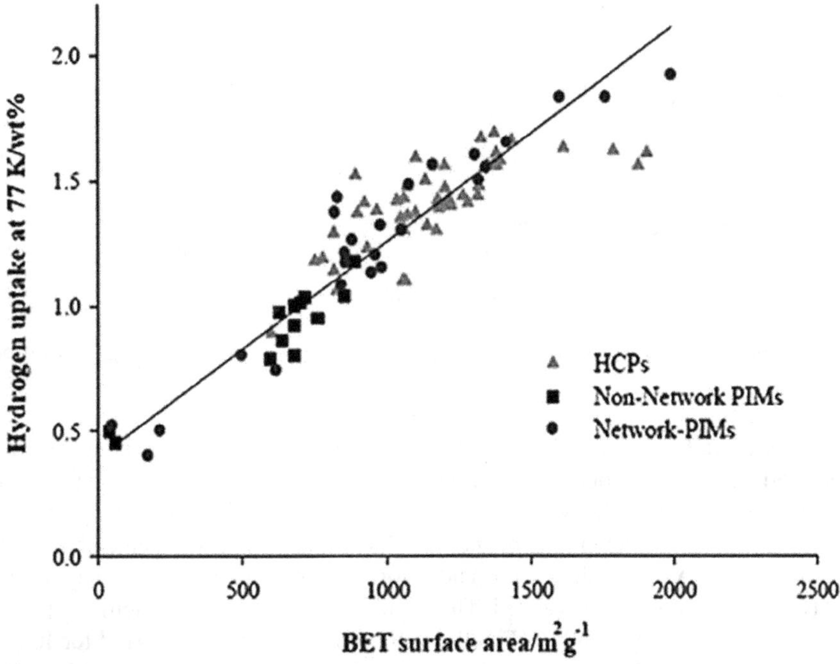

FIGURE 16.5 H_2 uptake versus surface area of BET for HCPs (at 77 K and 1.13 bar), non-network and network PIMs (at 77 K and 1 bar). (Ramimoghadam et al., 2016, with permission.)

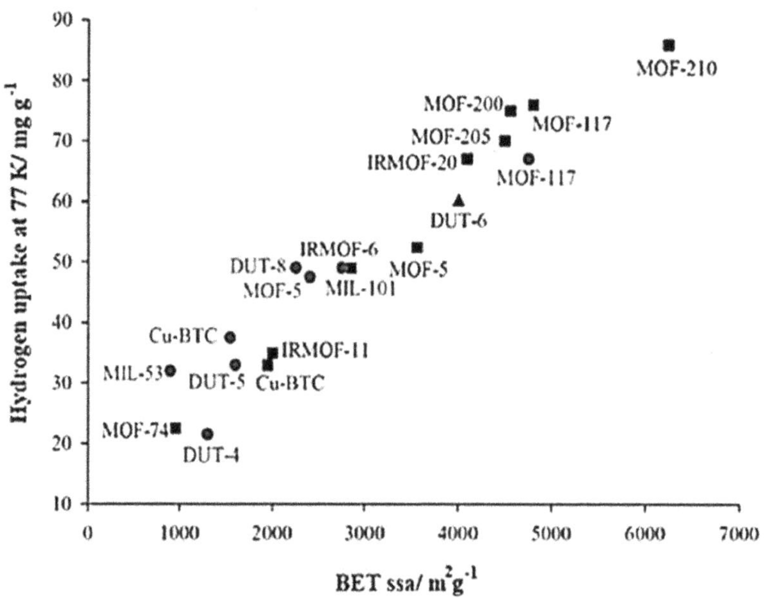

FIGURE 16.6 BET specific surface area dependence of H_2 uptake at 77 K for many high-porosity metal organic frameworks. The symbols explain measurements reported by various works: triangles at 60 bar (Klein et al., 2009), squares at 50 bar (Wong-Foy et al., 2006), and circles at 20 bar (Panella et al., 2006). (Hirscher, 2011, with permission.)

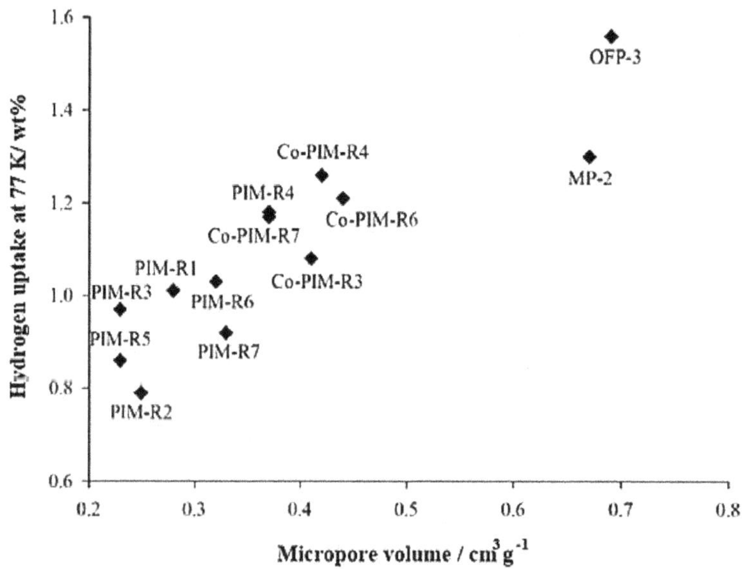

FIGURE 16.7 Micropore volume dependence of H_2 uptake for PIMs at 77 K and1 bar. (Ramimoghadam et al., 2016, with permission.)

suspension polymerization, a mixture of divinyl benzene, vinyl benzyl chloride, and 2, 2′-azobisisobutyronitrile is prepared, that is, the first precursor (poly (vinyl benzyl chloride) cross-linked with divinylbenzene). In the next step, the prepared mixture enters the multiple reactor, and then the aqueous poly (vinyl alcohol) solution is added. The system is then stirred for 24 h and heated to 80 °C. Finally, the product is washed in water and methanol several times before being extracted in a Soxhlet apparatus with diethyl ether and methanol and then dried.

The second precursor is synthesized from a polymerization mixture of divinyl benzene, vinyl benzyl chloride, toluene, and 2, 2′-azobisisobutyronitril, that is, in the form of a suspension. The organic phase is added, and the reactor's contents (a Buchi reactor is charged with aqueous poly (vinyl alcohol) solution) are purged with nitrogen for 10 min. They are then stirred for 10 min at room temperature before being heated for 24 h to 70 °C. The last step is similar to that mentioned in the synthesis of the first precursor.

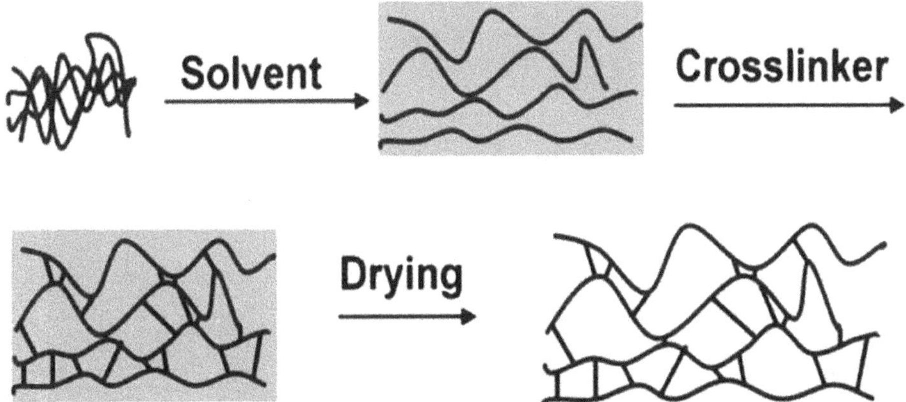

FIGURE 16.8 Schematic representation of the hyper-cross-linking process. (Germain et al., 2007, with permission.)

TABLE 16.2
H_2 Uptake Capacities and Surface Properties of Selected Polymers. (Yuan et al., 2009, with permission)

	BET Surface Area (m^2/g)	Langmuir Surface Area (m^2/g)	Micro-pore Surface Area (m^2/g)	Total Pore Volume (cm^3/g)	Micro-Pore Volume (cm^3/g)	H-K Median pore Diameter (nm)	H_2 Gravimetric Uptake (%) at 77 K (kg H_2/kg Adsorbent + H_2 (ads))[a]	H_2 Gravimetric Uptake (%) at 298 K (kg H_2/kg Adsorbent + H_2 (ads))[b]
PS4AC1	769	1122	492	0.427	0.249	0.62	2.8	0.44
PT4AC	762	1114	448	0.425	0.226	0.64	2.2	0.50
PS4AC2	1043	1412	496	0.477	0.251	0.67	3.7	0.43
PS4TH	971	1439	443	0.757	0.223	0.62	3.6	0.45

[a] H_2 uptake capacities at 77 K were measured at H_2 equilibrium pressure of 60 bar.
[b] H_2 uptake capacities at 298 K were measured at H_2 equilibrium pressure of 70 bar.

The post-cross-linking reaction is carried out using a polymeric precursor that has been pre-swollen for 2 h in 1, 2- dichloroethane. Then, the Friedel–Crafts catalyst is added to the slurry cooled in an ice bath. After homogeneously dispersing the catalyst in the mixture at room temperature, the cross-linking reaction is continued for 24 h at 80 °C. Before drying, the hyper-cross-linked polymer is separated and washed with methanol, HCl in acetone, and methanol again (Germain et al., 2006).

16.3.2.2 Characterization

Characterizations of the structures and H_2 storage capacities of nanoporous organic polymers are investigated by various techniques. Yuan et al. (2009) studied the preparation of numerous nanoporous polymers containing stereo-contorted cores for H_2 storage. The building blocks of these cross-linked polymers were spiro-bifluorene and tetraphenylmethane cores.

The prepared polymers had a surface area up to 1000 m^2 g^{-1} and a narrow pore size distribution. Also, the size of the nanopores and the surface area are affected by the reaction conditions. DSC (differential scanning calorimetry) and TGA (thermogravimetric analysis) were used to determine the thermal properties of polymers. The results showed that these polymers were thermally stable in a range of 300–380 °C. BET nitrogen adsorption methods were used for the investigation of surface properties. Table 16.2 shows the critical structural properties resulting from the isotherm data. A fundamental observation is that the polymers have a common pore width of 0.6 to 0.8 nm and relatively narrow pore size distribution.

A Sievert-type isotherm apparatus was used to measure excess hydrogen adsorption capacities (Zhou et al., 2007). At 60 bar and liquid nitrogen temperature, the adsorption capacity of PS4TH was 3.6 wt.%. For PT4AC, PS4AC1, PS4TH, and PS4AC2, the adsorption capacity at 7 bar and 77 K was 1.2, 1.7, 2.1, and 2.2 wt.%, respectively.

In another work, Germain et al. (2008) studied the synthesis of nanoporous size-selective hyper-cross-linked polymer networks with 37–92% of pores small enough for hydrogen adsorption but too small for nitrogen penetration. Buchwald and Ullman synthesis methods were used to combine diaminobenzene and polyaniline with tribromobenzene and diiodobenzene, and the Buchwald reaction appears to be more successful than the Ullman synthesis for the development of such materials. Figure 16.9 shows the low-pressure H_2 adsorption isotherms. Interestingly, hyper-cross-linked polystyrenes with surface areas seven

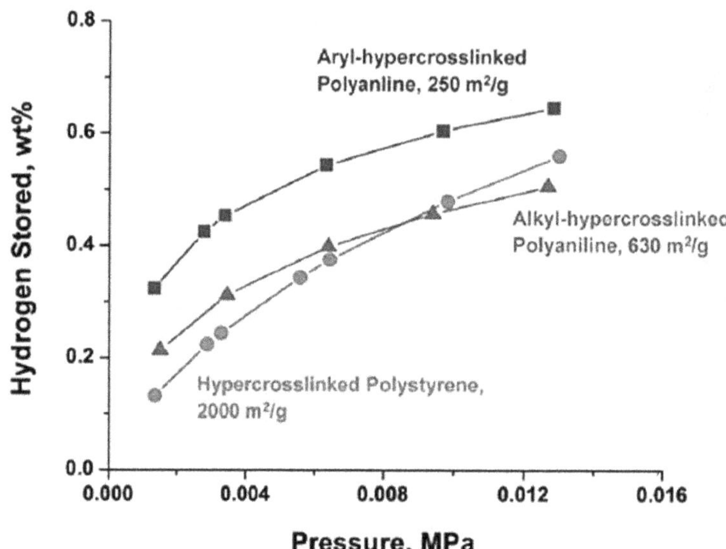

FIGURE 16.9 Low-pressure H_2 adsorption by hyper-cross-linked polystyrene with a surface area of 2000 m^2 g^{-1} as determined by applying the BET equation to N_2 adsorption isotherms (circles), a polymer network of aromatic rings with 250 m^2 g^{-1} (squares), and alkyl-hyper-cross-linked polyanilines with 630 m^2 g^{-1} (triangles). (Germain et al., 2008, with permission.)

times larger than the aromatic ring network adsorb less H_2 at low pressures. Moreover, the Langmuir surface areas measured by H_2 adsorption isotherms are higher than those calculated by the BET equation and nitrogen adsorption isotherms in these materials. Usually, because of capillary condensation, BET surface areas are greater than Langmuir surface areas (McKeown and Budd, 2006). The high H_2–polymer interaction energies have led to these interesting results for a network of aromatic rings.

Nitrogen adsorption isotherms are often used to ascertain the pore volume accessible by N_2 molecule adsorption, but some researchers have used H_2 adsorption isotherms to determine the pore volume available to H_2 (Züttel et al., 2004). The results show that a significant portion of the pore volume in nanoporous polymers consists of pores that are large enough to hold H_2 while eliminating N_2. The actual H_2 storage capacity for most of the aromatic rings' networks exceeds the capacity that could be realized in nitrogen-accessible nanoporous. For this purpose, some researchers have measured the diameter of adsorbed H_2 (Negri and Saendig, 2007) and N_2 (Lushington and Chabalowski, 2001) and believe that the size of the pores is in the range 0.29 to 0.36 nm.

16.3.2.3 Hydrogen Uptake and BET Surface Area

It is generally believed that larger pore volumes and higher BET surface areas will enhance H_2 storage capacities (Ren et al., 2017). As can be seen in Table 16.2, Yuan et al. showed that at 77 K, even with the differences in the molecular structures of the polymer, the adsorption capacity of these materials followed the so-called "Chahine rule" and is generally proportional to the BET surface area.

This is understandable because the kinetic energy of the H_2 molecule is defeated by the heat of adsorption at the polymer surface at 77 K, and H_2 will adsorb and occupy every accessible surface.

On the other hand, in the study of different types of commercial polymers, such as a poly (styrene-codivinylbenzene) resin Amberlite XAD4 and a hyper-cross-linked polystyrene-based Hypersol-Macronet MN200, Germain et al. (2008) found that the adsorption capacity of these materials is not directly proportional to their BET surface area and that the polymer with a lower surface area had a higher H_2 adsorption capacity than the polymer with a higher surface area (see Table 16.3).

16.3.3 Soluble Polymers

It is notable that a polymer without a network structure that has a specific quantity of free volume can behave like a microporous material due to the joining of voids.

This type of microporous polymer may be soluble and enable an easy solution-based processing that is not possible with other microporous materials. Consequently, they have attracted a favorable reception as important materials for heterogeneous catalysis, adsorption of small molecules, gas storage, and gas separation (Makhseed et al., 2012). The following are some of the soluble polymers used in H_2 storage.

- Soluble MP1 is a fluoropolymer with intrinsic microporosity, which also can be considered as a PIM. The H_2 adsorption of MP1 is 0.80 wt.% at 1 bar and 77 K, which is relatively significant. (Budd, Elabas, et al., 2004a).
- Ritter et al. (2009) studied the synthesis of a soluble microporous polyimide. These synthesized polyimides behaved similarly to PA1 (Weber et al., 2008); no microporosity was measured by nitrogen adsorption.

TABLE 16.3
List of Commercial Resins, Their Properties, and Their Capacity for H_2 Storage. (Germain et al., 2006, with permission)

Trade Name	Composition	Surface Area (m^2g-1)[a] *a*	*b*	H_2 Capacity (wt%)[b]
Amberlite XAD4	poly (styrene-*co*-divinylbenzene)	1060	425	0.8
Amberlite XAD16	poly (styrene-*co*-divinylbenzene)	770	336	0.6
Hayesep N	poly (divinylbenzene-*co*-ethylenedimethacrylate)	460	247	0.5
Hayesep B	polydivinylbenzene modified with polyethyleneimine	570	290	0.5
Hayesep S	poly (divinylbenzene-*co*-4-vinylpyridine)	510	254	0.5
Wofatit Y77	poly (styrene-*co*-divinylbenzene)	940	573	1.2
Lewatit EP63	poly (styrene-*co*-divinylbenzene)	1206	664	1.3
Lewatit VP OC 1064	poly (styrene-*co*-divinylbenzene)	810	377	0.7
Hyper-sol-Macro-net MN200	hyper-cross-linked polystyrene	840	576	1.3
Hyper-sol-Macro-net MN100	amine functionalized hyper-cross-linked polystyrene	600	477	1.1
Hyper-sol-Macro-net MN500	sulfonated hyper-cross-linked polystyrene	370	266	0.7

[a] Calculated from nitrogen adsorption using (a) the BET equation and (b) H_2 adsorption using the Langmuir equation.
[b] H_2 storage capacity at a pressure of 0.12 MPa.

On the other hand, the acceptable uptake of H_2 and carbon dioxide by these materials indicates that it is not enough to use one technique to prove the microporosity of polymers. Also, among the factors that affect pore availability, can be mentioned the smaller kinetic diameter and higher kinetic energy of CO_2 and H_2.

- The preparation of some soluble phthalimide-based microporous polymers has been reported by Makhseed et al. (2012). These materials have a good surface area, a microporosity like the high surface area PIMs, and a narrow pore diameter distribution that consequently may be able to store H_2.

16.3.3.1 Synthesis

Due to the easier processing of soluble microporous polymers, there is a lot of interest in them, although only a few of these materials are known (Ritter et al., 2009).

Ritter et al. (2009) prepared binaphthalene-based soluble polyimides by two different methods. In the first method, which is a single-step path, isoquinoline and m-cresol were used as a catalyst and a solvent, respectively. Benzene-1, 2, 4, 5-tetracarboxylic anhydride and (±) 2, 2′-diamino-1, 1′-binaphthalene, were dissolved in identical amounts of m-cresol under an N_2 atmosphere. Then, isoquinoline was added and the combination stirred for 1 h. In the final step, the mixture was heated first at 80 °C for 2 h and afterward at 200 °C for 6 h. The products generally have an M_w of approximately 5000 g mol^{-1}.

The second method for the preparation of these polymers is a two-stage process in which a polyamic acid is synthesized and then converted to a polyimide by condensation. As a result of the reaction between the equivalent quantities of benzene-1, 2, 4, 5-tetracarboxylic anhydride and (±)-2, 2′-diamino-1, 1′-binaphthalene, in a dry NMP (N-Methyl-2-pyrrolidone) under an N_2 atmosphere and stirred for 24 h, polyamic acid is formed. Then, the resulting polyamic acid is vacuum-dried at 40 °C and cured for 2 h at 300 °C under an N_2 atmosphere. The crude polyimide typically had an M_w of about 8000 g mol^{-1}. In general, the second method has advantages such as products with higher molar mass and less use of dangerous catalysts and solvents.

16.3.3.2 Characterization

Various analyses have been performed to characterize the properties and structure of soluble polymers. Makhseed et al. (2012) studied the synthesis of phthalimide-based microporous polymers that were synthesized using a traditional nucleophilic substitution reaction. FTIR, ^{1}H NMR, and elemental analyses confirmed the structures of the synthesized polymers. SEM was used to determine the particle size and morphology of the samples. Figure 16.10 shows the combination of spherical and non-spherical particles with nanoscale and microscale sizes. Moreover, GPC analysis shows that the synthesized samples have a high molecular weight. Based on the results obtained from TGA (thermogravimetric analysis), the stability of the synthesized polymers was determined up to 300 °C. The porous nature of the polymers was analyzed using nitrogen sorption at 77 K. Based on the porosity analysis, the prepared polymers are clearly comparable to high surface area PIMs (500–900 m^2 g^{-1}). The t-plot analysis revealed that the micropore surface area with the narrow size distribution of the ultra-micropores contributes significantly to the specific surface area, as indicated using the Horváth–Kawazoe analysis. The volumetric H_2 adsorption method was used for H_2 uptake measurement. The H_2 storage capacity of the synthesized CO-PIM (3, 4, 6, 7) and PIM-R (1–7) were promising (up to 1.26 wt.%, at 1.13 bar and 77 K) with high isoteric heats of H_2 adsorption (8.5 kJ mol^{-1}).

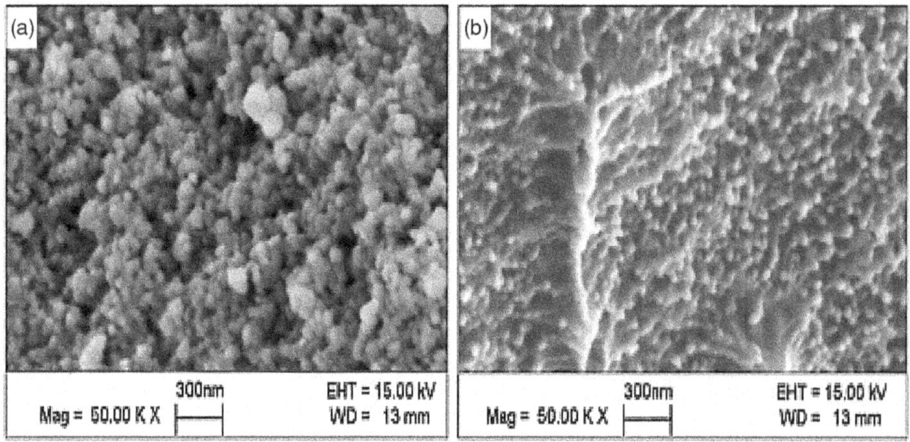

FIGURE 16.10 SEM images of CO-PIM (3, 4, 6, 7): (a) powder and (b) thin film. (Makhseed et al., 2012, with permission.)

16.3.3.3 Hydrogen Uptake and BET Surface Area

Many studies showed that there is an obvious relationship between H_2 adsorption density and the total surface area of porous materials (Dinca et al., 2006).

Moreover, the results show that small micropores in the soluble polymers can efficiently adsorb H_2, probably because of the fact that H_2 has a much smaller kinetic diameter than larger gas molecules like N_2. Makhseed et al. (2012) showed that the H_2 adsorption capacities by various samples usually have a linear relationship with a surface area of BET at low pressures. For instance, the H_2 uptake capacities of PIM-R2 (595 $m^2\ g^{-1}$) and PIM-R4 (889 $m^2\ g^{-1}$) at 1 bar and 77 K are 0.79 wt.% and 1.18 wt.%, respectively.

16.3.4 Polymer-Based Composites

Despite all the advantages, the materials mentioned so far are in particle or powder form, which limits their useability and mechanical properties. That is why polymer-based composites with properties such as high H_2 adsorption capacity, flexibility, and strength have attracted a lot of attention (Rochat et al., 2017). Safety, handling, and the potential for H_2 storage application are some of the advantages of polymer-based composites over powders (Tian et al., 2019).

16.3.4.1 Synthesis

The following methods can be used to prepare polymer-based composites. Tian et al. (2019) prepared PIM-1 based MIL-101 and AX21 composites as follows: PIM-1 solution and suspensions of AX21 and MIL-101 were prepared individually in chloroform to form 2 wt.% mixtures and stirred at room temperature for 24 h. In order to achieve the required gravimetric proportions of PIM-1 and (AX21 and MIL-101), the suspensions were stirred for another 24 h before being cast into Petri dishes of different sizes, then covered and allowed to evaporate slowly. In the next step, the prepared samples were dried under vacuum conditions for 8 h at 353 K. Eventually, the composite films were separated from the Petri dishes and prepared for analysis. All products had to be degassed before any gas adsorption measurement.

In another method for the synthesis of polymer-based composites, Molefe et al. (2019) prepared a PIM-1/MIL-101(Cr) monolithic composite. According to this study, cross-linked PIM-1/MIL-101(Cr) composites were prepared by the physical mixing of MIL-101(Cr) powder with PIM-1 in a tetrachloroethane solvent. This was obtained by making separate PIM-1 tetrachloroethane-based solutions and a suspension of different loadings of predried MIL-101(Cr) powder. To obtain the desired homogeneous mixture of PIM-1/MIL- 101(Cr), the solutions were combined in a 20 mL cylindrical glass vial and intensely stirred overnight. Finally, by heating the product up to 150 °C, the solvent evaporated and a cross-linked PIM-1/MIL-101(Cr) monolith was obtained.

Figure 16.11 shows a summary of the methods applied for the preparation of MIL-101(Cr) and PIM-1, together with the MOF composite.

16.3.4.2 Characterization

As mentioned before, polymer-based composites are desired for their appropriate mechanical properties and high H_2 adsorption, so these properties should be studied. Rochat et al. (2017) studied the preparation of a series of composites based on a PIM-1 matrix containing PAF-1 as a filler. These composites have a combination of properties obtained from PIM-1 and PAF-1, including enhanced surface area and processability.

The mechanical strength of the composite decreases as the amount of PAF-1 increases, but its processability is maintained. These composites show potential for useful H_2 storage, particularly for applications with pressures up to 7.5 MPa. By increasing the amount of PAF-1, the pore volume and surface area of the samples were increased (measured at 77.4 K by N_2 adsorption isotherms).

The skeletal density and porosity of the composites were revealed via helium pycnometry and N_2 adsorption,

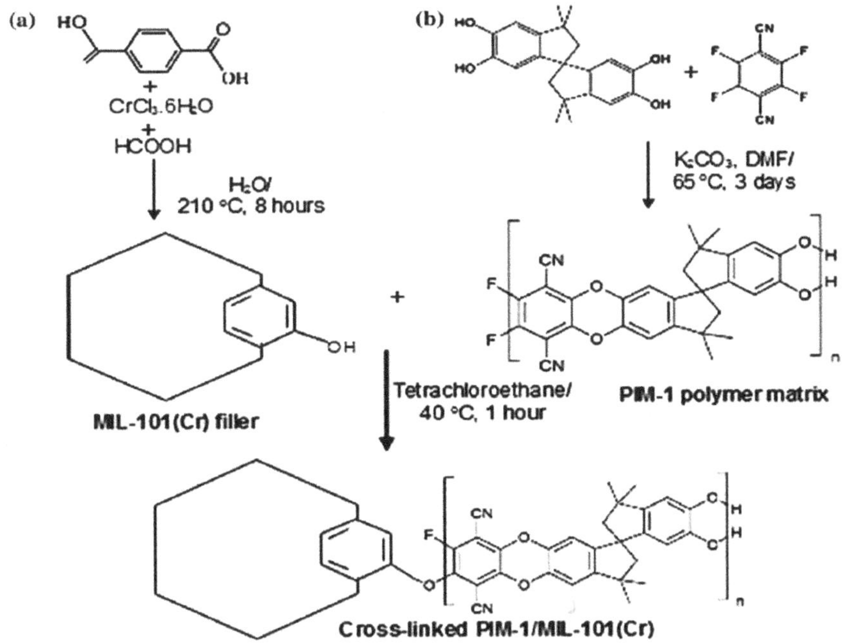

FIGURE 16.11 Schematic method for the preparation of molecular cross-linked PIM-1/MIL-101(Cr) monoliths. (Molefe et al., 2019, with permission.)

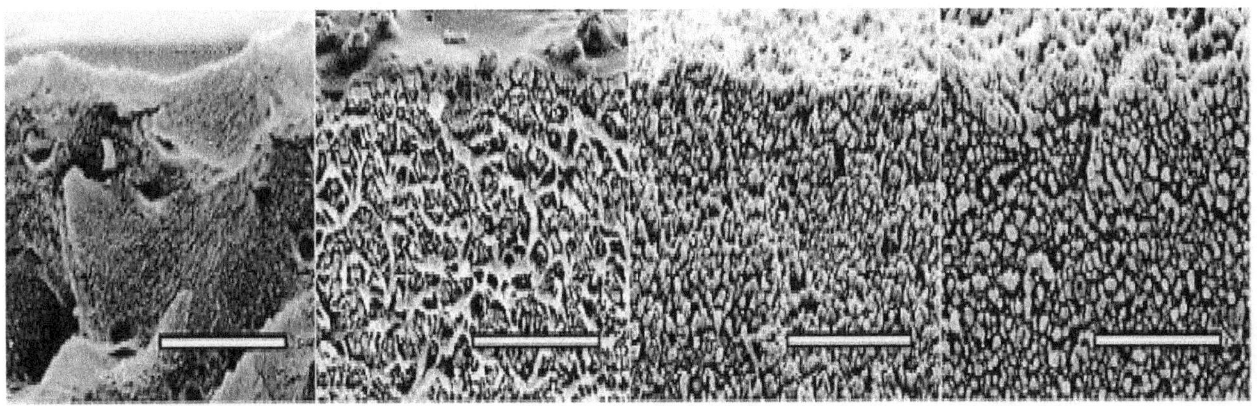

FIGURE 16.12 SEM images of PIM-1/PAF-1 composite films. (Rochat et al., 2017, with permission.)

respectively. Pore size distributions evaluated through analysis of the CO_2 isotherms is a combination of the ultra-micro-porosity of PIM-1 (in the 0.35–0.8 nm region) and the increased porosity (1.2–1.7 nm), due to PAF-1's significant contribution in that range.

To determine the PAF-1 particle dispersion in the PIM-1 polymer matrix as well as the prepared composite microstructure, SEM (Figure 16.12) was used. Since the particle distribution was mainly uniform, imaged fragments were reflective of the bulk samples. By increasing the amount of PAF-1, the number of particulates observed on the samples' surface were increased. In other words, conglomerated PAF-1 particles seemed to be immersed in the PIM-1 polymer matrix at a concentration of 7.5 wt.%, while the structure showed a granule-dominated character at concentrations of 37.5 wt.% and 22.5 wt.%. Fiber-like structures were obvious in cross-sectional pictures when PAF-1 concentration reached 7.5 wt.%, which can be attributed to the PIM-1 structure, while PAF-1 particles predominated in films with higher amounts.

The H_2 uptake of membranes at 0.1 MPa and 77.4 K increases with the increasing concentration of PAF-1. At low pressure, the H_2 uptake of PAF-1 was 1.43 wt.%, whereas samples prepared solely of PIM-1 had a 0.78 wt.% uptake. The low-pressure H_2 uptake amounts for composite films (Figure 16.13) are the same or even a little higher than the amounts predicted by the mixtures' rule. The amount of H_2 uptake (0.1 MPa at 77.4 K) for the composite film with a PAF-1 concentration of 30 wt.% was 0.96 wt.%.

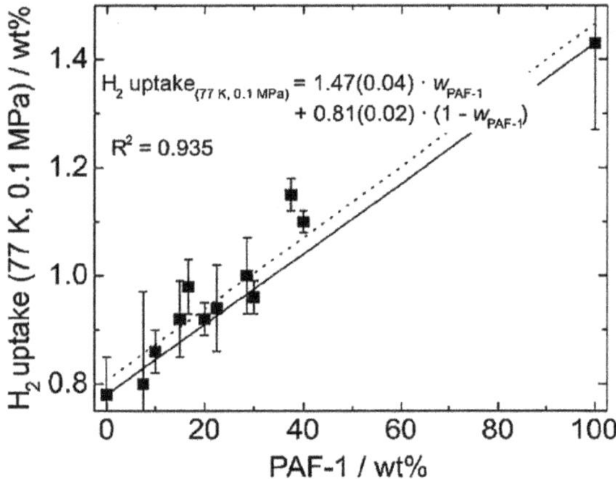

FIGURE 16.13 Excess H_2 uptakes for PIM-1/PAF-1 films (77.4 K, 0.1 MPa) as a function of PAF-1 weight proportion (average of three independent measurements). Dotted and solid lines show experimental and theoretical "rule of mixtures," respectively, and the standard errors are included in the fitting equation. (Rochat et al., 2017, with permission.)

16.3.4.3 Hydrogen Uptake and BET Surface Area

Much research has been done to describe the correlation between H_2 uptake and BET surface area in polymer-based composites. Rochat et al. in another work, investigated the direct relationship of H_2 storage in polymer-based composites at 77 K and low pressure up to 0.1 MPa with the surface area of materials. With a surface area of 1200 m^2 g^{-1}, the AX21 composite film adsorbed 1.12 wt.% H_2, while the uptake of MIL-101 and PAF-1 composite with the same surface area was 0.94 and 1.00 wt.%, respectively. Since the PIM-1/AX21 composite has a higher micropore volume than the PIM-1/MIL-101 composite with the same surface area, this shows that AX21 has more favorable shapes and pore sizes for increasing H_2 adsorption.

16.4 CONCLUSION

Briefly, the need to use H_2 as an energy carrier has led to studying H_2 storage with the goal of providing materials tailored to the specific desires of H_2 physisorption. Microporous polymers are materials that show significant versatile properties for H_2 storage and have attracted the attention of many researchers. In this study, the synthesis methods and characterization data of various polymers as solid-state materials for H_2 storage have been discussed in detail. Importantly, the storage capacity of H_2 was highlighted for the types of polymers that have been studied recently. Also, the advantages, issues, and limitations of each have been identified for H_2 physisorption, and instances of storage systems or prototypes manufactured have been provided to illustrate their feasibility and performance. It is noteworthy that the results of H_2 adsorption using polymers are comparable to organic and inorganic porous materials such as MOFs, activated carbon, and zeolites with identical surface areas. Despite the significant advantages of polymers for H_2 storage and acceptable results, their lower surface area than other materials has led to extensive research in this area to improve H_2 storage capacity. Moreover, the challenge of improving a polymer adsorbent with the objective of high storage capacity at room temperature instead of cryogenic temperature and high volumetric and gravimetric density is still prominent.

REFERENCES

Ahn, J.-H., Jang, J.-E., Oh, C.-G., Ihm, S.-K., Cortez, J., and Sherrington, D. C. 2006. Rapid generation and control of microporosity, bimodal pore size distribution, and surface area in Davankov-type hyper-cross-linked resins. *Macromolecules* 39(2): 627–632.

Bénard, P., and Chahine, R. 2001. Modeling of adsorption storage of hydrogen on activated carbons. *International Journal of Hydrogen Energy* 26(8): 849–855.

Broom, D., Webb, C., Hurst, K., Parilla, P., Gennett, T., Brown, C. M., Zacharia, R., Tylianakis, E., Klontzas, E., and Froudakis, G. 2016. Outlook and challenges for hydrogen storage in nanoporous materials. *Applied Physics A* 122(3): 151.

Broom, D. P. 2011. *Hydrogen Storage Materials: The Characterisation of Their Storage Properties*. Berlin: Springer Science and Business Media.

Budd, P. M., Butler, A., Selbie, J., Mahmood, K., McKeown, N. B., Ghanem, B., Msayib, K., Book, D., and Walton, A. 2007. The potential of organic polymer-based hydrogen storage materials. *Physical Chemistry Chemical Physics* 9(15): 1802–1808.

Budd, P. M., Elabas, E. S., Ghanem, B. S., Makhseed, S., McKeown, N. B., Msayib, K. J., Tattershall, C. E., and Wang, D. 2004a. Solution-processed, organophilic membrane derived from a polymer of intrinsic microporosity. *Advanced Materials* 16(5): 456–459.

Budd, P. M., Ghanem, B., Msayib, K., McKeown, N. B., and Tattershall, C. 2003. A nanoporous network polymer derived from hexaazatrinaphthylene with potential as an adsorbent and catalyst support. *Journal of Materials Chemistry* 13(11): 2721–2726.

Budd, P. M., Ghanem, B. S., Makhseed, S., McKeown, N. B., Msayib, K. J., and Tattershall, C. E. 2004b. Polymers of intrinsic microporosity (PIMs): Robust, solution processable, organic nanoporous materials. *Chemical Communications* (2): 230–231.

Carter, K. R., DiPietro, R. A., Sanchez, M. I., and Swanson, S. A. 2000. Nanoporous polyimides derived from highly fluorinated polyimide/poly (propylene oxide) copolymers. *Chemistry of Materials* 13(1): 213–221.

Cheng, H., Chen, L., Cooper, A. C., Sha, X., and Pez, G. P. 2008. Hydrogen spillover in the context of hydrogen storage using solid-state materials. *Energy and Environmental Science* 1(3): 338–354.

Cote, A. P., Benin, A. I., Ockwig, N. W., O'Keeffe, M., Matzger, A. J., and Yaghi, O. M. 2005. Porous, crystalline, covalent organic frameworks. *Science* 310(5751): 1166–1170.

David, E. 2005. An overview of advanced materials for hydrogen storage. *Journal of Materials Processing Technology* 162: 169–177.

Dinca, M., Yu, A. F., and Long, J. R. 2006. Microporous metal–organic frameworks incorporating 1, 4-benzeneditetrazolate: Syntheses, structures, and hydrogen storage properties. *Journal of the American Chemical Society* 128(27): 8904–8913.

Ding, S.-Y., and Wang, W. 2013. Covalent organic frameworks (COFs): From design to applications. *Chemical Society Reviews* 42(2): 548–568.

Germain, J., Fréchet, J. M., and Svec, F. 2007. Hypercrosslinked polyanilines with nanoporous structure and high surface area: Potential adsorbents for hydrogen storage. *Journal of Materials Chemistry* 17(47): 4989–4997.

Germain, J., Hradil, J., Fréchet, J. M., and Svec, F. 2006. High surface area nanoporous polymers for reversible hydrogen storage. *Chemistry of Materials* 18(18): 4430–4435.

Germain, J., Svec, F., and Fréchet, J. M. 2008. Preparation of size-selective nanoporous polymer networks of aromatic rings: Potential adsorbents for hydrogen storage. *Chemistry of Materials* 20(22): 7069–7076.

Ghanem, B. S., Msayib, K. J., McKeown, N. B., Harris, K. D., Pan, Z., Budd, P. M., Butler, A., Selbie, J., Book, D., and Walton, A. 2007. A triptycene-based polymer of intrinsic microposity that displays enhanced surface area and hydrogen adsorption. *Chemical Communications* (1): 67–69.

Gogotsi, Y., Dash, R. K., Yushin, G., Yildirim, T., Laudisio, G., and Fischer, J. E. 2005. Tailoring of nanoscale porosity in carbide-derived carbons for hydrogen storage. *Journal of the American Chemical Society* 127(46): 16006–16007.

Hirscher, M. 2011. Hydrogen storage by cryoadsorption in ultra-high-porosity metal–organic frameworks. *Angewandte Chemie International Edition* 50(3): 581–582.

Horváth, G., and Kawazoe, K. 1983. Method for the calculation of effective pore size distribution in molecular sieve carbon. *Journal of Chemical Engineering of Japan* 16(6): 470–475.

Iulianelli, A., and Basile, A. 2014. *Advances in Hhydrogen Production, Storage and Distribution.* Amsterdam: Woodhead Publishing, Elsevier.

Jena, P. 2011. Materials for hydrogen storage: Past, present, and future. *The Journal of Physical Chemistry Letters* 2(3): 206–211.

Jenkins, S. 2013. Hydrogen storage technology: Materials and applications. *Chemical Engineering* 120(5): 9–10.

Kaneko, K., and Ishii, C. (1992). Superhigh surface area determination of microporous solids. *Colloids and Surfaces* 67: 203–212.

Keller, J. U., and Staudt, R. 2005. *Gas Adsorption Equilibria: Experimental Methods and Adsorptive Isotherms.* Berline: Springer Science and Business Media.

Klein, N., Senkovska, I., Gedrich, K., Stoeck, U., Henschel, A., Mueller, U., and Kaskel, S. 2009. A mesoporous metal–organic framework. *Angewandte Chemie International Edition* 48(52): 9954–9957.

Lee, J.-Y., Wood, C. D., Bradshaw, D., Rosseinsky, M. J., and Cooper, A. I. 2006. Hydrogen adsorption in microporous hypercrosslinked polymers. *Chemical Communications* (25): 2670–2672.

Lim, K. L., Kazemian, H., Yaakob, Z., and Daud, W. W. 2010. Solid-state materials and methods for hydrogen storage: A critical review. *Chemical Engineering and Technology: Industrial Chemistry-Plant Equipment-Process Engineering-Biotechnology* 33(2): 213–226.

Long, T. M., and Swager, T. M. 2001. Minimization of free volume: Alignment of triptycenes in liquid crystals and stretched polymers. *Advanced Materials* 13(8): 601–604.

Lushington, G. H., and Chabalowski, C. F. 2001. Ab initio simulation of physisorption: N2 on pregraphitic clusters. *Journal of Molecular Structure: THEOCHEM* 544(1-3): 221–235.

Makhseed, S., Ibrahim, F., and Samuel, J. 2012. Phthalimide based polymers of intrinsic microporosity. *Polymer* 53(14): 2964–2972.

Makowski, P., Thomas, A., Kuhn, P., and Goettmann, F. 2009. Organic materials for hydrogen storage applications: From physisorption on organic solids to chemisorption in organic molecules. *Energy and Environmental Science* 2(5): 480–490.

Mason, C. R., Maynard-Atem, L., Heard, K. W., Satilmis, B., Budd, P. M., Friess, K., Lanč, M., Bernardo, P., Clarizia, G., and Jansen, J. C. 2014. Enhancement of CO_2 affinity in a polymer of intrinsic microporosity by amine modification. *Macromolecules* 47(3), 1021–1029.

McKeown, N. B. 2012. Polymers of intrinsic microporosity. *International Scholarly Research Notices* 2012: 1–16.

McKeown, N. B., and Budd, P. M. 2006. Polymers of intrinsic microporosity (PIMs): Organic materials for membrane separations, heterogeneous catalysis and hydrogen storage. *Chemical Society Reviews* 35(8): 675–683.

McKeown, N. B., Budd, P. M., and Book, D. 2007. Microporous polymers as potential hydrogen storage materials. *Macromolecular Rapid Communications* 28(9): 995–1002.

McKeown, N. B., Gahnem, B., Msayib, K. J., Budd, P. M., Tattershall, C. E., Mahmood, K., Tan, S., Book, D., Langmi, H. W., and Walton, A. 2006. Towards polymer-based hydrogen storage materials: Engineering ultramicroporous cavities within polymers of intrinsic microporosity. *Angewandte Chemie* 118(11): 1836–1839.

Molefe, L. Y., Musyoka, N. M., Ren, J., Langmi, H. W., Ndungu, P. G., Dawson, R., and Mathe, M. 2019. Synthesis of porous polymer-based metal–organic frameworks monolithic hybrid composite for hydrogen storage application. *Journal of Materials Science* 54(9): 7078–7086.

Negri, F., and Saendig, N. 2007. Tuning the physisorption of molecular hydrogen: Binding to aromatic, hetero-aromatic and metal-organic framework materials. *Theoretical Chemistry Accounts* 118(1): 149–163.

Panella, B., Hirscher, M., Pütter, H., and Müller, U. 2006. Hydrogen adsorption in metal–organic frameworks: Cu-MOFs and Zn-MOFs compared. *Advanced Functional Materials* 16(4): 520–524.

Polak-Kraśna, K., Dawson, R., Holyfield, L. T., Bowen, C. R., Burrows, A. D., and Mays, T. J. 2017. Mechanical characterisation of polymer of intrinsic microporosity PIM-1 for hydrogen storage applications. *Journal of Materials Science* 52(7): 3862–3875.

Principi, G., Agresti, F., Maddalena, A., and Russo, S. L. 2009. The problem of solid state hydrogen storage. *Energy* 34(12): 2087–2091.

Ramimoghadam, D., Gray, E. M., and Webb, C. 2016. Review of polymers of intrinsic microporosity for hydrogen storage applications. *International Journal of Hydrogen Energy* 41(38): 16944–16965.

Ren, J., Musyoka, N. M., Langmi, H. W., Mathe, M., and Liao, S. 2017. Current research trends and perspectives on materials-based hydrogen storage solutions: A critical review. *International Journal of Hydrogen Energy* 42(1): 289–311.

Ritter, N., Antonietti, M., Thomas, A., Senkovska, I., Kaskel, S., and Weber, J. 2009. Binaphthalene-based, soluble polyimides: The limits of intrinsic microporosity. *Macromolecules* 42(21): 8017–8020.

Rochat, S., Polak-Kraśna, K., Tian, M., Holyfield, L. T., Mays, T. J., Bowen, C. R., and Burrows, A. D. 2017. Hydrogen storage in polymer-based processable microporous composites. *Journal of Materials Chemistry A* 5(35): 18752–18761.

Rouquerol, J., Rouquerol, F., Llewellyn, P., Maurin, G., and Sing, K. S. 2013. *Adsorption by Powders and Porous Solids: Principles, Methodology and Applications*, 2nd edition. Academic Press, Elsevier.

Schlapbach, L., and Züttel, A. 2011. Hydrogen-storage materials for mobile applications. *Materials for Sustainable Energy: A Collection of Peer-Reviewed Research and Review Articles from Nature Publishing Group*, 414: 265–270.

Song, J., Du, N., Dai, Y., Robertson, G. P., Guiver, M. D., Thomas, S., and Pinnau, I. 2008. Linear high molecular weight ladder polymers by optimized polycondensation of tetrahydroxytetramethylspirobisindane and 1, 4-dicyanotetrafluorobenzene. *Macromolecules* 41(20): 7411–7417.

Tedds, S., Walton, A., Broom, D. P., and Book, D. 2011. Characterisation of porous hydrogen storage materials: Carbons, zeolites, MOFs and PIMs. *Faraday Discussions* 151: 75–94.

Texier-Mandoki, N., Dentzer, J., Piquero, T., Saadallah, S., and David, P. 2004. Hydrogen storage in activated carbon materials: Role of the nanoporous texture. *Carbon (New York, NY)* 42(12–13): 2744–2747.

Thomas, K. M. 2009. Adsorption and desorption of hydrogen on metal–organic framework materials for storage applications: Comparison with other nanoporous materials. *Dalton Transactions* (9): 1487–1505.

Thommes, M., Kaneko, K., Neimark, A. V., Olivier, J. P., Rodriguez-Reinoso, F., Rouquerol, J., and Sing, K. S. 2015. Physisorption of gases, with special reference to the evaluation of surface area and pore size distribution (IUPAC Technical Report). *Pure and Applied Chemistry* 87(9-10): 1051–1069.

Tian, M., Rochat, S., Polak-Kraśna, K., Holyfield, L. T., Burrows, A. D., Bowen, C. R., and Mays, T. J. 2019. Nanoporous polymer-based composites for enhanced hydrogen storage. *Adsorption* 25(4): 889–901.

Van Mullekom, H., Vekemans, J., Havinga, E., and Meijer, E. 2001. Developments in the chemistry and band gap engineering of donor–acceptor substituted conjugated polymers. *Materials Science and Engineering: R: Reports* 32(1): 1–40.

Walton, K. S., and Snurr, R. Q. 2007. Applicability of the BET method for determining surface areas of microporous metal– organic frameworks. *Journal of the American Chemical Society* 129(27): 8552–8556.

Webb, C., and Gray, E. M. 2014a. Analysis of the uncertainties in gas uptake measurements using the Sieverts method. *International Journal of Hydrogen Energy* 39(1): 366–375.

Webb, C., and Gray, E. M. 2014b. Analysis of uncertainties in gas uptake measurements using the gravimetric method. *International Journal of Hydrogen Energy* 39(13): 7158–7164.

Weber, J., Antonietti, M., and Thomas, A. 2008. Microporous networks of high-performance polymers: Elastic deformations and gas sorption properties. *Macromolecules* 41(8): 2880–2885.

Weber, J., Su, Q., Antonietti, M., and Thomas, A. 2007. Exploring polymers of intrinsic microporosity–microporous, soluble polyamide and polyimide. *Macromolecular Rapid Communications* 28(18–19): 1871–1876.

Wong-Foy, A. G., Matzger, A. J., and Yaghi, O. M. 2006. Exceptional H2 saturation uptake in microporous metal–organic frameworks. *Journal of the American Chemical Society* 128(11): 3494–3495.

Wood, C. D., Tan, B., Trewin, A., Niu, H., Bradshaw, D., Rosseinsky, M. J., Khimyak, Y. Z., Campbell, N. L., Kirk, R., and Stöckel, E. 2007. Hydrogen storage in microporous hypercrosslinked organic polymer networks. *Chemistry of Materials* 19(8): 2034–2048.

Wu, D., Xu, F., Sun, B., Fu, R., He, H., and Matyjaszewski, K. 2012. Design and preparation of porous polymers. *Chemical Reviews* 112(7): 3959–4015.

Xiang, Z., and Cao, D. 2013. Porous covalent–organic materials: Synthesis, clean energy application and design. *Journal of Materials Chemistry A* 1(8): 2691–2718.

Xu, S., and Tan, B. 2014. Microporous organic polymers: Synthesis, types, and applications. In *Synthesis and Applications of Copolymers*. New Jersey: Wiley Online Library.

Yu, X., Tang, Z., Sun, D., Ouyang, L., and Zhu, M. 2017. Recent advances and remaining challenges of nanostructured materials for hydrogen storage applications. *Progress in Materials Science* 88: 1–48.

Yuan, S., Kirklin, S., Dorney, B., Liu, D.-J., and Yu, L. 2009. Nanoporous polymers containing stereocontorted cores for hydrogen storage. *Macromolecules* 42(5): 1554–1559.

Zhou, W., Wu, H., Hartman, M. R., and Yildirim, T. 2007. Hydrogen and methane adsorption in metal– organic frameworks: A high-pressure volumetric study. *The Journal of Physical Chemistry C* 111(44): 16131–16137.

Züttel, A. 2003. Materials for hydrogen storage. *Materials Today* 6(9): 24–33.

Züttel, A., Sudan, P., Mauron, P., and Wenger, P. 2004. Model for the hydrogen adsorption on carbon nanostructures. *Applied Physics A* 78(7): 941–946.

17 Polymers and Their Composites for Water-Splitting Applications

Zahra Pezeshki and Zahra Heidari
Shahrood University of Technology, Shahrood, Semnan, Iran

Mashallah Rezakazemi
Shahrood University of Technology, Shahrood, Semnan, Iran

Mohammad Younas
University of Engineering and Technology, Peshawar, Pakistan

CONTENTS

17.1 INTRODUCTION

Today, human beings are looking for sustainable sources of energy, one of which is hydrogen energy. In fact, hydrogen energy is one of the few sources that can reduce the risk of environmental and air pollution to zero due to its high density and zero carbon emissions. But the question is: how can this energy be produced?

Water splitting is a chemical reaction for the decomposition of water (H_2O) into oxygen (O_2) and hydrogen (H_2). It is a kind of power-to-X technology considered as a sustainable production of H_2 as a coming fuel and universal energy carrier for the promotion of decarbonization and its market [1] as well as mathanation from suitable sources [2]. The produced H_2 and O_2 can be utilized to generate electricity as clean transportation fuel for electric vehicles (EVs) [3,4]. The chemical reactions of water splitting are classified into four groups: electrolysis, photocatalysis, radiolysis, and thermolysis. In these reactions, the catalysts must be designed to improve half-cell reactions called hydrogen evolution reactions (HERs) where hydrogen is

DOI: 10.1201/9781003169727-17

produced at the cathode, and oxygen evolution reactions (OERs) where oxygen is produced at the anode. Most of the time, a HER is a two-electron-transfer reaction while an OER is a four-proton-coupling reaction:

$$\text{HER: } 2H^+ + 2e^- \rightarrow H_2 \quad (17.1)$$

$$\text{OER: } 2H_2O \rightarrow O_2 + 4H^+ + 4e^- \quad (17.2)$$

Hydrogen is considered as a valuable energy for industries due to the fact that: (i) it is a harmless product, (ii) it can be generated by many energy sources especially renewable energies (REs), (iii) it can work with fuel cells, and (iv) it is pervasive everywhere from continents to nations [5].

Today, polymer materials and their composites are mostly used to design catalysts. These materials have many applications from daily necessities to high uses such as healthcare, automobiles, and aerospace due to their having light weight, less cost, and adaptive mechanical properties such as flexibility, stretchability, and easy manufacture. Other features related to them are associated with their complexity and integration where scalability, sustainability, and efficient resources are needed [6,7]. Furthermore, the addition of these capabilities to catalysts is now provided by polymers and their composites. This chapter after introducing the applications of water splitting methods, presents the polymers and their composites in these methods and how to use them in water splitting to generate hydrogen and oxygen gases.

17.2 ELECTROLYSIS

Electrolysis of water is the decomposition of H_2O into O_2 and H_2 due to an electric current being passed through the water. This process requires the decomposition voltage of water at 25 °C to be related to the theoretical electrical energy consumption of H_2 [8]. Thus, electrolysis application is limited to the cases where electrical generation is cheap [5]. Currently, this technology has been developed and accelerated by the fabrication of polymer materials and their composites to decrease energy demands and reduce investment and maintenance costs to ensure sufficient durability, safety, reliability, and flexibility [9]. There are four processes in electrolysis decomposition: high-temperature steam (HTS) electrolysis, proton exchange membrane (PEM) electrolysis, alkaline water (AW) electrolysis, and photoelectrochemical (PE) electrolysis (see Table 17.1). In the following, after a description of these four processes, we will introduce the polymers and their composites utilized for this chemical reaction during these processes.

17.2.1 HTS Electrolysis

This technology is known to have reduced electrical energy consumption compared to conventional low temperature water electrolysis [8,14]. Using this technology, we can have

TABLE 17.1
Comparing Electrolysis Processes. [8–11]

Process Name	Energy Conversion Efficiency (%)	Gap between Electrodes	Electrolysis Voltage (V)	Reference
HTS	100	—	Small	[10]
PEM	80–90	Very small	1.6–1.8	[11]
AW	60–82	Large	Less than 1	[12]
PE	40–87	Too large	1.47–3	[13]

the hydrogen without any CO_2 emissions, provided that the electricity is obtained from RE resources such as wind or solar power. But it needs a high temperature between 700 and 1000 °C compared to PEM and AW electrolyses, so its overvoltage is smaller [15]. Therefore, platinum-group metals are not needed as a catalyst. This technology works with steam and is called high-temperature steam electrolysis (HTSE) where two electrolytes, known as an oxide ion conductor and a proton conductor, can be used.

When an oxide ion conductor is used as the anode, steam is used as the cathode. So, hydrogen is collected, and oxygen ions are moved towards the anode, which is an air electrode, through the electrolyte, and oxidized into oxygen gas. So, the electrode equations are as follows (see also Figure 17.1):

$$\text{HER at cathode: } H_2O + 2e^- \rightarrow H_2 + O^{2-} \quad (17.3)$$

$$\text{OER at anode: } O^{2-} \rightarrow O_2 + 2e^- \quad (17.4)$$

In this kind of electrolysis reaction, the electrode structure is porous so as to facilitate gas diffusion. Also, durability and stability are considered as significant challenges [16–20].

Instead of an oxide ion conductor, a proton conductor is also used as the electrolyte, known as solid oxide (SO) electrolysis [21]. In this case, a proton conductor is used as the

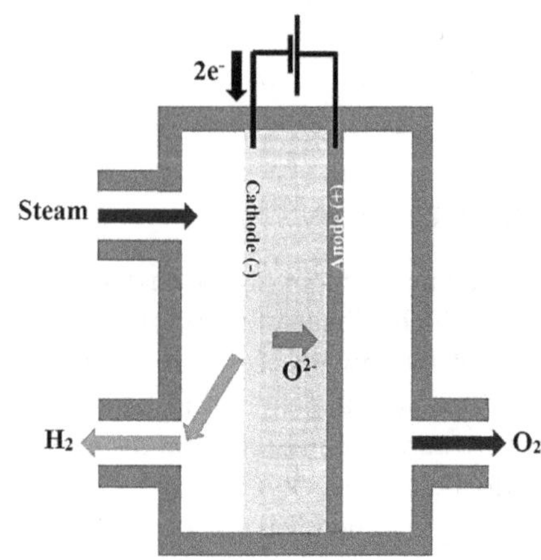

FIGURE 17.1 HTSE with oxide ion conductor.

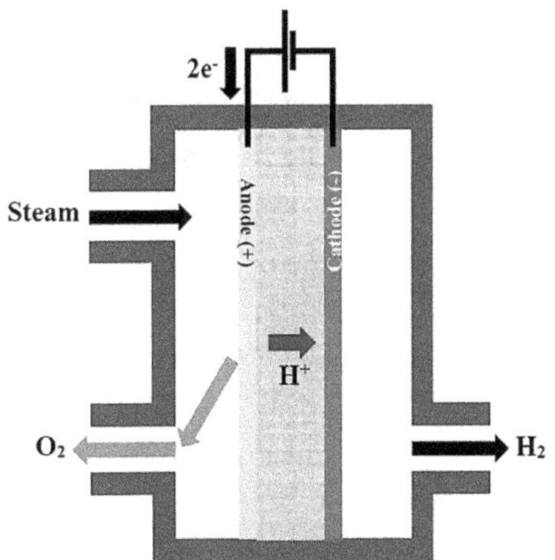

FIGURE 17.2 HTSE with proton conductor known as SO electrolysis.

cathode, and steam is used as the anode. So, the half-cell reactions are performed as follows (see also Figure 17.2):

$$\text{HER at cathode: } 2H^+ + 2e^- \rightarrow H_2 \tag{17.5}$$

$$\text{OER at anode: } H_2O \rightarrow 1/2O_2 + 2H^+ + 2e^- \tag{17.6}$$

Using a proton conductor is beneficial because the derived hydrogen needs no separation [19]. For improving HTS electrolysis technology, a few studies have been carried out for the improvement of the conductors by polymers and their composites.

17.2.1.1 Polymer-Obtained Boron-Doped Bismuth Oxide Nanocomposites

According to the research, bismuth oxide, Bi_2O_3, has many applications for solid oxide fuel cells (SOFCs) and oxygen sensors. A Bi_2O_3 nanocomposite-based ion conductor can improve the efficiency and longevity of a cell because it has high ionic conductivity. Polymers obtained from Bi_2O_3 nanocomposites can have stability in high temperatures from 730–825 °C, but they crack [22]. To eliminate this problem, these polymers can be stabilized with REs, but their ionic conductivity decreases. In this regard, doping boron to this kind of polymer is very effective because of its low melting point of 460°C [23,24].

17.2.2 PEM Electrolysis

This technology is known for its high efficiency, good corrosion stability, and flexible electrolysis in comparison to high temperature steam electrolysis. But the installation cost of this technology is high. This technology is also called solid polymer electrolysis. As shown in Figure 17.3, in this technology PEM acts as an acid solution, thereby conducting protons. It can also have higher efficiency over the alkaline system as well as good flexibility and corrosion stability, more than water splitting [5]. The half-cell

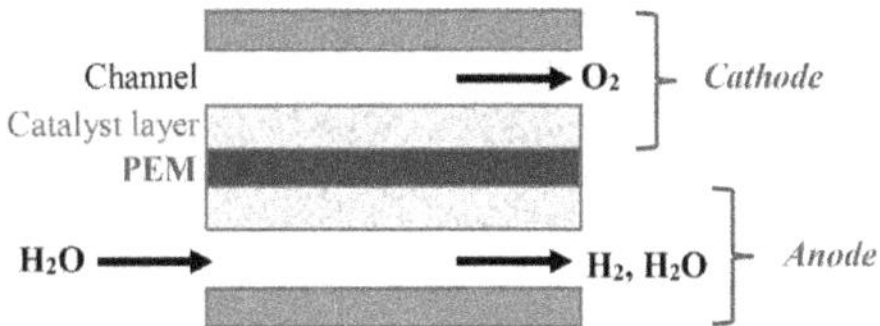

FIGURE 17.3 Schematic diagram of a PEM water electrolyzer.

reactions plus the overall reaction is defined in Equations (17.7–17.9).

$$\text{OER at anode: } H_2O \rightarrow 2H^+ + 1/2O_2 + 2e^- \tag{17.7}$$

$$\text{HER at cathode: } 2H^+ + 2e^- \rightarrow H_2 \tag{17.8}$$

$$\text{Overall reaction: } H_2O \rightarrow H_2 + 1/2O_2 \tag{17.9}$$

The Faradic efficiency of PEM electrolysis is higher than AW electrolysis, because the dense PEM is used to separate electrodes. But in AW electrolysis, it is used as the porous diaphragm. Also, in PEM, the gaps between the electrodes are very small, that is, of several hundred microns, resulting in ohmic losses, lower electrolysis voltage, and higher voltage efficiency. In order to improve voltage efficiency and higher current density in PEMs, the operating temperature must be higher, from 60 to 120°C [11], sometimes reaching to 200 °C [10]. Polymers and their composites have been used as electrocatalysts in PEM electrolysis for several decades to accelerate the water splitting reaction. The catalyst layer coated PEM, known as a catalyst coated membrane (CCM), comprises polymers and their composites called polymer electrolytes (PEs) [2]. The PEs ensure the separation of the produced gases and the conduction of protons while providing gas separation and electrical conduction between the electrodes. The goal of development is to decrease the cost and increase the durability and performance [25]. Table 17.2 shows the technological gaps and goals of PEs.

The next section introduces the polymers and their composites utilized for electrolysis chemical reaction during this process.

17.2.2.1 Ir and Ru Modified PANI Polymers

Studies have shown that iridium (Ir) presents high activity in OER reactions among other materials [26]. Using this property of Ir, scientists were able to create a new polymer called Ir and ruthenium (Ru) modified polyaniline (PANI) polymer. This polymer can accelerate the electrolysis reaction of water splitting exactly at the thermoneutral cell voltage of 1.45 V at 25 °C in 0.5 mole of H_2SO_4 with a satisfactory persistence of 96 h (see Figure 17.4) [27].

By using H_2SO_4, half-cell reactions—i.e., HERs and OERs—happen during the electrochemical kinetic of PAN-Ir and PANI-Ru, respectively, due to the combination of chemical polymerization and proper thermal treatment

TABLE 17.2
Technological Gaps and Goals for PE Development and Water Splitting Applications [2]

Gap/Goal	Materials	Operating Conditions	Durability	Economics
Higher temperature	×	×	×	×
Higher current density	×	×	×	×
Effects on safety and efficiency	×	×		×
Less catalyst loadings	×		×	×
Material properties relationships due to transport loss	×	×		×
Defined accelerated stress tests (ASTs) for durability assessments			×	
Replacing titanium (Ti) as structural material	×			×
Techno-economic analysis of the system level				×
Effects of intermittent and dynamic operation	×	×	×	

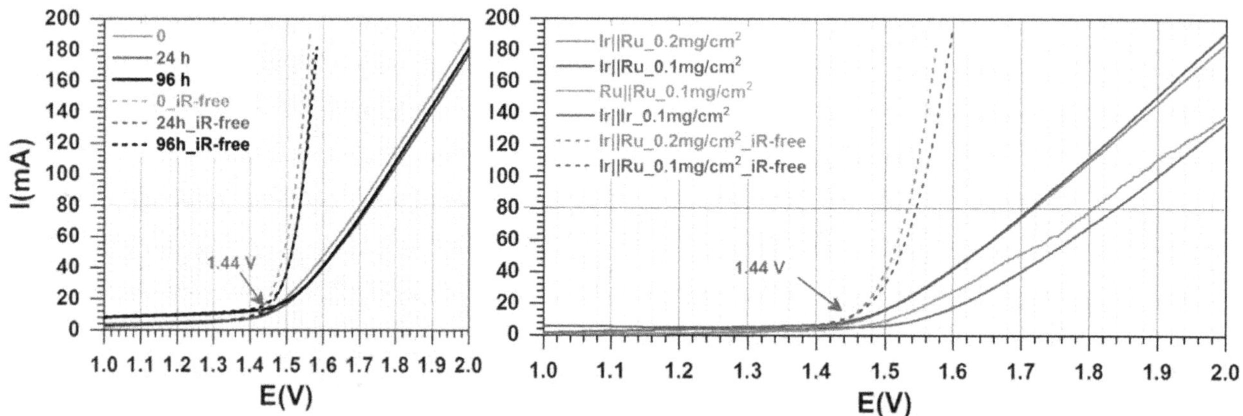

FIGURE 17.4 Polarization curves at anode and cathode during a satisfactory durability of 96 h at 10 mA/cm² [27].

to enhance the performance of electrocatalysts to produce sustainable H_2 (see Figure 17.5) [27].

The combination of polymerization and thermal treatment produces the metal nanostructured or oxide electrocatalysts. The presence of Ir(+III) cations causes the Ir metal nanoparticles to form during the polymerization, while Ru(+III) anions cause the RuO_2 oxide nanoparticles to form by thermal treatment. This reaction has demonstrated that the deposited electrocatalysts do not increase the ohmic resistance of the electrode, so the cell voltage does not increase [27].

17.2.2.2 Polymeric Nanofibers

Currently polymeric nanofibers can be used as electrodes for water splitting reactions in electrolysis. Among the nanofibers, cobalt (Co) catalysts such as SFCNF/$Co_{1-x}S$ @ CoN containing an S-doped flexible carbon nanofiber (SFCNF) matrix obtained by an electrospinning technique, $Co_{1-x}S$ nanoparticles, and CoN coatings are stable and highly active. Generally, Co_xS or CoN_x along with various compositions are considered as impressive electrocatalysts due to the rich variable valence high corrosion resistance for half-cell reactions [28–30]. Figure 17.6 shows the way to prepare an SFCNF/$Co_{1-x}S$ @ CoN catalyst.

As shown in Figure 17.6, sulfur is doped in activate adjacent carbon atoms to improve the conductivity. Then, active $Co_{1-x}S$ nanoparticles are formed under certain conditions between cobalt reagents and doped sulfur. Afterwards, depending on the surface area, uniform CoN is deposited on $Co_{1-x}S$ or carbon nanofibers by atomic layer deposition (ALD). At the end, the final electrocatalyst, that is SFCNF/$Co_{1-x}S$ @ CoN, can be used for half-cell reactions. In the following, the HER and OER activities of SFCNF is performed at a potential voltage of 1.58 V by the incorporation of $Co_{1-x}S$ @ CoN in 1 mole of KOH electrolyte with a satisfactory persistence of 20 h (see Figure 17.7) [31].

17.2.2.3 Nanocages Obtained from Polymer/Co Complexes

Nanocages obtained from polymer/Co complexes such as CoP@NCHNCs are the novel polymer composites which can be used for water splitting. These CoP@NCHNCs are made up of copolymer nanocomposites and N-doped carbon nanocages obtained from polydopamine (PDA)/Co nanocontainers through a pyrolysis/phosphidation technique (see Figure 17.8). In fact, in this polymer composite, PDA is selected as a carbon precursor to facilitate electrocatalyst activity.

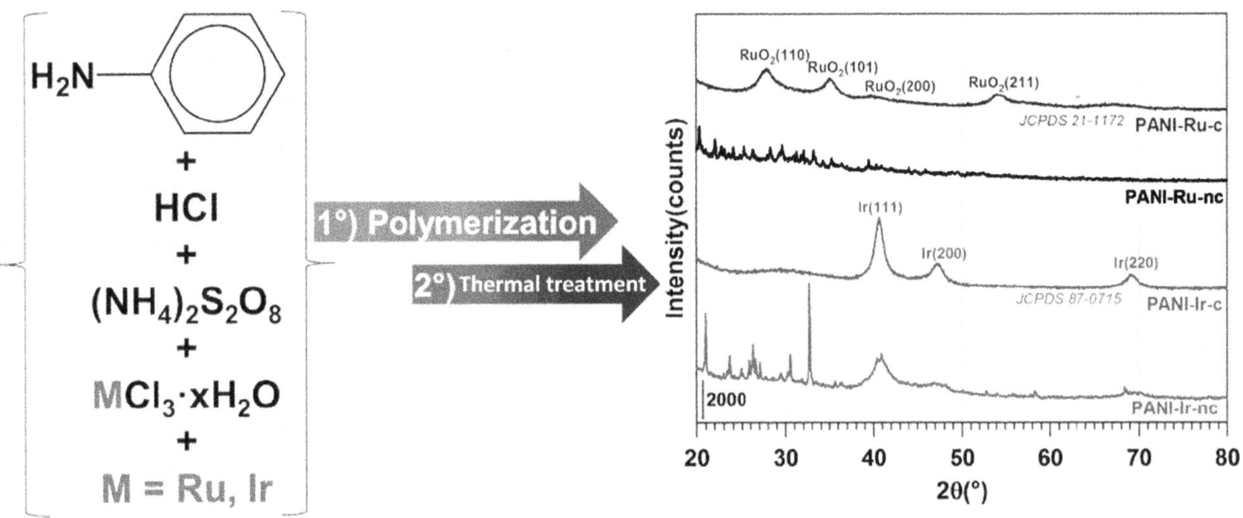

FIGURE 17.5 Attempt to produce nanostructured materials of PANI-Ir and PANI-Ru [27].

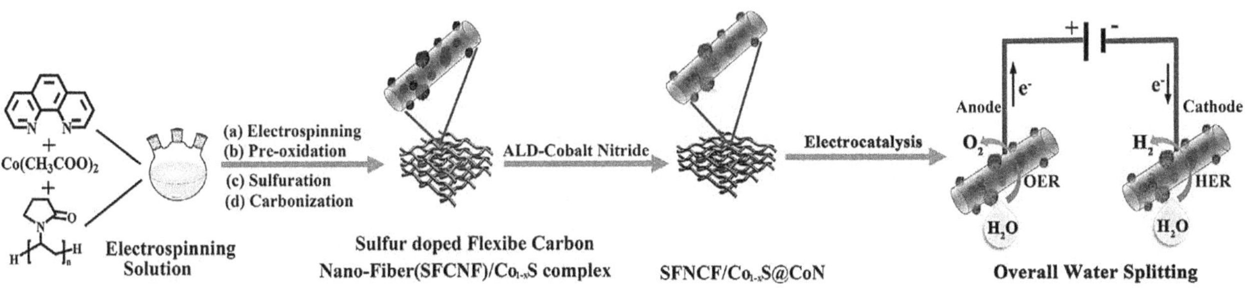

FIGURE 17.6 Preparation of SFCNF/Co_{1-x}S @ CoN catalyst for water splitting reactions [31].

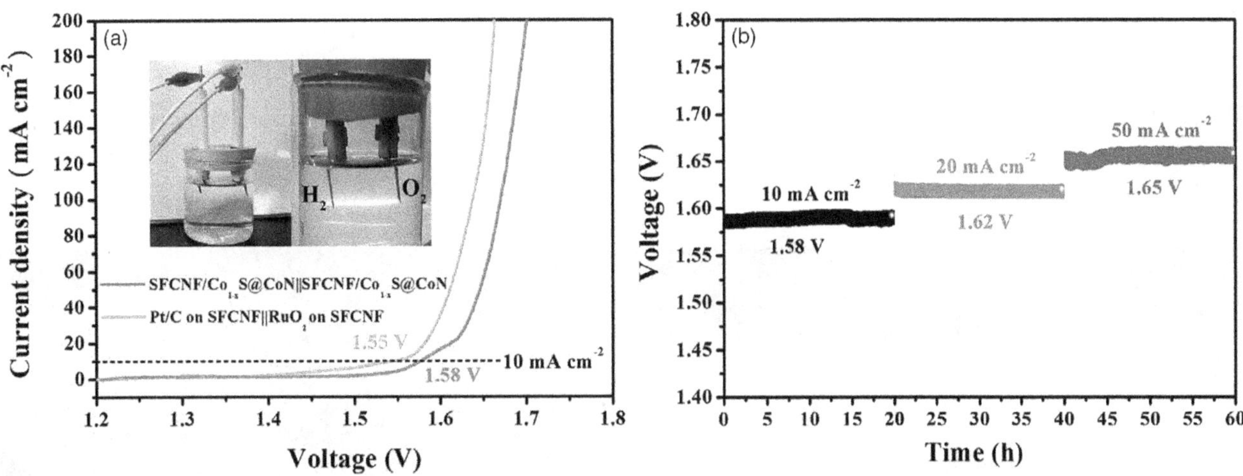

FIGURE 17.7 Overall water splitting (a) in 1 mole of KOH, (b) during 20–60 h at different current densities [31].

For an overall water splitting reaction, CoP@NCHNCs are used as the anode and cathode and half-cell reactions are done at a potential voltage of 1.62 V in 1 mole of KOH electrolyte with a satisfactory persistence and stability of 12 h (Figure 17.9) [32].

17.2.2.4 Polymeric Binders

Polymeric binders, such as ethyleneco-methacrylic acid and linear polyethylene, are usually used under certain conditions in PEM reactions (see Figure 17.10). They can show the highest stability performance along with anion-exchange

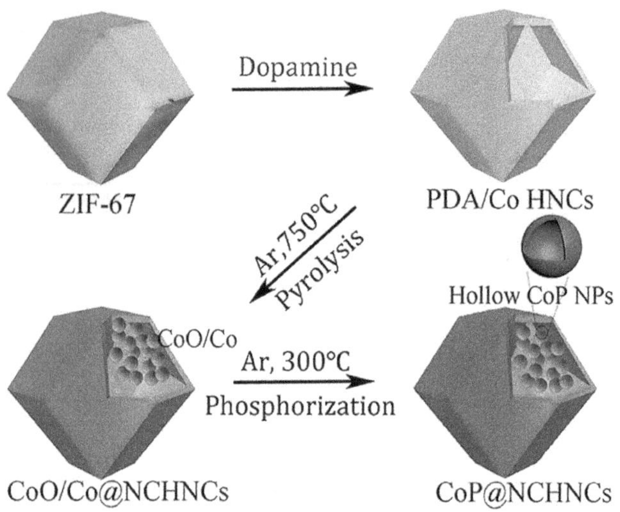

FIGURE 17.8 Schematic of CoP@NCHNCs preparation [32].

membranes prepared with trimethylbenzylammonium in an alkaline environment. In fact, the presence of this kind of binder enhances the mechanical strength of the membranes up to a temperature of 70°C due to the creation of good contact between ion selective particles filled with a water solution and the absence of β-hydrogen in the chain, even at high potassium hydroxide concentrations [5,33–36].

As the conductivity difference is more important than the difference in the ion exchange membrane (IEC), experiments have shown that linear polyethylene binders have the highest conductivity (see Figure 17.11) [5].

17.2.3 AW Electrolysis

This technology has been utilized in the chemical industry for many years because of its simplicity. The installation costs of this technology are less than PEM electrolysis technology due to the use of an alkaline environment rather than

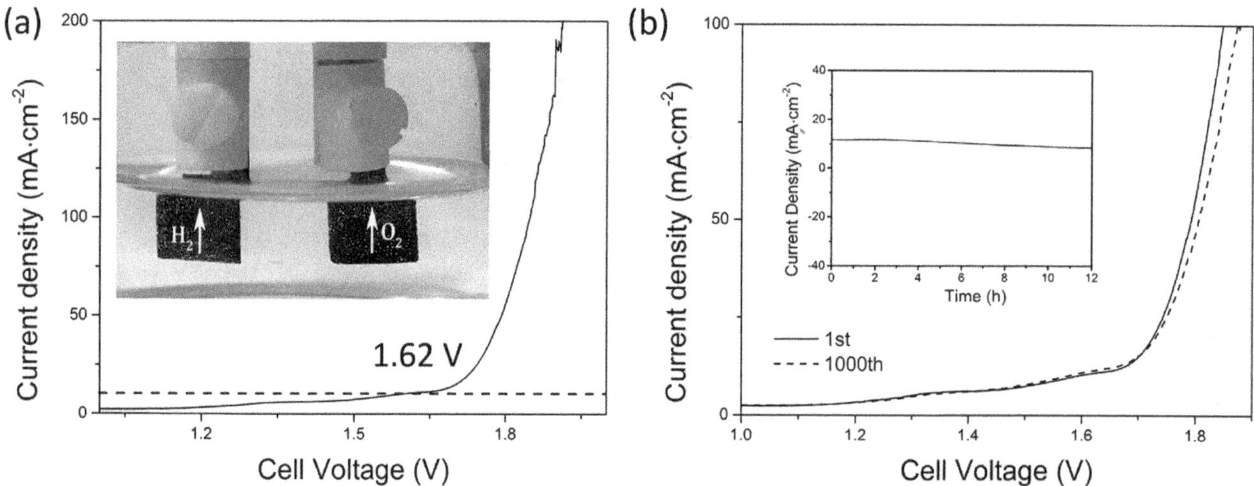

FIGURE 17.9 Overall water splitting (a) in 1 mole of KOH, (b) stability test over 12 h at 10 mA/cm^2 current density [32].

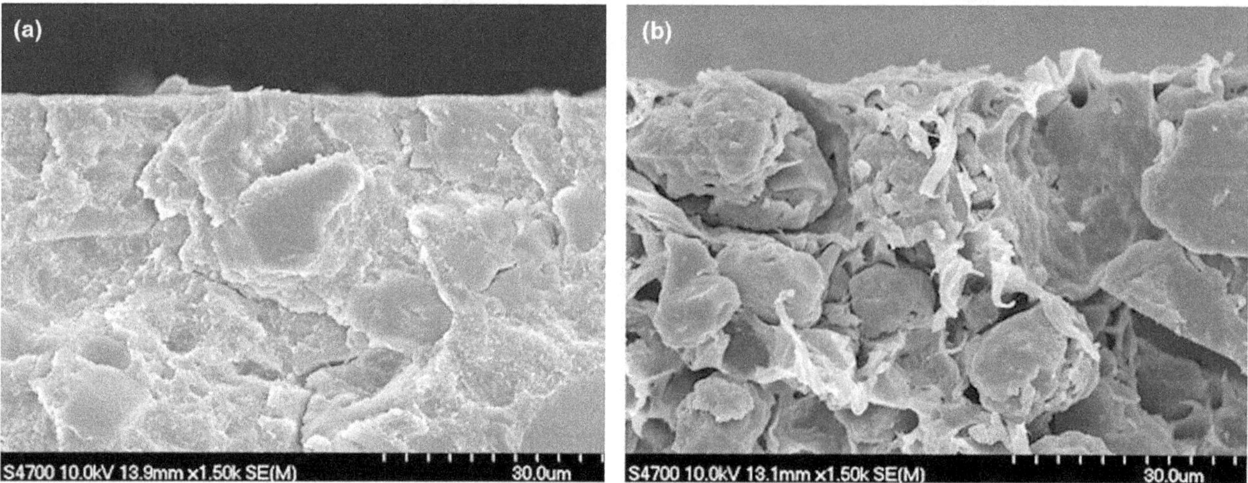

FIGURE 17.10 Cross-section of the polymer binders used: (a) ethyleneco-methacrylic acid and (b) linear polyethylene [5].

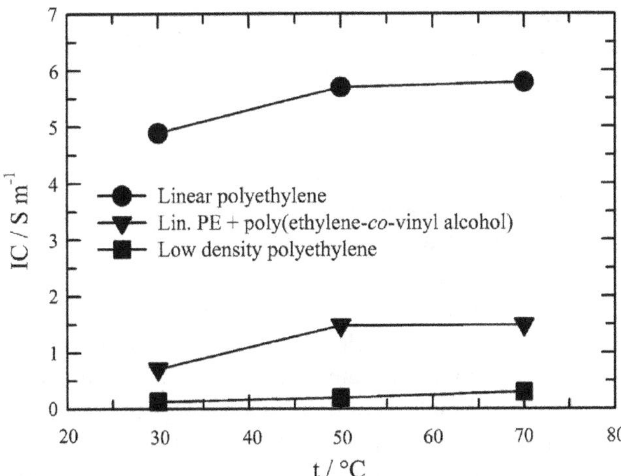

FIGURE 17.11 Dependence of ionic conductivity on temperature for three different polymer binders. The linear polyethylene binder has the highest conductivity [5].

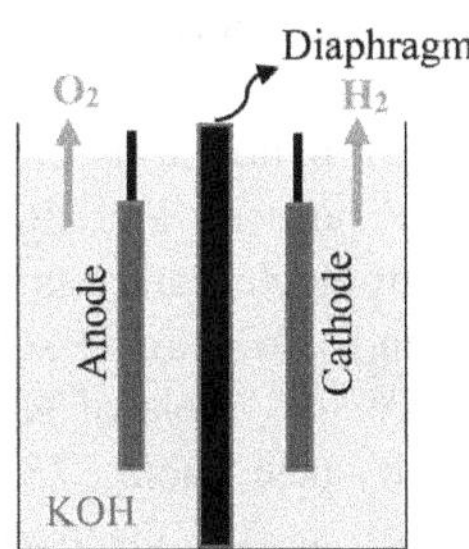

FIGURE 17.12 Schema of AW water electrolyzer.

expensive acid resistant materials. The anode and cathode are generally nickel-based materials or nickel-coated iron. In fact, this technology is known as the most advanced and cost-efficiency technology [5,12]. As shown in Figure 17.12, the electrolyte is an alkaline solution of KOH or NaOH, but an acidic solution such as H_2SO_4 can also be used. The half-cell reactions plus overall reaction is defined as in Equations (17.10–17.12).

$$\text{OER at anode: } 2OH^- \rightarrow H_2O + 1/2O_2 + 2e^- \quad (17.10)$$

$$\text{HER at cathode: } 2H_2O + 2e^- \rightarrow H_2 + 2OH^- \quad (17.11)$$

$$\text{Overall reaction: } H_2O \rightarrow H_2 + 1/2O_2 \quad (17.12)$$

Since OH^- is more easily oxidized, the OER is performed at the anode. So, the HER is done by the cathode and K^+ or Na^+ are more easily reduced from H_2O. In AW eletrolyses, the fraction of energy lost due to overpotential is defined as the voltage efficiency and is less than 1. So, AW electrolysis has a higher current density [12].

17.2.3.1 Organic Polymers

Organic polymers can be used to design electrocatalysts with thermal stability as well as chemical and physical characteristics. Electrocatalysts designed using organic polymers are used in three ways in water splitting: carbon materials obtained from organic polymers, ion-conductive polymers, and conjugated conductive polymers. These polymers have two main characteristics: dispersing active species and speeding up electronic transmission to improve water splitting reaction efficiency [37]. In the following, we will describe the role of these polymers in water decomposition.

17.2.3.1.1 Ion-Conductive Polymers

The ion-conductive polymers containing styrene and protonated imidazolium segments, or an amorphous molybdenum sulfide polymer, have the electronic capability to split water. In this case, they utilize active site exposure and conductivity to create a directional diffusion movement and act as an electrocatalyst for synthesis under an electric field action. Among this kind of polymer, an amorphous molybdenum sulfide polymer, MoS_x-polymer, is a HER electrocatalyst with higher conductivity and lower resistance. After combination of this polymer with highly oriented pyrolytic graphite (HOPG), oxidation occurs due to the ion exchange of the iodide anion in poly(2-(methacryloyloxy) ethyl ammonium iodide) with MoS_4^{2-}, which results in a HER precatalyst by an assembly involving MoS_3. Then, the best polymerization can create the best HER performance and the smallest overpotential of 200–250 mV in 1 mole of H_2SO_4 over 20 min (Figure 17.13) [37].

17.2.3.1.2 Conjugated Polymers

Conjugated polymers, a type of polymer semiconductor, are well-known because of their π-electron delocalization structure on the *p* orbital [38,39]. This structure makes the charge transfer faster by providing free electrons by conductivity. From this group of polymers, we should mention covalent-organic framework (COF) materials such as tris(4-formylphenyl)amine and 1,4-dialinobenzene 2-COF2, IISERP-COF2, 1, 3, 5-triformylphloroglucinol and bibyridine (Co-TpBpy), CoII-NDI-PG, CoII-BD-PG, and also conductive polymers containing polyoxometalates (POMs) such as polypyrrole, polyacetylene, polythiophene, polyphenylene, polyaniline, vinylene, and polydiacetylene [39–46], utilized as OER electrocatalysts. In fact, POMs are earth-abundant metals which are very usable because of their crystalline characteristics, porosity, and chemical tenability. Among this kind of polymer, CoII-NDI-PG has the best OER activity with a lower overpotential of

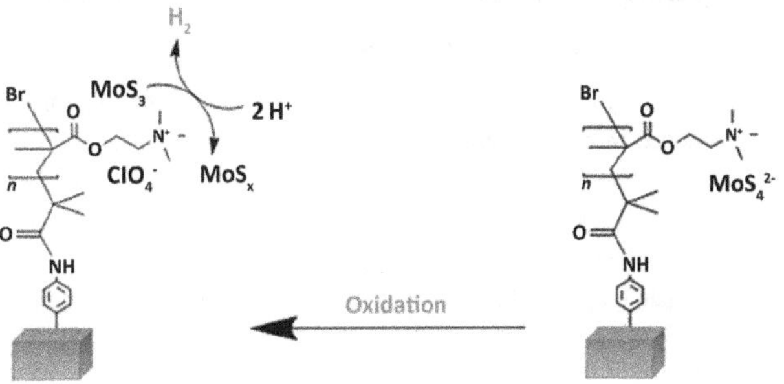

FIGURE 17.13 Schematic diagram of HER performance [37].

340 mV in 1 mole of KOH. Other polymers, such as THT (1,2,5,6,9,10-triphenylenehexathiol) and a nickel salt which form a 2D supramolecular polymer (2DSP), called THTNi 2DSP, are utilized as HER electrocatalysts. THTNi 2DSP has the better activity with an overpotential of 333 mV in 0.5 mole of H_2SO_4 (see Figure 17.14) [37,46].

17.2.3.1.3 Carbon Materials Obtained from Organic Polymers

Carbon materials obtained from organic polymers known as inorganic carbon materials are a type of material with improved migration efficiency. So, they can improve water splitting reactions. Among these materials, nitrogen-doped carbon materials can decrease the overpotential and enhance catalyst stability [47–51]. They are good HER and OER electrocatalysts in water splitting reactions due to their good chemical structure and heteroatom distribution. Instead of nitrogen, phosphorous can be doped in carbon materials. These materials are utilized in zinc-air batteries due to their having good activity. So, during the HER reaction, after generating H_2 desorbing from the electrode surface, the cathode accepts electrons to produce absorbed hydrogen atoms on the electrode surface. These batteries completely decompose water at 25°C with an overpotential of 0.24 V in 0.1 mole of KOH to produce H_2 and O_2 at a rate of 280 and 140 μmol/h [51] (see Figure 17.15). An assembly containing Fe/Co bimetallic as precursors to the materials can promote an OER reaction.

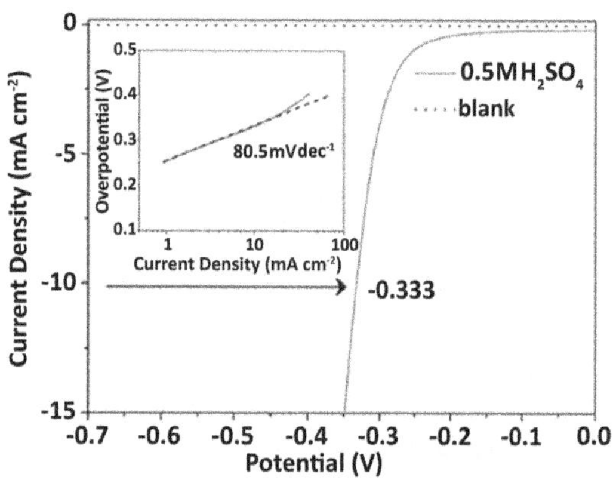

FIGURE 17.14 Polarization plot of a HER [37].

17.2.3.2 Carbonized Porous Conducting Polymers

Porous conducting polymers, called copolymers, such as n polythiophene (PTh) and polyparaphenylene (PPP), not only create a high surface area for electrocatalysts but also can absorb light. Today nanoporous conducting polymers due to having tunable porosity and light absorbancy are considered as excellent electrocatalysts for HER reactions with little deterioration in an alkaline solution and with an overpotential of 77 mV in 1 mole of KOH to obtain a 10 mA/cm^2 current density (see Table 17.3). Since electrodes play an important role in AW electrolysis for hydrogen

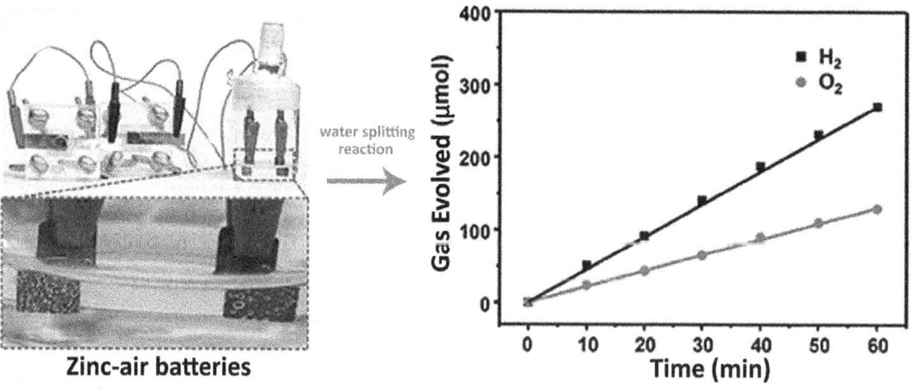

FIGURE 17.15 Water splitting of zinc-air batteries by inorganic carbon materials [37].

TABLE 17.3
Comparing Porous and Nanoporous Copolymers in 1 Mole of KOH

Carbonized (Non)Porous Copolymer	Method	Overpotential (V)	Reference
PTh:PPP copolymer	Electropolymerization + Pyrolysis	0.077	[53]
N, P codoped carbon network	Pyrolysis	0.16	[54]
3D porous carbon	Polymerization + pyrolysis	0.446	[55]
3D carbon network	Pyrolysis	0.131	[56]
N-doped graphene	Pyrolysis	0.432	[57]
N, S codoped graphitic sheet	Lyophilization and pyrolysis	0.31[a]	[58]

[a] In 1 mole of KOH

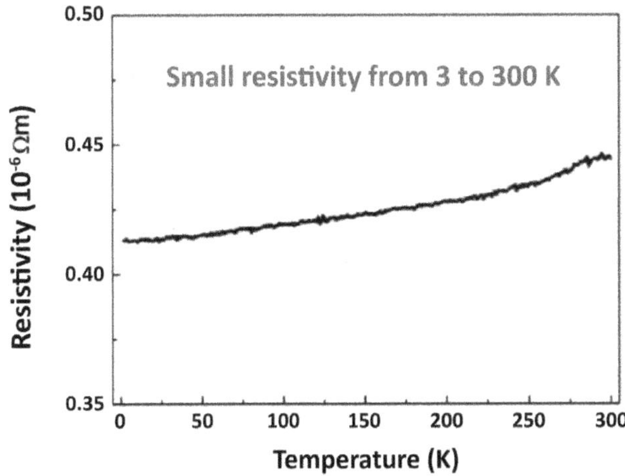

FIGURE 17.16 Nickel-phosphorous resistivity over a wide range of temperatures [6].

production, carbon-based materials due to providing a cheaper approach for half-cell reactions and being tolerant to both alkaline and acidic environments can be a suitable option. So nanoporous conducting polymers along with carbon-based materials can be easily synthesized with a thermal decomposition such as pyrolysis [52,53].

17.2.3.3 3D Printed Polymers

3D printed polymeric structures are well-known due to their high uniformity, though they usually lack conductivity. A novel 3D printed polymeric structure prepared by the high quality and low resistivity of a 0.45 μΩ nickel-phosphorous (NiP) coating (see Figure 17.16) has shown an effective conductivity of 4.7×10^4 S/m while maintaining its flexibility during conductivity. It has an excellent OER performance and the smallest overpotential of 197 mV in 1 mole of KOH with a satisfactory persistence of 48 h [6].

As shown in Figure 17.17, each part is made by 3D printing and then assembled. One of the benefits of 3D printing is that it minimizes the separation between HER and OER electrodes, decreases the electrical resistance, and increases the space usage. This can take place effectively in half-cell reactions at electrodes for the collection of H_2 and O_2 gases. The current 3D printed polymeric structure can work well with three 1.5 V batteries without mixing the gas bubbles [6].

17.2.4 PE Electrolysis

PE water splitting is an electrolysis method for breaking down water into H_2 and O_2 gases by photovoltaic systems/solar energy (see Figure 17.18). In this kind of water splitting, the electrical energy is obtained by photoelectrochemical cells (PECs) [59,60]. To have high conversion efficiency in a large system, the PECs are connected in series [61]. Over the years, several metal oxides such as TiO_2 [62], Cu_2O [63], Fe_2O_3 [64], $BiVO_4$ [65], WO_3 [66], NiO, and ZnO [67–70] have been used as photoelectrodes in PECs. Among them, ZnO is widely used due to its strong oxidizing ability, non-toxicity, and excellent electrical transportability [71,72]. For promoting the current density, the oxides can be prepared with each other and other materials as a heteroepitaxial system [73,74]. A novel method for preparing photoelectrodes utilizes the deposition of the thin layer of these oxides on the polymers and their composites to enhance the efficiency of PEC photoelectrodes by using flexibility, highly efficient light harvesting, and low cost. The half-cell reactions plus overall reaction is defined in Equations (17.13–17.15).

$$\text{OER at anode: } H_2O + 2H^+ \rightarrow 2H^+ + 1/2O_2 \quad (17.13)$$

$$\text{HER at cathode: } 2H_2O + 2e^- \rightarrow H_2 + 2OH^- \quad (17.14)$$

$$\text{Overall reaction: } H_2O \rightarrow H_2 + 1/2O_2 \quad (17.15)$$

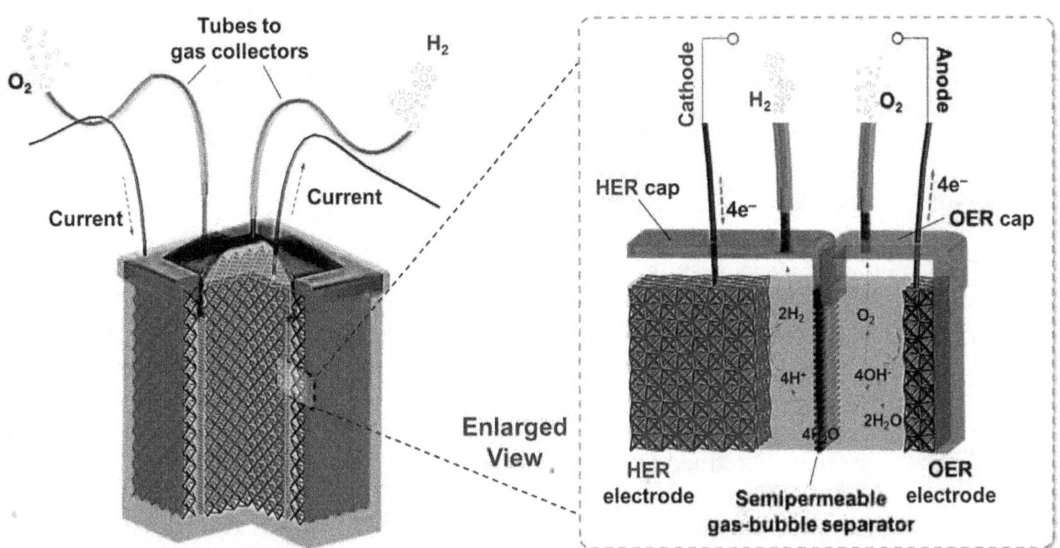

FIGURE 17.17 Schematic diagram of 3D printed polymeric structure water splitting cell [6].

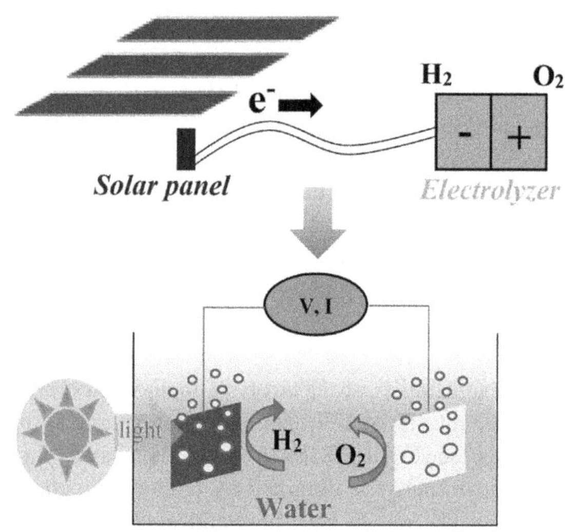

FIGURE 17.18 Schematic diagram of PE water electrolyzer.

TABLE 17.4
PEC Performance with Different ZnO Photoelectrodes

Photoelectrode Structure	Photocurrent Density (mA/cm²)	Reference
Vertically aligned ZnO nanowires (NWs)	0.1	[76]
ZnO nanorads	0.14	[77]
ZnO nanorads	0.18	[78]
ZP_F	0.17	[75]
Branched ZnO nanotetrapods	0.19	[79]
Vertically aligned ZnO nanowires (NWs)	0.2	[80]
ZnO nanorads	0.3	[81]
Caterpillar-like branched ZnO nanofibers (BZNs)	0.35	[80]
ZnO nanotrees	0.38	[78]
Textured porous ZnO plates	0.4	[82]
ZnO nanorads	0.45	[83]
ZnO nanorad clusters	0.5	[78]
ZP_M	0.52	[75]

17.2.4.1 Poly Urethane Acrylate

Poly urethane acrylate (PUA) is one of the polymers used for efficient PE water splitting application. It is prepared by depositing ZnO such as flat-PUA (ZP_F) or micropatterned-PUA (ZP_M), though ZP_M is better. For the preparation of the photoelectrodes, a thin layer of ZnO deposits on the PUA by a radio frequency (RF)-magnetron spurting technique. During the process, the photocurrent density (1.23 V vs. the reversible hydrogen electrode (RHE)) of ZP_M is enhanced by the increase of the collection ability, light capture capacity, and electrochemically active surface area; the charge transfer resistance decreases [75] (see Table 17.4).

17.2.4.2 Polymer-Based Dye-Sensitized Cells

Dye-sensitized photoelectrochemical cells (DS-PECs) are a type of PEC with their range of solar radiation improved (Figure 17.19). They have been organized based on the dye-sensitized photoelectrodes containing a photocathode with a passive anode, a photoanode with a passive cathode, or both photocathode and photoanode. Since the dye sensitization plays an essential role in light harvesting due to its large visible spectrum absorption, irradiation stability over a long time, and fast charge transfer while avoiding charge remix [84,85], polymers such as perylene derivatives, porphyrins, and donor-acceptor D-π-A molecules, that is two linear perylene-based photosensitizers and thienyl π-spacers, can enhance these features.

17.2.4.3 Polymer Electrolytes

Instead of conventional liquid electrolytes in PECs, we can use polymer electrolytes, such as an alkaline polymer electrolyte, as an anion-exchange membrane polymer electrolyte for water splitting to increase device stability by avoiding corrosion, especially in the photocathode, and maximize the performance of the electrolysis process. The photoanode and photocathode of this type of PEC can be made of phosphorus-modified α-Fe_2O_3 or an iron-modified CuO, respectively (Figure 17.20). Using a polymer electrolyte prevents the recombination of photogenerated oxygen and hydrogen while allowing well-suited ion transport for efficient oxygen and hydrogen separation [87–90].

17.2.4.4 Polymer-Templated Nanospiders

Polymer-templated nanospiders are the other materials utilized for photoelectrodes to have highly efficient water splitting in PECs for generating hydrogen. This is a cheap and facile approach in which organic polymers, such as benzene-swollen poly(ethylene glycol), are used for templating [91] semiconductor thin films such as TiO_2 thin films [92,93], called nanospiders (see Figure 17.21), to increase the liquid-semiconductor junction interaction and enhance water splitting performance at more than twice that of conventional thin films. Also, the photocurrent density of nanospider photoelectrodes is two times more than that of conventional thin films.

17.3 PHOTOCATALYSIS

In chemistry, photocatalysis is the acceleration of a photoreaction in the presence of a catalyst called a photocatalyst. In a water splitting reaction, photocatalysis technology is divided into three technologies—photolysis, photosynthesis, and cocatalysis processes—to generate H_2 and O_2. These processes are similar but with slight differences. For example, the selection of photocatalysts for photosynthesis should be markedly expanded compared to photolysis due to their having many thermodynamic constraints during the overall water splitting [95,96]. Here, after a description

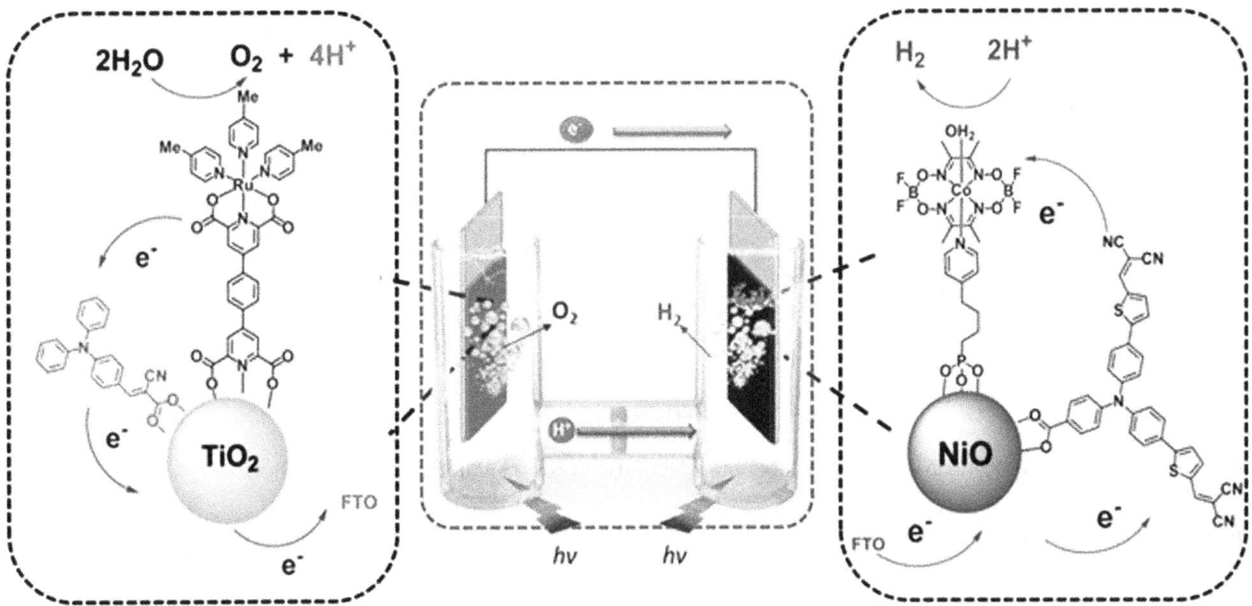

FIGURE 17.19 Schematic diagram of polymer based DS-PECS [86].

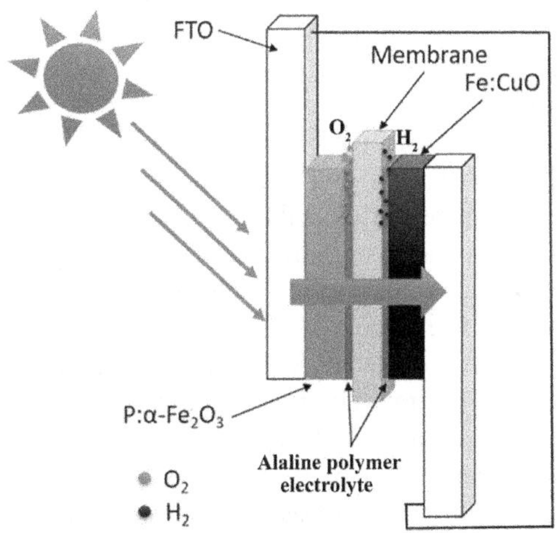

FIGURE 17.20 Using polymer electrolytes in PECs.

FIGURE 17.21 Top-view SEM images of a TiO_2 nanospider thin film [94].

of these processes, we will introduce the polymers and their composites utilized as photocatalysts for this chemical reaction.

17.3.1 Photolysis

Photolysis is a water splitting reaction with a one-step excitation process. It is a technology for converting radiant sunlight or artificial light into a two-electron hydrogen evolution reaction (0V vs. normal hydrogen electrode (NHE), pH = 0) and a four-electron oxygen evolution reaction (+1.23V vs. NHE, pH = 0) as a storable energy using photocatalysts:

$$\text{HER: } 2H^+ + 2e^- \rightarrow H_2 \quad (17.16)$$

$$\text{OER: } 4OH^- + 4h^+ \rightarrow O_2 + 2H_2O \quad (17.17)$$

In half-cell reactions, the thermal equilibrium of excitons occurs rapidly (usually less than 10 ps) with the moving of the electrons from the highest excited state to the lowest excited state; the fast electron-hole remix is associated with energy losses. So, the good stability and high efficiency photocatalysts require a bandgap larger than 1.8 eV and smaller than 2.2 eV so as to perform water splitting [95,97–100]. Moreover, they must be ecofriendly and have less cost, though they usually do not have a reasonable conversion efficiency. As a photocatalyst, polymers can absorb protons with enough energy for generating electron-hole pairs to diffuse free charge carriers. They can enhance and amplify photocatalytic performance. Note that most of the

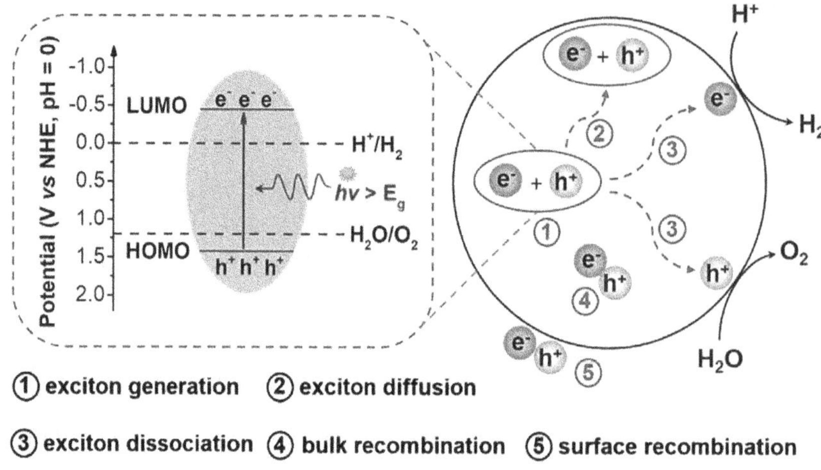

FIGURE 17.22 Schematic diagram of photolysis [103].

photocatalysts generate hydrogen in the water splitting process. So, during the process of photolysis for the generation of both hydrogen and oxygen gases, while the lowest unoccupied molecular orbital (LUMO) is lower than the potential of H_2/H^+, the highest occupied molecular orbital (HOMO) must be higher than the potential of O_2/H_2O (Figure 17.22) [101–103].

17.3.1.1 Conjugated Polymers

The conjugated polymers, a type of polymer semiconductor, widely used as photocatalysts for water splitting are divided into three groups. The first group are those that generate only H_2. The second group generate only O_2, and the third group are those which can generate both. Linear conjugated polymers such as linear poly(phenylene) derivatives [104] or linear polymer-(in)organic Z-schemes [105], cross-linked conjugated polymers such as N-rich covalent organic polymer [106], supramolecular self-assembled conjugated polymers shown in Table 17.5 [107], porous conjugated polymers (PCPs) [108], and microporous conjugated polymers such as donor–acceptor conjugated microporous polymers [109] fall into the first category. They can generate hydrogen with different performances based on their acceptor (co)monomers.

From the first group, donor–acceptor conjugated microporous polymers containing tetraphenylethylene (TPE) as donors and fluorenone (F) as acceptors are the most promising materials due to having a large potential for solar energy harvesting during the photolysis water splitting process. They are widely used for HER reactions. They usually have adjustable band gaps and display intramolecular charge transfer (ICT) absorption at the wavelength of 480 nm from 2.8–2.1 eV by increasing the acceptors (see Figure 17.23). So, the emission color of the polymers can change from green to red. They also exhibit charge transfer emissions in the wavelength range of 540–580 nm with the help of donors [109]. Overall, controlling the donor–acceptor ratio in this kind of polymer is considered a benefit where the energy transfer efficiency can be tuned.

The second group consist of microporous conjugated polymers such as photoactive Ag(I)-based coordination

TABLE 17.5
Examples of Supramolecular Self-Assembled Conjugated Polymers of the Second Group [107]

Name	Year of Creation	Reference
Methyl viologen (MV^{2+}) attached to the polymers through electrostatic interaction	2004	[110]
Perylene derivative supramolecular ribbons self-assembled through $\pi - \pi$ stacking	2015	[111]
Triazine-based 2D/2D heterostructures formed with van der Waals interaction	2017	[112]
Fluorescein supramolecular nanosheets self-assembled by hydrogen bonding	2018	[113]

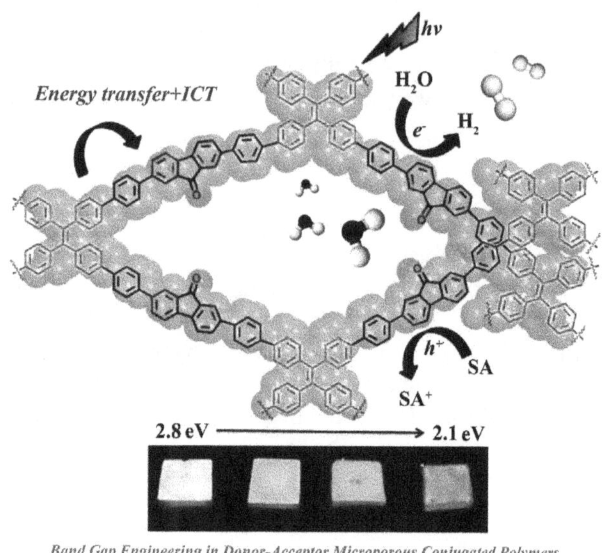

FIGURE 17.23 Schematic diagram of HER performance [109].

polymers [114,115] and supramolecular self-assembled conjugated polymers such as molecular perylene-3,4,9,10-tetracarboxylic diimide (PDI) as well as its derivatives [107]. The third group of conjugated polymers including conjugated polyelectrolytes, polymer dots, polymer nanocomposites such as polyacrylonitrile (PAN)/graphene nanofibers embedded with $Sr/TiO_3/NiO$ nanoparticles [116], polymer nanosheets such as porphyrin conjugated polymers grafted onto $BiVO_4$ [117], $N_3(C_3N_3S_3)_2$ coordination polymers [118], or Z-scheme-type conductive-polymer-P3HT/$KTa(Zr)O_3$ heterojunction composites [119] are called conjugated polymer nanomaterials (CPNMs). They are one of the promising and enhanced materials used as photocatalysts due to having tunable optoelectronic properties, easy molecular functionalization, good chemical stability, and a large contact area with water (Figure 17.24). They not only have been utilized for hydrogen production, but also for oxygen production [103].

17.3.2 Photosynthesis

For overcoming the drawback of conversion efficiency, a two-step excitation process called photosynthesis has been proposed [120–123]. This is a method for converting radiant sunlight or artificial light into fuel by photocatalysts. This technology is like the photosynthesis reaction that occurs in green plants to synthesize carbohydrates and split water in photosystems. As seen in Equations (17.18–17.21), in this technology, photon energy resulting from light sources generates electrons in the conduction band and holes in the valence band of photocatalysts, which leads to O_2-evolving and H_2-evolving, respectively, and all these reactions are recombined with each other via redox/solid-state mediators. Unlike the photolysis process that needs four photons to be completed, the photosynthesis process needs eight photons to be completed [124,125]. In half-cell light-dependent reactions, some energy is utilized to strip electrons from water producing oxygen gas. The hydrogen freed by the splitting of water and electrons reduced by hydrogen ions are used in the creation of hydrogen gas that serves as a short-term store of energy (Figure 17.25) [126].

$$2hv \rightarrow 2e^- + 2h^+ \tag{17.18}$$

$$\text{OER: } H_2O + 2h^+ \rightarrow 1/2O_2 + 2H^+ \tag{17.19}$$

$$\text{HER: } 2H^+ + 2e^- \rightarrow H_2 \tag{17.20}$$

$$\text{Overall reaction: } 2hv + H_2O \rightarrow 1/2O_2 + H_2 \tag{17.21}$$

Polymer semiconductors such as graphitic carbon nitride ($g\text{-}C_3N_4$) [127], Fe-NTiO_2 improved by polyethylene glycol (PEG) [126], polymer coated carbon quantum dot sensitizers [128], and donor–acceptor-based π-conjugated polymers [129] are the promising photocatalysts for photosynthesis

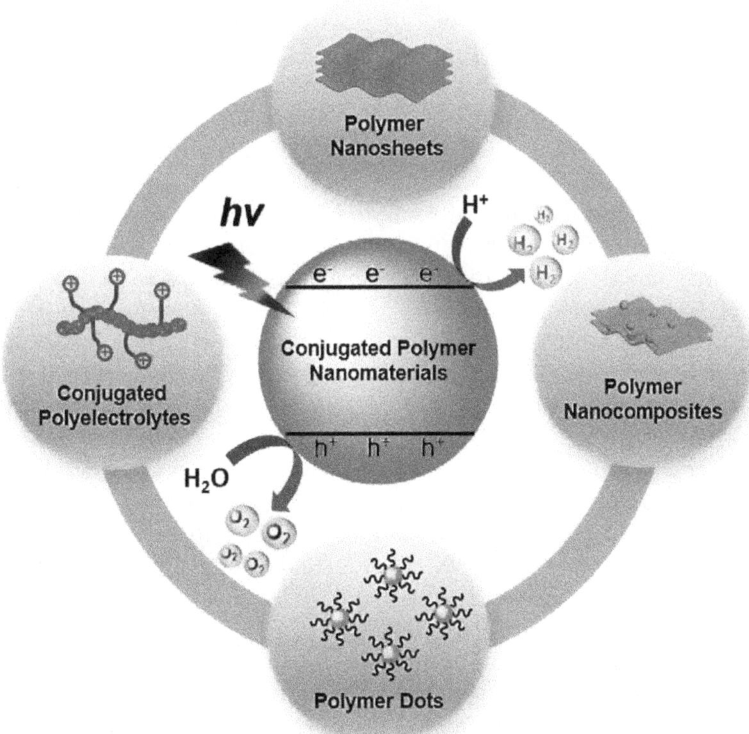

FIGURE 17.24 Types of CNPM for photolysis use [103].

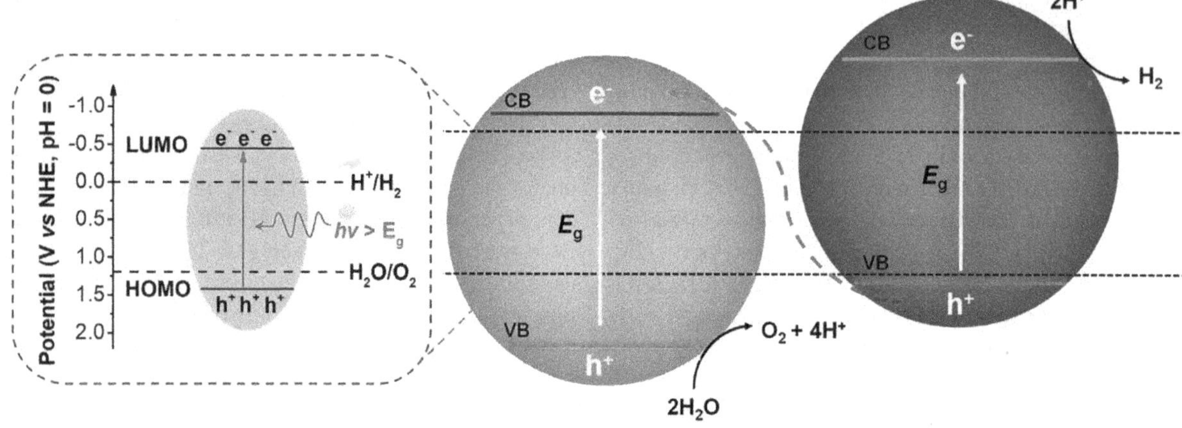

FIGURE 17.25 Schematic diagram of photosynthesis.

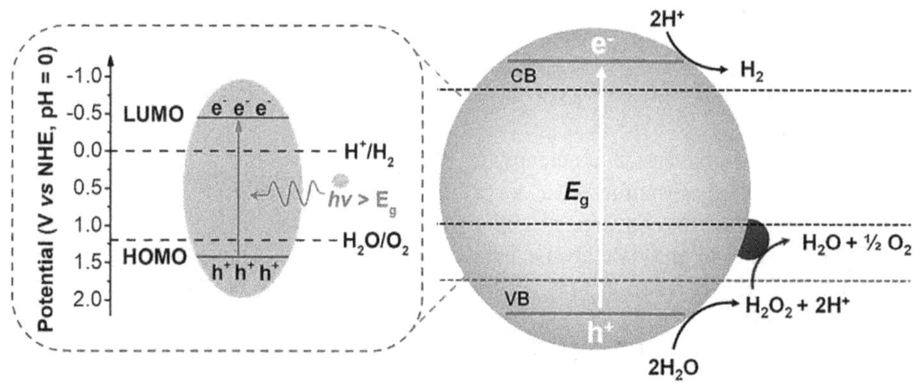

FIGURE 17.26 Schematic diagram of cocatalysis.

reaction due to their being benign, abundant, and having an adjustable band structure.

17.3.3 Cocatalysis

Since the photocatalysis OER reaction suffers from a very high overpotential, so this reaction can proceed with cocatalysts. Cocatalysis is a reaction with a stepwise two-electron/two-electron pathway process. This process is considered to be a notable technology for the decomposition of pure water. The first step of this process begins with photoexcitation by the oxidation of H_2O into H_2O_2 via photocatalysts followed by the dissociation of H_2O_2 into H_2 and O_2 and improved by cocatalysts (see Figure 17.26). In this regard, polymers and their composites can act better as photocatalysts, because they have a sufficiently low valence band to make the energy level of H_2O_2/H_2O (+1.23 V vs. NHE, pH = 0) more positive than that of O_2/H_2O (0 V vs. NHE, pH = 0) for a water-splitting reaction. Note that the lower valence band leads to a larger bandgap to maintain the mediator potential for proton reduction, resulting in a lower efficiency to absorb photons. So, further improvement is required here [130–133].

To be used, cocatalysts must be deposited onto polymers to create a rich catalytic coordination site on the polymer surface to facilitate charge migration and separation. In the cocatalysis process, Pt/Co or MnO_2 cocatalysts are among the most widely used materials [95,134].

The second and third groups of conjugated polymers are widely used in the cocatalysis process. The conjugated polymers used in the second group are 2D conjugated polymers [135] known as supramolecular self-assembled conjugated polymers such as 2DSP and microporous conjugated polymers. Those used in the third group are polymer dots [136–138] and polymer nanosheets. Depositing cocatalysts on the surface of them not only lowers the activation energy, but also forms a charge trap by a new midgap state within the polymer bandgap to extract the photogenerated electrons from the conduction band to suppress electron-hole remix. Using conjugated polymers improves the electrical conductivity and charge transfer for the overall water splitting activity for generating H_2 and O_2 by the decomposition of H_2O_2 (see Figure 17.27) [130,131,139–142].

17.4 RADIOLYSIS

Radiolysis can be used as a process to break down water into H_2 and O_2 gases with the help of fuel or nuclear waste. This process is divided into three stages: physical, physicochemical, and chemical. During this process, water goes

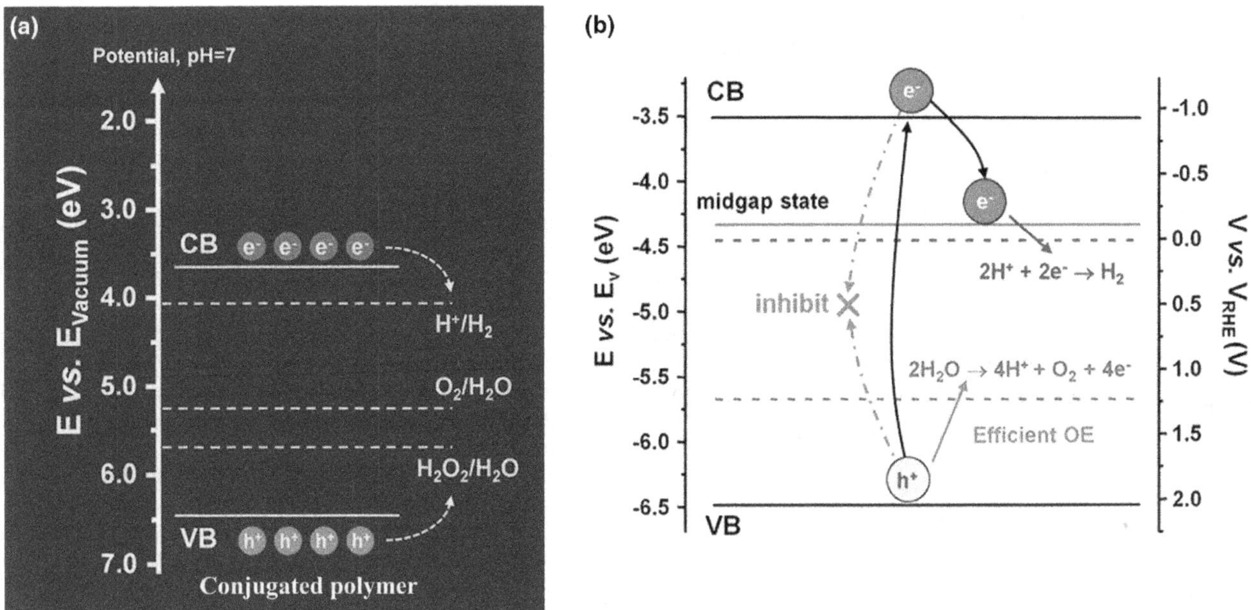

FIGURE 17.27 Schematic diagram of (a) the band energy diagram of conjugated polymers as photocatalysts in the cocatalysis process, (b) the electronic band structure for photocatalysts [95].

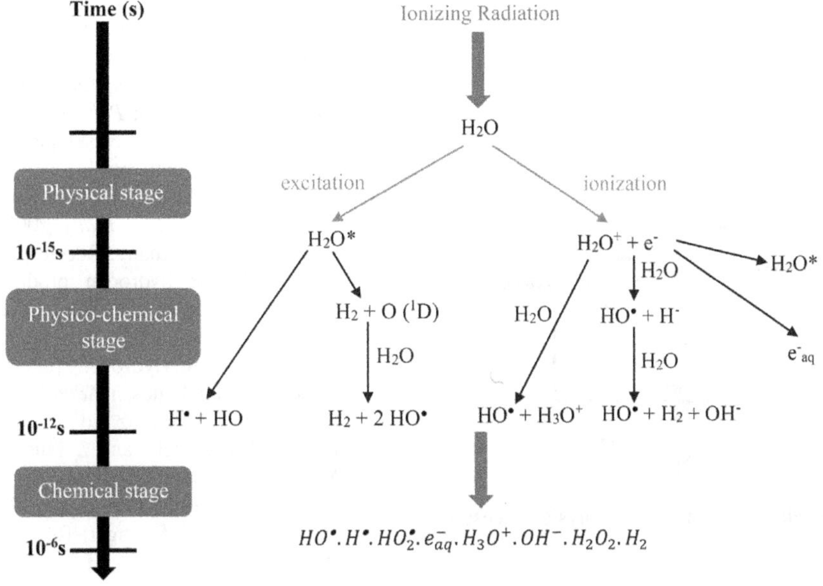

FIGURE 17.28 The stages of radiolysis for water splitting.

through a breakdown sequence into hydrogen peroxide, hydrogen radicals, and assorted oxygen compounds, such as ozone, which when converted back into oxygen releases great amounts of energy (Figure 17.28) [143,144]. Some of these releases are explosive. This decomposition is generated mostly by the α particles, which can be thoroughly absorbed by very thin layers of water. The radiolysis reaction can be written as:

$$H_2O \rightarrow HO^{\bullet}.H^{\bullet}.HO_2^{\bullet}.e_{aq}^{-}.H_3O^{+}.OH^{-}.H_2O_2.H_2 \quad (17.22)$$

Cross-linkable polymers such as poly(vinyl alcohol) (PVA) is widely utilized for the radiation–chemical synthesis of hydrogels. During this process, due to the combination of macroradicals produced by the reaction of polymer macromolecules with OH radicals upon γ-irradiation of aqueous polymer solutions, hydrogels are formed. The radiation-chemical results of the transformations of the aqueous PVA solution irradiated with X-rays at 77 K is as follows:

$$H_2O \rightarrow H^{\bullet}.e_{aq}^{-}.H_3O^{+}.OH^{-}.H_2O_2.H_2 \quad (17.23)$$

During this process most of the water creates polycrystalline ice. So, the chemical yield of the trapped electrons is very low. Some of the OH radicals produced are consumed before stabilization and others react with polymer chains to increase the mobility of the molecules at temperatures above 115 K in the second reaction, that is stage 2. The polymer macromolecules are stable up to 200 K [145].

17.5 THERMOLYSIS

Thermolysis can be used as a process for splitting water. During this process, water can be split up to some degree at temperatures starting at 500 °C but which can reach 2000 °C and generate H_2 and O_2 as follows (see Figure 17.29):

$$H_2O \rightleftharpoons H_2 + \frac{1}{2}O_2 \tag{17.24}$$

The temperature is obtained from concentrated solar power or the waste heat of nuclear power, whose reactions require temperatures above 1800 °C. Thermolysis is always performed by the PEM electrolysis process and an H_2SO_4 electrolyzer (see Figure 17.30; see also Section 17.2.2) [146].

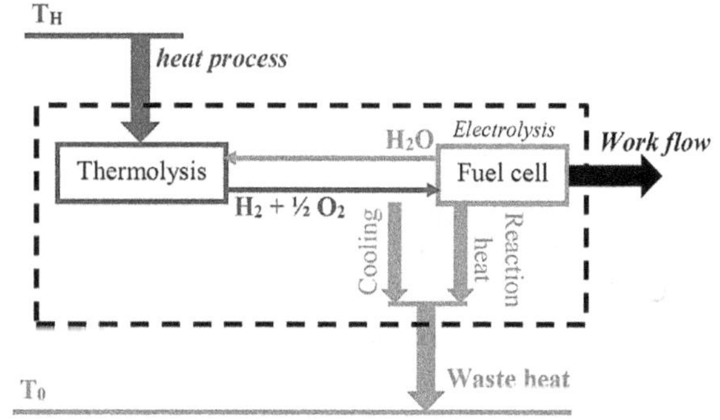

FIGURE 17.29 Schematic diagram of thermolysis process for water splitting.

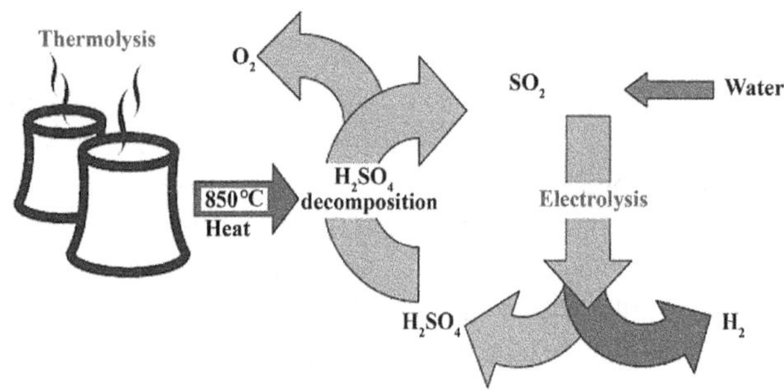

FIGURE 17.30 Schematic diagram of the thermolysis process.

REFERENCES

1. Pezeshki, Z., M. Younas, and M. Rezakazemi, Market Prospects of Membrane Contactors, in *Membrane Contactor Technology*, M. Rezakazemi and M. Younas, Editors. 2022, John Wiley and Sons. pp. 305–336. DOI: 10.1002/9783527831036.ch10
2. Babic, U., et al., critical review—identifying critical gaps for polymer electrolyte water electrolysis development. *Journal of The Electrochemical Society*, 2017. **164**(4): pp. F387–F399.
3. Naidu, K.C.B., et al., Recent advances in Ni-Fe batteries as electrical energy storage devices, in *Rechargeable Batteries*, Rajender Boddula, Inamuddin, Ramyakrishna Pothu, and Abdullah M. Asiri, Editors. 2020, Scrivener: Hoboken, NJ. pp. 115–130.
4. Pezeshki, Z., Classification, Modeling, and Requirements for Separators in Rechargeable Batteries, in *Rechargeable Batteries: History, Progress and Applications*, R. Boddula, Inamuddin, R. Pothu, and A. M. Asiri, Editors, 2020, John Wiley and Sons. pp. 265–314. DOI: 10.1002/9781119714774.ch13
5. Hnát, J., et al., Polymer anion selective membranes for electrolytic splitting of water. Part I: Stability of ion-exchange groups and impact of the polymer binder. *Journal of Applied Electrochemistry*, 2011. **41**(9): p. 1043.
6. Su, X., et al., Metallization of 3D printed polymers and their application as a fully functional water-splitting system. *Advanced Science*, 2019. **6**(6): p. 1801670.
7. McGrath, J.E., M.A. Hickner, and R. Höfer, 10.01 - Introduction: Polymers for a Sustainable Environment and Green Energy, in *Polymer Science: A Comprehensive Reference*, K. Matyjaszewski and M. Möller, Editors. 2012, Elsevier: Amsterdam. pp. 1–3.
8. Holladay, J.D., et al., An overview of hydrogen production technologies. *Catalysis Today*, 2009. **139**(4): pp. 244–260.
9. Zeng, K. and D. Zhang, Recent progress in alkaline water electrolysis for hydrogen production and applications. *Progress in Energy and Combustion Science*, 2010. **36**(3): pp. 307–326.
10. Fadhzir, M., et al., Hydrogen Production by Membrane Water Splitting Technologies, in *Advances In Hydrogen Generation Technologies*, M. Eyvaz, Editor. 2018, IntechOpen.
11. Ito, K., T. Sakaguchi, and Y. Tsuchiya, Polymer Electrolyte Membrane Water Electrolysis, in *Hydrogen Energy Engineering. Green Energy and Technology*, K. Sasaki, et al., Editors. 2016, Springer: Tokyo. p. 143–149.
12. Ito, K., H. Li, and Y.M. Hao, Alkaline Water Electrolysis, in *Hydrogen Energy Engineering. Green Energy and Technology*, K. Sasaki, et al., Editors. 2016, Springer: Tokyo. p. 137–142.
13. James, B.D., et al., Technoeconomic analysis of photoelectrochemical (PEC) hydrogen *Production* 2009. https://www.energy.gov/sites/prod/files/2014/03/f12/pec_technoeconomic_analysis.pdf
14. Brisse, A., J. Schefold, and M. Zahid, High temperature water electrolysis in solid oxide cells. *International Journal of Hydrogen Energy*, 2008. **33**(20): pp. 5375–5382.
15. Doenitz, W., et al., Hydrogen production by high temperature electrolysis of water vapour. *International Journal of Hydrogen Energy*, 1980. **5**(1): pp. 55–63.
16. Badwal, S.P.S., Zirconia-based solid electrolytes: Microstructure, stability and ionic conductivity. *Solid State Ionics*, 1992. **52**(1): pp. 23–32.

17. Ishihara, T., H. Matsuda, and Y. Takita, Doped LaGaO3 perovskite type oxide as a new oxide ionic conductor. *Journal of the American Chemical Society*, 1994. **116**(9): pp. 3801–3803.
18. Iwahara, H., et al., Proton conduction in sintered oxides and its application to steam electrolysis for hydrogen production. *Solid State Ionics*, 1981. **3–4**: pp. 359–363.
19. Matsumoto, H. and K. Leonard, Steam electrolysis, in *Hydrogen Energy Engineering. Green Energy and Technology*, Kazunari Sasaki, et al., Editors. 2016, Springer: Tokyo. https://doi.org/10.1007/978-4-431-56042-5_11
20. Mougin, J., 8 – Hydrogen production by high-temperature steam electrolysis, in *Compendium of Hydrogen Energy*, V. Subramani, A. Basile, and T.N. Veziroğlu, Editors. 2015, Woodhead Publishing: Oxford. pp. 225–253.
21. Laguna-Bercero, M.A., Recent advances in high temperature electrolysis using solid oxide fuel cells: A review. *Journal of Power Sources*, 2012. **203**: pp. 4–16.
22. Aytimur, A., et al., Synthesis and characterization of boron doped bismuth–calcium–cobalt oxide nanoceramic powders via polymeric precursor technique. *Ceramics International*, 2013. **39**(2): pp. 911–916.
23. Tunç, T., et al., Preparation of gadolina stabilized bismuth oxide doped with boron via electrospinning technique. *Journal of Inorganic and Organometallic Polymers and Materials*, 2012. **22**(1): pp. 105–111.
24. Aytimur, A., et al., Synthesis and characterization of boron-doped bismuth oxide-erbium oxide fiber derived nanocomposite precursor. *Journal of Composite Materials*, 2014. **48**(19): pp. 2317–2324.
25. Je, C.-H. and H.-M. Kim, A study on the water splitting using polymer electrolyte membrane for producing hydrogen and oxygen. *International Journal of Electrochemical Science*, 2019. **14**: pp. 6948–6975.
26. Wang, C., et al., Iridium-based catalysts for solid polymer electrolyte electrocatalytic water splitting. *ChemSusChem*, 2019. **12**(8): pp. 1576–1590.
27. Djara, R., et al., Iridium and ruthenium modified polyaniline polymer leads to nanostructured electrocatalysts with high performance regarding water splitting. *Polymers*, 2021. **13**(2): p. 190. https://doi.org/10.3390/polym13020190
28. Kong, D., et al., First-row transition metal dichalcogenide catalysts for hydrogen evolution reaction. *Energy & Environmental Science*, 2013. **6**(12): pp. 3553–3558.
29. Zhang, J., et al., Activating and optimizing activity of CoS2 for hydrogen evolution reaction through the synergic effect of N dopants and S vacancies. *ACS Energy Letters*, 2017. **2**(5): pp. 1022–1028.
30. Chen, P., et al., Metallic CO4N porous nanowire arrays activated by surface oxidation as electrocatalysts for the oxygen evolution reaction. *Angewandte Chemie International Edition*, 2015. **54**(49): pp. 14710–14714. https://doi.org/10.1002/anie.201506480
31. Guo, D., et al., Strategic atomic layer deposition and electrospinning of cobalt sulfide/nitride composite as efficient bifunctional electrocatalysts for overall water splitting. *Small*, 2020. **16**(35): p. 2002432.
32. Chen, Y., et al., Hierarchical hollow nanocages derived from polymer/cobalt complexes for electrochemical overall water splitting. *ACS Sustainable Chemistry & Engineering*, 2019. **7**(12): pp. 10912–10919.
33. Bauer, B., H. Strathmann, and F. Effenberger, Anion-exchange membranes with improved alkaline stability. *Desalination*, 1990. **79**(2): pp. 125–144.
34. Neagu, V., I. Bunia, and I. Plesca, Ionic polymers VI. Chemical stability of strong base anion exchangers in aggressive media. *Polymer Degradation and Stability*, 2000. **70**(3): pp. 463–468.
35. Shah, B.G., et al., Comparative studies on performance of interpolymer and heterogeneous ion-exchange membranes for water desalination by electrodialysis. *Desalination*, 2005. **172**(3): pp. 257–265.
36. Pezeshki, Z. and A. Kettab, Desalination Battery, in *Sustainable Materials and Systems for Water Desalination*, Inamuddin and A. Khan, Editors, 2021, Springer. pp. 137–162. DOI: 10.1007/978-3-030-72873-1_9
37. Niu, L., et al., Electrocatalytic water splitting using organic polymer materials-based hybrid catalysts. *MRS Bulletin*, 2020. **45**(7): pp. 562–568.
38. Günes, S., H. Neugebauer, and N.S. Sariciftci, Conjugated polymer-based organic solar cells. *Chemical Reviews*, 2007. **107**(4): pp. 1324–1338.
39. Heinze, J., B.A. Frontana-Uribe, and S. Ludwigs, Electrochemistry of conducting polymers—persistent models and new concepts. *Chemical Reviews*, 2010. **110**(8): pp. 4724–4771.
40. Gangopadhyay, R. and A. De, Conducting polymer nanocomposites: A brief overview. *Chemistry of Materials*, 2000. **12**(3): pp. 608–622.
41. Guimard, N.K., N. Gomez, and C.E. Schmidt, Conducting polymers in biomedical engineering. *Progress in Polymer Science*, 2007. **32**(8): pp. 876–921.
42. Ko, J.M., et al., Morphology and electrochemical properties of polypyrrole films prepared in aqueous and nonaqueous solvents. *Journal of The Electrochemical Society*, 1990. **137**(3): pp. 905–909.
43. Dalla Corte, D.A., et al., The hydrogen evolution reaction on nickel-polyaniline composite electrodes. *International Journal of Hydrogen Energy*, 2012. **37**(4): pp. 3025–3032.
44. Ibanez, J.G., et al., conducting polymers in the fields of energy, environmental remediation, and chemical–chiral sensors. *Chemical Reviews*, 2018. **118**(9): pp. 4731–4816.
45. Liao, G., Q. Li, and Z. Xu, The chemical modification of polyaniline with enhanced properties: A review. *Progress in Organic Coatings*, 2019. **126**: pp. 35–43.
46. Blasco-Ahicart, M., J. Soriano-López, and J.R. Galán-Mascarós, Conducting organic polymer electrodes with embedded polyoxometalate catalysts for water splitting. *ChemElectroChem*, 2017. **4**(12): pp. 3296–3301.
47. Antonietti, M. and M. Oschatz, The concept of "noble, heteroatom-doped carbons," their directed synthesis by electronic band control of carbonization, and applications in catalysis and energy materials. *Advanced Materials*, 2018 30(21):p. 1706836.
48. Han, S., et al., Metal-phosphide-containing porous carbons derived from an ionic-polymer framework and applied as highly efficient electrochemical catalysts for water splitting. *Advanced Functional Materials*, 2015. **25**(25): pp. 3899–3906.
49. Hua, S., et al., Highly dispersed few-layer MoS2 nanosheets on S, N co-doped carbon for electrocatalytic H2 production. *Chinese Journal of Catalysis*, 2017. **38**(6): pp. 1028–1037.
50. Yan, X.-H., et al., recent progress of metal carbides encapsulated in carbon-based materials for electrocatalysis of oxygen reduction reaction. *Small Methods*, 2020. **4**(1): p. 1900575.
51. Wang, H.-F., C. Tang, and Q. Zhang, A review of precious-metal-free bifunctional oxygen electrocatalysts: Rational design and applications in Zn–air batteries. *Advanced Functional Materials*, 2018. **28**(46): p. 1803329.

52. Zou, X. and Y. Zhang, Noble metal-free hydrogen evolution catalysts for water splitting. *Chemical Society Reviews*, 2015. **44**(15): p. 5148–5180.
53. Lahiri, A., G. Li, and F. Endres, Highly efficient electrocatalytic hydrogen evolution reaction on carbonized porous conducting polymers. *Journal of Solid State Electrochemistry*, 2020. **24**(11): pp. 2763–2771.
54. Zhang, J., et al., N,P-codoped carbon networks as efficient metal-free bifunctional catalysts for oxygen reduction and hydrogen evolution reactions. *Angewandte Chemie International Edition*, 2016. **55**(6): pp. 2230–2234.
55. Lai, J., et al., Unprecedented metal-free 3D porous carbonaceous electrodes for full water splitting. *Energy & Environmental Science*, 2016. **9**(4): pp. 1210–1214.
56. Peng, Z., et al., Homologous metal-free electrocatalysts grown on three-dimensional carbon networks for overall water splitting in acidic and alkaline media. *Journal of Materials Chemistry A*, 2016. **4**(33): pp. 12878–12883.
57. Zhang, B., et al., Nitrogen-doped graphene microtubes with opened inner voids: Highly efficient metal-free electrocatalysts for alkaline hydrogen evolution reaction. *Nano Research*, 2016. **9**(9): pp. 2606–2615.
58. Hu, C. and L. Dai, Multifunctional carbon-based metal-free electrocatalysts for simultaneous oxygen reduction, oxygen evolution, and hydrogen evolution. *Advanced Materials*, 2017. **29**(9): p. 1604942.
59. Naimi, Y. and A. Antar, *Hydrogen Generation by Water Electrolysis*, in *Advances In Hydrogen Generation Technologies*, M. Eyvaz, Editor. 2018, IntechOpen.
60. *Photoelectrochemical Water Splitting*. Vol. 71. 2020: Materials Research Foundations. https://doi.org/10.21741/9781644900734
61. Esiner, S., et al., Water splitting with series-connected polymer solar cells. *ACS Applied Materials & Interfaces*, 2016. **8**(40): pp. 26972–26981.
62. Cho, I.S., et al., Branched tio2 nanorods for photoelectrochemical hydrogen production. *Nano Letters*, 2011. **11**(11): pp. 4978–4984.
63. Yang, Y., et al., Cu 2 o/cuo bilayered composite as a high-efficiency photocathode for photoelectrochemical hydrogen evolution reaction. *Scientific Reports*, 2016. **6**(1): p. 35158.
64. Hsu, Y. K., Y.-C. Chen, and Y.-G. Lin, Novel ZnO/Fe2O3 Core–shell nanowires for photoelectrochemical water splitting. *ACS Applied Materials & Interfaces*, 2015. **7**(25): p. 14157–14162.
65. Yan, L., W. Zhao, and Z. Liu, 1D ZnO/BiVO4 heterojunction photoanodes for efficient photoelectrochemical water splitting. *Dalton Transactions*, 2016. **45**(28): pp. 11346–11352.
66. Su, J., et al., Nanostructured WO3/BiVO4 heterojunction films for efficient photoelectrochemical water splitting. *Nano Letters*, 2011. **11**(5): p. 1928–1933.
67. Lee, M.G., J.S. Park, and H.W. Jang, Solution-processed metal oxide thin film nanostructures for water splitting photoelectrodes: A review. *Journal of the Korean Ceramic Society*, 2018. **55**(3): pp. 185–202.
68. Zhang, S., et al., Enhanced piezoelectric-effect-assisted photoelectrochemical performance in ZnO modified with dual cocatalysts. *Applied Catalysis B: Environmental*, 2020. **262**: p. 118279.
69. Wei, S., et al., Fabrication of CuO/ZnO composite films with cathodic co-electrodeposition and their photocatalytic performance. *Journal of Molecular Catalysis A: Chemical*, 2010. **331**(1): pp. 112–116.
70. Ghobadi, A., et al., Angstrom Thick ZnO Passivation Layer to Improve the Photoelectrochemical Water Splitting Performance of a TiO 2 Nanowire Photoanode: The Role of Deposition Temperature. *Scientific Reports*, 2018. **8**(1): p. 16322.
71. Janotti, A. and C.G. Van de Walle, Fundamentals of zinc oxide as a semiconductor. *Reports on Progress in Physics*, 2009. **72**(12): p. 126501.
72. Soudi, A., P. Dhakal, and Y. Gu, Diameter dependence of the minority carrier diffusion length in individual ZnO nanowires. *Applied Physics Letters*, 2010. **96**(25).
73. Quynh, L.T., et al., flexible heteroepitaxy photoelectrode for photo-electrochemical water splitting. *ACS Applied Energy Materials*, 2018. **1**(8): pp. 3900–3907.
74. Singh, S. and N. Khare, Flexible PVDF/Cu/PVDF-NaNbO3 photoanode with ferroelectric properties: An efficient tuning of photoelectrochemical water splitting with electric field polarization and piezophototronic effect. *Nano Energy*, 2017. **42**: pp. 173–180.
75. Hou, T.-F., et al., Flexible, polymer-supported, ZnO nanorod array photoelectrodes for PEC water splitting applications. *Materials Science in Semiconductor Processing*, 2021. **121**: p. 105445.
76. Yaw, C.S., et al., A Type II n-n staggered orthorhombic V2O5/monoclinic clinobisvanite BiVO4 heterojunction photoanode for photoelectrochemical water oxidation: Fabrication, characterisation and experimental validation. *Chemical Engineering Journal*, 2019. **364**: pp. 177–185.
77. Liu, Y., et al., Synergistic effect of surface plasmonic particles and surface passivation layer on ZnO nanorods array for improved photoelectrochemical water splitting. *Scientific Reports*, 2016. **6**(1): pp. 29907.
78. Ren, X., et al., Photoelectrochemical water splitting strongly enhanced in fast-grown ZnO nanotree and nanocluster structures. *Journal of Materials Chemistry A*, 2016. **4**(26): pp. 10203–10211.
79. Qiu, Y., et al., Secondary branching and nitrogen doping of ZnO nanotetrapods: Building a highly active network for photoelectrochemical water splitting. *Nano Letters*, 2012. **12**(1): pp. 407–413.
80. Li, Q., et al., Facile and scalable synthesis of "Caterpillar-like" ZnO nanostructures with enhanced photoelectrochemical water-splitting effect. *The Journal of Physical Chemistry C*, 2014. **118**(25): pp. 13467–13475.
81. Wu, M., et al., In situ growth of matchlike ZnO/Au plasmonic heterostructure for enhanced photoelectrochemical water splitting. *ACS Applied Materials & Interfaces*, 2014. **6**(17): pp. 15052–15060.
82. Emin, S., et al., Photoelectrochemical properties of cadmium chalcogenide-sensitized textured porous zinc oxide plate electrodes. *ACS Applied Materials & Interfaces*, 2013. **5**(3): pp. 1113–1121.
83. Wang, T., et al., Au nanoparticle sensitized ZnO nanopencil arrays for photoelectrochemical water splitting. *Nanoscale*, 2015. **7**(1): pp. 77–81.
84. Kirner, J.T. and R.G. Finke, Water-oxidation photoanodes using organic light-harvesting materials: A review. *Journal of Materials Chemistry A*, 2017. **5**(37): pp. 19560–19592.
85. Gibson, E.A., Dye-sensitized photocathodes for H2 evolution. *Chemical Society Reviews*, 2017. **46**(20): pp. 6194–209.
86. Decavoli, C., et al., Molecular organic sensitizers for photoelectrochemical water splitting. *European Journal of Inorganic Chemistry*, 2020. **2020**(11–12): pp. 978–999.

87. Peter, L.M. and K.G. Upul Wijayantha, Photoelectrochemical water splitting at semiconductor electrodes: fundamental problems and new perspectives. *ChemPhysChem*, 2014. **15**(10): pp. 1983–1995.
88. Siracusano, S., et al., Enhanced performance and durability of low catalyst loading PEM water electrolyser based on a short-side chain perfluorosulfonic ionomer. *Applied Energy*, 2017. **192**: pp. 477–489.
89. Aricò, A.S., et al., Polymer electrolyte membranes for water photo-electrolysis. *Membranes*, 2017. **7**(2).
90. Cots, A., et al., Toward tandem solar cells for water splitting using polymer electrolytes. *ACS Applied Materials & Interfaces*, 2018. **10**(30): pp. 25393–25400.
91. Tian, B., et al., Self-adjusted synthesis of ordered stable mesoporous minerals by acid–base pairs. *Nature Materials*, 2003. **2**(3): pp. 159–163.
92. O'Regan, B. and M. Grätzel, A low-cost, high-efficiency solar cell based on dye-sensitized colloidal TiO 2 films. *Nature*, 1991. **353**(6346): pp. 737–740.
93. Fujishima, A. and K. Honda, Electrochemical photolysis of water at a semiconductor electrode. *Nature*, 1972. **238**(5358): pp. 37–38.
94. Fei, H., et al., Polymer-templated nanospider TiO2 thin films for efficient photoelectrochemical water splitting. *ACS Applied Materials & Interfaces*, 2010. **2**(4): pp. 974–979.
95. Wang, L., et al., 2d polymers as emerging materials for photocatalytic overall water splitting. *Advanced Materials*, 2018. **30**(48): p. 1801955.
96. Fisher, D.J., *Non-Electrolytic Water Splitting*. Materials Research Foundations Series, 2020, Materials Research Forum LLC. DOI: 10.21741/9781644900895
97. Fu, C.-F., X. Wu, and J. Yang, Material design for photocatalytic water splitting from a theoretical perspective. *Advanced Materials*, 2018. **30**(48): p. 1802106.
98. Le Bahers, T., M. Rérat, and P. Sautet, Semiconductors used in photovoltaic and photocatalytic devices: Assessing fundamental properties from DFT. *The Journal of Physical Chemistry C*, 2014. **118**(12): pp. 5997–6008.
99. Lin, S., et al., Photocatalytic oxygen evolution from water splitting. *Advanced Science*, 2021. **8**(1): p. 2002458.
100. Guiglion, P., C. Butchosa, and M.A. Zwijnenburg, Polymer photocatalysts for water splitting: Insights from computational modeling. *Macromolecular Chemistry and Physics*, 2016. **217**(3): pp. 344–353.
101. Li, L., et al., Rational design of porous conjugated polymers and roles of residual palladium for photocatalytic hydrogen production. *Journal of the American Chemical Society*, 2016. **138**(24): pp. 7681–7686.
102. Coropceanu, V., et al., Charge transport in organic semiconductors. *Chemical Reviews*, 2007. **107**(4): pp. 926–952.
103. Dai, C., Y. Pan, and B. Liu, Conjugated polymer nanomaterials for solar water splitting. *Advanced Energy Materials*, 2020. **10**(42): p. 2002474.
104. Sprick, R.S., et al., Nitrogen containing linear poly(phenylene) derivatives for photo-catalytic hydrogen evolution from water. *Chemistry of Materials*, 2018. **30**(16): pp. 5733–5742.
105. Bai, Y., et al., Photocatalyst Z-scheme system composed of a linear conjugated polymer and BiVO4 for overall water splitting under visible light. *Journal of Materials Chemistry A*, 2020. **8**(32): pp. 16283–16290.
106. Zhou, Z., P. Peng, and Z. Xiang, N-rich covalent organic polymer in situ modified TiO2 for highly efficient photocatalytic hydrogen evolution. *Science Bulletin*, 2018. **63**(6): pp. 369–375.
107. Zhao, C., et al., Recent advances in conjugated polymers for visible-light-driven water splitting. *Advanced Materials*, 2020. **32**(28): p. 1907296.
108. Li, L., et al., Donor–acceptor porous conjugated polymers for photocatalytic hydrogen production: The importance of acceptor comonomer. *Macromolecules*, 2016. **49**(18): pp. 6903–6909.
109. Mothika, V.S., et al., Rregulating charge-transfer in conjugated microporous polymers for photocatalytic hydrogen evolution. chemistry – a *European Journal*, 2019. **25**(15): pp. 3867–3874.
110. Jiang, D.-L., et al., Photosensitized hydrogen evolution from water using conjugated polymers wrapped in dendrimeric electrolytes. *Journal of the American Chemical Society*, 2004. **126**(38): pp. 12084–12089.
111. Weingarten, A.S., et al., Supramolecular packing controls H2 photocatalysis in chromophore amphiphile hydrogels. *Journal of the American Chemical Society*, 2015. **137**(48): pp. 15241–15246.
112. Schwarz, D., et al., Twinned growth of metal-free, triazine-based photocatalyst films as mixed-dimensional (2D/3D) van der Waals Heterostructures. *Advanced Materials*, 2017. **29**(40): p. 1703399.
113. Zhang, G.-Q., W. Ou, and Y.-S. Xu, Fluorescein supramolecular nanosheets: A novel organic photocatalyst for visible-light-driven H2 evolution from water. *Science China Materials*, 2018. **61**(7): pp. 1001–1006.
114. Xu, C., et al., Porous organic polymers: an emerged platform for photocatalytic water splitting. *Frontiers in Chemistry*, 2018. **6**: p. 592.
115. Mandal, S., et al., Photoactive Ag(I)-based coordination polymer as a potential semiconductor for photocatalytic water splitting and environmental remediation: Experimental and theoretical approach. *The Journal of Physical Chemistry C*, 2019. **123**(39): pp. 23940–23950.
116. Alharbi, A.R., et al., Synthesis and characterization of electrospun polyacrylonitrile/graphene nanofibers embedded with SrTiO3/NiO nanoparticles for water splitting. *Journal of Nanoscience and Nanotechnology*, 2017. **17**(8): pp. 5294–5302.
117. Wang, J., et al., Porphyrin conjugated polymer grafted onto BiVO4 nanosheets for efficient Z-scheme overall water splitting via cascade charge transfer and single-atom catalytic sites. *Advanced Energy Materials*, 2021. **n/a**(n/a): p. 2003575.
118. Guo, F., et al., Ni3(C3N3S3)2 coordination polymer as a novel broad spectrum-driven photocatalyst for water splitting into hydrogen. *Applied Catalysis B: Environmental*, 2017. **210**: pp. 205–211.
119. Koganemaru, Y., et al., Z-scheme-type conductive-polymer-P3HT/KTa(Zr)O3 heterojunction composites for enhancing the photocatalytic activity of water splitting. *Applied Catalysis A: General*, 2020. **602**: p. 117737.
120. Maeda, K., Z-scheme water splitting using two different semiconductor photocatalysts. *ACS Catalysis*, 2013. **3**(7): pp. 1486–1503.
121. Li, H., et al., Z-scheme photocatalytic systems for promoting photocatalytic performance: Recent progress and future challenges. *Advanced Science*, 2016. **3**(11): p. 1500389.
122. Zhou, P., J. Yu, and M. Jaroniec, All-solid-state z-scheme photocatalytic systems. *Advanced Materials*, 2014. **26**(29): pp. 4920–4935.
123. Bard, A.J., Photoelectrochemistry and heterogeneous photocatalysis at semiconductors. *Journal of Photochemistry*, 1979. **10**(1): pp. 59–75.

124. Yuan, Q., et al., Noble-metal-free janus-like structures by cation exchange for Z-scheme photocatalytic water splitting under broadband light irradiation. *Angewandte Chemie International Edition*, 2017. **56**(15): pp. 4206–4210.
125. Iwase, A., et al., Reduced graphene oxide as a solid-state electron mediator in Z-scheme photocatalytic water splitting under visible light. *Journal of the American Chemical Society*, 2011. **133**(29): pp. 11054–11057.
126. Priatmoko, S., et al., The influence of polyethylene glycol on characteristic of Fe-N-TiO2 photocatalyst and its activity for water splitting. *Journal of Physics: Conference Series*, 2020. **1567**: p. 022022.
127. Ruan, Q., *The Investigation into Polymer Semiconductor Photoelectrodes for Lightdriven Water Splitting*. 2020, UCL (University College London): London.
128. Virca, C.N., et al., Photocatalytic water reduction using a polymer coated carbon quantum dot sensitizer and a nickel nanoparticle catalyst. *Nanotechnology*, 2017. **28**(19): p. 195402.
129. Jayakumar, J. and H.-H. Chou, Recent advances in visible-light-driven hydrogen evolution from water using polymer photocatalysts. *ChemCatChem*, 2020. **12**(3): pp. 689–704.
130. Liu, J., et al., Metal-free efficient photocatalyst for stable visible water splitting via a two-electron pathway. *Science*, 2015. **347**(6225): p. 970.
131. Zhang, K., et al., Tunable bandgap energy and promotion of H_2O_2 oxidation for overall water splitting from carbon nitride nanowire bundles. *Advanced Energy Materials*, 2016. **6**(11): p. 1502352.
132. Wu, X., et al., Carbon dot and BiVO4 quantum dot composites for overall water splitting via a two-electron pathway. *Nanoscale*, 2016. **8**(39): pp. 17314–17321.
133. Wu, X., et al., Control strategy on two-/four-electron pathway of water splitting by multidoped carbon based catalysts. *ACS Catalysis*, 2017. **7**(3): pp. 1637–1645.
134. Liu, J., et al., A critical study of the generality of the two step two electron pathway for water splitting by application of a C3N4/MnO2 photocatalyst. *Nanoscale*, 2016. **8**(23): pp. 11956–11961.
135. Wang, L., et al., Conjugated microporous polymer nanosheets for overall water splitting using visible light. *Advanced Materials*, 2017. **29**(38): p. 1702428.
136. Li, H., et al., Carbon nanodots: Synthesis, properties and applications. *Journal of Materials Chemistry*, 2012. **22**(46): pp. 24230–24253.
137. Zheng, D.-W., et al., Carbon-dot-decorated carbon nitride nanoparticles for enhanced photodynamic therapy against hypoxic tumor via water splitting. *ACS Nano*, 2016. **10**(9): pp. 8715–8722.
138. Gu, Z.-G., et al., MOF-templated synthesis of ultrasmall photoluminescent carbon-nanodot arrays for optical applications. *Angewandte Chemie International Edition*, 2017. **56**(24): pp. 6853–6858.
139. Zhang, G., et al., Overall water splitting by Pt/g-C3N4 photocatalysts without using sacrificial agents. *Chemical Science*, 2016. **7**(5): pp. 3062–3066.
140. Guo, F., et al., CoO and g-C3N4 complement each other for highly efficient overall water splitting under visible light. *Applied Catalysis B: Environmental*, 2018. **226**: pp. 412–420.
141. Pan, Z., et al., Decorating CoP and Pt nanoparticles on graphitic carbon nitride nanosheets to promote overall water splitting by conjugated polymers. *ChemSusChem*, 2017. **10**(1): pp. 87–90.
142. Ming, H., et al., Large scale electrochemical synthesis of high quality carbon nanodots and their photocatalytic property. *Dalton Transactions*, 2012. **41** (31): pp. 9526–9531.
143. LaVerne, J.A. and L. Tandon, H2 production in the radiolysis of water on CeO2 and ZrO2. *The Journal of Physical Chemistry B*, 2002. **106**(2): pp. 380–386.
144. Le Caër, S., Water radiolysis: influence of oxide surfaces on H2 production under ionizing radiation. *Water*, 2011. **3**(1).
145. Zakurdaeva, O.A., S.V. Nesterov, and V.I. Feldman, Radiolysis of aqueous solutions of poly(vinyl alcohol) at 77K. *Radiation Physics and Chemistry*, 2010. **79**(8): pp. 876–879.
146. Mittelsteadt, C.K. and J.A. Staser, 10.42 – electrolyzer membranes, in *Polymer Science: A Comprehensive Reference*, K. Matyjaszewski and M. Möller, Editors. 2012, Elsevier: Amsterdam. pp. 849–871.

Index

Page numbers in *italics* refer to figures and **bold** refer to tables

D

E

F

G

H

T

For Product Safety Concerns and Information please contact our EU
representative GPSR@taylorandfrancis.com
Taylor & Francis Verlag GmbH, Kaufingerstraße 24, 80331 München, Germany

www.ingramcontent.com/pod-product-compliance
Lightning Source LLC
Chambersburg PA
CBHW080903220326
41598CB00034B/5462